VIBRATION
Fundamentals and Practice

VIBRATION
Fundamentals and Practice

Clarence W. de Silva

CRC Press
Boca Raton   London   New York   Washington, D.C.

Library of Congress Cataloging-in-Publication Data

De Silva, Clarence W.
　　Vibration : fundamentals and practice / Clarence W. de Silva
　　　　p.　cm.
　　Includes bibliographical references and index.
　　ISBN 0-8493-1808-4 (alk. paper)
　　1. Vibration.　I. Title.
TA355.D384　　1999
620.3—dc21　　　　　　　　　　　　　　　　　　　　　　　　　　　　　　　99-16238
　　CIP

　　This book contains information obtained from authentic and highly regarded sources. Reprinted material is quoted with permission, and sources are indicated. A wide variety of references are listed. Reasonable efforts have been made to publish reliable data and information, but the author and the publisher cannot assume responsibility for the validity of all materials or for the consequences of their use.
　　Neither this book nor any part may be reproduced or transmitted in any form or by any means, electronic or mechanical, including photocopying, microfilming, and recording, or by any information storage or retrieval system, without prior permission in writing from the publisher.
　　The consent of CRC Press LLC does not extend to copying for general distribution, for promotion, for creating new works, or for resale. Specific permission must be obtained in writing from CRC Press LLC for such copying.
　　Direct all inquiries to CRC Press LLC, 2000 Corporate Blvd., N.W., Boca Raton, Florida 33431.

　　Trademark Notice: Product or corporate names may be trademarks or registered trademarks, and are only used for identification and explanation, without intent to infringe.

Cover art is the U.S. Space Shuttle and the International Space Station. (Courtesy of NASA Langley Research Center, Hampton, VA. With permission.)
© 2000 by CRC Press LLC

No claim to original U.S. Government works
International Standard Book Number 0-8493-1808-4
Library of Congress Card Number 99-16238
Printed in the United States of America　1　2　3　4　5　6　7　8　9　0
Printed on acid-free paper

Preface

This book provides the background and techniques that will allow successful modeling, analysis, monitoring, testing, design, modification, and control of vibration in engineering systems. It is suitable as both a course textbook for students and instructors, and a practical reference tool for engineers and other professionals. As a textbook, it can be used in a single-semester course for third-year (junior) and fourth-year (senior) undergraduate students, or for Master's level graduate students in any branch of engineering such as aeronautical and aerospace, civil, mechanical, and manufacturing engineering. But, in view of the practical considerations, design issues, experimental techniques, and instrumentation that are presented throughout the book, and in view of the simplified and snapshot-style presentation of fundamentals and advanced theory, the book will also serve as a valuable reference tool for engineers, technicians, and other professionals in industry and in research laboratories.

The book is an outgrowth of the author's experience in teaching undergraduate and graduate courses in Dynamics, Mechanical Vibration, Dynamic System Modeling, Instrumentation and Design, Feedback Control, Modern Control Engineering, and Modal Analysis and Testing in the U.S. and Canada (Carnegie Mellon University and the University of British Columbia) for more than 20 years. The industrial experience and training that he received in product testing and qualification, analysis, design, and vibration instrumentation at places like Westinghouse Electric Corporation in Pittsburgh, IBM Corporation in Boca Raton, NASA's Langley and Lewis Research Centers, and Bruel and Kjaer in Denmark enabled the author to provide a realistic and practical treatment of the subject.

Design for vibration and control of vibration are crucial in maintaining a high performance level and production efficiency, and prolonging the useful life of machinery, structures, and industrial processes. Before designing or controlling an engineering system for good vibratory performance, it is important to understand, represent (i.e., model), and analyze the vibratory characteristics of the system. Suppression or elimination of undesirable vibrations and generation of required forms and levels of desired vibrations are general goals of vibration engineering. In recent years, researchers and practitioners have devoted considerable effort to studying and controlling vibration in a range of applications in various branches of engineering. With this book, designers, engineers, and students can reap the benefits of that study and experience, and learn the observation, instrumentation, modeling, analysis, design, modification, and control techniques that produce mechanical and aeronautical systems, civil engineering structures, and manufacturing processes that are optimized against the effects of vibration.

The book provides the background and techniques that will allow successful modeling, analysis, design, modification, testing, and control of vibration in engineering systems. This knowledge will be useful in the practice of vibration, regardless of the application area or the branch of engineering. A uniform and coherent treatment of the subject is presented, by introducing practical applications of vibration, through examples, in the very beginning of the book, along with experimental techniques and instrumentation, and then integrating these applications, design, and control considerations into fundamentals and analytical methods throughout the text. To maintain clarity and focus and to maximize the usefulness of the book, an attempt has been made to describe and illustrate industry-standard and state-of-the-art instrumentation, hardware, and computational techniques related to the practice of vibration. As its main features, the book:

- Introduces practical applications, design, and experimental techniques in the very beginning, and then uniformly integrates them throughout the book
- Provides 36 "Summary Boxes" that present key material covered in the book, in point form, within each chapter, for easy reference and recollection (these items are particularly suitable for use by instructors in their presentations)
- Outlines mathematics, dynamics, modeling, fast Fourier transform (FFT) techniques, and reliability analysis in appendices
- Provides over 60 worked examples and case studies, and over 300 problems
- Will be accompanied by an Instructor's Manual, for instructors, that contains complete solutions to all the end-of-chapter problems
- Describes sensors, transducers, filters, amplifiers, analyzers, and other instrumentation that is useful in the practice of vibration
- Describes industry-standard computer techniques, hardware, and tools for analysis, design, and control of vibratory systems, with examples
- Provides a comprehensive coverage of vibration testing and qualification of products
- Offers analogies of mechanical and structural vibration, to other oscillatory behavior such as in electrical and fluid systems, and contrasts these with thermal systems.

A Note to Instructors

The book is suitable as the text for a standard undergraduate course in Mechanical Vibration or for a specialized course for final-year undergraduate students and Master's level graduate students. Three typical course syllabuses are outlined below.

A. *A Standard Undergraduate Course*

As the textbook for an undergraduate (3rd year or 4th year) course in Mechanical Vibration, it may be incorporated into the following syllabus for a 12 week course consisting of 36 hours of lectures and 12 hours of laboratory experiments:

Lectures

Chapter 1 (1 hour)
Sections 8.1, 8.2, 8.4, 9.1, 9.2, 9.8 (3 hours)
Chapter 2 (6 hours)
Chapter 3 (6 hours)
Section 11.4 (2 hours)
Chapter 5 (6 hours)
Chapter 6 (6 hours)
Sections 12.1, 12.2, 12.3, 12.4, 12.5 (6 hours)

Laboratory Experiments

The following four laboratory experiments, each of 3-hour duration, may be incorporated.

1. Experiment on modal testing (hammer test and other transient tests) and damping measurement in the time domain (see Section 11.4)
2. Experiment on shaker testing and damping measurement in the frequency domain (see Section 11.4)
3. Experiment on single-plane and two-plane balancing (see Section 12.3)
4. Experiment on modal testing of a distributed-parameter system (see Section 11.4)

B. A Course in Industrial Vibration

Chapter 1 (1 hour)
Chapter 4 (3 hours)
Chapter 7 (5 hours)
Chapter 8 (5 hours)
Chapter 9 (4 hours)
Chapter 10 (6 hours)
Chapter 11 (6 hours)
Chapter 12 (6 hours)

A project may be included in place of a final examination.

C. A Course in Modal Analysis and Testing

Chapter 1 (1 hour)
Chapter 4 (3 hours)
Chapter 5 (6 hours)
Chapter 6 (6 hours)
Chapter 7 (5 hours)
Chapter 10 (6 hours)
Chapter 11 (6 hours)
Section 12.6 (hours)

A project may be included in place of a final examination.

Clarence W. de Silva
Vancouver, Canada

The Author

Clarence W. de Silva, Fellow ASME and Fellow IEEE, is Professor of Mechanical Engineering at the University of British Columbia, Vancouver, Canada, and has occupied the NSERC Chair in Industrial Automation since 1988. He obtained his first Ph.D. from the Massachusetts Institute of Technology in 1978 and, 20 years later, another Ph.D. from the University of Cambridge, England. De Silva has served as a consultant to several companies, including IBM and Westinghouse in the U.S., and has led the development of many industrial machines. He is recipient of the Education Award of the Dynamic Systems and Control Division of the American Society of Mechanical Engineers, the Meritorious Achievement Award of the Association of Professional Engineers of British Columbia, the Outstanding Contribution Award of the IEEE Systems, Man, and Cybernetics Society, the Outstanding Large Chapter Award of the IEEE Industry Applications Society, and the Outstanding Chapter Award from the IEEE Control Systems Society. He has authored 14 technical books, 10 edited volumes, over 120 journal papers, and a similar number of conference papers and book chapters. He has served on the editorial boards of 12 international journals, and is the Editor-in-Chief of the *International Journal of Knowledge-Based Intelligent Engineering Systems,* Senior Technical Editor of *Measurements and Control*, and Regional Editor, North America, of the *International Journal of Intelligent Real-Time Automation.* He has been a Lilly Fellow, Senior Fulbright Fellow to Cambridge University, ASI Fellow, and a Killam Fellow.

Acknowledgment

Preparation of this book would not have been possible if not for the support of many individuals and organizations, but it is not possible to list all of them here. I wish to recognize the following specific contributions:

Financial assistance for my research and professional activities has been provided primarily by:

- Ministry of Advanced Education, Training and Technology, Province of British Columbia, for the Network of Centres of Excellence Program
- Natural Sciences and Engineering Research Council of Canada (NSERC)
- Network of Centres of Excellence (Institute of Robotics and Intelligent Systems)
- Advanced Systems Institute of British Columbia
- Science Council of British Columbia
- Ministry of Environment of British Columbia
- British Columbia Hydro and Power Authority
- National Research Council
- Killam Memorial Faculty Fellowship Program
- B.C. Packers, Ltd.
- Neptune Dynamics, Ltd.
- Garfield Weston Foundation

Special acknowledgment should be made here of the Infrastructure Grant from the Ministry of Advanced Education, Training and Technology, Province of British Columbia, which made part of the secretarial support for my work possible. The Department of Mechanical Engineering at the University of British Columbia provided me with an excellent environment to carry out my educational activities, including the preparation of this book. My graduate students, research associates, teaching assistants, and office staff have contributed directly and indirectly to the success of the book. Particular mention should be made of the following people:

- Ricky Min-Fan Lee for systems assistance
- Hassan Bayoumi and Jay Choi for graphics assistance
- Yuan Chen, Scott Gu, Iwan Kurnianto, Farag Omar, and Roya Rahbari for teaching assistance
- Marje Lewis and Laura Gawronski for secretarial assistance.

I wish to thank the staff of CRC Press, particularly the Associate Editor, Felicia Shapiro and the Project Editor, Sara Seltzer, for their fine effort in the production of the book. Encouragement of various authorities in the field of engineering — particularly, Professor Devendra Garg of Duke University, Professor Mo Jamshidi of the University of New Mexico, and Professor Arthur Murphy (DuPont Fellow Emeritus) — is gratefully acknowledged. Finally, my family deserves an apology for the unintentional "neglect" that they may have suffered during the latter stages of production of the book.

Source Credits

The sources of several photos, figures, and tables are recognized and given credit, as follows:

- Figure 1.1: Courtesy of Ms. Kimberly Land, NASA Langley Research Center, Hampton, Virginia.
- Figure 1.3: Courtesy of Professor Carlos E. Ventura, Department of Civil Engineering, the University of British Columbia, Vancouver, Canada.
- Figure 1.4: Courtesy of Ms. Heather Conn of BC Transit, Vancouver, Canada. Photo by Mark Van Manen.
- Figure 1.5: Courtesy of Ms. Jeana Dugger, Key Technologies, Inc., Walla Walla, Washington.
- Figure 1.8: Courtesy of *Mechanical Engineering* magazine, from article "Semiactive Cone Suspension Smooths the Ride" by Bill Siuru, Vol. 116, No. 3, page 106. Copyright, American Society of Mechanical Engineers International, New York.
- Table 7.5: Reprinted from ASME BPVC, Section III-Division 1, Appendices, by permission of The American Society of Mechanical Engineers, New York. All rights reserved.
- Figure 8.8: Courtesy of Bruel & Kjaer, Naerum, Denmark.
- Figure 9.36 and Figure 9.38: Courtesy of Ms. Beth Daniels. Copyright 1999 Tektronix, Inc. All rights reserved. Reproduced by permission.
- Figure 11.6, Figure 11.8, and Figure 12.15: Experimental setups used by the author for teaching a fourth-year course in the Undergraduate Vibrations Laboratory, Department of Mechanical Engineering, the University of British Columbia, Vancouver, Canada.
- Figure 12.34, Figure 12.35, and Table 12.2: Courtesy of Dr. George Wang. Extracted from the report "Active Control of Vibration in Wood Machining for Wood Recovery" by G. Wang, J. Xi, Q. Zhong, S. Abayakoon, K. Krishnappa, and F. Lam, National Research Council, Integrated Manufacturing Technologies Institute, Vancouver, Canada, pp. 5, 8, 25-28, May 1998.

Dedication

Professor David N. Wormley.

"For the things we have to learn before we can do them, we learn by doing them."

— Aristotle (Author of *Mechanics and Acoustics,* 384–322 B.C.)

Table of Contents

Chapter 1   Vibration Engineering
1.1 Study of Vibration 2
1.2 Application Areas 5
1.3 History of Vibration 10
1.4 Organization of the Book 11
Problems 14
References and Further Reading 14
    Author's Work 14
    Other Useful Publications 15

Chapter 2   Time Response
2.1 Undamped Oscillator 18
    2.1.1 Energy Storage Elements 19
        Inertia 19
        Spring 20
        Gravitational Potential Energy 21
    2.1.2 Conservation of Energy 21
        System 1 (Translatory) 21
        System 2 (Rotatory) 22
        System 3 (Flexural) 22
        System 4 (Swinging) 22
        System 5 (Liquid Slosh) 24
        System 6 (Electrical) 24
        Capacitor 25
        Inductor 25
    2.1.3 Free Response 26
    Example 2.1 28
    Solution 29
2.2 Heavy Springs 32
    2.2.1 Kinetic Energy Equivalence 32
    Example 2.2 33
    Solution 34
2.3 Oscillations in Fluid Systems 34
    Example 2.3 35
    Solution 36
2.4 Damped Simple Oscillator 38
    2.4.1 Case 1: Underdamped Motion 40
        Initial Conditions 40
    2.4.2 Logarithmic Decrement Method 41
    2.4.3 Case 2: Overdamped Motion 42
    2.4.4 Case 3: Critically Damped Motion 43
    2.4.5 Justification for the Trial Solution 44

		First-Order System	44
		Second-Order System	44
		Repeated Roots	44
	2.4.6	Stability and Speed of Response	45
	Example 2.4		48
	Solution		48
2.5	Forced Response		52
	2.5.1	Impulse Response Function	52
	2.5.2	Forced Response	54
	2.5.3	Response to a Support Motion	55
		Impulse Response	58
		The Riddle of Zero Initial Conditions	59
		Step Response	60
		Liebnitz's Rule	62
Problems			63

Chapter 3   Frequency Response

3.1	Response to Harmonic Excitations		85
	3.1.1	Response Characteristics	87
		Case 1	87
		Case 2	88
		Case 3	90
		Particular Solution (Method 1)	90
		Particular Solution (Method 2): Complex Function Method	91
		Resonance	93
	3.1.2	Measurement of Damping Ratio (Q-Factor Method)	95
	Example 3.1		97
	Solution		97
3.2	Transform Techniques		101
	3.2.1	Transfer Function	101
	3.2.2	Frequency-Response Function (Frequency-Transfer Function)	104
		Impulse Response	106
		Case 1 ($\zeta < 1$)	106
		Case 2 ($\zeta > 1$)	107
		Case 3 ($\zeta = 1$)	107
		Step Response	107
	3.2.3	Transfer Function Matrix	108
	Example 3.2		109
	Example 3.3		111
	Example 3.4		113
	Solution		113
3.3	Mechanical Impedance Approach		116
		Mass Element	117
		Spring Element	117
		Damper Element	118
	3.3.1	Interconnection Laws	118
	Example 3.5		119
	Example 3.6		120
3.4	Transmissibility Functions		123
	3.4.1	Force Transmissibility	124

	3.4.2	Motion Transmissibility 124
		System Suspended on a Rigid Base (Force Transmissibility).....................124
		System with Support Motion (Motion Transmissibility)............................126
	3.4.3	General Case 127
	Example 3.7..128
	3.4.4	Peak Values of Frequency-Response Functions 130
3.5	Receptance Method 130
	3.5.1	Application of Receptance 130
		Undamped Simple Oscillator..131
		Dynamic Absorber 131
Problems..134

Chapter 4   Vibration Signal Analysis

4.1	Frequency Spectrum 151
	4.1.1 Frequency 152
	4.1.2 Amplitude Spectrum 152
	4.1.3 Phase Angle 153
	4.1.4 Phasor Representation of Harmonic Signals 154
	4.1.5 RMS Amplitude Spectrum 155
	4.1.6 One-Sided and Two-Sided Spectra 156
	4.1.7 Complex Spectrum 157
4.2	Signal Types 158
4.3	Fourier Analysis...160
	4.3.1 Fourier Integral Transform (FIT) 160
	4.3.2 Fourier Series Expansion (FSE) 161
	4.3.3 Discrete Fourier Transform (DFT) 162
	4.3.4 Aliasing Distortion 164
	Sampling Theorem..165
	Aliasing Distortion in the Time Domain..167
	Anti-Aliasing Filter 168
	Example 4.1..168
	4.3.5 Another Illustration of Aliasing 168
	Example 4.2..171
4.4	Analysis of Random Signals 171
	4.4.1 Ergodic Random Signals..171
	4.4.2 Correlation and Spectral Density..172
	4.4.3 Frequency Response Using Digital Fourier Transform....................173
	4.4.4 Leakage (Truncation Error)..174
	4.4.5 Coherence 175
	4.4.6 Parseval's Theorem..176
	4.4.7 Window Functions..177
	4.4.8 Spectral Approach to Process Monitoring 179
	4.4.9 Cepstrum..180
4.5	Other Topics of Signal Analysis 182
	4.5.1 Bandwidth..182
	4.5.2 Transmission Level of a Bandpass Filter 182
	4.5.3 Effective Noise Bandwidth 182
	4.5.4 Half-Power (or 3 dB) Bandwidth 183
	4.5.5 Fourier Analysis Bandwidth...183

4.6 Resolution in Digital Fourier Results 184
4.7 Overlapped Processing 184
      Example 4.3 186
      4.7.1 Order Analysis 186
            Speed Spectral Map 187
            Time Spectral Map 187
            Order Tracking 188
Problems 189

Chapter 5   Modal Analysis

5.1 Degrees of Freedom and Independent Coordinates 197
      5.1.1 Nonholonomic Constraints 198
      Example 5.1 198
      Example 5.2 199
5.2 System Representation 200
      5.2.1 Stiffness and Flexibility Matrices 202
      5.2.2 Inertia Matrix 204
      5.2.3 Direct Approach for Equations of Motion 206
5.3 Modal Vibrations 207
      Example 5.3 209
5.4 Orthogonality of Natural Modes 212
      5.4.1 Modal Mass and Normalized Modal Vectors 213
5.5 Static Modes and Rigid Body Modes 213
      5.5.1 Static Modes 213
      5.5.2 Linear Independence of Modal Vectors 214
      5.5.3 Modal Stiffness and Normalized Modal Vectors 214
      5.5.4 Rigid Body Modes 215
      Example 5.4 215
            Equation of Heave Motion 215
            Equation of Pitch Motion 216
      Example 5.5 218
            First Mode (Rigid Body Mode) 219
            Second Mode 220
      5.5.5 Modal Matrix 222
      5.5.6 Configuration Space and State Space 223
            State Vector 223
5.6 Other Modal Formulations 223
      5.6.1 Non-Symmetric Modal Formulation 224
      5.6.2 Transformed Symmetric Modal Formulation 224
      Example 5.6 226
            Approach 2 226
            Approach 3 228
5.7 Forced Vibration 230
      Example 5.7 232
            First Mode (Rigid Body Mode) 234
            Second Mode (Oscillatory Mode) 234
5.8 Damped Systems 235
      5.8.1 Proportional Damping 236
      Example 5.8 237

5.9 State-Space Approach..240
    5.9.1 Modal Analysis...241
    5.9.2 Mode Shapes of Nonoscillatory Systems...242
    5.9.3 Mode Shapes of Oscillatory Systems...242
    Example 5.9..243
Problems..246

Chapter 6   Distributed-Parameter Systems

6.1 Transverse Vibration of Cables 268
    6.1.1 Wave Equation..270
    6.1.2 General (Modal) Solution 271
    6.1.3 Cable with Fixed Ends 272
    6.1.4 Orthogonality of Natural Modes..275
    Example 6.1..276
    Solution..276
    6.1.5 Application of Initial Conditions 279
    Example 6.2..280
    Solution..280
6.2 Longitudinal Vibration of Rods 284
    6.2.1 Equation of Motion 284
    6.2.2 Boundary Conditions..285
    Example 6.3..286
    Solution..286
6.3 Torsional Vibration of Shafts 290
    6.3.1 Shaft with Circular Cross Section 290
    6.3.2 Torsional Vibration of Noncircular Shafts...293
    Example 6.4..294
    Solution..295
    Example 6.5..296
    Solution..296
6.4 Flexural Vibration of Beams 299
    6.4.1 Governing Equation for Thin Beams..299
        Moment-Deflection Relation 299
        Rotatory Dynamics (Equilibrium)...301
        Transverse Dynamics..301
    6.4.2 Modal Analysis...302
    6.4.3 Boundary Conditions..304
    6.4.4 Free Vibration of a Simply Supported Beam 305
        Normalization of Mode Shape Functions 307
        Initial Conditions 308
    6.4.5 Orthogonality of Mode Shapes...309
        Case of Variable Cross Section 309
    6.4.6 Forced Bending Vibration 310
    Example 6.6..313
    Solution..313
    Example 6.7..316
    Solution..316
    6.4.7 Bending Vibration of Beams with Axial Loads 318
    6.4.8 Bending Vibration of Thick Beams 322

		6.4.9	Use of the Energy Approach	324
		6.4.10	Orthogonality with Inertial Boundary Conditions	326
			Rotatory Inertia	328
	6.5	Damped Continuous Systems		328
		6.5.1	Modal Analysis of Damped Beams	329
		Example 6.8		330
		Solution		330
	6.6	Vibration of Membranes and Plates		332
		6.6.1	Transverse Vibration of Membranes	332
		6.6.2	Rectangular Membrane with Fixed Edges	333
		6.6.3	Transverse Vibration of Thin Plates	334
		6.6.4	Rectangular Plate with Simply Supported Edges	335
Problems				337

Chapter 7   Damping

7.1	Types of Damping		349
	7.1.1	Material (Internal) Damping	350
		Viscoelastic Damping	351
		Hysteretic Damping	352
	Example 7.1		353
	Solution		353
	7.1.2	Structural Damping	354
	7.1.3	Fluid Damping	356
	Example 7.2		357
	Solution		357
7.2	Representation of Damping in Vibration Analysis		359
	7.2.1	Equivalent Viscous Damping	360
	7.2.2	Complex Stiffness	362
	Example 7.3		364
	Solution		364
	7.2.3	Loss Factor	366
7.3	Measurement of Damping		368
	7.3.1	Logarithmic Decrement Method	368
	7.3.2	Step-Response Method	370
	7.3.3	Hysteresis Loop Method	371
	Example 7.4		372
	Solution		373
	7.3.4	Magnification-Factor Method	374
	7.3.5	Bandwidth Method	376
	7.3.6	General Remarks	378
7.4	Interface Damping		380
	Example 7.5		382
	Solution		382
	7.4.1	Friction In Rotational Interfaces	385
	7.4.2	Instability	386
Problems			386

Chapter 8   Vibration Instrumentation

8.1   Vibration Exciters 400
       8.1.1   Shaker Selection 403
                 Force Rating 404
                 Power Rating 405
                 Stroke Rating 405
       Example 8.1 406
       Solution 406
                 Hydraulic Shakers 407
                 Inertial Shakers 408
                 Electromagnetic Shakers 410
       8.1.2   Dynamics of Electromagnetic Shakers 411
                 Transient Exciters 414
8.2   Control System 415
       8.2.1   Components of a Shaker Controller 416
                 Compressor 416
                 Equalizer (Spectrum Shaper) 416
                 Tracking Filter 417
                 Excitation Controller (Amplitude Servo-Monitor) 417
       8.2.2   Signal-Generating Equipment 417
                 Oscillators 418
                 Random Signal Generators 419
                 Tape Players 419
                 Data Processing 420
8.3   Performance Specification 421
       8.3.1   Parameters for Performance Specification 422
                 Time-Domain Specifications 422
                 Frequency-Domain Specifications 423
       8.3.2   Linearity 423
       8.3.3   Instrument Ratings 425
                 Rating Parameters 425
       8.3.4   Accuracy and Precision 426
8.4   Motion Sensors and Transducers 428
       8.4.1   Potentiometer 428
                 Potentiometer Resolution 429
                 Optical Potentiometer 430
       8.4.2   Variable-Inductance Transducers 431
                 Mutual-Induction Transducers 432
                 Linear-Variable Differential Transformer (LVDT) 432
                 Signal Conditioning 434
       Example 8.2 434
       Solution 435
       8.4.3   Mutual-Induction Proximity Sensor 438
       8.4.4   Self-Induction Transducers 438
       8.4.5   Permanent-Magnet Transducers 438
       8.4.6   AC Permanent-Magnet Tachometer 441
       8.4.7   AC Induction Tachometer 442
       8.4.8   Eddy Current Transducers 443
       8.4.9   Variable-Capacitance Transducers 444
                 Capacitive Displacement Sensors 445

		Capacitive Angular Velocity Sensor	446
		Capacitance Bridge Circuit	447
	8.4.10	Piezoelectric Transducers	449
		Sensitivity	450
	Example 8.3		450
	Solution		450
		Piezoelectric Accelerometer	451
		Charge Amplifier	453
8.5	Torque, Force, and Other Sensors		456
	8.5.1	Strain-Gage Sensors	456
		Equations for Strain-Gage Measurements	456
		Bridge Sensitivity	459
		The Bridge Constant	460
	Example 8.4		461
	Solution		461
		The Calibration Constant	461
	Example 8.5		462
	Solution		464
		Data Acquisition	466
		Accuracy Considerations	467
		Semiconductor Strain Gages	467
		Force and Torque Sensors	468
		Strain-Gage Torque Sensors	470
		Deflection Torque Sensors	473
		Variable-Reluctance Torque Sensor	474
		Reaction Torque Sensors	474
	8.5.2	Miscellaneous Sensors	476
		Stroboscope	476
		Fiber-Optic Sensors and Lasers	477
		Fiber-Optic Gyroscope	478
		Laser Doppler Interferometer	478
		Ultrasonic Sensors	480
		Gyroscopic Sensors	481
8.6	Component Interconnection		482
	8.6.1	Impedance Characteristics	482
		Cascade Connection of Devices	484
		Impedance-Matching Amplifiers	485
		Operational Amplifiers	485
		Voltage Followers	487
		Charge Amplifiers	489
	8.6.2	Instrumentation Amplifier	491
		Ground Loop Noise	494
Problems			494

Chapter 9 Signal Conditioning and Modification

9.1	Amplifiers		509
	9.1.1	Operational Amplifier	510
	Example 9.1		511
	Solution		511

	9.1.2	Use of Feedback in Op-amps 512
	9.1.3	Voltage, Current, and Power Amplifiers 513
	9.1.4	Instrumentation Amplifiers 516
		Differential Amplifier 516
		Common Mode 518
		Amplifier Performance Ratings 519
		Example 9.2 519
		Solution 520
		Common-Mode Rejection Ratio (CMRR) 521
		AC-Coupled Amplifiers 521
9.2	Analog Filters 521
	9.2.1	Passive Filters and Active Filters 524
		Number of Poles 525
	9.2.2	Low-Pass Filters 525
		Example 9.3 527
		Solution 527
		Low-Pass Butterworth Filter 529
		Example 9.4 529
		Solution 529
	9.2.3	High-Pass Filters 531
	9.2.4	Bandpass Filters 533
		Resonance-Type Bandpass Filters 535
		Example 9.5 536
		Solution 536
	9.2.5	Band-Reject Filters 538
9.3	Modulators and Demodulators 539
	9.3.1	Amplitude Modulation 543
		Modulation Theorem 543
		Side Frequencies and Side Bands 544
	9.3.2	Application of Amplitude Modulation 545
		Fault Detection and Diagnosis 547
	9.3.3	Demodulation 547
9.4	Analog/Digital Conversion 548
	9.4.1	Digital-to-Analog Conversion (DAC) 549
		Weighted-Resistor DAC 550
		Ladder DAC 551
		DAC Error Sources 554
	9.4.2	Analog-to-Digital Conversion (ADC) 555
		Successive-Approximation ADC 555
		Dual-Slope ADC 557
		Counter ADC 559
	9.4.3	ADC Performance Characteristics 560
		Resolution and Quantization Error 560
		Monotonicity, Nonlinearity, and Offset Error 561
		ADC Conversion Rate 561
	9.4.4	Sample-and-Hold (S/H) Circuitry 563
	9.4.5	Multiplexers (MUX) 564
		Analog Multiplexers 564
		Digital Multiplexers 565
	9.4.6	Digital Filters 567

		Software Implementation and Hardware Implementation	568

9.5 Bridge Circuits..568
    9.5.1 Wheatstone Bridge 569
    9.5.2 Constant-Current Bridge 571
    9.5.3 Bridge Amplifiers 572
           Half-Bridge Circuits 572
    9.5.4 Impedance Bridges 574
           Owen Bridge 575
           Wien-Bridge Oscillator 577
9.6 Linearizing Devices 577
    9.6.1 Linearization by Software 580
    9.6.2 Linearization by Hardware Logic 581
    9.6.3 Analog Linearizing Circuitry 581
    9.6.4 Offsetting Circuitry 582
    9.6.5 Proportional-Output Circuitry 584
           Curve-Shaping Circuitry 585
9.7 Miscellaneous Signal-Modification Circuitry 586
    9.7.1 Phase Shifter..586
    9.7.2 Voltage-to-Frequency Converter (VFC)...............................589
    9.7.3 Frequency-to-Voltage Converter (FVC)...............................591
    9.7.4 Voltage-to-Current Converter (VCC)...................................592
    9.7.5 Peak-Hold Circuit..593
9.8 Signal Analyzers and Display Devices 594
    9.8.1 Signal Analyzers..595
    9.8.2 Oscilloscopes...597
           Triggering...599
           Lissajous Patterns..599
           Digital Oscilloscopes 600
Problems..601

Chapter 10   Vibration Testing

10.1 Representation of a Vibration Environment....................................611
    10.1.1 Test Signals 611
           Stochastic versus Deterministic Signals....................611
    10.1.2 Deterministic Signal Representation...................................612
           Single-Frequency Signals 612
           Sine Sweep..612
           Sine Dwell...614
           Decaying Sine 615
           Sine Beat 615
           Sine Beat with Pauses...616
           Multifrequency Signals...616
           Actual Excitation Records 616
           Simulated Excitation Signals 617
    10.1.3 Stochastic Signal Representation 617
           Ergodic Random Signals 618
           Stationary Random Signals 618
           Independent and Uncorrelated Signals......................619
           Transmission of Random Excitations........................620

- 10.1.4 Frequency-Domain Representations 623
  - Fourier Spectrum Method 623
  - Power Spectral Density Method 624
- 10.1.5 Response Spectrum 625
  - Displacement, Velocity, and Acceleration Spectra 627
  - Response-Spectra Plotting Paper 629
  - Zero-Period Acceleration 631
  - Uses of Response Spectra 632
- 10.1.6 Comparison of Various Representations 634

10.2 Pretest Procedures 636
- 10.2.1 Purpose of Testing 637
- 10.2.2 Service Functions 638
  - Active Equipment 638
  - Passive Equipment 638
  - Functional Testing 638
- 10.2.3 Information Acquisition 639
  - Interface Details 639
  - Effect of Neglecting Interface Dynamics 642
  - Effects of Damping 643
  - Effects of Inertia 643
  - Effect of Natural Frequency 643
  - Effect of Excitation Frequency 644
  - Other Effects of Interface 645
- 10.2.4 Test-Program Planning 645
  - Testing of Cabinet-Mounted Equipment 647
- 10.2.5 Pretest Inspection 649

10.3 Testing Procedures 650
- 10.3.1 Resonance Search 650
- 10.3.2 Methods of Determining Frequency-Response Functions 651
  - Fourier Transform Method 651
  - Spectral Density Method 651
  - Harmonic Excitation Method 651
- 10.3.3 Resonance-Search Test Methods 652
  - Hammer (Bump) Test and Drop Test 652
  - Pluck Test 653
  - Shaker Tests 654
- 10.3.4 Mechanical Aging 656
  - Equivalence for Mechanical Aging 656
  - Excitation-Intensity Equivalence 657
  - Dynamic-Excitation Equivalence 657
  - Cumulative Damage Theory 658
- 10.3.5 TRS Generation 658
- 10.3.6 Instrument Calibration 659
- 10.3.7 Test-Object Mounting 659
- 10.3.8 Test-Input Considerations 660
  - Test Nomenclature 660
  - Testing with Uncorrelated Excitations 662
  - Symmetrical Rectilinear Testing 664
  - Geometry versus Dynamics 664
  - Some Limitations 665
  - Testing of Black Boxes 666

	Phasing of Excitations	666
	Testing a Gray or White Box	667
	Overtesting in Multitest Sequences	667
10.4	Product Qualification Testing	668
10.4.1	Distribution Qualification	668
	Drive-Signal Generation	669
	Distribution Spectra	670
	Test Procedures	671
10.4.2	Seismic Qualification	673
	Stages of Seismic Qualification	674
10.4.3	Test Preliminaries	675
	Single-Frequency Testing	677
	Multifrequency Testing	679
10.4.4	Generation of RRS Specifications	680
Problems		682

Chapter 11   Experimental Modal Analysis

11.1	Frequency-Domain Formulation	629
11.1.1	Transfer Function Matrix	692
11.1.2	Principle of Reciprocity	696
Example 11.1		696
11.2	Experimental Model Development	698
11.2.1	Extraction of the Time-Domain Model	700
11.3	Curve-Fitting of Transfer Functions	702
11.3.1	Problem Identification	703
11.3.2	Single-Degree-of-Freedom and Multi-Degree-of-Freedom Techniques	704
11.3.3	Single-Degree-of-Freedom Parameter Extraction in the Frequency Domain	704
	Circle-Fit Method	704
	Peak Picking Method	708
11.3.4	Multi-Degree-of-Freedom Curve Fitting	709
	Formulation of the Method	709
11.3.5	A Comment on Static Modes and Rigid Body Modes	711
11.3.6	Residue Extraction	711
11.4	Laboratory Experiments	713
11.4.1	Lumped-Parameter System	713
	Frequency-Domain Test	715
	Time-Domain Test	716
11.4.2	Distributed-Parameter System	716
11.5	Commercial EMA Systems	718
11.5.1	System Configuration	719
	FFT Analysis Options	719
	Modal Analysis Components	720
Problems		720

Chapter 12   Vibration Design and Control

	Shock and Vibration	732
12.1	Specification of Vibration Limits	732
12.1.1	Peak Level Specification	733
12.1.2	RMS Value Specification	733

- 12.1.3 Frequency-Domain Specification 733
- 12.2 Vibration Isolation 735
  - Example 12.1 739
  - Solution 739
  - 12.2.1 Design Considerations 743
  - Example 12.2 745
  - Solution 745
  - 12.2.2 Vibration Isolation of Flexible Systems 745
- 12.3 Balancing of Rotating Machinery 749
  - 12.3.1 Static Balancing 749
    - Balancing Approach 750
  - 12.3.2 Complex Number/Vector Approach 751
  - Example 12.3 753
  - Solution 753
  - 12.3.3 Dynamic (Two-Plane) Balancing 755
  - Example 12.4 757
  - Solution 757
  - 12.3.4 Experimental Procedure of Balancing 760
- 12.4 Balancing of Reciprocating Machines 762
  - 12.4.1 Single-Cylinder Engine 763
  - 12.4.2 Balancing the Inertia Load of the Piston 764
  - 12.4.3 Multicylinder Engines 766
    - Two-Cylinder Engine 766
    - Six-Cylinder Engine 768
  - Example 12.5 769
  - Solution 769
  - 12.4.4 Combustion/Pressure Load 770
- 12.5 Whirling of Shafts 771
  - 12.5.1 Equations of Motion 772
  - 12.5.2 Steady-State Whirling 774
  - Example 12.6 776
  - Solution 776
  - 12.5.3 Self-Excited Vibrations 779
- 12.6 Design Through Modal Testing 779
  - 12.6.1 Component Modification 780
  - Example 12.7 782
  - Solution 783
  - 12.6.2 Substructuring 784
- 12.7 Passive Control of Vibration 787
  - 12.7.1 Undamped Vibration Absorber 788
  - Example 12.8 792
  - Solution 793
  - 12.7.2 Damped Vibration Absorber 793
    - Optimal Absorber Design 797
  - Example 12.9 798
  - Solution 798
  - 12.7.3 Vibration Dampers 801
- 12.8 Active Control of Vibration 805
  - 12.8.1 Active Control System 805
  - 12.8.2 Control Techniques 807
    - State-Space Models 807

    Example 12.10 809
    Solution 809
        Position and Velocity Feedback 811
        Linear Quadratic Regulator (LQR) Control 811
        Modal Control 812
   12.8.3 Active Control of Saw Blade Vibration 813
 12.9 Control of Beam Vibrations 816
   12.9.1 State-Space Model of Beam Dynamics 817
   12.9.2 Control Problem 820
   12.9.3 Use of Linear Dampers 821
        Design Example 822
Problems 824

Appendix A Dynamic Models and Analogies
A.1 Model Development 840
A.2 Analogies 840
A.3 Mechanical Elements 842
   A.3.1 Mass (Inertia) Element 842
   A.3.2 Spring (Stiffness) Element 843
A.4 Electrical Elements 843
   A.4.1 Capacitor Element 843
   A.4.2 Inductor Element 844
A.5 Thermal Elements 845
   A.5.1 Thermal Capacitor 845
   A.5.2 Thermal Resistance 845
A.6 Fluid Elements 846
   A.6.1 Fluid Capacitor 846
   A.6.2 Fluid Inertor 847
   A.6.3 Fluid Resistance 848
   A.6.4 Natural Oscillations 848
A.7 State-Space Models 848
   A.7.1 Linearization 849
   A.7.2 Time Response 850
   A.7.3 Some Formal Definitions 853
   A.7.4 Illustrative Example 853
   A.7.5 Causality and Physical Realizability 855

Appendix B Newtonian and Lagrangian Mechanics
B.1 Vector Kinematics 857
   B.1.1 Euler's Theorem 857
        Important Corollary 857
        Proof 857
   B.1.2 Angular Velocity and Velocity at a Point of a Rigid Body 858
        Theorem 858
        Proof 858
   B.1.3 Rates of Unit Vectors Along Axes of Rotating Frames 859
        General Result 859
        Cartesian Coordinates 859
        Polar Coordinates (2-D) 860

| | | Spherical Polar Coordinates ....861 |
| | | Tangential-Normal (Intrinsive) Coordinates (2-D)................................861 |

	B.1.4	Acceleration Expressed in Rotating Frames.....................................862
		Spherical Polar Coordinates 862
		Tangential-Normal Coordinates (2-D)..862

B.2 Newtonian (Vector) Mechanics..864
    B.2.1 Frames of Reference Rotating at Angular Velocity ω 864
    B.2.2 Newton's Second Law for a Particle of Mass m 865
    B.2.3 Second Law for a System of Particles — Rigidly or Flexibly Connected...........866
    B.2.4 Rigid Body Dynamics — Inertia Matrix and Angular Momentum......................867
    B.2.5 Manipulation of Inertia Matrix 869
        Parallel Axis Theorem — Translational Transformation of $[I]$ 869
        Rotational Transformation of $[I]$ 869
        Principal Directions (Eigenvalue Problem)..869
        Mohr's Circle 871
    B.2.6 Euler's Equations (for a Rigid Body Rotating at ω).................................871
    B.2.7 Euler's Angles 871

B.3 Lagrangian Mechanics...873
    B.3.1 Kinetic Energy and Kinetic Coenergy 873
    B.3.2 Work and Potential Energy 874
        Examples 875
    B.3.3 Holonomic Systems, Generalized Coordinates, and Degrees of Freedom 875
    B.3.4 Hamilton's Principle..875
    B.3.5 Lagrange's Equations 876
    Example 876
        Generalized Coordinates..876
        Generalized Nonconservative Forces 876
        Lagrangian 877
        Lagrange's Equations..877

Appendix C   Review of Linear Algebra

C.1 Vectors and Matrices 879
C.2 Vector-Matrix Algebra..881
    C.2.1 Matrix Addition and Subtraction 882
    C.2.2 Null Matrix 882
    C.2.3 Matrix Multiplication 883
    C.2.4 Identity Matrix...884
C.3 Matrix Inverse...884
    C.3.1 Matrix Transpose...885
    C.3.2 Trace of a Matrix 885
    C.3.3 Determinant of a Matrix 886
    C.3.4 Adjoint of a Matrix 887
    C.3.5 Inverse of a Matrix 888
C.4 Vector Spaces..889
    C.4.1 Field ($\mathscr{F}$)..889
    C.4.2 Vector Space ($\mathscr{L}$) 889
        Properties 889
        Special Case 890
    C.4.3 Subspace $\mathscr{S}$ of $\mathscr{L}$..890

		C.4.4	Linear Dependence	890

	C.4.5	Basis and Dimension of a Vector Space	891
	C.4.6	Inner Product	891
	C.4.7	Norm	892
		Properties	892
	C.4.8	Gram-Schmidt Orthogonalization	892
	C.4.9	Modified Gram-Schmidt Procedure	892
C.5	Determinants		893
	C.5.1	Properties of Determinant of a Matrix	893
	C.5.2	Rank of a Matrix	893
C.6	System of Linear Equations		894
References			894

Appendix D Digital Fourier Analysis and FFT

D.1	Unification of the Three Fourier Transform Types		895
	D.1.1	Relationship Between DFT and FIT	895
	D.1.2	Relationship Between DFT and FSE	897
D.2	Fast Fourier Transform (FFT)		898
	D.2.1	Development of the Radix-Two FFT Algorithm	898
	D.2.2	The Radix-Two FFT Procedure	901
	D.2.3	Illustrative Example	901
D.3	Discrete Correlation and Convolution		902
	D.3.1	Discrete Correlation	902
		Discrete Correlation Theorem	903
		Discrete Convolution Theorem	903
D.4	Digital Fourier Analysis Procedures		904
	D.4.1	Fourier Transform Using DFT	904
	D.4.2	Inverse DFT Using DFT	905
	D.4.3	Simultaneous DFT of Two Real Data Records	905
	D.4.4	Reduction of Computation Time for a Real Data Record	906
	D.4.5	Convolution of Finite Duration Signals Using DFT	907
		Wraparound Error	907
		Data-Record Sectioning in Convolution	908

Appendix E Reliability Considerations for Multicomponent Units

E.1	Failure Analysis		911
	E.1.1	Reliability	911
	E.1.2	Unreliability	911
	E.1.3	Inclusion–Exclusion Formula	912
		Example	912
E.2	Bayes' Theorem		913
	E.2.1	Product Rule for Independent Events	913
	E.2.2	Failure Rate	914
	E.2.3	Product Rule for Reliability	916

Answers to Numerical Problems ... 919

Index ... 921

1 Vibration Engineering

Vibration is a repetitive, periodic, or oscillatory response of a mechanical system. The rate of the vibration cycles is termed "frequency." Repetitive motions that are somewhat clean and regular, and that occur at relatively low frequencies, are commonly called oscillations, while any repetitive motion, even at high frequencies, with low amplitudes, and having irregular and random behavior falls into the general class of vibration. Nevertheless, the terms "vibration" and "oscillation" are often used interchangeably, as is done in this book.

Vibrations can naturally occur in an engineering system and may be representative of its free and natural dynamic behavior. Also, vibrations may be forced onto a system through some form of excitation. The excitation forces may be either generated internally within the dynamic system, or transmitted to the system through an external source. When the frequency of the forcing excitation coincides with that of the natural motion, the system will respond more vigorously with increased amplitude. This condition is known as resonance, and the associated frequency is called the resonant frequency. There are "good vibrations," which serve a useful purpose. Also, there are "bad vibrations," which can be unpleasant or harmful. For many engineering systems, operation at resonance would be undesirable and could be destructive. Suppression or elimination of bad vibrations and generation of desired forms and levels of good vibration are general goals of vibration engineering.

This book deals with

1. Analysis
2. Observation
3. Modification

of vibration in engineering systems. Applications of vibration are found in many branches of engineering such as aeronautical and aerospace, civil, manufacturing, mechanical, and even electrical. Usually, an analytical or computer model is needed to analyze the vibration in an engineering system. Models are also useful in the process of design and development of an engineering system for good performance with respect to vibrations. Vibration monitoring, testing, and experimentation are important as well in the design, implementation, maintenance, and repair of engineering systems. All these are important topics of study in the field of vibration engineering, and the book will cover pertinent

1. Theory and modeling
2. Analysis
3. Design
4. Experimentation
5. Control

In particular, practical applications and design considerations related to modifying the vibrational behavior of mechanical devices and structures will be studied. This knowledge will be useful in the practice of vibration regardless of the application area or the branch of engineering; for example, in the analysis, design, construction, operation, and maintenance of complex structures such as the Space Shuttle and the International Space Station. Note in Figure 1.1 that long and flexible components, which would be prone to complex "modes" of vibration, are present. The structural design should take this into consideration. Also, functional and servicing devices such as robotic manipu-

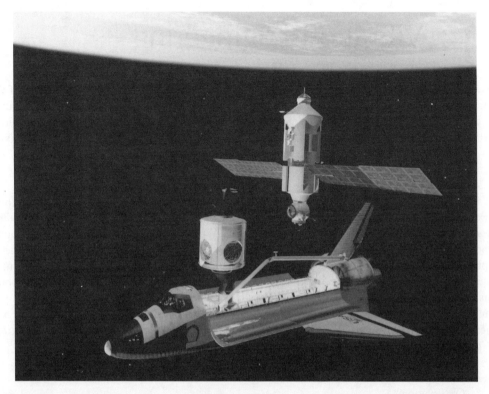

FIGURE 1.1 The U.S. Space Shuttle and the International Space Station with the Canadarm. (Courtesy of NASA Langley Research Center, Hampton, VA. With permission.)

lators (e.g., Canadarm) can give rise to vibration interactions that need to be controlled for accurate performance. The approach used in the book is to introduce practical applications of vibration in the very beginning, along with experimental techniques, and then integrate these applications and design considerations into fundamentals and analytical methods throughout the text.

1.1 STUDY OF VIBRATION

Natural, free vibration is a manifestation of the oscillatory behavior in mechanical systems, as a result of repetitive interchange of kinetic and potential energies among components in the system. Such natural oscillatory response is not limited, however, to purely mechanical systems, and is found in electrical and fluid systems as well, again due to a repetitive exchange of two types of energy among system components. But, purely thermal systems do not undergo free, natural oscillations, primarily because of the absence of two forms of reversible energy. Even a system that can hold two reversible forms of energy may not necessarily display free, natural oscillations. The reason for this would be the strong presence of an energy dissipation mechanism that could use up the initial energy of the system before completing a single oscillation cycle (energy interchange). Such dissipation is provided by damping or friction in mechanical systems, and resistance in electrical systems. Any engineering system (even a purely thermal one) is able to undergo forced oscillations, regardless of the degree of energy dissipation. In this case, the energy necessary to sustain the oscillations will come from the excitation source, and will be continuously replenished.

Proper design and control are crucial in maintaining a high performance level and production efficiency, and prolonging the useful life of machinery, structures, and industrial processes. Before designing or controlling an engineering system for good vibratory performance, it is important to understand, represent (model), and analyze the vibratory characteristics of the system. This can be

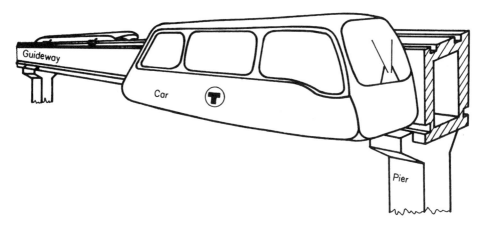

FIGURE 1.2(a) An elevated guideway transit system.

accomplished through purely analytical means, computer analysis of analytical models, testing and analysis of test data, or a combination of these approaches. As an example, a schematic diagram of an innovative elevated guideway transit system is shown in Figure 1.2(a). This is an automated transit system that is operated without drivers. The ride quality, which depends on the vibratory motion of the vehicle, can be analyzed using an appropriate model. Usually, the dynamics (inertia, flexibility, and energy dissipation) of the guideway, as well as the vehicle, must be incorporated into such a model. A simplified model is shown in Figure 1.2(b). It follows that modeling, analysis, testing, design, and control are all important aspects of study in mechanical vibration.

The analysis of a vibrating system can be done either in the time domain or in the frequency domain. In the time domain, the independent variable of a vibration signal is time. In this case, the system itself can be modeled as a set of differential equations with respect to time. A model of a vibrating system can be formulated by applying either force-momentum rate relations (Newton's second law) or the concepts of kinetic and potential energies. Both Newtonian (force-motion) and Lagrangian (energy) approaches will be utilized in this book.

In the frequency domain, the independent variable of a vibration signal is frequency. In this case, the system can be modeled by input-output transfer functions which are algebraic, rather than differential, models. Transfer function representations such as mechanical impedance, mobility, receptance, and transmissibility can be conveniently analyzed in the frequency domain, and effectively used in vibration design and evaluation. Modeling and vibration-signal analysis in both time and frequency domains will be studied in this book. The two domains are connected by the Fourier transformation, which can be treated as a special case of the Laplace transformation. These transform techniques will be studied, first in the purely analytical and analog measurement situation of continuous time. In practice, however, digital electronics and computers are commonly used in signal analysis, sensing, and control. In this situation, one needs to employ concepts of discrete time, sampled data, and digital signal analysis in the time domain. Correspondingly, then, concepts of discrete or digital Fourier transformation and techniques of fast Fourier transform (FFT) will be applicable in the frequency domain. These concepts and techniques are also studied in this book.

An engineering system, when given an initial disturbance and allowed to execute free vibrations without a subsequent forcing excitation, will tend to do so at a particular "preferred" frequency and maintaining a particular "preferred" geometric shape. This frequency is termed a "natural frequency" of the system, and the corresponding shape (or motion ratio) of the moving parts of the system is termed a "mode shape." Any arbitrary motion of a vibrating system can be represented in terms of its natural frequencies and mode shapes. The subject of modal analysis primarily concerns determination of natural frequencies and mode shapes of a dynamic system. Once the

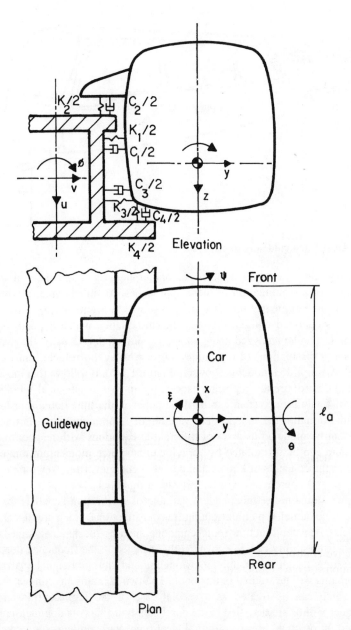

FIGURE 1.2(b) A model for determining the ride quality of the elevated guideway transit system.

modes are determined, they can be used in understanding the dynamic nature of the systems, and also in design and control. Modal analysis is extremely important in vibration engineering, and will be studied in this book. Natural frequencies and mode shapes of a vibrating system can be determined experimentally through procedures of modal testing. In fact, a dynamic model (an experimental model) of the system can be determined in this manner. The subject of modal testing, experimental modeling (or model identification), and associated analysis and design is known as *experimental modal analysis*. This subject will also be treated in this book.

Energy dissipation (or damping) is present in any mechanical system. It alters the dynamic response of the system, and has desirable effects such as stability, vibration suppression, power transmission (e.g., in friction drives), and control. Also, it has obvious undesirable effects such as energy wastage, reduction of the process efficiency, wear and tear, noise, and heat generation. For these reasons, damping is an important topic of study in the area of vibration, and will be covered in this book. In general, energy dissipation is a nonlinear phenomenon. But, in view of well-known difficulties of analyzing nonlinear behavior, and because an equivalent representation of the overall energy dissipation is often adequate in vibration analysis, linear models are primarily used to represent damping in the analyses herein. However, nonlinear representations are discussed as well; and how equivalent linear models can be determined for nonlinear damping are described.

Properties such as mass (inertia), flexibility (spring-like effect), and damping (energy dissipation) are continuously distributed throughout practical mechanical devices and structures to a large extent. This is the case with distributed components such as cables, shafts, beams, membranes, plates, shells, and various solids, as well as structures made of such components. Representation (i.e., modeling) of these distributed-parameter (or continuous) vibrating systems will require independent variables in space (spatial coordinates) in addition to time; these models are partial differential equations in time and space. The analysis of distributed-parameter models will require complex procedures and special tools. This book studies vibration analysis, particularly modal analysis, of several types of continuous components, as well as how approximate lumped-parameter models can be developed for continuous systems, using procedures such as modal analysis and energy equivalence.

Vibration testing is useful in a variety of stages in the development and utilization of a product. In the design and development stage, vibration testing can be used to design, develop, and verify the performance of individual components of a complex system before the overall system is built (assembled) and evaluated. In the production stage, vibration testing can be used for screening of selected batches of products for quality control. Another use of vibration testing is in product qualification. Here, a product of good quality is tested to see whether it can withstand various dynamic environments that it may encounter in a specialized application. An example of a large-scale shaker used for vibration testing of civil engineering structures is shown in Figure 1.3. The subject of vibration testing is addressed in some detail in this book.

Design is a subject of paramount significance in the practice of vibration. In particular, mechanical and structural design for acceptable vibration characteristics will be important. Modification of existing components and integration of new components and devices, such as vibration dampers, isolators, inertia blocks, and dynamic absorbers, can be incorporated into these practices. Furthermore, eliminations of sources of vibration — for example, through component alignment and balancing of rotating devices — is a common practice. Both passive and active techniques are used in vibration control. In passive control, actuators that require external power sources are not employed. In active control, vibration is controlled by means of actuators (which need power) to counteract vibration forces. Monitoring, testing, and control of vibration will require devices such as sensors and transducers, signal conditioning and modification hardware (e.g., filters, amplifiers, modulators, demodulators, analog-digital conversion means), and actuators (e.g., vibration exciters or shakers). The underlying subject of vibration instrumentation will be covered in this book. Particularly, within the topic of signal conditioning, both hardware and software (numerical) techniques will be presented.

1.2 APPLICATION AREAS

The science and engineering of vibration involve two broad categories of applications:

1. Elimination or suppression of undesirable vibrations
2. Generation of the necessary forms and quantities of useful vibrations

Undesirable and harmful types of vibration include structural motions generated due to earthquakes, dynamic interactions between vehicles and bridges or guideways, noise generated by construction equipment, vibration transmitted from machinery to its supporting structures or environment, and

FIGURE 1.3 A multi-degree-of-freedom hydraulic shaker used in testing civil engineering structures. (Courtesy of Prof. C.E. Ventura, University of British Columbia. With permission.)

damage, malfunction, and failure due to dynamic loading, unacceptable motions, and fatigue caused by vibration. As an example, dynamic interactions between an automated transit vehicle and a bridge (see Figure 1.4) can cause structural problems as well as degradation in ride quality. Rigorous analysis and design are needed, particularly with regard to vibration, in the development of these ground transit systems. Lowering the levels of vibration will result in reduced noise and improved work environment, maintenance of a high performance level and production efficiency, reduction in user/operator discomfort, and prolonging the useful life of industrial machinery. Desirable types of vibration include those generated by musical instruments, devices used in physical therapy and medical applications, vibrators used in industrial mixers, part feeders and sorters, and vibratory material removers such as drills and polishers (finishers). For example, product alignment for

Vibration Engineering 7

FIGURE 1.4 The SkyTrain in Vancouver, Canada, a modern automated transit system. (Photo by Mark Van Manen, courtesy of BC Transit. With permission.)

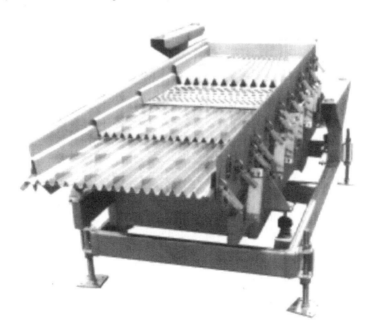

FIGURE 1.5 An alignment shaker. (Key Technology, Inc., of Walla Walla, WA. With permission.)

industrial processing or grading can be carried out by means of vibratory conveyors or shakers, as shown in Figure 1.5.

Concepts of vibration have been used for many centuries in practical applications. Recent advances of vibration are quite significant, and the corresponding applications are numerous. Many of the recent developments in the field of vibration were motivated perhaps for two primary reasons:

1. The speeds of operation of machinery have doubled over the past 50 years and, consequently, the vibration loads generated due to rotational excitations and unbalances would have quadrupled if proper actions of design and control were not taken.
2. Mass, energy, and efficiency considerations have resulted in lightweight, optimal designs of machinery and structures consisting of thin members with high strength. Associated structural flexibility has made the rigid-structure assumption unsatisfactory, and given rise to the need for sophisticated procedures of analysis and design that govern distributed-parameter flexible structures.

One can then visualize several practical applications where modeling, analysis, design, control, monitoring, and testing, related to vibration are important.

A range of applications of vibration can be found in various branches of engineering: particularly civil, mechanical, aeronautical and aerospace, and production and manufacturing. Modal analysis and design of flexible civil engineering structures such as bridges, guideways, tall buildings, and chimneys directly incorporate theory and practice of vibration. A fine example of an elongated building where vibration analysis and design are crucial is the Jefferson Memorial Arch, shown in Figure 1.6.

In the area of ground transportation, vehicles are designed by incorporating vibration engineering, not only to ensure structural integrity and functional operability, but also to achieve required levels of ride quality and comfort. Specifications such as the one shown in Figure 1.7, where limits on root-mean-square (rms) levels of vibration (expressed in units of acceleration due to gravity, g) for different frequencies of excitation (expressed in cycles per second, or hertz, or Hz) and different trip durations, are used to specify ride quality requirements in the design of transit systems. In particular, the design of suspension systems, both active and passive, falls within the field of vibration engineering. Figure 1.8 shows a test setup used in the development of an automotive suspension system. In the area of air transportation, mechanical and structural components of aircraft are designed for good vibration performance. For example, proper design and balancing can reduce helicopter vibrations caused by imbalance in their rotors. Vibrations in ships can be suppressed through structural design, propeller and rudder design, and control. Balancing of internal combustion engines is carried out using principles of design for vibration suppression.

Oscillation of transmission lines of electric power and communication signals (e.g., overhead telephone lines) can result in faults, service interruptions, and sometimes major structural damage. Stabilization of transmission lines involves direct application of the principles of vibration in cables and the design of vibration dampers and absorbers.

In the area of production and manufacturing engineering, mechanical vibration has direct implications of product quality and process efficiency. Machine tool vibrations are known to not only degrade the dimensional accuracy and the finish of a product, but also will cause fast wear and tear and breakage of tools. Milling machines, lathes, drills, forging machines, and extruders, for example, should be designed for achieving low vibration levels. In addition to reducing the tool life, vibration will result in other mechanical problems in production machinery, and will require more frequent maintenance. Associated downtime (production loss) and cost can be quite significant. Also, as noted before, vibrations in production machinery will generate noise problems and also will be transmitted to other operations through support structures, thereby interfering with their performance as well. In general, vibration can degrade performance and production efficiency of

Vibration Engineering

FIGURE 1.6 Jefferson Memorial Arch in St. Louis, MO.

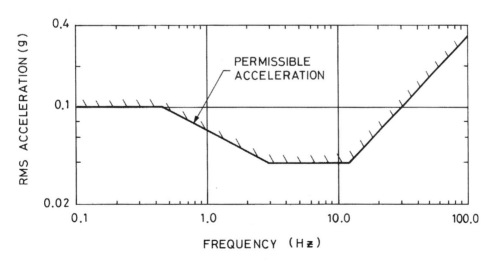

FIGURE 1.7 A typical specification of vehicle ride quality for a specified trip duration.

FIGURE 1.8 Cone suspension system installed on a Volvo 480ES automobile for testing. (Copyright *Mechanical Engineering* magazine; the American Society of Mechanical Engineers International. With permission.)

manufacturing processes. Proper vibration isolation (e.g., mountings) will be needed to reduce these transmissibility problems.

Heavy machinery in the construction industry (e.g., cranes, excavators, pile drivers, impacting and compacting machinery, and bulldozers) rely on structural integrity, reliability, and safety. Their design must be based on sound principles of engineering. Although the dynamic loading in these machines is generally random, it is also quite repetitive from the point of view of both the excitation generated by the engine and the functional operation of the tasks performed. Design based on vibration and fatigue is an important requirement for these machines: for maintaining satisfactory performance, prolonging the useful life, and reducing the cost and frequency of maintenance.

1.3 HISTORY OF VIBRATION

The origins of the theory of vibration can be traced back to the design and development of musical instruments (good vibration). It is known that drums, flutes, and stringed instruments existed in China and India for several millennia B.C. Also, ancient Egyptians and Greeks explored sound and vibration from both practical and analytical points of view. For example, while Egyptians had known of a harp since at least 3000 B.C., the Greek philosopher, mathematician, and musician Pythagoras (of the Pythagoras theorem fame) who lived during 582 to 502 B.C., experimented on sounds generated by blacksmiths and related them to music and physics. The Chinese developed a mechanical seismograph (an instrument to detect and record earthquake vibrations) in the 2nd century A.D.

The foundation of the modern-day theory of vibration was probably laid by scientists and mathematicians such as Robert Hooke (1635–1703) of the Hooke's law fame, who experimented on the vibration of strings; Sir Isaac Newton (1642–1727), who gave us calculus and the laws of motion for analyzing vibrations; Daniel Bernoulli (1700–1782) and Leonard Euler (1707–1783), who studied beam vibrations (Bernoulli-Euler beam) and also explored dynamics and fluid mechanics; Joseph Lagrange (1736–1813), who studied vibration of strings and also explored the energy approach to formulating equations of dynamics; Charles Coulomb (1736–1806), who studied torsional vibrations and friction; Joseph Fourier (1768–1830), who developed the theory of frequency analysis of signals; and Simeon-Dennis Poisson (1781–1840), who analyzed vibration of membranes and also analyzed elasticity (Poisson's ratio). As a result of the industrial revolution and associated developments of steam turbines and other rotating machinery, an urgent need was felt for developments in the analysis, design, measurement, and control of vibration. Motivation for many aspects of the existing techniques of vibration can be traced back to related activities since the industrial revolution.

Much credit should go to scientists and engineers of more recent history, as well. Among the notable contributors are Rankine (1820–1872), who studied critical speeds of shafts; Kirchhoff (1824–1887), who analyzed vibration of plates; Rayleigh (1842–1919), who made contributions to the theory of sound and vibration and developed computational techniques for determining natural vibrations; de Laval (1845–1913), who studied the balancing problem of rotating disks; Poincaré (1854–1912), who analyzed nonlinear vibrations; and Stodola (1859–1943), who studied vibrations of rotors, bearings, and continuous systems. Distinguished engineers who made significant contributions to the published literature and also to the practice of vibration include Timoshenko, Den Hartog, Clough, and Crandall.

1.4 ORGANIZATION OF THE BOOK

This book provides the background and techniques for modeling, analysis, design, instrumentation and monitoring, modification, and control of vibration in engineering systems. This knowledge will be useful in the practice of vibration, regardless of the application area or the branch of engineering. A uniform and coherent treatment of the subject is given by introducing practical applications of vibration in the very beginning of the book, along with experimental techniques and instrumentation, and then integrating these applications, design and experimental techniques, and control considerations into fundamentals and analytical methods throughout the text.

The book consists of 12 chapters and 5 appendices. The chapters have summary boxes for easy reference and recollection. Many worked examples and problems (over 300) are included. Some background material is presented in the appendices, rather than in the main text, in order to avoid interference with the continuity of the subject matter.

The present introductory chapter provides some background material on the subject of vibration engineering, and sets the course for the study. It gives the objectives and motivation of the study and indicates key application areas. A brief history of the field of vibration is given as well.

Chapter 2 provides the basics of time response analysis of vibrating systems. Both undamped and damped systems are studied. Also, analysis of both free (unforced) and forced response is given. The concept of a state variable is introduced. Some analogies of purely mechanical and structural vibrating systems — specifically, translatory, flexural, and torsional; to electrical and fluid oscillatory systems — are introduced. An energy-based approximation of a distributed-parameter system (a heavy spring) to a lumped-parameter system is developed in detail. The logarithmic decrement method of damping measurement is developed. Although the chapter primarily considers single-degree-of-freedom systems, the underlying concepts can be easily extended to multi-degree-of-freedom systems.

Chapter 3 concerns frequency response analysis of vibrating systems. First, the response of a vibrating system to harmonic (sinusoidal) excitation forces (inputs) is analyzed, primarily using the time-domain concepts developed in Chapter 2. Then, its interpretation in the frequency domain is given. The link between the time domain and the frequency domain, through Fourier transform, is highlighted. In particular, Fourier transform is interpreted as a special case of Laplace transform. The response analysis using transform techniques is presented, along with the associated basic ideas of convolution integral, and the impulse response function whose Laplace transform is the transfer function, and Fourier transform is the frequency response function. The half-power bandwidth approach of measuring damping is given. Special types of frequency transfer functions — specifically, force transmissibility, motion transmissibility, and receptance — are studied and their complementary relationships are highlighted. Their use in the practice of vibration, particularly in vibration isolation, is discussed.

Chapter 4 presents the fundamentals of analyzing vibration signals. First, the idea of frequency spectrum of a time signal is given. Various types and classifications of signals encountered in vibration engineering are discussed. The technique of Fourier analysis is formally introduced and linked to the concepts presented in Chapter 3. The idea of random signals is introduced, and useful analytical techniques for these signals are presented. Practical issues pertaining to vibration signal analysis are raised. Computational techniques of signal analysis are given and various sources of error, such as aliasing and truncation, are indicated; and ways of improving the accuracy of digital signal analysis are given.

Chapter 5 deals with the modal analysis of lumped-parameter vibrating systems. The basic assumption made is that distributed effects of inertia and flexibility in a vibrating system can be represented by an interconnected set of lumped inertia and spring elements. The total number of possible independent, incremental motions of these inertia elements is the number of degrees of freedom of the system. For holonomic systems, this is also equal to the total number of independent coordinates needed to represent an arbitrary configuration of the system; but for non-holonomic systems, the required number of coordinates will be larger. For this reason, the concepts of holonomic and non-holonomic systems and the corresponding types of constraints are discussed. The representation of a general lumped-parameter vibrating system by a differential equation model is given, and methods of obtaining such a model are discussed. Apart from the Newtonian and Lagrangian approaches, the influence coefficient approach is given for determining the mass and stiffness matrices. The concepts of natural frequencies and mode shapes are discussed, and the procedure for determining these characteristic quantities, through modal analysis, is developed. The orthogonality property of natural modes is derived. The ideas of static modes and rigid body modes are explored, and the causes of these conditions will be indicated. In addition to the standard formulation of the modal analysis problem, two other modal formulations are developed. The analysis of the problem of forced vibration, using modal analysis, is given. Damped lumped-parameter vibrating systems are studied from the point of view of modal analysis. The conditions of existence of real modes for damped systems are explored, with specific reference to proportional damping. The state-space approach of representing and analyzing a vibrating system is presented. Practical problems of modal analysis are presented.

Chapter 6 studies distributed-parameter vibrating systems such as cables, rods, shafts, beams, membranes, and plates. Practical examples of associated vibration problems are indicated. Vibration of continuous systems is treated as a generalization of lumped-parameter systems, discussed in Chapter 5. In particular, the modal analysis of continuous systems is addressed in detail. The issue of orthogonality of modes is studied. The influence of system boundary conditions on the modal problem in general and the orthogonality in particular is discussed, with special emphasis on "inertial" boundary conditions (e.g., continuous systems with lumped masses at the boundaries). The influence of damping on the modal analysis problem is discussed. The analysis of response to a forcing excitation is performed.

Chapter 7 exclusively deals with the problem of energy dissipation or damping in vibrating systems. Various types of damping present in mechanical and structural systems are discussed, with practical examples, and particular emphasis on interface damping. Methods of representation or modeling of damping in the analysis of vibrating systems are indicated. Techniques and principles of measurement of damping are given, with examples.

Chapter 8 studies instrumentation issues in the practice of vibration. Applications range from monitoring and fault diagnosis of industrial processes, to product testing for quality assessment and qualification, experimental modal analysis for developing experimental models and for designing of vibrating systems, and control of vibration. Instrumentation types, basics of operation, industrial practices pertaining to vibration exciters, control systems, motion sensors and transducers, torque and force sensors, and other types of transducers are addressed. Performance specification of an instrumented system is discussed. Issues and implications of component interconnection in the practical use of instrumentation are addressed.

Chapter 9 addresses signal conditioning and modification for practical vibration systems. These considerations are closely related to the subject of instrumentation discussed in Chapter 8 and signal analysis discussed in Chapter 4. Particular emphasis is given to commercial instruments and hardware that are useful in monitoring, analyzing, and control of vibration. Specific devices considered include amplifiers, analog filters, modulators and demodulators, analog-to-digital converters, digital-to-analog converters, bridge circuits, linearizing devices, and other types of signal modification circuitry. Commercial spectrum analyzers and digital oscilloscopes commonly employed in the practice of vibration are discussed as well.

Chapter 10 deals with vibration testing. This is a practical topic that is directly applicable to product design and development, experimental modeling, quality assessment and control, and product qualification. Various methods of representing a vibration environment in a test program are discussed. Procedures that need to be followed prior to testing an object (i.e., pre-test procedures) are given. Available testing procedures are presented, with a discussion of appropriateness, advantages, and disadvantages of various test procedures. The topic of product qualification testing is addressed in some length.

Chapter 11 studies experimental modal analysis, which is directly related to vibration testing (Chapter 10), experimental modeling, and design. It draws from the analytical procedures presented in previous chapters, particularly Chapters 5 and 6. Frequency domain formulation of the problem is given. The procedure of developing a complete experimental model of a vibrating system is presented. Procedures of curve fitting of frequency transfer functions, which are essential in model parameter extraction, are discussed. Several laboratory experiments in the area of vibration testing (modal testing) are described, giving details of the applicable instrumentation. Features and capabilities of several commercially available experimental modal analysis systems are described, and a comparative evaluation is given.

Chapter 12 addresses practical and analytical issues of vibration design and control. The emphasis here is in the ways of designing, modifying, or controlling a system for good performance with regard to vibration. Ways of specification of vibration limits for proper performance of an engineering system are discussed. Techniques and practical considerations of vibration isolation are described, with an emphasis on the use of transmissibility concepts developed in Chapter 3. Static and dynamic balancing of rotating machinery is studied by presenting both analytical and practical procedures. The related topic of balancing multi-cylinder reciprocating machines is addressed in some detail. The topic of whirling of rotating components and shafts is studied. The subject of design through modal testing, which is directly related to the material in chapters 10 and 11, is discussed. Both passive control and active control of vibration are studied, giving procedures and practical examples.

The background material that is not given in the main body of the text, but is useful in comprehending the underlying procedures, is given in the appendices. Reference is made in the main text to these appendices, for further reading. Appendix A deals with dynamic models and analogies. Main steps of developing analytical models for dynamic systems are indicated. Analogies between mechanical, electrical, fluid, and thermal systems are presented, with particular emphasis on the cause of free natural oscillations. Development procedure of state-space models for these systems is indicated. Appendix B summarizes Newtonian and Lagrangian approaches to writing equations of motion for dynamic systems. Appendix C reviews the basics of linear algebra. Vector-matrix techniques that are useful in vibration analysis and practice are summarized. Appendix D further explores the topic of digital Fourier analysis, with a special emphasis on the computational procedure of fast Fourier transform (FFT). As the background theory, the concepts of Fourier series, Fourier integral transform, and discrete Fourier transform are discussed and integrated, which leads the digital computation of these quantities using FFT. Practical procedures and applications of digital Fourier analysis are given. Appendix E addresses reliability considerations for multicomponent devices. These considerations have a direct relationship to vibration monitoring and testing, failure diagnosis, product qualification, and design optimization.

PROBLEMS

1.1 Explain why mechanical vibration is an important area of study for engineers. Mechanical vibrations are known to have harmful effects as well as useful ones. Briefly describe five practical examples of *good* vibrations and also five practical examples of *bad* vibrations.

1.2 Under some conditions it may be necessary to modify or redesign a machine with respect to its performance under vibrations. What are possible reasons for this? What are some of the modifications that can be carried out on a machine in order to suppress its vibrations?

1.3 On the one hand, modern machines are designed with sophisticated procedures and computer tools, and should perform better than the older designs, with respect to mechanical vibration. On the other hand, modern machines have to operate under more stringent specifications and requirements in a somewhat optimal fashion. In general, design for satisfactory performance under vibration takes an increased importance for modern machinery. Indicate some reasons for this.

1.4 Dynamic modeling — both analytical and experimental (e.g., experimental modal analysis) — is quite important in the design and development of a product, for good performance with regard to vibration. Indicate how a dynamic model can be utilized in the vibration design of a device.

1.5 Outline one practical application of mechanical vibration in each of the following branches of engineering:

1. Civil engineering
2. Aeronautical and aerospace engineering
3. Mechanical engineering
4. Manufacturing engineering
5. Electrical engineering

REFERENCES AND FURTHER READING

The book has relied on many publications, directly and indirectly, in its evolution and development. The author's own work as well as other excellent books have provided a wealth of knowledge. Although it is not possible or useful to list all such material, some selected publications are listed below.

AUTHOR'S WORK

1. De Silva, C.W., *Dynamic Testing and Seismic Qualification Practice*, D.C. Heath and Co., Lexington, MA, 1983.
2. De Silva C.W. and Wormley, D.N., *Automated Transit Guideways: Analysis and Design*, D.C. Heath and Co., Lexington, MA, 1983.
3. De Silva, C.W., *Control Sensors and Actuators*, Prentice-Hall, Englewood Cliffs, NJ, 1989.
4. De Silva, C.W., *Control System Modeling*, Measurements and Data Corp., Pittsburgh, PA, 1989.
5. De Silva, C.W., A technique to model the simply supported timoshenko beam in the design of mechanical vibrating systems, *International Journal of Mechanical Sciences*, 17, 389-393, 1975.
6. Van de Vegte, J. and de Silva, C.W., Design of passive vibration controls for internally damped beams by modal control techniques, *Journal of Sound and Vibration*, 45(3), 417-425, 1976.
7. De Silva, C.W., Optimal estimation of the response of internally damped beams to random loads in the presence of measurement noise, *Journal of Sound and Vibration*, 47(4), 485-493, 1976.

8. De Silva, C.W., Dynamic beam model with internal damping, rotatory inertia and shear deformation, *AIAA Journal*, 14(5), 676-680, 1976.
9. De Silva, C.W. and Wormley, D.N., Material optimization in a torsional guideway transit system, *Journal of Advanced Transportation*, 13(3), 41-60, 1979.
10. De Silva, C.W., Buyukozturk, O., and Wormley, D.N., Postcracking compliance of RC beams, *Journal of the Structural Division, Trans. ASCE*, 105(ST1), 35-51, 1979.
11. De Silva, C.W., Seismic qualification of electrical equipment using a uniaxial test, *Earthquake Engineering and Structural Dynamics*, 8, 337-348, 1980.
12. De Silva, C.W., Loceff, F., and Vashi, K.M., Consideration of an optimal procedure for testing the operability of equipment under seismic disturbances, *Shock and Vibration Bulletin*, 50(5), 149-158, 1980.
13. De Silva, C.W. and Wormley, D.N., Torsional analysis of cutout beams, *Journal of the Structural Division*, Trans. ASCE, 106(ST9), 1933-1946, 1980.
14. De Silva, C.W., An algorithm for the optimal design of passive vibration controllers for flexible systems, *Journal of Sound and Vibration*, 74(4), 495-502, 1982.
15. De Silva, C.W., Matrix eigenvalue problem of multiple-shaker testing, *Journal of the Engineering Mechanics Division*, Trans. ASCE, 108(EM2), 457-461, 1982.
16. De Silva, C.W., Selection of shaker specifications in seismic qualification tests, *Journal of Sound and Vibration*, 91(2), 21-26, 1983.
17. De Silva, C.W., Shaker test-fixture design, *Measurements and Control*, 17(6), 152-155, 1983.
18. De Silva, C.W., On the modal analysis of discrete vibratory systems, *International Journal of Mechanical Engineering Education*, 12(1), 35-44, 1984.
19. De Silva, C.W. and Palusamy, S.S., Experimental modal analysis — A modeling and design tool, *Mechanical Engineering*, ASME, 106(6), 56-65, 1984.
20. De Silva, C.W., A dynamic test procedure for improving seismic qualification guidelines, *Journal of Dynamic Systems, Measurement, and Control*, Trans. ASME, 106(2), 143-148, 1984.
21. De Silva, C.W., Hardware and software selection for experimental modal analysis, *The Shock and Vibration Digest*, 16(8), 3-10, 1984.
22. De Silva, C.W., Computer-automated failure prediction in mechanical systems under dynamic loading, *The Shock and Vibration Digest*, 17(8), 3-12, 1985.
23. De Silva, C.W., Henning, S.J., and Brown, J.D., Random testing with digital control — Application in the distribution qualification of microcomputers, *The Shock and Vibration Digest*, 18(9), 3-13, 1986.
24. De Silva, C.W., The digital processing of acceleration measurements for modal analysis, *The Shock and Vibration Digest*, 18(10), 3-10, 1986.
25. De Silva, C.W., Price, T.E., and Kanade, T., A torque sensor for direct-drive manipulators, *Journal of Engineering for Industry*, Trans. ASME, 109(2), 122-127, 1987.
26. De Silva, C.W., Optimal input design for the dynamic testing of mechanical systems, *Journal of Dynamic Systems, Measurement, and Control*, Trans. ASME, 109(2), 111-119, 1987.
27. De Silva, C.W., Singh, M., and Zaldonis, J., Improvement of response spectrum specifications in dynamic testing, *Journal of Engineering for Industry*, Trans. ASME, 112(4), 384-387, 1990.
28. De Silva, C.W., Schultz, M., and Dolejsi, E., Kinematic analysis and design of a continuously-variable transmission, *Mechanism and Machine Theory*, 29(1), 149-167, 1994.
29. Bussani, F. and de Silva, C.W., Use of finite element method to model machine processing of fish, *Finite Element News*, 5, 36-42, 1994.
30. Caron, M., Modi, V.J., Pradhan, S., de Silva, C.W., and Misra, A.K., Planar dynamics of flexible manipulators with slewing deployable links, *Journal of Guidance, Control, and Dynamics*, 21(4), 572-580, 1998.

OTHER USEFUL PUBLICATIONS

1. Beards, C.F., *Engineering Vibration Analysis with Application to Control Systems*, Halsted Press, New York, 1996.
2. Bendat, J.S. and Piersol, A.G., *Random Data: Analysis and Measurement Procedures*, Wiley-Interscience, New York, 1971.
3. Blevins, R.D., *Flow-Induced Vibration*, Van Nostrand Reinhold, New York, 1977.
4. Brigham, E.O., *The Fast Fourier Transform*, Prentice-Hall, Englewood Cliffs, NJ, 1974.

5. Broch, J.T., *Mechanical Vibration and Shock Measurements*, Bruel and Kjaer, Naerum, Denmark, 1980.
6. Buzdugan, G., Mihaiescu, E., and Rades, M., *Vibration Measurement*, Martinus Nijhoff Publishers, Dordrecht, The Netherlands, 1986.
7. Crandall, S.H., Karnopp, D.C., Kurtz, E.F., and Prodmore-Brown, D.C., *Dynamics of Mechanical and Electromechanical Systems*, McGraw-Hill, New York, 1968.
8. Den Hartog, J.P., *Mechanical Vibrations*, McGraw-Hill, New York, 1956.
9. Dimarogonas, A., *Vibration for Engineers*, 2nd edition, Prentice-Hall, Upper Saddle River, NJ, 1996.
10. Ewins, D.J., *Modal Testing: Theory and Practice*, Research Studies Press Ltd., Letchworth, England, 1984.
11. Inman, D.J., *Engineering Vibration*, Prentice-Hall, Englewood Cliffs, NJ, 1996.
12. Irwin, J.D. and Graf, E.R., *Industrial Noise and Vibration Control*, Prentice-Hall, Englewood Cliffs, NJ, 1979.
13. McConnell, K.G., *Vibration Testing*, John Wiley & Sons, New York, 1995.
14. Meirovitch, L., *Computational Methods in Structural Dynamics*, Sijthoff & Noordhoff, Rockville, MD, 1980.
15. Meirovitch, L., *Elements of Vibration Analysis*, 2nd edition, McGraw-Hill, New York, 1986.
16. Randall, R.B., *Application of B&K Equipment to Frequency Analysis*, Bruel and Kjaer, Naerum, Denmark, 1977.
17. Rao, S.S., *Mechanical Vibrations*, 3rd edition, Addison-Wesley, Reading, MA, 1995.
18. Shearer, J.L. and Kulakowski, B.T., *Dynamic Modeling and Control of Engineering Systems*, MacMillan Publishing, New York, 1990.
19. Shearer, J.L., Murphy, A.T., and Richardson, H.H., *Introduction to System Dynamics*, Addison-Wesley, Reading, MA, 1971.
20. Steidel, R.F., *An Introduction to Mechanical Vibrations*, 2nd edition, John Wiley & Sons, New York, 1979.
21. Volterra, E. and Zachmanoglou, E.C., *Dynamics of Vibrations*, Charles E. Merrill Books, Columbus, OH, 1965.

2 Time Response

Vibrations are oscillatory responses of dynamic systems. Natural vibrations occur in these systems due to the presence of two modes of energy storage. Specifically, when the stored energy is converted from one form to the other, repeatedly back and forth, the resulting time response of the system is oscillatory in nature. In a mechanical system, natural vibrations can occur because kinetic energy, which is manifested as velocities of mass (inertia) elements, can be converted into potential energy (which has two basic types: elastic potential energy due to the deformation in spring-like elements, and gravitational potential energy due to the elevation of mass elements against the Earth's gravitational pull) and back to kinetic energy, repetitively, during motion. Similarly, natural oscillations of electrical signals occur in circuits due to the presence of electrostatic energy (of the electric charge storage in capacitor-like elements) and electromagnetic energy (due to the magnetic fields in inductor-like elements). Fluid systems can also exhibit natural oscillatory responses as they possess two forms of energy. But purely thermal systems do not produce natural oscillations because they, as far as anyone knows, have only one type of energy. These ideas are summarized in Appendix A. Note, however, that an oscillatory forcing function is able to make a dynamic system respond with an oscillatory motion (usually at the same frequency as the forcing excitation) even in the absence of two forms of energy storage. Such motions are forced responses rather than natural or free responses. This book concerns vibrations in mechanical systems. Nevertheless, clear analogies exist with electrical and fluid systems as well as mixed systems such as electromechanical systems.

Mechanical vibrations can occur as both free (natural) responses and forced responses in numerous practical situations. Some of these vibrations are desirable and useful, and others are undesirable and should be avoided or suppressed. The sound that is generated after a string of a guitar is plucked is a free vibration, while the sound of a violin is a mixture of both free and forced vibrations. These sounds are generally pleasant and desirable. The response of an automobile after it hits a road bump is an undesirable free vibration. The vibrations felt while operating a concrete drill are desirable for the drilling process itself, but are undesirable forced vibrations for the human who operates the drill. In the design and development of a mechanical system, regardless of whether it is intended for generating desirable vibrations or for operating without vibrations, an analytical model of the system can serve a very useful function. The model will represent the dynamic system, and can be analyzed and modified more quickly and cost effectively than one could build and test a physical prototype. Similarly, in the control or suppression of vibrations, it is possible to design, develop, and evaluate vibration isolators and control schemes through analytical means before they are physically implemented. It follows that analytical models (see Appendix A) are useful in the analysis, control, and evaluation of vibrations in dynamic systems, and also in the design and development of dynamic systems for desired performance in vibration environments.

An analytical model of a mechanical system is a set of equations, and can be developed either by the Newtonian approach where Newton's second law is explicitly applied to each inertia element, or by the Lagrangian or Hamiltonian approach, which is based on the concepts of energy (kinetic and potential energies). These approaches are summarized in Appendix B. A time-domain analytical model is a set of differential equations, with respect to the independent variable time (t). A frequency-domain model is a set of input-output transfer functions with respect to the independent variable frequency (ω). The time response will describe how the system moves (responds) as a function of time. Both free and forced responses are useful. The frequency response will describe the way the system moves when excited by a harmonic (sinusoidal) forcing input, and is a function of the frequency of excitation. This chapter introduces some basic concepts of vibration analysis

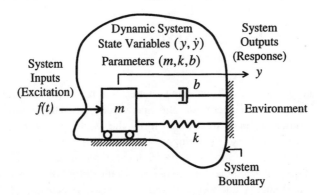

FIGURE 2.1 A mechanical dynamic system.

using time-domain methods. The frequency-domain analysis will be studied in subsequent chapters (Chapters 3 and 4, in particular).

2.1 UNDAMPED OSCILLATOR

Consider the mechanical system that is schematically shown in Figure 2.1. The inputs (or excitation) applied to the system are represented by the force $f(t)$. The outputs (or response) of the system are represented by the displacement y. The system boundary demarcates the region of interest in this analysis. This boundary could be an imaginary one. What is outside the system boundary is the environment in which the system operates. An analytical model of the system can be given by one or more equations relating the outputs to the inputs. If the rates of changes of the response (outputs) are not negligible, the system is a *dynamic* system. In this case, the analytical model in the time domain becomes one or more differential equations rather than algebraic equations. System parameters (e.g., mass, stiffness, damping constant) are represented in the model, and their values should be known in order to determine the response of the system to a particular excitation. State variables are a minimum set of variables that completely represent the dynamic state of a system at any given time t. These variables are not unique (more than one choice of a valid set of state variables is possible). The concepts of state variables and state models are introduced in Appendix A and also in this chapter. For a simple oscillator (a single-degree-of-freedom mass-spring-damper system as in Figure 2.1), an appropriate set of state variables would be the displacement y and the velocity $\dot{y}$. An alternative set would be $\dot{y}$ and the spring force.

This chapter provides an introduction to the response analysis of mechanical vibrating systems in the time domain. In this introductory chapter, single-degree-of-freedom systems that require only one coordinate (or one independent displacement variable) in their model, are considered almost exclusively. Higher-degree-of-freedom systems will be analyzed elsewhere in the book (e.g., Chapter 5). Mass (inertia) and spring are the two basic energy storage elements in a mechanical vibrating system. A mass can store gravitational potential energy as well when located against a gravitational force. These elements are analyzed first. In a practical system, mass and stiffness properties can be distributed (continuous) throughout the system. But in this present analysis, lumped-parameter models are employed where inertia, flexibility, and damping effects are separately lumped into single parameters, with a single geometric coordinate used to represent the location of each lumped inertia.

This chapter section first shows that many types of oscillatory systems can be represented by the equation of an undamped simple oscillator. In particular, mechanical, electrical, and fluid systems are considered. Please refer to Appendix A for some foundation material on this topic. The conservation of energy is a straightforward approach for deriving the equations of motion for undamped oscillatory systems (or conservative systems). The equations of motion for mechanical systems can

Time Response

FIGURE 2.2 A mass element.

be derived using the free-body diagram approach with the direct application of Newton's second law. An alternative and rather convenient approach is the use of Lagrange equations, as described in Appendix B. The natural (free) response of an undamped simple oscillator is a *simple harmonic motion*. This is a periodic, sinusoidal motion. This simple time response is also discussed.

2.1.1 ENERGY STORAGE ELEMENTS

Mass (inertia) and spring are the two basic energy storage elements in mechanical systems. The concept of state variables can be introduced as well through these elements (see Appendix A for details), and will be introduced along with associated energy and state variables.

Inertia (*m*)

Consider an inertia element of lumped mass m, excited by force f, as shown in Figure 2.2. The resulting velocity is v.

Newton's second law gives

$$m\frac{dv}{dt} = f \tag{2.1}$$

Kinetic energy stored in the mass element is equal to the work done by the force f on the mass. Hence,

$$\text{Energy} \qquad E = \int f dx = \int f \frac{dx}{dt} dt = \int f v dt = \int \frac{mdv}{dt} v dt = m \int v dv$$

or

$$\text{Kinetic energy} \qquad KE = \frac{1}{2} mv^2 \tag{2.2}$$

Note: v is an appropriate state variable for a mass element because it can completely represent the energy of the element.

Integrate equation (2.1) from a time instant immediately before $t = 0$ (i.e., $t = 0^-$).

$$v(t) = v(0^-) + \frac{1}{m} \int_{0^-}^{t} f dt \tag{2.3}$$

Hence, with $t = 0^+$, for a time instant immediately after $t = 0$, one obtains

$$v(0^+) = v(0^-) + \frac{1}{m} \int_{0^-}^{0^+} f dt \tag{2.4}$$

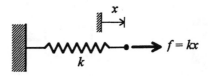

FIGURE 2.3 A spring element.

Since the integral of a finite quantity over an almost zero time interval is zero, these results imply that a finite force will not cause an instantaneous change in velocity in an inertia element. In particular, for a mass element subjected to finite force, since the integral on the RHS of equation (2.4) is zero, one obtains

$$v(0^+) = v(0^-) \tag{2.5}$$

Spring (k)

Consider a massless spring element of lumped stiffness k, as shown in Figure 2.3. One end of the spring is fixed and the other end is free. A force f is applied at the free end, which results in a displacement (extension) x in the spring.

Hooke's law gives

$$f = kx \quad \text{or} \quad \frac{df}{dt} = kv \tag{2.6}$$

Elastic potential energy stored in the spring is equal to the work done by the force on the spring. Hence,

$$\text{Energy}: \quad E = \int f \, dx = \int kx \, dx = \tfrac{1}{2} kx^2$$

$$= \int f \frac{dx}{dt} dt = \int f v \, dt = \int f \frac{1}{k} \frac{df}{dt} dt = \frac{1}{k} \int f \, df = \frac{1}{2k} f^2$$

or

$$\text{Elastic potential energy } PE = \frac{1}{2} kx^2 = \frac{1}{2} \frac{f^2}{k} \tag{2.7}$$

Note: f is an appropriate state variable for a spring, and so is x, because they can completely represent the energy in the spring.

Integrate equation (2.6).

$$f(t) = f(0^-) + \frac{1}{k} \int_{0^-}^{t} v \, dt \tag{2.8}$$

Set $t = 0^+$. Then,

$$f(0^+) = f(0^-) + \frac{1}{k} \int_{0^-}^{0^+} v \, dt \tag{2.9}$$

Time Response

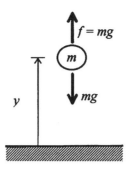

FIGURE 2.4 A mass element subjected to gravity.

From these results, it follows that at finite velocities, there cannot be an instantaneous change in the force of a spring. In particular, from equation (2.9) one sees that at finite velocities of a spring

$$f(0^+) = f(0^-) \tag{2.10}$$

Also, it follows that

$$x(0^+) = x(0^-) \tag{2.11}$$

Gravitational Potential Energy

The work done in raising an object against the gravitational pull is stored as gravitational potential energy of the object. Consider a lumped mass m, as shown in Figure 2.4, that is raised to a height y from some reference level. The work done gives

$$\text{Energy}: E = \int f dy = \int mg dy$$

Hence,

$$\text{Gravitational potential energy}: PE = mgy \tag{2.12}$$

2.1.2 Conservation of Energy

There is no energy dissipation in undamped systems, which contain energy storage elements only. In other words, energy is conserved in these systems, which are known as conservative systems. For mechanical systems, conservation of energy gives

$$KE + PE = const \tag{2.13}$$

These systems tend to be oscillatory in their natural motion, as noted before. Also, as discussed in Appendix A, analogies exist with other types of systems (e.g., fluid and electrical systems). Consider the six systems sketched in Figure 2.5.

System 1 (Translatory)

Figure 2.5 (a) shows a translatory mechanical system (an undamped oscillator) that has just one degree of freedom x. This can represent a simplified model of a rail car that is impacting against a snubber. The conservation of energy (equation (2.13)) gives

$$\frac{1}{2}m\dot{x}^2 + \frac{1}{2}kx^2 = const \tag{2.14}$$

Here, m is the mass and k is the spring stiffness. Differentiate equation (2.14) with respect to time t to obtain

$$m\dot{x}\ddot{x} + kx\dot{x} = 0$$

Since $\dot{x} \neq 0$ at all t, in general, one can cancel it out. Hence, by the method of conservation of energy, one obtains the equation of motion

$$\ddot{x} + \frac{k}{m}x = 0 \tag{2.15}$$

System 2 (Rotatory)

Figure 2.5(b) shows a rotational system with the single degree of freedom θ. It may represent a simplified model of a motor drive system. As before, the conservation energy gives

$$\frac{1}{2}J\dot{\theta}^2 + \frac{1}{2}K\theta^2 = const \tag{2.16}$$

In this equation, J is the moment of inertia of the rotational element and K is the torsional stiffness of the shaft. Then, by differentiating equation (2.16) with respect to t and canceling $\dot{\theta}$, one obtains the equation of motion

$$\ddot{\theta} + \frac{K}{J}\theta = 0 \tag{2.17}$$

System 3 (Flexural)

Figure 2.5(c) is a lateral bending (flexural) system, which is a simplified model of a building structure. Again, a single degree of freedom x is assumed. Conservation of energy gives

$$\frac{1}{2}m\dot{x}^2 + \frac{1}{2}kx^2 = const \tag{2.18}$$

Here, m is the lumped mass at the free end of the support and k is the lateral bending stiffness of the support structure. Then, as before, the equation of motion becomes

$$\ddot{x} + \frac{k}{m}x = 0 \tag{2.19}$$

System 4 (Swinging)

Figure 2.5(d) shows a simple pendulum. It may represent a swinging-type building demolisher or a skilift and has a single-degree-of-freedom θ. Thus,

$$KE = \frac{1}{2}m(l\dot{\theta})^2$$

Gravitational $\quad PE = E_{ref} - mgl\cos\theta$

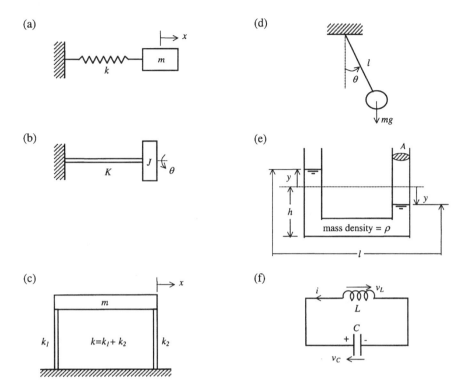

FIGURE 2.5 Six examples of single D.O.F. oscillatory systems: (a) translatory, (b) rotatory, (c) flexural, (d) swinging, (e) liquid slosh, and (f) electrical.

Here, m is the pendulum mass, l is the pendulum length, g is the acceleration due to gravity, and E_{ref} is the *PE* at the reference point, which is a constant. Hence, conservation of energy gives

$$\frac{1}{2}ml^2\dot{\theta}^2 - mgl\cos\theta = const \tag{2.20}$$

Differentiate with respect to t after canceling the common ml:

$$l\dot{\theta}\ddot{\theta} + g\sin\theta\,\dot{\theta} = 0$$

Since $\dot{\theta} \neq 0$ at all t, the equation of motion becomes

$$\ddot{\theta} + \frac{g}{l}\sin\theta = 0 \tag{2.21}$$

This system is nonlinear, in view of the term $\sin\theta$. For small θ, $\sin\theta$ is approximately equal to θ. Hence, the linearized equation of motion is

$$\ddot{\theta} + \frac{g}{l}\theta = 0 \tag{2.22}$$

System 5 (Liquid Slosh)

Consider a liquid column system shown in Figure 2.5(e). It may represent two liquid tanks linked by a pipeline. The system parameters are

Area of cross section of each column = A
Mass density of liquid = ρ
Length of liquid mass = l

Then,

$$KE = \frac{1}{2}(\rho l A)\dot{y}^2$$

$$\text{Gravitational} \quad PE = \rho A(h+y)\frac{g}{2}(h+y) + \rho A(h-y)\frac{g}{2}(h-y)$$

Note that the center of gravity of each column is used in expressing the gravitational *PE*. Hence, conservation of energy gives

$$\frac{1}{2}\rho l A \dot{y}^2 + \frac{1}{2}\rho A g(h+y)^2 + \frac{1}{2}\rho A g(h-y)^2 = const \qquad (2.23)$$

Differentiate:

$$l\dot{y}\ddot{y} + g(h+y)\dot{y} - g(h-y)\dot{y} = 0$$

But, one has

$$\dot{y} \neq 0 \text{ for all } t.$$

Hence,

$$\ddot{y} + g(h+y) - g(h-y) = 0$$

or

$$\ddot{y} + \frac{2g}{l}y = 0 \qquad (2.24)$$

System 6 (Electrical)

Figure 2.5(f) shows an electrical circuit with a single capacitor and a single inductor. Again, conservation of energy can be used to derive the equation of motion. First, an alternative is given.
Voltage balance gives

$$v_L + v_C = 0 \qquad (2.25)$$

where v_L and v_C are the voltages across the inductor and the capacitor, respectively.

Time Response

The constitutive equation for the inductor is

$$L\frac{di}{dt} = v_L \tag{2.26}$$

The constitutive equation for the capacitor is

$$C\frac{dv_C}{dt} = i \tag{2.27}$$

Hence, by differentiating equation (2.26), substituting equation (2.25), and using equation (2.27), one obtains

$$L\frac{d^2i}{dt^2} = \frac{dv_L}{dt} = -\frac{dv_C}{dt} = -\frac{i}{C}$$

or

$$LC\frac{d^2i}{dt^2} + i = 0 \tag{2.28}$$

Now consider the energy conservation approach for this electrical circuit, which will give the same result. Note that power is given by the product vi.

Capacitor

$$\text{Electrostatic energy}: E = \int vi\,dt = \int vC\frac{dv}{dt}dt = C\int v\,dv = \frac{Cv^2}{2} \tag{2.29}$$

Here, v denotes v_C. Also,

$$v = \frac{1}{C}\int i\,dt \tag{2.30}$$

Since the current i is finite for a practical circuit, then $\int_{0^-}^{0^+} i\,dt = 0$.

Hence, in general, the voltage across a capacitor cannot change instantaneously. In particular,

$$v(0^+) = v(0^-) \tag{2.31}$$

Inductor

$$\text{Electromagnetic energy } E = \int vi\,dt = \int L\frac{di}{dt}i\,dt = L\int i\,di = \frac{Li^2}{2}$$

Here, v denotes v_L. Also,

$$i = \frac{1}{L}\int v\,dt \tag{2.32}$$

Since v is finite in a practical circuit, then $\int_{0^-}^{0^+} v\,dt = 0$.

Hence, in general, the current through an inductor cannot change instantaneously. In particular,

$$i(0^+) = i(0^-) \tag{2.33}$$

Since the circuit in Figure 2.5(f) does not have a resistor, there is no energy dissipation. As a result, conservation energy gives

$$\frac{Cv^2}{2} + \frac{Li^2}{2} = const \tag{2.34}$$

Differentiate equation (2.34) with respect to t.

$$Cv\frac{dv}{dt} + Li\frac{di}{dt} = 0$$

Note that $v = v_c$ in this equation.

Substitute the capacitor constitutive equation (2.27).

$$iv + Li\frac{di}{dt} = 0$$

Since $i \neq 0$ in general, one can cancel it. Now, by differentiating equation (2.27), one has $\frac{di}{dt} = C\frac{d^2v}{dt^2}$.
Substitute this in the above equation to obtain

$$LC\frac{d^2v}{dt^2} + v = 0 \tag{2.35}$$

Similarly, one obtains

$$LC\frac{d^2i}{dt^2} + i = 0 \tag{2.36}$$

2.1.3 Free Response

Note that the equation of free (i.e., no excitation force) motion of the six linear systems considered above (Figure 2.5) is of the same general form

$$\ddot{x} + \omega_n^2 x = 0 \tag{2.37}$$

Time Response

This is the equation of an undamped, simple oscillator. For a mechanical system of mass m and stiffness k,

$$\omega_n = \sqrt{\frac{k}{m}} \tag{2.38}$$

To determine the time response x of this system, use the trial solution

$$x = A \sin(\omega_n t + \phi) \tag{2.39}$$

in which A and ϕ are unknown constants, to be determined by the initial conditions (for x and $\dot{x}$); say,

$$x(0) = x_o, \quad \dot{x}(0) = v_o \tag{2.40}$$

Substitute the trial solution into equation (2.37) and obtain

$$\left(-A\omega_n^2 + A\omega_n^2\right)\sin(\omega_n t + \phi) = 0$$

This equation is identically satisfied for all t. Hence, the general solution of equation (2.37) is indeed equation (2.39), which is periodic and sinusoidal.

This response is sketched in Figure 2.6. Note that this sinusoidal oscillatory motion has a *frequency* of oscillation of ω rad/s). Hence, a system that provides this type of natural motion is called a *simple oscillator*. In other words, the response exactly repeats itself in time periods of T, corresponding to a cyclic frequency $f = \dfrac{1}{T}$ (Hz). The frequency ω is in fact the *angular frequency* given by $\omega = 2\pi f$. Also, the response has an *amplitude A*, which is the peak value of the sinusoidal response. Now, suppose that the response curve is shifted to the right through ϕ/ω. Consider the resulting curve to be the reference signal (with signal value = 0 at $t = 0$, and increasing). It should be clear that the response shown in Figure 2.6 leads the reference signal by a time period of ϕ/ω. This can be verified from the fact that the value of the reference signal at time t is the same as that of the signal in Figure 2.6 at time $t - \phi/\omega$. Hence, ϕ is termed the *phase angle* of the response, and it represents a phase lead.

The left-hand portion of Figure 2.6 is the phasor representation of a sinusoidal response. In this representation, an arm of length A rotates in the counterclockwise direction at angular speed ω. This is the phasor. The arm starts at an angular position ϕ from the horizontal axis, at time $t = 0$. The projection of the arm onto the vertical (x) axis is the time response. In this manner, the phasor representation can conveniently indicate the amplitude, frequency, phase angle, and the actual time response (at any time t) of a sinusoidal motion.

As noted previously, a repetitive (periodic) motion of the type (2.39) is called *simple harmonic motion*, meaning it is a pure sinusoidal oscillation at a single frequency.

Next, it is shown that the amplitude A and the phase angle ϕ both depend on the initial conditions. Substitute the ICs (2.40) into equation (2.39) and its time derivative to get

$$x_o = A \sin \phi \tag{2.41}$$

$$v_o = A\omega_n \cos \phi \tag{2.42}$$

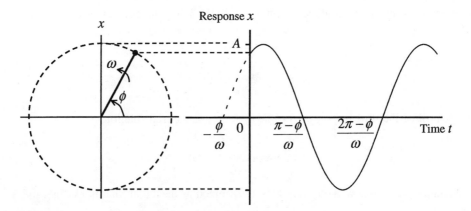

FIGURE 2.6 Free response of an undamped simple oscillator.

Now divide equation (2.41) by (2.42), and also use the fact that $\sin^2\phi + \cos^2\phi = 1$ and obtain

$$\tan\phi = \omega_n \frac{x_o}{v_o}$$

$$\left(\frac{x_o}{A}\right)^2 + \left(\frac{v_o}{A\omega_n}\right)^2 = 1$$

Hence,

$$\text{Amplitude}: A = \sqrt{x_o^2 + \frac{v_o^2}{\omega_n^2}} \qquad (2.43)$$

$$\text{Phase}: \phi = \tan^{-1}\frac{\omega_n x_o}{v_o} \qquad (2.44)$$

EXAMPLE 2.1

A simple model for a tracked gantry conveyor system in a factory is shown in Figure 2.7.

The carriage of mass (m) moves on a frictionless track. The pulley is supported on frictionless bearings, and its axis of rotation is fixed. Its moment of inertia about this axis is J. The motion of the carriage is restrained by a spring of stiffness k_1, as shown. The belt segment that drives the carriage runs over the pulley without slip, and is attached at the other end to a fixed spring of stiffness k_2. The displacement of the mass is denoted by x and the corresponding rotation of the pulley is denoted by θ. When $x = 0$ (and $\theta = 0$), the springs k_1 and k_2 have an extension of x_{10} and x_{20}, respectively, from their unstretched (free) configurations. Assume that the springs will remain in tension throughout the motion of the system.

   a. Using Newton's second law, first principles, and free-body diagrams, develop an equivalent equation of motion for this system in terms of the response variable x. What is the equivalent mass, and what is the equivalent stiffness of the system?
   b. Verify the result in part (a) using the energy method.

Time Response

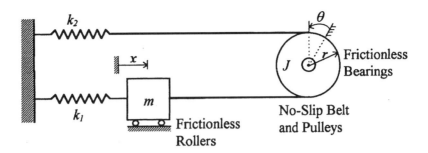

FIGURE 2.7 A tracked conveyor system.

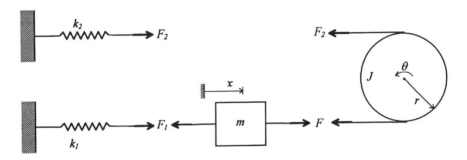

FIGURE 2.8 A free-body diagram for the conveyor system.

c. What is the natural frequency of vibration of the system?
d. Express the equation of the system in terms of the rotational response variable θ.

What is the natural frequency of vibration corresponding to this rotational form of the system equation?

What is the equivalent moment of inertia and the equivalent torsional stiffness of the rotational form of the system?

Solution

A free-body diagram for the system is shown in Figure 2.8.

a. Hooke's law for the spring elements:

$$F_1 = k_1(x_{10} + x) \tag{i}$$

$$F_2 = k_2(x_{20} - x) \tag{ii}$$

Newton's second law for the inertia elements:

$$m\ddot{x} = F - F_1 \tag{iii}$$

$$J\ddot{\theta} = rF_2 - rF \tag{iv}$$

Compatibility:

$$x = r\theta \quad (v)$$

Straightforward elimination of F_1, F_2, F, and θ in (i) to (v), using algebra, gives

$$\left(m + \frac{J}{r^2}\right)\ddot{x} + (k_1 + k_2)x = k_2 x_{20} - k_1 x_{10} \quad (vi)$$

It follows that

$$\text{Equivalent mass}: m_{eq} = m + \frac{J}{r^2}$$

$$\text{Equivalent stiffness}: k_{eq} = k_1 + k_2$$

b. Total energy in the system:

$$E = \frac{1}{2}m\dot{x}^2 + \frac{1}{2}J\dot{\theta}^2 + \frac{1}{2}k_1(x_{10} + x)^2 + \frac{1}{2}k_2(x_{20} - x)^2 = \text{constant}$$

Differentiate w.r.t. time:

$$m\dot{x}\ddot{x} + J\dot{\theta}\ddot{\theta} + k_1(x_{10} + x)\dot{x} - k_2(x_{20} - x)\dot{x} = 0$$

Substitute the compatibility relation, $\dot{x} = r\dot{\theta}$, to get

$$m\dot{x}\ddot{x} + \frac{J}{r^2}\dot{x}\ddot{x} + k_1(x_{10} + x)\dot{x} - k_2(x_{20} - x)\dot{x} = 0$$

Eliminate the common velocity variable $\dot{x}$ (which cannot be zero for all t). Obtain

$$\left(m + \frac{J}{r^2}\right)\ddot{x} + (k_1 + k_2)x = k_2 x_{20} - k_1 x_{10}$$

which is the same result as before.

c. The natural frequency (undamped) of the system is

$$\omega_n = \sqrt{k_{eq}/m_{eq}} = \sqrt{(k_1 + k_2)\bigg/\left(m + \frac{J}{r^2}\right)} \quad (vii)$$

d. Substitute for x and its derivatives into (vi) using the compatibility condition (v) to obtain

$$(r^2 m + J)\ddot{\theta} + r^2(k_1 + k_2)\theta = rk_2 x_{20} - rk_1 x_{10}$$

The natural frequency

Time Response

BOX 2.1 Approaches for Developing Equations of Motion

---

1. **Conservative Systems (No Nonconservative Forces/No Energy Dissipation):**
   Kinetic energy = T
   Potential energy = V
   Conservation of energy: $T + V = const$
   Differentiate with respect to time t

2. **Lagrange's Equations:**
   Lagrangian $L = T - V$

   $$\frac{d}{dt}\frac{\partial L}{\partial \dot{q}_i} - \frac{\partial L}{\partial q_i} = Q_i \quad \text{for } i = 1, 2, \ldots, n$$

   n = number of degrees of freedom
   Q_i = generalized force corresponding to generalized coordinate q_i.

   Find Q_i using: $\delta W = \Sigma\, Q_i\, \delta q_i$

   where δW = work done by nonconservative forces in a general incremental motion $(\delta q_1, \delta q_2, \ldots, \delta q_n)$.

3. **Newtonian Approach:**

   $$\Sigma \text{ Forces} = \frac{d}{dt} \Sigma \text{ Linear momentum}$$

   $$\Sigma \text{ Torques} = \frac{d}{dt} \Sigma \text{ Angular momentum} \quad \text{(About centriod or a fixed point)}$$

$$\omega_n = \sqrt{r^2(k_1 + k_2)/(r^2 m + J)} = \sqrt{(k_1 + k_2)/\left(m + \frac{J}{r^2}\right)} \qquad (viii)$$

is identical to the previous answer *(vii)*. This is to be expected, as the system has not changed (only the response variable was changed).

$$\text{Equivalent moment of inertia}: J_{eq} = r^2 m + J$$

$$\text{Equivalent torsional stiffness}: K_{eq} = r^2(k_1 + k_2)$$

□

Common approaches of developing equations of motion for mechanical systems are summarized in Box 2.1.

2.2 HEAVY SPRINGS

A heavy spring has its mass and flexibility properties continuously distributed throughout its body. In that sense, it has an infinite number of degrees of freedom, and a single coordinate cannot represent its motion. However, for many practical purposes, a lumped-parameter approximation with just one lumped mass to represent the inertial characteristics of the spring may be sufficient. Such an approximation can be obtained using the energy approach. Here, the spring is represented by a lumped-parameter "model" such that the original spring and the model have the same net kinetic energy and potential energy. This *energy equivalence* is used in deriving a lumped mass parameter for the model. Although damping (energy dissipation) is neglected in the present analysis, it is not difficult to incorporate that as well in the model.

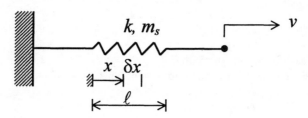

FIGURE 2.9 A uniform heavy spring.

2.2.1 KINETIC ENERGY EQUIVALENCE

Consider the uniform, heavy spring shown in Figure 2.9, with one end fixed and the other end moving at velocity v. Note that:

$$k = \text{stiffness of spring}$$

$$m_s = \text{mass of spring}$$

$$l = \text{length of spring}$$

Local speed of element δx of the spring is given by $\dfrac{x}{l} v$.

Element mass $= \dfrac{m_s}{l} \cdot \delta x$.

Hence, element $KE = \dfrac{1}{2} \dfrac{m_s}{l} \cdot \delta x \left(\dfrac{x}{l} v \right)^2$.

In the limit, $\delta x \to dx$. Then,

$$\text{Total } KE = \int_0^l \frac{1}{2} \frac{m_s}{l} dx \left(\frac{x}{l} v \right)^2 = \frac{1}{2} \frac{m_s v^2}{l^3} \int_0^l x^2 dx = \frac{1}{2} \frac{m_s v^2}{3} \tag{2.45}$$

Hence,

Equivalent lumped mass concentrated at the free end $= \dfrac{1}{3} \times$ spring mass.

Note: This derivation assumes that one end of the spring is fixed and, furthermore, that the conditions are *uniform* along the spring.

An example of utilizing this result is shown in Figure 2.10. Here, a system with a heavy spring and a lumped mass is approximated by a light spring (having the same stiffness) and a lumped mass.

Time Response

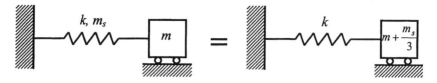

FIGURE 2.10 Lumped-parameter approximation for an oscillator with heavy spring.

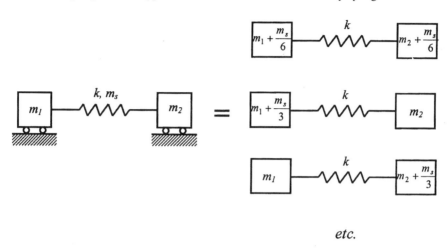

etc.

FIGURE 2.11 An example where the lumped-parameter approximation for a spring is ambiguous.

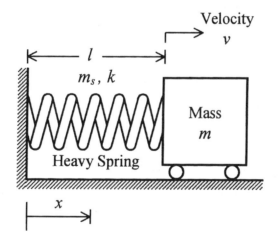

FIGURE 2.12 A heavy spring connected to a rolling stock.

Another example is shown in Figure 2.11. In this case, it is not immediately clear which of the approximations shown on the right-hand side is most appropriate.

EXAMPLE 2.2

A uniform heavy spring of mass m_s and stiffness k is attached at one end to a mass m that is free to roll on a frictionless horizontal plane. The other end is anchored to a vertical post. A schematic diagram of this arrangement is shown in Figure 2.12.

The unstretched length of the spring is l. Assume that when the velocity of the connected mass is v, the velocity distribution along the spring is given by

$$v_s(x) = v \sin \frac{\pi x}{2l}$$

where x is the distance of a point along the spring, as measured from the fixed end. Determine an equivalent lumped mass located at the moving end of the spring (i.e., at the moving mass m) to represent the inertia effects of the spring. What are the limitations of your result?

SOLUTION

Consider an element of length δx at location x of the spring. Since the spring is uniform, the element mass $= \frac{m_s}{l} \delta x$. Also, according to the given assumption, the element velocity $= v \sin \frac{\pi x}{2l}$. Hence, the kinetic energy of the spring is

$$\int_0^l \frac{1}{2} \left(v \sin \frac{\pi x}{2l} \right)^2 \frac{m_s}{l} dx = \frac{1}{2} \frac{m_s}{l} v^2 \int_0^l \sin^2 \frac{\pi x}{2l} dx = \frac{1}{4} \frac{m_s}{l} v^2 \int_0^l \left[1 - \cos \frac{\pi x}{l} \right] dx$$

$$= \frac{1}{4} \frac{m_s}{l} v^2 \left[x - \frac{l}{\pi} \sin \frac{\pi x}{l} \right]_0^l = \frac{1}{4} \frac{m_s}{l} v^2 l = \frac{1}{2} \frac{m_s}{2} v^2$$

It follows that the equivalent lumped mass to be located at the moving end of the spring is $\frac{m_s}{2}$. This result is valid only for the assumed velocity distribution, and corresponds to the first mode of motion only. In fact, a linear velocity distribution would be more realistic in this low-frequency (quasi-static motion) region, which will give an equivalent lumped mass of $\frac{1}{3} m_s$, as seen before. Such approximations will not be valid for high frequencies (say, higher than $\sqrt{\frac{k}{m_s}}$).

□

2.3 OSCILLATIONS IN FLUID SYSTEMS

As discussed in Appendix A, fluid systems can undergo oscillations (vibrations) quite analogous to mechanical and electrical systems. Again, the reason for their natural oscillation is the ability to store and repeatedly interchange two types of energy — kinetic energy and potential energy. The kinetic energy comes from the velocity of fluid particles during motion. The potential energy arises primarily from the following three main sources.

1. Gravitational potential energy
2. Compressibility of the fluid volume
3. Flexibility of the fluid container

A detailed analysis of these three effects is not undertaken here. However, one sees from the example in Figure 2.5(e) how a liquid column can oscillate due to repeated interchange between kinetic energy and gravitational potential energy. Now consider another example.

EXAMPLE 2.3

A university laboratory has developed a procedure for optimal cutting (portion control) of fish for can filling, with the objective of minimizing the wastage (overfill) and regulatory violations (underfill). The procedure depends on the knowledge of the volumetric distribution of a dressed (cleaned) fish. In fact, a group of volumeteric models is developed through off-line experimentation so that extensive measurements need not be made on-line, during processing. One set of such off-line experiments consists of dipping a fish into a tank of water in fixed increments and measuring the volume of water that is displaced. An illustration of the experimental setup is given in Figure 2.13(a).

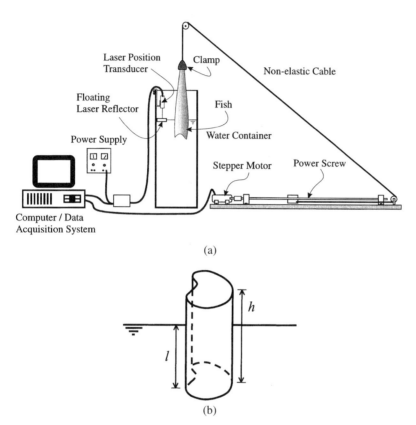

FIGURE 2.13 (a) An experimental system for determining the volumetric distribution of a fish body. (b) Buoyancy experiment.

One day, an adventurous student decided to try a different test with the experimental system. Instead of a fish, he used a cylindrical wooden peg of uniform cross section and height h. Realizing that the object could not be completely immersed in water, he pushed it down by hand, in the upright orientation (see Figure 2.13(b)). The object oscillated up and down while floating in the tank. Let ρ_b and ρ_l be the mass densities of the body (peg) and the liquid (water), respectively.

a. Clearly stating the assumptions that are made, obtain an expression for the natural frequency of oscillations.
b. If the object is slightly tilted to one side would it return to its upright configuration? Explain.

SOLUTION

a. Suppose that, under equilibrium in the upright position of the body, the submersed length is l. The mass of the body is

$$m = Ah\rho_b \qquad (i)$$

where A is the area of cross section (uniform).

By Archimedes principle, the buoyancy force R is equal to the weight of the liquid displaced by the body. Hence,

$$R = Al\rho_l g \qquad (ii)$$

For equilibrium, one has

$$R = mg \qquad (iii)$$

or

$$Al\rho_l g = Ah\rho_b g$$

Hence,

$$l = \frac{\rho_b}{\rho_l} h \qquad (iv)$$

For a vertical displacement y from the equilibrium position, the equation of motion is (Figure 2.14(a))

$$m\ddot{y} = mg - A(l+y)\rho_l g$$

Substitute equations (ii) and (iii) to obtain

$$m\ddot{y} = -A\rho_l g y$$

Substitute equation (i):

$$Ah\rho_b \ddot{y} + A\rho_l g y = 0$$

or

$$\ddot{y} + \frac{\rho_l g}{\rho_b h} y = 0$$

The natural frequency of oscillations is

$$\omega_n = \sqrt{\frac{\rho_l g}{\rho_b h}}$$

Note that this result is independent of the area of cross section of the body.

Assumptions:
1. The tank is very large compared to the body. The change in liquid level is negligible as the body is depressed into the water.
2. Fluid resistance (viscous effects, drag, etc.) is negligible.
3. Dynamics of the liquid itself are negligible. Hence, "added inertia" due to liquid motion is neglected.

The buoyancy force R acts through the centroid of the volume of displaced water [Figure 2.14(b)]. Its line of action passes through the central axis of the body at point M. The point is known as the *metacenter*. Let C be the centroid of the body.

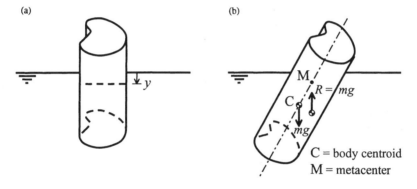

FIGURE 2.14 (a) Upright oscillations of the body. (b) Restoring buoyancy couple due to a stable metacenter.

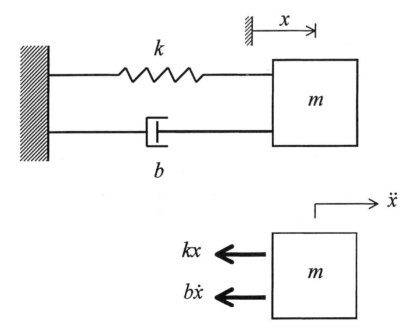

FIGURE 2.15 A damped simple oscillator and its free-body diagram.

If M is above C, then, when tilted, there will be a restoring couple that will tend to restore the body to its upright position. Otherwise, the body will be in an unstable situation, and the buoyancy couple will tend to tilt it further toward a horizontal configuration.

□

2.4 DAMPED SIMPLE OSCILLATOR

Now consider the free (natural) response of a simple oscillator in the presence of energy dissipation (damping).

Assume viscous damping, and consider the oscillator shown in Figure 2.15. The free-body diagram of the mass is shown separately.

The following notation is used in this book.

ω_n = undamped natural frequency
ω_d = damped natural frequency
ω_r = resonant frequency
ω = frequency of excitation.

The concept of resonant frequency will be addressed in Chapter 3.

The viscous damping constant is denoted by b (but sometimes c will be used instead of b, as done in some literature).

Apply Newton's second law. From the free-body diagram in Figure 2.15, one has the equation of motion

$$m\ddot{x} = -kx - b\dot{x}$$

or

$$m\ddot{x} + b\dot{x} + kx = 0 \tag{2.46}$$

or

$$\ddot{x} + 2\zeta\omega_n\dot{x} + \omega_n^2 x = 0 \tag{2.47}$$

This is a free (or unforced, or *homogeneous*) equation of motion. Its solution is the free (natural) response of the system and is also known as the *homogeneous solution*. Note that $\omega_n = \sqrt{\dfrac{k}{m}}$, which is the natural frequency when there is no damping, and

$$2\zeta\omega_n = \frac{b}{m} \tag{2.48}$$

Hence, $\zeta = \dfrac{1}{2}\sqrt{\dfrac{m}{k}}\dfrac{b}{m}$
or

Time Response

$$\zeta = \frac{1}{2}\frac{b}{\sqrt{km}} \tag{2.49}$$

Also note that ζ is called the *damping ratio*. The formal definition and the rationale for this terminology will be discussed later.

Assume an exponential solution:

$$x = Ce^{\lambda t} \tag{2.50}$$

This is justified by the fact that linear systems have exponential or oscillatory (i.e., complex exponential) free responses. A more detailed justification will be provided later.

Substitute equation (2.50) into (2.47) to obtain

$$\left[\lambda^2 + 2\zeta\omega_n\lambda + \omega_n^2\right]Ce^{\lambda t} = 0$$

Note that $Ce^{\lambda t}$ is not zero in general. It follows that when λ satisfies the equation

$$\lambda^2 + 2\zeta\omega_n\lambda + \omega_n^2 = 0 \tag{2.51}$$

then equation (2.50) will represent a solution of equation (2.47).

Equation (2.51) is called the *characteristic equation* of the system. This equation depends on the natural dynamics of the system, not forcing excitation or initial conditions.

Solution of equation (2.51) gives the two roots:

$$\begin{aligned}\lambda &= -\zeta\omega_n \pm \sqrt{\zeta^2-1}\,\omega_n \\ &= \lambda_1 \text{ and } \lambda_2\end{aligned} \tag{2.52}$$

These are called *eigenvalues* or *poles* of the system.

When $\lambda_1 \neq \lambda_2$, the general solution is

$$x = C_1 e^{\lambda_1 t} + C_2 e^{\lambda_2 t} \tag{2.53}$$

The two unknown constants C_1 and C_2 are related to the integration constants, and can be determined by two initial conditions, which should be known.

If $\lambda_1 = \lambda_2 = \lambda$, one has the case of repeated roots. In this case, the general solution (2.53) does not hold because C_1 and C_2 would no longer be independent constants, to be determined by two initial conditions. The repetition of the roots suggests that one term of the homogenous solution should have the multiplier t (a result of the double-integration of zero). Then, the general solution is

$$x = C_1 e^{\lambda t} + C_2 t e^{\lambda t} \tag{2.54}$$

One can identify three categories of damping level, as discussed below, and the nature of the response will depend on the particular category of damping.

2.4.1 CASE 1: UNDERDAMPED MOTION ($\zeta < 1$)

In this case, it follows from equation (2.52) that the roots of the characteristic equation are

$$\lambda = -\zeta\omega_n \pm j\sqrt{1-\zeta^2}\,\omega_n = -\zeta\omega_n \pm j\omega_d = \lambda_1 \text{ and } \lambda_2 \quad (2.55)$$

where, the *damped natural frequency* is given by

$$\omega_d = \sqrt{1-\zeta^2}\,\omega_n \quad (2.56)$$

Note that λ_1 and λ_2 are complex conjugates. The response (2.53) in this case can be expressed as

$$x = e^{-\zeta\omega_n t}\left[C_1 e^{j\omega_d t} + C_2 e^{-j\omega_d t}\right] \quad (2.57)$$

The term within the square brackets of equation (2.57) has to be real because it represents the time response of a real physical system. It follows that C_1 and C_2, as well, have to be complex conjugates.

Note: $e^{j\omega_d t} = \cos\omega_d t + j\sin\omega_d t$

$e^{-j\omega_d t} = \cos\omega_d t - j\sin\omega_d t$

Thus, an alternative form of the general solution would be

$$x = e^{-\zeta\omega_n t}\left[A_1 \cos\omega_d t + A_2 \sin\omega_d t\right] \quad (2.58)$$

Here, A_1 and A_2 are the two unknown constants. By equating the coefficients, it can be shown that

$$A_1 = C_1 + C_2$$
$$A_2 = j(C_1 - C_2) \quad (2.59)$$

Hence,

$$C_1 = \frac{1}{2}(A_1 - jA_2)$$
$$C_2 = \frac{1}{2}(A_1 + jA_2) \quad (2.60)$$

which are complex conjugates, as required.

Initial Conditions:
Let, $x(0) = x_o$, $\dot{x}(0) = v_o$ as before. Then,

$$x_o = A_1$$

and (2.61)

$$v_o = -\zeta\omega_n A_1 + \omega_d A_2$$

or

$$A_2 = \frac{v_o}{\omega_d} + \frac{\zeta\omega_n x_o}{\omega_d} \qquad (2.62)$$

Yet, another form of the solution would be:

$$x = Ae^{-\zeta\omega_n t}\sin(\omega_d t + \phi) \qquad (2.63)$$

Here, A and ϕ are the unknown constants with

$$A = \sqrt{A_1^2 + A_2^2} \text{ and } \sin\phi = \frac{A_1}{\sqrt{A_1^2 + A_2^2}}. \qquad (2.64)$$

Also, $\cos\phi = \dfrac{A_2}{\sqrt{A_1^2 + A_2^2}}$ and $\tan\phi = \dfrac{A_1}{A_2}$ (2.65)

Note that the response $x \to 0$ as $t \to \infty$. This means the system is *asymptotically stable*.

2.4.2 Logarithmic Decrement Method

The damping ratio ζ can be experimentally determined from the free response by the logarithmic decrement method. To illustrate this approach, note from equation (2.63) that the period of damped oscillations is

$$T = \frac{2\pi}{\omega_d} \qquad (2.66)$$

Also, from equation (2.63),

$$\frac{x(t)}{x(t+nT)} = \frac{Ae^{-\zeta\omega_n t}\sin(\omega_d t + \phi)}{Ae^{-\zeta\omega_n(t+nT)}\sin[\omega_d(t+nT) + \phi]}$$

But, $\sin[\omega_d(t+nT) + \phi] = \sin(\omega_d t + \phi + 2n\pi) = \sin(\omega_d + \phi)$

Hence,

$$\frac{x(t)}{x(t+nT)} = \frac{e^{-\zeta\omega_n t}}{e^{-\zeta\omega_n(t+nT)}} = e^{\zeta\omega_n nT} \qquad (2.67)$$

Take the natural logarithm of equation (2.67), the logarithmic decrement:

$$\zeta\omega_n nT = \ln\left[\frac{x(t)}{x(t+nT)}\right]$$

But, $\omega_n T = \omega_n \frac{2\pi}{\omega_d} = \frac{\omega_n 2\pi}{\sqrt{1-\zeta^2}\,\omega_n} = \frac{2\pi}{\sqrt{1-\zeta^2}}$. Hence, with $x(t)/x(t+nT) = r$, one has the logarithmic decrement

$$\frac{2\pi n\zeta}{\sqrt{1-\zeta^2}} = \ln r$$

Note that $\frac{1}{n}\ln r$ is the "per-cycle" logarithmic decrement, and $\frac{1}{2\pi n}\ln r$ is the "per-radian" logarithmic decrement. The latter is

$$\frac{\zeta}{\sqrt{1-\zeta^2}} = \frac{1}{2\pi n}\ln r = \alpha \tag{2.68}$$

Then, one has

$$\zeta = \sqrt{\frac{\alpha^2}{1+\alpha^2}} \tag{2.69}$$

This is the basis of the logarithmic decrement method of measuring damping. Start by measuring a point $x(t)$ and another point $x(t+nT)$ at n cycles later. For high accuracy, pick the peak points of the response curve for the measurement of $x(t)$ and $x(t+nT)$. From equation (2.68), it is clear that for small damping, $\zeta = \alpha$ = per-radian logarithmic decrement.

2.4.3 Case 2: Overdamped Motion ($\zeta > 1$)

In this case, roots λ_1 and λ_2 of the characteristic equation (2.51) are real. Specifically,

$$\lambda_1 = -\zeta\omega_n + \sqrt{\zeta^2-1}\,\omega_n < 0 \tag{2.70}$$

$$\lambda_2 = -\zeta\omega_n - \sqrt{\zeta^2-1}\,\omega_n < 0 \tag{2.71}$$

and the response (2.53) is nonoscillatory. Also, it should be clear from equations (2.70) and (2.71) that both λ_1 and λ_2 are negative. Hence, $x \to 0$ as $t \to \infty$. This means the system is asymptotically stable.

From the initial conditions

$$x(0) = x_o, \quad \dot{x}(0) = v_o$$

we get

$$x_o = C_1 + C_2 \tag{i}$$

Time Response

and

$$v_o = \lambda_1 C_1 + \lambda_2 C_2 \qquad (ii)$$

Multiply the first IC(i) by λ_1 to obtain $\lambda_1 x_o = \lambda_1 C_1 + \lambda_1 C_2$. $\qquad (iii)$

Then subtract (iii) from (ii) to obtain $v_o - \lambda_1 x_o = C_2(\lambda_2 - \lambda_1)$

and
$$C_2 = \frac{v_o - \lambda_1 x_o}{\lambda_2 - \lambda_1} \qquad (2.72)$$

Similarly, multiply the first IC(i) by λ_2 and subtract from (ii). One obtains

$$v_o - \lambda_2 x_o = C_1(\lambda_1 - \lambda_2)$$

Hence,

$$C_1 = \frac{v_o - \lambda_2 x_o}{\lambda_1 - \lambda_2} \qquad (2.73)$$

2.4.4 CASE 3: CRITICALLY DAMPED MOTION ($\zeta = 1$)

Here, we have repeated roots, given by

$$\lambda_1 = \lambda_2 = -\omega_n \qquad (2.74)$$

The response for this case is given by (see equation (2.54))

$$x = C_1 e^{-\omega_n t} + C_2 t e^{-\omega_n t} \qquad (2.75)$$

Since the term $e^{-\omega_n t}$ goes to zero faster than t goes to infinity, one has

$$t e^{-\omega_n t} \to 0 \text{ as } t \to \infty. \text{ Hence, the system is asymptotically stable.}$$

Now use the initial conditions $x(0) = x_o$, $\dot{x}(0) = v_o$. One obtains

$$x_o = C_1$$
$$v_o = -\omega_n C_1 + C_2$$

Hence,

$$C_1 = x_o \qquad (2.76)$$

$$C_2 = v_o + \omega_n x_o \qquad (2.77)$$

Note: When $\zeta = 1$, one has the critically damped response because below this value, the response is oscillatory (underdamped), and above this value, the response is nonoscillatory (overdamped). It follows that one can define the damping ratio as

$$\zeta = \text{Damping ratio} = \frac{\text{Damping constant}}{\text{Damping constant for critically damped condition}} \tag{2.78}$$

2.4.5 Justification for the Trial Solution

In the present analysis, the trial solution (2.50) has been used for the response of a linear system having constant parameter values. A justification for this is provided now.

First-Order System

Consider a first-order linear system given by (homogeneous, no forcing input)

$$\left(\frac{d}{dt} - \lambda\right)x = \dot{x} - \lambda x = 0 \tag{2.79}$$

This equation can by written as

$$\frac{dx}{x} = \lambda dt$$

Integrate:

$$\ln x = \lambda t + \ln C$$

Here, $\ln C$ is the constant of integration. Hence,

$$x = Ce^{\lambda t} \tag{2.80}$$

This is then the general form of the free response of a first-order system. It incorporates one constant of integration and, hence, will need one initial condition.

Second Order System

One can write the equation of a general second-order (homogenous, unforced) system in the operational form

$$\left(\frac{d}{dt} - \lambda_1\right)\left(\frac{d}{dt} - \lambda_2\right)x = 0 \tag{2.81}$$

By reasoning as before, the general solution would be of the form $x = C_1 e^{\lambda_1 t} + C_2 e^{\lambda_2 t}$. Here, C_1 and C_2 are the constants of integration, which are determined using two initial conditions.

Repeated Roots

The case of repeated roots deserves a separate treatment. First consider

$$\frac{d^2 x}{dt^2} = 0 \tag{2.82}$$

Time Response

Integrate twice:
$$\frac{dx}{dt} = C; \quad x = Ct + D \tag{2.83}$$

Note the term with t in this case. Hence, a suitable trial solution for the system

$$\left(\frac{d}{dt} - \lambda\right)\left(\frac{d}{dt} - \lambda\right) x = 0 \tag{2.84}$$

would be $x = C_1 e^{\lambda t} + C_2 t e^{\lambda t}$.

The main results for free (natural) response of a damped oscillator are given in Box 2.2.

2.4.6 Stability and Speed of Response

The free response of a dynamic system (particularly a vibrating system) can provide valuable information concerning the natural characteristics of the system. The free (unforced) excitation can be obtained, for example, by giving an initial-condition excitation to the system and then allowing it to respond freely. Two important characteristics that can be determined in this manner are:

1. Stability
2. Speed of response

The stability of a system implies that the response will not grow without bounds when the excitation force itself is finite. This is known as bounded-input-bounded-output (BIBO) stability. In particular, if the free response eventually decays to zero, in the absence of a forcing input, the system is said to be asymptotically stable. It was shown that a damped simple oscillator is asymptotically stable, but an undamped oscillator, while being stable in a general (BIBO) sense, is not asymptotically stable. It is marginally stable.

Speed of response of a system indicates how fast the system responds to an excitation force. It is also a measure of how fast the free response (1) rises or falls if the system is oscillatory; or (2) decays, if the system is non-oscillatory. Hence, the two characteristics — stability and speed of response — are not completely independent. In particular, for non-oscillatory (overdamped) systems, these two properties are very closely related. It is clear then, that stability and speed of response are important considerations in the analysis, design, and control of vibrating systems.

The level of stability of a linear dynamic system depends on the real parts of the eigenvalues (or poles, which are the roots of the characteristic equations). Specifically, if all the roots have real parts that are negative, then the system is stable. Also, the more negative the real part of a pole, the faster the decay of the free response component corresponding to that pole. The inverse of the negative real part is the time constant. Hence, the smaller the time constant, the faster the decay of the corresponding free response and, hence, the higher the level of stability associated with that pole. One can summarize these observations as follows:

Level of stability: Depends on decay rate of free response (and hence on time constants or real parts of poles)
Speed of response: Depends on natural frequency and damping for oscillatory systems and decay rate for non-oscillatory systems
Time constant: Determines stability and decay rate of free response (and speed of response in non-oscillatory systems)

Now consider the specific case of a damped simple oscillator given by equation (2.47).

BOX 2.2 Free (Natural) Response of a Damped Simple Oscillator

---

System Equation: $m\ddot{x} + b\dot{x} + kx = 0$ or $\ddot{x} + 2\zeta\omega_n\dot{x} + \omega_n^2 x = 0$

Undamped natural frequency $\omega_n = \sqrt{\dfrac{k}{m}}$

Damping ratio $\zeta = \dfrac{b}{2\sqrt{km}}$

Characteristic equation: $\lambda^2 + 2\zeta\omega_n\lambda + \omega_n^2 = 0$

Roots (eigenvalues or poles): λ_1 and $\lambda_2 = -\zeta\omega_n \pm \sqrt{\zeta^2 - 1}\,\omega_n$

Response: $x = C_1 e^{\lambda_1 t} + C_2 e^{\lambda_2 t}$ for unequal roots $(\lambda_1 \neq \lambda_2)$

$x = (C_1 + C_2 t) e^{\lambda t}$ for equal roots $(\lambda_1 = \lambda_2 = \lambda)$

Initial conditions: $x(0) = x_o$, $\dot{x}(0) = v_o$

Case 1: Underdamped ($\zeta < 1$)

Poles are complex conjugates: $-\zeta\omega_n \pm j\omega_d$

Damped natural frequency $\omega_d = \sqrt{1-\zeta^2}\,\omega_n$

$$x = e^{-\zeta\omega_n t}\left[C_1 e^{j\omega_d t} + C_2 e^{-j\omega_d t}\right]$$
$$= e^{-\zeta\omega_n t}\left[A_1 \cos\omega_d t + A_2 \sin\omega_d t\right]$$
$$= A e^{-\zeta\omega_n t}\sin(\omega_d t + \phi)$$

$A_1 = C_1 + C_2$ and $A_2 = j(C_1 - C_2)$

$C_1 = \dfrac{1}{2}(A_1 - jA_2)$ and $C_2 = \dfrac{1}{2}(A_1 + jA_2)$

$A = \sqrt{A_1^2 + A_2^2}$ and $\tan\phi = \dfrac{A_1}{A_2}$

ICs give: $A_1 = x_o$ and $A_2 = \dfrac{v_o + \zeta\omega_n x_o}{\omega_d}$

Logarithmic decrement per radian: $\alpha = \dfrac{1}{2\pi n}\ln r = \dfrac{\zeta}{\sqrt{1-\zeta^2}}$ or $\zeta = \sqrt{\dfrac{\alpha^2}{1+\alpha^2}}$

where $r = \dfrac{x(t)}{x(t+nT)}$ = decay ratio over n complete cycles.

For small ζ: $\zeta \cong \alpha$

Case 2: Overdamped ($\zeta > 1$)

Poles are real and negative: $\lambda_1, \lambda_2 = -\zeta\omega_n \pm \sqrt{\zeta^2 - 1}\,\omega_n$

$$x = C_1 e^{\lambda_1 t} + C_2 e^{\lambda_2 t}$$

$$C_1 = \dfrac{v_0 - \lambda_2 x_0}{\lambda_1 - \lambda_2} \quad \text{and} \quad C_2 = \dfrac{v_0 - \lambda_1 x_0}{\lambda_2 - \lambda_1}$$

Case 3: Critically Damped ($\zeta = 1$)

Two identical poles: $\lambda_1 = \lambda_2 = \lambda = -\omega_n$

$x = (C_1 + C_2 t)e^{-\omega_n t}$ with $C_1 = x_0$ and $C_2 = v_0 + \omega_n x_0$

Time Response

Case 1 ($\zeta < 1$): The free response is given by $x = Ae^{-\zeta\omega_n t}\sin(\omega_d t + \phi)$

$$\text{Time constant } \tau = \frac{1}{\zeta\omega_n} \tag{2.85}$$

The system is asymptotically stable. The larger the $\zeta\omega_n$ the more stable the system. Also, the speed of response increases with both ω_d and $\zeta\omega_n$.

Case 2 ($\zeta > 1$): The response is non-oscillatory, and is given by

$$x = \underset{(\text{decays slower})}{A_1 e^{\lambda_1 t}} + \underset{(\text{decays faster})}{A_2 e^{\lambda_2 t}}$$

where, $\lambda_1 = -\zeta\omega_n + \sqrt{\zeta^2 - 1}\,\omega_n$ and $\lambda_2 = -\zeta\omega_n - \sqrt{\zeta^2 - 1}\,\omega_n$

This system has two time constants:

$$\tau_1 = \frac{1}{|\lambda_1|} \quad \text{and} \quad \tau_2 = \frac{1}{|\lambda_2|} \tag{2.86}$$

Note that τ_1 is the dominant (slower) time constant. The system is also asymptotically stable. The larger the $|\lambda_1|$, the faster and more stable the system.

Consider an underdamped system and an overdamped system with damping ratios ζ_u and ζ_o, respectively. One can show that the underdamped system is more stable than the overdamped system if and only if:

$$\zeta_o - \sqrt{\zeta_o^2 - 1} < \zeta_u \tag{2.87a}$$

or equivalently,

$$\zeta_o > \frac{\zeta_u^2 + 1}{2\zeta_u} \tag{2.87b}$$

where $\zeta_o > 1 > \zeta_u > 0$ by definition.

Proof

To be more stable, one should have the underdamped pole located farther away than the dominant overdamped pole, from the imaginary axis of the pole plane; thus,

$$\zeta_u \omega_n > \zeta_o \omega_n - \sqrt{\zeta_o^2 - 1}\,\omega_n$$

Hence,

$$\zeta_u > \zeta_o - \sqrt{\zeta_o^2 - 1}$$

Now, bring the square-root term to the LHS and square it.

$$\zeta_o^2 - 1 > (\zeta_o - \zeta_u)^2 = \zeta_o^2 - 2\zeta_o\zeta_u + \zeta_u^2$$

Hence,

$$2\zeta_o\zeta_u > \zeta_u^2 + 1$$

or

$$\zeta_o > \frac{\zeta_u^2 + 1}{2\zeta_u}$$

This completes the proof.

To explain this result further, consider an undamped ($\zeta = 0$) simple oscillator of natural frequency ω_n. Its poles are at $\pm j\omega_n$ (on the imaginary axis of the pole plane). Now add damping and increase ζ from 0 to 1. Then the complex conjugates poles $-\zeta\omega_n \pm j\omega_d$ will move away from the imaginary axis as ζ increases (because $\zeta\omega_n$ increases) and, hence, the level of stability will increase. When ζ reaches the value 1 (critical damping), one obtains two identical and real poles at $-\omega_n$. When ζ is increased beyond 1, the poles will be real and unequal, with one pole having a magnitude smaller than ω_n and the other having a magnitude larger than ω_n. The former (closer to the "origin" of zero) is the dominant pole, and will determine both stability and the speed of response of the overdamped system. It follows that as ζ increases beyond 1, the two poles will branch out from the location $-\omega_n$, one moving toward the origin (becoming less stable) and the other moving away from the origin. It is now clear that as ζ is increased beyond the point of critical damping, the system becomes less stable. Specifically, for a given value of $\zeta_u < 1.0$, there is a value of $\zeta_o > 1$, governed by (2.87), above which the overdamped system is less stable and slower than the underdamped system.

EXAMPLE 2.4

Consider the simple oscillator shown in Figure 2.15, with parameters $m = 4$ kg, $k = 1.6 \times 10^3$ N m^{-1}, and the two cases of damping:

1. $b = 80$ N m^{-1} s^{-1}
2. $b = 320$ N m^{-1} s^{-1}

Study the nature of the free response in each case.

SOLUTION

The undamped natural frequency of the system is

$$\omega_n = \sqrt{\frac{k}{m}} = \sqrt{\frac{1.6 \times 10^3}{4}} = 20.0 \text{ rad s}^{-1}$$

Case 1:
$$2\zeta\omega_n = \frac{b}{m} \quad \text{or} \quad 2\zeta \times 20 = \frac{80}{4}$$

Then,
$$\zeta_u = 0.5$$

The system is underdamped in this case.

Case 2:
$$2\zeta \times 20 = \frac{320}{4}$$

Then,
$$\zeta_o = 2.0$$

The system is overdamped in this case.

Case 1: The characteristic equation is
$$\lambda^2 + 2 \times 0.5 \times 20\lambda + 20^2 = 0$$

or
$$\lambda^2 + 20\lambda + 20^2 = 0$$

The roots (eigenvalues or poles) are
$$\lambda = -10 \pm j\sqrt{20^2 - 10^2} = -10 \pm j10\sqrt{3}$$

The free (no force) response is given by
$$x = Ae^{-10t}\sin\left(10\sqrt{3}t + \phi\right)$$

The amplitude A and the phase angle ϕ can be determined using initial conditions.

$$\text{Time constant}: \tau = \frac{1}{10} = 0.1 \text{ s}$$

Case 2: The characteristic equation is
$$\lambda^2 + 2 \times 2 \times 20\lambda + 20^2 = 0$$

or
$$\lambda^2 + 80\lambda + 20^2 = 0$$

The roots are
$$\lambda = -40 \pm \sqrt{40^2 - 20^2} = -40 \pm 20\sqrt{3}$$
$$= -5.36, \ -74.64$$

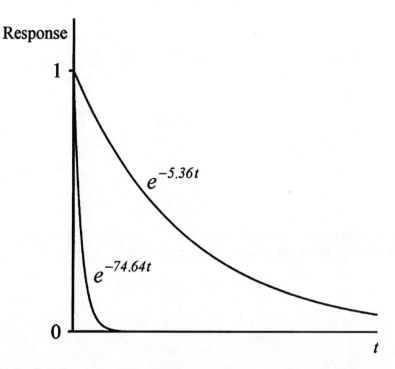

FIGURE 2.16 The free (homogeneous) response components of an overdamped system.

The free response is given by

$$x = C_1 e^{-5.36t} + C_2 e^{-74.64t}$$

The constants C_1 and C_2 can be determined using initial conditions. The second term on the RHS goes to zero much faster than the first term, as shown in Figure 2.16. Hence, the first term will dominate and will determine the dominant time constant, level of stability, and speed of response. Specifically, the response may be approximated as

$$x \cong C_1 e^{-5.36t}$$

TABLE 2.1
Natural Characteristics of a Damped Oscillator

Damping ratio	Level of damping	Oscillatory response	Stability	Speed of response	Time constant
$\zeta < 1$	Underdamped	Yes	Asymptotically stable (less stable than $\zeta = 1$ case but not necessarily less stable than the overdamped case)	Better than overdamped	$1/(\zeta \omega_n)$
$\zeta > 1$	Overdamped	No	Asymptotically stable; less stable than the critically damped case	Lower than critical	$1/\left(\zeta \omega_n \pm \sqrt{\zeta^2 - 1}\, \omega_n\right)$
$\zeta = 1$	Critically damped	No	Asymptotically stable; most stable	Good	$1/\omega_n$

Hence,

$$\text{Time constant } \tau = \frac{1}{5.36} = 0.19 \text{ s}$$

This value is double that of Case 1. Consequently, it is clear that the underdamped system (Case 1) decays faster than the overdamped system (Case 2). In fact, according to equation (2.87b), with $\zeta_u = 0.5$, we have $\zeta_o > \frac{0.5^2 + 1}{2 \times 0.5} = 1.25$. Hence, an overdamped system of damping ratio greater than 1.25 will be less stable than the underdamped system of damping ratio 0.5.

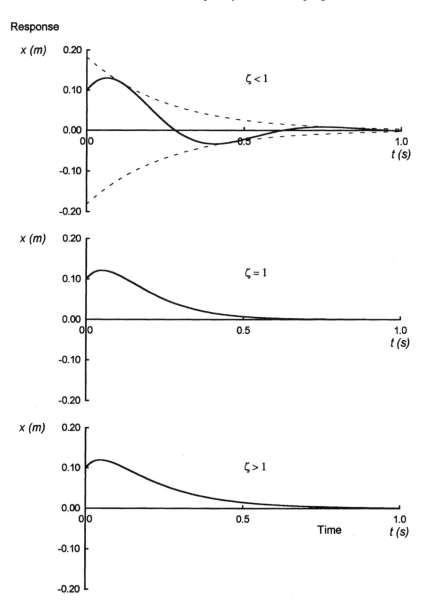

FIGURE 2.17 Free response of a damped oscillator: (a) underdamped, (b) critically damped, (c) overdamped.

Table 2.1 summarizes some natural characteristics of a damped simple oscillator under three different levels of damping. The nature of the natural response for these three cases is sketched in Figure 2.17.

□

2.5 FORCED RESPONSE

Thus far only the "free response" of a vibratory system has been studied. This is the response to some initial excitation and in the absence of any subsequent forcing input. This corresponds to the "natural" response of the system. Mathematically, it is the homogeneous solution because it is obtained by solving the homogeneous equation of the system (i.e., without the input terms). The natural response, the free response, and the homogeneous solution are synonymous in the absence of a forcing input to the system. But when there is a forcing excitation (i.e., an input), the equation of motion will be non-homogeneous (i.e., the right-hand side will not be zero). Then, the total solution (total response T) will be given by the sum of the homogeneous solution (H) and the particular integral (P), subject to the system initial conditions. This is a mathematical solution of the equation of motion. This total response can be separated into the terms that depend on the initial conditions (X) and the terms that depend on the forcing excitation (F). This is the physical interpretation of the total solution. Note that X is called the "free response," "initial-condition response," or the "zero-input response." The term F is called the "forced response," the "zero-initial condition response," or the "zero-state response." In general, H is not identical to X, and P is not identical to F. But when there is no forcing excitation (no input), then by definition H and X will be identical. Furthermore, under steady state, the homogeneous part or initial-condition response will die down (assuming that the system is stable). Then, F will become equal to P. Note that, even when the initial conditions are zero, F and P may not be identical because F may contain a natural response term that is excited by the forcing input. This term will die out with time, however.

This section will look at the forced response of a dynamic (vibratory) system. This is the response of the system to a forcing input. The total response will depend on the natural characteristics of the system (as for the free response) and also on the nature of the forcing excitation. Mathematically, then, the total response will be determined by both the homogeneous solution and the particular solution. The complete solution will require a knowledge of the input (forcing excitation) and the initial conditions.

The behavior of a dynamic system when subjected to a certain forcing excitation can be studied by analyzing a model of the system. This is commonly known as system-response analysis. System response can be studied either in the time domain, where the independent variable of system response is time, or in the frequency domain, where the independent variable of system response is frequency. Time-domain analysis and frequency-domain analysis are equivalent. Variables in the two domains are connected through Fourier (integral) transform. The preference of one domain over the other depends on such factors as the nature of the excitation input, the type of analytical model available, the time duration of interest, and the quantities that need to be determined. The frequency-domain analysis will be addressed in detail in Chapters 3 and 4. The present section concentrates only on the time-domain analysis of the forced response. In particular, the impulse-response approach will be presented.

2.5.1 IMPULSE RESPONSE FUNCTION

Consider a linear dynamic (vibratory) system; then, the principle of superposition holds. More specifically, if y_1 is the system response to excitation $u_1(t)$ and y_2 is the response to excitation $u_2(t)$, then $\alpha y_1 + \beta y_2$ is the system response to input $\alpha u_1(t) + \beta u_2(t)$ for any constants α and β and any

Time Response

excitation functions $u_1(t)$ and $u_2(t)$. This is true for both time-variant-parameter linear systems and constant-parameter linear systems.

A unit pulse of width $\Delta\tau$ starting at time $t = \tau$ is shown in Figure 2.18(a). Its area is unity. A unit impulse is the limiting case of a unit pulse when $\Delta\tau \to 0$. Unit impulse acting at time $t = \tau$ is denoted by $\delta(t - \tau)$ and is graphically represented as in Figure 2.18(b). In mathematical analysis, this is known as the *Dirac delta function*. It is mathematically defined by the two conditions:

$$\delta(t - \tau) = 0 \quad \text{for} \quad t \neq \tau \qquad (2.88)$$
$$\to \infty \quad \text{at} \quad t = \tau$$

and

$$\int_{-\infty}^{\infty} \delta(t - \tau) dt = 1 \qquad (2.89)$$

The Dirac delta function has the following well-known and useful properties:

$$\int_{-\infty}^{\infty} f(t)\delta(t - \tau) dt = f(\tau) \qquad (2.90)$$

and

$$\int_{-\infty}^{\infty} \frac{d^n f(t)}{dt^n} \delta(t - \tau) dt = \left. \frac{d^n f(t)}{dt^n} \right|_{t=\tau} \qquad (2.91)$$

for any well-behaved time function $f(t)$.

The system response (output) to a unit-impulse excitation (input) acted at time $t = 0$ is known as the *impulse-response function* and is denoted by $h(t)$.

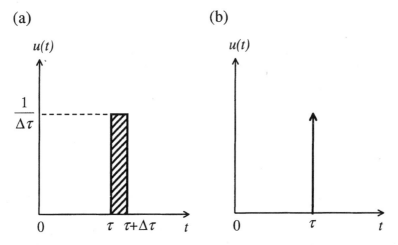

FIGURE 2.18 Illustrations of (a) unit pulse and (b) unit impulse.

2.5.2 FORCED RESPONSE

The system output to an arbitrary input can be expressed in terms of its impulse-response function. This is the essence of the impulse-response approach to determining the forced response of a dynamic system.

Without loss of generality, assume that the system input $u(t)$ starts at $t = 0$; that is,

$$u(t) = 0 \quad \text{for} \quad t < 0 \quad (2.92)$$

For physically realizable systems, the current response does not depend on the future values of the input. Consequently,

$$y(t) = 0 \quad \text{for} \quad t < 0 \quad (2.93)$$

and

$$h(t) = 0 \quad \text{for} \quad t < 0 \quad (2.94)$$

where $y(t)$ is the response of the system to any general excitation $u(t)$.

Furthermore, if the system is a constant-parameter system, then the response does not depend on the time origin used for the input. Mathematically, this is stated as follows: if the response to input $u(t)$ satisfying equation (2.92) is $y(t)$, which satisfies equation (2.93), then the response to input $u(t - \tau)$, which satisfies,

$$u(t - \tau) = 0 \quad \text{for} \quad t < \tau \quad (2.95)$$

is $y(t - \tau)$, and it satisfies

$$y(t - \tau) = 0 \quad \text{for} \quad t < \tau \quad (2.96)$$

This situation is illustrated in Figure 2.19. It follows that the delayed-impulse input $\delta(t - \tau)$, having time delay τ, produces the delayed response $h(t - \tau)$.

A given input $u(t)$ can be divided approximately into a series of pulses of width $\Delta\tau$ and magnitude $u(\tau)\Delta\tau$. In Figure 2.20, for $\Delta\tau \to 0$, the pulse shown by the shaded area becomes an impulse acting at $t = \tau$, having the magnitude $u(\tau)d\tau$. This impulse is given by $\delta(t - \tau)u(\tau)d\tau$. In a linear, constant-parameter system, it produces the response $h(t - \tau)u(\tau)d\tau$. By integrating over the entire time duration of the input $u(t)$, the overall response $y(t)$ is obtained as

$$y(t) = \int_0^\infty h(t - \tau)u(\tau)d\tau \quad (2.97)$$

Equation (2.97) is known as the *convolution integral*. This is, in fact, the forced response under zero initial conditions. In view of equation (2.94), it follows that $h(t - \tau) = 0$ for $\tau > t$. Consequently, the upper limit of integration in equation (2.97) could be made equal to t without affecting the result. Similarly, in view of equation (2.92), the lower limit of integration in equation (2.97) could be made $-\infty$. Furthermore, by introducing the change of variable $\tau \to t - \tau$, an alternative version of the convolution integral is obtained. Several valid versions of the convolution integral (or response equation) for linear, constant-parameter systems are as follows:

$$y(t) = \int_0^\infty h(\tau)u(t-\tau)d\tau \qquad (2.97\text{a})$$

$$y(t) = \int_{-\infty}^\infty h(t-\tau)u(\tau)d\tau \qquad (2.97\text{b})$$

$$y(t) = \int_{-\infty}^\infty h(\tau)u(t-\tau)d\tau \qquad (2.97\text{c})$$

$$y(t) = \int_{-\infty}^t h(t-\tau)u(\tau)d\tau \qquad (2.97\text{d})$$

$$y(t) = \int_{-\infty}^t h(\tau)u(t-\tau)d\tau \qquad (2.97\text{e})$$

$$y(t) = \int_0^t h(t-\tau)u(\tau)d\tau \qquad (2.97\text{f})$$

$$y(t) = \int_0^t h(\tau)u(t-\tau)d\tau \qquad (2.97\text{g})$$

In fact, the lower limit of integration in the convolution integral could be any value satisfying $\tau \leq 0$, and the upper limit could be any value satisfying $\tau \geq t$. The use of a particular pair of integration limits depends on whether the functions $h(t)$ and $u(t)$ implicitly satisfy the conditions given by equations (2.93) and (2.94) or these conditions have to be imposed on them by means of the proper integration limits. It should be noted that the two versions given by equations (2.97f) and (2.97g) explicitly take these conditions into account and therefore are valid for all inputs and impulse-response functions.

It should be emphasized that the response given by the convolution integral assumes a zero initial state, and is known as the *zero-state response* because the impulse response itself assumes a zero initial state. As stated, this is not necessarily equal to the "particular solution" in mathematical analysis. Also, as t increases ($t \to \infty$), this solution approaches the *steady-state response* denoted by y_{ss}, which is typically the particular solution. The impulse response of a system is the inverse Laplace transform of the transfer function. Hence, it can be determined using Laplace transform techniques. This aspect will be addressed in Chapter 3. Some useful concepts of forced response are summarized in Box 2.3.

2.5.3 Response to a Support Motion

An important consideration in vibration analysis and testing of machinery and equipment is the response to a support motion. To illustrate the method of analysis, consider the linear, single-degree-of-freedom system consisting of mass m, spring constant k, and damping constant b, subjected to

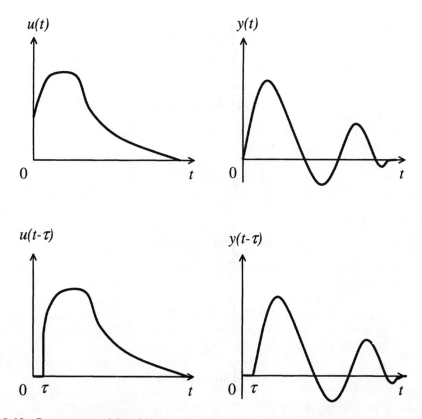

FIGURE 2.19 Response to a delayed input.

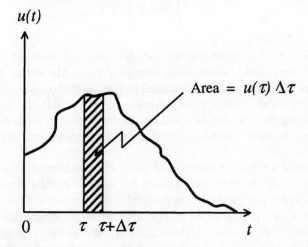

FIGURE 2.20 General input treated as a continuous series of impulses.

support motion (displacement) $u(t)$. Vertical and horizontal configurations of this system are shown in Figure 2.21. Both configurations possess the same equation of motion, provided the support motion $u(t)$ and the mass response (displacement) y are measured from the fixed points that correspond to the initial, static-equilibrium position of the system. In the vertical configuration, the compressive force in the spring exactly balances the weight of the mass when it is in static equilibrium. In the horizontal configuration, the spring is unstretched when in static equilibrium. It may be easily verified that the equation of motion is given by

BOX 2.3 Concepts of Forced Response

Total Response (T) = Homogeneous solution + Particular integral
$\qquad\qquad\qquad\qquad\qquad (H) \qquad\qquad\qquad\qquad (P)$

$\qquad\qquad\quad$ = Free response + Forced response
$\qquad\qquad\qquad\quad (X) \qquad\qquad\qquad (F)$

$\qquad\qquad\quad$ = Initial-condition response + Zero-initial-condition response
$\qquad\qquad\qquad\qquad (X) \qquad\qquad\qquad\qquad\qquad (F)$

$\qquad\qquad\quad$ = Zero-input response + Zero-state response
$\qquad\qquad\qquad\qquad (X) \qquad\qquad\qquad\qquad (F)$

Note: In general, $H \neq X$ and $P \neq F$
With no input (no forcing excitation), by definition, $H \equiv X$
At steady state, F becomes equal to P.

Convolution Integral: Response $y = \int_0^t h(t-\tau)u(\tau)d\tau = \int_0^t h(\tau)u(t-\tau)d\tau$

where u = excitation (input) and h = impulse response function (response to a unit impulse input).

Damped Simple Oscillator: $\ddot{y} + 2\zeta\omega_n\dot{y} + \omega_n^2 y = \omega_n^2 u(t)$

Poles (eigenvalues) $\lambda_1, \lambda_2 = -\zeta\omega_n \pm \sqrt{\zeta^2 - 1}\,\omega_n$ for $\zeta \geq 1$

$\qquad\qquad\qquad\qquad = -\zeta\omega_n \pm j\omega_d$ for $\zeta < 1$

ω_n = undamped natural frequency, ω_d = damped natural frequency

ζ = damping ratio. *Note*: $\omega_d = \sqrt{1-\zeta^2}\,\omega_n$

Impulse Response Function:
(zero ICs) $\quad h(t) = \dfrac{\omega_n}{\sqrt{1-\zeta^2}} \exp(-\zeta\omega_n t) \sin\omega_d t$ for $\zeta < 1$

$\qquad\qquad\qquad = \dfrac{\omega_n}{2\sqrt{\zeta^2 - 1}} \left[\exp\lambda_1 t - \exp\lambda_2 t\right]$ for $\zeta > 1$

$\qquad\qquad\qquad = \omega_n^2 t \exp(-\omega_n t)$ for $\zeta = 1$

Unit Step Response:
(zero ICs) $\quad y(t)_{step} = 1 - \dfrac{1}{\sqrt{1-\zeta^2}} \exp(-\zeta\omega_n t)\sin(\omega_d t + \phi)$ for $\zeta < 1$

$\qquad\qquad\qquad = 1 - \dfrac{1}{2\sqrt{\zeta^2-1}\,\omega_n}\left[\lambda_1 \exp\lambda_2 t - \lambda_2 \exp\lambda_1 t\right]$ for $\zeta > 1$

$\qquad\qquad\qquad = 1 - (\omega_n t + 1)\exp(-\omega_n t)$ for $\zeta = 1$

$\qquad\qquad \cos\phi = \zeta$

Note: Impulse response = $\dfrac{d}{dt}$ (Step response).

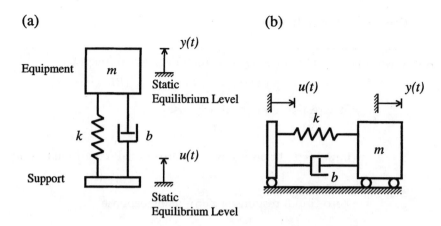

FIGURE 2.21 A system subjected to support motion: (a) vertical configuration, (b) horizontal configuration.

$$m\ddot{y} + b\dot{y} + ky = ku(t) + b\dot{u}(t) \tag{2.98}$$

in which $(\dot{\ }) = d/dt$ and $(\ddot{\ }) = d^2/dt^2$. The two parameters ω_n and ζ are undamped natural frequency and damping ratio, respectively, given by $\omega_n = \sqrt{\dfrac{k}{m}}$ and $2\zeta\omega_n = \dfrac{b}{m}$, as usual. This results in the equivalent equation of motion

$$\ddot{y} + 2\zeta\omega_n \dot{y} + \omega_n^2 y = \omega_n^2 u(t) + 2\zeta\omega_n \dot{u}(t) \tag{2.99}$$

There are several ways to determine the response y from equation (2.99) once the excitation function $u(t)$ is specified. The procedure adopted here is to first solve the modified equation

$$\ddot{y} + 2\zeta\omega_n \dot{y} + \omega_n^2 y = \omega_n^2 u(t) \tag{2.100}$$

This can be identified as the equation of motion of the single-degree-of-freedom system shown in Figure 2.15. Once this response is known, the response of the system (2.99) is obtained by the principle of superposition.

Impulse Response

Many important characteristics of a system can be studied by analyzing the system response to an impulse or a step-input excitation. Such characteristics include system stability, speed of response, time constants, damping properties, and natural frequencies. In this way, an idea of the system response to an arbitrary excitation is gained. A unit impulse or a unit step are baseline inputs or test inputs. Responses to such inputs can also serve as the basis for system comparison. In particular, it is usually possible to determine the degree of nonlinearity in a system by exciting it with two input intensity levels, separately, and checking whether the proportionality is retained at the output; or whether limit cycles are encountered by the response when the excitation is harmonic.

The response of the system (2.100) to a unit impulse $u(t) = \delta(t)$ can be conveniently determined by the Laplace transform approach (See Chapter 3). In the present section, a time-domain approach is used instead. First integrate equation (2.100) over the almost zero interval from $t = 0^-$ to $t = 0^+$. One obtains

$$\dot{y}(0^+) = \dot{y}(0^-) - 2\zeta\omega_n\left[y(0^+) - y(0^-)\right] - \omega_n^2\int_{0^-}^{0^+} y\,dt + \omega_n^2\int_{0^-}^{0^+} u(t)\,dt \qquad (2.101)$$

Suppose that the system starts from rest. Hence, $y(0^-) = 0$ and $\dot{y}(0^-) = 0$. Also, when an impulse is applied over an infinitesimal time period $[0^-,0^+]$, the system will not be able to move through a finite distance during that time. Hence, $y(0^+) = 0$ as well, and furthermore, the integral of y on the RHS of equation (2.101) will also be zero. Now, by definition of a unit impulse, the integral of u on the RHS of equation (2.101) will be unity. Hence, $\dot{y}(0^+) = \omega_n^2$. It follows that as soon as a unit impulse is applied to the system (2.100), the initial conditions will become

$$y(0^+) = 0 \quad \text{and} \quad \dot{y}(0^+) = \omega_n^2 \qquad (2.102)$$

Also, beyond $t = 0^+$, the excitation $u(t) = 0$, according to the definition of an impulse. Hence, the impulse response of the system (2.100) is obtained by its homogeneous solution (as carried out before, under free response), but with the initial conditions (2.102). The three cases of damping ratio ($\zeta < 1$, $\zeta > 1$, and $\zeta = 1$) should be considered separately. Then, one can conveniently obtain the following results:

$$y_{\text{impulse}}(t) = h(t) = \frac{\omega_n}{\sqrt{1-\zeta^2}}\exp(-\zeta\omega_n t)\sin\omega_d t \qquad \text{for } \zeta < 1 \qquad (2.103a)$$

$$y_{\text{impulse}}(t) = h(t) = \frac{\omega_n}{\sqrt{\zeta^2-1}}\left[\exp\lambda_1 t - \exp\lambda_2 t\right] \qquad \text{for } \zeta > 1 \qquad (2.103b)$$

$$y_{\text{impulse}}(t) = h(t) = \omega_n^2 t\exp(-\omega_n t) \qquad \text{for } \zeta = 1 \qquad (2.103c)$$

An explanation concerning the dimensions of $h(t)$ is appropriate at this juncture. Note that $y(t)$ has the same dimensions as $u(t)$. Since $h(t)$ is the response to a unit impulse $\delta(t)$, it follows that they have the same dimensions. The magnitude of $\delta(t)$ is represented by a unit area in the $u(t)$ vs. t plane. Consequently, $\delta(t)$ has the dimensions of time^{-1}, or frequency. It follows that $h(t)$ also has the dimensions of time^{-1} or frequency.

The Riddle of Zero Initial Conditions

For a second-order system, zero initial conditions correspond to $y(0) = 0$ and $\dot{y}(0) = 0$. It is clear from equations (2.103) that $h(0) = 0$, but $\dot{h}(0) \neq 0$, which appears to violate the zero-initial-conditions assumption. This situation is characteristic in system response to impulses and their derivatives and can be explained as follows. When an impulse is applied to a system at rest (zero initial state), the highest derivative of the system differential equation becomes infinity momentarily. As a result, the next lower derivative becomes finite (nonzero) at $t = 0^+$. The remaining lower derivatives maintain their zero values at that instant. When an impulse is applied to the system given by equation (2.100), for example, the acceleration $\ddot{y}(t)$ becomes infinity, and the velocity $\dot{y}(t)$ takes a non-zero (finite) value shortly after its application ($t = 0^+$). The displacement $y(t)$, however, would not have sufficient time to change at $t = 0^+$. The impulse input is therefore equivalent

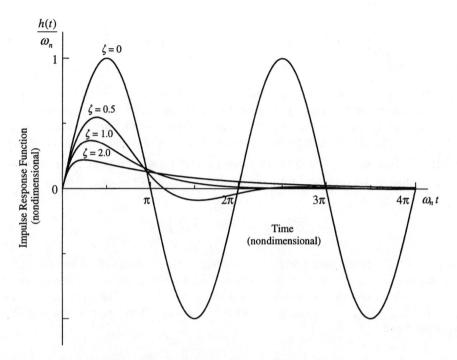

FIGURE 2.22 Impulse-response functions of a damped oscillator.

to a velocity initial condition in this case. This initial condition is determined using the integrated version (2.101) of the system equation (2.100), as has been done.

The impulse-response functions given by equations (2.103) are plotted in Figure 2.22 for some representative values of damping ratio. It should be noted that, for $0 < \zeta < 1$, the angular frequency of damped vibrations is ω_d, which is smaller than the undamped natural frequency ω_n.

Step Response

A unit step excitation is defined by

$$\begin{aligned}\mathcal{U}(t) &= 1 \quad \text{for} \quad t > 0 \\ &= 0 \quad \text{for} \quad t \leq 0\end{aligned} \tag{2.104}$$

Unit impulse excitation $\delta(t)$ can be interpreted as the time derivative of $\mathcal{U}(t)$:

$$\delta(t) = \frac{d\mathcal{U}(t)}{dt} \tag{2.105}$$

Note that equation (2.105) re-establishes the fact that for nondimensional $\mathcal{U}(t)$, the dimension of $\delta(t)$ is time^{-1}. Then, because a unit step is the integral of a unit impulse, the step response can be obtained directly as the integral of the impulse response; thus,

$$y_{\text{step}}(t) = \int_0^t h(\tau)d\tau \tag{2.106}$$

Time Response

This result also follows from the convolution integral (2.97g) because, for a delayed unit step, one has

$$\mathcal{U}(t-\tau) = 1 \quad \text{for} \quad \tau < t$$
$$= 0 \quad \text{for} \quad \tau \geq t \tag{2.107}$$

Thus, by integrating equations (2.103) with zero initial conditions, the following results are obtained for step response:

$$y_{\text{step}}(t) = 1 - \frac{1}{\sqrt{1-\zeta^2}} \exp(-\zeta\omega_n t)\sin(\omega_d t + \phi) \text{ for } \zeta < 1 \tag{2.108a}$$

$$y_{\text{step}} = 1 - \frac{1}{2\sqrt{1-\zeta^2}\,\omega_n} \left[\lambda_1 \exp\lambda_2 t - \lambda_2 \exp\lambda_1 t\right] \text{ for } \zeta > 1 \tag{2.108b}$$

$$y_{\text{step}} = 1 - (\omega_n t + 1)\exp(-\omega_n t) \text{ for } \zeta = 1 \tag{2.108c}$$

$$\text{with } \cos\phi = \zeta \tag{2.109}$$

The step responses given by equations (2.108) are plotted in Figure 2.23 for several values of damping ratio.

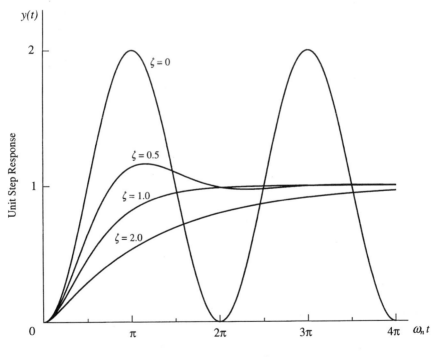

FIGURE 2.23 Unit step response of a damped simple oscillator.

Note that, because a step input does not cause the highest derivative of the system equation to approach infinity at $t = 0^+$, the initial conditions required to solve the system equation remain unchanged at $t = 0^+$, provided there are no derivative terms on the input side of the system equation. If there are derivative terms in the input, then, for example, a step can become an impulse and the situation changes.

Now the response of the system in Figure 2.21, when subjected to a unit step of support excitation (see equation (2.99)), is obtained using the principle of superposition, as the sum of the unit step response and $(2\zeta/\omega_n)$ times the unit impulse response of equation (2.100). Thus, from equations (2.103) and (2.108), one obtains the step response of the system in Figure 2.21 as

$$y(t) = 1 - \frac{\exp(-\zeta\omega_n t)}{\sqrt{1-\zeta^2}}\left[\sin(\omega_d t + \phi) - 2\zeta \sin\omega_d t\right] \text{ for } \zeta < 1 \tag{2.110a}$$

$$y(t) = 1 + \frac{1}{2\sqrt{1-\zeta^2}\,\omega_n}\left[\lambda_2 \exp\lambda_2 t - \lambda_1 \exp\lambda_1 t\right] \text{ for } \zeta > 1 \tag{2.110b}$$

$$y(t) = 1 + (\omega_n t - 1)\exp(-\omega_n t) \text{ for } \zeta = 1 \tag{2.110c}$$

Liebnitz's Rule

The time derivative of an integral for which the limits of integration are also functions of time can be obtained using Liebnitz's rule. It is expressed as

$$\frac{d}{dt}\int_{a(t)}^{b(t)} f(\tau,t)d\tau = f[b(t),t]\frac{db(t)}{dt} - f[a(t),t]\frac{da(t)}{dt} + \int_{a(t)}^{b(t)} \frac{\partial f}{\partial t}(\tau,t)d\tau \tag{2.111}$$

By repeated application of Liebnitz's rule to equation (2.97g), one can determine the ith derivative of the response variable; thus,

$$\frac{d^i y(t)}{dt^i} = \left[h(t) + \frac{dh(t)}{dt} + \cdots + \frac{d^{i-1}h(t)}{dt^{i-1}}\right]u(0)$$

$$+ \left[h(t) + \frac{dh(t)}{dt} + \cdots + \frac{d^{i-2}h(t)}{dt^{i-2}}\right]\frac{du(0)}{dt} + \cdots \tag{2.112}$$

$$+ h(t)\frac{d^{i-1}u(0)}{dt^{i-1}} + \int_0^t h(\tau)\frac{d^i u(t-\tau)}{dt^i}d\tau$$

From this result, it follows that the zero-state response to input $[d^i u(t)]/dt^i$ is $[d^i y(t)]/dt^i$, provided that all lower-order derivatives of $u(t)$ vanish at $t = 0$. This result verifies the fact, for instance, that the first derivative of the unit step response gives the impulse-response function.

It should be emphasized that the convolution integral (2.97) gives the forced response of a system, assuming that the initial conditions are zero. For non-zero initial conditions, the homogeneous solution (e.g., equation (2.54) or (2.58)) should be added to this zero-IC response and then the unknown constants should be evaluated by using the initial conditions. Care should be exercised

Time Response

in the situation where there is an initial velocity in the system and then an impulsive excitation is applied. In this case, one approach would be to first determine the velocity at $t = 0^+$ by adding to the initial velocity at $t = 0^-$, the velocity change in the system due to the impulse. The initial displacement will not change, however, due to the impulse. Once the initial conditions at $t = 0^+$ are determined in this manner, the complete solution can be obtained as usual.

PROBLEMS

2.1 From the point of view of energy, explain the phenomenon of natural mechanical vibration. Compare this with natural oscillations in electrical circuits by giving the electromechanical analogy, associated variables, and parameters.

2.2 Consider the undamped, simple oscillator given by

$$\ddot{x} + \omega_n^2 x = 0$$

It is known that $x = A \sin(\omega_n t + \phi)$ represents the complete solution to this system equation.

a. What are the physical meanings of the parameters ω_n and ϕ? Show that the velocity $\dot{x}$ leads the displacement x by an angle of $\pi/2$.

b. Explain why x alone does not represent a complete state of this system, but the pair x and $\dot{x}$ does.

c. With $x(0) = x_0$ and $\dot{x}(0) = v_0$ as initial conditions of the system, one can show that

$$A = x_0 \sqrt{1 + \left(\frac{v_0}{\omega_n x_0}\right)^2} \quad \text{and} \quad \tan \phi = \frac{\omega_n x_0}{v_0}.$$

Using these results, explain how the amplitude and the phase angle of the motion are affected by the initial conditions and the natural frequency of the system. Why are these observations intuitively clear as well?

2.3 It is claimed that "the natural variable for representing the response of a simple mass (inertia) element is its velocity and not displacement." Can you justify this statement?

Consider a mass m moving in a straight line on a horizontal, frictionless plane. If the velocity of the mass is v and the displacement is x, what is its equation of motion? Show that x alone cannot completely represent the "state" of the mass, but either v alone or x and v taken together can. What is the kinetic energy of the mass? Can it be represented in terms of x?

2.4 Discuss how mechanical vibrations in a robot arm could adversely affect its performance.

Consider the simplified model of a single-degree-of-freedom robot arm (single link) shown in Figure P2.4. The link is driven by a DC motor through a light shaft of torsional stiffness k_s.

The moments of inertia of the motor rotor and the robot arm (link) are J_m and J_l, respectively, about the common axis of rotation. The rotations of the motor and the link are denoted by θ_m and θ_l, respectively, as shown in the figure.

a. Write a single equation representing the differential motion $\theta_m - \theta_l$ of the robot.

b. What is the natural frequency of vibration of the robot joint?

c. Practically, would you design the robot to have a high natural frequency or a low natural frequency? Why? Explain how the natural frequency of vibration depends on the parameters k_s, J_m, and J_l. Which of these parameters can be adjusted in order to obtain the required (design) natural frequency?

2.5 Discuss how the mass of a spring can affect the natural frequency of vibration of a system of which the spring is a component.

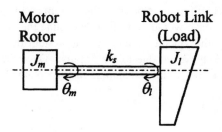

FIGURE P2.4 A single-link robot arm.

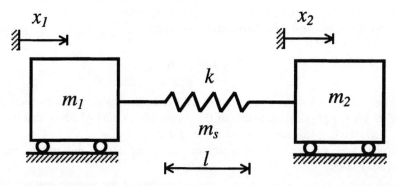

FIGURE P2.5 A two-car train.

Consider two rail cars of mass m_1 and m_2 linked by a spring of mass m_s and stiffness k_s, as shown by the simplified model in Figure P2.5.

Assume that the spring is uniform with a uniform velocity distribution along its length (l). The displacements of the two cars are denoted by x_1 and x_2, as shown, such that $x_1 = 0$ and $x_2 = 0$ correspond to the unstretched configuration of the spring.

a. Obtain expressions for the total kinetic energy and the elastic potential energy of the system.
b. Clearly providing justification, suggest an approximate lumped-parameter model of the system where the mass of the spring is lumped at one or more locations. You must give all the mass elements, spring elements, and their parameters in terms of m_1, m_2, m_s, and k_s.

2.6 Define the terms:
   a. Undamped natural frequency
   b. Damped natural frequency
   Consider the damped second-order system given by

$$\ddot{x} + 2\zeta\omega_n \dot{x} + \omega_n^2 x = 0$$

Define the parameters ω_n and ζ, giving their physical meanings within the context of mechanical vibration. Give an expression for damped natural vibration ω_d of this system.

Describe the free response of the system for the three cases:
   (i) $\zeta < 1$
   (ii) $\zeta > 1$
   (iii) $\zeta = 1$

2.7 a. Consider a heavy uniform spring of mass m_s and stiffness k, with one end fixed and the other end free to move. Clearly showing all the steps and stating the necessary

Time Response

assumptions, show that this distributed-parameter system can be approximated by a massless spring of stiffness k, with a lumped mass $m_s/3$ at the free end.

Hint: Obtain the kinetic energy (KE) and the elastic potential energy (PE) of the system and then establish a lumped-parameter system having the same KE and PE.

b. Explain why the lumped-parameter approximation obtained in part (a) essentially corresponds to the first mode of the heavy (distributed-parameter) spring, and not a higher mode. No analysis is needed.

Hint: Consider the assumptions that were made in obtaining the lumped-parameter approximation.

c. A heavy spring of mass m_s was fixed at one end. The free end was pressed through a distance of A_1 from the static equilibrium position, held stationary, and released. At the completion of the first cycle of vibration, from this starting time, the free end was found to be deflected through A_2 from the static equilibrium position (*Note*: $A_2 < A_1$). Also, the time period of the cycle was found to be T. Obtain expressions for the following, in terms of the measured parameters m_s, A_1, A_2, and T:
  (i) Damping ratio ζ
  (ii) Undamped natural frequency ω_n
  (iii) Spring stiffness k

2.8 Energy method is useful in determining the equivalent mass and equivalent stiffness of a vibrating system. Consider the example shown in Figure P2.8.

First consider the lumped-parameter system shown in Figure P2.8(a). A light, yet rigid beam is hinged at one end using a frictionless pivot, and restrained using a torsional spring, with torsional stiffness k at the hinged end. Two point masses m_1 and m_2 are attached at distances l_1 and l_2, respectively, from the hinged end. Under static equilibrium conditions, the beam remains in a horizontal configuration.

Suppose that the beam is excited by an initial push and left to vibrate in a small angular displacement θ. As a result, the mass m_2 undergoes a lateral-displacement vibration y. An equivalent vibratory system is shown in Figure P2.8(b). Here, m_{eq} is an equivalent mass assumed to be present at the location of m_2 and restrained by an equivalent linear spring there with stiffness k_{eq}.

a. Explain why gravity effects do not enter the equations of motion of Figure P2.8, assuming that θ and y are measured from the static equilibrium configuration.

b. Obtain expressions for m_{eq} and k_{eq} in terms of the parameters of the original system shown in Figure P2.8(a).

c. What is the natural frequency of the system in Figure P2.8(a)?
What is the natural frequency of the system in Figure P2.8(b)?
Comment on these results.

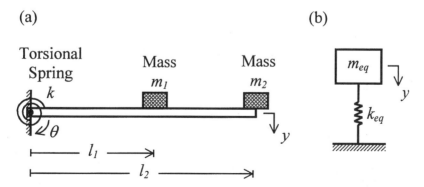

FIGURE P2.8 (a) A lumped-parameter mechanical system; (b) an equivalent system.

2.9 Consider the second-order, nonlinear, autonomous, dynamic system that is represented by the state-space model.

$$\dot{q}_1 = q_1 q_2 - r(t)$$
$$\dot{q}_2 = q_1 - q_2$$

where q_1 and q_2 are the state variables and $r(t)$ is the excitation (input) variable.

a. Linearize this system about an operating point where a steady excitation $\bar{r}$ is applied, and the resulting state of the system is $[\bar{q}_1, \bar{q}_2]^T$. Identify the matrices A and B of the linear state-space model (see Appendix A).

b. Discuss the stability of the linearized model.

2.10 Consider a test setup for a delicate instrument enclosed in a massless casing. The system is modeled as in Figure P2.10. A velocity excitation $u(t)$ is applied to the shaker table using a linear, electromagnetic actuator. Determine a state-space model for the system that can be used to study the velocity (v) of the instrument.

2.11 A heavy engine was placed on a flexible mount of sufficient stiffness and negligible damping. The mount was displaced through a vertical distance of y_0 from its relaxed position, due to the weight of the engine. A sketch of the system is shown in Figure P2.11.

What is the natural frequency (undamped) of the engine-mount system? If the damping in the engine mount is such that the damping ratio is ζ, what is the true (damped) natural frequency?

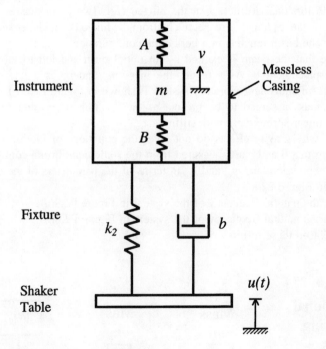

FIGURE P2.10 Model for an instrument test setup.

2.12 A compound pendulum of mass m is suspended from a smooth pivot at the point O, as shown in Figure P2.12.

The radius of gyration of the pendulum about O is k_0. Suppose that the centroid G of the pendulum is at a distance l from O.

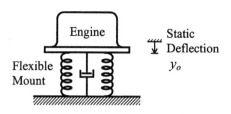

FIGURE P2.11 An engine placed on a flexible mount.

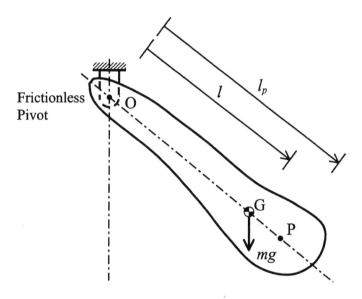

FIGURE P2.12 A compound pendulum.

a. Write an equation of motion for the pendulum in terms of the angle of swing θ about the vertical equilibrium configuration. What is the natural frequency of motion of the pendulum? Show that this is less than the natural frequency of a simple pendulum of length l.
b. The center of precession P of the pendulum is defined, with $OP = l_p$, such that the natural frequency of the compound pendulum is equal to that of a simple pendulum of length l_p. Obtain an expression for l_p in terms of l and k_0.
c. Show that if the compound pendulum is hung at P instead of O, its natural frequency of motion will remain the same as before. (*Note*: This is a defining property of the center of precession of a compound pendulum.)

2.13 The inverted pendulum is a simple model that is used in the stability study of inherently unstable systems such as rockets. Consider an inverted pendulum of point mass m and length l that is restrained at its pivot (smooth) by a torsional spring of stiffness k. This arrangement is sketched in Figure P2.13.
  a. Derive an equation of motion for this system?
  b. Obtain an expression for the natural frequency of small oscillations θ about the vertical configuration. Under what conditions would such oscillations not be possible (i.e., the system become unstable)?

2.14 Some types of industrial conveyors have sequentially placed holding pockets (pans) for the objects that are transported on these conveyors. Each pocket is appropriately curved for stable holding of an object. Consider the idealized case where a holding pocket has a circular curvature of radius R. Three types of uniform objects are placed in the pockets

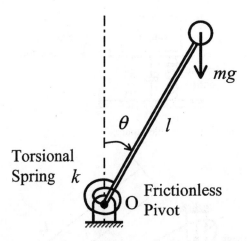

FIGURE P2.13 Spring-restrained inverted pendulum used in stability studies.

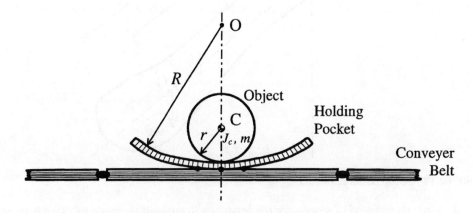

FIGURE P2.14 A conveyor with holding pans (pockets) for curved objects.

$$\text{Disk} \quad \left(J_c = 1/2 \; mr^2\right)$$

$$\text{Cylinder} \quad \left(J_c = 1/2 \; mr^2\right)$$

$$\text{Sphere} \quad \left(J_c = 3/10 \; mr^2\right)$$

where J_c is the moment of inertia about the rolling axis through the center of the object, m is the mass, and r is the radius of the object. This arrangement is illustrated in Figure P.2.14.

Suppose that the object rolls up slightly from its equilibrium position as a result of an initial jerk in the conveyor. Determine the natural frequency of the rolling motion that ensues, assuming that there is no slip between the object and the pan.

2.15 A simple analysis can be carried out to study the stability of rolling (or pitch) motions of a ship. It is known that the requirement for stability is that the metacenter M should fall above the centroid C of the ship. The Archimedes principle states that the buoyancy force R on an object immersed in a liquid is equal to the weight of the liquid displaced by the object. Furthermore, R acts upward through the centroid of the liquid mass that is displaced. Its line of action will intersect the upright axis of the body that passes through the centroid C. This point of intersection is the metacenter M, as shown in Figure P2.15.

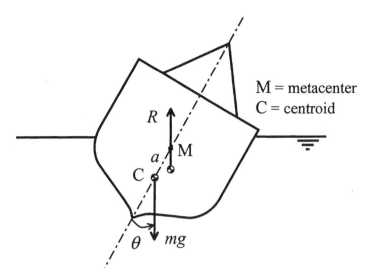

FIGURE P2.15 Stable rolling oscillations of a ship.

Consider a ship of mass m and rolling moment of inertia J about its centroid. Let $CM = a$ as in Figure P2.15. Obtain an equation for rolling motion θ of the ship for the stable configuration. What is the natural frequency of small oscillations?

2.16 A reciprocating carriage system of a photocopier is sketched in Figure P2.16. It consists of a carriage of mass M driven by a spring-loaded linkage mechanism. The four-bar linkage is symmetric with each bar (which is assumed light and rigid) having a length l. The cross spring has a stiffness k, and has two end masses m as shown.

Consider a general configuration where each linkage bar makes an angle θ with the cross spring axis. What is the equivalent mass m_{eq} and the equivalent stiffness k_{eq} of the system, with respect to the location of the carriage M? What is the natural frequency of motion of the carriage system in the close neighborhood of this configuration? Neglect energy dissipation and consider the free (i.e., no drive force) motion. You may assume that the plane of motion is horizontal.

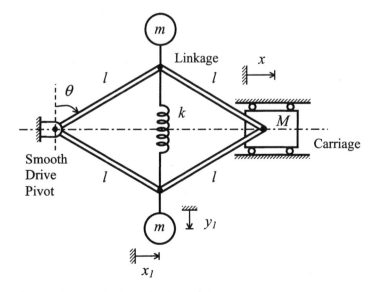

FIGURE P2.16 The carriage mechanism of a photocopier.

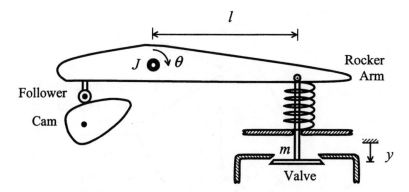

FIGURE P2.17 A vibrating cam-follower mechanism.

Would the natural frequency of motion be different if the plane of motion is vertical with the carriage (M) moving
a. Horizontally?
b. Vertically?

2.17 Cam-follower mechanisms are commonly used to realize timed, periodic motions having some desired characteristics. For example, they are used in synchronized opening and closing of valves in internal combustion (IC) engines. A schematic representation of such an arrangement is shown in Figure P2.17.

The rocker arm is supported on a smooth pivot. One end of it carries a spring-loaded valve. The other end (drive end) has the follower (roller-type) for which the input motion is determined by the shape profile and the rotatory speed of the cam that is in intimate contact with the follower. The following parameters are given:

J = moment of inertia of the rocker arm and follower combination about the supporting pivot
m = mass of the valve and stem combination (not included in J)
l = lever arm length of the valve weight from the pivot point
k = stiffness of the valve spring

a. Determine expressions for the equivalent mass m_{eq} and the equivalent stiffness k_{eq} of the entire system as located at the valve.
b. What is the undamped natural frequency of rocking motions? What is the significance of this frequency in the proper operation of the cam-follower system?
c. If the equivalent (linear, viscous) damping constant at the valve location is b, what is the damped natural frequency and the damping ratio of the system?

2.18 Consider a heavy coil of helical spring, as shown in Figure P2.18. The force (f) vs. deflection (x) relationship of the spring is given by

$$f = \frac{Gd^4}{8nD^3} x$$

where
d = diameter of the coil wire of the spring
D = mean diameter of the spring (coil)
n = number of (active) turns in the spring
G = shear modulus of the coil wire

Assuming that the mass density of the spring material is ρ, derive an expression for the first undamped natural frequency of oscillation of the spring in the fixed-free configuration of end conditions shown in the figure.

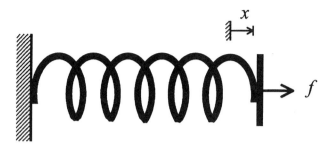

FIGURE P2.18 A heavy helical spring.

2.19 An excavator boom/stick along with its bucket is modeled as a light rigid rod of length l with a lumped end mass m, as shown in Figure P2.19. The flexibility in the system is modeled as a torsional spring of stiffness k located at the base O. Write an equation of motion for the boom for small rotations α from the static equilibrium configuration where the boom is inclined at an angle θ with the vertical. What is the corresponding natural frequency of oscillation? Determine the condition of stability. Neglect damping effects.

2.20 Consider a uniform elastic post of mass m, length l, and area of cross section A that is vertically mounted on a rigid concrete floor. There is a mss M attached to the top end of the post, as shown in Figure P2.20(a). In studying longitudinal vibrations of the post, we wish to obtain an equivalent model as shown in Figure P2.20(b), where m_e is the equivalent mass of the post as concentrated at the top end and k_e is the equivalent stiffness for longitudinal motions. Assuming a linear variation of longitudinal deflection and velocity along the post, determine k_{eq} and m_{eq}. What is the natural frequency of undamped longitudinal vibrations? The Young's modulus of the post material is E.

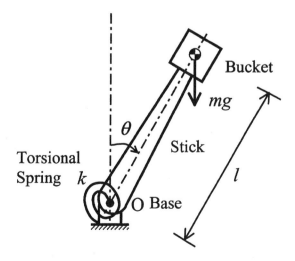

FIGURE P2.19 An excavator stick (boom) with the bucket.

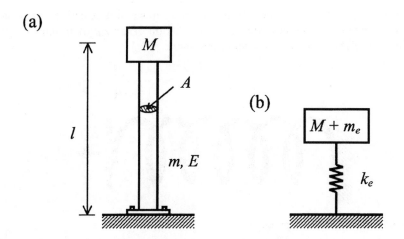

Hint:

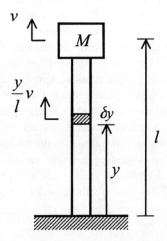

FIGURE P2.20 (a) A vertical post with an end mass.; (b) an equivalent model for longitudinal vibrations.

2.21 This problem deals with resolving the direction of a spring force in vibration studies. Consider the spring-loaded carriage mechanism sketched in Figure P2.21. The carriage unit of mass m is supported by two springs (with restraining guide plates), one vertical having stiffness k and the other inclined at θ to the horizontal, with stiffness k_a.

Show that the equivalent stiffness k_{eq} on the carriage M in the vertical direction is $k + k_a/\sin^2\theta$. What is the undamped natural frequency of vibration of the carriage system?

2.22 A centrifugal water pump is to be located at the free end of an overhung beam, as shown in Figure P2.22. It is required that the operating speed of the pump does not correspond to a natural frequency (strictly, a resonant frequency) of the structural system. Given that:

M = mass of the pump
m = mass of the beam
l = length of the beam
I = 2nd moment of area of the beam cross section about the horizontal neutral axis of bending
E = Young's modulus of the beam material

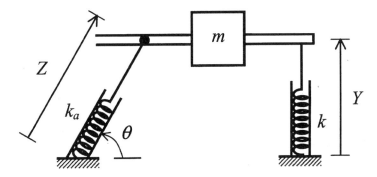

FIGURE P2.21 A carriage with bi-directional spring restraints.

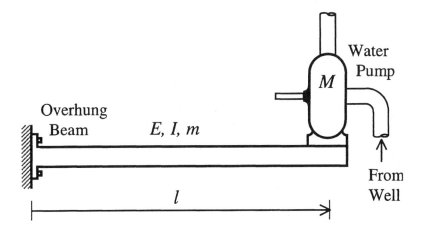

Hint:

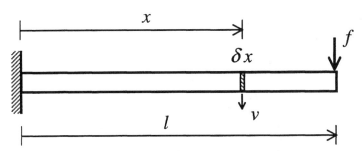

FIGURE P2.22 A water pump mounted on an overhung beam.

determine an expression for the equivalent linear stiffness k_{eq} and mass m_{eq} of the beam, as located at the free end. What is the corresponding natural frequency of vibration of the pump-beam system?

2.23 A gear transmission has a meshed pair of gear wheels of moments of inertia J_d and J_l about their axes of rotation, and a step-down gear ratio r. The load gear wheel is connected to a purely torsional load of stiffness k_l as shown in Figure P2.23. Suppose that the angle of rotation of the drive gear wheel is θ and that of the load gear wheel is α in the opposite direction, so that $r = \theta/\alpha$. Determine the equivalent moment of inertia and the equivalent torsional stiffness of the system, both with respect to the drive side and the load side of the transmission. For each case, what is the natural frequency of torsional vibration? Justify the results.

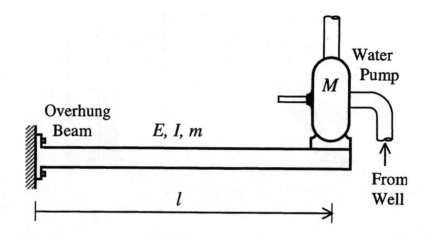

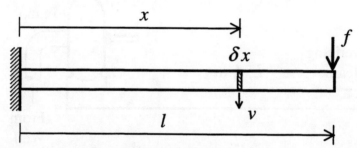

FIGURE P2.22 A water pump mounted on an overhung beam.

Hint:

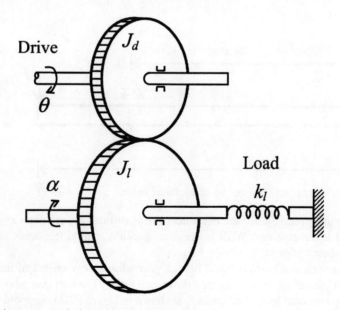

FIGURE P2.23 A gear transmission with a torsional load.

2.24 The handle of a hoist is modeled as a rigid light rod pivoted at the bottom and restrained by a torsional spring of stiffness k, along with a uniform circular disk of mass m and radius r attached to the top end, as shown in Figure P2.24. The distance from the bottom pivot to the center of the disk is l. Initially, the handle is in its vertical configuration, where the spring is in its relaxed position. Obtain an equation for angular motion θ of the handle with respect to this configuration. What is the natural frequency of small vibrations of the handle about its upright position? Under what conditions would such vibrations be not possible? Neglect energy dissipation.

2.25 A simplified model of an elevator is shown in Figure P2.25.

Note that:

J = moment of inertia of the cable pulley
r = radius of the pulley
k = stiffness of the cable
m = mass of the car and occupants.

a. Which system parameters are variable? Explain.
b. Suppose that the damping torque $T_d(\omega)$ at the bearings of the pulley is a *nonlinear* function of the angular speed ω of the pulley. Taking the state vector x as

$$x = [\omega \ f \ v]^T$$

in which

f = tension force in the cable
v = velocity of the car (taken positive upwards)

the input vector u as

$$u = [T_m]$$

in which

T_m = torque applied by the motor to the pulley
(positive in the direction indicated in figure)

and, the output vector y as

$$y = [v]$$

obtain a complete, *nonlinear* state-space model for the system.

c. With T_m as the input and v as the output, convert the state-space model into the nonlinear input-output differential equation model. What is the order of the system?
d. Give an equation for which the solution provides the steady-state operating speed $\bar{v}$ of the elevator car.
e. Linearize the nonlinear input/output differential-equation model obtained in part (c), for small changes $\hat{T}_m$ of the input and $\hat{v}$ of the output, about an operating point. *Note*: $\bar{T}_m$ = steady-state operating-point torque of the motor (assumed known).

Hint: Denote $\dfrac{d}{d\omega} T_d(\omega)$ as $b(\omega)$.

f. Linearize the state-space model obtained in part (b) and give the model matrices A, B, C, and D in the usual notation. Obtain the linear input/output differential equation from this state-space model and verify that it is identical to what was obtained in part (e).

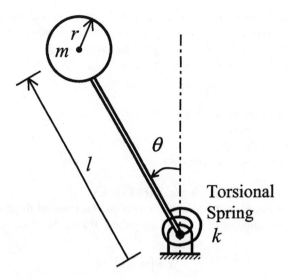

FIGURE P2.24 A handle of a mechanical hoist.

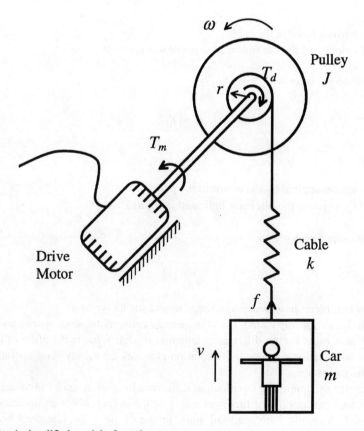

FIGURE P2.25 A simplified model of an elevator.

2.26 An automated wood cutting system contains a cutting unit that consists of a DC motor and a cutting blade, linked by a flexible shaft and coupling. The purpose of the flexible shaft is to locate the blade unit at any desirable configuration, away from the motor itself. A simplified, lumped-parameter, dynamic model of the cutting unit is shown in Figure P2.26.

Time Response

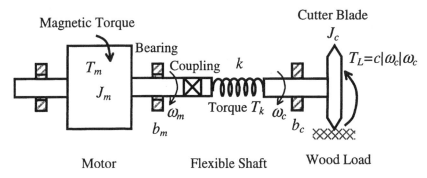

FIGURE P2.26 A wood cutting machine.

The following parameters and variables are shown in the figure:

J_m = axial moment of inertia of the motor rotor
b_m = equivalent viscous damping constant of the motor bearings
k = torsional stiffness of the flexible shaft
J_c = axial moment of inertia of the cutter blade
b_c = equivalent viscous damping constant of the cutter bearings
T_m = magnetic torque of the motor
ω_m = motor speed
T_k = torque transmitted through the flexible shaft
ω_c = cutter speed
T_L = load torque on the cutter from the workpiece (wood)

In comparison with the flexible shaft, the coupling unit is assumed rigid, and is also assumed light. The cutting load is given by

$$T_L = c|\omega_c|\omega_c$$

The parameter c, which depends on factors such as the depth of cut and the material properties of the workpiece, is assumed to be constant in the present analysis.

a. Using T_m as the input, T_L as the output, and $x = \left[\hat{\omega}_m, \hat{T}_k, \hat{\omega}_c\right]^T$ as the state vector, develop a complete (nonlinear) state model for the system shown in Figure P2.26. What is the order of the system?
b. Using the state model derived in part (a), obtain a single input-output differential equation for the system, with T_m as the input and ω_c as the output.
c. Consider the steady operating conditions, where $T_m = \overline{T}_m$, $\omega_m = \overline{\omega}_m$, $T_k = \overline{T}_k$, $\omega_c = \overline{\omega}_c$, and $T_L = \overline{T}_L$ are all constants. Express the operating point values $\overline{\omega}_m$, $\overline{T}_k$, $\overline{\omega}_c$, and $\overline{T}_L$ in terms of $\overline{T}_m$ and the model parameters. You must consider both cases: $\overline{T}_m > 0$ and $\overline{T}_m < 0$.
d. Now consider an incremental change $\hat{T}_m$ in the motor torque and the corresponding changes $\hat{\omega}_m$, $\hat{T}_k$, $\hat{\omega}_c$, and $\hat{T}_L$ in the system variables. Determine a linear state model (A, B, C, D) for the incremental dynamics of the system in this case, using $x = \left[\hat{\omega}_m, \hat{T}_k, \hat{\omega}_c\right]^T$ as the state vector, $u = \left[\hat{T}_m\right]$ as the input and $y = \left[\hat{T}_L\right]$ as the output.

e. In the nonlinear model (see part (a)), if the twist angle of the flexible shaft (i.e., $\theta_m - \theta_c$) is used as the output what would be a suitable state model? What is the system order then?

f. In the nonlinear model, if the angular position θ_c of the cutter blade is used as the output variable, explain how the state model obtained in part (a) should be modified. What is the system order in this case?

Hint for Part (b):

$$\frac{d}{dt}\left(|\omega_c|\omega_c\right) = 2|\omega_c|\dot{\omega}_c$$

$$\frac{d^2}{dt^2}\left(|\omega_c|\omega_c\right) = 2|\omega_c|\ddot{\omega}_c + 2\dot{\omega}_c^2 \operatorname{sgn}(\omega_c)$$

2.27 It is required to study the dynamic behavior of an automobile during the very brief period of a sudden start from rest. Specifically, the vehicle acceleration a in the direction of primary motion, as shown in Figure P2.27(a), is of interest and should be considered as the system output. The equivalent force $f(t)$ of the engine, applied in the direction of primary motion, is considered as the system input. A simple dynamic model that can be used for the study is shown in Figure P2.27(b).

Note that k is the equivalent stiffness, primarily due to tire flexibility, and b is the equivalent viscous damping constant, primarily due to energy dissipations at the tires and other moving parts of the vehicle, taken in the direction of a. Also, m is the mass of the vehicle.

a. Discuss advantages and limitations of the proposed model for the specific purpose.

b. Using force f_k of the spring (stiffness k) and velocity v of the vehicle as the state variables, engine force $f(t)$ as the input, and the vehicle acceleration a as the output, develop a complete state-space model for the system.
(*Note*: You must derive the matrices **A**, **B**, **C**, and **D** for the model.)

c. Obtain the input/output differential equation of the system.

d. Discuss the characteristics of this model by observing the nature of matrix **D**, and the input and output orders of the input-output differential equation.

2.28 a. Briefly explain why a purely thermal system typically does not have a free oscillatory response, whereas a fluid system can.

b. Figure P2.28 shows a pressure-regulated system that can provide a high-speed jet of liquid. The system consists of a pump, a spring-loaded accumulator, and a fairly long section of piping that ends with a nozzle. The pump is considered as a flow source of value Q_s. The following parameters are important:

A = area of cross section (uniform) of the accumulator cylinder
k = spring stiffness of the accumulator piston wall
L = length of the section of piping from the accumulator to the nozzle
A_p = area of cross section (uniform, circular) of the piping
A_o = exit area of the nozzle
C_d = discharge coefficient of the nozzle
ρ = mass density of the liquid

Assume that the liquid is incompressible. The following variables are important:

$P_{1r} = P_1 - P_r$ = pressure at the inlet of the accumulator with respect to the ambient reference P_r

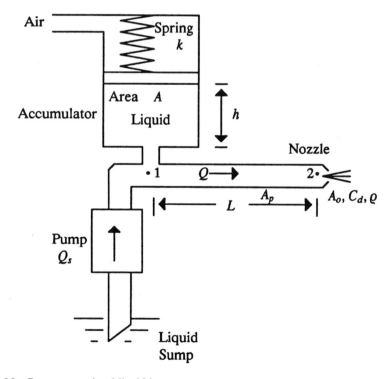

FIGURE P2.27 (a) A vehicle suddenly accelerating from rest; (b) a simplified model.

FIGURE P2.28 Pressure regulated liquid jet system.

Q = volume flow rate through the nozzle
h = height of the liquid column in the accumulator

Note that the piston (wall) of the accumulator can move against the spring, thereby varying h.

i. Considering the effects of the movement of the spring-loaded wall and the gravity head of the liquid, obtain an expression for the equivalent fluid capacitance C_a of the accumulator in terms of k, A, ρ, and g. Are the two capacitances that contribute to C_a (i.e., wall stretching and gravity) connected in parallel or in series?
Note: Neglect the effect of bulk modulus of the liquid.

ii. Considering the capacitance C_a, the inertance I of the fluid volume in the piping (length L and cross-sectional area A_p), and the resistance of the nozzle only, develop a nonlinear state-space model for the system. The state vector $x = [P_{1r}, Q]^T$ and the input $u = [Q_s]$.

For flow in the (circular) pipe with a parabolic velocity profile, the inertance is given by

$$I = \frac{2\rho L}{A_p}$$

and the discharge through the nozzle is given by

$$Q = A_0 C_d \sqrt{\frac{2 P_{2r}}{\rho}}$$

in which P_{2r} is the pressure inside the nozzle with respect to the outside reference pressure P_r.

2.29 Give reasons for the common experience that in the flushing tank of a household toilet, some effort is needed to move the handle for the flushing action but virtually no effort is needed to release the handle at the end of the flush.

A simplified model for the valve movement mechanism of a household flushing tank is shown in Figure P2.29. The overflow tube on which the handle lever is hinged is assumed rigid. Also, the handle rocker is assumed light, and the rocker hinge is assumed frictionless. The following parameters are indicated in the figure:

$r = \dfrac{l_v}{l_h}$ = the lever arm ratio of the handle rocker

m = equivalent lumped mass of the valve flapper and the lift rod
k = stiffness of the spring action on the valve flapper

The damping force f_{NLD} on the valve is assumed quadratic and is given by

$$f_{NLD} = a |v_{NLD}| v_{NLD}$$

where the positive parameter

$a = a_u$ for upward motion of the flapper ($v_{NLD} \geq 0$)
$ = a_d$ for downward motion of the flapper ($v_{NLD} \leq 0$)

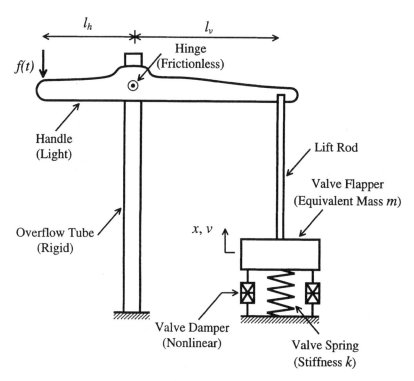

FIGURE P2.29 Simplified model of a toilet flushing mechanism.

with

$$a_u \gg a_d$$

The force applied at the handle is $f(t)$, as shown.

We are interested in studying the dynamic response of the flapper valve. Specifically, the valve displacement x and the valve speed v are considered outputs, as shown in Figure P2.29. Note that x is measured from the static equilibrium point of the spring where the weight mg is balanced by the spring force.

a. Using valve speed (v) and the spring force (f_k) as the state variables, develop a (nonlinear) state-space model for the system.
b. Linearize the state-space model about an operating point where the valve speed is $\bar{v}$. For the linearized model, obtain the model matrices **A**, **B**, **C**, and **D** in the usual notation. Note that the incremental variables $\hat{x}$ and $\hat{v}$ are the outputs in the linear model, and the incremental variable $\hat{f}(t)$ is the input.
c. From the linearized state-space model, derive the input-output model relating $\hat{f}(t)$ and $\hat{x}$.
d. Give expressions for the undamped natural frequency and the damping ratio of the linear model in terms of the parameters a, $\bar{v}$, m, and k. Show that the damping ratio increases with the operating speed.

2.30 The electrical circuit shown in Figure P2.30 has two resistors R_1 and R_2, an inductor L, a capacitor C, and a voltage source $u(t)$. The voltage across the capacitor is considered the output y of the circuit.

a. What is the order of the system and why?
b. Show that the input-output equation of the circuit is given by

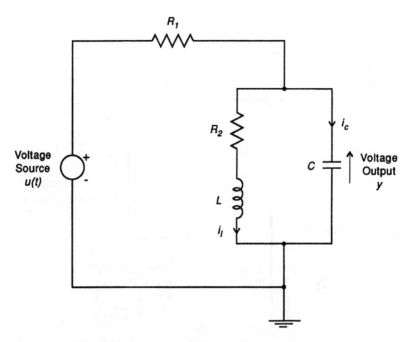

FIGURE P2.30 An RLC circuit driven by a voltage source.

$$a_2 \frac{d^2y}{dt^2} + a_1 \frac{dy}{dt} + a_0 y = b_1 \frac{du}{dt} + b_0 u$$

Express the coefficients a_0, a_1, b_0, and b_1 in terms of the circuit parameters R_1, R_2, L, and C. What is the undamped natural frequency? What is the damping ratio?

c. Starting with the auxiliary differential equation

$$a_2 \ddot{x} + a_1 \dot{x} + a_0 x = u$$

and using $x = [x \; \dot{x}]^T$ as the state vector, obtain a complete state-space model for the system in Figure P2.30. Note that this is the "superposition method" of developing a state model.

d. Clearly explain why, for the system in Figure P2.30, neither the current i_c through the capacitor, nor the time derivative of the output (i.e., $\dot{y}$) can be chosen as a state variable.

2.31 Consider two water tanks joined by a horizontal pipe with an on/off valve. With the valve closed, the water levels in the two tanks were initially maintained unequal. When the valve is suddenly opened, some oscillations were observed in the water levels of the tanks. Suppose that the system is modeled as two gravity-type capacitors linked by a fluid resistor. Would this model exhibit oscillations in the water levels when subjected to an initial-condition excitation? Clearly explain your answer.

A centrifugal pump is used to pump water from a well into an overhead tank. This fluid system is schematically shown in Figure P2.31(a). The pump is considered as a pressure source $P_s(t)$ and the water level h in the overhead tank is the system output. The ambient pressure is denoted by P_a. The following parameters are given:

L_v, d_v = length and the internal diameter of the vertical segment of the pipe

(a)

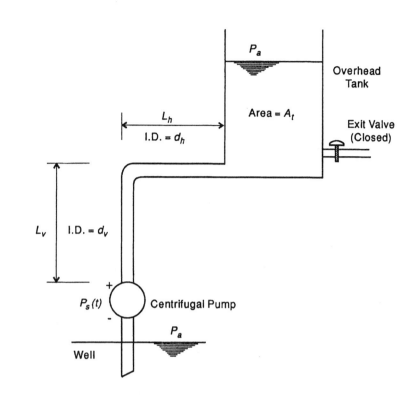

(b)

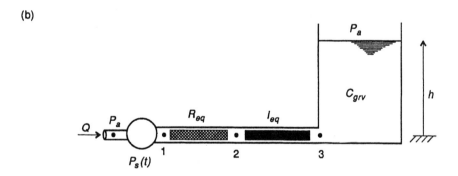

FIGURE P2.31 (a) A system for pumping water from a well into an overhead tank; (b) lumped parameter model of the fluid system.

L_h, d_h = length and the internal diameter of the horizontal segment of the pipe
A_t = area of cross section of the overhead tank (uniform)
ρ = mass density of water
μ = dynamic viscosity of water
g = acceleration due to gravity

Suppose that this fluid system may be approximated by the lumped parameter model shown in Figure P2.31(b).

a. Give expressions for the equivalent linear fluid resistance of the overall pipe (i.e., combined vertical and horizontal segments) R_{eq}, the equivalent fluid inertance within the overall pipe I_{eq}, and the gravitational fluid capacitance of the overhead tank C_{grv} in terms of the system parameters defined above.

b. Treating $x = \begin{bmatrix} P_{3a} & Q \end{bmatrix}^T$ as the state vector,
where

P_{3a} = pressure head of the overhead tank
Q = volume flow rate through the pipe

develop a complete state-space model for the system. Specifically, obtain the matrices A, B, C, and D.

c. Obtain the input-output differential equation of the system. What is the characteristic equation of this system?

d. Using the following numerical values for the system parameters:
$L_v = 10.0$ m, $L_h = 4.0$ m, $d_v = 0.025$ m, $d_h = 0.02$ m
$\rho = 1000.0$ kg m^{-3}, $\mu = 1.0 \times 10^{-3}$ N s m^{-2}, and tank diameter = 0.5 m,
compute the undamped natural frequency ω_n and the damping ratio ζ of the system. Will this system provide an oscillatory natural response? If so, what is the corresponding frequency? If not, explain why.

2.32 a. Define the following terms with reference to the response of a dynamic system:
   i. Homogeneous solution
   ii. Particular solution (or particular integral)
   iii. Zero-input (or free) response
   iv. Zero-state (or forced) response
   v. Steady-state response

b. Consider the first-order system

$$\tau \frac{dy}{dt} + y = u(t)$$

in which u is the input, y is the output, and τ is a system constant.
   i. Suppose that the system is initially at rest with $u = 0$ and $y = 0$, and suddenly a unit step input is applied. Obtain an expression for the ensuing response of the system. Into which of the above five categories does this response fall? What is the corresponding steady-state response?
   ii. If the step input in part (i) above is of magnitude A, what is the corresponding response?
   iii. If the input in part (i) above was an impulse of magnitude P, what would be the response?

2.33 Consider a mechanical system that is modeled by a simple mass-spring-damper unit with a forcing excitation. Its equation of motion is given by the normalized form

$$\ddot{y} + 2\zeta\omega_n \dot{y} + \omega_n^2 y = \omega_n^2 u(t)$$

in which $u(t)$ is the forcing excitation and y is the resulting displacement response. The system is assumed to be underdamped ($\zeta < 1$).

Suppose that a unit step excitation is applied to the system. At a subsequent time (which can be assumed $t = 0$, without loss of generality) when the displacement is y_0 and the velocity is v_0, a unit impulse is applied to the mass of the system. Obtain an expression for y that describes the subsequent response of the system, with proper initial conditions.

3 Frequency Response

In many vibration problems, the primary excitation force typically has a repetitive periodic nature and in some cases this periodic forcing function may be even purely sinusoidal. Examples are excitations due to mass eccentricity and misalignments in rotational components, tooth meshing in gears, and electromagnetic devices excited by ac or periodic electrical signals. In basic terms, the frequency response of a dynamic system is the response to a pure sinusoidal excitation. As the amplitude and frequency of the excitation are changed, the response also changes. In this manner, the response of the system over a range of excitation frequencies can be determined and this represents the frequency response. In this case, frequency (ω) is the independent variable and hence one is dealing with the *frequency domain*. In contrast, in Chapter 2 for response consideration in the time domain, the independent variable is time (t).

Frequency-domain considerations are applicable even when the signals are not periodic. In fact, a time signal can be transformed into its frequency spectrum through the Fourier transform. This subject will be studied in more detail in Chapter 4. For the time being, it is adequate to realize that for a given time signal, an equivalent Fourier spectrum, which contains all the frequency (sinusoidal) components of the signal, can be determined either analytically or computationally. Hence, a time-domain representation and analysis has an equivalent frequency-domain representation and analysis, at least for linear dynamic systems. For this reason, and also because of the periodic nature of typical vibration signals, frequency response analysis is extremely useful in the subject of mechanical vibrations. This chapter considers the topic of frequency response analysis of dynamic (and vibratory) systems. Because the response to a particular form of "excitation" is what is considered here, one is specifically dealing with the subject of "forced response" analysis — albeit in the frequency domain.

3.1 RESPONSE TO HARMONIC EXCITATIONS

Consider a simple oscillator with an excitation force $f(t)$, as shown in Figure 3.1. The equation of motion is given by

$$m\ddot{x} + b\dot{x} + kx = f(t) \tag{3.1}$$

Suppose that $f(t)$ is sinusoidal (i.e., harmonic). Pick the time reference such that:

$$f(t) = f_o \cos \omega t \tag{3.2}$$

where ω = excitation frequency
f_o = forcing excitation amptitude

From Chapter 2, for a system subjected to a forcing excitation,

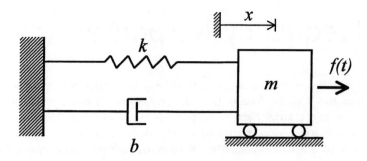

FIGURE 3.1 A forced simple oscillator.

Total response	=	Homogeneous response x_h	+	Particular response x_p
T		H		P
		(Natural response)	+	(Enforced response)
		X		F
		(Depends only on ICs)		(Depends only on f_o)
		Does not contain enforced response and depends entirely on the natural / homogeneous response		But contains a natural / homogeneous component

With this in mind, proceed to analyze the problem. Thus,

$$\ddot{x} + \frac{b}{m}\dot{x} + \frac{k}{m}x = \frac{f_o}{m}\cos\omega t = u(t) \tag{3.3}$$

or

$$\ddot{x} + 2\zeta\omega_n\dot{x} + \omega_n^2 x = a\cos\omega t = u(t) \tag{3.4}$$

where $u(t)$ is the modified excitation.
Total response is

$$x = x_h + x_p \tag{3.5}$$

with

$$x_h = C_1 e^{\lambda_1 t} + C_2 e^{\lambda_2 t} \tag{3.6}$$

as obtained in Chapter 2. The particular solution x_p, by definition, is one solution that satisfies equation (3.4). It should be intuitively clear that this will be of the form

$$x_p = a_1\cos\omega t + a_2\sin\omega t \quad \{\text{Except for the case}: \zeta = 0 \text{ and } \omega = \omega_n\} \tag{3.7}$$

Frequency Response

where the constants a_1 and a_2 are determined by substituting equation (3.7) into the system equation (3.4) and equating the like coefficient — *the method of undetermined coefficients.*

Now consider several important cases.

3.1.1 Response Characteristics

Case 1: Undamped Oscillator with Excitation Frequency ≠ Natural Frequency

In this case,

$$\ddot{x} + \omega_n^2 x = a\cos\omega t \quad \text{with} \quad \omega \neq \omega_n \tag{3.8}$$

Homogeneous solution:
$$x_h = A_1 \cos\omega_n t + A_2 \sin\omega_n t \tag{3.9}$$

Particular solution:
$$x_p = \frac{a}{(\omega_n^2 - \omega^2)} \cos\omega t \tag{3.10}$$

It can be easily verified that x_p given by equation (3.10) satisfies the forced system equations (3.8) or (3.4), with $\zeta = 0$. Hence, it is a particular solution.

Complete solution:

$$x = \underbrace{A_1 \cos\omega_n t + A_2 \sin\omega_n t}_{H} + \underbrace{\frac{a}{(\omega_n^2 - \omega^2)} \cos\omega t}_{P} \tag{3.11}$$

Satisfies the homogeneous equation   Satisfies the equation with input

Now A_1 and A_2 are determined using the initial conditions (ICs):

$$x(0) = x_o \quad \text{and} \quad \dot{x}(0) = v_o \tag{3.12}$$

Specifically, one obtains

$$x_o = A_1 + \frac{a}{\omega_n^2 - \omega^2} \tag{3.13a}$$

$$v_o = A_2 \omega_n \tag{3.13b}$$

Hence, the complete response is

$$x = \underbrace{\left[x_o - \frac{a}{(\omega_n^2 - \omega^2)}\right]\cos\omega_n t + \frac{v_o}{\omega_n}\sin\omega_n t}_{H} + \underbrace{\frac{a}{(\omega_n^2 - \omega^2)}\cos\omega t}_{P} \tag{3.14a}$$

Homogeneous solution   Particular solution

$$= \underbrace{x_o \cos\omega_n t + \frac{v_o}{\omega_n}\sin\omega_n t}_{X} \; + \; \underbrace{\frac{a}{(\omega_n^2 - \omega^2)}\underbrace{[\cos\omega t - \cos\omega_n t]}_{2\sin\frac{(\omega_n+\omega)}{2}t \sin\frac{(\omega_n-\omega)}{2}t}}_{F}$$

Free response
(depends only on ICs)
Comes from x_h.
*Sinusodal at ω_n.

Forced response (depends on input)
Comes from both x_h and x_p.
*Will exhibit a beat phenomenon for small $\omega_n - \omega$; i.e.,
$\frac{(\omega_n + \omega)}{2}$ wave *modulated* by $\frac{(\omega_n - \omega)}{2}$ wave.

(3.14b)

This is a *stable* response in the sense of bounded-input-bounded-output (BIBO) stability, as it is bounded and does not increase steadily.

Note: If there is no forcing excitation, the homogeneous solution H and the free response X will be identical. With a forcing input, the natural response (the homogeneous solution) will be influenced by it in general, as discussed in Chapter 2, and as clear from equation (3.14a).

Case 2: Undamped Oscillator with $\omega = \omega_n$ (Resonant condition)

In this case, the x_p that was used before is no longer valid (this is the degenerate case), because otherwise the particular solution cannot be distinguished from the homogeneous solution and the former will be completely absorbed into the latter. Instead, in view of the double-integration nature of the forced system equation when $\omega = \omega_n$, use the particular solution (*P*):

$$x_p = \frac{at}{2\omega}\sin\omega t \qquad (3.15)$$

This choice of particular solution is strictly justified by the fact that it satisfies the forced system equation.

Complete solution:

$$x = A_1 \cos\omega t + A_2 \sin\omega t + \frac{at}{2\omega}\sin\omega t \qquad (3.16)$$

ICs:

$$x(0) = x_o \quad \text{and} \quad \dot{x}(0) = v_o.$$

One obtains

$$x_o = A_1 \qquad (3.17a)$$

$$v_o = \omega A_2 \qquad (3.17b)$$

The total response is

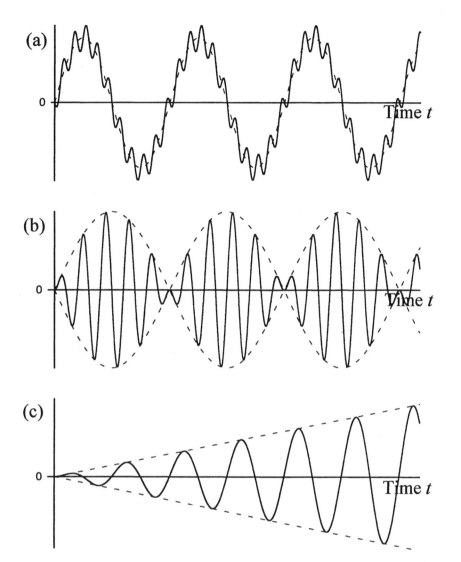

FIGURE 3.2 Forced response of a harmonically excited undamped simple oscillator: (a) for a large frequency difference, (b) for a small frequency difference (beat phenomenon), and (c) response at resonance.

$$x = \underbrace{x_o \cos \omega t + \frac{v_o}{\omega} \sin \omega t}_{\substack{X=H \\ \text{Free response (depends on ICs)} \\ \text{*Sinusodal at } \omega}} + \underbrace{\frac{at}{2\omega} \sin \omega t}_{\substack{F=P \\ \text{Forced response (depends on input)} \\ \text{*Amplitude increases linearly}}} \quad (3.18)$$

Since the forced response increases steadily, this is an *unstable* response in the bounded-input-bounded-output (BIBO) sense. Furthermore, the homogeneous solution H and the free response X are identical, and the particular solution P is identical to the forced response F in this case.

Note that the same system (undamped oscillator) gives a bounded response for some excitations while producing an unstable (steady linear increase) response when the excitation frequency is equal to its natural frequency. Hence, the system is not quite unstable, but is not quite stable either. In fact, the undamped oscillator is said to be marginally stable. When the excitation frequency is equal to the natural frequency, it is reasonable for the system to respond in a complementary and steadily increasing manner because this corresponds to the most "receptive" excitation. Specifically, in this case, the excitation complements the natural response of the system. In other words, the system is "in resonance" with the excitation, and the condition is called a resonance. This aspect for the more general case of a damped oscillator is addressed in Case 3.

Figure 3.2 shows typical forced responses of an undamped oscillator for a large difference between the excitation and the natural frequencies (Case 1); for a small difference between the excitation and the natural frequencies (also Case 1), where a beat phenomenon is clearly manifested; and for the resonant case (Case 2).

Case 3: Damped Oscillator

The equation of forced motion is

$$\ddot{x} + 2\zeta\omega_n \dot{x} + \omega_n^2 x = a\cos\omega t \tag{3.19}$$

Particular Solution (Method 1)

Since derivatives of both odd order and even order are present in this equation, the particular solution should have terms corresponding to odd and even derivatives of the forcing function (i.e., $\sin\omega t$ and $\cos\omega t$). Hence, the appropriate particular solution will be of the form

$$x_p = a_1 \cos\omega t + a_2 \sin\omega t \tag{3.20}$$

Substitute equation (3.20) in (3.19) to obtain

$$-\omega^2 a_1 \cos\omega t - \omega^2 a_2 \sin\omega t + 2\zeta\omega_n\left[-\omega a_1 \sin\omega t + \omega a_2 \cos\omega t\right] +$$
$$\omega_n^2\left[a_1 \cos\omega t + a_2 \sin\omega t\right] = a\cos\omega t$$

Equate like coefficients:

$$-\omega^2 a_1 + 2\zeta\omega_n \omega a_2 + \omega_n^2 a_1 = a$$
$$-\omega^2 a_2 - 2\zeta\omega_n \omega a_1 + \omega_n^2 a_2 = 0$$

Hence,

$$\left(\omega_n^2 - \omega^2\right)a_1 + 2\zeta\omega_n \omega a_2 = a \tag{3.21a}$$

$$-2\zeta\omega_n \omega a_1 + \left(\omega_n^2 - \omega^2\right)a_2 = 0 \tag{3.21b}$$

This can be written in the matrix-vector form:

$$\begin{bmatrix} \left(\omega_n^2 - \omega^2\right) & 2\zeta\omega_n\omega \\ -2\zeta\omega_n\omega & \left(\omega_n^2 - \omega^2\right) \end{bmatrix} \begin{bmatrix} a_1 \\ a_2 \end{bmatrix} = \begin{bmatrix} a \\ 0 \end{bmatrix} \tag{3.21c}$$

Solution is:

Frequency Response

$$\begin{bmatrix} a_1 \\ a_2 \end{bmatrix} = \frac{1}{D} \begin{bmatrix} (\omega_n^2 - \omega^2) & -2\zeta\omega_n\omega \\ 2\zeta\omega_n\omega & (\omega_n^2 - \omega^2) \end{bmatrix} \begin{bmatrix} a \\ 0 \end{bmatrix} \quad (3.22)$$

with the determinant

$$D = (\omega_n^2 - \omega^2)^2 + (2\zeta\omega_n\omega)^2 \quad (3.23)$$

On simplification,

$$a_1 = \frac{(\omega_n^2 - \omega^2)}{D} a \quad (3.24a)$$

$$a_2 = \frac{2\zeta\omega_n\omega}{D} a \quad (3.24b)$$

This is the method of undetermined coefficients.

Particular Solution (Method 2): Complex Function Method

Consider

$$\ddot{x} + 2\zeta\omega_n\dot{x} + \omega_n^2 x = ae^{j\omega t} \quad (3.25)$$

where the excitation is complex. (*Note*: $e^{j\omega t} = \cos\omega t + j\sin\omega t$).
The resulting "complex" particular solution is

$$x_p = X(j\omega)e^{j\omega t}. \quad (3.26)$$

Note that one should take the "real part" of this solution as the true particular solution.
First substitute equation (3.26) into (3.25):

$$X\left[-\omega^2 + 2\zeta\omega_n j\omega + \omega_n^2\right]e^{j\omega t} = ae^{j\omega t}$$

Hence (since $e^{j\omega t} \neq 0$ in general),

$$X = \frac{a}{\left[-\omega^2 + 2\zeta\omega_n j\omega + \omega_n^2\right]} \quad (3.27)$$

It is known (see Chapter 2) that the *characteristic polynomial* of the system is

$$\Delta(\lambda) = \lambda^2 + 2\zeta\omega_n\lambda + \omega_n^2 \quad (3.28a)$$

or, with the Laplace variable s,

$$\Delta(s) = s^2 + 2\zeta\omega_n s + \omega_n^2 \qquad (3.28b)$$

If $s = j\omega$, one obtains

$$\Delta(j\omega) = -\omega^2 + 2\zeta\omega_n j\omega + \omega_n^2 \qquad (3.28c)$$

Note that equation (3.28c) is indeed the denominator of (3.27). Hence, equation (3.27) can be written as

$$X = \frac{a}{\Delta(j\omega)} \qquad (3.29)$$

It follows from equation (3.26) that the complex particular solution is

$$x_p = \frac{a}{\Delta(j\omega)} e^{j\omega t} \qquad (3.30)$$

Next, let

$$\Delta(j\omega) = |\Delta| e^{j\phi} \qquad (3.31)$$

Then, by substituting equation (3.31) in (3.30), obtain

$$x_p = \frac{a}{|\Delta|} e^{j(\omega t - \phi)} \qquad (3.32)$$

where it is clear from equation (3.31) that

$$|\Delta| = \text{magnitude of } \Delta(j\omega)$$

$$\phi = \text{phase angle of } \Delta(j\omega)$$

The actual (real) particular solution is the real part of equation (3.32) and is given by

$$x_p = \frac{a}{|\Delta|} \cos(\omega t - \phi) \qquad (3.33)$$

It can be easily verified that this result is identical to that obtained previously (by Method 1), as given by equation (3.20) together with (3.23) and (3.24).

In passing, note here that the frequency-domain transfer function (i.e., response/excitation in the frequency domain) of the system (3.19) is:

$$G(j\omega) = \frac{1}{\Delta} = \left[\frac{1}{s^2 + 2\zeta\omega_n s + \omega_n^2} \right]_{s=j\omega} \qquad (3.34)$$

Frequency Response

This frequency transfer function (also known as the frequency response function) is obtained from the Laplace transfer function $G(s)$ by setting $s = j\omega$; this aspect is discussed in more detail later.

As observed in Chapter 2, the particular solution (P) is equal to the steady-state solution because the homogeneous solution dies out due to damping. The particular solution (3.33) has the following characteristics:

1. Frequency is same as the excitation frequency ω.
2. Amplitude is amplified by the magnitude $\dfrac{1}{|\Delta|} = |G(j\omega)|$.
3. Response is *lagged* by the phase angle ϕ of Δ (or *led* by the phase angle of $G(j\omega)$, denoted by $\angle G(j\omega)$).
4. Because the homogenous solution of a stable system decays to zero, the particular solution is also the *steady-state* solution.

Resonance

The amplification $\left(|G(j\omega)| = \dfrac{1}{|\Delta|}\right)$ is maximum (i.e., resonance) when $|\Delta|$ is a minimum or $|\Delta|^2$ is a minimum. As noted earlier, this condition of peak amplification of a system when excited by a sinusoidal input is called *resonance*, and the associated frequency of excitation is called *resonant frequency*. One can determine the resonance of the system (3.19) as follows.

Equation (3.28c) is: $\Delta = \omega_n^2 - \omega^2 + 2\zeta\omega_n\omega j$
Hence,

$$|\Delta|^2 = \left(\omega_n^2 - \omega^2\right)^2 + \left(2\zeta\omega_n\omega\right)^2 = D \tag{3.35}$$

The resonance corresponds to a minimum value of D; or

$$\frac{dD}{d\omega} = 2\left(\omega_n^2 - \omega^2\right)(-2\omega) + 2\left(2\zeta\omega_n\right)^2\omega = 0 \quad \text{(for a minimum)} \tag{3.36}$$

Hence, with straightforward algebra, the required condition for resonance is

$$-\omega_n^2 + \omega^2 + 2\zeta^2\omega_n^2 = 0$$

or

$$\omega^2 = \left(1 - 2\zeta^2\right)\omega_n^2$$

or

$$\omega = \sqrt{1 - 2\zeta^2}\,\omega_n$$

This is the *resonant frequency*, and is denoted as

$$\omega_r = \sqrt{1 - 2\zeta^2}\,\omega_n \tag{3.37}$$

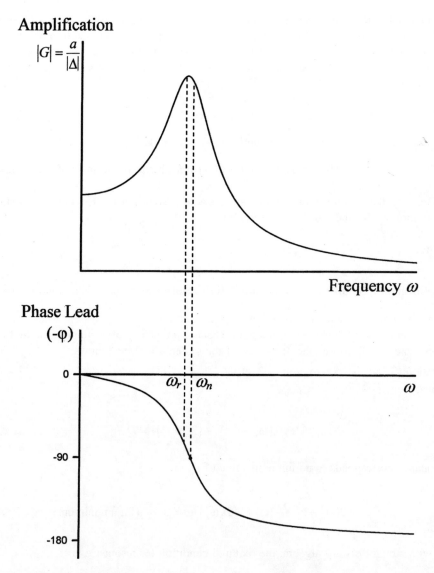

FIGURE 3.3 Magnitude and phase angle curves of a simple oscillator (a Bode plot).

Note that $\omega_r \leq \omega_d \leq \omega_n$, where ω_d is the damped natural frequency given by $\omega_d = \sqrt{1-\zeta^2}\,\omega_n$, as discussed in Chapter 2. These three frequencies (resonant frequency, damped natural frequency, and undamped natural frequency) are almost equal for small ζ (i.e., for light damping).

The magnitude and the phase angle plots of $G(j\omega)$ are shown in Figure 3.3. These curves correspond to the amplification and the phase change of the particular response (or the steady-state response) with respect to the excitation input. This pair of magnitude and phase angle plots of a transfer function with respect to frequency is termed a *Bode plot*. Usually, logarithmic scales are used for both magnitude (e.g., decibels) and frequency (e.g., decades). In summary, the steady-state response of a linear system to a sinusoidal excitation is completely determined by the frequency transfer function of the system. The total response is determined by adding H to P and substituting ICs, as usual.

For an undamped oscillator ($\zeta = 0$), note from equation (3.34) that the magnitude of $G(j\omega)$ becomes infinity when the excitation frequency is equal to the natural frequency (ω_n) of the oscillator. This frequency (ω_n) is clearly the resonant frequency (as well as natural frequency) of the oscillator. This fact has been further supported by the nature of the corresponding time response [see equation (3.18) and Figure 3.2(c)], which grows (linearly) with time.

3.1.2 Measurement of Damping Ratio (Q-Factor Method)

The frequency transfer function of a simple oscillator (3.19) can be used to determine the damping ratio. This frequency-domain method is also termed the *half-power point* method, for reasons that should be clear from the following development.

First assume that $\zeta < \frac{1}{\sqrt{2}}$. Strictly speaking, one should assume that $\zeta < \frac{1}{2\sqrt{2}}$.

Without loss of generality, consider the normalized (or nondimensionalized) transfer function

$$G(j\omega) = \left[\frac{\omega_n^2}{s^2 + 2\zeta\omega_n s + \omega_n^2}\right]_{s=j\omega} = \frac{\omega_n^2}{\omega_n^2 - \omega^2 + 2\zeta\omega_n\omega j} \quad (3.38)$$

As noted before, the transfer function $G(s)$, where s is the Laplace variable, can be converted into the corresponding frequency transfer function by simply setting $s = j\omega$. Its value at the undamped natural frequency is

$$G(j\omega)\big|_{\omega=\omega_n} = \frac{1}{2\zeta j} \quad (3.39a)$$

Hence, the magnitude of $G(j\omega)$ (amplification) at $\omega = \omega_n$ is

$$|G(j\omega)|_{\omega=\omega_n} = \frac{1}{2\zeta} \quad (3.39b)$$

For small ζ, $\omega_r \cong \omega_n$. Hence, $\frac{1}{2\zeta}$ is approximately the peak magnitude at resonance (resonant peak). The actual peak is slightly larger.

It is clear from equation (3.39a) that the phase angle of $G(j\omega)$ at $\omega = \omega_n$ is $-\pi/2$.

When the amplification is $\frac{1}{\sqrt{2}}$ of the peak value (i.e., when power is $\frac{1}{2}$ of the peak power value; because, for example, the displacement squared is proportional to potential energy, the velocity squared is proportional to kinetic energy, and power is the rate of change of energy), the half-power points are given by:

$$\frac{1}{\sqrt{2}}\frac{1}{2\zeta} = \left|\frac{\omega_n^2}{\omega_n^2 - \omega^2 + 2j\zeta\omega_n\omega}\right| = \left|\frac{1}{1-\left(\frac{\omega}{\omega_n}\right)^2 + 2j\zeta\left(\frac{\omega}{\omega_n}\right)}\right| \quad (3.40)$$

Square equation (3.40):

$$\frac{1}{2\times 4\zeta^2} = \frac{1}{\left[1-\left(\frac{\omega}{\omega_n}\right)^2\right]^2 + 4\zeta^2\left(\frac{\omega}{\omega_n}\right)^2}$$

Hence,

$$\left(\frac{\omega}{\omega_n}\right)^4 - 2\left(\frac{\omega}{\omega_n}\right)^2 + 1 + 4\zeta^2\left(\frac{\omega}{\omega_n}\right)^2 = 8\zeta^2$$

or

$$\left(\frac{\omega}{\omega_n}\right)^4 - 2(1-2\zeta^2)\left(\frac{\omega}{\omega_n}\right)^2 + \underbrace{(1-8\zeta^2)}_{>0} = 0 \tag{3.41}$$

Now assume that $\zeta^2 < \frac{1}{8}$ or, $\zeta < \frac{1}{2\sqrt{2}}$. Otherwise, one will not get two positive roots for $\left(\frac{\omega}{\omega_n}\right)^2$. Solve for $\left(\frac{\omega}{\omega_n}\right)^2$, which will give two roots ω_1^2 and ω_2^2 for ω^2. Next, assume $\omega_2^2 > \omega_1^2$. Compare $(\omega^2 - \omega_1^2)(\omega^2 - \omega_2^2) = 0$ with equation (3.41).

$$\text{Sum of roots}: \frac{\omega_2^2 + \omega_1^2}{\omega_n^2} = 2(1-2\zeta^2) \tag{3.42}$$

$$\text{Product of roots}: \frac{\omega_2^2 \omega_1^2}{\omega_n^4} = (1-8\zeta^2) \tag{3.43}$$

Hence,

$$\left(\frac{\omega_2 - \omega_1}{\omega_n}\right)^2 = \left(\frac{\omega_2^2 + \omega_1^2 - 2\omega_2\omega_1}{\omega_n^2}\right)$$

$$= 2(1-2\zeta^2) - 2\sqrt{1-8\zeta^2}$$

$$= 2 - 4\zeta^2 - 2\left[1 - \frac{1}{2}\times 8\zeta^2 + O(\zeta^4)\right]$$

$$\cong 2 - 4\zeta^2 - 2 + 8\zeta^2 \quad \left(\text{because } O(\zeta^4) \to 0 \text{ for small } \zeta\right)$$

$$\cong 4\zeta^2$$

or

$$\frac{\omega_2 - \omega_1}{\omega_n} \cong 2\zeta$$

Frequency Response

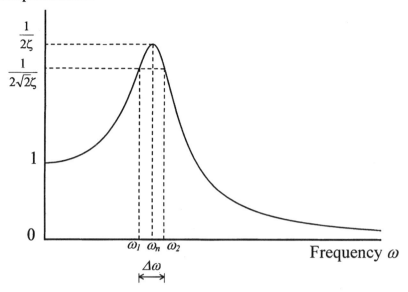

FIGURE 3.4 The Q-factor method of damping measurement.

Hence, the damping ratio

$$\zeta \cong \frac{(\omega_2 - \omega_1)}{2\omega_n} = \frac{\Delta\omega}{2\omega_n} \cong \frac{\omega_2 - \omega_1}{\omega_2 + \omega_1} \quad (3.44)$$

It follows that, once the magnitude of the frequency response function $G(j\omega)$ is experimentally determined, the damping ratio can be estimated from equation (3.44), as illustrated in Figure 3.4. The Q-factor, which measures the sharpness of resonant peak, is defined by

$$\text{Q-factor} = \frac{\omega_n}{\Delta\omega} = \frac{1}{2\zeta} \quad (3.45)$$

The term originated from the field of electrical tuning circuits where sharpness of the resonant peak is a desirable thing (*quality factor*). Some useful results on the frequency response of a simple oscillator are summarized in Box 3.1.

EXAMPLE 3.1

A dynamic model of a fluid coupling system is shown in Figure 3.5. The fluid coupler is represented by a rotatory viscous damper with damping constant b. It is connected to a rotatory load of moment of inertia J, restrained by a torsional spring of stiffness k, as shown. Obtain the frequency transfer function of the system relating the restraining torque τ of the spring to the angular displacement excitation $\alpha(t)$ that is applied at the free end of the fluid coupler. If $\alpha(t) = \alpha_o \sin\omega t$, what is the magnitude (i.e., amplitude) of τ at steady state?

SOLUTION

Newton's second law gives

$$J\ddot{\theta} = b(\dot{\alpha} - \dot{\theta}) - k\theta$$

Hence,

$$J\ddot{\theta} + b\dot{\theta} + k\theta = b\dot{\alpha} \qquad (i)$$

Motion transfer function is

$$\frac{\theta}{\alpha} = \frac{bs}{Js^2 + bs + k} \qquad (ii)$$

Note that the frequency transfer function is obtained simply by setting $s = j\omega$.
Restraining torque of the spring is $\tau = k\theta$. Hence,

$$\frac{\tau}{\alpha} = \frac{k\theta}{\alpha} = \frac{kbs}{Js^2 + bs + k} = G(s) \qquad (iii)$$

Then, the corresponding frequency response function (frequency transfer function) is

$$G(j\omega) = \frac{kbj\omega}{(k - J\omega^2) + bj\omega} \qquad (iv)$$

For a harmonic excitation of

$$\alpha = \alpha_o e^{j\omega t} \qquad (v)$$

one has

$$\tau = |G(j\omega)|\alpha_o e^{(j\omega t + \phi)} \qquad (vi)$$

at steady state.
Here, the phase "lead" of τ with respect to α is

$$\phi = \angle G(j\omega)$$
$$= \angle j\omega - \angle(k - J\omega^2 + bj\omega) \qquad (vii)$$
$$= \frac{\pi}{2} - \tan^{-1}\frac{b\omega}{(k - J\omega^2)}$$

The magnitude of the restraining torque at steady state is

$$\tau_o = \alpha_o |G(j\omega)|$$
$$= \alpha_o \frac{kb\omega}{|k - J\omega^2 + bj\omega|} = \frac{\alpha_o kb\omega}{\sqrt{(k - J\omega^2)^2 + b^2\omega^2}}$$

Hence,

Frequency Response

BOX 3.1 Harmonic Response of a Simple Oscillator

Undamped Oscillator: $\ddot{x} + \omega_n^2 x = a\cos\omega t;\ x(0) = x_0,\ \dot{x}(0) = v_0$

$$x = \underbrace{x_0 \cos\omega_n t + \frac{v_o}{\omega_n}\sin\omega_n t}_{X} + \underbrace{\frac{a}{\omega_n^2 - \omega^2}\left[\cos\omega t - \cos\omega_n t\right]}_{F} \quad \text{for } \omega \neq \omega_n$$

$$x = \text{Same } X + \frac{at}{2\omega}\sin\omega t \quad \begin{array}{l}\text{for } \omega = \omega_n \\ \text{(resonance)}\end{array}$$

Damped Oscillator: $\ddot{x} + 2\zeta\omega_n \dot{x} + \omega_n^2 x = a\cos\omega t$

$$x = H + \underbrace{\frac{a}{\left|\omega_n^2 - \omega^2 + 2j\zeta\omega_n\omega\right|}\cos(\omega t - \phi)}_{P}$$

where $\tan\phi = \dfrac{2\zeta\omega_n\omega}{\omega_n^2 - \omega^2};\ \phi = $ phase lag.

Particular solution P is also the steady-state response.

Homogeneous solution $H = A_1 e^{\lambda_1 t} + A_2 e^{\lambda_2 t}$

where λ_1 and λ_2 are roots of $\lambda^2 + 2\zeta\omega_n\lambda + \omega_n^2 = 0$
(characteristic equation)

A_1 and A_2 are determined from ICs: $x(0) = x_o,\ \dot{x}(0) = v_0$

Resonant Frequency: $\omega_r = \sqrt{1 - 2\zeta^2}\ \omega_n$

The magnitude of P will peak at resonance.

Damping Ratio: $\zeta = \dfrac{\Delta\omega}{2\omega_n} = \dfrac{\omega_2 - \omega_1}{\omega_2 + \omega_1}$ for low damping

where, $\Delta\omega = $ half-power bandwidth $= \omega_2 - \omega_1$

Note: Q-factor $= \dfrac{\omega_n}{\Delta\omega} = \dfrac{1}{2\zeta}$ for low damping.

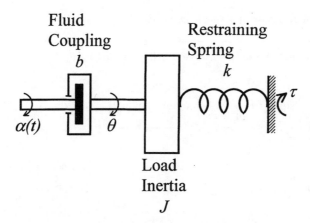

FIGURE 3.5 A fluid coupling system.

$$\tau_o = \frac{J\alpha_o \omega_n^2 \omega \cdot 2\zeta\omega_n}{\sqrt{\left(\omega_n^2 - \omega^2\right)^2 + \left(2\zeta\omega_n\omega\right)^2}} \quad (viii)$$

or

$$\tau_o = \frac{\alpha_o k 2\zeta\omega_n \omega}{\sqrt{\left(\omega_n^2 - \omega^2\right)^2 + \left(2\zeta\omega_n\omega\right)^2}} \quad (ix)$$

with $\frac{k}{J} = \omega_n^2$; $\frac{b}{J} = 2\zeta\omega_n$; and ω_n = undamped natural frequency of the load.
Now define the normalized frequency

$$r = \frac{\omega}{\omega_n} \quad (x)$$

Then, from (*ix*), one obtains

$$\tau_o = \frac{2k\alpha_o \zeta r}{\sqrt{\left(1 - r^2\right)^2 + \left(2\zeta r\right)^2}} \quad (xi)$$

For $r = 1$: $\tau_o = \frac{2k\alpha_o \zeta}{2\zeta} = k\alpha_o \quad (xii)$

This means, at resonance, the applied twist is directly transmitted to the load spring.

For small r: $\tau_o = \frac{2k\alpha_o \zeta r}{1} \quad (xiii)$

which is small, and becomes zero at $r = 0$. Hence, at low frequencies, the transmitted torque is small.

Frequency Response

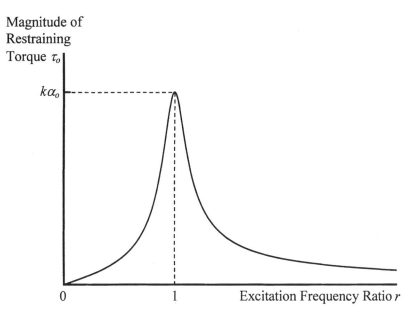

FIGURE 3.6 Variation of the steady-state transmitted torque with frequency.

For large r: $\tau_o = \dfrac{2k\alpha_o \zeta r}{r^2} = \dfrac{2k\alpha_o \zeta}{r}$ (xiv)

which is small, and goes to zero. Hence, at high frequencies as well, the transmitted torque is small. The variation of τ_o with the frequency ratio r is sketched in Figure 3.6.

□

3.2 TRANSFORM TECHNIQUES

Concepts of frequency-response analysis originate from the nature of the response of a dynamic system to a sinusoidal (i.e., harmonic) excitation. These concepts can be generalized because the time-domain analysis, where the independent variable is time (t) and the frequency-domain analysis, where the independent variable is frequency (ω) are linked through the *Fourier transformation*. Analytically, it is more general and versatile to use the Laplace transformation, where the independent variable is the Laplace variable (s) which is complex (non-real). This is true because analytical Laplace transforms may exist even for time functions that do not have "analytical" Fourier transforms. But with compatible definitions, the Fourier transform results can be obtained form the Laplace transform results simply by setting $s = j\omega$. This chapter section formally introduces the Laplace transformation and the Fourier transformation, and illustrates how these techniques are useful in the response analysis of vibrating systems. Fourier analysis and techniques will be discussed further in Chapter 4. The preference of one domain over another will depend on such factors as the nature of the excitation input, the type of the analytical model available, the time duration of interest, and the quantities that need to be determined.

3.2.1 TRANSFER FUNCTION

The Laplace transform of a piecewise-continuous function $f(t)$ is denoted here by $F(s)$ and is given by the Laplace transformation

$$F(s) = \int_0^\infty f(t)\exp(-st)dt \tag{3.46}$$

in which s is a complex independent variable known as the Laplace variable, expressed as

$$s = \sigma + j\omega \tag{3.47}$$

and $j = \sqrt{-1}$. Laplace transform operation is denoted as $\mathcal{L}f(t) = F(s)$. The inverse Laplace transform operation is denoted by $f(t) = \mathcal{L}^{-1}F(s)$ and is given by

$$f(t) = \frac{1}{2\pi j} \int_{\sigma-j\infty}^{\sigma+j\infty} F(s)\exp(st)ds \tag{3.48}$$

The integration is performed along a vertical line parallel to the imaginary (vertical) axis, located at σ from the origin in the complex Laplace plane (s-plane). For a given piecewise-continuous function $f(t)$, the Laplace transform exists if the integral in equation (3.46) converges. A sufficient condition for this is

$$\int_0^\infty |f(t)|\exp(-\sigma t)dt < \infty \tag{3.49}$$

Convergence is guaranteed by choosing a sufficiently large and positive σ. This property is an advantage of the Laplace transformation over the Fourier transformation (For a more complete discussion of the Fourier transformation, see later in this chapter and Chapter 4).

By use of Laplace transformation, the convolution integral equation can be converted into an algebraic relationship. To illustrate this, consider the *convolution integral* that gives the response $y(t)$ of a dynamic system to an excitation input $u(t)$, with zero ICs, as discussed in Chapter 2. By definition (3.46), its Laplace transform is written as

$$Y(s) = \int_0^\infty \int_0^\infty h(\tau)u(t-\tau)d\tau \exp(-st)dt \tag{3.50}$$

Note that $h(t)$ is the *impulse response function* of the system. Because the integration with respect to t is performed while keeping τ constant, $dt = d(t - \tau)$. Consequently,

$$Y(s) = \int_{-\tau}^\infty u(t-\tau)\exp[-s(t-\tau)]d(t-\tau)\int_0^\infty h(\tau)\exp(-s\tau)d\tau$$

The lower limit of the first integration can be made equal to zero, in view of the fact that $u(t) = 0$ for $t < 0$. Again, using the definition of Laplace transformation, the foregoing relation can be expressed as

$$Y(s) = H(s)U(s) \tag{3.51}$$

Frequency Response

in which

$$H(s) = \mathcal{L}h(t) = \int_0^\infty h(t)\exp(-st)dt \quad (3.52)$$

Note that, by definition, the transfer function of a system, denoted by $H(s)$, is given by equation (3.51). More specifically, system transfer function is given by the ratio of the Laplace-transformed output and the Laplace-transformed input, with zero initial conditions. In view of equation (3.52), it is clear that the system transfer function can be expressed as the Laplace transform of the impulse-response function of the system. Transfer function of a linear and constant-parameter system is a unique function that completely represents the system. A physically realizable, linear, constant-parameter system possesses a unique transfer function, even if the Laplace transforms of a particular input and the corresponding output do not exist. This is clear from the fact that the transfer function is a system model and does not depend on the system input itself. *Note*: The transfer function is also commonly denoted by $G(s)$. But in the present context, we use $H(s)$ in view of its relation to $h(t)$. Some useful Laplace transform relations are given in Table 3.1.

Consider the nth-order linear, constant-parameter dynamic system given by

TABLE 3.1
Important Laplace Transform Relations

$\mathcal{L}^{-1}F(s) = f(t)$	$\mathcal{L}f(t) = F(s)$
$\dfrac{1}{2\pi j}\displaystyle\int_{\sigma-j\infty}^{\sigma+j\infty} F(s)\exp(st)ds$	$\displaystyle\int_0^\infty f(t)\exp(-st)dt$
$k_1 f_1(t) + k_2 f_2(t)$	$k_1 F_1(s) + k_2 F_2(s)$
$\exp(-at)f(t)$	$F(s+a)$
$f^{(n)}(t) = \dfrac{d^n f(t)}{dt^n}$	$s^n F(s) - s^{n-1}f(0^+) - s^{n-2}f'(0^+)$ $\cdots - f^{n-1}(0^+)$
$\displaystyle\int_{-\infty}^t f(t)dt$	$\dfrac{F(s)}{s} + \dfrac{\int_{-\infty}^0 f(t)dt}{s}$
Impulse function $\delta(t)$	1
Step function $\mathcal{U}(t)$	$\dfrac{1}{s}$
t^n	$\dfrac{n!}{s^{n+1}}$
$\exp(-at)$	$\dfrac{1}{s+a}$
$\sin \omega t$	$\dfrac{\omega}{s^2+\omega^2}$
$\cos \omega t$	$\dfrac{s}{s^2+\omega^2}$

$$a_n \frac{d^n y}{dt^n} + a_{n-1} \frac{d^{n-1} y}{dt^{n-1}} + \cdots + a_o y = b_o u + b_1 \frac{du(t)}{dt} + \cdots + b_m \frac{d^m u(t)}{dt^m} \quad (3.53)$$

For physically realizable systems, $m \leq n$. By applying Laplace transformation and then integrating by parts, it may be verified that

$$\mathcal{L} \frac{d^k f(t)}{dt^k} = s^k F(s) - s^{k-1} f(0) - s^{k-2} \frac{df(0)}{dt} - \cdots + \frac{d^{k-1} f(0)}{dt^{k-1}} \quad (3.54)$$

By definition, the initial conditions are set to zero in obtaining the transfer function. This results in

$$H(s) = \frac{b_0 + b_1 s + \cdots + b_m s^m}{a_0 + a_1 s + \cdots + a_n s^n} \quad (3.55)$$

for $m \leq n$. Note that equation (3.55) contains all the information that is contained in equation (3.53). Consequently, transfer function is an analytical model of a system. The transfer function may be employed to determine the total response of a system for a given input and any initial conditions, although it is defined in terms of the response under zero initial conditions. This is quite logical because the analytical model of a system is independent of the system's initial conditions.

The denominator polynomial of a transfer function is the system's *characteristic polynomial*. Its roots are the poles or the eigenvalues of the system. If all the eigenvalues have negative real parts, the system is stable. Response of a stable system is bounded (i.e., remains finite) when the input is bounded (which is the BIBO stability). The zero-input response of an asymptotically stable system approaches zero with time.

3.2.2 Frequency-Response Function (Frequency-Transfer Function)

The Fourier integral transform of the impulse-response function is given by

$$H(f) = \int_{-\infty}^{\infty} h(t) \exp(-j2\pi ft) dt \quad (3.56)$$

where f is the *cyclic frequency* (measured in cycles per second or hertz). This is known as the frequency-response function (or frequency transfer function) of a system. Fourier transform operation is denoted as $\mathcal{F}h(t) = H(f)$. In view of the fact that $h(t) = 0$ for $t < 0$, the lower limit of integration in equation (3.56) could be made zero. Then, from equation (3.52), it is clear that $H(f)$ is obtained simply by setting $s = j2\pi f$ in $H(s)$. Hence, strictly speaking, one should use the notation $H(j2\pi f)$ and not $H(f)$. But for the notational simplicity, denote $H(j2\pi f)$ by $H(f)$. Furthermore, since the angular frequency $\omega = 2\pi f$, one can express the frequency response function by $H(j\omega)$, or simply by $H(\omega)$ for the notational convenience. It should be noted that the frequency-response function, like the (Laplace) transfer function, is a complete representation of a linear, constant-parameter system. In view of the fact that both $u(t) = 0$ and $y(t) = 0$ for $t < 0$, one can write the Fourier transforms of the input and the output of a system directly by setting $s = j2\pi f = j\omega$ in the corresponding Laplace transforms. Specifically, according to the notation used here,

$$U(f) = U(j2\pi f) = U(j\omega) \text{ and}$$

$$Y(f) = Y(j2\pi f) = Y(j\omega)$$

Then, from equation (3.51),

$$Y(f) = H(f)U(f) \qquad (3.57)$$

Note: Sometimes, for notational convenience, the same lowercase letters are used to represent the Laplace and Fourier transforms as well as the original time-domain variables.

If the Fourier integral transform of a function exists, then its Laplace transform also exists. The converse is not generally true, however, because of poor convergence of the Fourier integral in comparison to the Laplace integral. This arises from the fact that the factor $\exp(-\sigma t)$ is not present in the Fourier integral. For a physically realizable, linear, constant-parameter system, $H(f)$ exists even if $U(f)$ and $Y(f)$ do not exist for a particular input. The experimental determination of $H(f)$, however, requires system stability. For the nth-order system given by equation (3.53), the frequency-response function is determined by setting $s = j2\pi f$ in equation (3.55) as

$$H(f) = \frac{b_0 + b_1 j2\pi f + \cdots + b_m (j2\pi f)^m}{a_0 + a_1 j2\pi f + \cdots + a_n (j2\pi f)^n} \qquad (3.58)$$

This generally is a complex function of f that has a magnitude denoted by $|H(f)|$ and a phase angle denoted by $\angle H(f)$.

A further interpretation of the frequency-response function can be given in view of the developments in Section 3.1. Consider a harmonic input having cyclic frequency f, expressed by

$$u(t) = u_0 \cos 2\pi f t \qquad (3.59a)$$

In analysis, it is convenient to use the complex input

$$u(t) = u_0 (\cos 2\pi f t + j \sin 2\pi f t) = u_0 \exp(j2\pi f t) \qquad (3.59b)$$

and take only the real part of the final result. Note that equation (3.59b) does not implicitly satisfy the requirement of $u(t) = 0$ for $t < 0$. Therefore, an appropriate version of the convolution integral where the limits of integration automatically accounts for this requirement should be used. For example, one can write

$$y(t) = \mathrm{Re}\left[\int_{-\infty}^{t} h(\tau) u_0 \exp[j2\pi f(t - \tau)] d\tau\right] \qquad (3.60a)$$

or

$$y(t) = \mathrm{Re}\left[u_0 \exp(j2\pi f t) \int_{-\infty}^{t} h(\tau) \exp(-j2\pi f \tau) d\tau\right] \qquad (3.60b)$$

in which Re[ ] denotes the real part. As $t \to \infty$, the integral term in equation (3.60b) becomes the frequency-response function $H(f)$, and the response $y(t)$ becomes the steady-state response y_{ss}. Accordingly,

$$y_{ss} = \text{Re}\left[H(f)u_0 \exp(j2\pi ft)\right] \qquad (3.61a)$$

or

$$y_{ss} = u_0 |H(f)| \cos(2\pi ft + \phi) \qquad (3.61b)$$

for a harmonic excitation, in which the phase lead angle $\phi = \angle H(f)$. It follows from equation (3.61b) that, when a harmonic excitation is applied to a stable, linear, constant-parameter dynamic system having frequency-response function $H(f)$, its steady-state response will also be harmonic at the same frequency, but with an amplification factor of $|H(f)|$ in its amplitude and a phase lead of $\angle H(f)$. This result has been established previously, in Section 3.1. Consequently, the frequency-response function of a stable system can be experimentally determined using a sine-sweep test or a sine-dwell test (see Chapter 10). With these methods, a harmonic excitation is applied as the system input, and the amplification factor and the phase-lead angle in the corresponding response are determined at steady state. The frequency of excitation is varied continuously for a sine sweep, and in steps for a sine dwell. The sweep rate should be slow enough, and the dwell times should be long enough to guarantee steady-state conditions at the output (see Chapter 10). The pair of plots of $|H(f)|$ and $\angle H(f)$ against f completely represents the complex frequency-response function, and is the Bode plot or the Bode diagram, as noted before. In Bode plots, logarithmic scales are normally used for both frequency f and magnitude $|H(f)|$.

Impulse Response

The impulse-response function of a system can be obtained by taking the inverse Laplace transform of the system transfer function. For example, consider the damped simple oscillator given by the normalized transfer function:

$$H(s) = \frac{\omega_n^2}{s^2 + 2\zeta\omega_n s + \omega_n^2} \qquad (3.62)$$

The characteristic equation of this system is given by

$$s^2 + 2\zeta\omega_n s + \omega_n^2 = 0 \qquad (3.63)$$

The eigenvalues (poles) are given by its roots. Three possible cases exist, as discussed below and also in Chapter 2.

Case 1 ($\zeta < 1$)

This is the case of complex eigenvalues λ_1 and λ_2. Because the coefficients of the characteristic equation are real, the complex roots should occur in conjugate pairs. Hence,

$$\lambda_1, \lambda_2 = -\zeta\omega_n \pm j\omega_d \qquad (3.64)$$

in which

$$\omega_d = \sqrt{1-\zeta^2}\,\omega_n \qquad (3.65)$$

is the damped natural frequency.

Frequency Response

Case 2 ($\zeta > 1$)

This case corresponds to real and unequal eigenvalues

$$\lambda_1, \lambda_2 = -\zeta\omega_n \pm \sqrt{\zeta^2 - 1}\,\omega_n \qquad (3.66)$$
$$= -a, -b$$

with $a \neq b$, in which

$$ab = \omega_n^2 \qquad (3.67)$$

and

$$a + b = 2\zeta\omega_n \qquad (3.68)$$

Case 3 ($\zeta = 1$)

In this case, the eigenvalues are real and equal:

$$\lambda_1 = \lambda_2 = -\omega_n \qquad (3.69)$$

In all three cases, the real parts of the eigenvalues are negative. Consequently, these second-order systems of a damped simple oscillator are always stable.

The impulse-response functions $h(t)$ corresponding to the three cases are determined by taking the inverse Laplace transform (Table 3.1) of equation (3.62) for $\zeta < 1$, $\zeta > 1$, and $\zeta = 1$, respectively. The following results are obtained:

$$y_{\text{impulse}}(t) = h(t) = \frac{\omega_n}{\sqrt{1-\zeta^2}} \exp(-\zeta\omega_n t) \sin\omega_d t \qquad \text{for } \zeta < 1 \qquad (3.70a)$$

$$y_{\text{impulse}}(t) = h(t) = \frac{ab}{(b-a)}\left[\exp(-at) - \exp(-bt)\right] \qquad \text{for } \zeta > 1 \qquad (3.70b)$$

$$y_{\text{impulse}}(t) = h(t) = \omega_n^2 t \exp(-\omega_n t) \qquad \text{for } \zeta = 1 \qquad (3.70c)$$

These results are identical to what was obtained in Chapter 2.

Step Response

Unit step function is defined by

$$\mathcal{U}(t) = 1 \qquad \text{for } t > 0$$
$$= 0 \qquad \text{for } t \leq 0 \qquad (3.71)$$

Unit impulse function $\delta(t)$ can be interpreted as the time derivative of $\mathcal{U}(t)$; thus,

$$\delta(t) = \frac{d\mathcal{U}(t)}{dt} \qquad (3.72)$$

Note that equation (3.72) reestablishes the fact that for nondimensional $\mathcal{U}(t)$, the dimension of $\delta(t)$ is (time)$^{-1}$. Because $\mathcal{L}\mathcal{U}(t) = 1/s$ (see Table 3.1), the unit step response of the dynamic system (3.62) can be obtained by taking the inverse Laplace transform of

$$Y_{step}(s) = \frac{1}{s} \frac{\omega_n^2}{\left(s^2 + 2\zeta\omega_n s + \omega_n^2\right)} \tag{3.73}$$

which follows from equation (3.73).

To facilitate using Table 3.1, partial fractions of equation (3.73) are determined in the form

$$\frac{a_1}{s} + \frac{a_2 + a_3 s}{\left(s^2 + 2\zeta\omega_n s + \omega_n^2\right)}$$

in which the constants a_1, a_2, and a_3 are determined by comparing the numerator polynomial; thus,

$$\omega_n^2 = a_1\left(s^2 + 2\delta\omega_n s + \omega_n^2\right) + s(a_2 + a_3 s)$$

Then $a_1 = 1$, $a_2 = -2\zeta\omega_n$, and $a_3 = 1$.
The following results are obtained:

$$y_{step}(t) = 1 - \frac{1}{\sqrt{1-\zeta^2}} \exp(-\zeta\omega_n t) \sin(\omega_d t + \phi) \qquad \text{for } \zeta < 1 \tag{3.74a}$$

$$y_{step} = 1 - \frac{1}{(b-a)}\left[b\exp(-at) - a\exp(-bt)\right] \qquad \text{for } \zeta > 1 \tag{3.74b}$$

$$y_{step} = 1 - (\omega_n t + 1)\exp(-\omega_n t) \qquad \text{for } \zeta = 1 \tag{3.74c}$$

In equation (3.74c),

$$\cos\phi = \zeta \tag{3.75}$$

These results are the same as those obtained in Chapter 2.

3.2.3 Transfer Function Matrix

Consider again the state-space model of a linear dynamic system as discussed in Chapter 2 and Appendix A. It is given by

$$\dot{x} = Ax + Bu \tag{3.76a}$$

$$y = Cx + Du \tag{3.76b}$$

where, x = nth order state vector, u = rth order input vector, y = mth order output vector, A = system matrix, B = input gain matrix, C = output (measurement) gain matrix, and D = feedforward

Frequency Response

gain matrix. One can express the input-output relation between u and y, in the Laplace domain, by a transfer function matrix of the order $m \times r$.

To obtain this relation, Laplace transform the equations (3.76a) and (3.76b) and use zero initial conditions for x; thus,

$$sX(s) = AX(s) + BU(s) \tag{3.77a}$$

$$Y(s) = CX(s) + DU(s) \tag{3.77b}$$

From equation (3.77a), it follows that,

$$X(s) = (sI - A)^{-1} BU(s) \tag{3.78}$$

in which, I is the nth order identity matrix. By substituting equation (3.78) into (3.77b), one obtains the transfer relation

$$Y(s) = \left[C(sI - A)^{-1} B + D \right] U(s) \tag{3.79a}$$

or

$$Y(s) = G(s)U(s) \tag{3.79b}$$

The transfer-function matrix $G(s)$ is an $m \times r$ matrix given by

$$G(s) = C(sI - A)^{-1} B + D \tag{3.80}$$

In practical systems with dynamic delay, the excitation $u(t)$ is not fed forward into the response y. Consequently, $D = 0$ for systems that are normally encountered. For such systems,

$$G(s) = C(sI - A)^{-1} B \tag{3.81}$$

Several examples are given now to illustrate the approaches of obtaining transfer function models when the time domain differential equation models are given and to indicate some uses of a transfer function model. Some useful results in the frequency domain are summarized in Box 3.2.

EXAMPLE 3.2

Consider the simple oscillator equation given by

$$m\ddot{y} + b\dot{y} + ky = ku(t) \tag{i}$$

Note that $u(t)$ can be interpreted as a displacement input (e.g., support motion) or $ku(t)$ can be interpreted as the input force applied to the mass. Take the Laplace transform of the system equation (i) with zero initial conditions; thus,

$$\left(ms^2 + bs + k \right) Y(s) = kU(s) \tag{ii}$$

BOX 3.2 Useful Frequency-Domain Results

Laplace Transform ($\mathcal{L}$): $F(s) = \int_0^\infty f(t)\exp(-st)dt$

Fourier Transform ($\mathcal{F}$): $F(j\omega) = \int_{-\infty}^{\infty} f(t)\exp(-j\omega t)dt$

Note: May use $F(\omega)$ to denote $F(j\omega)$
Note: Set $s = j\omega = j2\pi f$ to convert Laplace results into Fourier results.

ω = angular frequency (radians per second)

f = cyclic frequency (cycles per second or Hz)

Transfer function $H(s) = \dfrac{\text{Output}}{\text{Input}}$ in Laplace domain, with zero ICs.

Frequency transfer function (or frequency response function) = $H(j\omega)$
Note: Notation $G(s)$ is also used to denote a system transfer function
Note: $H(s) = \mathcal{L}h(t)$
$h(t)$ = impulse response function = response to a unit impulse input.

Frequency Response:

$Y(j\omega) = H(j\omega)U(j\omega)$

where, $U(j\omega)$ = Fourier spectrum of input $u(t)$
$Y(j\omega)$ = Fourier spectrum of output $y(t)$
Note: $|H(j\omega)|$ = response amplification for a harmonic excitation of frequency ω
$\angle H(j\omega)$ = response phase "lead" for a harmonic excitation

Multivariable Systems:

State-space model: $\dot{x} = Ax + Bu$

$y = Cx + Du$

Transfer-matrix model: $Y(s) = G(s)U(s)$

where $G(s) = C(sI - A)^{-1}B + D$

The corresponding transfer function is

$$G(s) = \frac{Y(s)}{U(s)} = \frac{k}{ms^2 + bs + k} \quad\quad (iii)$$

Frequency Response

or, in terms of the undamped natural frequency ω_n and the damping ratio ζ, where, $\omega_n^2 = k/m$ and $2\zeta\omega_n = b/m$, the transfer function is given by

$$G(s) = \frac{\omega_n^2}{s^2 + 2\zeta\omega_n s + \omega_n^2} \tag{iv}$$

This is the transfer function corresponding to the displacement output. It follows that the output velocity transfer function is

$$\frac{sY(s)}{U(s)} = sG(s) = \frac{s\omega_n^2}{s^2 + 2\zeta\omega_n s + \omega_n^2} \tag{v}$$

and the output acceleration transfer function is

$$\frac{s^2 Y(s)}{U(s)} = s^2 G(s) = \frac{s^2 \omega_n^2}{s^2 + 2\zeta\omega_n s + \omega_n^2} \tag{vi}$$

In the output acceleration transfer function, $m = n = 2$. This means that if the acceleration of the mass that is caused by an applied force is measured, the input (applied force) is instantly felt by the acceleration. This corresponds to a feedforward action of the input excitation or a lack of dynamic delay. For example, this is the primary mechanism through which road disturbances are felt inside a vehicle that has very hard suspensions.

EXAMPLE 3.3

Again consider the simple oscillator differential equation

$$\ddot{y} + 2\zeta\omega_n \dot{y} + \omega_n^2 y = \omega_n^2 u(t) \tag{i}$$

By defining the state variables as

$$x = [x_1, x_2]^T = [y, \dot{y}]^T \tag{ii}$$

a state model for this system can be expressed as

$$\dot{x} = \begin{bmatrix} 0 & 1 \\ -\omega_n^2 & -2\zeta\omega_n \end{bmatrix} x + \begin{bmatrix} 0 \\ \omega_n^2 \end{bmatrix} u(t) \tag{iii}$$

If one considers both displacement and velocity as outputs, then

$$y = x \tag{iv}$$

Note that the output gain matrix C is the identity matrix in this case. From equations (3.79b) and (3.81), it follows that:

$$Y(s) = \begin{bmatrix} s & -1 \\ \omega_n^2 & s+2\zeta\omega_n \end{bmatrix}^{-1} \begin{bmatrix} 0 \\ \omega_n^2 \end{bmatrix} U(s)$$

$$= \frac{1}{[s^2+2\zeta\omega_n s+\omega_n^2]} \begin{bmatrix} s+2\zeta\omega_n & 1 \\ -\omega_n^2 & s \end{bmatrix} \begin{bmatrix} 0 \\ \omega_n^2 \end{bmatrix} \qquad (v)$$

$$= \frac{1}{[s^2+2\zeta\omega_n s+\omega_n^2]} \begin{bmatrix} \omega_n^2 \\ s\omega_n^2 \end{bmatrix} U(s) \qquad (vi)$$

The transfer function matrix is

$$G(s) = \begin{bmatrix} \omega_n^2/\Delta(s) \\ s\omega_n^2/\Delta(s) \end{bmatrix} \qquad (vii)$$

in which $\Delta s = s^2 + 2\zeta\omega_n s + \omega_n^2$ is the characteristic polynomial of the system. The first element in the only column in $G(s)$ is the *displacement-response* transfer function and the second element is the *velocity-response* transfer function. These results agree with the expressions obtained in the previous example.

Now consider the acceleration $\ddot{y}$ as an output and denote it by y_3. It is clear from the system equation (*i*) that,

$$y_3 = \ddot{y} = -2\zeta\omega_n \dot{y} - \omega_n^2 y + \omega_n^2 u(t) \qquad (viii)$$

or, in terms of the state variables,

$$y_3 = -2\zeta\omega_n x_2 - \omega_n^2 x_1 + \omega_n^2 u(t) \qquad (ix)$$

Note that this output explicitly contains the input variable. The feedforward situation implies that the matrix D is non-zero for the output y_3. Now,

$$Y_3(s) = -2\zeta\omega_n X_2(s) - \omega_n^2 X_1(s) + \omega_n^2 U(s)$$

$$= -2\zeta\omega_n \frac{s\omega_n^2}{\Delta(s)} U(s) - \frac{\omega_n^2}{\Delta(s)} U(s) + \omega_n^2 U(s)$$

which simplifies to

$$Y_3(s) = \frac{s^2 \omega_n^2}{\Delta(s)} U(s) \qquad (x)$$

This again confirms the result for the acceleration output transfer function that was obtained in the previous example.

Frequency Response

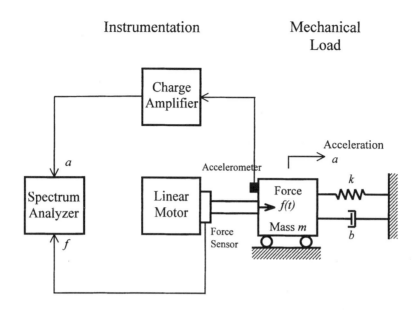

FIGURE 3.7 Measurement of the acceleration spectrum of a mechanical system.

Example 3.4

a. Briefly explain an approach that one could use to measure the resonant frequency of a mechanical system. Do you expect this measured frequency to depend on whether displacement, velocity or acceleration is used as the response variable? Justify your answer.
b. A vibration test setup is schematically shown in Figure 3.7.

In this experiment, a mechanical load is excited by a linear motor and its acceleration response is measured by an accelerometer and charge amplifier combination. The force applied to the load by the linear motor is also measured, using a force sensor (strain-gage type). The frequency response function [acceleration/force] is determined from the sensor signals, using a spectrum analyzer.

Suppose that the mechanical load is approximated by a damped oscillator with mass m, stiffness k, and damping constant b, as shown in Figure 3.7. If the force applied to the load is $f(t)$ and the displacement in the same direction is y, show that the equation of motion of the system is given by

$$m\ddot{y} + b\dot{y} + ky = f(t)$$

Obtain an expression for the acceleration frequency response function $G(j\omega)$ in the frequency domain, with excitation frequency ω as the independent variable. Note that the applied force f is the excitation input and the acceleration a of the mass is the response, in this case.

Express $G(j\omega)$ in terms of (normalized) frequency ratio $r = \dfrac{\omega}{\omega_n}$, where ω_n is the undamped natural frequency.

Giving all the necessary steps, determine an expression for r at which the acceleration frequency response function will exhibit a resonant peak. What is the corresponding peak magnitude of $|G|$?

For what range of values of damping ratio ζ would such a resonant peak be possible?

SOLUTION

(a) For a single-degree-of-freedom (dof) system, apply a sinusoidal forcing excitation at the dof and measure the displacement response at the same location. Vary the excitation frequency ω in small steps and, for each frequency at steady state, determine the amplitude ratio of the [displacement response/forcing excitation]. The peak amplitude ratio will correspond to the resonance. For a multi-dof system, several tests may be needed, with excitations applied at different locations of freedom and the response measured at various locations as well. (See Chapters 10 and 11). In the frequency domain:

$$|\text{Velocity response spectrum}| = \omega \times |\text{Displacement response spectrum}|$$

$$|\text{Acceleration response spectrum}| = \omega \times |\text{Velocity response spectrum}|$$

It follows that the shape of the frequency response function will depend on whether the displacement, velocity, or acceleration is used as the response variable. Hence, it is likely that the frequency at which the peak amplification occurs (i.e., resonance) will also depend on the type of response variable that is used.

(b) A free-body diagram of the mass element is shown in Figure 3.8.
Newton's second law gives

$$m\ddot{y} = f(t) - b\dot{y} - ky \tag{i}$$

Hence, the equation of motion is

$$m\ddot{y} + b\dot{y} + ky = f(t) \tag{ii}$$

The displacement transfer function is

$$\frac{y}{f} = \frac{1}{\left(ms^2 + bs + k\right)} \tag{iii}$$

Note that, for notational convenience, the same lowercase letters are used to represent the Laplace transforms as well as the original time-domain variables (y and f). The acceleration transfer function is obtained by multiplying equation (iii) by s^2. (From Table 3.1, the Laplace transform of $\frac{d}{dt}$ is s, with zero ICs). Hence,

$$\frac{a}{f} = \frac{s^2}{\left(ms^2 + bs + k\right)} = G(s) \tag{iv}$$

In the frequency domain, the corresponding frequency response function is obtained by substituting $j\omega$ for s. Hence,

$$G(j\omega) = \frac{-\omega^2}{\left(-m\omega^2 + bj\omega + k\right)} \tag{v}$$

Frequency Response

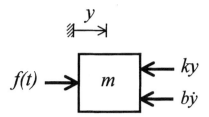

FIGURE 3.8 Free-body diagram.

Divide throughout by m and use $\dfrac{b}{m} = 2\zeta\omega_n$ and $\dfrac{k}{m} = \omega_n^2$, where ω_n = undamped natural frequency, and ζ = damping ratio.
Then,

$$G(j\omega) = \dfrac{-\omega^2/m}{\left(\omega_n^2 - \omega^2 + 2j\zeta\omega_n\omega\right)} \qquad (vi)$$

$$= \dfrac{-r^2/m}{1 - r^2 + 2j\zeta r}$$

where $r = \dfrac{\omega}{\omega_n}$. The magnitude of $G(j\omega)$ gives the amplification of the acceleration signal with respect to the forcing excitation:

$$|G(j\omega)| = \dfrac{r^2/m}{\sqrt{(1-r^2)^2 + (2\zeta r)^2}} \qquad (vii)$$

Its peak value corresponds to the peak value of

$$p(r) = \dfrac{r^4}{(1-r^2)^2 + (2\zeta r)^2} \qquad (viii)$$

and gives the resonance. This occurs when $\dfrac{dp}{dr} = 0$. Hence,

$$\left[(1-r^2)^2 + (2\zeta r)^2\right]4r^3 - r^4\left[2(1-r^2)(-2r) + 8\zeta^2 r\right] = 0$$

The solution is

$$r = 0 \text{ or } \left[(1-r^2)^2 + 4\zeta^2 r^2\right] + r^2\left[1 - r^2 - 2\zeta^2\right] = 0$$

The first result ($r = 0$) corresponds to static conditions, and is ignored. Hence, the resonant peak occurs when

$$(1-r^2)^2 + 4\zeta^2 r^2 + r^2 - r^4 - 2\zeta^2 r^2 = 0$$

which has the valid root

$$r = \frac{1}{\sqrt{1-2\zeta^2}} \qquad (ix)$$

Note that r has to be real and positive. It follows that, for a resonance to occur,

$$0 < \zeta < \frac{1}{\sqrt{2}}$$

Substitute in (*vii*) the resonant value of r to get

$$|G|_{peak} = \frac{\frac{1}{m(1-2\zeta^2)}}{\sqrt{\left(1-\frac{1}{1-2\zeta^2}\right)^2 + \frac{4\zeta^2}{1-2\zeta^2}}}$$

or

$$|G|_{peak} = \frac{1}{2m\zeta\sqrt{1-\zeta^2}} \qquad (x)$$

□

3.3 MECHANICAL IMPEDANCE APPROACH

Any type of force or motion variable can be used as input and output variables in defining a system transfer function. In vibration studies, three particular choices are widely used. The corresponding frequency transfer functions are named *impedance functions*, *mobility functions*, and *transmissibility functions*. These are described in the present section and in the subsequent section, and their use is illustrated.

Through variables (force) and across variables (velocity), when expressed in the frequency domain (as Fourier spectra), are used in defining the two important frequency transfer functions: *mechanical impedance* and *mobility*. In the case of impedance function, velocity is considered the input variable and the force is the output variable; whereas in the case of mobility function, the converse applies. Specifically,

$$M = \frac{1}{Z} \qquad (3.82)$$

It is clear that mobility (M) is the inverse of impedance (Z). Either transfer function can be used in a given problem. One can define several other versions of frequency transfer functions that might be useful in the modeling and analysis of mechanical systems. Some of the relatively common ones are listed in Table 3.2. Note that, in the frequency domain, since Acceleration = $j\omega$ × Velocity; and Displacement = Velocity/($j\omega$), the alternative types of transfer functions as defined in Table 3.2 are related to *Mechanical Impedance* and *Mobility* through a factor of $j\omega$; specifically,

Frequency Response

TABLE 3.2
Definitions of Useful Mechanical Transfer Functions

Transfer Function	Definition (in the frequency domain)
Dynamic stiffness	Force/displacement
Receptance, dynamic flexibility, or compliance	Displacement/force
Impedance (Z)	Force/velocity
Mobility (M)	Velocity/force
Dynamic inertia	Force/acceleration
Accelerance	Acceleration/force
Force transmissibility (T_f)	Transmitted force/applied force
Motion transmissibility (T_m)	Transmitted velocity/applied velocity

$$\text{Dynamic inertia} = \text{Force/acceleration} = \text{Impedance}/(j\omega)$$

$$\text{Accelerance} = \text{Acceleration/force} = \text{Mobility} \times j\omega$$

$$\text{Dynamic stiffness} = \text{Force/displacement} = \text{Impedance} \times j\omega$$

$$\text{Receptance} = \text{Displacement/force} = \text{Mobility}/(j\omega)$$

In these definitions, the variables — force, acceleration, and displacement — should be interpreted as the corresponding Fourier spectra.

The time-domain constitutive relations for the mass, spring, and the damper elements are well known. The corresponding transfer relations are obtained by replacing the derivative operator d/dt by the Laplace operator s. The frequency transfer functions are obtained by substituting $j\omega$ or $j2\pi f$ for s. These results are derived below.

Mass Element:

$$m \frac{dv}{dt} = f$$

In the frequency domain,

$$mj\omega v = f \quad \text{or} \quad \frac{f}{v} = mj\omega$$

Hence,

$$Z_m = mj\omega \quad \text{and} \quad M_m = \frac{1}{mj\omega} \qquad (3.83 \text{ a and b})$$

Spring Element:

$$f = kx \quad \text{or} \quad \frac{df}{dt} = kv$$

TABLE 3.3
Impedance and Mobility Functions of Basic Mechanical Elements

	Frequency Transfer Function (Set $s = j\omega = j2\pi f$)		
Element	Impedance	Mobility	Receptance
Mass, m	$Z_m = ms$	$M_m = \dfrac{1}{ms}$	$R_m = \dfrac{1}{ms^2}$
Spring, k	$Z_k = \dfrac{k}{s}$	$M_k = \dfrac{s}{k}$	$R_k = \dfrac{1}{k}$
Damper, b	$Z_b = b$	$M_b = \dfrac{1}{b}$	$R_b = \dfrac{1}{bs}$

In the frequency domain,

$$j\omega f = kv \quad \text{or} \quad \frac{f}{v} = \frac{k}{j\omega}$$

Hence,

$$Z_k = \frac{k}{j\omega} \quad \text{and} \quad M_k = \frac{j\omega}{k} \qquad (3.84 \text{ a and b})$$

Damper Element:

$$f = bv \quad \text{and} \quad \frac{f}{v} = b$$

Then,

$$Z_b = b \quad \text{or} \quad M_b = \frac{1}{b} \qquad (3.85 \text{ a and b})$$

These results are summarized in Table 3.3.

3.3.1 Interconnection Laws

Any general impedance element or a mobility element can be interpreted as a *two-port element* in which, under steady conditions, energy (or power) transfer into the device takes place at the *input port* and energy (or power) transfer out of the device takes place at the *output port*. Each port of a two-port element has a *through variable*, such as force or current, and an *across variable*, such as velocity or voltage, associated with it. Through variables are called *flux variables*, and across variables are called *potential variables*. Through variables are not always the same as *flow variables* (velocity and current). Similarly, across variables are not the same as *effort variables* (force and voltage). For example, force is an effort variable, but it is also a through variable. Similarly, velocity is a flow variable and is also an across variable. The concept of effort and flow variables is useful in giving unified definitions for electrical and mechanical impedance. But in component interconnecting and circuit analysis, mechanical impedance is not analogous to electrical impedance. The

TABLE 3.4
Interconnection Laws for Impedance and Mobility

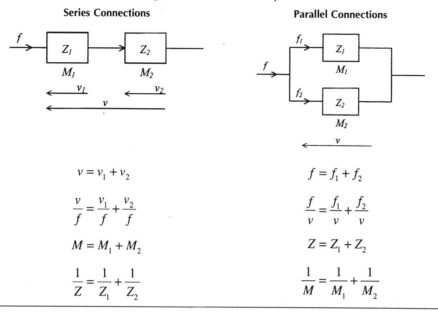

$$v = v_1 + v_2$$

$$\frac{v}{f} = \frac{v_1}{f} + \frac{v_2}{f}$$

$$M = M_1 + M_2$$

$$\frac{1}{Z} = \frac{1}{Z_1} + \frac{1}{Z_2}$$

$$f = f_1 + f_2$$

$$\frac{f}{v} = \frac{f_1}{v} + \frac{f_2}{v}$$

$$Z = Z_1 + Z_2$$

$$\frac{1}{M} = \frac{1}{M_1} + \frac{1}{M_2}$$

definition of mechanical impedance is force/velocity in the frequency domain. This is a ratio of (through variable)/(across variable), whereas electrical impedance, defined as voltage/current in the frequency domain, is a ratio of (across variable)/(through variable). Because both force and voltage are "effort" variables and velocity and current are flow variables, it is then convenient to use the definition

$$\text{Impedance (electrical or mechanical)} = \frac{\text{Effort}}{\text{Flow}}$$

In other words, impedance measures how much effort is needed to drive a system at unity flow. Nevertheless, this definition does not particularly help in analyzing interconnected systems with mechanical impedance, because mechanical impedance cannot be manipulated using the rules for electrical impedance. For example, if two electric components are connected in series, the current (through variable) will be the same for both components, and the voltage (across variable) will be additive. Accordingly, impedance of a series-connected electrical system is just the sum of the impedances of the individual components. Now consider two mechanical components connected in series. Here, the force (through variable) will be the same for both components, and velocity (across variable) will be additive. Hence, it is mobility, not impedance, that is additive in the case of series-connected mechanical components. It can be concluded that, in circuit analysis, mobility behaves like electrical impedance, and mechanical impedance behaves like electrical admittance. Hence, the "generalized series element" is electrical impedance or mechanical mobility, and the "generalized parallel element" is electrical admittance or mechanical impedance. The corresponding interconnection laws are summarized in Table 3.4.

Now, two examples are given to demonstrate the use of impedance and mobility methods in frequency-domain problems.

EXAMPLE 3.5

Consider the simple oscillator shown in Figure 3.9(a). A schematic mechanical circuit is given in Figure 3.9(b). Note that in this circuit, the broken line from the mass to the ground represents how the "inertia force" of the mass is felt by (or virtually transmitted to) the ground. This is the case because the net force that generates the acceleration in the mass (i.e., the inertia force) must be transmitted to the ground at the reference point of the force source. This is the same reference with respect to which the velocity of the mass is expressed. If the input is the force $f(t)$, the source element is a force source. The corresponding response is the velocity v, and in this situation the transfer function $V(f)/F(f)$ is a mobility function. On the other hand, if the input is the velocity $v(t)$, the source element is a velocity source. Then, f is the output, and the transfer function $F(f)/V(f)$ is an impedance function.

Suppose that a known forcing function is applied to this system (with zero initial conditions) using a force source, and the velocity is measured. Now, if one were to move the mass exactly at this predetermined velocity (using a velocity source), the force generated at the source would be identical to the originally applied force. In other words, mobility is the reciprocal (inverse) of impedance, as noted earlier. This reciprocity should be intuitively clear because one is dealing with the same system and same initial conditions. Due to this property, one can use either the impedance representation or the mobility representation, depending on whether the elements are connected in parallel or in series, irrespective of whether the input is a force or a velocity. Once the transfer function is determined in one form, its reciprocal gives the other form.

In the present example, the three elements are connected in parallel, as clear from the mechanical circuit shown in Figure 3.9(b). Hence, the impedance representation is appropriate. The overall impedance function of the system is

$$Z(f) = \frac{F(f)}{V(f)} = Z_m + Z_k + Z_b = ms + \frac{k}{s} + b \bigg|_{s=j2\pi f}$$

$$= \frac{ms^2 + bs + k}{s} \bigg|_{s=j2\pi f}$$

(3.86)

Then, the mobility function is

$$M(f) = \frac{V(f)}{F(f)} = \frac{s}{ms^2 + bs + k} \bigg|_{s=j2\pi f}$$

(3.87)

Note that if in fact the input is the force, the mobility function will govern the system behavior. In this case, the characteristic polynomial of the system is $s^2 + bs + k$, which corresponds to a simple oscillator and, accordingly, the (dependent) velocity response of the system would be governed by this. If, on the other hand, the input is the velocity, the impedance function will govern the system behavior. The characteristic polynomial of the system in this case is s, which corresponds to a simple integrator. This is a physically non-realizable system because the numerator order (2) is greater than the denominator order (1). The (dependent) force response of the system would be governed by an integrator-type behavior. To explore this behavior further, suppose that the velocity source has a constant value. The inertia force will be zero. The damping force will be constant. The spring force will increase linearly. Hence, the net force will have an integration (linearly increasing) effect. If the velocity source provides a linearly increasing velocity (constant acceleration), the inertia force will be constant, the damping force will increase linearly, and the spring force will increase quadratically.

Frequency Response

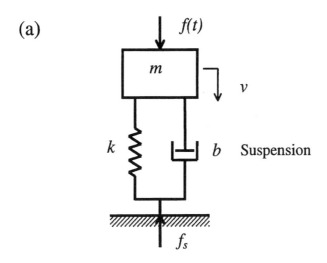

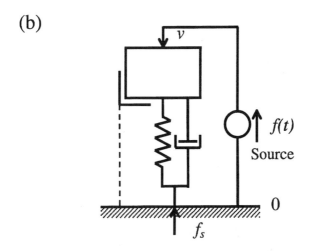

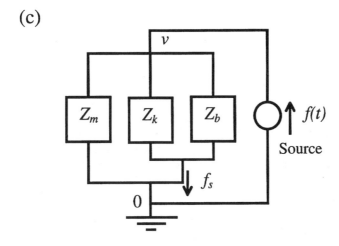

FIGURE 3.9 (a) A ground-based mechanical oscillator; (b) schematic mechanical circuit; and (c) impedance circuit.

EXAMPLE 3.6

Consider the system shown in Figure 3.10(a). In this example, the motion of the mass m is not associated with an external force. The support motion v, however, is associated with a force f.

The schematic mechanical circuit representation shown in Figure 3.10(b) and the corresponding impedance circuit shown in Figure 3.10(c) indicate that the spring and the damper are connected in parallel and the mass is connected in series with this pair. By impedance addition for parallel elements and mobility addition for series elements, it follows that the mobility function is

$$\frac{V(f)}{F(f)} = M_m + \frac{1}{(Z_k + Z_b)} = \frac{1}{ms} + \frac{1}{(k/s + b)}\bigg|_{s=j2\pi f}$$

$$= \frac{ms^2 + bs + k}{ms(bs + k)}\bigg|_{s=j2\pi f} \tag{3.88}$$

It also follows that when the support force is the input (force source) and the support velocity is the output, the system characteristic polynomial is $ms(bs + k)$, which is known to be inherently unstable due to the presence of a free integrator, and has a nonoscillatory transient response.

The impedance function that corresponds to support velocity input (velocity source) is the reciprocal of the previous mobility function; thus,

$$\frac{F(f)}{V(f)} = \frac{ms(bs + k)}{ms^2 + bs + k}\bigg|_{s=j2\pi f} \tag{3.89}$$

and

$$\frac{V_m(f)}{F(f)} = \frac{1}{ms}\bigg|_{s=j2\pi f} \tag{3.90}$$

The resulting impedance function $F(f)/V_m(f)$ is not admissible and physically nonrealizable since V_m cannot be an input because the associated force can become infinity. This is confirmed by the fact that the corresponding transfer function is a differentiator, which is not physically realizable. The mobility function $V_m(f)/F(f)$ corresponds to a simple integrator. Physically, when a force f is applied to the support, it is transmitted to the mass, unchanged, through the parallel spring-damper unit. Accordingly, when f is constant, a constant acceleration is produced at the mass, causing its velocity to increase linearly (an integration behavior).

Maxwell's principle of reciprocity can be demonstrated by noting that the mobility function $V_m(f)/F(f)$ obtained in this example will be identical to the mobility function when the locations of f and v_m are reversed (i.e., when a force f is applied to the mass m and the resulting motion v_m of the support, which is not restrained by a force, is measured), with the same initial conditions. The reciprocity property is valid for linear, constant-parameter systems in general, and is particularly useful in vibration analysis and testing of multi-degree-of-freedom systems; for example, to determine a transfer function that is difficult to measure, by measuring its symmetrical counterpart in the transfer function matrix.

3.4 TRANSMISSIBILITY FUNCTIONS

Transmissibility functions are transfer functions that are particularly useful in the analysis of vibration isolation in machinery and other mechanical systems. Two types of transmissibility

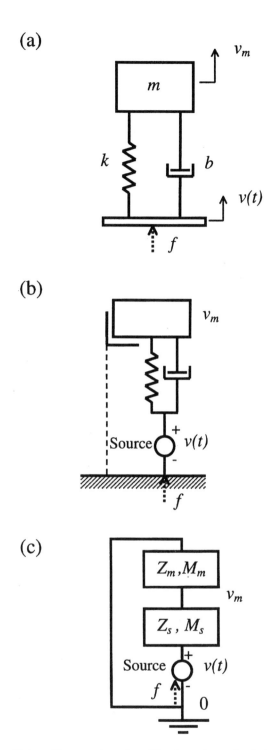

FIGURE 3.10 (a) An oscillator with support motion; (b) schematic mechanical circuit; and (c) impedance circuit.

functions — force transmissibility and motion transmissibility — can be defined as given in Table 3.2. Due to a reciprocity characteristic in linear systems, it can be shown that these two transfer functions are equal and, consequently, it is sufficient to consider only one of them. However, consider both types first to show their equivalence.

3.4.1 Force Transmissibility

Consider a mechanical system that is supported on a rigid foundation through a suspension system. If a forcing excitation is applied to the system, the force is not directly transmitted to the foundation. The suspension system acts as a vibration isolation device (see Chapter 12). Force transmissibility determines the fraction of the forcing excitation that is transmitted to the support structure (foundation) through the suspension, at different excitation frequencies, and is defined as:

$$\text{Force transmissibility } T_f = \frac{\text{Force transmitted to support } F_s}{\text{Applied force } F} \quad (3.91)$$

Note that this is defined in the frequency domain and, accordingly, F_s and F should be interpreted as the Fourier spectra of the corresponding forces.

A schematic representation of the force transmissibility mechanism is shown in Figure 3.11(a). The reason for the suspension force f_s being not equal to the applied force f is attributed to the inertia paths (broken line in the figure) that are present in a mechanical system.

3.4.2 Motion Transmissibility

Consider a mechanical system that is supported through a suspension on a structure that may be subjected to undesirable motions (e.g., guideway deflections, vehicle motions, seismic disturbances). Motion transmissibility determines the fraction of the support motion that is transmitted to the system through its suspension at different frequencies. It is defined as

$$\text{Motion transmissibility } T_m = \frac{\text{System motion } V_m}{\text{Support motion } V} \quad (3.92)$$

The velocities V_m and V are expressed in the frequency domain as Fourier spectra.

A schematic representation of the motion transmissibility mechanism is shown in Figure 3.11(b). Typically, the motion of the system is taken as the velocity of one of its critical masses. Different transmissibility functions are obtained when different mass points (or degrees of freedom) of the system are considered.

Next, two examples are given to show a reciprocity property that makes the force transmissibility and the motion transmissibility functions identical in complementary (reciprocal) systems.

System Suspended on a Rigid Base (Force Transmissibility)

Consider the system suspended on a rigid base and excited by force $f(t)$, as shown earlier in Figure 3.9(a). Here the *system* is the inertia element m, and the *suspension* is the parallel spring and damper combination. Note that the three elements m, k, and b are all in parallel.

Here, $Z_m = mj\omega$; $Z_b = b$; and $Z_k = \frac{k}{j\omega}$, as given in Table 3.3. Now, because the elements are connected in parallel, one has (see Table 3.4)

$$\frac{f}{v} = Z_m + Z_b + Z_k \quad (3.93)$$

Frequency Response

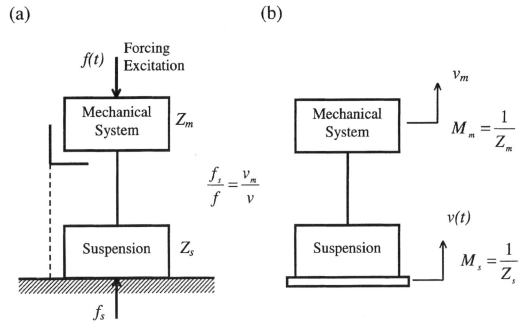

FIGURE 3.11 (a) An inertial system with ground-based suspension, and (b) the counterpart (reciprocal) system with support motion.

Hence,

$$\frac{v}{f} = \frac{1}{Z_m + Z_b + Z_d} \tag{3.94}$$

Also, suspension impedance is

$$\frac{f_s}{v} = Z_b + Z_k = Z_s \tag{3.95}$$

where, f_s is the force transmitted to the support structure (foundation). Then,

$$\frac{f_s}{f} = \frac{Z_b + Z_k}{Z_m + Z_b + Z_k} = \frac{Z_s}{Z_m + Z_s} \tag{3.96}$$

This result should be immediately clear because force is divided among parallel branches in proportion to their impedance values (because the velocity is common). Now,

$$\text{Force transmissibility magnitude } |T_f| = \left|\frac{\text{Force transmitted to support}}{\text{Applied force to system}}\right| = \left|\frac{Z_s}{Z_m + Z_s}\right| \tag{3.97}$$

Substitute parameters:

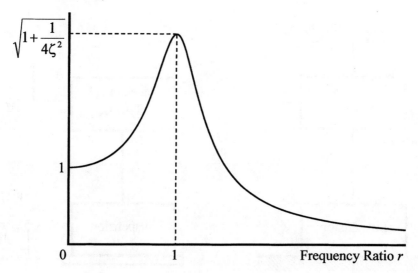

FIGURE 3.12 Transmissibility curve for a simple mechanical oscillator.

$$|T_f| = \left|\frac{f_s}{f}\right| = \left|\frac{b + \dfrac{k}{j\omega}}{mj\omega + b + \dfrac{k}{j\omega}}\right| = \left|\frac{bj\omega + k}{k - m\omega^2 + bj\omega}\right|$$

$$= \sqrt{\frac{b^2\omega^2 + k^2}{(k - m\omega^2)^2 + b^2\omega^2}} = \sqrt{\frac{(2\zeta\omega_n\omega)^2 + \omega_n^4}{(\omega_n^2 - \omega^2)^2 + (2\zeta\omega_n\omega)^2}}$$

On simplification one obtains

$$|T_f| = \sqrt{\frac{1 + 4\zeta^2 r^2}{(1 - r^2)^2 + 4\zeta^2 r^2}} \tag{3.98}$$

where the normalized frequency

$$\frac{\omega}{\omega_n} = r$$

At $r = 0$, $\quad |T_f| = 1$

At $r = 1$, $\quad |T_f| = \sqrt{1 + \dfrac{1}{4\zeta^2}} \tag{3.99}$

This transmissibility magnitude curve is shown in Figure 3.12.

Frequency Response

System with Support Motion (Motion Transmissibility)

Consider again the system suspended on a moving platform as shown in Figure 3.10(a).
For this system,

$$Z_m = mj\omega \quad \text{and} \quad M_m = \frac{1}{mj\omega}$$

for the mass element. Since, the damper and the spring are connected in parallel, the corresponding impedances are additive. Hence,

$$Z_s = Z_b + Z_k = b + \frac{k}{j\omega} \quad \text{and} \quad M_s = \frac{1}{b + \frac{k}{j\omega}}$$

$$\text{Motion transmissibility magnitude } T_m = \left|\frac{\text{Applied support motion}}{\text{Motion of system inertia}}\right| = \left|\frac{M_m}{M_m + M_s}\right| \quad (3.100)$$

This directly follows from the fact that the velocity is divided among series elements in proportion to their mobilities (because the force is common).
However,

$$\frac{M_m}{M_m + M_s} = \frac{1}{1 + \frac{M_s}{M_m}}$$

$$= \frac{1}{1 + \frac{Z_m}{Z_s}} = \frac{Z_s}{Z_m + Z_s} \quad (3.101)$$

Hence,

$$|T_m| = \left|\frac{Z_s}{Z_m + Z_s}\right| \quad (3.102)$$

It follows that

$$T_f = T_m \quad (3.103)$$

This establishes the reciprocity property.

3.4.3 GENERAL CASE

Consider an inertial system with a ground-based suspension, as shown in Figure 3.11(a) and its counterpart with a moving support, as shown in Figure 3.11(b). The corresponding impedance circuits are shown in Figure 3.13.
For system (a):

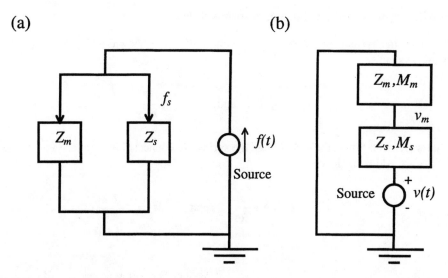

FIGURE 3.13 Impedance circuits: (a) inertial system with ground-based suspension; and (b) system and suspension with support motion.

$$\frac{f_s}{f} = \frac{Z_s}{Z_m + Z_s}$$

For system (b):

$$\frac{v_m}{v} = \frac{M_m}{M_m + M_s} = \frac{1}{1 + \frac{M_s}{M_m}} = \frac{1}{1 + \frac{Z_m}{Z_s}} = \frac{Z_s}{Z_s + Z_m}$$

It follows that

$$\frac{f_s}{f} = \frac{v_m}{v} \quad \text{and} \quad T_f = T_m \tag{3.104}$$

for this more general situation.

EXAMPLE 3.7

Consider again the problem of fluid coupling system shown in Figure 3.5 and studied in Example 3.1. A schematic mechanical circuit for the system is shown in Figure 3.14(a). The corresponding impedance circuit is shown in Figure 3.14(b). Use the impedance method to solve the same problem.

For this problem: Input angular velocity = $\dot{\alpha}(t)$; Angular velocity of the load = $\dot{\theta}$; and Load impedance $Z_l = Z_m + Z_k$. In view of the series connection of Z_b and Z_l, one has

$$\frac{\dot{\theta}}{\dot{\alpha}} = \frac{M_l}{M_l + M_b} = \frac{\frac{1}{M_b}}{\frac{1}{M_b} + \frac{1}{M_l}} = \frac{Z_b}{Z_b + Z_m + Z_k} \tag{i}$$

Frequency Response

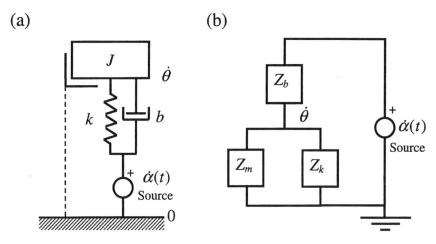

FIGURE 3.14 (a) Schematic mechanical circuit of the fluid coupling system, and (b) impedance circuit.

For the torsional spring,

$$\frac{\tau}{\theta} = Z_k \qquad (ii)$$

Multiply the equations (*i*) and (*ii*) together.

$$\frac{\tau}{\dot{\alpha}} = \frac{Z_k Z_b}{(Z_b + Z_m + Z_k)} \qquad (iii)$$

Because time derivative corresponds to multiplication by $j\omega$ in the frequency domain, one can write (*iii*) in the form

$$\frac{\tau}{\alpha} = j\omega \frac{Z_k Z_b}{(Z_b + Z_m + Z_k)} \qquad (iv)$$

Substituting $Z_b = b$, $Z_m = j\omega J$, and $Z_k = \dfrac{k}{j\omega}$, one obtains

$$\frac{\tau}{\alpha} = \frac{kbj\omega}{(k - J\omega^2) + bj\omega}$$

which is identical to what we obtained in Example 3.1.

3.4.4 Peak Values of Frequency-Response Functions

The peak values of a frequency transfer function correspond to the resonances. The frequencies at these points are called resonant frequencies. Because a transfer function is the ratio of a response variable to an input variable, it is reasonable to get different peak frequencies for the same excitation input, if the response variable that is considered is different. Some results obtained for a damped oscillator model are summarized in Table 3.5.

TABLE 3.5
Some Practical Frequency Response Functions and Their Peaks

System	Response/Excitation	Frequency Response Function (Normalized)	Normalized Frequency at Peak (r_p)	Peak Magnitude (Normalized)
Simple oscillator	Displacement/force	$\dfrac{1}{(1-r^2)+2j\zeta r}$	$\sqrt{1-2\zeta^2}$	$\dfrac{1}{2\zeta\sqrt{1-\zeta^2}}$
Simple oscillator with velocity response	Velocity/force	$\dfrac{jr}{(1-r^2)+2j\zeta r}$	1	$\dfrac{1}{2\zeta}$
Simple oscillator with acceleration response	Acceleration/force	$\dfrac{-r^2}{(1-r^2)+2j\zeta r}$	$\dfrac{1}{\sqrt{1-2\zeta^2}}$	$\dfrac{1}{2\zeta\sqrt{1-\zeta^2}}$
Fluid coupling system	Torque/displacement	$\dfrac{2j\zeta r}{(1-r^2)+2j\zeta r}$	1	1
Force transmissibility	Force/force	$\dfrac{1+2j\zeta r}{(1-r^2)+2j\zeta r}$	$\dfrac{\sqrt{\sqrt{1-8\zeta^2}-1}}{2\zeta}$	$\sqrt{1+\dfrac{1}{4\zeta^2}}$ (for small ζ)
Motion transmissibility	Velocity/velocity	Same	Same	Same

3.5 RECEPTANCE METHOD

Receptance is another name for dynamic flexibility or compliance and is given by the transfer function Output displacement/Input force, in the frequency domain (see Table 3.2). Also, as mentioned previously, it is directly related to mobility through

$$\text{Receptance} = \frac{\text{Mobility}}{j\omega} \quad (3.105)$$

and this relationship should be clear due to the fact that Velocity = $j\omega$ × Displacement, with zero initial conditions, in the frequency domain. Hence, the receptance functions for the basic elements mass (m), spring (k), and damper (b) can be derived from the mobility functions of these elements, as given in Table 3.3. Furthermore, as a result, the interconnection laws given for mobility (M) will be valid for receptance (R) as well. Specifically, for two receptance elements R_1 and R_2 connected in series, the combined receptance is

$$\text{Series:} \quad R = R_1 + R_2 \quad (3.106)$$

because the displacements are additive and the force is common. For two receptance elements connected in parallel, the combined receptance R is given by

$$\text{Parallel:} \quad \frac{1}{R} = \frac{1}{R_1} + \frac{1}{R_2} \quad (3.107)$$

because the forces are additive and the displacement is common. The inverse of receptance is *dynamic stiffness*.

3.5.1 Application of Receptance

The receptance method is widely used in the frequency-domain analysis of multi-degree-of-freedom systems. This is true particularly because the receptance of a multi-component system can be expressed in a convenient form in terms of the receptances of its constituent components. In deriving such relations, one can use the conditions of continuity (force balance at points of interconnection, or nodes) and compatibility (relative displacements in a loop add to zero). In fact, equations (3.106) and (3.107) are special cases of receptance relations for multi-component systems.

It should be clear from Table 3.3 that the receptance R_m of an inertia element ($-1/\omega^2 m$)) and the receptance R_k of a spring element ($1/k$) are real quantities, unlike the corresponding mobility functions. The receptance R_b of a damper ($1/(bj\omega)$) is imaginary, however. It follows that receptance functions of undamped systems are real, and one will have to deal only with real quantities in the receptance analysis of undamped systems. This makes the analysis quite convenient. Also, because the displacement response of an undamped system becomes infinite when excited by a harmonic force at its natural frequency, one can see that the receptance function of an undamped system goes to infinity (or its inverse becomes zero) at its natural frequencies. This property can be utilized in determining an undamped natural frequency (say, the fundamental natural frequency) of a system using the receptance method. In particular, the characteristic equations for a system with two interconnected components are

$$\text{Series:} \quad \frac{1}{R_1 + R_2} = 0 \tag{3.108}$$

$$\text{Parallel:} \quad R_1 + R_2 = 0 \tag{3.109}$$

and their solutions will give the undamped natural frequencies of the combined system. Now consider two examples to illustrate the application of receptance techniques.

Undamped Simple Oscillator

Consider the simple oscillator shown in Figure 3.9, but assume that the damper is not present. As has been noted before, the mass and the spring elements are connected in parallel. Hence, the characteristic equation of the undamped system is

$$R_m + R_k = 0 \tag{3.110}$$

or

$$-\frac{1}{\omega^2 m} + \frac{1}{k} = 0 \tag{3.111}$$

or

$$-k + \omega^2 m = 0$$

whose positive solution is

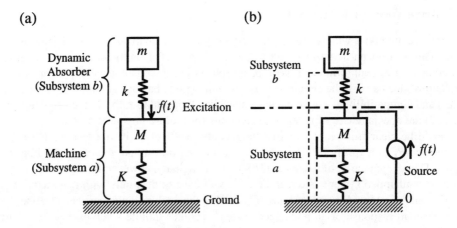

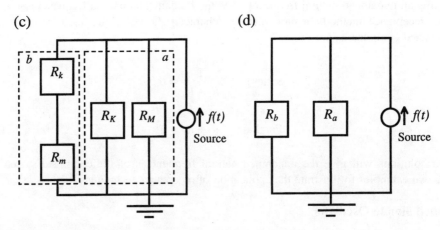

FIGURE 3.15 (a) A machine with a vibration absorber; (b) schematic mechanical circuit; (c) component receptance circuit; and (d) subsystem receptance circuit.

$$\omega = \omega_n = \sqrt{\frac{k}{m}}$$

which gives the undamped natural frequency.

Dynamic Absorber

Dynamic absorbers are commonly used in practice for vibration suppression in machinery over narrow frequency ranges. Specifically, a dynamic absorber can "absorb" the vibration energy from the main system (machine) at a specific frequency (tuned frequency) and thereby completely balance the vibration excitation in the system. Details are found in Chapter 12.

Consider a machine of equivalent mass M and equivalent stiffness K that is mounted on a rigid foundation, as modeled in Figure 3.15(a). A dynamic absorber, which is a lightly damped oscillator, of mass m and stiffness k is mounted on the machine. The damping is neglected in the model. The machine receives a vibration excitation $f(t)$, and the objective of the absorber is to counteract this excitation. A schematic mechanical circuit of the system is shown in Figure 3.15(b). The overall system can be considered to consist of the two subsystems: the subsystem a, representing the

Frequency Response

BOX 3.3 Concepts of Reception

$$\frac{\text{Receptance } R}{(\text{Compliance, Dynamic flexibility})} = \frac{\text{Displacement}}{\text{Force}}$$

Receptance $= \dfrac{\text{Mobility}}{j\omega}$

Series connection: $R = R_1 + R_2$

Parallel connection: $\dfrac{1}{R} = \dfrac{1}{R_1} + \dfrac{1}{R_2}$

Note: R is real for undamped systems.
Natural frequency: $R \to \infty$ (undamped case)

Characteristic Equation:

For system with series components: $\dfrac{1}{\Sigma R_i} = 0$

For system with parallel components: $\Sigma \dfrac{1}{R_i} = 0$

(e.g., for two parallel components: $R_1 + R_2 = 0$ gives natural frequency.)

---

machine, has M and K connected in parallel with the excitation source; and the subsystem b, representing the vibrating absorber, has m and k connected in series, as is clear from Figure 3.15(b). The corresponding receptance circuit, indicating the two subsystems with receptances R_a and R_b, is shown in Figure 3.15(d).

Because M and K are connected in parallel, from equation (3.107) one obtains

$$\frac{1}{R_a} = -\omega^2 M + K \tag{3.112}$$

Because m and k are connected in series, from equation (3.106) one obtains

$$R_b = -\frac{1}{\omega^2 m} + \frac{1}{k} \tag{3.113}$$

Now, because the subsystems a and b are connected in parallel, from equation (3.109), the characteristic equation of the overall system is given by

$$R_a + R_b = 0 \tag{3.114}$$

Substituting equations (3.112) and (3.113) into (3.114), one obtains

$$\frac{1}{-\omega^2 M + K} + \frac{1}{-\omega^2 m} + \frac{1}{k} = 0 \qquad (3.115)$$

On simplification, after multiplying throughout by the common denominator, one obtains the characteristic equation

$$mM\omega^4 - (kM + Km + km)\omega^2 + kK = 0 \qquad (3.116)$$

This will give two positive roots for ω, which are the two undamped natural frequencies of the system. Typically, the natural frequency of the vibration absorber must be tuned to the frequency of excitation in order to achieve effective vibration suppression, as discussed in Chapter 12.

Here, one has only considered *direct receptance* functions, where the considered excitation and response are both for the same node. For more complex, multi-component, multi-degree-of-freedom systems, one needs to consider *cross receptance* functions, where the response is considered at a node other than where the excitation force is applied. Such situations are beyond the scope of the present introductory material. Some concepts of receptance are summarized in Box 3.3.

PROBLEMS

3.1 What is meant by the frequency spectrum of a signal?
  a. Define the following terms:
   i. Mechanical impedance
   ii. Mobility
   iii. RMS value of a signal
  b. Using sketches, show how the following specifications can be represented in the frequency domain:
   i. Ride quality of a ground transit vehicle for short trips and for long trips
   ii. Specification for a vibration (shaker) test

3.2 a. Define the following terms:
   i. Undamped natural frequency
   ii. Damped natural frequency
   iii. Resonant frequency
   iv. Damping constant
   v. Damping ratio
   vi. Half-power frequencies
   vii. Q-factor

For the mass-spring-damper system described by

$$m\ddot{x} + b\dot{x} + kx = f(t)$$

with the usual notation, give expressions for each of the six quantities that you have defined, in terms of the system parameters m, b, and k. Full credit will be given if you can correctly and completely describe the procedure for deriving expressions for these quantities.

Four tests are carried out on the same mass-spring-damper system, as schematically shown in Figure P3.2. Assume that the damping ratio of the test object is less than 0.7. The experiments (i) and (ii) are "hammar tests," and (iii) and (iv) are "sine tests." In a hammar test, a single quick impact is given to the mass and the response (displacement) of the mass is measured. From that, the most significant frequency of oscillation is determined; for example, using a spectrum analyzer. In a sine test, a sinusoidal forcing function $f_0 \sin\omega t$ is applied to the test object and its steady-state response is measured for

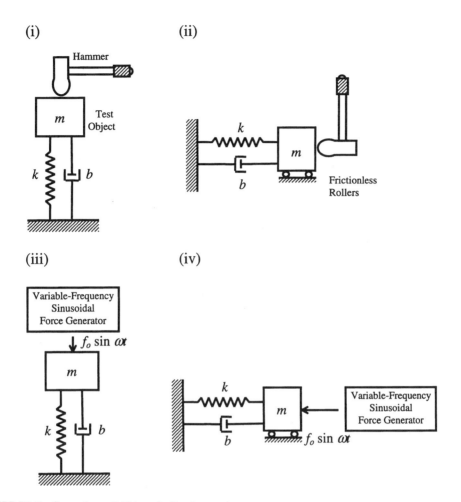

FIGURE P3.2 Several possibilities of vibration testing.

a given frequency ω. The test is repeated for a range of frequencies, and from the results the excitation frequency that corresponds to the highest amplification of the object response is observed.

Note that the tests (i) and (iii) are carried out for the vertical configuration of the object in the presence of gravity. Tests (ii) and (iv) are done in the horizontal configuration, so that the gravity effects can be neglected.

By carrying out the four tests, the four frequency values are determined as explained in the above procedure. Are the four frequencies obtained from the four tests identical or different? Clearly justify your answer.

3.3 a. In relation to a mechanical system that executes forced vibrations, define, compare, and contrast the following pairs of terms:
  i. Homogeneous solution and Free response
  ii. Particular solution and Forced response
 b. Vibrations in the flexible coupling unit of a motor-driven mechanical device is to be investigated. A simple schematic diagram of the test arrangement that is being considered is shown in Figure P.3.3(a). A simplified model is shown in Figure P.3.3b. Assume that all forms of dissipation are negligible. The motor is harmonically excited so that the magnetic torque generated is given by

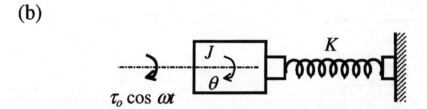

FIGURE P3.3 (a) Vibration investigation of a flexible coupling system; and (b) a simplified model.

$$\tau(t) = \tau_0 \cos \omega t$$

i. Show that the equation of motion of the simplified model is given by

$$J\ddot{\theta} + K\theta = \tau_0 \cos \omega t$$

in the usual notation.

ii. Suppose that the excitation is applied at $t = 0$, when the system has the initial conditions:

$$\theta(0) = \theta_0 \quad \text{and} \quad \dot{\theta}(0) = \omega_0$$

Obtain expressions for the total (complete) response θ of the system for the two cases of

$$\omega \neq \omega_n \quad \text{and} \quad \omega = \omega_n$$

where ω_n is the natural frequency of the system.

Sketch the behavior of the time response for the following three cases:
1. ω very close to ω_n
2. ω quite different from ω_n
3. ω equal to ω_n

3.4 a. Define the terms: Mechanical impedance; Mobility; Force transmissibility; and Motion transmissibility.

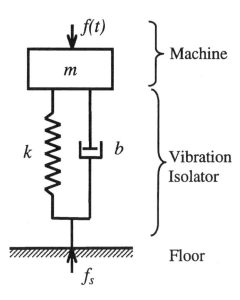

FIGURE P3.4 Model of a machine mounted on a vibration isolator.

In a vibration test, a harmonic forcing excitation is applied to the test object by means of a shaker and the velocity response at some other location is measured by using an accelerometer.
i. Strictly speaking, is this a mobility test or an impedance test? Justify your answer.
ii. How could the velocity response be measured using an accelerometer?

b. A machine of mass m has a rotating device that generates a harmonic forcing excitation $f(t)$ in the vertical direction. The machine is mounted on the factory floor using a vibration isolator of stiffness k and damping constant b. The harmonic component of the force that is transmitted to the floor, due to the forcing excitation, is $f_s(t)$. A corresponding model is shown in Figure P3.4. Giving all necessary steps, obtain an expression for the force transmissibility magnitude $|T_f|$ from f to f_s in terms of r and ζ where

$$r = \frac{\omega}{\omega_n}; \quad \zeta = \text{damping ratio}; \quad \omega_n = \text{undamped natural frequency of the system; and}$$

ω = excitation frequency (of $f(t)$).

c. Suppose that in part (b), $m = 100$ kg and $k = 1.0 \times 10^6$ N m^{-1}. Also, the frequency of the excitation force $f(t)$ in the operating range of the machine is known to be 200 rad s^{-1} or higher. Determine the damping constant b of the vibration isolator so that the force transmissibility magnitude is 0.5 or less.

3.5 Piezoelectric accelerometers are commonly used in vibration testing. A schematic diagram of a typical version of uniaxial piezoelectric accelerometer is shown in Figure P3.5(a). The piezoelectric element is sandwiched between a seismic mass and the base of the accelerometer housing, and an intimate contact is maintained by means of a high-stiffness spring that is placed between the seismic mass and the housing. The accelerometer typically has a magnetic base, with screw-in threading as well. It is mounted on an object whose acceleration needs to be measured. As the object accelerates, the seismic mass will accelerate together. There will be some energy dissipation (damping) due to the relative movement between the seismic mass and the accelerometer housing. This dissipation is present primarily in the piezoelectric element, but also can take place in

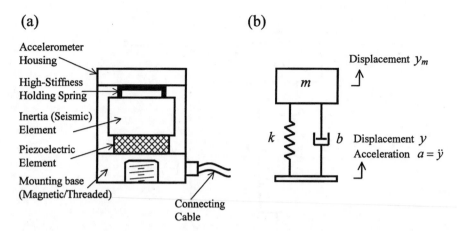

FIGURE P3.5 (a) Schematic diagram of a piezoelectric accelerometer; and (b) a simplified model.

the spring element and attachments. Strain (stress) in the piezoelectric element, due to its deflection, creates an electric charge and it is sensed through associated signal conditioning hardware (e.g., charge amplifier). This charge signal is known to represent the base acceleration of the accelerometer.

A simplified model of the accelerometer is shown in Figure P3.5(b). In this model, m is the equivalent moving mass within the accelerometer housing and it primarily represents the seismic mass. The stiffness k represents the equivalent spring restraint between the seismic mass and the accelerometer housing and is contributed by both the high-stiffness holding spring and the piezoelectric element. The linear viscous damping element with damping constant b represents the overall energy dissipation, as noted before. Also,

y = motion (displacement) of the accelerometer base

y_m = motion (displacement) of the seismic mass

Base acceleration, $a = \ddot{y}$, is the variable that needs to be measured. Assume that the charge generated in the piezoelectric element is given by

$$q = g(y - y_m)$$

in which g is a constant of the piezoelectric element, and $y - y_m$ is the net deflection of the piezoelectric element along its axis of sensitivity. We are interested in studying the relation between the sensed (measured) signal (q) and the quantity that is measured (a).

a. Write a differential equation in y and y_m to describe the motion of the system in Figure P3.5(b). Assume that the displacements are measured from the static equilibrium configuration so that gravity effects do not enter into the equation of motion.

b. Obtain a frequency transfer function $G(j\omega)$ for the ratio $\dfrac{\text{Measurement}}{\text{Measurand}}$, which is equal to $\dfrac{\text{Piezoelectric charge } q}{\text{Base acceleration } a}$.

c. Determine the frequency at which the magnitude of the accelerometer transfer function $G(j\omega)$ will be a maximum. Suggest a frequency range in which the accelerometer should be used, in order to obtain an acceptable measurement accuracy.

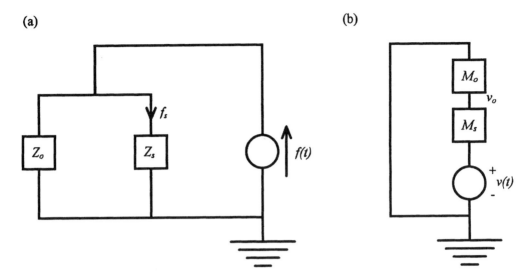

FIGURE P3.6 (a) A mechanical circuit with parallel-connected impedances; and (b) a mechanical circuit with series-connected impedances.

3.6 Define the following terms:
  a. Mechanical impedance
  b. Mechanical mobility
  c. Force transmissibility
  d. Motion transmissibility

Compare and contrast mechanical impedance and electrical impedance.

Consider an object (say, a piece of equipment) having mechanical impedance Z_o and mobility M_o. It is suspended on a supporting structure of mechanical impedance Z_s and mobility M_s. Two types of structural arrangements, as shown by the circuits in Figures P3.6(a) and (b) are of interest here.

Show that the force transmissibility $\dfrac{F_s(j\omega)}{F(j\omega)}$ of the system in Figure P3.6(a) is identical to the motion transmissibility $\dfrac{V_o(j\omega)}{V(j\omega)}$ of the system in Figure P3.6(b).

Interpret (i.e., give the meanings of) the variables F_s, F, V_o, and V of these expressions in the frequency domain.

Sketch a mechanical system consisting of mass, spring, and damper elements, that would be in the form of Figure P3.6(a) and give expressions for its Z_o and Z_s. Sketch the corresponding Figure P3.6(b) for a mechanical system and give the expressions for M_o and M_s for that system.

3.7 A machine tool and its supporting structure are modeled as the simple mass-spring-damper system shown in Figure P3.4.
  a. Draw a mechanical-impedance circuit for this system in terms of the impedances of the three elements: mass (m), spring (k), and viscous damping (b). Show that the magnitude force transmissibility $|T_f|$ of the system is given by

$$\frac{|F_s(j\omega)|}{|F(j\omega)|} = \sqrt{\frac{1+4\zeta^2 r^2}{\left(1-r^2\right)^2 + 4\zeta^2 r^2}}$$

in which

$r = \dfrac{\omega}{\omega_n}$ = ratio of the excitation frequency (ω) to the undamped natural frequency (ω_n)

ζ = damping ratio

b. Determine the exact value of r, in terms of ζ, at which the transmissibility magnitude will peak. Show that for small ζ, this value is $r = 1$.
c. Plot $|T_f|$ vs. r, for the interval $r = [0,5]$, with one curve for each of the five ζ values 0.0, 0.3, 0.7, 1.0, and 2.0 on the same plane. Discuss the behavior of these transmissibility curves.
d. From part (c), determine for each of the five ζ values, the excitation frequency range, with respect to ω_n, for which the transmissibility magnitude is
   i. Less than 1.05
   ii. Less than 0.5.
e. Suppose that the device in Figure P3.4 has a primary, undamped natural frequency of 6 Hz and a damping ratio of 0.2. It is required that the system has a force transmissibility magnitude of less than 0.5 for operating frequency values greater than 12 Hz.

   Does the existing system meet this requirement? If not, explain how you should modify the system to meet the requirement.

3.8 a. A vibrating system has a frequency-domain transfer function of $G(j\omega)$. If a forcing excitation with a frequency spectrum $F(j\omega)$ is applied to this system, the resulting motion response $x(t)$ can be given by its frequency spectrum $X(j\omega)$. Customarily we use the relation

$$X(j\omega) = G(j\omega)F(j\omega)$$

   i. Give the main assumptions that are made in arriving at this relation.
   ii. What are the advantages and disadvantages of using this relation?
b. An overhead transport device that is used for the transfer of parts and tools in a factory is schematically shown in Figure P3.8(a). The cart is driven by rubber tires on a supporting structure and is suspended from the overhead guiding track by means of a guiding wheel. Assume that the damping effects in the tires dominate over their elastic effects. Also, the overhead suspension mechanism has some flexibility but negligible damping

   A simplified model that may be used to study the vertical force transmitted to the overhead guiding track due to surface irregularities of the driven-on surface, is shown in Figure P3.8(b). Specifically, one is interested in determining the force f transmitted to the guiding track due to a harmonic component $u = u_0 \sin\omega t$ of the vertical displacement of the tires.
   i. What does ω depend on?
   ii. Under steady conditions, express the forcing response f in terms of $r = \omega/\omega_n$, ζ, k, and u_0, where $\omega_n = \sqrt{k/m}$ and $2\zeta\omega_n = b/m$.
   iii. Determine the value of r at which the amplitude of f will peak.

3.9 The Laplace transform of a signal $x(t)$ is given by

$$x(s) = \dfrac{2s^4 + 11s^3 + 14s^2 - 12s - 25}{(s+2)^2(s+3)}$$

What is the signal?

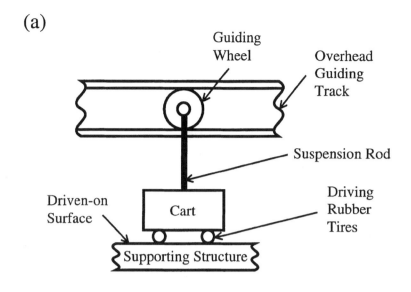

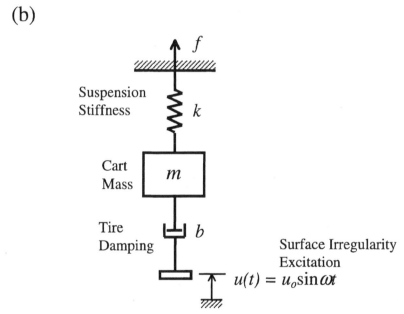

FIGURE P3.8 (a) Schematic representation of an overhead transport system in a factory; and (b) an approximate model for studying the vertical force-transfer characteristics.

3.10 Consider one joint (one degree of freedom) of a direct-drive robotic manipulator as shown in Figure P3.10. The joint is driven by an armature-controlled DC motor. Since the electrical time constant is negligible, the magnetic torque T_m of the motor is proportional to the input voltage v_a of the armature circuit; thus,

$$T_m = k_m v_a$$

The following parameters are defined:

J_m = motor rotor inertia (kg m²)

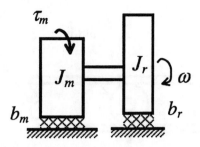

FIGURE P3.10 A joint of a direct-drive robotic arm.

b_m = motor mechanical damping constant (N m rad^{-1} s^{-1})
J_r = inertia of the robot arm
b_r = damping constant at the robot arm

Assume that the motor rotor is directly and rigidly linked to the robot arm. Neglect the effects of other joints (i.e., assume a one-degree-of-freedom robot arm).
  i. Can the system produce an oscillatory motion under (a) free conditions? (b) forced conditions? Explain your answer.
  ii. Write the differential equation relating the motor (magnetic) torque T_m to the arm speed ω.
  iii. Obtain the transfer function of the system with v_a as the input and ω as the output.
  iv. A unit step voltage input is applied to the arm that is initially at rest. Obtain an expression for the subsequent speed $\omega(t)$ of the arm as a function of time, in terms of k_m, b_m, b_r, J_m, and J_r. *Hint*: Use the Laplace transform method.

3.11 A second order mechanical device (with transfer function $G_p(s)$) has the following characteristics:

When a step input is applied to the device, the subsequent response appeared to be damped oscillations with frequency 2 rad s^{-1} and a time constant of 0.5 s. Also, at steady state, the output was found to be equal to the input (i.e., steady-state error = 0). What is the transfer function $G_p(s)$ of the device?

3.12 A mechanical system that is at rest is subjected to a unit step input $\mathcal{U}(t)$. Its response is given by

$$y = \left[2e^{-t} \sin t\right]\mathcal{U}(t)$$

  a. Write the input-output differential equation of the system.
  b. What is the transfer function?
  c. Determine the damped natural frequency, undamped natural frequency, and the damping ratio.
  d. Write the response of the system to a unit impulse and find the corresponding initial condition $y(0^+)$.
  e. What is the steady-state response for a unit step input?

3.13 A mechanical dynamic system is represented by the transfer function

$$\frac{Y(s)}{U(s)} = G(s) = \frac{\omega_n^2}{s^2 + 2\zeta\omega_n s + \omega_n^2}$$

with positive parameters ζ and ω_n.
  a. Is the system asymptotically stable?

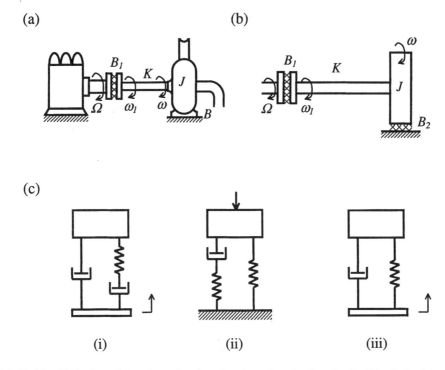

FIGURE P3.15 (a) A pump driven by a diesel engine through a clutch and a flexible shaft.; (b) schematic diagram of the system; and (c) possibilities of equivalent translatory mechanical systems.

  b. If the system is given an impulse input, at what frequency will it oscillate? What is the steady-state value of the output?
  c. If the system is given a unit step input, what is the frequency of the resulting output oscillations? What is its steady-state value?
  d. The system is given the sinusoidal input $u(t) = a \sin \omega t$. Determine an expression for the output $y(t)$ at steady state in terms of a, ω, ω_n, and ζ. At what value of the excitation frequency ω will the amplitude of the output $y(t)$ be maximum at steady state?

3.14 Consider the dynamic system represented by the transfer function

$$\frac{Y(s)}{U(s)} = \frac{(s+4)}{(s^2 + 3s + 2)}$$

  a. Indicate the correct statement among the following:
     i. The system is stable
     ii. The system is unstable
     iii. System stability depends on the input
     iv. None of the above
  b. Obtain the system differential equation.
  c. Using the Laplace transform technique determine the system response $y(t)$ to a unit step input with zero initial conditions.
  d. Determine the steady-state value of the system for a unit step input.

3.15 The rigid output shaft of a diesel engine prime mover is running at a known angular velocity $\Omega(t)$. It is connected through a friction clutch to a flexible shaft that, in turn, drives a hydraulic pump (see Figure P3.15(a)).

A *linear* model for this system is shown schematically in Figure P3.15(b). The clutch is represented by a viscous rotatory damper of damping constant B_1 (units: torque/angular velocity). The stiffness of the flexible shaft is K (units: torque/angular rotation). The pump is represented by a rotor of moment of inertia J (units: torque/ angular acceleration) and viscous damping constant B_2.

a. Write the two state equations relating the state variables T and ω to the input Ω, where T = torque in flexible shaft and ω = pump speed.

Hints:
1. Use a free-body diagram for the shaft, with ω_1 as the angular speed at the left end of the shaft.
2. Write the "torque balance" and "constitutive" relations for the shaft and eliminate ω_1.
3. Draw the free-body diagram for the rotor J and use D'Alembert's principle (Newton's second Law).

b. Obtain the system transfer functions, with Ω as the input, and T and ω as the outputs.
c. Which one of the translatory systems shown in Figure P3.15(c) is analogous to the system in Figure P3.15(b)?

3.16 A dynamic system that is at rest is subjected to a unit step input $\mathcal{U}(t)$. Its response is given by

$$y = 2e^{-t}(\cos t - \sin t)\mathcal{U}(t)$$

a. Write the input-output differential equation for the system.
b. What is its transfer function?
c. Determine the damped natural frequency, undamped natural frequency, and the damping ratio.
d. Write the response of the system to a unit impulse.

3.17 An air circulation fan system of a building is shown in Figure P3.17(a); and a simplified model of the system has been developed, as represented in Figure P3.17(b). The induction motor is represented as a torque source $\tau(t)$. The speed ω of the fan, which determines the volume flow rate of air, is of interest. The moment of inertia of the fan impeller is J. The energy dissipation in the fan is modeled by a linear viscous component (of damping constant b) and a quadratic aerodynamic component (of coefficient d).

a. Show that the system equation can be given by

$$J\dot{\omega} + b\omega + d|\omega|\omega = \tau(t)$$

b. Suppose that the motor torque is given by

$$\tau(t) = \bar{\tau} + \hat{\tau}_a \sin \Omega t$$

in which $\bar{\tau}$ is the steady torque and $\hat{\tau}_a$ is a very small amplitude (compared to $\bar{\tau}$) of the torque fluctuations at frequency Ω. Determine the steady-state operating speed $\bar{\omega}$, which is assumed positive, of the fan.

c. Linearize the model about the steady-state operating conditions and express it in terms of the speed fluctuations $\hat{\omega}$. From this, estimate the amplitude of the speed fluctuations.

3.18 When body flexibilities are neglected, an aircraft has six major degrees of freedom. These can be defined as three translatory motions (longitudinal, heave, and lateral) and three angular motions (roll, yaw, and pitch). Usually, all six degrees of freedom are coupled.

Frequency Response

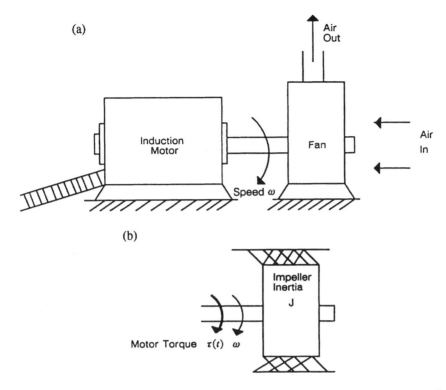

FIGURE P3.17 (a) A motor/fan combination of a building ventilation system; and (b) a simplified model of the ventilation fan.

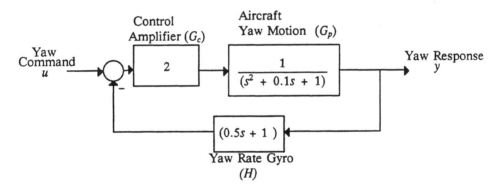

FIGURE P3.18 An uncoupled control system for yaw motion of an aircraft.

For a very preliminary analysis, however, each degree-of-freedom can be modeled separately (i.e., an uncoupled model).

Suppose that the yaw motion of an aircraft is given by the transfer function:

$$G_p(s) = \frac{1}{\left(s^2 + 0.1s + 1\right)}$$

A yaw rate gyro with transfer function

$$H(s) = 0.5s + 1$$

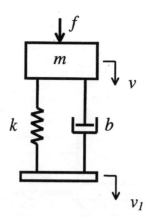

FIGURE P3.19 A degenerate mechanical system.

is used as the feedback sensor in the control of the yaw motion. The controller is approximated by a simple gain $G_c(s) = 2$. Accordingly, the yaw control loop is represented by the block diagram shown in Figure P3.18.

a. Compute the eigenvalues (poles) of the closed loop system (transfer function = y/u). Is the control system stable?

   Hint: $y = G_c G_p (u - Hy)$

b. Suppose that a sinusoidal yaw command

$$u = 2\cos 2t$$

is applied to the system. What is the yaw response y at steady state?

3.19 Consider the intuitively degenerate example of a mechanical system as shown in Figure P3.19. Note that the support motion v_1 is not associated with an external force. The mass m has an external force f and velocity v. The mass is supported on a spring of stiffness k and a linear damper of damping constant b.

   a. Draw a schematic mechanical circuit and an impedance circuit for the system. Can the velocity v be considered as an input to the system? Explain.
   b. Obtain various possibilities of impedance and mobility functions for this system.

3.20 Consider the two-degree-of-freedom systems shown in Figure P3.20(a) and (b). The mechanical elements m_1, m_2, k_1, k, b_1, and b are the same for both systems. The first system is supported on a rigid foundation and a force $f(t)$ is applied to mass m_1. The second system is supported on a light platform that is moved at velocity $v(t)$.

   a. Draw an impedance circuit for the system (a) and a mobility circuit for the system (b).
   b. Determine the force transmissibility f_s/f for system (a) where f_s is the force transmitted to the foundation.
   c. Determine the motion transmissibility v_m/v for system (b), where v_m is the velocity transmitted to mass m_1.
   d. Show that the transmissibility functions obtained in parts (b) and (c) above are identical.

3.21 The tachometer is a velocity-measuring device (passive) that uses the principle of electromagnetic generation. A DC tachometer is shown schematically in Figure P3.21(a). The field windings are powered by DC voltage v_f. The across variable at the input port is the measured angular speed ω_i. The corresponding torque T_i is the through variable at the input port. The output voltage v_o of the armature circuit is the across variable at the output port. The corresponding current i_o is the through variable at the output port. A

Frequency Response

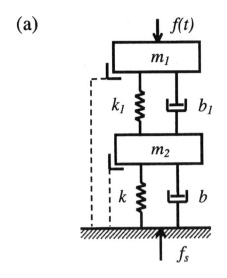

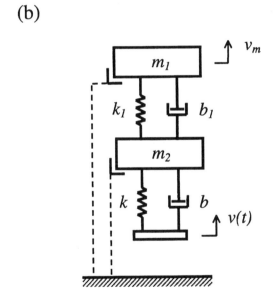

FIGURE P3.20 A two-degree-of-freedom mechanical system supported on a (a) rigid foundation, (b) light moving platform.

free-body diagram for the motor armature is shown in Figure P3.21(b). Obtain a transfer-function model for this device.

3.22 Consider the linear state-space model of a mechanical system as given by

$$\dot{x} = Ax + Bu$$

$$y = Cx$$

in the usual notation. Obtain a frequency transfer function relationship between the system output y and the input u.

3.23 Consider again, the model of an instrument test setup, as schematically shown in Figure P3.23. The springs A and B have a combined stiffness of k_1. Show that the transfer function between the excitation (support displacement) u and the response (velocity of m) v is given by

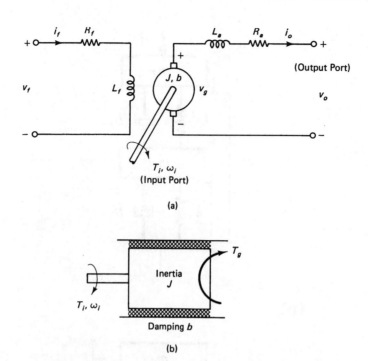

FIGURE P3.21 A DC tachometer example: (a) equivalent circuit; and (b) free-body diagram.

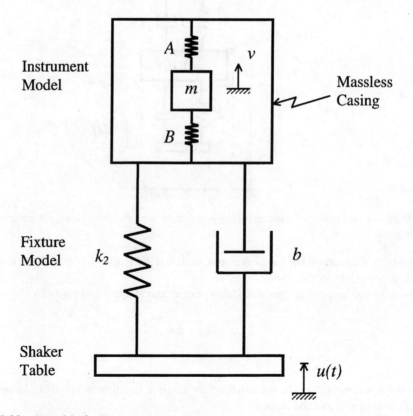

FIGURE P3.23 A model of an instrument test setup.

$$H(s) = \frac{k_1 b(bs + k_2)}{ms^2[b^2 s + b(k_1 + k_2)] + kb(bs + k_2)}$$

3.24 Consider a mass-spring-damper system (a simple oscillator). What is its force transmissibility function? What is its motion transmissibility function? Approximate them for the case of light damping. Indicate a use of this approximate result.

3.25 A two-degree-of-freedom, undamped, mechanical system is shown in Figure P3.25(a). Verify that the receptance circuit of the system is as given in Figure P3.25(b). Show that the receptance relations for this system can be expressed in the vector-matrix form as

$$\begin{bmatrix} \dfrac{1}{R_{mk1}} + \dfrac{1}{R_k} & -\dfrac{1}{R_k} \\ -\dfrac{1}{R_k} & \dfrac{1}{R_{mk2}} + \dfrac{1}{R_k} \end{bmatrix} \begin{bmatrix} X_1 \\ X_2 \end{bmatrix} = \begin{bmatrix} F_1 \\ F_2 \end{bmatrix}$$

where

$$\frac{1}{R_{mki}} = \frac{1}{R_{mi}} + \frac{1}{R_{ki}} \quad \text{for} \quad k = 1, 2$$

and X_i and F_i are the frequency spectra of the variables x_i and f_i, respectively, for $i = 1, 2$. Using this result, obtain the characteristic equation of the system, whose roots give the natural frequencies.

(a)

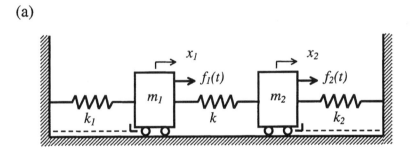

(b)

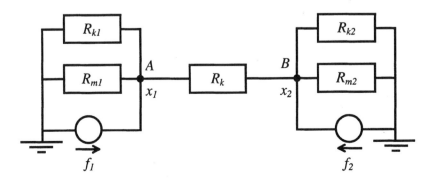

FIGURE P3.25 (a) A two-degree-of-freedom mechanical system; and (b) impedance circuit.

4 Vibration Signal Analysis

Numerous examples can be drawn from engineering applications for vibrating (dynamic) systems — a steam generator of a nuclear power plant that undergoes flow-induced vibration; a high-rise building subjected to seismic motions at its foundation; an incinerator tower subjected to aerodynamic disturbances; an airplane excited by atmospheric turbulence, a gate valve under manual operation; and a heating, ventilating, and air conditioning (HVAC) control panel stressed due to vibrations in its support structure are such examples.

Consider an aircraft in flight, as schematically shown in Figure 4.1. There are many excitations on this dynamic system. For example, jet engine forces and control surface movements are intentional excitations, whereas aerodynamic disturbances are unintentional (and unwanted) excitations. The primary response of the aircraft to these excitations will be the motions in various degrees of freedom, including rigid-body and flexible (vibratory) mode motions.

Although the inputs and outputs (excitations and responses) are functions of time, they can also be represented as functions of frequency, through Fourier transformation. The resulting *Fourier spectrum* of a signal can be interpreted as the set of frequency components that the original signal contains. As noted in Chapter 3, this *frequency-domain* representation of a signal can highlight many salient characteristics of the signal and also those of the corresponding system. For this reason, frequency-domain methods, particularly Fourier analysis, are used in a wide variety of applications such as data acquisition and interpretation, experimental modeling and modal analysis, diagnostic techniques, signal/image processing and pattern recognition, acoustics and speech research, signal detection, telecommunications, and dynamic testing for design development, quality control, and qualification of products. Many such applications involve the study of mechanical vibrations. This chapter discusses the nature and analysis of vibration signals.

4.1 FREQUENCY SPECTRUM

Excitations (inputs) to a dynamic system progress with time, thereby producing responses (outputs) which themselves vary with time. These are signals that can be recorded or measured. A measured signal is a *time history*. Note that in this case, the independent variable is *time* and the signal is represented in the *time domain*. A limited amount of information can be extracted by examining a time history. As an example, consider the time history record that is shown in Figure 4.2. It can be characterized by parameters such as the following.

a_p = peak amplitude
T_p = period in the neighborhood of the peak
   = 2 × interval between successive zero crossings near the peak
T_e = duration of the record
T_s = duration of strong response (i.e., the time interval beyond which no peaks occur that are larger than $a_p/2$)
N_z = number of zero crossings within T_s (N_z = 14 in Figure 4.2)

It is obviously cumbersome to keep track of so many parameters and, furthermore, not all of them are equally significant in a given application. Note, however, that all the parameters listed above are directly or indirectly related to either the *amplitude* or the *frequency* of zero crossings within

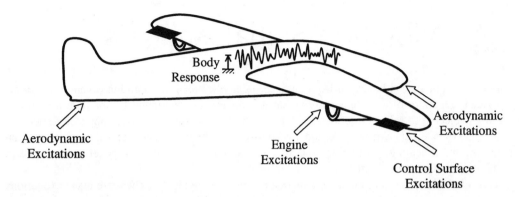

FIGURE 4.1 In-flight excitations and responses of an aircraft.

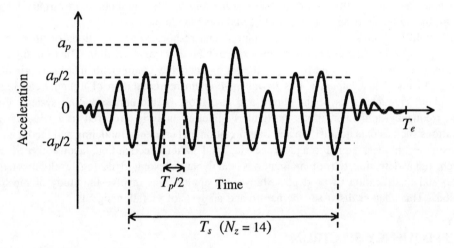

FIGURE 4.2 A time history record.

a given time interval. This signifies the importance of a frequency variable in representing a time signal. This is probably the fundamental motivation for using frequency-domain representations. In this context, more rigorous definitions are needed, however, for the parameters amplitude and frequency. A third parameter, known as *phase angle*, is also needed for unique representation of a signal in the frequency domain.

4.1.1 FREQUENCY

Let us further examine the basis of frequency domain analysis. Consider the *periodic* signal of period T that is formed by combining two harmonic (or, sinusoidal) components of periods T and $T/2$ and amplitudes a_1 and a_2, as shown in Figure 4.3. The *cyclic frequency* (cycles per second, or hertz, Hz) of the two components are $f_1 = 1/T$ and $f_2 = 2/T$. Note that in order to obtain the *angular frequency* (radians per second), the cyclic frequency must be multiplied by 2π.

4.1.2 AMPLITUDE SPECTRUM

An alternative graphical representation of the periodic signal shown in Figure 4.3 is given in Figure 4.4. In this representation, the amplitude of each harmonic component of the signal is plotted against the corresponding frequency. This is known as the *amplitude spectrum* of the signal, and it forms the basis of the frequency domain representation. Note that this representation is often

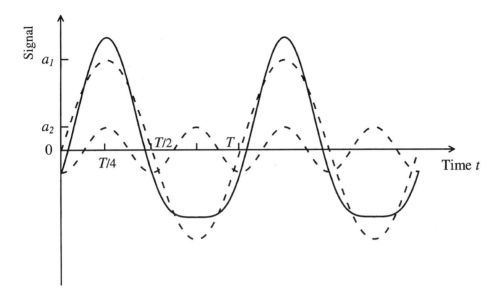

FIGURE 4.3 Time domain representation of a periodic signal.

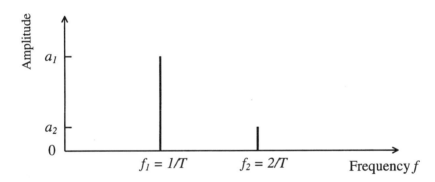

FIGURE 4.4 The amplitude spectrum of a periodic signal.

more compact and can be far more useful than the time domain representation. Note further that in the frequency domain representation, the independent variable is frequency.

4.1.3 PHASE ANGLE

In its present form, Figure 4.4 does not contain all the information of Figure 4.3. For example, if the high-frequency component in Figure 4.3 is shifted through half its period ($T/4$), the resulting signal is shown in Figure 4.5; this signal is quite different from that in Figure 4.3. But because the amplitudes and the frequencies of the two harmonic components are identical for both signals, they possess the same amplitude spectrum. Then, what is lacking in Figure 4.3 in order to make it a unique representation of a signal, is the information concerning the exact location of the harmonic components with respect to the time reference or origin ($t = 0$). This is known as the *phase information*. As an example, the distance of the first positive peak of each harmonic component from the time origin can be expressed as an angle (in radians) by multiplying it by $2\pi/T$; this is termed the *phase angle* of the particular component. In both signals (shown in Figures 4.3 and 4.5), the phase angle of the first harmonic component is the same and equals $\pi/2$ according to the present convention. The phase angle of the second harmonic component is $\pi/2$ in Figure 4.3 and zero (0) in Figure 4.5.

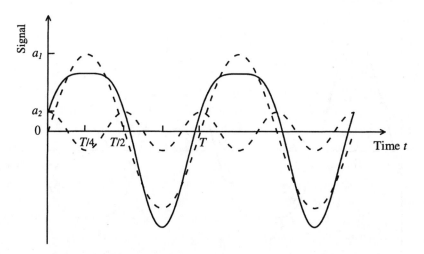

FIGURE 4.5 A periodic signal having an identical amplitude spectrum as for Figure 4.3.

4.1.4 Phasor Representation of Harmonic Signals

A convenient geometric representation of a harmonic signal of the form

$$y(t) = a\cos(\omega t + \phi) \tag{4.1}$$

is possible by means of a *phasor*. This representation is illustrated in Figure 4.6. Specifically, consider a rotating arm of radius a, rotating in the counterclockwise (ccw) direction at an angular speed of ω rad s^{-1}. Suppose that the arm starts (i.e., at $t = 0$) at an angular position ϕ with respect to the y-axis (vertical axis) in the ccw sense; then, it is clear from Figure 4.6(a) that the projection of the rotating arm on the y-axis gives the time signal $y(t)$. This is the phasor representation, where

   Signal amplitude = Length of the phasor

   Signal frequency = Angular speed of the phasor

   Signal phase angle = Initial position of the phasor with respect to the y-axis

It should be clear that a phase angle makes practical sense only when two or more signals are compared. This is so because, for a given harmonic signal, one can pick any point as the time reference ($t = 0$); but when two harmonic signals are compared, as in Figure 4.6(b), for example, one may consider one of those signals that starts (at $t = 0$) at its position peak as the reference signal. This will correspond to a phasor whose initial configuration coincides with the positive y-axis. As is clear from Figure 4.6(b), for this reference signal, $\phi = 0$. Then the phase angle ϕ of any other harmonic signal will correspond to the angular position of its phasor with respect to the reference phasor. Note that, in this example, the time shift between the two signals is ϕ/ω, which is also a direct representation of the phase. It should be clear, then, that the phase difference between two signals is also a representation of the *time lead* or *time lag* (delay) of one signal with respect to the other. Specifically, the phase that is ahead of the reference phasor is considered to "lead" the reference signal. In other words, the signal $a\cos(\omega t + \phi)$ has a phase "lead" of ϕ or a time "lead" of ϕ/ω with respect to the signal of $a\cos\omega t$.

Another important observation can be made with regard to the phasor representation of a harmonic signal. A phasor can be expressed as the complex quantity

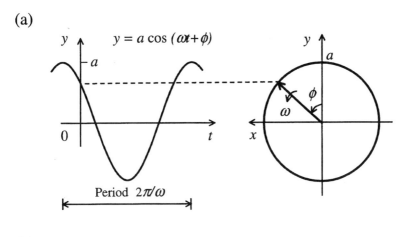

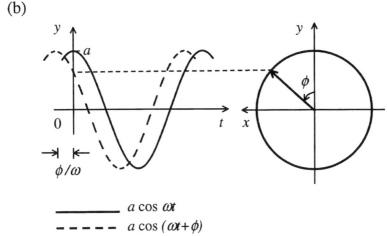

FIGURE 4.6 Phasor representation of a harmonic signal: (a) a phasor and the corresponding signal, and (b) representation of a phase angle (phase lead) ϕ.

$$y(t) = ae^{j(\omega t+\phi)} = a\cos(\omega t + \phi) + ja\sin(\omega t + \phi) \qquad (4.2)$$

whose real part is $a\cos(\omega t + \phi)$ — which is in fact the signal of interest. It is clear from Figure 4.6 that if one takes the y-axis to be real and the x-axis to be imaginary, then the complex representation (4.2) is indeed a complete representation of a phasor. By using the complex representation (4.2) for a harmonic signal, significant benefits of mathematical convenience can be derived in vibration analysis. It suffices to remember that practical vibrations are "real" signals, and regardless of the type of mathematical analysis that is used, only the real part of a complex signal of the form (4.2) will make physical sense.

4.1.5 RMS Amplitude Spectrum

If a harmonic signal $y(t)$ is averaged over one period T, the negative portion cancels out with the positive portion, giving zero. Consider a harmonic signal of angular frequency ω (or cyclic frequency f), phase angle ϕ and amplitude a, as given by

$$y(t) = a\cos(\omega t + \phi) = a\cos(2\pi f t + \phi) \qquad (4.3)$$

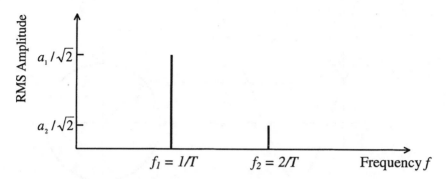

FIGURE 4.7 The rms amplitude spectrum of a periodic signal.

Its *average* (*mean*) value is

$$y_{mean} = \frac{1}{T}\int_0^T y(t)dt = 0 \qquad (4.4)$$

which can be verified by direct integration, while noting that

$$T = 1/f = 2\pi/\omega \qquad (4.5)$$

For this reason, the mean value is not a measure of the "strength" of a signal in general. Now define the *root-mean-square* (rms) value of a signal. This is the square root of the mean value of the square of the signal. By direct integration, it can be shown that for a sinusoidal (or harmonic) signal, the rms value is given by

$$y_{rms} = \left[\frac{1}{T}\int_0^T y^2(t)dt\right]^{1/2} = \frac{a}{\sqrt{2}} \qquad (4.6)$$

It follows that the *rms amplitude spectrum* is obtained by dividing the amplitude spectrum by $\sqrt{2}$. For example, for the periodic signal formed by combining two harmonic components as in Figure 4.3, the rms amplitude spectrum is shown in Figure 4.7. This again is a frequency domain representation of a signal, and the independent variable is frequency.

4.1.6 ONE-SIDED AND TWO-SIDED SPECTRA

The mean squared amplitude spectrum of a signal (sometimes called *power spectrum* because the square of a variable such as voltage and velocity is a measure of quantities such as power and energy, although it is not strictly the spectrum of power in the conventional sense) is obtained by plotting the mean squared amplitude of the signal against frequency. Note that these are one-sided spectra because only the positive frequency band is considered. This is a realistic representation because one cannot talk about negative frequencies for a real system. However, from a mathematical point of view, one can consider negative frequencies as well. In a spectral representation, it is at times convenient to consider the entire frequency band (consisting of both negative and positive values of frequency) — then it becomes a two-sided spectrum. In this case, the spectral component at each frequency value should be equally divided between the positive and the negative frequency values (hence, the spectrum is *symmetric*) such that the overall mean squared amplitude (or power, or energy) remains the same.

Vibration Signal Analysis

It has been seen that, for a harmonic signal component of amplitude a and frequency f (e.g., $a\cos(2\pi f + \phi)$), the rms amplitude is $a^2/2$ at frequency f; whereas, the two-side spectrum has a magnitude of $a^2/4$ at both the frequency values $-f$ and $+f$.

Note that, although it is possible to interpret the meaning of a negative time (which represents the past, previous to the starting point), it is not possible to give a realistic meaning to a negative frequency. This concept is introduced primarily for analytical convenience, and may be interpreted as clockwise rotation of a phasor.

4.1.7 COMPLEX SPECTRUM

It has been shown that, for unique representation of a signal in the frequency domain, both amplitude and phase information should be provided for each frequency component. Alternatively, the spectrum can be expressed as a *complex* function of frequency, having a *real part* and an *imaginary part*. For example, for a harmonic component given by $a\cos(2\pi f_i + \phi)$, the two-sided complex spectrum can be expressed as

$$Y(f_i) = \frac{a_i}{2}(\cos\phi_i + j\sin\phi_i) = \frac{a_i}{2}e^{-j\phi_i}$$

and

$$Y(-f_i) = \frac{a_i}{2}(\cos\phi_i - j\sin\phi_i) = \frac{a_i}{2}e^{-j\phi_i} \qquad (4.7)$$

in which j is the *imaginary unity* as given by $j = \sqrt{-1}$. Note that the spectral component at the negative frequency is the *complex conjugate* of that at the positive frequency. This concept of complex spectrum is the basis of (complex) *Fourier series expansion* that will be considered in detail in a later section. It should be clear that the complex conjugate of a spectrum is obtained by changing either j to $-j$ or ω to $-\omega$ (or f to $-f$).

4.2 SIGNAL TYPES

Signals can be classified into different types, depending on their characteristics. Note that the signal itself is a time function, but its frequency domain representation can bring up some of its salient features. Signals particularly important in the present study are the excitations and responses of vibrating systems. These can be divided into two broad classes: *deterministic signals* and *random signals*, depending on whether one is dealing with deterministic vibrations or random vibrations. Consider a damped cantilever beam that is subjected to a sinusoidal base excitation of frequency ω and amplitude u_o in the lateral direction (Figure 4.8). In the steady state, the tip of the beam will also oscillate at the same frequency, but with a different amplitude y_o; and furthermore, there will be a phase shift by an angle ϕ. For a given frequency and known beam properties, the quantities y_o and ϕ can be completely determined. Under these conditions, the tip response of the cantilever is a deterministic signal in the sense that when the experiment is repeated, the same response is obtained. Furthermore, the response can be expressed as a mathematical relationship in terms of parameters for which the values are determined with 100% certainty, and probabilities are not associated with these parameters (such parameters are termed *deterministic parameters*). Random signals are nondeterministic (or *stochastic*) signals. Their mathematical representation requires probability considerations. Furthermore, if the process were to repeat, there would always be some uncertainty as to whether an identical response signal could be obtained again.

Deterministic signals can be classified as *periodic*, *quasi-periodic*, or *transient*. Periodic signals repeat exactly at equal time periods. The frequency (Fourier) spectrum of a periodic signal consti-

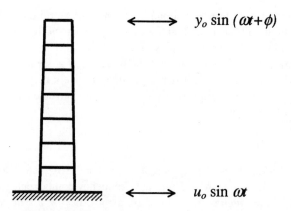

FIGURE 4.8 Response to base excitations of a tall structure (cantilever).

TABLE 4.1
Deterministic Signals

Primary Classification	Nature of the Fourier Spectrum	An Example
Periodic	Discrete and equally spaced	$y_o \sin \omega t + y_1 \sin\left(\frac{5}{3}\omega t + \phi\right)$
Quasi-periodic	Discrete and irregularly spaced	$y_o \sin \omega t + y_1 \sin\left(\sqrt{2}\omega t + \phi\right)$
Transient	Continuous	$y_o \exp(-\lambda t)\sin(\omega t + \phi)$

tutes a series of equally spaced impulses. Furthermore, a periodic signal will have a Fourier series representation. This implies that a periodic signal can be expressed as a sum of sinusoidal components for which the frequency ratios are rational numbers (not necessarily integers). Quasi-periodic (or almost periodic) signals, as well, have discrete Fourier spectra but the spectral lines are not equally spaced. Typically, a quasi-periodic signal can be generated by combining two or more sinusoidal components, provided that at least two of the components have as their frequency ratio an irrational number. Transient signals have continuous Fourier spectra. These types of signals cannot be expressed as a sum of sinusoidal components (or a Fourier series). All signals that are not periodic or quasi-periodic can be classified as transient. Most often, highly damped (overdamped) signals with exponentially decaying characteristics are termed "transient," although various other forms of signals such as exponentially increasing (unstable) responses, sinusoidal decays (underdamped responses), and sine sweeps (sinewaves with variable frequency) also fall into this category. Table 4.1 gives examples for these three types of deterministic signals. The corresponding amplitude spectra are sketched in Figure 4.9. A general classification of signals, with some examples, is given in Box 4.1.

4.3 FOURIER ANALYSIS

Fourier analysis is the key to frequency analysis of vibration signals. The frequency domain representation of a time signal is obtained through the Fourier transform. One immediate advantage of the Fourier transform is that, through its use, differential operations (differentiation and integration) in the time domain are converted into simpler algebraic operations (multiplication and division). Transform techniques are quite useful in mathematical applications. For example, a simple,

Vibration Signal Analysis

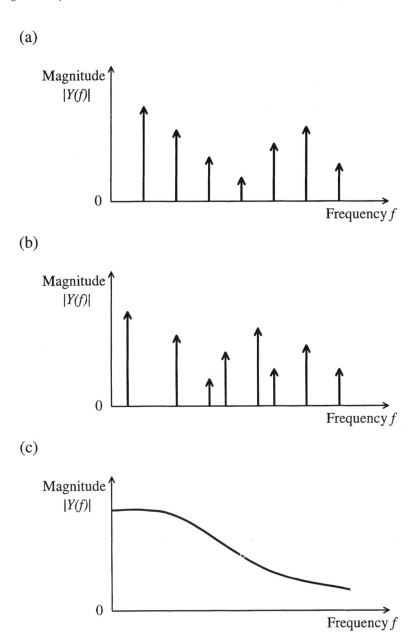

FIGURE 4.9 Magnitude spectra for three types of deterministic signals: (a) periodic; (b) quasi-periodic; and (c) transient.

yet versatile, transformation from products into sums is accomplished through the use of the logarithm. Three versions of Fourier transform are important: the Fourier integral transform can be applied to any general signal, the Fourier series expansion is applicable only to periodic signals, and the discrete Fourier transform is used for discrete signals. As shall be seen, all three versions of transform are interrelated. In particular, one must use the discrete Fourier transform in digital computation of both Fourier integral transform and Fourier series expansion.

BOX 4.1 Signal Classification

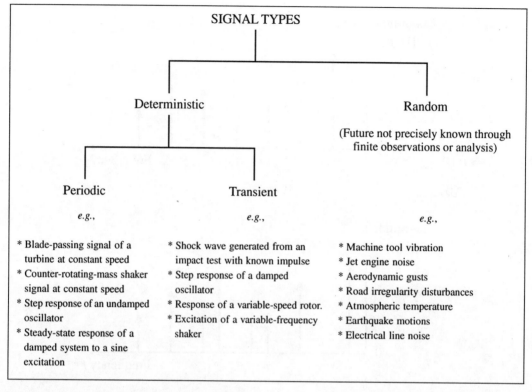

4.3.1 Fourier Integral Transform (FIT)

The *Fourier spectrum X(f)* of a time signal *x(t)* is given by the *forward transform* relation

$$X(f) = \int_{-\infty}^{\infty} x(t)\exp(-j2\pi ft)dt \qquad (4.8)$$

with $j = \sqrt{-1}$ and f the cyclic frequency variable. When equation (4.8) is multiplied by $\exp(j2\pi f\tau)$ and integrated with respect to f using the orthogonal property (which can be considered as a definition of the Dirac delta function δ)

$$\int_{-\infty}^{\infty} \exp[j2\pi f(t-\tau)]dt = \delta(t-\tau) \qquad (4.9)$$

one obtains the inverse transform relation

$$x(t) = \int_{-\infty}^{\infty} X(f)\exp(2j\pi ft)df \qquad (4.10)$$

The forward transform is denoted by the operator $\mathscr{F}$, and the inverse transform by $\mathscr{F}^{-1}$. Hence, the Fourier transform pair is given by

Vibration Signal Analysis

$$X(f) = \mathscr{F} x(t) \quad \text{and} \quad x(t) = \mathscr{F}^{-1} X(f) \tag{4.11}$$

Note that for *real* systems, $x(t)$ is a real function but $X(f)$ is a complex function in general. Hence, the Fourier spectrum of a signal can be represented by the *magnitude* $|X(f)|$ and the *phase angle* $\angle X(f)$ of the (complex) Fourier spectrum $X(f)$. Alternatively, the *real part* Re $X(f)$ and the *imaginary part* Im$X(f)$ together can be used to represent the Fourier spectrum.

According to the present definition, the Fourier spectrum is defined for negative frequency values as well as positive frequencies (i.e., a two-sided spectrum). The *complex conjugate* of a complex value is obtained by simply reversing the sign of the imaginary part; in other words, by replacing j by $-j$. By noting that replacing j by $-j$ in the forward transform relation is identical to replacing f by $-f$ it should be clear that the Fourier spectrum (of real signals) for negative frequencies is given by the complex conjugate $X^*(f)$ of the Fourier spectrum for positive frequencies. As a result, only the positive-frequency spectrum needs to be specified and the negative-frequency spectrum can be conveniently derived from it — through complex conjugation.

The Laplace transform, which was introduced in Chapter 3, is similar to the Fourier integral transform. The Laplace transform is defined by the forward and inverse relations

$$X(s) = \int_0^\infty x(t) \exp(-st) dt \tag{4.12}$$

and

$$x(t) = \frac{1}{2\pi j} \int_{\sigma - j\infty}^{\sigma + j\infty} X(s) \exp(st) ds \tag{4.13}$$

Since the signal itself is zero for $t < 0$, it is seen that for all practical purposes, Fourier transform results can be deduced from the Laplace transform analysis, simply by substituting $s = j2\pi f = j\omega$ and $\sigma = 0$.

4.3.2 FOURIER SERIES EXPANSION (FSE)

For a periodic signal $x(t)$ of period T, the *Fourier series expansion* (FSE) is given by

$$x(t) = \Delta F \sum_{n=-\infty}^{\infty} A_n \exp(j2\pi nt/T) \tag{4.14}$$

with $\Delta F = 1/T$. Strictly speaking (see FIT relations), this is the inverse transform relation. The scaling factor ΔF is not essential, but is introduced so that the Fourier coefficients A_n will have the same units as the Fourier spectrum. The Fourier coefficients are obtained by multiplying the inverse transform relation by $\exp(-j2\pi mt/T)$ and integrating with respect to t, from 0 to T, using the orthogonality condition

$$\frac{1}{T} \int_0^T \exp[j2\pi(n-m)t/T] dt = \delta_{mn} \tag{4.15}$$

Note that the *Kronecker delta* δ_{mn} is defined as

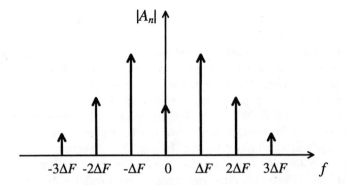

FIGURE 4.10 Fourier spectrum of a periodic signal and its relation to Fourier series.

$$\delta_{mn} = 1 \quad \text{for } m = n$$
$$\phantom{\delta_{mn}} = 0 \quad \text{for } m \neq n \tag{4.16}$$

for integer values of m and n. The forward transform that results is given by

$$A_n = \int_0^T x(t) \exp(-j2\pi nt/T) dt \tag{4.17}$$

Note that A_n are complex quantities in general.

It can be shown that for periodic signals, FSE is a special case of FIT — as expected. Consider a Fourier spectrum consisting of a sum of equidistant impulses separated by the frequency interval $\Delta F = 1/T$;

$$X(f) = \Delta F \sum_{n=-\infty}^{\infty} A_n \delta(f - n \cdot \Delta F) \tag{4.18}$$

This is shown in Figure 4.10 (only the magnitudes $|A_n|$ can be plotted in this figure because A_n is complex in general). Substituting this spectrum into the inverse FIT relation given earlier, one obtains the inverse FSE relation (4.14). Furthermore, this shows that the Fourier spectrum of a periodic signal is a series of equidistant impulses.

4.3.3 Discrete Fourier Transform (DFT)

The discrete Fourier transform relates an N-element sequence of sampled (discrete) data signal

$$\{x_m\} = [x_o, x_1, \ldots, x_{N-1}] \tag{4.19}$$

to an N-element sequence of spectral results

$$\{X_n\} = [X_o, X_1, \ldots, X_{N-1}] \tag{4.20}$$

through the forward transform relation

$$X_n = \Delta T \sum_{m=0}^{N-1} x_m \exp(-j2\pi mn/N) \qquad (4.21)$$

with $n = 0, 1, \ldots, N - 1$. The values X_n are called the spectral lines. It can be shown that these quantities approximate the values of the Fourier spectrum (continuous) at the corresponding discrete frequencies. Identify ΔT as the sampling period (i.e., the time step between two adjacent points of sampled data).

The inverse transform relation is obtained by multiplying the forward transform relation by $\exp(j2\pi nr/N)$ and summing over $n = 0$ to $N - 1$, using the orthogonality property

$$\frac{1}{N} \sum_{n=0}^{N-1} \exp[j2\pi n(r-m)/N] = \delta_{rm} \qquad (4.22)$$

Note that this orthogonality relation can be considered as a definition of *Kronecker delta*. The inverse transform is

$$x_m = \Delta F \sum_{n=0}^{N-1} X_n \exp(j2\pi mn/N) \qquad (4.23)$$

The data record length is given by

$$T = N\Delta T = 1/\Delta F \qquad (4.24)$$

The DFT is a transform, in its own right, independent of the FIT. It is possible, however, to interpret this transformation as the *trapezoidal integration* approximation of FIT. (The author has deliberately chosen appropriate scaling factors ΔT and ΔF in order to maintain this equivalence, and it is very useful in computing the Fourier spectrum of a general signal or the Fourier coefficients of a periodic signal using a digital computer.) Proper interpretation of the digital results is crucial, however, in using DFT to compute (approximate) Fourier spectra of a (continuous) signal. In particular, two types of error — *aliasing* and *leakage* (or *truncation error*) — should be considered. This subject will be treated later. The three transform relations, corresponding inverse transforms, and the othogonality relations are summarized in Table 4.2.

The link between the time domain signals and models, and the corresponding frequency domain equivalents is the Fourier integral transform. Table 4.3 provides some important properties of the FIT and the corresponding time-domain relations that are useful in the analysis of signals and system models. These properties can be easily derived from the basic FIT relations (4.8) through (4.10). It should be noted that, inherent in the definition of the DFT given in Table 4.2 is the N-point periodicity of the two sequences; that is, $X_n = X_{n+iN}$ and $x_m = x_{m+iN}$, for $i = \pm 1, \pm 2, \ldots$.

The definitions given in Table 4.2 may differ from the versions available in the literature by a multiplicative constant. However, it is observed that according to the present definitions, the DFT may be interpreted as the trapezoidal integration of the FIT. The close similarity between the definitions of the FSE and the DFT is also noteworthy. Furthermore, according to the last column in Table 4.2, the FSE can be expressed as a special FIT consisting of an equidistant set of impulses of magnitude A_n/T located at $f = n/T$.

TABLE 4.2
Unified Definitions for Three Fourier Transform Types

Relation Name	Fourier Integral Transform	Discrete Fourier Transform (DFT)	Fourier Series Expansion (FSE)
Forward transform	$X(f) = \int_{-\infty}^{\infty} x(t)\exp(-j2\pi ft)dt$	$X_n = \Delta T \sum_{m=0}^{N-1} x_m \exp(-j2\pi nm/N)$ $n = 0, 1, \ldots, N-1.$	$A_n = \int_0^T x(t)\exp(-j2\pi nt/T)dt$ $n = 0, 1, \ldots$
Inverse transform	$x(t) = \int_{-\infty}^{\infty} X(f)\exp(j2\pi ft)df$	$x_m = \Delta F \sum_{n=0}^{N-1} X_n \exp(j2\pi nm/N)$ $m = 0, 1, \ldots, N-1$	$x(t) = \Delta F \sum_{n=-\infty}^{\infty} A_n \exp(j2\pi nt/T)$
Orthogonality	$\int_{-\infty}^{\infty} \exp[j2\pi f(\tau - t)]df = \delta(\tau - t)$	$\frac{1}{N}\sum_{n=0}^{N-1} \exp[j2\pi n(r-m)/N] = \delta_{rm}$	$\frac{1}{T}\int_0^T \exp[j2\pi(r-n)t/T]dt = \delta_{rn}$
Notes:	$T = N\Delta T$	$\Delta F = 1/T$	$X(f) = \Delta F \sum_{n=-\infty}^{\infty} A_n \delta(f - n/T)$

TABLE 4.3
Important Properties of the Fourier Transform

Function of Time	Fourier Spectrum
$x(t)$	$X(f)$
$k_1 x_1(t) + k_2 x_2(t)$	$k_1 X_1(f) + k_2 X_2(f)$
$x(t)\exp(-j2\pi ta)$	$X(f + a)$
$x(t + \tau)$	$X(f)\exp(j2\pi f\tau)$
$\dfrac{d^n x(t)}{dt^n}$	$(j2\pi f)^n X(f)$
$\int_{-\infty}^t x(t)dt$	$\dfrac{X(f)}{j2\pi f}$

4.3.4 ALIASING DISTORTION

Recalling that the primary task of digital Fourier analysis is to obtain a discrete approximation to the FIT of a piecewise continuous function, it is advantageous to interpret the DFT as a discrete (digital computer) version of the FIT, rather than an independent discrete transform. Accordingly, the results from a DFT must be consistent with the exact results obtained if the FIT were used. The definitions given in Table 4.2 are consistent in this respect because the DFT is given as the trapezoidal integration of the FIT. However, it should be clear that if $X(f)$ is the FIT of $x(t)$, the sequence of sampled values $\{X(n \cdot \Delta F)\}$ is not exactly the DFT of the sampled data sequence $\{x(m \cdot \Delta T)\}$. Only an approximate relationship exists.

A further advantage of the definitions given in Table 4.2 is apparent when dealing with the Fourier series expansion. As previously noted, the FIT of a periodic function is a set of impulses. One can avoid dealing with impulses by relating the complex Fourier coefficients to the DFT sequence of sampled data from the periodic function, via the present definitions.

TABLE 4.4
Unified Fourier Transform Relationships

Description	Relationship	
	DFT and FIT	DFT and FSE
Given	$x(t) \xrightarrow{\text{FIT}} X(f)$	$x(t) \xrightarrow{\text{FSE}} \{A_n\}$
Form	$\tilde{x}(t) = \sum_{k=-\infty}^{\infty} x(t+kT)$	$\tilde{A}_n = \sum_{k=-\infty}^{\infty} A_{n+kN}$
	$\tilde{X}(f) = \sum_{k=-\infty}^{\infty} X(f+kF)$	
Then	$\{\tilde{x}_m\} \xrightarrow{\text{DFT}} \{\tilde{X}_n\}$	$\{x_m\} \xrightarrow{\text{DFT}} \{\tilde{A}_n\}$
Where	$\tilde{x}_m = \tilde{x}(m \cdot \Delta T), \quad \tilde{X}_n = \tilde{X}(n \cdot \Delta F)$	$x_m = x(m \cdot \Delta T)$
	$F = 1/\Delta T, \quad T = 1/\Delta F$	$N = T/\Delta T$

Aliasing distortion is an important consideration when dealing with sampled data from a continuous signal. This error can enter into computation in both the time domain and the frequency domain, depending on the domain in which the results are presented. This issue will now be addressed.

Sampling Theorem

The basic relationships among the FIT, the DFT, and the FSE are summarized in Table 4.4. By means of straightforward mathematical procedures, the relationship between the FIT and the DFT can be established. Although $\{X(n \cdot \Delta F)\}$ is not the DFT of $\{x(m \cdot \Delta T)\}$, the results in Table 4.4 show that $\{\tilde{X}(n \cdot \Delta F)\}$ is the DFT of $\{\tilde{x}(m \cdot \Delta T)\}$ where the periodic functions $\tilde{X}(f)$ and $\tilde{x}(t)$ are as defined in Table 4.4. This situation is illustrated in Figure 4.11. It should be recalled that $X(f)$ is a complex function in general and as such it cannot be displayed as a single curve in a two-dimensional coordinate system. Both the magnitude and the phase angle variations with respect to frequency f are needed. For brevity, only the magnitude $|X(f)|$ is shown in Figure 4.11(a). Nevertheless, the argument presented applies to the phase angle $\angle X(f)$ as well.

It is obvious that in the time interval $[0, T]$, $x(t) = \tilde{x}(t)$ and $x_m = \tilde{x}_m$. However, $\tilde{X}(n \cdot \Delta F)$ is only approximately equal to $X(n \cdot \Delta F)$ in the frequency interval $[0, F]$. This is known as the aliasing distortion in the frequency domain. As ΔT decreases (i.e., as F increases), $\tilde{X}(f)$ will become closer to $X(f)$ in the frequency interval $[0, F/2]$, as is clear from Figure 4.11(c). Furthermore, due to the F-periodicity of $\tilde{X}(f)$, its value in the frequency range $[F/2, F]$ will approximate $X(f)$ in the frequency range $[-F/2, 0]$.

It is clear from the preceding discussion that if a time signal $x(t)$ is sampled at equal steps of ΔT, no information regarding its frequency spectrum $X(f)$ is obtained for frequencies higher than $f_c = 1/(2\Delta T)$. This fact is known as *Shannon's sampling theorem*, and the limiting (cutoff) frequency is called the *Nyquist frequency*. In vibration signal analysis, a sufficiently small sample step ΔT should be chosen in order to reduce aliasing distortion in the frequency domain, depending on the highest frequency of interest in the analyzed signal. This, however, increases the signal processing time and the computer storage requirements, which is undesirable particularly in real-time analysis. It also can result in stability problems in numerical computations. The Nyquist sampling criterion requires that the sampling rate $(1/\Delta T)$ for a signal should be at least twice the highest frequency of interest. Instead of making the sampling rate very high, a moderate value that satisfies the Nyquist

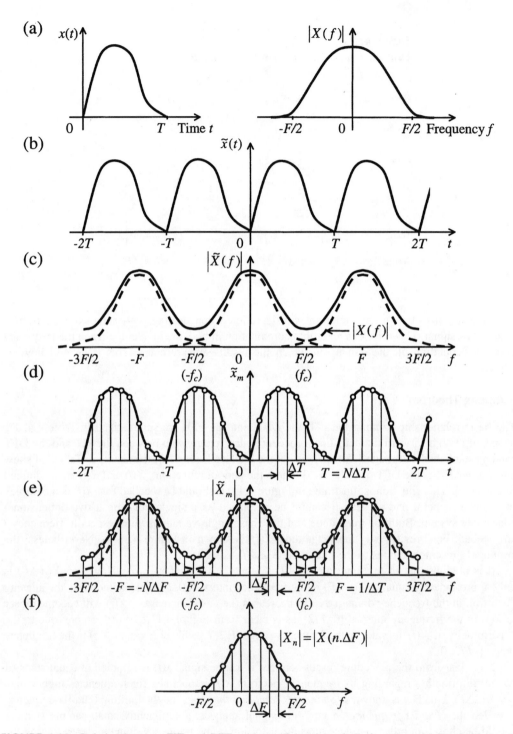

FIGURE 4.11 Relation between FIT and DFT, with an illustration of aliasing error: (a) Fourier integral transformation (FIT) of a signal; (b) periodically arranged time signal; (c) periodicity of the frequency spectrum; (d) sampled time signal; (e) sampled frequency spectrum (with aliasing error); and (f) sampled original spectrum (no aliasing error).

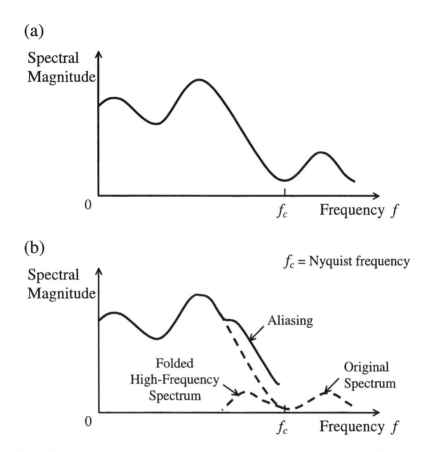

FIGURE 4.12 Aliasing distortion of frequency spectrum: (a) original spectrum, and (b) distorted spectrum due to aliasing.

sampling criterion is used in practice, together with an *anti-aliasing filter* to remove the distorted frequency components. It should be noted that the DFT results in the frequency interval $[f_c, 2f_c]$ are redundant because they merely approximate the frequency spectrum in the negative frequency interval $[-f_c, 0]$, which is known for real signals. This fact is known as the *Hermitian property*.

The last column of Table 4.4 presents the relationship between the FSE and the DFT. It is noted that the sequence $\{\tilde{A}_n\}$ rather than the sequence of complex Fourier series coefficients $\{A_n\}$ represents the DFT of the sampled data sequence $\{x(m \cdot \Delta T)\}$. In practice, however, $A_n \to 0$ as $n \to \infty$. Consequently, $\tilde{A}_n$ is a good approximator to A_n in the range $[-N/2 \leq n \leq N/2]$ for sufficiently large N. This basic result is useful in determining the Fourier coefficients of a periodic signal using discrete data that are sampled at time steps of $\Delta T = 1/F$, in which F is the fundamental frequency of the periodic signal. Again, the aliasing error ($\tilde{A}_n - A_n$) can be reduced by increasing the sampling rate (i.e., by decreasing ΔT or increasing N).

Aliasing Distortion in the Time Domain

In vibration applications, it is sometimes required to reconstruct the signal from its Fourier spectrum. Inverse DFT is used for this purpose and is particularly applicable in digital equalizers in vibration testing. Due to sampling in the frequency domain, the signal becomes distorted. The aliasing error ($\tilde{x}_m - x(m\Delta T)$) is reduced by decreasing the sample period ΔF. It should be noted that no information regarding the signal for times greater than $T = 1/\Delta F$ is obtained from the analysis.

By comparing Figure 4.11(a) with (c), or (e) with (f), it should be clear that the aliasing error in $\tilde{X}$ in comparison with the original spectrum X, is caused by "folding" of the high-frequency segment of X beyond the Nyquist frequency into the low-frequency segment of X. This is illustrated in Figure 4.12(b).

Anti-Aliasing Filter

It should be clear from Figure 4.12 that if the original spectrum is low-pass filtered at a cutoff frequency equal to the Nyquist frequency, then the aliasing distortion would not occur due to sampling. A filter of this type is call an anti-aliasing filter. In practice, it is not possible to achieve perfect filtering. Hence, some aliasing could remain even after using an anti-aliasing filter. Such residual errors can be reduced using a filter cutoff frequency that is slightly less than the Nyquist frequency. Then the resulting spectrum would only be valid up to this filter cutoff frequency (and not up to the theoretical limit of Nyquist frequency).

EXAMPLE 4.1

Consider 1024 data points from a signal, sampled at 1 millisecond (ms) intervals.

$$\text{Sample rate } f_s = 1/0.001 \text{ samples/s} = 1000 \text{ Hz} = 1 \text{ kHz}$$

$$\text{Nyquist frequency} = 1000/2 \text{ Hz} = 500 \text{ Hz}$$

Due to aliasing, approximately 20% of the spectrum (i.e., spectrum beyond 400 Hz) will be distorted. Here, one can use an anti-aliasing filter.

Suppose that a digital Fourier transform computation provides 1024 frequency points of data up to 1000 Hz. Half of this number is beyond the Nyquist frequency and will not give any new information about the signal.

$$\text{Spectral line separation} = 1000/1024 \text{ Hz} = 1 \text{ Hz (approx.)}$$

Keep only the first 400 spectral lines as the useful spectrum.

Note: Almost 500 spectral lines can be retained if an accurate anti-aliasing filter is used.

Some useful information of signal sampling is summarized in Box 4.2.

4.3.5 Another Illustration of Aliasing

A simple illustration of aliasing is given in Figure 4.13. Here, two sinusoidal signals of frequency $f_1 = 0.2$ Hz and $f_2 = 0.8$ Hz are shown (Figure 4.13(a)). Suppose that the two signals are sampled at the rate of $f_s = 1$ sample per second. The corresponding Nyquist frequency is $f_c = 0.5$ Hz. It is seen that, at this sampling rate, the data samples from the two signals are identical. In other words, the high-frequency signal cannot be distinguished from the low-frequency signal. Hence, a high-frequency signal component of frequency 0.8 Hz will appear as a low-frequency signal component of frequency 0.2 Hz. This is aliasing, as clear from the signal spectrum shown in Figure 4.13. Specifically, the spectral segment of the signal beyond the Nyquist frequency (f_c) cannot be recovered.

It is apparent from Figure 4.11(e) that the aliasing error becomes more and more prominent for frequencies of the spectrum closer to the Nyquist frequency. With reference to the expression for $\tilde{X}(f)$ in Table 4.4, it should be clear that when the true Fourier spectrum $X(f)$ has a steep roll-off prior to $F/2$ ($=f_c$), the influence of the $X(f - nF)$ segments for $n \geq 2$ and $n \leq -1$ is negligible in the discrete spectrum in the frequency range $[0, F/2]$. Hence, the aliasing distortion in the frequency band $[0, F/2]$ comes primarily from $X(f - F)$, which is the true spectrum shifted to the right through F. Therefore, a reasonably accurate expression for the aliasing error is

Vibration Signal Analysis

BOX 4.2 Signal Sampling Considerations

- The maximum useful frequency in digital Fourier results is half the sampling rate.

Nyquist Frequency or Cutoff Frequency or Computational Bandwidth:

$$f_c = \frac{1}{2} \times \text{Sampling rate}$$

Aliasing Distortion:
High-frequency spectrum beyond Nyquist frequency folds onto the useful spectrum, thereby distorting it.

Summary:
1. Pick a sufficiently small sample step ΔT in the time domain to reduce the aliasing distortion in the frequency domain.
2. The highest frequency for which the Fourier transform (frequency-spectrum) information would be valid is the Nyquist frequency $f_c = 1/(2\Delta T)$
3. DFT results that are computed for the frequency range $[f_c, 2f_c]$ merely approximate the frequency spectrum in the negative frequency range $[-f_c, 0]$.

$$e_n = X\bigl(-(F - n \cdot \Delta/F)\bigr) - X(n \cdot \Delta F - F) = X_{-(N-n)} - X_{n-N} \quad (4.25)$$

for $n = 0, 1, 2, \ldots, N/2$

Note from equation (4.8) that the spectral value obtained when f in the complex exponential is replaced by $-f$ is the same as the spectral value obtained when j is replaced by $-j$. Since the signal $x(t)$ is real, it follows that the Fourier spectrum for the negative frequencies is simply the complex conjugate of the Fourier spectrum for the positive frequencies; thus,

$$X(-f) = X^*(f) \quad (4.26)$$

or, in the discrete case,

$$X_{n-N} = X^*_{N-n} \quad (4.27)$$

It follows from equation (4.25) that the aliasing distortion is given by

$$e_n = X^*_{N-n} \quad \text{for } n = 0, 1, 2, \ldots, N/2 \quad (4.28)$$

This result confirms that aliasing can be interpreted as folding of the complex conjugate of the true spectrum beyond the Nyquist frequency $f_c(=F/2)$ over to the original spectrum. In other words, due to aliasing, frequency components higher than the Nyquist frequency appear as lower frequency components (due to folding). These aliasing components enter into the digital Fourier results in the useful frequency range $[0, f_c]$.

Aliasing reduces the valid frequency range in digital Fourier results. Typically, the useful frequency limit is $f_c/1.28$ so that the last 20% of the spectral points near the Nyquist frequency should be neglected. It should be clear that if a low-pass filter with its cutoff frequency set at f_c is used on the time signal prior to sampling and digital Fourier analysis, the aliasing distortion can

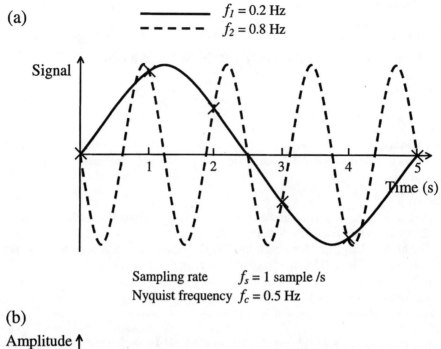

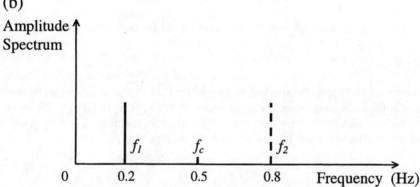

FIGURE 4.13 A simple illustration of aliasing: (a) two harmonic signals with identical sampled data; and (b) frequency spectra of the two harmonic signals.

be virtually eliminated. Analog hardware filters can be used for this purpose. They are the *anti-aliasing filters*. Note that sometimes $f_c/1.28 (\cong 0.8 f_c)$ is used as the filter cutoff frequency. In this case, the computed spectrum is accurate up to $0.8 f_c$ and not up to f_c.

The buffer memory of a typical commercial Fourier analyzer can store $N = 2^{10} = 1024$ samples of data from the time signal. This is the size of the data block analyzed in each digital Fourier transform calculation. This will result in $N/2 = 512$ spectral points (spectral lines) in the frequency range $[0, f_c]$. Out of this, only the first 400 spectral lines (approx. 80%) are considered free of aliasing distortion.

Example 4.2

Suppose that the frequency range of interest in a particular vibration signal is 0 to 200 Hz. One is interested in determining the sampling rate (digitization speed) and the cutoff frequency for the anti-aliasing (low-pass) filter.

The Nyquist frequency f_c is given by $f_c/1.28 = 200$
Hence, $f_c = 256$ Hz.

The sampling rate (or digitization speed) for the time signal that is needed to achieve this range of analysis is $F = 2f_c = 512$ Hz. With this sampling frequency, the cutoff frequency for the anti-aliasing filter could be set at a value between 200 and 256 Hz.

4.4 ANALYSIS OF RANDOM SIGNALS

Random (stochastic) signals are generated by some random mechanism. Each time the mechanism is operated, a new time history (sample function) is generated. The likelihood of any two sample functions becoming identical is governed by some probabilistic law. If all sample functions are identical (with unity probability), then the corresponding signal is a deterministic signal. A random process is denoted by $\tilde{X}(t)$, while any sample function of it is denoted by $x(t)$. No numerical computations can be performed on $\tilde{X}(t)$ because it is not known for certainty. Its Fourier transform, for example, can be written down as an analytical expression, but cannot be numerically computed. However, once the signal is generated, numerical computations can be performed on that sample function $x(t)$ because it is a completely known function of time.

4.4.1 Ergodic Random Signals

At any given time t_1, $\tilde{X}(t_1)$ is a random variable that has a certain probability distribution. Consider a well-behaved function $f\{\tilde{X}(t)\}$ of this random variable (which is also a random variable). Its expected value (statistical mean) is $E[f\{\tilde{X}(t)\}]$. This is also known as the *ensemble average* because it is equivalent to the average value at t of a collection (ensemble) of a large number of sample functions $x(t)$.

Now consider the function $f\{x(t)\}$ of *one* sample function $x(t)$. Its temporal (time) mean is expressed by

$$\lim_{T \to \infty} \frac{1}{2T} \int_{-T}^{T} f\{x(t)\} dt$$

Now, if

$$E\left[f\{\tilde{X}(t_1)\}\right] = \lim_{T \to \infty} \frac{1}{2T} \int_{-T}^{T} f\{x(t)\} dt \qquad (4.29)$$

then the random signal is said to be *ergodic*. It should be noted that the right-hand side of equation (4.29) does not depend on time. Consequently, the left-hand side also should be independent of the time point t_1.

For analytical convenience, random vibration signals are usually assumed to be ergodic (the *ergodic hypothesis*). Using this hypothesis, the properties of a random signal could be determined by performing computations on a sufficiently long record (sample function) of the signal. Because the ergodic hypothesis is not exactly satisfied for vibration signals, and because it is impossible to analyze infinitely long data records, the accuracy of the numerical results depends on various factors such as the record length, sampling rate, frequency range of interest, and the statistical nature of the random signal (e.g., closeness to a deterministic signal, frequency content, periodicity, damping characteristics). Accuracy can be improved, in general, by averaging the results for more than one data record.

4.4.2 Correlation and Spectral Density

If for a random signal $\tilde{X}(t)$, the joint statistical properties of $\tilde{X}(t_1)$ and $\tilde{X}(t_2)$ depend on the time difference $(t_2 - t_1)$ and not on t_1 itself, then the signal is said to be *stationary*. Consequently, the statistical properties of a stationary $\tilde{X}(t)$ will be independent of t. It is noted from equation (4.29) that ergodic random signals are necessarily stationary. However, the converse is not true in general.

The cross-correlation function of two random signals $\tilde{X}(t)$ and $\tilde{Y}(t)$ is given by $E[\tilde{X}(t)\tilde{Y}(t+\tau)]$. If the signals are stationary, this expected value is a function of τ (not t) and is denoted by $\phi_{xy}(\tau)$. In view of the ergodic hypothesis, the cross-correlation function can be expressed as

$$\phi_{xy}(\tau) = \lim_{T \to \infty} \left[\frac{1}{T} \int_0^T x(t)y(t+\tau)dt \right] \tag{4.30}$$

The FIT of $\phi_{xy}(\tau)$ is the cross-spectral density function, which is denoted by $\Phi_{xy}(f)$. When the two signals are identical, one obtains the auto-correlation function $\phi_{xx}(\tau)$ in the time domain and the power spectral density (psd) $\Phi_{xx}(f)$ in the frequency domain. The continuous and discrete versions of the correlation theorem are given in the first row of Table 4.5. It follows that the cross-spectral density can be estimated using the DFT (FFT) of the two signals as, $[X_n]^* Y_n / T$, in which T is the record length and $[X_n]^*$ is the complex conjugate of $[X_n]$.

Parseval's theorem (second row of Table 4.5) follows directly from the correlation theorem. Consequently, the mean square value of a random signal can be obtained from the area under the psd curve. This suggests a hardware-based method of estimating the psd, as illustrated by the functional diagram in Figure 4.14(a). Alternatively, a software-based digital Fourier analysis could be used (Figure 4.14(b)). A single sample function would not give the required accuracy, and averaging is usually needed. In real-time digital analysis, the running average as well as the current estimate are usually computed. In the running average, it is desirable to give a higher weighting to the more recent estimates. The fluctuations in the psd estimate about the local average could be reduced by selecting a large filter bandwidth Δf and a large record length T. A measure of this fluctuations is given by

$$\varepsilon = \frac{1}{\sqrt{\Delta f T}} \tag{4.31}$$

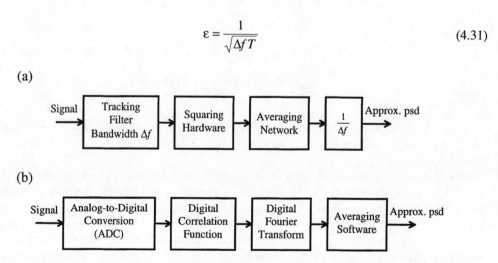

FIGURE 4.14 Power spectral density computation: (a) narrow-band filtering method, and (b) correlation and fourier transformation method.

Vibration Signal Analysis

TABLE 4.5
Some Useful Fourier Transform Results

Description		Continuous	Discrete
Correlation theorem	If	$z(\tau) = \int_{-\infty}^{\infty} x(t)y(t+\tau)dt$	$Z_m = \Delta T \sum_{r=0}^{N-1} x_r y_{r+m}$
	Then	$Z(f) = [X(f)]^* Y(f)$	$Z_n = [X_n]^* Y_n$
Parseval's theorem	If	$y(t) \xrightarrow{FIT} Y(f)$	$\{y_m\} \xrightarrow{DFT} \{Y_n\}$
	Then	$\int_{-\infty}^{\infty} y^2(t)dt = \int_{-\infty}^{\infty} |Y(f)|^2 df$	$\Delta T \sum_{m=0}^{N-1} y_m^2 = \Delta F \sum_{n=0}^{N-1} |Y_n|^2$
Convolution theorem	If	$y(t) = \int_{-\infty}^{\infty} h(\tau)u(t-\tau)d\tau$	$y_m = \Delta T \sum_{r=0}^{N-1} h_r u_{m-r}$
		$= \int_{-\infty}^{\infty} h(t-\tau)u(\tau)d\tau$	$= \Delta T \sum_{r=0}^{N-1} h_{m-r} u_r$
	Then	$Y(f) = H(f)U(f)$	$Y_n = H_n U_n$

It should be noted that a large Δf results in reduction of the precision of the estimates while improving the appearance. To offset this, T has to be increased further.

4.4.3 Frequency Response Using Digital Fourier Transform

Vibration test programs usually require a resonance search-type pretesting. In order to minimize the damage potential, it is carried out at a much lower intensity than the main test. The objective of such exploratory tests is to determine the significant frequency-response functions of the test specimen. These provide the natural frequencies, damping ratios, and mode shapes of the test specimen. Such frequency response data are useful in planning and conducting the main test. For example, more attention is required when testing in the vicinity of the resonance points (slower sweep rates, larger dwell periods, etc.). Also, the frequency-response data are useful in determining the most desirable test input directions and intensities. The degree of nonlinearity and time variance of the test object can be determined by conducting more than one frequency-response test at different input intensities. If the deviation of the frequency-response function is sufficiently small, then linear, time-invariant analysis is considered to be adequate. Often, frequency-response tests are conducted at full test intensity. In such cases, it is considered as a part of the main test rather than a prescreening test. Other uses of the frequency-response function include the following: it can be employed as a system model (experimental model) for further analysis of the test specimen (experimental modal analysis). Most desirable frequency range and sweep rates for vibration testing can be estimated by examining frequency-response functions.

The time response $h(t)$ to a unit impulse is known as the *impulse response function*. For each pair of input and output locations (A, B) of the test specimen, a corresponding single response function would be obtained (assuming linearity and time invariance). The entire collection of these responses would determine the response of the test specimen to an arbitrary input signal. The response $y(t)$ at B to an arbitrary input $u(t)$ applied at A is given by

$$y(t) = \int_{-\infty}^{\infty} h(\tau)u(t-\tau)d\tau$$

$$= \int_{-\infty}^{\infty} h(t-\tau)u(\tau)d\tau \qquad (4.32)$$

The right-hand side of equation (4.32) is the convolution integral of $h(t)$ and $u(t)$, and is denoted by $h(t) * u(t)$. By substituting the inverse FIT relations (Table 4.2) in equation (4.32), the frequency-response function (frequency-transfer function) $H(f)$ is obtained as the ratio of the (complex) FITs of the output and the input. It exists for physically realizable (causal) systems even when the individual FITs of the input and output signals do not converge. The continuous convolution theorem and the discrete counterpart are given in the last row of Table 4.5. The discrete convolution can be interpreted as the trapezoidal integration of equation (4.32). The frequency-response function is a valid representation (model) for linear, time-invariant systems. It is related to the system transfer function $G(s)$ (ratio of the Laplace transforms of output and input with zero initial conditions) through

$$H(f) = \frac{Y(f)}{U(f)} = G(2\pi f j) \qquad (4.33)$$

But for notational convenience, the frequency-response function corresponding to $G(s)$ can be denoted by either $G(f)$ or $G(\omega)$, where ω is the angular frequency and f is the cyclic frequency.

Using Fourier transform theory, three methods of determining $H(f)$ can be established. First, using any transient excitation signal to a system at rest and the corresponding output, $H(f)$ is determined from their FITs (Table 4.5). Second, if the input is sinusoidal, the signal amplification of the steady-state output is the magnitude $|H(f)|$ at the input frequency, and the phase lead of the steady-state output is the corresponding phase angle $\angle H(f)$. Third, using a random input signal and the corresponding input and output spectral density functions, $H(f)$ is determined as the ratio

$$H(f) = \Phi_{uy}(f)/\Phi_{uu}(f) \qquad (4.34)$$

4.4.4 Leakage (Truncation Error)

In digital processing of vibration signals (e.g., accelerometer signals), sampled data are truncated to eliminate less significant parts. This is of course essential in real-time processing because, in that case, only sufficiently short segments of continuously acquisitioned data are processed at one time. The computer memory (and buffer) limitations, the speed and cost of processing, the frequency range of importance, sampling rate, and the nature of the signal (level of randomness, periodicity, decay rate, etc.) should be taken into consideration in selecting the truncation point of data.

The effect of direct truncation of a signal $x(t)$ on its Fourier spectrum is shown in Figure 4.15. In the time domain, truncation is accomplished by multiplying $x(t)$ by the boxcar function $b(t)$. This is equivalent to a convolution $(X(f) * B(f))$ in the frequency domain. This procedure introduces ripples (side lobes) into the true spectrum. The resulting error $(X(f) - X(f) * B(f))$ is known as leakage or truncation error. Similar leakage effects arise in the time domain, as a result of truncation of the frequency spectrum. The truncation error can be reduced by suppressing the side lobes, which requires modification of the truncation function (window) from the boxcar shape $b(t)$ to a more desirable shape. Commonly used windows are the Hanning, Hamming, Parzen, and Gaussian windows.

Vibration Signal Analysis

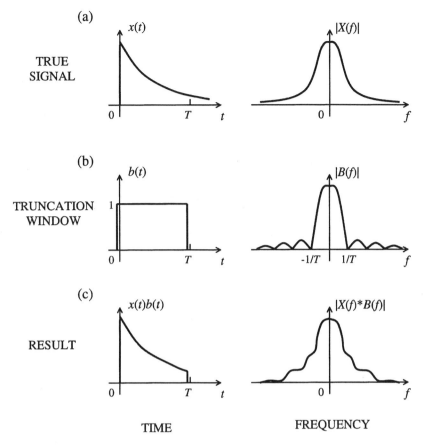

FIGURE 4.15 Illustration of truncation error: (a) signal and its frequency spectrum; (b) a rectangular (boxcar) window and its frequency spectrum; and (c) truncated signal and its frequency spectrum.

4.4.5 COHERENCE

Random vibration signals $\tilde{X}(t)$ and $\tilde{Y}(t)$ are said to be statistically independent if their joint probability distribution is given by the product of the individual distributions. A special case of this is the uncorrelated signals that satisfy

$$E\left[\tilde{X}(t_1)\tilde{Y}(t_2)\right] = E\left[\tilde{X}(t_1)\right]E\left[\tilde{Y}(t_2)\right] \tag{4.35}$$

In the stationary case, the means $\mu_x = E\left[\tilde{X}(t)\right]$ and $\mu_y = E\left[\tilde{Y}(t)\right]$ are time independent. The autocovariance functions are given by

$$\psi_{xx}(\tau) = E\left[\left\{\tilde{X}(t) - \mu_x\right\}\left\{\tilde{X}(t+\tau) - \mu_x\right\}\right] = \phi_{xx}(\tau) - \mu_x^2 \tag{4.36}$$

$$\psi_{yy}(\tau) = E\left[\left\{\tilde{Y}(t) - \mu_y\right\}\left\{\tilde{Y}(t+\tau) - \mu_x\right\}\right] = \phi_{yy}(\tau) - \mu_y^2 \tag{4.37}$$

and the cross-covariance function is given by

$$\psi_{xy}(\tau) = E\left[\left\{\tilde{X}(t) - \mu_x\right\}\left\{\tilde{Y}(t+\tau) - \mu_y\right\}\right] = \phi_{xy}(\tau) - \mu_x \mu_y \quad (4.38)$$

For uncorrelated signals, $\phi_{xy}(\tau) = \mu_x \mu_y$ and $\psi_{xy}(\tau) = 0$. The correlation function coefficient is defined by

$$\rho_{xy}(\tau) = \frac{\psi_{xy}(\tau)}{\sqrt{\psi_{xx}(0)\psi_{yy}(0)}} \quad (4.39)$$

which satisfies $-1 \leq \rho_{xy}(\tau) \leq 1$.

For uncorrelated signals, $\rho_{xy}(\tau) = 0$. This function measures the degree of correlation of the two signals. In the frequency domain, the correlation is determined by its (ordinary) coherence function

$$\gamma_{xy}^2(f) = \frac{\left|\Phi_{xy}(f)\right|^2}{\Phi_{xx}(f)\Phi_{yy}(f)} \quad (4.40)$$

which satisfies the condition $0 \leq \gamma_{xy}^2(f) \leq 1$. In this definition, the signals are assumed to have zero means. Alternatively, the FIT of the covariance functions can be used. If the signals are uncorrelated (or better, independent), the coherence function vanishes. On the other hand, if $\tilde{Y}(t)$ is the response of a linear, time-invariant system to an input $\tilde{X}(t)$, then

$$\Phi_{xy}(f) = \Phi_{xx}(f)H(f) \quad (4.41)$$

$$\Phi_{yy}(f) = \Phi_{xx}(f)\left|H(f)\right|^2 \quad (4.42)$$

Consequently, the coherence function becomes unity for this ideal case. In practice, however, the coherence function of an excitation and the corresponding response are usually less than unity. This is due to deviations such as measurement noise, system nonlinearities, and time-variant effects. Consequently, the coherence function is commonly used as a measure of the accuracy of frequency response estimates.

4.4.6 Parseval's Theorem

For a pair of rapidly decaying (aperiodic) deterministic signals $x(t)$ and $y(t)$, the cross-correlation function is given by

$$\phi_{xy}(\tau) = \int_{-\infty}^{\infty} x(t)y(t+\tau)dt \quad (4.43)$$

This is equivalent to equation (4.30), for a pair of ergodic, random (stochastic) signals $x(t)$ and $y(t)$. Using the definition of the inverse FIT (see Table 4.2) in equation (4.43) and following straightforward mathematical manipulation, it can be shown that

$$\Phi_{xy}(f) = \left[X(f)\right]^* Y(f) \quad (4.44)$$

in which the cross-spectral density $\Phi_{xy}(f)$ is the FIT of $\phi_{xy}(\tau)$, as given by

$$\Phi_{xy}(f) = \int_{-\infty}^{\infty} \phi_{xy}(\tau)\exp(-j2\pi f\tau)d\tau \tag{4.45}$$

and [ ]* denotes the complex conjugation operation. This result, which is known as the correlation theorem (see Table 4.5), has applications in the evaluation of the correlation functions and psd functions of finite-record-length data.

The inverse FIT relation corresponding to equation (4.45) is

$$\phi_{xy}(\tau) = \int_{-\infty}^{\infty} \Phi_{xy}(f)\exp(j2\pi f\tau)df \tag{4.46}$$

From equation (4.44),

$$\phi_{xy}(\tau) = \int_{-\infty}^{\infty} [X(f)]^* Y(f)\exp(j2\pi f\tau)df \tag{4.47}$$

If $\tau = 0$ and $x = y$ in equation (4.47), one obtains

$$\phi_{xy}(0) = \int_{-\infty}^{\infty} |Y(f)|^2 df \tag{4.48}$$

Similarly, from equation (4.43), one gets

$$\phi_{xy}(0) = \int_{-\infty}^{\infty} y^2(t)dt \tag{4.49}$$

By comparing equations (4.48) and (4.49), one obtains Parseval's theorem; thus,

$$\int_{-\infty}^{\infty} y^2(t)dt = \int_{-\infty}^{\infty} |Y(f)|^2 df \tag{4.50}$$

Using the discrete correlation theorem in an analogous manner, one can establish the discrete version of equation (4.50):

$$\Delta T \sum_{m=0}^{N-1} y_m^2 = \Delta F \sum_{n=0}^{N-1} |Y_n|^2 \tag{4.51}$$

These results are listed in the second row of Table 4.5.

4.4.7 Window Functions

Consider the unit boxcar window function $w(t)$, defined as

TABLE 4.6
Some Common Window Functions

Function Name	Time-Domain Representation [$w(t)$]	Frequency-Domain Representation [$W(f)$]
Boxcar	$= 1 \quad$ for $0 \leq t < T$ $= 0 \quad$ otherwise	$\dfrac{1}{j2\pi f}\left[1 - \cos 2\pi fT + j\sin 2\pi fT\right]$
Hanning	$= \dfrac{1}{2} + \dfrac{1}{2}\cos\dfrac{\pi t}{T} \quad$ for $\|t\| < T$ $= 0$ otherwise	$\dfrac{T\sin 2\pi fT}{2\pi fT\left[1 - (2fT)^2\right]}$
Parzen	$= 1 - 6\left[\dfrac{t}{T}\right]^2 + 6\left[\dfrac{\|t\|}{T}\right]^3 \quad$ for $\|t\| < \dfrac{T}{2}$ $= 2\left[1 - \dfrac{\|t\|}{T}\right]^3 \quad$ for $\dfrac{T}{2} < \|t\| \leq T$ $= 0$ otherwise	$\dfrac{3}{4}T\left[\dfrac{\sin \pi fT/2}{\pi fT/2}\right]^4$
Bartlett	$= 1 - \dfrac{\|t\|}{T} \quad$ for $\|t\| \leq T$ $= 0$ otherwise	$T\left[\dfrac{\sin \pi fT}{\pi fT}\right]^4$

$$w(t) = 1 \quad \text{for } 0 \leq t < T$$
$$= 0 \quad \text{otherwise} \tag{4.52}$$

This is shown in Figure 4.15(b). The FIT of $w(t)$ is

$$W(f) = \frac{1}{j2\pi f}\left[1 - \cos 2\pi fT + j\sin 2\pi fT\right] \tag{4.53}$$

Clearly, this (rectangular) window function produces side lobes (leakage) in the frequency domain.

In the spectral analysis of vibration signals, one is often required to segment the time history into several parts, and then perform spectral analysis on the individual results to observe the time development of the spectrum. If segmenting is done by simple truncation (multiplication by the boxcar window), the process would introduce rapidly fluctuating side lobes into the spectral results. Window functions, or smoothing functions other than the boxcar function, are widely used to suppress the side lobes (leakage error). Some common smoothing functions are defined in Table 4.6.

A graphical comparison of these four window types is given in Figure 4.16. Hanning windows are very popular in practical applications. A related window is the Hamming window, which is simply a Hanning window with rectangular cutoffs at the two ends. A Hamming window will have characteristics similar to those of a Hanning window, except that the side lobe fall-off rate at higher frequencies is less in the Hamming window.

From Figure 4.16(b), one observes that the frequency-domain weight of each window varies with the frequency range of interest. Obviously, the boxcar window is the worst. In practical applications, the performance of any window could be improved by simply increasing the window length T.

Characteristics of the signal that is being analyzed and also the nature of the system that generates the signal should be considered in choosing an appropriate truncation window. In particular, the Hamming window is recommended for signals generated by heavily damped systems

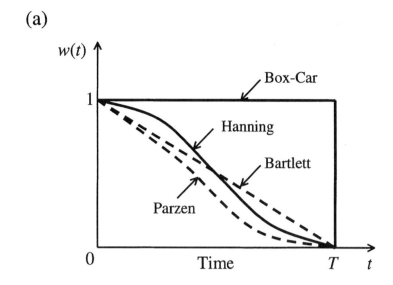
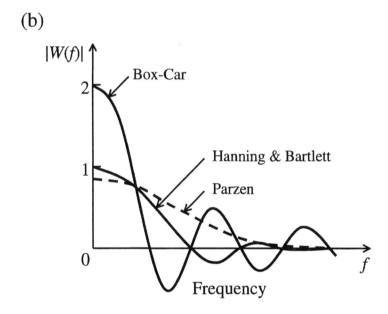

FIGURE 4.16 Some common window functions: (a) time-domain function, and (b) frequency spectrum.

and the Hamming window is recommended for use with lightly damped system. Table 4.7 lists some useful signal types and appropriate window functions.

4.4.8 SPECTRAL APPROACH TO PROCESS MONITORING

In mechanical systems, component degradation can be caused by vibration excitations, which can result in malfunction or failure. In this sense, continuous monitoring during testing of mechanical deterioration in various critical components of a vibratory system is of prime importance. This usually cannot be done by simple visual observation, unless malfunction is detected by operability monitoring of the system. Because mechanical degradation is always associated with a change in vibration level, however, by continuously monitoring the development of Fourier spectra in time (during system operation) at various critical locations of the system, it is possible to conveniently

TABLE 4.7
Signal Types and Appropriate Windows

Signal Type	Window
Periodic with period = T	Rectangular
Rapid transients within [0, T]	
Periodic with period ≠ T	Flat-top Cosine
Quasi-periodic	
Slow transients beyond [0, T]	
Nonstationary random	
Beats-like signals with period ≈ T	Bartlett (Triangular)
Narrow-band random	Hanning (Cosine)
Stationary random	
Important low-level components mixed with widely spaced high-level spectral components	Parzen
Broad-band random (white noise, pink noise, etc.)	

detect any mechanical deterioration and impending failure. In this respect, real-time Fourier analysis is very useful in process monitoring and failure detection and prediction. Special-purpose real-time analyzers with the capability of spectrum comparison (often done by an external command) are available for this purpose.

Various mechanical deteriorations manifest themselves at specific frequency values. A change in spectrum level at a particular frequency (and its multiples) would indicate a specific type of mechanical degradation or component failure. An example is given in Figure 4.17, which compares the Fourier spectrum at a monitoring location of a vibratory system at the start of test with the Fourier spectrum after some mechanical degradation has taken place. To facilitate spectrum comparison within a narrow-frequency band, it is customary to plot such Fourier spectra on a linear frequency axis. It is seen that the overall spectrum levels have increased as a result of mechanical degradation. Also, a significant change has occurred near 30 Hz. This information is useful in diagnosing the cause of degradation or malfunction. Figure 4.17 might indicate, for example, impending failure of a component having resonant frequency close to 30 Hz.

4.4.9 CEPSTRUM

A function known as the *cepstrum* is sometimes used to facilitate the analysis of Fourier spectrum in detecting mechanical degradation. The cepstrum (complex) $C(\tau)$ of a Fourier spectrum $Y(f)$ is defined by

$$C(\tau) = \mathcal{F}^{-1} \log Y(f) \tag{4.54}$$

The independent variable τ is known as *quefrency*, and it has the units of time.

An immediate advantage of cepstrum arises from the fact that the logarithm of the Fourier spectrum is taken. From equation (4.33), it is clear that, for a system having frequency-transfer function $H(f)$, and excited by a signal having Fourier spectrum $U(f)$, the response Fourier spectrum $Y(f)$ can be expressed in the logarithmic form:

$$\log Y(f) = \log H(f) + \log U(f) \tag{4.55}$$

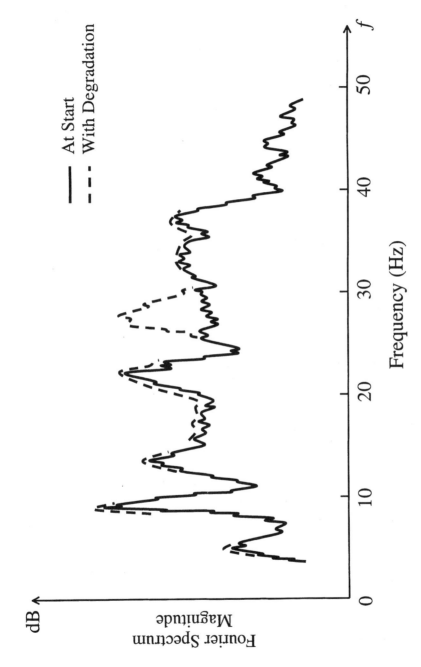

FIGURE 4.17 Effect of mechanical degradation on a monitored Fourier spectrum.

Because the right-hand-side terms are added rather than multiplied, any variation in $H(f)$ at a particular frequency will be less affected by a possible low spectrum level in the excitation $U(f)$ at that frequency, when considering log $Y(f)$ rather than $Y(f)$. Consequently, any degradation will be more conspicuous in the cepstrum than in the Fourier spectrum. Another advantage of cepstrum is that it is better capable of detecting variation in phenomena that manifest themselves as periodic components in the Fourier spectrum (e.g., harmonics and sidebands). Such phenomena that appear as repeated peaks in the Fourier spectrum occur as a single peak in the cepstrum, and so any variations can be detected more easily.

4.5 OTHER TOPICS OF SIGNAL ANALYSIS

This section briefly addresses some other important topics of signal analysis, starting with bandwidth, in different contexts. Then, several practically useful analysis procedures and results on vibration signals are presented.

4.5.1 BANDWIDTH

Bandwidth has different meanings, depending on the particular context and application. For example, when studying the response of a dynamic system, the bandwidth relates to the fundamental resonant frequency and, correspondingly, to the speed of response for a given excitation. In bandpass filters, the bandwidth refers to the frequency band within which the frequency components of the signal are allowed through the filter, the frequency components outside the band being rejected by it. With respect to measuring instruments, bandwidth refers to the range of frequencies within which the instrument accurately measures a signal. Note that these various interpretations of bandwidth are somewhat related. For example, if a signal passes through a bandpass filter, its frequency content is within the bandwidth of the filter; but one cannot determine the actual frequency content of the signal through such an observation. In this context, the bandwidth appears to represent a frequency uncertainty in the observation (i.e., the larger the bandwidth of the filter, the less certain one can be about the actual frequency content of a signal that is allowed through the filter).

4.5.2 TRANSMISSION LEVEL OF A BANDPASS FILTER

Practical filters can be interpreted as dynamic systems. In fact, all physical, dynamic systems (e.g., mechanical structures) are analog filters. It follows that the filter characteristic can be represented by the frequency transfer function $G(f)$ of the filter. A magnitude-squared plot of such a filter transfer function is shown in Figure 4.18. In a logarithmic plot, the magnitude-squared curve is obtained by simply doubling the corresponding magnitude (Bode plot) curve. Note that the actual filter transfer function [Figure 4.18(b)] is not flat like the ideal filter shown in Figure 4.18(a). The reference level G_r is the average value of the transfer function magnitude in the neighborhood of its peak.

4.5.3 EFFECTIVE NOISE BANDWIDTH

Effective noise bandwidth of a filter is equal to the bandwidth of an ideal filter that has the same reference level and that transmits the same amount of power from a white noise source. Note that white noise has a constant (flat) power spectral density (psd). Hence, for a noise source of unity psd, the power transmitted by the practical filter is given by

$$\int_0^\infty |G(f)|^2 df$$

Vibration Signal Analysis

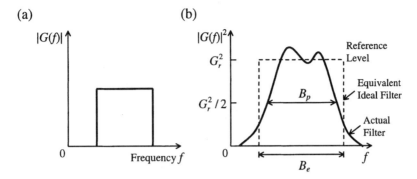

FIGURE 4.18 Characteristics of (a) an ideal bandpass filter; and (b) a practical bandpass filter.

which, by definition, is equal to the power $G_r^2 B_e$ transmitted by the equivalent ideal filter. Hence, the effective noise bandwidth B_e is given by

$$B_e = \int_0^\infty |G(f)|^2 df / G_r^2 \qquad (4.56)$$

4.5.4 Half-Power (or 3dB) Bandwidth

Half of the power from a unity-psd noise source as transmitted by an ideal filter, is $G_r^2 B_e / 2$. Hence, $G_r / \sqrt{2}$ is referred to as the *half-power level*. This is also known as a 3 dB level because $20 \log_{10} \sqrt{2}$ = $10 \log_{10} 2 = 3$ dB. (*Note*: 3 dB refers to a power ratio of 2 or an amplitude ratio of $\sqrt{2}$. Furthermore, 20dB corresponds to an amplitude ratio of 10 or a power ratio of 100). The 3 dB (or half-power) bandwidth corresponds to the width of the filter transfer function at the half-power level. This is denoted by B_p in Figure 4.18(b). Note that B_e and B_p are different in general. In an ideal case where the magnitude-squared filter characteristic has linear rise and fall-off segments, however, these two bandwidths are equal (see Figure 4.19).

4.5.5 Fourier Analysis Bandwidth

In Fourier analysis, bandwidth is interpreted, again, as the *frequency uncertainty* in the spectral results. In analytical Fourier integral transform (FIT) results, which assume that the entire signal is available for analysis, the spectrum is continuously defined over the entire frequency range $[-\infty, \infty]$ and the frequency increment df is infinitesimally small ($df \to 0$). There is no frequency uncertainty in this case, and the analysis bandwidth is infinitesimally narrow.

In digital Fourier transform, the discrete spectral lines are generated at frequency intervals of ΔF. This finite frequency increment ΔF, which is the frequency uncertainty, is therefore, the analysis bandwidth B for this analysis. Note that $\Delta F = 1/T$, where T is the record length (or window length for a rectangular window). It follows also that the minimum frequency that has a meaningful accuracy is the bandwidth. This interpretation for analysis bandwidth is confirmed by noting the fact that harmonic components of frequency less than ΔF (or period greater than T) cannot be studied by observing a signal record of length less than T. Analysis bandwidth carries information regarding distinguishable minimum frequency separation in computed results. In this sense, bandwidth is directly related to the frequency resolution of analyzed results. The accuracy of analysis increases by increasing the record length T (or decreasing the analysis bandwidth B).

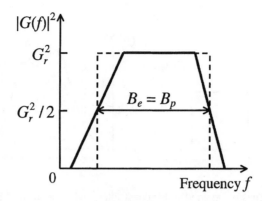

FIGURE 4.19 An idealized filter with linear segments.

When a time window other than the rectangular window is used to truncate a measured vibration signal, then reshaping of data occurs according to the shape of the window. This reshaping reduces leakage due to suppression of side lobes of the Fourier spectrum of the window. At the same time, however, an error is introduced due to the information lost through data reshaping. This error is proportional to the bandwidth of the window itself. The effective noise bandwidth of a rectangular window is only slightly less than $1/T$ because the main lobe of its Fourier spectrum is nearly rectangular. Hence, for all practical purposes, the effective noise bandwidth can be taken as the analysis bandwidth. Note that data truncation (multiplication in the time domain) is equivalent to convolution of the Fourier spectrum (in the frequency domain). The main lobe of the spectrum uniformly affects all spectral lines in the discrete spectrum of the data signal. It follows that a window main lobe having a broader bandwidth (effective noise bandwidth) introduces a larger error into the spectral results. Hence, in digital Fourier analysis, bandwidth is taken as the effective noise bandwidth of the time window that is employed.

4.6 RESOLUTION IN DIGITAL FOURIER RESULTS

In digital Fourier analysis results, resolution is the frequency separation between spectral lines. For a data record of length T, the resolution is $\Delta F = 1/T$, irrespective of the type of window used. There is a noteworthy distinction between analysis bandwidth and resolution. Suppose that one has a data record of length T. If one doubles the length by augmenting it with trailing zeros, digital Fourier analysis of the resulting record of length $2T$ will yield a spectral line separation of $1/(2T)$. Thus, the resolution is halved. But, unless the true signal value is also zero in the second time interval $t[T, 2T]$, no new information in presented in the augmented record of duration $[0, 2T]$ in comparison to the original record of duration $[0, T]$. Thus, the analysis bandwidth (a measure of accuracy) will remain unchanged. If, on the other hand, the signal itself was sampled over $[0, 2T]$ and the resulting $2N$ data points were used in digital Fourier analysis, the bandwidth as well as the resolution would be halved and the accuracy doubled.

Some relations that are useful in the digital computation of spectral results for signals are summarized in Box 4.3.

4.7 OVERLAPPED PROCESSING

Digital Fourier analysis is performed on blocks of sampled data (e.g., $2^{10} = 1024$ samples at a time). In overlapped processing, each data block is made to include part of the previous data block that was analyzed. After completing a computation, the overlapped data at the end of the computed block is moved to the beginning of the block, and the leading vacancy is filled with new data so

BOX 4.3 Useful Relations for Digital Spectral Computations

$$\mathscr{F}\, y(t) \xrightarrow[\text{(FFT)}]{\text{DFT}} Y(f) \qquad \text{Fourier spectrum}$$

$$\frac{1}{T} Y^*(f) Y(f) \qquad = \text{Power spectral density (psd)}$$

Power spectrum $\qquad = B \times$ Power spectral density $\qquad = \dfrac{B}{T} Y^*(f) Y(f)$

Energy spectrum $\qquad = T \times$ Power spectrum $\qquad = B Y^*(f) Y(f)$

Energy spectral density $\qquad = \dfrac{1}{B} \times$ Energy spectrum $\qquad = Y^*(f) Y(f)$

RMS spectra
(always shown for
+ve frequencies only)

$$= \left[\frac{2}{B} \int_{f}^{f+B} |Y(f')|^2 \, df' \right]^{\frac{1}{2}}$$

*One sided
*Like $|Y(f)|$ but smoother
*No phase information
*Increase $B \to$ higher bandwidth

Note:

T = record length
B = bandwidth of digital analysis
(Min. freq. for which meaningful results are obtained) $\to$ includes window effect.

Periodic or stationary signals: Use power spectra
(infinite energy)
Transient signals: Energy spectra can be used
(finite energy)

One sided spectrum = 2 × (+ve frequency part of two-sided spectrum)

Coherent output power = Coherence γ_{uy}^2 × Output power $\leftarrow$ could be power spectrum or
(Spectrum or spectral density) $\qquad\qquad\qquad\qquad\qquad\qquad\qquad$ psd of the output

---

that the end data in one block is identical to the beginning data in the next block, in the overlapped region. In other words, the overlapped portions of each data block (the two end portions) are processed twice. It follows that if there is 50% (or more) overlapping, then the entire data block is processed twice. Three main reasons can be given for using overlapped processing in digital Fourier analysis:

1. It is an effective means of averaging spectral results.
2. It reduces the waiting time for assembling the data buffer.
3. It reduces the error caused by the end shaping effect of time windows (when a window other than the rectangle window is used).

From reasons (1) and (2), it is clear that, due to overlapping, the statistical error of computations is reduced for the same speed of computation, and the computing power is more efficiently used.

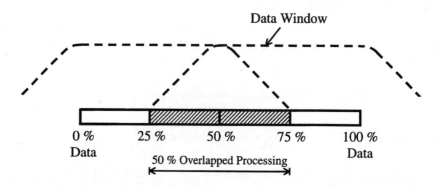

FIGURE 4.20 Overlapped processing of windowed signals.

To explain reason (3), examine Figure 4.20. This example shows a 50% overlap in data. It is seen that the window function can be assumed relatively flat at least over 50% of the window length (record length). Then the entire data block will correspond to the flat part of the window in three successive analyses. Consequently, the shaping error (or the error due to increased analysis bandwidth) that is caused by a nonrectangular time window is virtually eliminated by overlapped processing. The flatness of a time window is determined by its effective noise bandwidth B_e. The effective record length T_e is defined as

$$T_e = \frac{1}{B_e} \tag{4.57}$$

which provides a measure for the flat segment of the window. The percentage effective record length is given by T_e as a percentage of the actual record length T. The degree of overlapping is chosen using the relation

$$\% \text{ overlap} = 100\left(1 - \frac{T_e}{T}\right) \tag{4.58}$$

EXAMPLE 4.3

For a Hamming window, $B_e = 1.4/T$. Hence, a typical value for the percentage overlap is

$$100\left(1 - \frac{1}{1.4}\right) = 29\%$$

One might want to use a conservative overlap and even go up to 50% in this case because the window is not quite flat.

4.7.1 ORDER ANALYSIS

Speed-related vibrations in rotation machinery can be analyzed through order analysis. Machinery vibrations under startup (accelerating) and shutdown (decelerating) conditions are analyzed in this manner. Orders represent the rotating-speed-related frequency components in a response signal. The ratio of the response frequency to the rotating speed is termed "order."

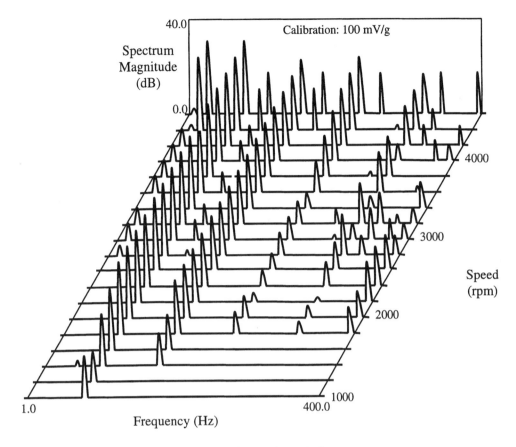

FIGURE 4.21 A speed spectral map obtained from order analysis.

Order analysis is done essentially through digital Fourier analysis of a rotating-speed-related response signal. Practically, this can be accomplished in many ways. The format in which the spectral results are presented will depend on the procedure used in order analysis. Some of the typical formats of data presentation are given below.

Speed Spectral Map

As the rotating speed of a machine is changing in a given range, the Fourier spectrum of the response signal is determined for equal increments of speed. The results are presented as a speed spectral map, which is a three-dimensional cascade diagram (or *waterfall display*). The two base axes of the plot are spectral frequency and rotating speed. The third axis gives the spectral magnitude (see Figure 4.21). These types of plots are useful in identifying order-related components during startup or coast-down conditions. Note that for each speed, the frequency band of digital Fourier analysis is kept the same (i.e., fixed sampling rate). Each distinct crest trace denotes an order-related resonance. The fact that these traces are almost straight lines indicates the significance of order (the ratio, frequency/rotating speed) in exciting these resonances.

Time Spectral Map

Under variable speed conditions (not necessarily accelerating or decelerating), the response signal is Fourier analyzed at equal increments of time. The results are plotted in a *cascade diagram*, with frequency and time as the base axes. The third axis again represents the magnitude of the Fourier

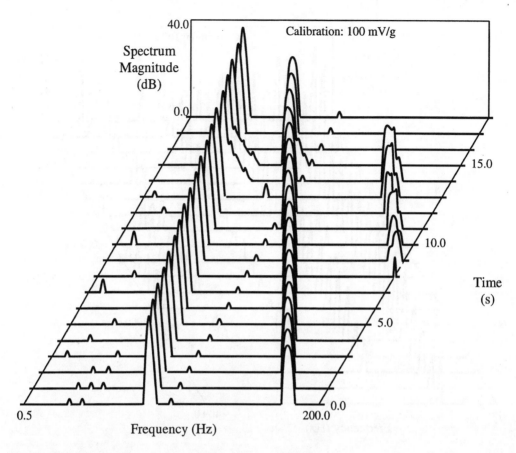

FIGURE 4.22 A time spectral map obtained from order analysis.

spectrum (see Figure 4.22). In this case, the crest traces are not necessarily straight, and can change their orientation arbitrarily. This variation in crest orientation is determined by the degree of speed variation.

Order Tracking

In order tracking, a "tracking frequency multiplier" monitors the rotating speed of the machine (as for a speed spectral map). But, in the present case, the sampling rate of the response signal (for Fourier analysis) is changed in proportion to the rotating speed. Note that, in this manner, the maximum useful frequency (approximately 400/512 × Nyquist frequency) is increased as the rotating speed increases, so that the aliasing effects are reduced. If the same sampling rate is used for high speeds (as in Case 1 above), aliasing error can be significant at high rotating speeds.

In presenting order tracking spectral results, the frequency axis is typically calibrated in orders. Both speed spectral maps and time spectral maps can be presented in this manner. Other types of data presentation can be used, as well, in order analysis. For example, instead of the Fourier spectrum of the response signal, the power spectrum or composite power spectrum (in which the total signal power is computed in specified frequency bands and presented as a function of the rotating speed) can be used in the schemes described in this section.

Order analysis provides information on most severe operating speeds with respect to vibration (and dynamic stress). For example, suppose that for a given speed of operation, two major resonances occur: one at 10 Hz and the other at 80 Hz. Then, the structure of the system (rotating machine and its support fixtures) should be modified to change and preferably damp out these

Vibration Signal Analysis

resonances. Furthermore, the most desirable operating speed can be chosen in terms of the lowest resonant peaks by observing a speed spectral map.

PROBLEMS

4.1 The Fourier transform of a position measurement $y(t)$ is $Y(j\omega)$. The Fourier transform of the corresponding velocity signal is:
  a. $Y(j\omega)$
  b. $j\omega Y(j\omega)$
  c. $Y(j\omega)/(j\omega)$
  d. $\omega Y(j\omega)$.

4.2 The Fourier transform of the acceleration signal in Problem 4.1 is:
  a. $Y(j\omega)$
  b. $\omega^2 Y(j\omega)$
  c. $-\omega^2 Y(j\omega)$
  d. $Y(j\omega)/(j\omega)$.

4.3 What is the Fourier integral transform (FIT) of a unit impulse excitation $\delta(t)$? The excitation in a *bump test* or a *hammer test* that is commonly used in structural dynamic testing can be approximated by an impulse. Discuss the implication of your answer in such tests.

4.4 The real part and the imaginary part of the Fourier spectrum of a (real) signal are shown in Figure P4.4. Complete these spectral curves by including the negative spectrum as well.

4.5 The frequency transfer function for a simple oscillator is given by

$$H(\omega) = \frac{\omega_n^2}{\left[\omega_n^2 - \omega^2 + 2j\zeta\omega_n\omega\right]}$$

Determine an expression for the half-power (3dB) bandwidth at low damping.

4.6 An approximate frequency transfer function of a system was determined by Fourier analysis of excitation-response data (measured) and fitting into an appropriate analytical expression (by curve fitting using a least-squares method). This was found to be

$$H(f) = \frac{5}{10 + j2\pi f}$$

What is its magnitude, phase angle, real part, and imaginary part at $f = 2$ Hz? If the reference frequency is taken as 1 Hz, what is the transfer function magnitude at 2 Hz, expressed in dB?

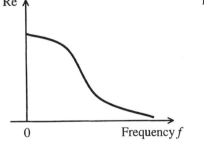 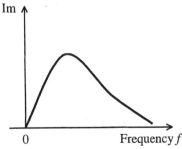

FIGURE P4.4 An example of one-sided spectra.

4.7 Using the definition and direct analytical integration, determine the Fourier spectrum of the signal

$$x(t) = ae^{-\lambda t} \quad t \geq 0$$
$$= 0 \quad t < 0$$

a. as a real part and an imaginary part
b. as a magnitude and a phase angle.

4.8 Consider the harmonic signal

$$x(t) = a\cos 2\pi f_0 t$$

defined over $t = [-\infty, \infty]$. Determine, using analysis, the Fourier integral transform of $x(t)$.

4.9 The discrete Fourier transform (DFT) of a sampled signal $\{x_m\}$ is denoted by $\{X_n\}$ and is given by

$$X_n = \Delta T \sum_{m=0}^{N-1} x_m \exp(-j2\pi nm/N)$$

for $n = 0, 1, \ldots, N-1$, where ΔT is the sample time step and N is the number of samples in the data record.

The inverse DFT is given by

$$x_m = \Delta F \sum_{n=0}^{N-1} X_n \exp(j2\pi nm/N)$$

for $m = 0, 1, \ldots, N-1$, with $\Delta F = \dfrac{1}{N\Delta T} = \dfrac{1}{T}$.

Describe a way to use a computer program that computes the forward DFT, and for computing the inverse DFT as well.

4.10 Two signals $y(t)$ and $z(t)$ are sampled at time period ΔT to generate the data sequences $\{y_n\}$ and $\{z_n\}$, each having N points. Explain how the discrete Fourier spectra sequences $\{Y_m\}$ and $\{Z_m\}$ of these data records could be computed using a single DFT operation.

4.11 Consider the two time signals $u(t)$ and $h(t)$. If the first signal is shifted through t_1 and the second signal is shifted through t_2 in the same direction, show that the convolution function

$$y(t) = \int_{-\infty}^{\infty} u(t)h(t-\tau)d\tau$$

will be shifted by $t_1 + t_2$ in the same direction. If $u(t)$ is a signal of short duration and $h(t)$ is a signal that is several times longer, describe an efficient way to compute the convolution function, by sectioning $h(t)$ into several segments and then computing the convolution of each segment with $u(t)$.

4.12 Give several interpretations of the term "bandwidth" in the frequency-domain analysis of signals and systems. In particular, discuss the "analysis bandwidth" in the context of

TABLE P4.12
Analysis Bandwidths for Some Common Time Windows

Window Name	Bandwidth B
Rectangular	$1/T$
Flat-top cosine	$1.01/T$
Bartlett (triangular)	$1.4/T$
Hamming	$1.4/T$
Hanning (cosine)	$1.5/T$
Parzen	$1.9/T$

digital frequency analysis of a signal. Note the analysis bandwidth (B) corresponding to several data window functions of length (truncation) T, as given in Table P4.12.

4.13 a. What is an anti-aliasing filter? If a sensor signal is sampled at f_s Hz, suggest a suitable cutoff frequency for an anti-aliasing filter to be used in this application.

b. Suppose that a sinusoidal signal of frequency f_1 Hz is sampled at the rate of f_s samples per second. Another sinusoidal signal of the same amplitude, but of a somewhat higher frequency f_2 Hz was found to yield the same data when sampled at f_s. What is the likely analytical relationship between f_1, f_2, and f_s?

c. Consider an inertial system (e.g., mass and damper combination) having the transfer function

$$G(s) = \frac{k}{(1+\tau s)}$$

What is the static gain of this system? Show that the magnitude of the transfer function reaches $1/\sqrt{2}$ of the static gain when the excitation frequency is $1/\tau$ rad s^{-1}. Note that the frequency, $\omega_b = 1/\tau$ rad s^{-1}, can be considered as the operating bandwidth of the system.

4.14 a. Define the following terms:
  i. Sampling theorem
  ii. Nyquist frequency
  iii. Aliasing distortion
  iv. Anti-aliasing filters
  v. Truncation error or leakage
  vi. Data windows.

b. Consider 1024 data points sampled at 0.1 s. Determine:
  i. Sampling rate
  ii. Data record length
  iii. The number of points obtained by DFT analysis
  iv. The number of spectral lines obtained
  v. Spectral line separation
  vi. Maximum useful frequency available in the spectrum
  vii. The number of accurate spectral lines
  viii. Maximum undistorted frequency.

4.15 A window function somewhat distorts a signal that is truncated and also changes the amplitude and energy characteristics of the signal. Hence, the window is often scaled by a multiplication factor in order to reduce the resulting error. The scaling factor that is sometimes used for a window in digital Fourier analysis, is simply the ratio of the main lobe heights (in the frequency domain) of the particular window and the rectangular window, both having the same window length T and the same peak value of unity.

A more logical approach is used in many hardware Fourier analyzers. The method is based on scaling the peak value of the window (which is equivalent to scaling the computed spectrum) such that the energy of the original signal in the interval $[0, T]$ is equal to the energy of the windowed signal. The instantaneous power of the original signal is $y(t)^2$, and that of the weighted signal is $[w(t)y(t)]^2 = w(t)^2 y(t)^2$, where $w(t)$ is the window function. Assuming that the signal is stationary (this is a necessary assumption for spectral density computations as well), a good estimate for the *energy reduction factor* F_e is given by

$$F_e = \frac{1}{T}\int_0^T w(t)^2 dt$$

In the discrete case, the corresponding relation is

$$F_e = \frac{1}{N}\sum_{m=0}^{N-1} w_n^2$$

TABLE P4.15
Common Window Functions

Window Name	Describing Equation	Scaling Factor $1/\sqrt{F_e}$
Rectangular	$w(t) = 1$ for $t = 0$ to T $= 0$ elsewhere	
Flat-top (cosine)	$w(t) = 0.5(1 - \cos 10\pi t/T)$ for $t = 0$ to $T/10$ and $t = 9T/10$ to T $= 1$ for $t = T/10$ to $9T/10$ $= 0$ elsewhere	
Bartlett (triangular)	$w(t) = 2t/T$ for $t = 0$ to $T/2$ $= -2t/T + 2$ for $t = T/2$ to T $= 0$ elsewhere	
Hamming	$w(t) = 0.8 + 0.46(\cos 2\pi t/T)$ for $t = 0$ to T $= 0$ elsewhere	
Hanning (cosine)	$w(t) = 0.5(1 - \cos 2\pi t/T)$ for $t = 0$ to T $= 0$ elsewhere	
Parzen	$w(t) = 1 - 6(2t/T - 1)^2 + 6\|2t/T - 1\|^3$ for $t = T/4$ to $3T/4$ $= 2(1 - \|2t/T - 1\|^3)$ for $t = 3T/4$ to T $= 0$ elsewhere	

Vibration Signal Analysis

in which w_n are the sampled values of window, sampled at the same instants as the original signal. The window amplitude must be scaled by the factor $1/\sqrt{F_e}$ in order to achieve the energy equivalence. One can compute F_e using the above result and, for example, using the typical value of $N = 2^{10} = 1024$. Note that once the window is scaled in this manner, one is not required to scale the results (rms spectra, psd, cross spectra, etc.). Compute the scaling factors for common window functions as listed in Table P4.15.

4.16 a. Starting with the discrete Fourier transform relation

$$Y_n = \Delta T \sum_{m=0}^{N-1} y_m \exp(-j2\pi mn/N)$$

for a sampled data sequence $\{y_m\}$ of a signal $y(t)$, provide consistent relations for digital computation of the following functions:
  i. Autocorrelation function of $y(t)$
  ii. Power spectral density of $y(t)$
  iii. Cross spectral density of $u(t)$ and $y(t)$
  iv. Power spectrum of $y(t)$
  v. Energy spectrum of $y(t)$
  vi. Energy spectral density of $y(t)$
  vii. Coherence function of $u(t)$ and $y(t)$
  viii. Coherent output power spectral density for input $u(t)$ and output $y(t)$
  ix. Coherent output power spectrum for input $u(t)$ and output $y(t)$.
  b. Consider an acceleration time signal given in units of acceleration due to gravity (g). What are the corresponding units of correlation, Fourier spectrum, and spectral density of the signal?

4.17 The transfer function of a dynamic system is the ratio:

$$\frac{\text{Output}}{\text{Input}}$$

in the frequency domain. Note that this is a model for the system, and it does not depend on the nature of the input itself.
Is the frequency-transfer-function complex or real?
Methods of experimental computation of a system transfer function are:
  1. Apply a transient input and measure the output
     FFT (output)/FFT (input)
  2. Apply a sinusoidal input
     Measure amplification and phase shift of the output for various frequencies of interest.
  3. Apply a random input
     Cross-spectral density/PSD of input
In methods 1 and 3, averaging would be necessary to improve accuracy. Give analytical relations used for the associated computations in these three methods.

4.18 A frequency-domain signal analysis package provides the following computational options:
  1. Power spectrum of a signal
  2. rms spectrum of a signal
  3. Power spectral density of a signal
  4. Cross spectrum of two signals
  5. Transfer function from an input signal and an output signal

6. Transmissibility functions from input and output signals
7. Coherence of two signals
8. Coherent output power of a signal with respect to another signal
9. rms velocity spectrum of a signal
10. rms displacement spectrum of a signal.

In testing this package for its performance, the issues addressed would be:
(a) Parameter selection for sample problems to be used for testing the package
(b) Generation of input time histories
(c) Generation of the output (frequency-domain) data for verification.

a. *Parameter Selection*:
- Buffer (block) length for input data = T
- Buffer size (number of data points in the buffer) = N (typically 1024)

Then,
- Ordinary resolution in the frequency domain (or bandwidth) $\Delta F = \dfrac{1}{T}$
- Sampling period (in the time domain) $\Delta T = \dfrac{T}{N}$
- Amplitude parameter for the input signal = a (typically 0.1)
- Time constant of the input signal = τ (typically a fraction of T)
- Damping ratio of the dynamic system = ζ (typically 0.05)
- Undamped natural frequency (Hz) of the dynamic system = f_n (typically within T, i.e., $Tf_n = 4$)

Then,
- Angular undamped natural frequency (rad s^{-1}) $\omega_n = 2\pi f_n$
- Damped natural frequency $\omega_d = \sqrt{1-\zeta^2}\,\omega_n$
- Phase angle $\phi = \tan^{-1}\left[\dfrac{\omega_d}{\dfrac{1}{\tau}-\zeta\omega_n}\right]$

b. *Input Time-History Generation*
Input signal $x(t) = ae^{-t/\tau}$
(acceleration)

Ouput signal $y(t) = a\omega_n^2\left[\dfrac{e^{-t/\lambda}}{\left(\dfrac{1}{\tau}-\zeta\omega_n\right)^2 + \omega_d^2} + \dfrac{e^{-\zeta\omega_n t}\sin(\omega_d t - \phi)}{\omega_d\left[\left(\dfrac{1}{\tau}-\zeta\omega_n\right)^2 + \omega_d^2\right]^{1/2}}\right]$
(acceleration)

Normally, one samples both signals at sampling period ΔT and generate data in multiples of N.

Outline a procedure for generating the output signal in the frequency domain and subsequent computation of the ten functions listed above.

4.19 Define the following terms:
a. Sample record
b. Ensemble
c. Ergodic signal.

Explain why erogodicty is usually assumed in digital processing of random signals.

4.20 a. Define:
i. Octave
ii. Decade
iii. One-third octave
iv. Decibel (dB).

b. How many decades are there in 100 Hz?
c. How many decades correspond to a frequency change from 0.1 rad s^{-1} to 10.0 rad s^{-1}?
d. How many octaves correspond to a frequency change from 1 Hz to $\sqrt{2}$ Hz?
e. What are the dimensions of a decade?
f. What are the advantages of using log-log plots?

4.21 a. Give frequency bandwidths that are appropriate in the frequency-domain analysis of the following situations:
   i.   Whole-body vibration of humans
   ii. Ride quality considerations
   iii. Transportation excitations
   iv. Sound perceived by human ear
   x.   Physical systems (structures, circuits, etc.).
b. Indicate the rationale for using frequency-domain methods in vibration monitoring and malfunction diagnosis of dynamic systems. Indicate examples.

5 Modal Analysis

Complex vibrating systems usually consist of components that possess distributed energy-storage and energy-dissipative characteristics. In these systems, inertial, stiffness, and damping properties vary (piecewise) continuously with respect to the spatial location. Consequently, partial differential equations, with spatial coordinates (e.g., Cartesian coordinates x, y, z) and time t as independent variables, are necessary to represent their vibration response.

A distributed (continuous) vibrating system can be approximated (modeled) by an appropriate set of lumped masses properly interconnected using discrete spring and damper elements. Such a model is called a *lumped-parameter model* or *discrete model*. An immediate advantage resulting from this lumped-parameter representation is that the system equations become ordinary differential equations. Often, linear springs and linear viscous damping elements are used in these models. The resulting linear ordinary differential equations can be solved by the modal analysis method. The method is based on the fact that these idealized systems (models) have preferred frequencies and geometric configurations (or *natural modes*), in which they tend to execute free vibration. An arbitrary response of the system can be interpreted as a linear combination of these *modal vibrations*; and as a result, its analysis can be conveniently done using modal techniques.

Modal analysis is an important tool in vibration analysis, diagnosis, design, and control. In some systems, mechanical malfunction or failure can be attributed to the excitation of their preferred motion such as modal vibrations and resonances. By modal analysis, it is possible to establish the extent and location of severe vibrations in a system. For this reason, it is an important diagnostic tool. For the same reason, modal analysis is also a useful method for predicting impending malfunctions or other mechanical problems. Structural modification and substructuring are techniques of vibration analysis and design, which are based on modal analysis. By sensitivity analysis methods using a "modal" model, it is possible to determine what degrees of freedom of a mechanical system are most sensitive to addition or removal of mass and stiffness elements. In this manner, a convenient and systematic method can be established for making structural modifications to eliminate an existing vibration problem or to verify the effects of a particular modification. A large and complex system can be divided into several subsystems that can be independently analyzed. By modal analysis techniques, the dynamic characteristics of the overall system can be determined from the subsystem information. This approach has several advantages, including: (1) subsystems can be developed by different methods such as experimentation, finite element method, or other modeling techniques and assembled to obtain the overall model; (2) the analysis of a high-order system can be reduced to several lower-order analyses; and (3) the design of a complex system can be done by designing and developing its subsystems separately. These capabilities of structural modification and substructure analysis possessed by the modal analysis method make it a useful tool in the design development process of mechanical systems. Modal control, a technique that employs modal analysis, is quite effective in the vibration control of complex mechanical systems.

5.1 DEGREES OF FREEDOM AND INDEPENDENT COORDINATES

The geometric configuration of a vibrating system can be completely determined by a set of independent coordinates. This number of independent coordinates, for most systems, is termed the number of *degrees of freedom* (dof) of the system. For example, a particle moving freely on a plane requires two independent coordinates to completely locate it (e.g., x and y Cartesian coordinates or r and θ polar coordinates); its motion has two degrees of freedom. A rigid body that is free to take any orientation in (the three-dimensional) space needs six independent coordinates to completely define its position. For example, its centroid is positioned using three independent Cartesian coordinates (x, y, z). Any axis fixed in the body and passing through its centroid can be oriented by two independent angles (θ, ϕ). The orientation of the body about this body axis can be fixed by a third independent angle (ψ). Altogether, six independent coordinates have been utilized; the system has six degrees of freedom.

Strictly speaking, the number of degrees of freedom is equal to the number of independent "incremental" generalized coordinates that are needed to represent a general motion. In other words, it is the number of "incremental independent motions" that are possible. For *holonomic systems* (i.e., systems possessing *holonomic constraints* only), the number of independent incremental generalized coordinates is equal to the number of independent generalized coordinates; hence, either definition can be used for the number of degrees of freedom. If, on the other hand, the system has nonholonomic constraints, the definition based on incremental coordinates should be used because in these systems the number of independent incremental coordinates is in general less than the number of independent coordinates required to completely position the system.

5.1.1 Nonholonomic Constraints

Constraints of a system that cannot be represented by purely algebraic equations in its generalized coordinates and time are termed "nonholonomic constraints." For a nonholonomic system, more coordinates than the number of degrees of freedom are required to completely define the position of the system. The number of excess coordinates is equal to the number of nonalgebraic relations that define the nonholonomic constraints in the system. Examples for nonholonomic systems are afforded by bodies rolling on surfaces (see Example 5.1), and bodies whose velocities are constrained in some manner (see Example 5.2).

Example 5.1

A good example for a nonholonomic system is provided by a sphere rolling, without slipping, on a plane surface. In Figure 5.1, the point O denotes the center of the sphere at a given instant, and P is an arbitrary point within the sphere. The instantaneous point of contact with the plane surface is denoted by Q, so that the radius of the sphere is $OQ = a$. This system requires five independent generalized coordinates to position it. For example, the center O is fixed by the Cartesian coordinates x and y. Because the sphere is free to roll along any arbitrary path on the plane and return to the starting point, the line OP can assume any arbitrary orientation for any given position for the center O. This line can be oriented by two independent coordinates θ and ϕ, defined as in Figure 5.1. Furthermore, because the sphere is free to spin about the z-axis and is also free to roll on any trajectory (and return to its starting point), it follows that the sphere can take any orientation about the line OP (for a specific location of point O and line OP). This position can be oriented by the angle ψ. These five generalized coordinates $x, y, \theta, \phi,$ and ψ are independent. The corresponding incremental coordinates $\delta x, \delta y, \delta \theta, \delta \phi,$ and $\delta \psi$ are, however, not independent as a result of the constraint of rolling without slipping. It can be shown that two independent differential equations can be written for this constraint and, consequently, there exist only three independent incremental coordinates; the system actually has only three degrees of freedom.

To establish the equations for the two nonholonomic constraints, note that the incremental displacements δx and δy of the center O about the instantaneous point of contact Q can be written:

$$\delta x = a \delta \beta$$

$$\delta y = -a \delta \alpha$$

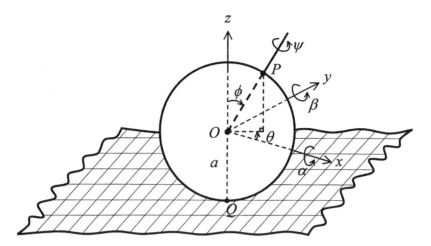

FIGURE 5.1 Rolling sphere on a plane (an example of a nonholonomic system).

in which the rotations of α and β are taken positive about the positive directions of x and y, respectively (Figure 5.1). Next, one can express $\delta\alpha$ and $\delta\beta$ in terms of the generalized coordinates. Note that $\delta\beta$ is directed along the z-direction and has no components along the x and y directions. On the other hand, $\delta\phi$ has the components $\delta\phi\cos\theta$ in the positive y-direction and $\delta\phi\sin\theta$ in the negative x-direction. Furthermore, the horizontal component of $\delta\psi$ is $\delta\psi\sin\phi$. This in turn has the components $(\delta\psi\sin\phi)\cos\theta$ and $(\delta\psi\sin\phi)\sin\theta$ in the positive x and y directions, respectively. It follows that

$$\delta\alpha = -\delta\phi\sin\theta + \delta\psi\sin\phi\cos\theta$$

$$\delta\beta = \delta\phi\cos\theta + \delta\psi\sin\phi\sin\theta$$

Consequently, the two nonholonomic constraint equations are

$$\delta x = a(\delta\phi\cos\theta + \delta\psi\sin\phi\sin\theta)$$

$$\delta y = a(\delta\phi\sin\theta - \delta\psi\sin\phi\cos\theta)$$

Note that these are differential equations that cannot be directly integrated to give algebraic equations. A particular choice for the three independent incremental coordinates associated with the three degrees of freedom in the present system of rolling sphere would be $\delta\theta$, $\delta\phi$, and $\delta\psi$. The incremental variables $\delta\alpha$, $\delta\beta$, and $\delta\theta$ will form another choice. The incremental variables δx, δy, and $\delta\theta$ will also form a possible choice. Once three incremented displacements are chosen in this manner, the remaining two incremental generalized coordinates are not independent and can be expressed in terms of these three incremented variables, using the constraint differential equations.

□

Example 5.2

A relatively simple example for a nonholonomic system is a single-dimensional rigid body (a straight line) moving on a plane such that its velocity is always along the body axis. Idealized motion of a ship in calm water is a practical situation representing such a system. This body needs three independent coordinates to completely define all possible configurations that it can take. For example, the centroid of the body can be fixed by two Cartesian coordinates x and y on the plane,

BOX 5.1 Some Definitions and Properties of Mechanical Systems

Holonomic constraints:	Constraints that can be represented by purely algebraic relations
Nonholonomic constraints:	Constraints that require differential relations for their representation
Holonomic system:	A system that possesses holonomic constraints only
Nonholonomic system:	A system that possesses one or more nonholonomic constraints
Number of degrees of freedom: (dof)	The number of independent incremental coordinates that are needed to represent a general incremental motion of a system; = Number of independent incremental motions
Order of a system	= 2 × number of dof (typically)
For a holonomic system: # independent incremental coordinates	= number of independent coordinates = number of dof
For a nonholonomic system: # independent incremental coordinates	< number of independent coordinates

and the orientation of the axis through the centroid can be fixed by a single angle θ. Note that for a given location (x, y) of the centroid, any arbitrary orientation (θ) for the body axis is feasible because, as in the previous example, any arbitrary trajectory can be followed by this body and return the centroid to the starting point, but with a different orientation of the axis of the body. Because the velocity is always directed along the body axis, a nonholonomic constraint exists and it is expressed as

$$\frac{dy}{dx} = \tan\theta$$

It follows that there are only two independent incremental variables; the system has only two degrees of freedom.

□

Some useful definitions and properties discussed in this section are summarized in Box 5.1.

5.2 SYSTEM REPRESENTATION

Some damped systems do not possess real modes. If a system does not possess real modes, modal analysis could still be used but the results would be only approximately valid. In modal analysis, it is convenient to first neglect damping and develop the fundamental results, and subsequently extend them to damped systems — for example, by assuming a suitable damping model that possesses real modes. Because damping is an energy dissipation phenomenon, it is usually possible to determine a model that possesses real modes and also has an energy dissipation capacity equivalent to that of the actual system.

Consider the three undamped system representations (models) shown in Figure 5.2. The motion of the system (a) consists of the *translatory* displacements y_1 and y_2 of the lumped masses m_1 and m_2. The masses are subjected to the external excitation forces (inputs) $f_1(t)$ and $f_2(t)$ and the restraining forces of the discrete, tensile-compressive stiffness (spring) elements k_1, k_2, and k_3. Only two independent incremental coordinates (δy_1 and δy_2) are required to completely define the

(a)

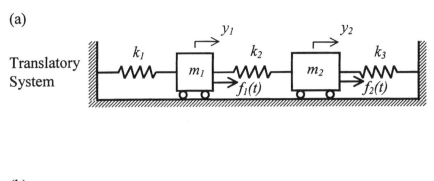

Translatory System

(b)

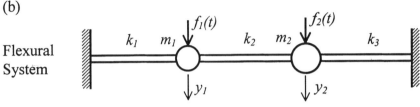

Flexural System

(c)

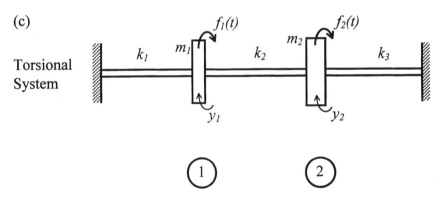

Torsional System

FIGURE 5.2 Three types of two-degree-of-freedom systems.

incremental motion of the system subject to its inherent constraints. It follows that the system has two degrees of freedom.

In system (b) shown in Figure 5.2, the elastic stiffness to the *transverse* displacements y_1 and y_2 of the lumped masses is provided by three bending (*flexural*) springs, which are considered massless. This flexural system is very much analogous to the translatory system (a) although the physical construction and the motion itself are quite different. The system (c) in Figure 5.2 is the analogous *torsional* system. In this case, the lumped elements m_1 and m_2 should be interpreted as polar moments of inertia about the shaft axis, and k_1, k_2, and k_3 as the torsional stiffness in the connecting shafts. Furthermore, the motion coordinates y_1 and y_2 are rotations, and the external excitations $f_1(t)$ and $f_2(t)$ are torques applied at the inertia elements. Practical examples where these three types of vibration system models may be useful are: (a) two-car train, (b) bridge with two separate vehicle loads, and (c) electric motor and pump combination.

The three systems shown in Figure 5.2 are analogous to each other in the sense that the dynamics of all three systems can be represented by similar equations of motion. For modal analysis, it is convenient to express the system equations as a set of coupled second-order differential equations in terms of the displacement variables (coordinates) of the inertia elements. Since in modal analysis one is concerned with *linear* systems, the system parameters can be given by a *mass matrix* and a *stiffness matrix* or a *flexibility matrix*. Lagrange's equations of motion directly yield these matrices. An intuitive method for identifying the stiffness and mass matrices is presented below.

The linear, lumped-parameter, undamped systems shown in Figure 5.2 satisfy the set of dynamic equations

$$\begin{bmatrix} m_{11} & m_{12} \\ m_{21} & m_{22} \end{bmatrix} \begin{bmatrix} \ddot{y}_1 \\ \ddot{y}_2 \end{bmatrix} + \begin{bmatrix} k_{11} & k_{12} \\ k_{21} & k_{22} \end{bmatrix} \begin{bmatrix} y_1 \\ y_2 \end{bmatrix} = \begin{bmatrix} f_1 \\ f_2 \end{bmatrix}$$

or

$$M\ddot{y} = Ky = f \qquad (5.1)$$

Here, M is the inertia matrix, which is the generalized case of mass matrix, and K is the stiffness matrix. There are many ways to derive equations (5.1). Below is described an approach, termed the *influence coefficient method*, that accomplishes the task by separately determining K and M.

5.2.1 STIFFNESS AND FLEXIBILITY MATRICES

In the systems shown in Figure 5.2, suppose the accelerations $\ddot{y}_1$ and $\ddot{y}_2$ both are zero at a particular instant, so that the inertia effects are absent. The stiffness matrix K is given under these circumstances, by the *constitutive relation* for the spring elements:

$$\begin{bmatrix} f_1 \\ f_2 \end{bmatrix} = \begin{bmatrix} k_{11} & k_{12} \\ k_{21} & k_{22} \end{bmatrix} \begin{bmatrix} y_1 \\ y_2 \end{bmatrix}$$

or

$$f = Ky \qquad (5.2)$$

in which f is the force vector $[f_1, f_2]^T$ and y is the displacement vector $[y_1, y_2]^T$. Both are column vectors. The elements of the stiffness matrix, in this two-degree-of-freedom (2 dof) case, are explicitly given by

$$K = \begin{bmatrix} k_{11} & k_{12} \\ k_{21} & k_{22} \end{bmatrix}$$

Suppose that $y_1 = 1$ and $y_2 = 0$ (i.e., give a unit displacement to m_1 while holding m_2 at its original position). Then, k_{11} and k_{21} are the forces needed at location 1 and location 2, respectively, to maintain this static configuration. For this condition, it is clear that $f_1 = k_1 + k_2$ and $f_2 = -k_2$. Accordingly,

$$k_{11} = k_1 + k_2$$

$$k_{21} = -k_2$$

Similarly, suppose that $y_1 = 0$ and $y_2 = 1$. Then, k_{12} and k_{22} are the forces needed at location 1 and location 2, respectively, to maintain the corresponding static configuration. It follows that

Modal Analysis

$$k_{12} = -k_2$$

$$k_{22} = k_2 + k_3$$

Consequently, the complete stiffness matrix can be expressed in terms of the stiffness elements in the system as

$$K = \begin{bmatrix} k_1 + k_2 & -k_2 \\ -k_2 & k_2 + k_3 \end{bmatrix}$$

From the foregoing development, it should be clear that the stiffness parameter k_{ij} represents what force is needed at the location i to obtain a unit displacement at location j. Hence, these parameters are called *stiffness influence coefficients*.

Observe that the stiffness matrix is symmetric. Specifically,

$$k_{ij} = k_{ji} \qquad \text{for } i \neq j$$

or

$$K^T = K \tag{5.3}$$

Note, however, that K is not diagonal in general ($k_{ij} \neq 0$ for at least two values of $i \neq j$). This means that the system is *statically coupled* (or *flexibly coupled*).

The flexibility matrix L is the inverse of the stiffness matrix;

$$L = K^{-1} \tag{5.4}$$

To determine the flexibility matrix using the influence coefficient approach, one must start with a constitutive relation of the form

$$y = Lf \tag{5.5}$$

assuming that there are no inertia forces at a particular instant, and then proceed as before. For the systems in Figure 5.2, for example, start with $f_1 = 1$ and $f_2 = 0$. In this manner, one can determine the elements l_{11} and l_{21} of the flexibility matrix

$$L = \begin{bmatrix} l_{11} & l_{12} \\ l_{21} & l_{22} \end{bmatrix}$$

But, here, the result is not as straightforward as in the previous case. For example, to determine l_{11}, one must find the flexibility contributions from either side of m_1. The flexibility of the stiffness element k_1 is $1/k_1$. The combined flexibility of k_2 and k_3, which are connected in series, is $1/k_2 + 1/k_3$ because the displacements (*across variables*) are additive in series. The two flexibilities on either side of m_1 are applied in *parallel* at m_1. Since the forces (*through variables*) are additive in parallel, the stiffness will also be additive. Consequently,

$$\frac{1}{l_{11}} = \frac{1}{(1/k_1)} + \frac{1}{(1/k_2 + 1/k_3)}$$

After some algebraic manipulation, one obtains

$$l_{11} = \frac{k_2 + k_3}{k_1 k_2 + k_2 k_3 + k_3 k_1}$$

Because there is no external force at m_2 in the assumed loading configuration, the deflections at m_2 and m_1 are proportioned according to the flexibility distribution along the path. Accordingly,

$$l_{21} = \left[\frac{1/k_3}{1/k_3 + 1/k_2}\right] l_{11}$$

or

$$l_{21} = \frac{k_2}{k_1 k_2 + k_2 k_3 + k_3 k_1}$$

Similarly, one can obtain

$$l_{12} = \frac{k_2}{k_1 k_2 + k_2 k_3 + k_3 k_1}$$

and

$$l_{22} = \frac{k_1 + k_2}{k_1 k_2 + k_2 k_3 + k_3 k_1}$$

Note that these results confirm the symmetry of flexibility matrices;

$$l_{ij} = l_{ji} \quad \text{for } i \neq j$$

or

$$\boldsymbol{L}^T = \boldsymbol{L} \tag{5.6}$$

Also, one can verify the fact that $\boldsymbol{L}$ is the inverse of $\boldsymbol{K}$. The series-parallel combination rules for stiffness and flexibility that are useful in the present approach are summarized in Table 5.1.

The flexibility parameters l_{ij} represent the displacement at the location i when a unit force is applied at location j. Hence, these parameters are called *flexibility influence coefficients*.

5.2.2 Inertia Matrix

Mass matrix, which is used in the case of translatory motions, can be generalized as *inertia matrix* $\boldsymbol{M}$ in order to include rotatory motions as well. To determine $\boldsymbol{M}$ for the systems shown in Figure 5.2, suppose the deflections y_1 and y_2 both are zero at a particular instant, so that the springs are in their static equilibrium configuration. Under these conditions, the equation of motion (5.1) becomes

$$\boldsymbol{f} = \boldsymbol{M}\ddot{\boldsymbol{y}} \tag{5.7}$$

TABLE 5.1
Combination Rules for Stiffness and Flexibility Elements

Connection	Graphical Representation	Combined Stiffness	Combined Flexibility
Series	k_1, l_1 k_2, l_2 (springs in series)	$\dfrac{1}{(1/k_1 + 1/k_2)}$	$l_1 + l_2$
Parallel	k_1, l_1 and k_2, l_2 (springs in parallel)	$k_1 + k_2$	$\dfrac{1}{(1/l_1 + 1/l_2)}$

For the present two-degree-of-freedom case, the elements of M are denoted by

$$M = \begin{bmatrix} m_{11} & m_{12} \\ m_{21} & m_{22} \end{bmatrix}$$

To identify these elements, first set $\ddot{y}_1 = 1$ and $\ddot{y}_2 = 0$. Then, m_{11} and m_{21} are the forces needed at the locations 1 and 2, respectively, to sustain the given accelerations; specifically, $f_1 = m_1$ and $f_2 = 0$. It follows that

$$m_{11} = m_1$$
$$m_{21} = 0$$

Similarly, by setting $\ddot{y}_1 = 0$ and $\ddot{y}_2 = 1$, one obtains

BOX 5.2 Influence Coefficient Method of Determining System Matrices (Undamped Case)

Stiffness Matrix (K):

1. Set $\ddot{y} = 0$

   $f = Ky$

2. Set $y_j = 1$ and $y_i = 0$ for all $i \neq j$
3. Determine f from the system diagram that is needed to main equilibrium = jth column of K.
   Repeat for all j.

Mass Matrix (M):

1. Set $y = 0$

   $f = M\ddot{y}$

2. Set $\ddot{y}_j = 1$ and $\ddot{y}_i = 0$ for all $i \neq j$
3. Determine f to maintain this condition = jth column of M.

   Repeat for all j.

$$m_{12} = 0$$

$$m_{22} = m_2$$

Then, the mass matrix is obtained as

$$M = \begin{bmatrix} m_1 & 0 \\ 0 & m_2 \end{bmatrix}$$

It should be clear now that the inertia parameter m_{ij} represents what force should be applied at the location i in order to produce a unit acceleration at location j. Consequently, these parameters are called *inertia influence coefficients*.

Note that the mass matrix is symmetric in general; specifically,

$$m_{ij} = m_{ji} \quad \text{for } i \neq j$$

or

$$M^T = M \tag{5.8}$$

Furthermore, when the independent displacements of the lumped inertia elements are chosen as the motion coordinates, as is typical, the inertia matrix becomes *diagonal*. If not, it can be made diagonal using straightforward algebraic substitutions, so that each equation contains the second derivative of just one displacement variable. Hence, one can assume that

$$m_{ij} = 0 \quad \text{for } i \neq j \tag{5.9}$$

Then the system is said to be *inertially* uncoupled. This approach to finding K and M is summarized in Box 5.2. It can be conveniently extended to damped systems for determining the damping matrix C.

5.2.3 Direct Approach for Equations of Motion

The influence coefficient approach that was described in the previous section is a rather indirect way of obtaining the equations of motion (5.1) for a multi-degree-of-freedom (multi-dof) system. The most straightforward approach, however, is to sketch a free-body diagram for the system, mark the forces or torques on each inertia element, and finally apply Newton's second law. This approach is now illustrated for the system shown in Figure 5.2(a). The equations of motion for the systems in Figures 5.2(b) and (c) will follow analogously.

The free-body diagram of the system in Figure 5.2(a) is sketched in Figure 5.3. Note that all the forces on each inertia element are marked. Application of Newton's second law to the two mass elements separately gives

$$m_1 \ddot{y}_1 = -k_1 y_1 + k_2 (y_2 - y_1) + f_1(t)$$

$$m_2 \ddot{y}_2 = -k_2 (y_2 - y_1) - k_3 y_2 + f_2(t)$$

The terms can be rearranged to obtain the following two coupled, second-order, linear, ordinary differential equations:

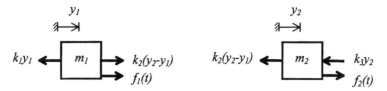

FIGURE 5.3 Free-body diagram of the two-degree-of-freedom system.

$$m_1\ddot{y}_1 + (k_1 + k_2)y_1 - k_2 y_2 = f_1(t)$$

$$m_2\ddot{y}_2 - k_2 y_1 + (k_2 + k_3)y_2 = f_2(t)$$

which can be expressed in the vector-matrix form as

$$\begin{bmatrix} m_1 & 0 \\ 0 & m_2 \end{bmatrix}\begin{bmatrix} \ddot{y}_1 \\ \ddot{y}_2 \end{bmatrix} + \begin{bmatrix} k_1+k_2 & -k_2 \\ -k_2 & k_2+k_3 \end{bmatrix}\begin{bmatrix} y_1 \\ y_2 \end{bmatrix} = \begin{bmatrix} f_1(t) \\ f_2(t) \end{bmatrix}$$

Observe that this result is identical to that obtained by the influence coefficient approach.

Another convenient approach that would provide essentially the same result is the *energy method* through the application of *Lagrange's equations*. This method, which is explained in Appendix B, should be studied carefully, and is left as an exercise for the student. Two common types of models used in vibration analysis and applications are summarized in Box 5.3.

5.3 MODAL VIBRATIONS

Among the infinite number of relative geometric configurations the lumped masses in a multi-degree-of-freedom system could assume under free motion (i.e., with $f(t) = 0$) when excited by an arbitrary initial state, there is a finite number of configurations that are naturally preferred by the system. Each such configuration will have an associated frequency of motion. These motions are termed *modal motions*. By choosing the initial displacement $y(0)$ proportional to a particular modal configuration, with zero initial velocity ($\dot{y}(0) = 0$), that particular mode can be excited at the associated natural frequency of motion. The displacements of different degrees of freedom retain this initial proportion at all times. This constant proportion in displacement can be expressed as a vector ψ for that mode, and represents the *mode shape*. Note that each modal motion is a harmonic motion executed at a specific frequency ω known as the *natural frequency* (undamped). In view of these general properties of modal motions, they can be expressed by

$$y = \psi\cos\omega t \tag{5.10}$$

or, in the complex form, for its ease of analysis, as

$$y = \psi e^{j\omega t} \tag{5.11}$$

When equation (5.11) is substituted into the equation of unforced (free) motion,

$$M\ddot{y} + Ky = 0 \tag{5.12}$$

as required by the definition of modal motion, the following *eigenvalue problem* results:

BOX 5.3 Model Types

Linear	**Nonlinear**

- Coupled second-order equations:

$$M\ddot{y} + C\dot{y} + Ky = f(t) \qquad\qquad M\ddot{y} = h(y, \dot{y}, f(t))$$

Response vector: $y = [y_1, y_2, ..., y_p]^T$; $p =$ # degrees of freedom (dof)

Excitation vector: $f(t) = [f_1, f_2, ..., f_p]^T$

M = mass matrix
C or B = damping matrix
K = stiffness matrix

- Coupled first-order equations:
  (state-space models)

$$\dot{x} = Ax + Bu \qquad\qquad\qquad \dot{x} = a(x, u)$$

$$y = Cx \qquad\qquad\qquad\qquad y = y(x)$$

State vector: $x = [x_1, x_2, ..., x_n]^T$; n = order of the system

Input (excitation) vector: $u = [u_1, u_2, ..., u_m]^T$

Output (response) vector: $y = [y_1, y_2, ..., y_p]^T$

Notes:
1. Definition of state: If $x(t_0)$, and u from t_0 to t_1, are known, $x(t_1)$ can be determined completely.
2. State vector x contains a minimum number (n) of elements.
3. State model is not unique (different state models are possible for the same system).
4. One approach to obtaining a state model is to use $x = \begin{bmatrix} y \\ \dot{y} \end{bmatrix}$

$$[\omega^2 M - K]\psi = 0 \qquad\qquad (5.13)$$

For this reason, natural frequencies are sometimes called *eigenfrequencies*, and mode shapes are termed *eigenvectors*. The feasibility of modal motions for a given system is determined by the existence of nontrivial solutions for ψ (i.e., $\psi \neq 0$). Specifically, nontrivial solutions for ψ are possible if and only if the determinant of the system of linear homogeneous equations (5.13) vanishes; thus,

$$\det[\omega^2 M - K] = 0 \qquad\qquad (5.14)$$

Equation (5.14) is known as the *characteristic equation* of the system. For an n-degree-of-freedom system, both M and K are $n \times n$ matrices. It follows that the characteristic equation has n roots for ω^2. For physically realizable systems, these n roots are all non-negative and they yield the n natural frequencies $\omega_1, \omega_2, \ldots, \omega_n$ of the system. For each natural frequency ω_i when substituted into equation (5.13) and solved for ψ, there results a mode shape vector ψ_i that is determined up to one unknown parameter which can be used as a scaling parameter. Extra care should be exercised, however, when determining mode shapes for zero natural frequencies (i.e., *rigid body modes*) and repeated natural frequencies (i.e., for systems with a *dynamic symmetry*). These considerations are discussed in later sections.

EXAMPLE 5.3

Consider a mechanical system modeled as in Figure 5.4. This is obtained from the systems in Figure 5.2 by setting $m_1 = m$, $m_2 = \alpha m$, $k_1 = k$, $k_2 = \beta k$, and $k_3 = 0$. The corresponding mass matrix and the stiffness matrix are

$$M = \begin{bmatrix} m & 0 \\ 0 & \alpha m \end{bmatrix} \quad K = \begin{bmatrix} (1+\beta)k & -\beta k \\ -\beta k & \beta k \end{bmatrix}$$

The natural frequencies are given by the roots of the characteristic equation

$$\det \begin{bmatrix} \omega^2 m - (1+\beta)k & \beta k \\ \beta k & \omega^2 \alpha m - \beta k \end{bmatrix} = 0$$

By expanding the determinant, this can be expressed as

$$\left[\omega^2 m - (1+\beta)k\right]\left[\omega^2 \alpha m - \beta k\right] - \beta^2 k^2 = 0$$

or

$$\omega^4 \alpha m^2 - \omega^2 km\left[\beta + \alpha(1+\beta)\right] + \beta k^2 = 0$$

One can define a frequency parameter $\omega_o = \sqrt{k/m}$. This parameter is identified as the natural frequency of an undamped simple oscillator (single-degree-of-freedom mass-spring system) with mass m and stiffness k. Consequently, the characteristic equation of the given two-degree-of-freedom system can be written as

$$\alpha \left(\frac{\omega}{\omega_o}\right)^4 - (\alpha + \beta + \alpha\beta)\left(\frac{\omega}{\omega_o}\right)^2 + \beta = 0$$

whose roots are

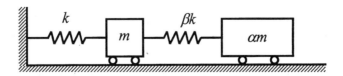

FIGURE 5.4 A modal vibration example.

$$\left(\frac{\omega_1}{\omega_o}\right)^2, \left(\frac{\omega_2}{\omega_o}\right)^2 = \frac{1}{2\alpha}\{\alpha+\beta+\alpha\beta\}\left\{1\pm\sqrt{1-\frac{4\alpha\beta}{(\alpha+\beta+\alpha\beta)^2}}\right\}$$

The mode shapes are obtained by solving for ψ in

$$\begin{bmatrix} \omega^2 m - (1+\beta)k & \beta k \\ \beta k & \omega^2 \alpha m - \beta k \end{bmatrix}\psi = 0$$

or

$$\begin{bmatrix} \left(\frac{\omega}{\omega_o}\right)^2 - (1+\beta) & \beta \\ \beta & \alpha\left(\frac{\omega}{\omega_o}\right)^2 - \beta \end{bmatrix}\psi = 0$$

In a mode shape vector, only the ratio of the elements is needed. This is because, in determining a mode shape, one is concerned about the relative motions of the lumped masses, not the absolute motions. From the above equation, it is clear that this ratio is given by

$$\frac{\psi_2}{\psi_1} = \frac{(1+\beta)-\left(\frac{\omega}{\omega_o}\right)^2}{\beta} = \frac{\beta}{\beta-\alpha\left(\frac{\omega}{\omega_o}\right)^2}$$

which is evaluated by substituting the appropriate value for (ω/ω_o), depending on the mode, into any one of the right-hand-side expressions above.

The dependence of the natural frequencies on the parameters α and β is illustrated by the curves in Figure 5.5. Some representative values of the natural frequencies and mode shape ratios are listed in Table 5.2.

Note that when $\beta = 0$, the spring connecting the two masses does not exist, and the system reduces to two separate systems: a simple oscillator of natural frequency ω_o and a single mass particle (of zero natural frequency). It is clear that in this case, $\omega_1/\omega_o = 0$ and $\omega_2/\omega_o = 1$. This fact can be established from the expressions for natural frequencies of the original system by setting $\beta = 0$. The mode corresponding to $\omega_1/\omega_o = 0$ is a *rigid body mode* in which the free mass moves indefinitely (zero frequency) and the other mass (restrained mass) stands still. It follows that the mode shape ratio $(\psi_2/\psi_1)_1 \to \infty$. In the second mode, the free mass stands still and the restrained mass moves. Hence, $(\psi_2/\psi_1)_1 = 0$. These results are also obtained from the general expressions for the mode shape ratios of the original system.

When $\beta \to \infty$, the spring connecting the two masses becomes rigid and the two masses act as a single mass $(1 + \alpha)m$ restrained by a spring of stiffness k. This simple oscillator has a squared natural frequency of $\omega_o^2/(1 + \alpha)$. This is considered the smaller natural frequency of the corresponding system: $(\omega_1/\omega_o)^2 = 1/(1 + \alpha)$. The larger natural frequency ω_2 approaches ∞ in this case and it corresponds to the natural frequency of a massless spring. These limiting results can be derived from the general expressions for natural frequencies of the original system by using the fact that for small

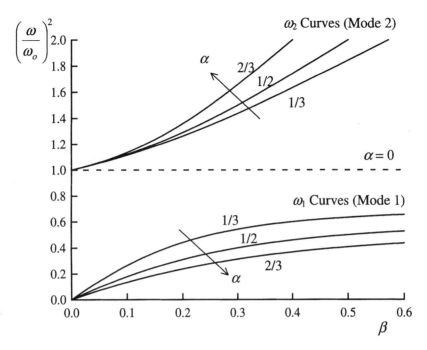

FIGURE 5.5 Dependence of natural frequencies (ω/ω_o) on mass ratio (α) and stiffness ratio (β).

TABLE 5.2
The Dependence of Natural Frequencies and Mode Shapes on Inertia and Stiffness

α	0.5				1.0				2.0			
β	ω_1/ω_o	ω_2/ω_o	$(\psi_2/\psi_1)_1$	$(\psi_2/\psi_1)_2$	ω_1/ω_o	ω_2/ω_o	$(\psi_2/\psi_1)_1$	$(\psi_2/\psi_1)_2$	ω_1/ω_o	ω_2/ω_o	$(\psi_2/\psi_1)_1$	$(\psi_2/\psi_1)_2$
0	0	1.0	∞	0	0	1.0	∞	0	0	1.0	∞	0
0.5	0.71	1.41	2.0	−1.0	0.54	1.31	2.41	−0.41	0.40	1.26	2.69	−0.19
1.0	0.77	1.85	1.41	−1.41	0.62	1.62	1.62	−0.62	0.47	1.51	1.78	−0.28
2.0	0.79	2.52	1.19	−1.69	0.66	2.14	1.28	−0.78	0.52	1.93	1.37	−0.37
5.0	0.81	3.92	1.07	−1.87	0.69	3.24	1.10	−0.91	0.55	2.86	1.14	−0.44
∞	0.82	∞	1.0	−2.0	0.71	∞	1.0	−1.0	0.57	∞	1.0	−0.5

$|x| \ll 1$, the expression $\sqrt{1-x}$ is approximately equal to $1 - \frac{1}{2}x$. (Proof: Use the Taylor series expansion.) In the first mode, the two masses move as one unit and hence the mode shape ratio $(\psi_2/\psi_1)_1 = 1$. In the second mode, the two masses move in opposite directions restrained by an infinitely stiff spring. This is considered the *static mode* that results from the redundant situation of associating two degrees of freedom to a system that actually has only one lumped mass $(1 + \alpha)m$. In this case, the mode shape ratio is obtained from the general result as follows. For large β, one can neglect α in comparison. Hence,

$$\left(\frac{\omega_2}{\omega_o}\right) = \frac{1}{2\alpha}\{\beta + \alpha\beta\}\{1+1\} = \{1+\alpha\}\frac{\beta}{\alpha}$$

By substituting this result in the expression for the mode shape ratio, one obtains

$$\left(\frac{\psi_2}{\psi_1}\right)_2 = \frac{\beta}{\beta - \alpha\left(\frac{\omega_2}{\omega_o}\right)^2} = \frac{\beta}{\beta - \alpha(1+\alpha)\frac{\beta}{\alpha}} = -\frac{1}{\alpha}$$

Finally, consider the case $\alpha = 0$ (with $\beta \neq 0$). In this case, only one mass m is present, which is restrained by a spring of stiffness k. The spring of stiffness βk has an open end. The first mode corresponds to a simple oscillator of natural frequency ω_o. Hence, $\omega_1/\omega_o = 1$. The open end has the same displacement as the point mass. Consequently, $(\psi_2/\psi_1)_1 = 1$. These results can be derived from the general expressions for the original system. In the second mode, the simple oscillator stands still and the open-ended spring oscillates (at an infinite frequency). Hence, $\omega_2/\omega_o = \infty$, and this again corresponds to a static mode situation that arises by assigning two degrees of freedom to a system that has only one degree of freedom associated with its inertia elements. Because the lumped mass stands still, one has $(\psi_2/\psi_1)_2 = \infty$.

Note that when $\alpha = 0$ and $\beta = 0$, the system reduces to a simple oscillator, and the second mode is completely undefined. Hence, the corresponding results cannot be derived from the general results for the original system.

□

5.4 ORTHOGONALITY OF NATURAL MODES

One can write equation (5.13) explicitly for the two distinct modes i and j. Distinct modes are defined as those having distinct natural frequencies (i.e., $\omega_i \neq \omega_j$). Therefore,

$$\omega_i^2 M\psi_i - K\psi_i = 0 \tag{5.15}$$

$$\omega_j^2 M\psi_j - K\psi_j = 0 \tag{5.16}$$

Premultiply equation (5.15) by ψ_j^T and equation (5.16) by ψ_i^T.

$$\omega_i^2 \psi_j^T M\psi_i - \psi_j^T K\psi_i = 0 \tag{5.17}$$

$$\omega_j^2 \psi_i^T M\psi_j - \psi_i^T K\psi_j = 0 \tag{5.18}$$

Take the transpose of equation (5.18), which is a scaler:

$$\omega_j^2 \psi_j^T M^T \psi_i - \psi_j^T K^T \psi_i = 0$$

This, in view of the symmetry of M and K (see equations (5.8) and (5.3)), becomes

$$\omega_j^2 \psi_j^T M\psi_i - \psi_j^T K\psi_i = 0$$

By subtracting this result from equation (5.17), one obtains

$$\left(\omega_i^2 - \omega_j^2\right)\psi_j^T M\psi_j = 0$$

Modal Analysis

Now, because $\omega_i \neq \omega_j$, it follows that

$$\psi_i^T M \psi_j = 0 \quad \text{for } i \neq j$$
$$\qquad\qquad = M_i \quad \text{for } i = j \tag{5.19}$$

Equation (5.19) is a useful *orthogonality condition* for natural modes.

Although the foregoing condition of *M-orthogonality* was proved for distinct (unequal) natural frequencies, it is generally true even if two or more modes have repeated (equal) natural frequencies. Indeed, if a particular natural frequency is repeated r times, there will be r arbitrary elements in the modal vector. As a result, one can determine r independent mode shapes that are orthogonal with respect to the M matrix. An example is given later in the problem of Figure 5.6. Note further that any such mode shape vector corresponding to a repeated natural frequency will also be M-orthogonal to the mode shape vector corresponding to any of the remaining distinct natural frequencies. Consequently, one concludes that the entire set of n mode shape vectors is M-orthogonal, even in the presence of various combinations of repeated natural frequencies.

5.4.1 MODAL MASS AND NORMALIZED MODAL VECTORS

Note that in equation (5.19), a parameter M_i has been defined to denote $\psi_i^T M \psi_i$. This parameter is termed *generalized mass* or *modal mass* for the ith mode. Because the modal vectors ψ_i are determined up to one unknown parameter, it is possible to set the value of M_i arbitrarily. The process of specifying the unknown scaling parameter in the modal vector, according to some convenient rule, is called the *normalization* of modal vectors. The resulting modal vectors are termed *normal modes*. A particularly useful method of normalization is to set each modal mass to unity ($M_i = 1$). The corresponding normal modes are said to be "normalized with respect to the mass matrix," or "M-normal." Note that if ψ_i is normal with respect to M, then it follows from equation (5.18) that $-\psi_i$ is also normal with respect to M. Specifically,

$$\left(-\psi_i\right)^T M \left(-\psi_i\right) = \psi_i^T M \psi_i = 1 \tag{5.20}$$

It follows that M-normal modal vectors are still arbitrary up to a multiplier of -1. A convenient practice for eliminating this arbitrariness is to make the first element of each normalized modal vector positive.

5.5 STATIC MODES AND RIGID BODY MODES

5.5.1 STATIC MODES

Modes corresponding to infinite natural frequencies are known as *static modes*. For these modes, the modal mass is zero; in the normalization process with respect to M, static modes cannot be included. If one assigns a degree of freedom for a massless location, the resulting mass matrix M becomes singular (det $M = 0$) and a static mode arises. Two similar situations were observed in a previous example: in one case, the stiffness of the spring connecting two masses is made infinite so that they act as a single mass in the limit; and in the other case, one of the two masses is made equal to zero so that only one mass is left. One should take extra precaution to avoid such situations by using proper modeling practices; the presence of a static mode amounts to assigning a degree of freedom to a system that it does not actually possess. In a static mode, the system behaves like a simple massless spring.

In the literature of experimental modal analysis, the static modes are represented by a *residual flexibility* term in the transfer functions. Note that in this case, modes of natural frequencies that are higher than the analysis bandwidth, or the maximum frequency of interest, are considered static modes. Such issues of experimental modal analysis will be discussed in Chapter 11.

5.5.2 Linear Independence of Modal Vectors

In the absence of static modes (i.e., modal masses $M_i \neq 0$), the inertia matrix M will be non-singular. Then the orthogonality condition (5.19) implies that the modal vectors are *linearly independent*, and consequently, they will form a *basis* for the n-dimensional space of all possible displacement trajectories y for the system. Any vector in this *configuration space* (or displacement space), therefore, can be expressed as a linear combination of the modal vectors.

Note the assumption made in the earlier development that the natural frequencies are distinct (or unequal). Nevertheless, linearly independent modal vectors are possessed by modes of equal natural frequencies as well. An example is the situation where these modes are physically uncoupled. These modal vectors are not unique, however; there will be arbitrary elements in the modal vector equal in number to the repeated natural frequencies. Any linear combination of these modal vectors can also serve as a modal vector at the same natural frequency. To explain this point further, without loss of generality, suppose that $\omega_1 = \omega_2$. Then, from equation (5.15), one obtains

$$\omega_1^2 M \psi_1 - K \psi_1 = 0$$
$$\omega_1^2 M \psi_2 - K \psi_2 = 0$$

Multiply the first equation by α, the second equation by β, and add the resulting equations to obtain

$$\omega_1^2 M (\alpha \psi_1 + \beta \psi_2) - K (\alpha \psi_1 + \beta \psi_2) = 0$$

This verifies that any linear combination $\alpha \psi_1 + \beta \psi_2$ of the two modal vectors ψ_1 and ψ_2 will also serve as a modal vector for the natural frequency ω_1. The physical significance of this phenomenon should be clear in Example 5.4.

5.5.3 Modal Stiffness and Normalized Modal Vectors

It is possible to establish an alternative version of the orthogonality condition given as equation (5.19) by substituting it into equation (5.18). This gives

$$\begin{aligned} \psi_i^T K \psi_j &= 0 \quad \text{for } i \neq j \\ &= K_i \quad \text{for } i = j \end{aligned} \tag{5.21}$$

This condition is termed K-*orthoganlity*.

Because the M-orthogonality condition (equation (5.19)) is true even for the case of repeated natural frequencies, it should be clear that the K-orthogonality condition (equation (5.21)) is also true in general, even with repeated natural frequencies. The newly defined parameter K_i represents the value of $\psi_i^T K \psi_i$ and is known as the *generalized stiffness* or *modal stiffness* corresponding to the ith mode.

Another useful way to *normalize* modal vectors is to choose their unknown parameters so that all modal stiffnesses are unity ($K_i = 1$ for all i). This process is known as normalization with respect

Modal Analysis

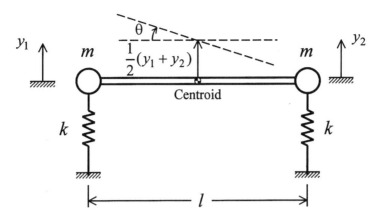

FIGURE 5.6 A simplified vehicle model for heave and pitch motions.

to the stiffness matrix. The resulting normal modes are said to be **K**-normal. These normal modes are still arbitrary up to a multiplier of −1. This can be eliminated by assigning positive values to the first element of all modal vectors.

Note that, in general, it is not possible to normalize a modal vector simultaneously with respect to both **M** and **K**. To understand this further, observe that $\omega_i^2 = K_i/M_i$ and, consequently, one is unable to pick both K_i and M_i arbitrarily. In particular, for the **M**-normal case, $K_i = \omega_i^2$; and for the **K**-normal case, $M_i = 1/\omega_i^2$.

5.5.4 Rigid Body Modes

Rigid body modes are those for which the natural frequency is zero. Modal stiffness is zero for rigid body modes; as a result, it is not possible to normalize these modes with respect to the stiffness matrix. Note that when rigid body modes are present, the stiffness matrix becomes *singular* (det $K = 0$). Physically, removing a spring and setting free an inertia element results in a rigid body mode. Example 5.3 provided a similar situation. In experimental modal analysis applications, low-stiffness connections or restraints that might be present in a test object could result in approximate rigid body modes that would become prominent at low frequencies.

Some important results of modal analysis discussed thus far are summarized in Table 5.3.

Example 5.4

Consider a light rod of length l having equal masses m attached to its ends. Each end is supported by a spring of stiffness k, as shown in Figure 5.6. Note that this system may represent a simplified model of a vehicle in heave and pitch motions. Gravity effects can be eliminated by measuring the displacements y_1 and y_2 of the two masses about their respective static equilibrium positions. Assume small front-to-back rotations θ in the pitch motion and small up-and-down displacements $\frac{1}{2}(y_1 + y_2)$ of the centroid in its heave motion.

Equation of Heave Motion

From Newton's second law for rigid body motion, one obtains

$$2m\frac{1}{2}(\ddot{y}_1 + \ddot{y}_2) = -ky_1 - ky_2$$

TABLE 5.3
Some Important Results of Modal Analysis

System	$M\ddot{y} + Ky = f(t)$
Symmetry	$M^T = M$ and $K^T = K$
Modal problem	$[\omega^2 M - K]\psi = 0$
Characteristic equation (gives natural frequencies)	$\det[\omega^2 M - K] = 0$
M-orthogonality	$\psi_i^T M \psi_j = 0$ for $i \neq j$ $ = M_i$ for $i = j$
K-orthogonality	$\psi_i^T K \psi_j = 0$ for $i \neq j$ $ = K_i$ for $i = j$
Modal mass (generalized mass)	M_i
Modal stiffness (generalized stiffness)	K_i
Natural frequency	$\omega_i = \sqrt{K_i/M_i}$
M-normal case	$M_i = 1$ $K_i = \omega_i^2$
K-normal case	$K_i = 1$ $M_i = 1/\omega_i^2$
Presence of rigid body modes	$\det K = 0$ $K_i = 0$ and $\omega_i = 0$
Presence of static modes	$\det M = 0$ $M_i = 0$ and $\omega_i \to \infty$

Equation of Pitch Motion

Note that for small angles of rotation, $\theta = (y_1 - y_2)/l$. The moment of inertia of the system about the centroid is $2 \times m(l/2)^2 = \frac{1}{2}ml^2$. Hence, by Newton's second law for rigid body rotation,

$$\frac{1}{2}ml^2\left(\frac{\ddot{y}_1 - \ddot{y}_2}{l}\right) = -\frac{l}{2}ky_1 + \frac{l}{2}ky_2$$

These two equations of motion can be written as

Modal Analysis

$$\ddot{y}_1 + \ddot{y}_2 + \omega_o^2(y_1 + y_2) = 0$$

$$\ddot{y}_1 - \ddot{y}_2 + \omega_o^2(y_1 - y_2) = 0$$

in which $\omega_o = \sqrt{k/m}$. By straightforward algebraic manipulation, a pair of completely uncoupled equations of motion are obtained; thus,

$$\ddot{y}_1 + \omega_o^2 y_1 = 0$$

$$\ddot{y}_2 - \omega_o^2 y_2 = 0$$

It follows that the resulting mass matrix and the stiffness matrix are both diagonal. In this case, there are an infinite number of choices for mode shapes, and any two linearly independent second-order vectors can serve as modal vectors for the system. Two particular choices are shown in Figure 5.7. Any of these mode shapes will correspond to the same natural frequency ω_o.

In each of these two choices, the mode shapes have been chosen so that they are orthogonal with respect to both M and K. This fact is verified below. Note that, in the present example,

$$M = \begin{bmatrix} 1 & 0 \\ 0 & 1 \end{bmatrix} \quad \text{and} \quad K = \begin{bmatrix} \omega_o^2 & 0 \\ 0 & \omega_o^2 \end{bmatrix}$$

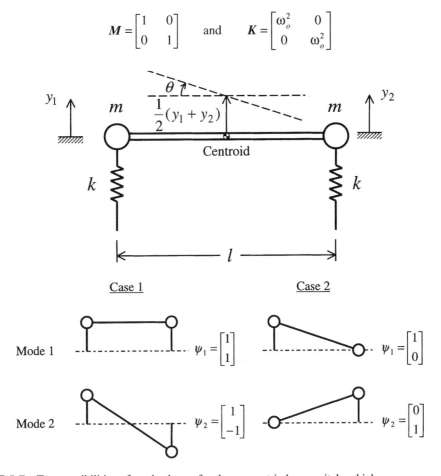

FIGURE 5.7 Two possibilities of mode shapes for the symmetric heave–pitch vehicle.

For Case 1:

$$[1 \quad 1]M = \begin{bmatrix} 1 \\ -1 \end{bmatrix} = 0 \quad \text{and} \quad [1 \quad 1]K = \begin{bmatrix} 1 \\ -1 \end{bmatrix} = 0$$

For Case 2:

$$[1 \quad 0]M = \begin{bmatrix} 0 \\ 1 \end{bmatrix} = 0 \quad \text{and} \quad [1 \quad 0]K = \begin{bmatrix} 0 \\ 1 \end{bmatrix} = 0$$

In general, because both elements of each eigenvector can be picked arbitrarily, one can write

$$\psi_1 = \begin{bmatrix} 1 \\ a \end{bmatrix} \quad \text{and} \quad \psi_2 = \begin{bmatrix} 1 \\ b \end{bmatrix}$$

where a and b are arbitrary, limited only by the orthogonality requirement for ψ_1 and ψ_2. The *M*-orthogonality requires that

$$[1 \quad a]\begin{bmatrix} 1 & 0 \\ 0 & 1 \end{bmatrix}\begin{bmatrix} 1 \\ b \end{bmatrix} = 0$$

and *K*-orthogonality requires that

$$[1 \quad a]\begin{bmatrix} \omega_o^2 & 0 \\ 0 & \omega_o^2 \end{bmatrix}\begin{bmatrix} 1 \\ b \end{bmatrix} = 0$$

Both conditions give $1 + ab = 0$, which corresponds to $ab = -1$. Note that Case 1 corresponds to $a = 1$ and $b = -1$, and Case 2 corresponds to $a = 0$ and $b \to \infty$. More generally, one can pick as modal vectors

$$\psi_1 = \begin{bmatrix} 1 \\ a \end{bmatrix} \quad \text{and} \quad \psi_2 = \begin{bmatrix} 1 \\ -1/a \end{bmatrix}$$

such that the two mode shapes are both *M*-orthogonal and *K*-orthogonal. In fact, if this particular system is excited by an arbitrary initial displacement, it will undergo free vibrations at frequency ω_o while maintaining the initial displacement ratio. Hence, if *M-orthogonality* and *K-orthogonality* are not required, any arbitrary second-order vector can serve as a modal vector to this system. □

EXAMPLE 5.5

An example for a system possessing a rigid body mode is shown in Figure 5.8. This system, a crude model of a two-car train, can be derived from the system shown in Figure 5.4 by removing the end spring (inertia restraint), and setting $\alpha = 1$ and $\beta = 1$. The equation for unforced motion of this system is

Modal Analysis

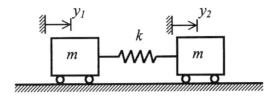

FIGURE 5.8 A simplified model of a two-car train.

$$\begin{bmatrix} m & 0 \\ 0 & m \end{bmatrix} \begin{bmatrix} \ddot{y}_1 \\ \ddot{y}_2 \end{bmatrix} + \begin{bmatrix} k & -k \\ -k & k \end{bmatrix} \begin{bmatrix} y_1 \\ y_2 \end{bmatrix} = \begin{bmatrix} 0 \\ 0 \end{bmatrix}$$

Note that, det $M = m^2 \neq 0$ and hence the system does not possess static modes. This should also be obvious from the fact that each degree of freedom (y_1 and y_2) has an associated, independent mass element. On the other hand, det $K = k^2 - k^2 = 0$, which signals the presence of rigid body modes.

The characteristic equation of the system is

$$\det \begin{bmatrix} \omega^2 m - k & k \\ k & \omega^2 m - k \end{bmatrix} = 0$$

or

$$\left(\omega^2 m - k\right)^2 - k^2 = 0$$

The two natural frequencies are given by the roots: $\omega_1 = 0$ and $\omega_2 = \sqrt{2k/m}$. Note that the zero natural frequency corresponds to the rigid body mode. The mode shapes can reveal further interesting facts.

First Mode (Rigid Body Mode)

In this case, $\omega_1 = 0$. Consequently, from equation (5.15), the mode shape is given by

$$\begin{bmatrix} -k & k \\ k & -k \end{bmatrix} \begin{bmatrix} \psi_1 \\ \psi_2 \end{bmatrix} = \begin{bmatrix} 0 \\ 0 \end{bmatrix}$$

which has the general solution $\psi_1 = \psi_2$, or

$$\begin{bmatrix} \psi_1 \\ \psi_2 \end{bmatrix}_1 = \begin{bmatrix} a \\ a \end{bmatrix}$$

The parameter a can be chosen arbitrarily. The corresponding modal mass is

$$M_1 = \begin{bmatrix} a & a \end{bmatrix} \begin{bmatrix} m & 0 \\ 0 & m \end{bmatrix} \begin{bmatrix} a \\ a \end{bmatrix} = 2ma^2$$

If the modal vector is normalized with respect to M, then $M_1 = 2ma^2 = 1$. Then, $a = \pm \dfrac{1}{\sqrt{2m}}$, and the corresponding normal mode vector would be

$$\begin{bmatrix} \psi_1 \\ \psi_2 \end{bmatrix}_1 = \begin{bmatrix} \dfrac{1}{\sqrt{2m}} \\ \dfrac{1}{\sqrt{2m}} \end{bmatrix} \quad \text{or} \quad \begin{bmatrix} -\dfrac{1}{\sqrt{2m}} \\ -\dfrac{1}{\sqrt{2m}} \end{bmatrix}$$

which is arbitrary up to a multiplier of -1. If the first element of the normal mode is restricted to be positive, the former vector (one with positive elements) should be used.

As previously noted, it is not possible to normalize a rigid body mode with respect to K. Specifically, the modal stiffness for the rigid body mode is

$$K_1 = \begin{bmatrix} a & a \end{bmatrix} \begin{bmatrix} k & -k \\ -k & k \end{bmatrix} \begin{bmatrix} a \\ a \end{bmatrix} = 0$$

for any choice for a, as expected.

Second Mode

For this mode, $\omega_2 = \sqrt{2k/m}$. By substituting into equation (5.15), one obtains

$$\begin{bmatrix} k & k \\ k & k \end{bmatrix} \begin{bmatrix} \psi_1 \\ \psi_2 \end{bmatrix}_2 = \begin{bmatrix} 0 \\ 0 \end{bmatrix}$$

the solution of which gives the corresponding modal vector (mode shape). The general solution is $\psi_2 = -\psi_1$, or

$$\begin{bmatrix} \psi_1 \\ \psi_2 \end{bmatrix}_2 = \begin{bmatrix} a \\ -a \end{bmatrix}$$

in which a is arbitrary. The corresponding modal mass is given by

$$M_2 = \begin{bmatrix} a & -a \end{bmatrix} \begin{bmatrix} m & 0 \\ 0 & m \end{bmatrix} \begin{bmatrix} a \\ -a \end{bmatrix} = 2ma^2$$

and the modal stiffness is given as

$$K_2 = \begin{bmatrix} a & -a \end{bmatrix} \begin{bmatrix} k & -k \\ -k & k \end{bmatrix} \begin{bmatrix} a \\ -a \end{bmatrix} = 4ka^2$$

Then, for M-normality, one must have $2ma^2 = 1$ or $a = \pm 1/\sqrt{2m}$.

It follows that the M-normal mode vector would be

Modal Analysis

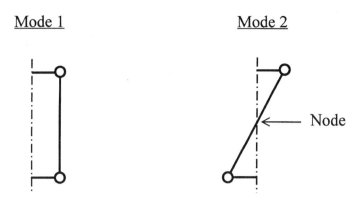

FIGURE 5.9 Mode shapes of the two-car train example.

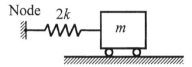

FIGURE 5.10 Equivalent system for mode 2 of the two-car train example.

$$\begin{bmatrix} \psi_1 \\ \psi_2 \end{bmatrix}_1 = \begin{bmatrix} \dfrac{1}{\sqrt{2m}} \\ -\dfrac{1}{\sqrt{2m}} \end{bmatrix} \quad \text{or} \quad \begin{bmatrix} -\dfrac{1}{\sqrt{2m}} \\ \dfrac{1}{\sqrt{2m}} \end{bmatrix}$$

The corresponding value of the modal stiffness is $K_2 = 2k/m$, which is equal to ω_2^2, as expected. Similarly, for K-normality, one must have $4ka^2 = 1$, or $a = \pm 1/\sqrt{4k}$. Hence, the K-normal modal vector would be

$$\begin{bmatrix} \psi_1 \\ \psi_2 \end{bmatrix}_2 = \begin{bmatrix} \dfrac{1}{\sqrt{4k}} \\ -\dfrac{1}{\sqrt{4k}} \end{bmatrix} \quad \text{or} \quad \begin{bmatrix} -\dfrac{1}{\sqrt{4k}} \\ \dfrac{1}{\sqrt{4k}} \end{bmatrix}$$

The corresponding value of the modal mass is $M_2 = m/(2k)$, which is equal to $1/\omega_2^2$, as expected.

The mode shapes of the system are shown in Figure 5.9. Note that in the rigid body mode, both masses move in the same direction through the same distance, with the connecting spring maintained in the unstretched configuration. In the second mode, the two masses move in opposite directions with equal amplitudes. This results in a *node point* halfway along the spring. A node is a point in the system that remains stationary under a modal motion. It follows that in the second mode, the system behaves like an identical pair of simple oscillators, each possessing twice the stiffness of the original spring (see Figure 5.10). The corresponding natural frequency is $\sqrt{2k/m}$, which is equal to ω_2.

Orthogonality of the two modes can be verified with respect to the mass matrix as

$$\begin{bmatrix} 1 & 1 \end{bmatrix} \begin{bmatrix} m & 0 \\ 0 & m \end{bmatrix} \begin{bmatrix} 1 \\ -1 \end{bmatrix} = 0$$

and, with respect to the stiffness matrix as

$$\begin{bmatrix} 1 & 1 \end{bmatrix} \begin{bmatrix} k & -k \\ -k & k \end{bmatrix} \begin{bmatrix} 1 \\ -1 \end{bmatrix} = 0$$

Because K is singular, due to the presence of the rigid body mode, the first orthogonality condition (equation (5.19)) — and not the second (equation (5.21)) — is the useful result for this system. In particular, because M is non-singular, the orthogonality of the modal vectors with respect to the mass matrix implies that they are linearly independent vectors by themselves. This is further verified by the non-singularity of the *modal matrix*; specifically,

$$\det[\psi_1, \psi_2] = \det \begin{bmatrix} 1 & 1 \\ 1 & -1 \end{bmatrix} \neq 0$$

Since M is a scalar multiple of the identity matrix, one sees that the modal vectors are in fact orthogonal, as is clear from

$$\psi_1^T \psi_2 = \begin{bmatrix} 1 & 1 \end{bmatrix} \begin{bmatrix} 1 \\ -1 \end{bmatrix} = 0$$

□

5.5.5 MODAL MATRIX

An n-degree-of-freedom system has n modal vectors $\psi_1, \psi_2, \ldots, \psi_n$, which are independent. The $n \times n$ square matrix Ψ, for which the columns are the modal vectors, is known as the *modal matrix*:

$$\Psi = [\psi_1, \psi_2, \ldots, \psi_n] \tag{5.22}$$

Because the mass matrix M can always be made non-singular through proper modeling practices (in choosing the degrees of freedom), it can be concluded that the modal matrix is non-singular:

$$\det \Psi \neq 0 \tag{5.23}$$

and the inverse Ψ^{-1} exists. Before showing this fact, note that the orthogonality conditions (5.19) and (5.21) can be written in terms of the modal matrix as

$$\Psi^T M \Psi = \text{diag}[M_1, M_2, \ldots, M_n] = \overline{M} \tag{5.24}$$

$$\Psi^T K \Psi = \text{diag}[K_1, K_2, \ldots, K_n] = \overline{K} \tag{5.25}$$

in which $\overline{M}$ and $\overline{K}$ are the *diagonal matrices* of modal masses and modal stiffnesses, respectively.

Next, the result from linear algebra that states that the determinant of the product of two square matrices is equal to the product of the determinants. Also, a square matrix and its transpose have the same determinent. Then, by taking the determinant of both sides of equation (5.24), it follows that

$$\det \Psi^T M \Psi = (\det \Psi)^2 \det M$$

$$= \det \overline{M} = M_1 M_2 \ldots M_n \tag{5.26}$$

Here, one has also used the fact that in equation (5.24), the RHS matrix is diagonal. Now, $M_i \neq 0$ for all i, since there are no static modes in a well-posed modal problem. It follows that

$$\det \Psi \neq 0 \qquad (5.27)$$

which implies that Ψ is nonsingular.

5.5.6 CONFIGURATION SPACE AND STATE SPACE

All solutions of the displacement response y span an Euclidean space known as the *configuration space*. This is an n-Euclidean space (L^n). This is also the *displacement space*.

The trace of the displacement vector y is not a complete representation of the dynamic response of a vibrating system because the same y can correspond to more than one *dynamic state* of the system. Hence, y is not a state vector; but $\begin{bmatrix} y \\ \dot{y} \end{bmatrix}_{2n}$ is a state vector because it includes both displacement and velocity, and completely represents the state of the system. This *state vector* spans the *state space* (L^{2n}), which is a $2n$-Euclidean space.

State Vector

This is a vector x consisting of a minimal set of response variables of a dynamic system such that, with the knowledge of the initial state $x(t_0)$ and the subsequent input $u[t_0, t_1]$ to the system over a finite time interval $[t_0, t_1]$, the end state $x(t_1)$ can be *uniquely* determined. Each point in a state space *uniquely* (and completely) determines the state of the dynamic system under these conditions.

Note: Configuration space can be thought of as a subspace of the state space that is obtained by projecting the state space into the subspace formed by the axes of the y vector.

For an n-degree-of-freedom vibrating system (see equation (5.1)), the displacement response vector y is of order n. If one knows the initial condition $y(0)$ and the forcing excitation $f(t)$, it is not possible to completely determine $y(t)$, in general. However, if one does know $y(0)$ and $\dot{y}(0)$ as well as $f(t)$, then it is possible to completely determine $y(t)$ and $\dot{y}(t)$. This says what was noted before: that y alone does not constitute a state vector, but y and $\dot{y}$ together do. In this case, the order of the state space is $2n$, which is twice the number of degrees of freedom.

5.6 OTHER MODAL FORMULATIONS

The modal problem (eigenvalue problem) studied in the previous sections consists of the solution of

$$\omega^2 M \psi = K \psi \qquad (5.28)$$

which is identical to equation (5.13). The natural frequencies (eigenvalues) are given by solving the characteristic equation (5.14). The corresponding mode shape vectors (eigenvectors) ψ_i are determined by substituting each natural frequency ω_i into equation (5.13) and solving for a nontrivial solution. This solution will have at least one arbitrary parameter. Hence, ψ represents the relative displacements at the various degrees of freedom of the vibrating system, and not the absolute displacements. Now, two other formulations are given for the modal problem.

The first alternative formulation given below involves solution of the eigenvalue problem of a nonsymmetric matrix ($M^{-1}K$). The other formulation given consists of first transforming the original problem into a new set of motion coordinates and then solving the eigenvalue problem of a symmetric matrix $\left(M^{-\frac{1}{2}} K M^{-\frac{1}{2}}\right)$, and then transforming the resulting modal vectors back to the original motion coordinates. Of course, all three formulations will give the same end result for the natural frequencies and mode shapes of the system — because the physical problem would remain the same regardless of what formulation and solution approach are employed. This fact will be illustrated using an example.

5.6.1 Non-Symmetric Modal Formulation

Consider the original modal formulation given by equation (5.28), which has been studied previously. Since the inertia matrix M is nonsingular, its inverse M^{-1} exists. The premultiplication of equation (5.28) by M^{-1} gives

$$\omega^2 \psi = M^{-1} K \psi \qquad (5.29)$$

This vector-matrix equation is of the form

$$\lambda \psi = S \psi \qquad (5.30)$$

where $\lambda = \omega^2$ and $S = M^{-1} K$. Equation (5.30) represents the standard matrix eigenvalue problem for matrix S. It follows that,

Squared natural frequencies = Eigenvalues of $M^{-1} K$

Mode shape vectors = Eigenvectors of $M^{-1} K$

5.6.2 Transformed Symmetric Modal Formulation

Now consider the free (unforced) system equations

$$M\ddot{y} + Ky = 0 \qquad (5.31)$$

whose modal problem needs to be solved. First, define the square-root of matrix M, as denoted by $M^{\frac{1}{2}}$ such that

$$M^{\frac{1}{2}} M^{\frac{1}{2}} = M \qquad (5.32)$$

Because M is symmetric, $M^{\frac{1}{2}}$ must also be symmetric. Next, define $M^{-\frac{1}{2}}$ as the inverse of $M^{\frac{1}{2}}$. Specifically,

$$M^{-\frac{1}{2}} M^{\frac{1}{2}} = M^{\frac{1}{2}} M^{-\frac{1}{2}} = I \qquad (5.33)$$

where I is the identity matrix. Note that $M^{-\frac{1}{2}}$ is also symmetric.

Once $M^{-\frac{1}{2}}$ is defined in this manner, transform the original problem (5.31) using the coordinate transformation

$$y = M^{-\frac{1}{2}} q \qquad (5.34)$$

Here, q denotes the transformed displacement vector, which is related to the actual displacement vector y through the matrix transformation using $M^{-\frac{1}{2}}$.

Modal Analysis

By differentiating equation (5.34) twice, one obtains

$$\ddot{y} = M^{-\frac{1}{2}}\ddot{q} \qquad (5.35)$$

Substitute equations (5.34) and (5.35) into (5.31). This gives

$$MM^{-\frac{1}{2}}\ddot{q} + KM^{-\frac{1}{2}}q = 0$$

Premultiply this result by $M^{-\frac{1}{2}}$ and use the fact that

$$M^{-\frac{1}{2}}MM^{-\frac{1}{2}} = M^{-\frac{1}{2}}M^{\frac{1}{2}}M^{\frac{1}{2}}M^{-\frac{1}{2}} = I$$

which follows from equations (5.32) and (5.33). One obtains

$$\ddot{q} + M^{-\frac{1}{2}}KM^{-\frac{1}{2}}q = 0 \qquad (5.36)$$

Equation (5.36) is the transformed problem, whose modal response can be given by

$$q = e^{j\omega t}\phi \qquad (5.37)$$

where ω represents a natural frequency and ϕ represents the corresponding modal vector, as usual. Then, in view of equation (5.34),

$$\begin{aligned}\psi &= e^{j\omega t}M^{-\frac{1}{2}}\phi \\ &= e^{j\omega t}\psi\end{aligned} \qquad (5.38)$$

It follows that the natural frequencies of the original problem (5.31) are identical to the natural frequencies of the transformed problem (5.36), and the modal vectors ψ of the original problem are related to the modal vectors ϕ of the transformed problem through

$$\psi = M^{-\frac{1}{2}}\phi \qquad (5.39)$$

Substitute the modal response (5.37) into (5.36) to obtain

$$\lambda\phi = P\phi \qquad (5.40)$$

where $\lambda = \omega^2$ and $P = M^{-\frac{1}{2}}KM^{-\frac{1}{2}}$

Equation (5.40), just like equation (5.30), represents a standard matrix eigenvalue problem. But now, matrix P is symmetric. As a result, its eigenvectors ϕ will not only be real but also orthogonal.

The solution steps for the present, transformed, and symmetric modal problem are:

1. Determine $M^{-\frac{1}{2}}$.
2. Solve for eigenvalues λ and eigenvectors ϕ of $M^{-\frac{1}{2}}KM^{-\frac{1}{2}}$. Eigenvalues are squares of the natural frequencies of the original system.
3. Determine the modal vectors ψ of the original system using $\psi = M^{-\frac{1}{2}}\phi$.

TABLE 5.4
Three Approaches of Modal Analysis

Approach	Standard	Non-symmetric Matrix Eigenvalue	Symmetric Matrix Eigenvalue
Modal formulation	$[\omega^2 M - K]\psi = 0$	$\omega^2 \psi = M^{-1} K \psi$	$\omega^2 \phi = M^{-\frac{1}{2}} K M^{-\frac{1}{2}} \phi$
Squared natural frequencies (ω_i^2)	Roots of $\det[\omega^2 M - K] = 0$	Eigenvalues of $M^{-1} K$	Eigenvalues of $M^{-\frac{1}{2}} K M^{-\frac{1}{2}}$
Mode-shape vectors (ψ_i)	Nontrivial Solutions of $[\omega_i^2 M - K]\psi = 0$	Eigenvectors of $M^{-1} K$	Determine eigenvectors ϕ_i of $M^{-\frac{1}{2}} K M^{-\frac{1}{2}}$ Then $\psi_i = M^{-\frac{1}{2}} \phi_i$

The three approaches of modal analysis that have been studied are summarized in Table 5.4.

EXAMPLE 5.6

Use the two-degree-of-freedom vibration problem given in Figure 5.4 (Example 5.3) to demonstrate the fact that all three approaches summarized in Table 5.4 will lead to the same results.

Consider the special case of $\alpha = 0.5$ and $\beta = 0.5$. Then one has:

$$M = \begin{bmatrix} m & 0 \\ 0 & \frac{m}{2} \end{bmatrix} \quad \text{and} \quad K = \begin{bmatrix} \frac{3}{2}k & -\frac{k}{2} \\ -\frac{k}{2} & \frac{k}{2} \end{bmatrix}$$

Using the standard approach, one obtains the modal results given in Table 5.3. Specifically, one obtains the natural frequencies (normalized with respect to $\omega_o = \sqrt{k/m}$)

$$\frac{\omega_1}{\omega_o} = \frac{1}{\sqrt{2}} \quad \text{and} \quad \frac{\omega_2}{\omega_o} = \sqrt{2}$$

and the mode shapes

$$\left(\frac{\psi_2}{\psi_1}\right)_1 = 2 \quad \text{and} \quad \left(\frac{\psi_2}{\psi_1}\right)_2 = -1$$

Now obtain these results using the other two approaches of modal analysis.

Approach 2

$$M^{-1} = \begin{bmatrix} \frac{1}{m} & 0 \\ 0 & \frac{2}{m} \end{bmatrix}$$

Modal Analysis

$$M^{-1}K = \begin{bmatrix} \dfrac{1}{m} & 0 \\ 0 & \dfrac{2}{m} \end{bmatrix} \begin{bmatrix} \dfrac{3}{2}k & -\dfrac{k}{2} \\ -\dfrac{k}{2} & \dfrac{k}{2} \end{bmatrix} = \begin{bmatrix} \dfrac{3}{2}\dfrac{k}{m} & -\dfrac{1}{2}\dfrac{k}{m} \\ -\dfrac{k}{m} & \dfrac{k}{m} \end{bmatrix}$$

$$= \omega_o^2 \begin{bmatrix} \dfrac{3}{2} & -\dfrac{1}{2} \\ -1 & 1 \end{bmatrix}$$

Note that this is not a symmetric matrix. Solve the eigenvalue problem of $\begin{bmatrix} \dfrac{3}{2} & -\dfrac{1}{2} \\ -1 & 1 \end{bmatrix}$. Eigenvalues λ are given by

$$\det \begin{bmatrix} \lambda - \dfrac{3}{2} & \dfrac{1}{2} \\ 1 & \lambda - 1 \end{bmatrix} = 0$$

or

$$\left(\lambda - \dfrac{3}{2}\right)(\lambda - 1) - \dfrac{1}{2} = 0$$

or

$$\lambda^2 - \dfrac{5}{2}\lambda + 1 = 0$$

the roots of which are

$$\lambda_1, \lambda_2 = \dfrac{1}{2},\ 2$$

It follows that $\dfrac{\omega_1}{\omega_o} = \dfrac{1}{\sqrt{2}}$ and $\dfrac{\omega_2}{\omega_o} = \sqrt{2}$, as before.

The eigenvector corresponding to λ_1 (Mode 1) is given by

$$\begin{bmatrix} \dfrac{1}{2} - \dfrac{3}{2} & \dfrac{1}{2} \\ 1 & \dfrac{1}{2} - 1 \end{bmatrix} \begin{bmatrix} \psi_1 \\ \psi_2 \end{bmatrix}_1 = \begin{bmatrix} 0 \\ 0 \end{bmatrix}$$

The solution is $\left(\dfrac{\psi_2}{\psi_1}\right)_1 = 2$, as before.

The eigenvector corresponding to λ_2 (Mode 2) is given by

$$\begin{bmatrix} 2 - \dfrac{3}{2} & \dfrac{1}{2} \\ 1 & 2 - 1 \end{bmatrix} \begin{bmatrix} \psi_1 \\ \psi_2 \end{bmatrix}_2 = \begin{bmatrix} 0 \\ 0 \end{bmatrix}$$

The solution is $\left(\dfrac{\psi_2}{\psi_1}\right)_2 = -1$, as before.

Approach 3

Since M is diagonal, it is easy to obtain $M^{\frac{1}{2}}$. Simply take the square root of the diagonal elements; thus,

$$M^{\frac{1}{2}} = \begin{bmatrix} \sqrt{m} & 0 \\ 0 & \sqrt{\dfrac{m}{2}} \end{bmatrix}$$

Its inverse is given by inverting the diagonal elements; thus,

$$M^{-\frac{1}{2}} = \begin{bmatrix} \dfrac{1}{\sqrt{m}} & 0 \\ 0 & \sqrt{\dfrac{2}{m}} \end{bmatrix}$$

Now,

$$M^{-\frac{1}{2}} K M^{-\frac{1}{2}} = \begin{bmatrix} \dfrac{1}{\sqrt{m}} & 0 \\ 0 & \sqrt{\dfrac{2}{m}} \end{bmatrix} \begin{bmatrix} \dfrac{3}{2}k & -\dfrac{k}{2} \\ -\dfrac{k}{2} & \dfrac{k}{2} \end{bmatrix} \begin{bmatrix} \dfrac{1}{\sqrt{m}} & 0 \\ 0 & \sqrt{\dfrac{2}{m}} \end{bmatrix}$$

$$= \begin{bmatrix} \dfrac{3}{2}\dfrac{k}{m} & -\dfrac{1}{\sqrt{2}}\dfrac{k}{m} \\ -\dfrac{1}{\sqrt{2}}\dfrac{k}{m} & \dfrac{k}{m} \end{bmatrix} = \omega_o^2 \begin{bmatrix} \dfrac{3}{2} & -\dfrac{1}{\sqrt{2}} \\ -\dfrac{1}{\sqrt{2}} & 1 \end{bmatrix}$$

Note that, as expected, this is a symmetric matrix. Solve for eigenvalues and eigenvectors of

$$\begin{bmatrix} \dfrac{3}{2} & -\dfrac{1}{\sqrt{2}} \\ -\dfrac{1}{\sqrt{2}} & 1 \end{bmatrix}$$

Eigenvalues are given by

$$\det \begin{bmatrix} \lambda - \dfrac{3}{2} & \dfrac{1}{\sqrt{2}} \\ \dfrac{1}{\sqrt{2}} & \lambda - 1 \end{bmatrix} = 0$$

or

$$\left(\lambda - \dfrac{3}{2}\right)(\lambda - 1) - \dfrac{1}{2} = 0$$

Modal Analysis

or

$$\lambda^2 - \frac{5}{2}\lambda + 1 = 0$$

which is identical to the characteristic equation obtained in the first two approaches. It follows that the same two natural frequencies are obtained by this method. The eigenvector ϕ_1 for Mode 1 is given by

$$\begin{bmatrix} \frac{1}{2} - \frac{3}{2} & \frac{1}{\sqrt{2}} \\ \frac{1}{\sqrt{2}} & \frac{1}{2} - 1 \end{bmatrix} \begin{bmatrix} \phi_1 \\ \phi_2 \end{bmatrix}_1 = \begin{bmatrix} 0 \\ 0 \end{bmatrix}$$

which gives $\begin{bmatrix} \phi_2 \\ \phi_1 \end{bmatrix}_1 = [\sqrt{2}]$

Accordingly, one can use

$$\begin{bmatrix} \phi_1 \\ \phi_2 \end{bmatrix}_1 = \begin{bmatrix} 1 \\ \sqrt{2} \end{bmatrix}$$

The eigenvector ϕ_2 for Mode 2 is given by

$$\begin{bmatrix} 2 - \frac{3}{2} & \frac{1}{\sqrt{2}} \\ \frac{1}{\sqrt{2}} & 2 - 1 \end{bmatrix} \begin{bmatrix} \phi_1 \\ \phi_2 \end{bmatrix}_2 = \begin{bmatrix} 0 \\ 0 \end{bmatrix}$$

which gives $\left(\dfrac{\phi_2}{\phi_1}\right)_2 = -\dfrac{1}{\sqrt{2}}$

Accordingly, use

$$\begin{bmatrix} \phi_1 \\ \phi_2 \end{bmatrix}_2 = \begin{bmatrix} 1 \\ -\frac{1}{\sqrt{2}} \end{bmatrix}$$

Now, transform these eigenvectors back to the original coordinate system using equation (5.39) to obtain

$$\begin{bmatrix} \psi_1 \\ \psi_2 \end{bmatrix}_1 = \begin{bmatrix} \frac{1}{\sqrt{m}} & 0 \\ 0 & \sqrt{\frac{2}{m}} \end{bmatrix} \begin{bmatrix} 1 \\ \sqrt{2} \end{bmatrix} = \begin{bmatrix} \frac{1}{\sqrt{m}} \\ \frac{2}{\sqrt{m}} \end{bmatrix}$$

which gives $\left(\dfrac{\psi_2}{\psi_1}\right)_1 = 2$, as before.

Also,

$$\begin{bmatrix} \psi_1 \\ \psi_2 \end{bmatrix}_2 = \begin{bmatrix} \frac{1}{\sqrt{m}} & 0 \\ 0 & \sqrt{\frac{2}{m}} \end{bmatrix} \begin{bmatrix} 1 \\ -\frac{1}{\sqrt{2}} \end{bmatrix} = \begin{bmatrix} \frac{1}{\sqrt{m}} \\ -\frac{1}{\sqrt{m}} \end{bmatrix}$$

which gives

$$\left(\frac{\psi_2}{\psi_1}\right)_2 = -1, \text{ as before.}$$

□

5.7 FORCED VIBRATION

The forced motion of a linear, n-degree-of-freedom undamped system is given by the nonhomogeneous equation of motion (5.1):

$$M\ddot{y} + Ky = f(t) \tag{5.41}$$

Although the following discussion is based on this undamped model, the results can be easily extended to the damped case.

It has been observed that the modal vectors form a basis for the *configuration space*. In other words, it is possible to express the response y as a linear combination of the modal vectors ψ_i:

$$y = q_1\psi_1 + q_2\psi_2 + \ldots + q_n\psi_n \tag{5.42}$$

The parameters q_i are a set of *generalized coordinates* and are functions of time t. Equation (5.42) is written in the vector-matrix form as

$$y = [\psi_1, \psi_2, \ldots, \psi_n] \begin{bmatrix} q_1 \\ q_2 \\ \vdots \\ q_n \end{bmatrix}$$

or

$$y = \Psi q \tag{5.43}$$

and can be viewed as a coordinate transformation from the trajectory space to the *canonical space* of generalized coordinates (principal coordinates or natural coordinates). Note that the inverse transformation exists because the modal matrix Ψ is non-singular. On substituting equation (5.43) into (5.41), one obtains

$$M\Psi\ddot{q} + K\Psi q = f(t)$$

This result is premultiplied by Ψ^T, and the orthogonality conditions (5.24) and (5.25) are substituted to obtain the canonical form of the system equation:

$$\overline{M}\ddot{q} + \overline{K}q = \bar{f}(t) \tag{5.44}$$

in which $\overline{M}$ and $\overline{K}$ are the diagonal matrices given by equations (5.24) and (5.25), and the transformed forcing vector is given by

$$\bar{f}(t) = \Psi^T f(t) \tag{5.45}$$

Since $\overline{M}$ and $\overline{K}$ are diagonal matrices, equation (5.44) corresponds to the set of n uncoupled *simple oscillator* equations:

$$M_i \ddot{q}_i + K_i q_i = \bar{f}_i(t) \quad \text{for } i = 1, 2, \ldots, n \tag{5.46}$$

In other words, the coordinate transformation (5.43) using the modal matrix has uncoupled the equations of forced motion. It follows from equation (5.46) that the natural frequencies of the system are given by

$$\omega_i^2 = K_i / M_i \quad \text{for } i = 1, 2, \ldots, n \tag{5.47}$$

as noted before.

It is particularly convenient to employ M-normal modal vectors. In this case, $\overline{M}$ becomes the nth-order identity matrix, and $\overline{K}$ the diagonal matrix having ω_i^2 as its diagonal elements; thus,

$$\overline{M} = I \tag{5.48}$$

$$\overline{K} = \mathrm{diag}[\omega_1^2, \omega_2^2, \ldots, \omega_n^2] \tag{5.49}$$

The corresponding uncoupled equations (5.46) take the form

$$\ddot{q}_i + \omega_i^2 q_i = \bar{f}_i(t) \quad \text{for } i = 1, 2, \ldots, n \tag{5.50}$$

in which, the forcing terms $\bar{f}_i(t)$ are given by equation (5.45), Ψ being M-normal.

Typically, the initial conditions for the original system are provided as the initial position $y(0)$ and the initial velocity $\dot{y}(0)$. The corresponding initial conditions for the transformed equations of motion are obtained using equation (5.43) as

$$q(0) = \Psi^{-1} y(0) \tag{5.51}$$

$$\dot{q}(0) = \Psi^{-1} \dot{y}(0) \tag{5.52}$$

The complete response of the original system can be conveniently obtained by first solving the simple oscillator equations (5.50) and then transforming the results back into the trajectory space using equation (5.43).

The complete solution to this linear system can be viewed as the sum of the initial condition response in the absence of the forcing function, and the forced response with zero initial conditions, as discussed in Chapter 2. For the simple oscillator equations (5.50), the initial condition response q_{iI} is given by

$$q_{iI} = q_i(0)\cos\omega_i t + \frac{\dot{q}_i(0)}{\omega_i}\sin\omega_i t \tag{5.53}$$

The impulse response function (i.e., response to a unit impulse excitation) for the undamped oscillator equation is

$$h_i(t) = \frac{1}{\omega_i}\sin\omega_i t \quad \text{for } t \geq 0 \tag{5.54}$$

The forced response q_{iF}, with zero initial conditions, is obtained using the *convolution integral* method (see Chapter 2); specifically,

$$q_{iF} = \int_0^t \bar{f}_i(\tau) h_i(t-\tau) d\tau \tag{5.55}$$

The complete solution in the canonical domain is

$$q_i = q_{iI} + q_{iF}$$

$$= q_i(0)\cos\omega_i t + \frac{\dot{q}_i(0)}{\omega_i}\sin\omega_i t + \frac{1}{\omega_i}\int_0^t \bar{f}_i(\tau)\sin\omega_i(t-\tau)d\tau \tag{5.56}$$

$$i = 1, 2, \ldots, n$$

Each of the n responses is a modal response which is the contribution from that mode to the actual response y. This approach is summarized in Box 5.4. Next, this approach of solving for the forced vibration is illustrated by means of an example.

EXAMPLE 5.7

Consider again the system shown in Figure 5.8. A step input force $f(t)$ given by

$$f(t) = f_o \quad \text{for } t \geq 0$$
$$= 0 \quad \text{for } t < 0$$

is applied to the left-hand mass (degree of freedom y_1). Assume that the system starts from rest $(y(0) = \mathbf{0}$ and $\dot{y}(0) = \mathbf{0})$.

As before, the M-normal modal matrix of the system is

$$\Psi = \begin{bmatrix} \dfrac{1}{\sqrt{2m}} & \dfrac{1}{\sqrt{2m}} \\ \dfrac{1}{\sqrt{2m}} & -\dfrac{1}{\sqrt{2m}} \end{bmatrix} = \frac{1}{\sqrt{2m}}\begin{bmatrix} 1 & 1 \\ 1 & -1 \end{bmatrix}$$

No forcing input is applied to the second degree of freedom (y_2). Hence, the overall forcing input vector is

Modal Analysis

BOX 5.4 Modal Approach to Forced Response

Forced system: $M\ddot{y} + Ky = f(t)$
Number of degrees of freedom $= n$

Modal transformation: $y = \Psi q$

where, modal matrix $\Psi = [\psi_1, \psi_2, ..., \psi_n]$ = matrix of mode shape vectors.
We get the diagonalized system:

$$\overline{M}\ddot{q} + \overline{K}q = \bar{f}(t)$$

or

$$M_i \ddot{q}_i + K_i q_i = \bar{f}_i(t);\ i = 1, 2, ..., n$$

where

$$\overline{M} = \Psi^T M \Psi = \text{diag}[M_1, M_2, ..., M_n]$$

$$\overline{K} = \Psi^T K \Psi = \text{diag}[K_1, K_2, ..., K_n]$$

$$\bar{f}(t) = \Psi^T f(t) = [\bar{f}_1, \bar{f}_2, ..., \bar{f}_n]^T$$

Initial conditions: $q(0) = \Psi^{-1} y(0)$

$$\dot{q}(0) = \Psi^{-1} \dot{y}(0)$$

Steps:
Use the *M*-normal case: $M_i = 1$, $K_i = \omega_i^2$
Then,

$$\ddot{q}_i + \omega_i^2 q_i = \bar{f}_i(t);\ i = 1, 2, ..., n$$

1. Free, initial-condition response (zero-input response):

$$q_{iI} = q_i(0)\cos\omega_i t + \frac{\dot{q}_i(0)}{\omega_i}\sin\omega_i t$$

2. Forced, zero IC response:

$$q_{iF} = \frac{1}{\omega_i}\int_0^t \bar{f}_i(\tau)\sin\omega_i(t - \tau)d\tau$$

3. $q_i = q_{iI} + q_{iF}$
4. Transform back to y using $y = \Psi q$

$$f(t) = \begin{bmatrix} f_o \\ 0 \end{bmatrix} \quad \text{for } t \geq 0$$

From equation (5.45), the transformed forcing input vector is obtained as

$$\bar{f}(t) = \Psi^T f(t) = \frac{1}{\sqrt{2m}} \begin{bmatrix} 1 & 1 \\ 1 & -1 \end{bmatrix} \begin{bmatrix} f_o \\ 0 \end{bmatrix}$$

$$= \frac{f_o}{\sqrt{2m}} \begin{bmatrix} 1 \\ 1 \end{bmatrix} \quad \text{for } t \geq 0$$

From equations (5.51) and (5.52), the initial conditions for the modal (canonical) variables are obtained as, $\dot{q}(0) = \mathbf{0}$ and $q(0) = \mathbf{0}$. The modal responses $q_1(t)$ and $q_2(t)$ are obtained using equation (5.56).

First Mode (Rigid Body Mode)

Note that: $\lim\limits_{\omega_i \to 0} \dfrac{\sin \omega_i (t-\tau)}{\omega_i} = t - \tau$

It follows that

$$q_1(t) = \int_0^t \frac{f_o(t-\tau)}{\sqrt{2m}} d\tau = \frac{f_o t^2}{2\sqrt{2m}}$$

Second Mode (Oscillatory Mode)

$$q_2(t) = \frac{1}{\omega_2} \int_0^t \frac{f_o}{\sqrt{2m}} \sin \omega_2 (t-\tau) d\tau$$

$$= \frac{f_o}{\omega_2^2 \sqrt{2m}} (1 - \cos \omega_2 t)$$

The overall response in the physical trajectory space is obtained by transforming the modal responses using equation (5.43); thus,

$$y = \frac{1}{\sqrt{2m}} \begin{bmatrix} 1 & 1 \\ 1 & -1 \end{bmatrix} \begin{bmatrix} \dfrac{f_o t^2}{2\sqrt{2m}} \\ \dfrac{f_o (1 - \cos \omega_2 t)}{\omega_2^2 \sqrt{2m}} \end{bmatrix}$$

$$= \frac{f_o}{2m} \begin{bmatrix} \dfrac{t^2}{2} + \dfrac{1}{\omega_2^2} (1 - \cos \omega_2 t) \\ \dfrac{t^2}{2} - \dfrac{1}{\omega_2^2} (1 - \cos \omega_2 t) \end{bmatrix}$$

with $\omega_2^2 = 2k/m$. The response of both masses grows (unstable) quadratically in an oscillatory manner, as shown in Figure 5.11.

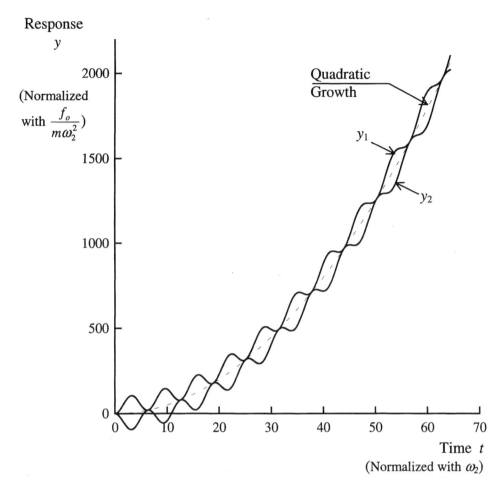

FIGURE 5.11 Forced response obtained through modal analysis.

5.8 DAMPED SYSTEMS

Now one can examine the possibility of extending the modal analysis results to damped systems. Damping is an energy dissipation phenomenon. In lumped-parameter models of vibrating systems, damping force can be represented by a resisting force at each lumped mass. In view of equation (5.1), then, the system equations for a damped system can be written as

$$M\ddot{y} + Ky = f(t) - d \tag{5.57}$$

in which d is the damping force vector. Modeling of damping is usually quite complicated. Often, a linear model, whose energy dissipation capacity is equivalent to that of the actual system, is employed. Such a model is termed an *equivalent damping model*. The most popular model for damping is the *linear viscous model*, in which the damping force is proportional to the relative velocity. In lumped-parameter dynamic models (linear), viscous damping elements can be assigned across pairs of degrees of freedom or across a degree of freedom and a fixed reference. The damping force for such a model can be expressed as

$$d = C\dot{y} \tag{5.58}$$

in which C is the *damping matrix*. The resulting damped system equation is

$$M\ddot{y} + C\dot{y} + Ky = f(t) \qquad (5.59)$$

To determine the elements of C for a given system model, by the influence coefficient approach, the same procedure outlined previously for obtaining the elements of K can be used, except that the velocities $\dot{y}_i$ should be used in place of the displacements y_i. The coefficients c_{ij} are called the *damping influence coefficients*.

5.8.1 Proportional Damping

Because the modal vectors are orthogonal with respect to both M and K, the transformation (5.43) will decouple the undamped system equation (5.41). It should be clear, however, that the same transformation will not diagonalize the C matrix in general. As observed in the previous section, decoupling, or *modal decomposition*, is a convenient tool in response analysis. This is because each uncoupled modal equation is a simple oscillator equation with a well-known solution that can be transformed back (or recombined) to obtain the total response. This simple procedure cannot be used for damped systems unless the modal vectors are orthogonal with respect to C as well.

In modal vibration, all degrees of freedom move in the same displacement proportion, as given by the modal vector. This type of "synchronous" motion may not be possible in damped systems. Another way to state this is that most damped systems do not possess real modes. If one tries to excite such a damped system at one of its natural frequencies, one will notice that the constant proportion given by the modal vector (for the undamped system) is violated during motion. Furthermore, if the undamped system has a (fixed) node point in the particular mode, that point would not remain fixed but would move in a cyclic manner when set to vibrate at the natural frequency of the mode.

Note that viscous damping is just a "model" for energy dissipation. If one becomes rather restrictive in choosing the parameters of viscous damping, one will be able to develop an equivalent damping matrix with respect to which, the modal vectors of the undamped system would be orthogonal. In other words, it is required that the transformed damping matrix $\overline{C}$ be a diagonal matrix; thus,

$$\Psi^T C \Psi = \text{diag}[C_1, C_2, \ldots, C_n] = \overline{C} \qquad (5.60)$$

In this case, the corresponding viscous damping model is termed *proportional damping* or *Rayleigh damping* (after the person who first identified this simplification).

Modal decomposition of equation (5.59), assuming proportional damping, using the transformation (5.43) results in the canonical form (uncoupled modal equation)

$$\overline{M}\ddot{q} + \overline{C}\dot{q} + \overline{K}q = \bar{f}(t) \qquad (5.61)$$

or

$$M_i \ddot{q}_i + C_i \dot{q}_i + K_i q_i = \bar{f}_i(t) \qquad \text{for } i = 1, 2, \ldots, n$$

Equation (5.61) can be written in the standard form for a damped simple oscillator:

$$\ddot{q}_i + 2\zeta_i \omega_i \dot{q}_i + \omega_i^2 q_i = \bar{f}_i(t) \qquad \text{for } i = 1, 2, \ldots, n \qquad (5.62)$$

in which the modal matrix is assumed M-normal (i.e., $M_i = 1$). It can be concluded that a proportionally damped system possesses real modal vectors that are identical to the modal vectors of its undamped counterpart. The damped natural frequency, however, is smaller than the undamped natural frequency and is given by

$$\omega_{di} = \sqrt{1-\zeta_i^2}\,\omega_i \tag{5.63}$$

One way to guarantee proportional damping is to pick a damping matrix that satisfies

$$C = c_m M + c_k K \tag{5.64}$$

This, however, is not the only way to achieve real modes in a damped system (equation (5.60)). The first term on the right-hand side (RHS) of equation (5.64) is termed the *inertial damping matrix*. The corresponding damping force on each lumped mass in the model will be proportional to the momentum. This term may represent the energy loss associated with a momentum and is termed *momentum damping*. Physically, this is incorporated by assigning a viscous damper between each degree of freedom and its fixed reference, with the damping constant proportional to the mass concentrated at that location.

The second term in equation (5.64) is termed the *stiffness damping matrix*. The corresponding damping force is proportional to the rate of change of the local deformation forces (stresses) in flexible structural members and joints. It may be interpreted as a simplified model for *structural damping*. Physically, this model is realized by assigning a viscous damper across every spring element in the model, with the damping constant proportional to the stiffness. It is known that rate of change of stresses or rate of change of strains will give rise to *viscoelastic damping*, which is associated with plasticity and viscoelasticity. This type of damping is known as *strain-rate damping*.

Usually, structural damping is most appropriately modeled as being present across (lumped) stiffness elements. Coulomb damping is modeled as acting between an inertia element and its fixed reference point. These intuitive observations also support the damping model given by equation (5.64). Some terminology and properties of damped systems are summarized in Box 5.5.

EXAMPLE 5.8

Now examine the lumped-parameter damped model shown in Figure 5.12 to determine whether it has real modes.

Note that if one modifies the model as in Figure 5.13, the damping matrix will become proportional to the stiffness matrix. This will be a case of proportional damping and the modified (damping) system will possess real modes with modal vectors that are identical to those for the corresponding undamped system. It should be verified that the equations of motion for this modified system (Figure 5.13) are

$$\begin{bmatrix} m & 0 \\ 0 & m \end{bmatrix}\ddot{y} + \begin{bmatrix} 2c & -c \\ -c & 2c \end{bmatrix}\dot{y} + \begin{bmatrix} 2k & -k \\ -k & 2k \end{bmatrix}y = \begin{bmatrix} 0 \\ f(t) \end{bmatrix}$$

Notice the similarity between the stiffness matrix and the damping matrix.

Another damped system that possesses classical modes is shown in Figure 5.14. In this model, the damping matrix is proportional to the mass (inertia) matrix. It can be easily verified that the system equations are

$$\begin{bmatrix} m & 0 \\ 0 & m \end{bmatrix}\ddot{y} + \begin{bmatrix} c & 0 \\ 0 & c \end{bmatrix}\dot{y} + \begin{bmatrix} 2k & -k \\ -k & 2k \end{bmatrix}y = \begin{bmatrix} 0 \\ f(t) \end{bmatrix}$$

Here, notice the similarity between the inertia matrix and the damping matrix.

BOX 5.5 Terminology and Properties of Damped Systems

Characteristic equation:

$$\det[Ms^2 + Cs + K] = 0$$

Roots s are the eigenvalues λ_i

$\text{Im}(\lambda_i) = \omega_{di}$ = damped natural frequencies

$|\lambda_i| = \omega_i$ = undamped natural frequencies

$-\text{Re}(\lambda_i)/\omega_i = \zeta_i$ = modal damping ratios

$\sqrt{1 - 2\zeta_i^2}\,\omega_i = \omega_{ri}$ = resonant frequencies

$\sqrt{1 - \zeta_i^2}\,\omega_i = \omega_{di}$

Existence of real modes:
- Condition for existence of modes that are identical to undamped modes:

$\Psi^T C \Psi$ = diagonal matrix

where

Ψ = modal matrix of undamped modes

- Another (equivalent) condition:

$M^{-1}CM^{-1}K = M^{-1}KM^{-1}C$ (Prove)

i.e., $M^{-1}C$ and $M^{-1}K$ must commute
- Special case:
Proportional damping $C = c_m M + c_k K$

Returning to the original system shown in Figure 5.12, the equations of motion can be written as

$$\begin{bmatrix} m & 0 \\ 0 & m \end{bmatrix} \ddot{y} + \begin{bmatrix} c_1 + c_2 & -c_2 \\ -c_2 & c_2 \end{bmatrix} \dot{y} + \begin{bmatrix} 2k & -k \\ -k & 2k \end{bmatrix} y = \begin{bmatrix} 0 \\ f(t) \end{bmatrix}$$

From these equations, it is not obvious whether this system possesses real modes. The undamped natural frequencies are given by the roots of the characteristic equation

$$\det \begin{bmatrix} \omega^2 m - 2k & k \\ k & \omega^2 m - 2k \end{bmatrix} = 0$$

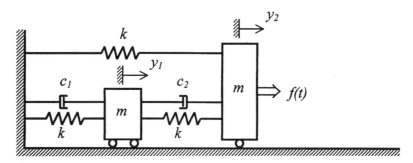

FIGURE 5.12 A system with linear viscous damping.

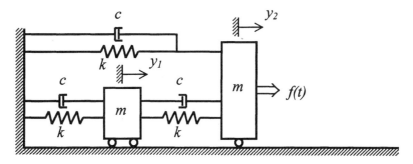

FIGURE 5.13 A system with proportional damping in proportion to stiffness.

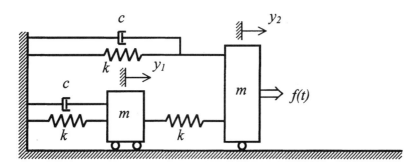

FIGURE 5.14 A system with proportional damping in proportion to inertia.

The natural frequencies are $\omega_1 = \sqrt{k/m}$ and $\omega_2 = \sqrt{3k/m}$. The corresponding mode shapes are given by the nontrivial solution of

$$\begin{bmatrix} \omega_i^2 m - 2k & k \\ k & \omega_i^2 m - 2k \end{bmatrix} \begin{bmatrix} \psi_1 \\ \psi_2 \end{bmatrix}_i = \begin{bmatrix} 0 \\ 0 \end{bmatrix} \quad \text{for } i = 1, 2$$

Also, one can normalize the modal vectors by choosing the first element of each vector to be unity. Then, by following the usual procedure of modal analysis, one obtains the normalized modal vectors

$$\psi_1 = \begin{bmatrix} 1 \\ 1 \end{bmatrix} \quad \text{and} \quad \psi_2 = \begin{bmatrix} 1 \\ -1 \end{bmatrix}$$

The modal matrix is

$$\Psi = [\psi_1, \psi_2] = \begin{bmatrix} 1 & 1 \\ 1 & -1 \end{bmatrix}$$

Consequently, one obtains

$$\Psi^T M \Psi = \overline{M} = \begin{bmatrix} 2m & 0 \\ 0 & 2m \end{bmatrix}$$

$$\Psi^T K \Psi = \overline{K} = \begin{bmatrix} 2k & 0 \\ 0 & 6k \end{bmatrix}$$

$$\Psi^T C \Psi = \overline{C} = \begin{bmatrix} c_1 & c_1 \\ c_1 & c_1 + 4c_2 \end{bmatrix}$$

and

$$\Psi^T f(t) = \bar{f}(t) = \begin{bmatrix} f(t) \\ -f(t) \end{bmatrix}$$

Notice that the transformed damping matrix $\overline{C}$ is not diagonal in general; consequently, real modes will not exist. This is to be expected in the absence of the condition of proportional damping. Proportional damping is realized in this model if $c_1 = 0$. Then, $\overline{C}$ will be diagonal and the transformed systems equations will be uncoupled. The uncoupled modal equations are

$$2m\ddot{q}_1 + 2kq_1 = f(t)$$

$$2m\ddot{q}_2 + 4c_2\dot{q}_2 + 6kq_2 = -f(t)$$

The first mode is always undamped for this choice of damping model. This confirms that, in the case of proportional damping, it is not generally possible to pick an arbitrary structure for the damping matrix. For this reason, proportional damping is sometimes an analytical convenience rather than a strict physical reality.

□

Modal analysis and response analysis of a system with general viscous damping can be accomplished using *state-space* concepts as well. In this case, modal analysis is carried out in terms of eigenvalues and eigenvectors of the *system matrix* of a suitable *state variable* model. These "eigen results," if complex, will occur in complex conjugate pairs, and then the modes are said to be *complex*. This approach is outlined next.

5.9 STATE-SPACE APPROACH

The state-space approach to modal analysis can be used for any linear dynamic system. The starting point is to formulate a state-space model of the system, which is a set of coupled first-order differential equations:

Modal Analysis

$$\dot{x} = Ax + Bu \tag{5.65}$$

where

x = state vector
u = input vector
A = system matrix
B = input gain matrix.

There are many approaches to formulating a vibration problem as a state-space model (5.65). One simple method is to first obtain the conventional, coupled, second-order differential equations:

$$M\ddot{y} + C\dot{y} + Ky = f(t) \tag{5.59}$$

Next, the state vector and the input vector are defined as

$$x = \begin{bmatrix} y \\ \dot{y} \end{bmatrix} \quad \text{and} \quad u = f(t) \tag{5.66}$$

Note that equation (5.59) can be written as

$$\ddot{y} = -M^{-1}Ky - M^{-1}C\dot{y} + M^{-1}f(t) \tag{5.67}$$

which is identical to

$$\ddot{y} = \begin{bmatrix} -M^{-1}K & -M^{-1}C \end{bmatrix} \begin{bmatrix} y \\ \dot{y} \end{bmatrix} + M^{-1}f(t) \tag{5.67*}$$

This, together with the identity $\dot{y} = \dot{y}$, can be expressed in the form

$$\begin{bmatrix} \dot{y} \\ \ddot{y} \end{bmatrix} = \begin{bmatrix} 0 & I \\ -M^{-1}K & -M^{-1}C \end{bmatrix} \begin{bmatrix} y \\ \dot{y} \end{bmatrix} + \begin{bmatrix} 0 \\ M^{-1} \end{bmatrix} f(t) \tag{5.68}$$

which is in the state-space form (5.65) where,

$$A = \begin{bmatrix} 0 & I \\ -M^{-1}K & -M^{-1}C \end{bmatrix} \quad \text{and} \quad B = \begin{bmatrix} 0 \\ M^{-1} \end{bmatrix} \tag{5.69}$$

Note that in equations (5.68) and (5.69), I denotes an identity matrix of an appropriate size.

5.9.1 Modal Analysis

Consider the free motion ($u = 0$) of the nth-order system given by equation (5.65). Its solution is given by

$$x = \Phi(t)x(0) \tag{5.70}$$

It is known that the state transition matrix $\Phi(t)$ is given by the matrix-exponential expansion equation.

$$\Phi(t) = \exp(At) = I + At + \frac{1}{2!}A^2 t^2 + \ldots \qquad (5.71)$$

To discuss the rationale for this exponential response further, begin by assuming a homogeneous solution of the form

$$x = X \exp(\lambda t) \qquad (5.72)$$

By substituting equation (5.72) into the homogeneous equation of motion (that is, equation (5.65) with $u = 0$), the following matrix-eigenvalue problem results:

$$(A - sI)X = 0 \qquad (5.73)$$

Assume that the n eigenvalues ($\lambda_1, \lambda_2, \ldots, \lambda_n$) of A are distinct. Then the corresponding eigenvectors $X_1, X_2, \ldots, X_n$ are linearly independent vectors; that is, any one eigenvector cannot be expressed as a linear combination of the rest of the eigenvectors in the set. Thus, the general solution for free dynamics is

$$x(t) = X_1 \exp(\lambda_1 t) + X_2 \exp(\lambda_2 t) + \ldots + X_n \exp(\lambda_n t) \qquad (5.74)$$

Each of the n eigenvectors has an unknown parameter. The total of n unknowns is determined using n initial conditions:

$$x(0) = x_0 \qquad (5.75)$$

5.9.2 Mode Shapes of Nonoscillatory Systems

Since the eigenvectors are independent, if the initial state is set at $x_0 = X_i$, then the subsequent motion should not have any X_j terms with $j \neq i$ in equation (5.74). Otherwise, when $t = 0$, X_i becomes a linear combination of the remaining eigenvectors, which contradicts the linear independence. Hence, the motion due to this eigenvector initial condition is given by $x(t) = X_i \exp(\lambda_i t)$, for which the vector is parallel to X_i throughout the motion. Thus, X_i gives the mode shape of the system corresponding to the eigenvalue λ_i.

5.9.3 Mode Shapes of Oscillatory Systems

The analysis in the preceding section is valid for real eigenvalues and eigenvectors. In vibratory systems, λ_i and X_i generally are complex. Let

$$\lambda_i = \sigma_i + j\omega_i \qquad (5.76)$$

$$X_i = R_i + jI_i \qquad (5.77)$$

Modal Analysis

For real systems, there exist corresponding complex conjugates:

$$\overline{\lambda}_i = \sigma_i - j\omega_i \tag{5.78}$$

$$\overline{X}_j = R_i - jI_i \tag{5.79}$$

Equations (5.76) through (5.79) represent the ith mode of the system. The corresponding damped natural frequency is ω_i, and the damping parameter is σ_i. The net contribution of the ith mode to the solution — equation (5.74) — is

$$\left(R_i \cos\omega_i t - I_i \sin\omega_i t\right) 2 \exp(\sigma_i t)$$

It should be clear, for instance from equation (5.66) that only some of the state variables in $x(t)$ correspond to displacements of the masses (or spring forces). These can be extracted through an output relationship of the form

$$y = Cx \tag{5.80}$$

The contribution of the ith mode to the displacement variables is

$$Y_i = C\left[R_i \cos\omega_i t - I_i \sin\omega_i t\right] 2 \exp(\sigma_i t) \tag{5.81}$$

If equation (5.81) can be expressed in the form

$$Y_i = S_i \sin(\omega_i t + \phi_i) \exp(\sigma_i t) \tag{5.82}$$

in which S_i is a constant vector that is defined up to one unknown, then it is possible to excite the system so that every independent mass element undergoes oscillations in phase (hence, passing through the equilibrium state simultaneously) at a specific frequency ω_i. It has been noted that this type of motion is known as *normal mode motion*. The vector S_i gives the mode shape corresponding to the (damped) natural frequency ω_i. A normal mode motion is possible for undamped systems and for certain classes of damped systems. The initial state that is required to excite the ith mode is $x_0 = R_i$. The corresponding displacement and velocity initial conditions are obtained from equation (5.81); thus,

$$Y_i(0) = CR_i \tag{5.83}$$

$$\dot{Y}_i(0) = C(R_i\sigma_i - I_i\omega_i) \tag{5.84}$$

Note that the constant factor 2 has been ignored because X_i is known up to one arbitrary complex parameter.

EXAMPLE 5.9

A torsional dynamic model of a pipeline segment is shown in Figure 5.15(a). Free-body diagrams in Figure 5.15(b) show internal torques acting at sectioned inertia junctions for free motion. A state

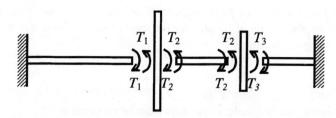

FIGURE 5.15 (a) Dynamic model of a pipeline segment, and (b) free-body diagrams.

model is obtained using the generalized velocities (angular velocities Ω_i) of the inertia elements and the generalized forces (torques T_i) as state variables. A minimum set that is required for complete representation determines the system order. There are two inertia elements and three spring elements — a total of five energy-storage elements. The three springs are not independent, however. The motion of two springs completely determines the motion of the third. This indicates that the system is a fourth-order system. One obtains the model as follows.

Newton's second law gives:

$$I_1 \dot{\Omega}_1 = -T_1 + T_2 \qquad \text{(i)}$$

$$I_2 \dot{\Omega}_2 = -T_2 - T_3 \qquad \text{(ii)}$$

Hooke's law gives:

$$\dot{T}_1 = k_1 \Omega_1 \qquad \text{(iii)}$$

$$\dot{T}_2 = k_2 (\Omega_2 - \Omega_1) \qquad \text{(iv)}$$

Torque T_3 is determined in terms of T_1 and T_2, using the displacement relation for the inertia I_2:

$$\frac{T_1}{k_1} + \frac{T_2}{k_2} = \frac{T_3}{k_3} \qquad \text{(v)}$$

This is in fact the motion compatibility condition.

The state vector is chosen as

Modal Analysis

$$x = [\Omega_1, \Omega_2, T_1, T_2]^T \tag{vi}$$

The corresponding system matrix is

$$A = \begin{bmatrix} 0 & 0 & -\dfrac{1}{I_1} & \dfrac{1}{I_1} \\ 0 & 0 & -\dfrac{1}{I_2}\left(\dfrac{k_3}{k_1}\right) & -\dfrac{1}{I_2}\left(1+\dfrac{k_3}{k_2}\right) \\ k_1 & 0 & 0 & 0 \\ -k_2 & k_2 & 0 & 0 \end{bmatrix} \tag{vii}$$

The output-displacement vector is

$$y = \left[\dfrac{T_1}{k_1}, \dfrac{T_1}{k_1}+\dfrac{T_2}{k_2}\right]^T \tag{viii}$$

which corresponds to the output-gain matrix,

$$C = \begin{bmatrix} 0 & 0 & \dfrac{1}{k_1} & 0 \\ 0 & 0 & \dfrac{1}{k_1} & \dfrac{1}{k_2} \end{bmatrix} \tag{ix}$$

For the special case given by $I_1 = I_2 = I$ and $k_1 = k_3 = k$, the system eigenvalues are

$$\lambda_1, \bar{\lambda}_1 = \pm j\omega_1 = \pm j\sqrt{\dfrac{k}{I}} \tag{x}$$

$$\lambda_2, \bar{\lambda}_2 = \pm j\omega_2 = \pm j\sqrt{\dfrac{k+2k_2}{I}} \tag{xi}$$

and the corresponding eigenvectors are

$$X_1, \bar{X}_1 = R_1 \pm jI_1 = \dfrac{\alpha_1}{2}[\omega_1, \omega_1, \mp jk_1, 0]^T \tag{xii}$$

$$X_2, \bar{X}_2 = R_2 \pm jI_2 = \dfrac{\alpha_1}{2}[\omega_2, -\omega_2, \mp jk_1, \pm 2jk_2]^T \tag{xiii}$$

In view of equation (5.81), the modal contributions to the displacement vector are

$$Y_1 = \begin{bmatrix} 1 \\ 1 \end{bmatrix} \alpha_1 \sin\omega_1 t \quad \text{and} \quad Y_2 = \begin{bmatrix} 1 \\ -1 \end{bmatrix} \alpha_2 \sin\omega_2 t \tag{xiv}$$

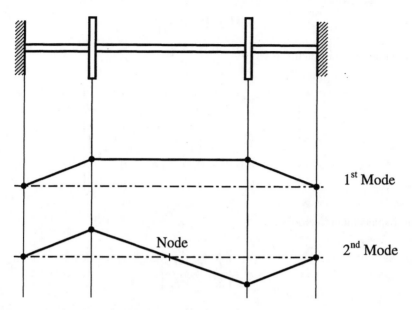

FIGURE 5.16 Mode shapes of the pipeline segment.

These equations (*xiv*) are of the form given by equation (5.82). The mode shapes are given by the vectors $S_1 = [1,1]^T$ and $S_2 = [1,-1]^T$, which are illustrated in Figure 5.16. In general, each modal contribution introduces two unknown parameters, a_i and ϕ_i, into the free response (homogeneous solution), where ϕ_i are the phase angles associated with the sinusoidal terms. For an n-degree-of-freedom (order-$2n$) system, this results in $2n$ unknowns, which require the $2n$ initial conditions $x(0)$.

□

PROBLEMS

5.1 A bird swing is sketched in Figure P5.1. It consists of a rigid slender bar hung with two wires from a branch of a tree. Assume that the wires are taut and the motions are small.
  a. How many degrees of freedom does the system have?
  b. Give two separate choices of complete and independent set of coordinates to describe the motion of the swing.

5.2 Two thin disks are connected together at their centers by a spring and rest upright on a horizontal plane. They are allowed to move along a straight line on the horizontal plane, while remaining in a common, fixed, vertical plane, as shown in Figure P5.2. Give a complete and independent set of coordinates to describe the motion of the system, assuming that:
  a. Slipping is allowed;
  b. Slipping is not allowed.

5.3 A disk rolls upright (i.e., its plane remains vertical) on a horizontal plane, as shown in Figure P5.3. Using clear discussions, indicate the number of degrees of freedom of the system and whether the system is holonomic if:
  a. Slipping is permitted between the disk and the plane;
  b. Slipping is not permitted.

5.4 Define the terms natural frequency, mode shape, rigid-body mode, and node.
A centrifugal pump driven by an induction motor is modeled as in Figure P5.4. The rotor inertia of the motor is represented by m_1, drive-shaft stiffness by k, and pump rotor inertia and equivalent fluid load by m_2. Qualitatively describe the modes of vibration of the system. Without analyzing the overall system equations, and by considering single-

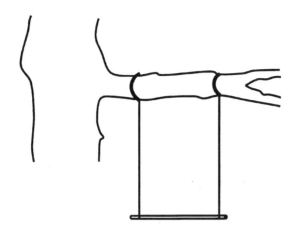

FIGURE P5.1 A bird swing.

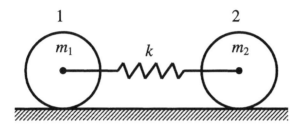

FIGURE P5.2 Two upright disks connected by a spring.

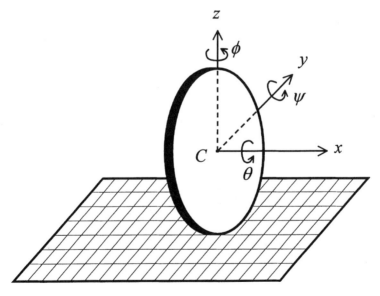

FIGURE P5.3 A disk rolling on a horizontal plane.

degree-of-freedom subsystems and the concept of "node," completely obtain expressions for the natural frequencies of the system in terms of m_1, m_2, and k. Completely determine the corresponding mode shapes.

5.5 a. Consider a linear, lumped-parameter, undamped mechanical system given by the vector-matrix form of equation of motion

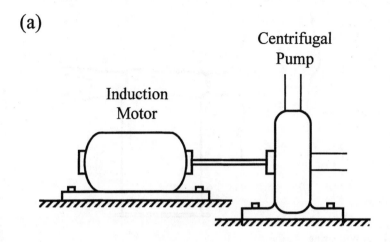

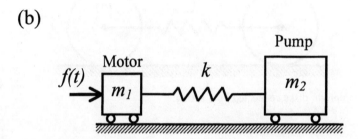

FIGURE P5.4 (a) A centrifugal pump driven by an induction motor, and (b) a simplified model for rotatory vibration.

$$M\ddot{y} + Ky = f(t)$$

What do $f(t)$ and y represent in these equations? Describe three ways of obtaining the stiffness matrix K and the mass matrix M of a system of this nature, assuming that a sketch of the system, showing its point masses and the stiffness elements, is available.

b. A sketch of a two-story steel structure is shown in Figure P5.5(a). A model that can be used to study its lateral planar motions is shown in Figure P5.5(b). Assume that the displacements y_1 and y_2 of the lumped masses m_1 and m_2 are measured from the static equilibrium position of the system. It is required to determine the system equations in the form

$$\begin{bmatrix} m_{11} & m_{12} \\ m_{21} & m_{22} \end{bmatrix} \begin{bmatrix} \ddot{y}_1 \\ \ddot{y}_2 \end{bmatrix} + \begin{bmatrix} k_{11} & k_{12} \\ k_{21} & k_{22} \end{bmatrix} \begin{bmatrix} y_1 \\ y_2 \end{bmatrix} = \begin{bmatrix} f_1(t) \\ f_2(t) \end{bmatrix}$$

Suppose that $\ddot{y}_1 = 0 = \ddot{y}_2$ and also $y_2 = 0$. What are the forces f_1 and f_2 that are needed to maintain a unit displacement $y_1 = 1$ at mass m_1?

Similarly, suppose that $\ddot{y}_1 = 0 = \ddot{y}_2$ and also $y_1 = 0$. What are the forces f_1 and f_2 that are needed to maintain a unit displacement $y_2 = 1$ at mass m_2?

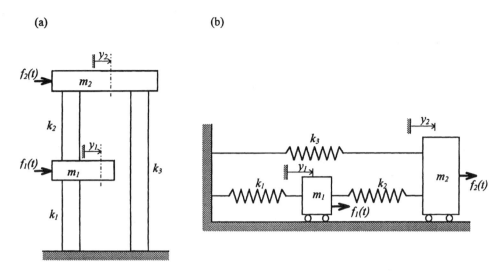

FIGURE P5.5 (a) A two-story building, and (b) a simplified model of lateral planar dynamics.

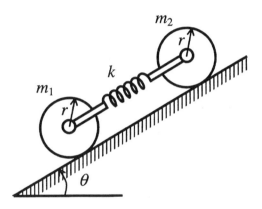

FIGURE P5.6 A cart rolling down a hill.

    By proceeding in this manner, obtain the mass matrix and the stiffness matrix of the system.

  c. List some properties of M and K of the given system that can be extended to similar, higher-order systems.

  d. Consider the special case of the system in Figure P5.5, with $k_1 = k_2 = k$, $k_3 = 3k$, $m_1 = m$, and $m_2 = 2m$. Determine the natural frequencies and mode shapes of the system. Normalize the mode shapes with respect to:
    i. the mass matrix;
    ii. the stiffness matrix.
  What are the modal masses and modal stiffnesses for each of these two normalized cases of mode shapes?
  Sketch the mode shapes of the systems and describe the system behavior in each of these modes.

  e. Sketch two rotational system that are analogous to the translational system that is shown in Figure P5.5(b).

5.6 A simplified (planar) model of a cart rolling down hill is shown in Figure P5.6. The wheels are approximated by rigid disks of radius r. The mass of the front wheels is m_1

and that of the rear wheels is m_2. There is a longitudinal flexibility in the cart that is represented by a spring of stiffness k. The inclination of the hill is θ. Neglect friction at the wheel axles and assume that there is no slip between the wheels and the ground surface.

Using Lagrange's equations, derive the equations of motion of the system. What is the mass matrix and what is the stiffness matrix?

5.7 A simplified planar model that can be used to study heave and pitch vibrations of a vehicle is shown in Figure P5.7. The vehicle body is modeled as a uniform rod of length L and mass m. The stiffness of the rear suspensions is k_1 and that of the front suspensions is k_2. Energy dissipation is neglected. Using Lagrange's equations, determine the heave (y) and pitch (θ) equations of free motion. Identify the mass matrix and the stiffness matrix.

5.8 An indicator mechanism of a centrifuge consists of a spring with one of its ends attached to the center of spin and the other end carrying the indicator mass. The spring-mass unit slides inside a smooth glass tube. A schematic diagram of the system is given in Figure P5.8. Neglect energy dissipation due to friction. The indicator is given a spin of angular velocity ω and then the power is turned off.
   a. Using Lagrange's equations, develop the equations of motion assuming constant ω.
   b. What are the steady-state equilibrium states of the device?
   c. Investigate the oscillations and stability of motion in the neighborhood of each equilibrium state.
   d. What are the modes of oscillation of the system?

5.9 A flexible shaft-rotor system is shown in Figure P5.9. The rotors are identical and have the same mass m and the same polar radius of gyration r. The shaft segments have the same torsional stiffness k.
   a. Formulate the modal vibration problem for this system and express it in terms of a mass matrix M and a stiffness matrix K.
   b. Determine the natural frequencies and mode shapes of the system.
   c. Construct the modal matrix Ψ and verify that both $\Psi^T M \Psi$ and $\Psi^T K \Psi$ are diagonal.
   d. Sketch the mode shapes of the system.

5.10 A simplified model of a three-car train is shown in Figure P5.10. Assume that the cars have the same mass m and the couplers have the same longitudinal stiffness k.
   a. Formulate the free vibration problem in the form $Kx + M\ddot{x} = 0$.
   b. Determine the natural frequencies and modes of vibration.
   c. Suppose that a sinusoidal force $F \sin\omega t$ is applied to Car 1. Determine and sketch the resulting frequency response at Car 1 (in a steady state).

5.11 Consider an identical pair of simple pendulums of length l and point mass m. They are mounted at the same horizontal level, and the two masses are linked using a spring of stiffness k. The pendulums rest in a vertical position (with the spring in its relaxed position), as shown in Figure P5.11. Applying Lagrange's equations, obtain the equations of motion of this connected system of two pendulums. Formulate the modal problem for this system assuming that the angles of swing θ_1 and θ_2 of the pendulums, from their static equilibrium configuration, are small. Solve the modal problem and determine the natural frequencies and the mode shapes. Discuss the nature of these two modes.

5.12 Consider an overhead gantry truck carrying a pendulous load, as found in a factory (manufacturing plant). We wish to study the vibrations when the truck is braking against a flexible coupler. A simplified model of this system is shown in Figure P5.12. The following parameters are given:

   m = mass of the gantry truck
   k = stiffness of the flexible coupler

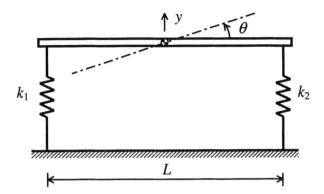

FIGURE P5.7 A simplified vehicle model.

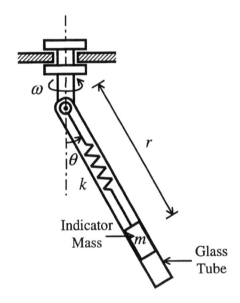

FIGURE P5.8 An indicator device of a centrifuge.

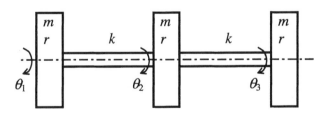

FIGURE P5.9 A flexible shaft-rotor system.

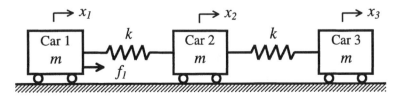

FIGURE P5.10 A three-car train.

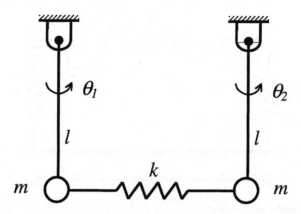

FIGURE P5.11 A pair of simple pendulums linked by a spring.

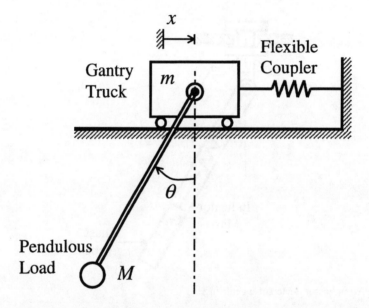

FIGURE P5.12 A gantry truck with a pendulous load and braking against a flexible coupler.

M = mass of the pendulous load
l = length of the pendulous arm.

Also, let x be the displacement of the truck from the relaxed position of the flexible coupler, and θ be the angle of swing of the pendulous load from the vertical static configuration.
  a. Neglecting energy dissipation and any external excitation forces, obtain the equations of motion of the system using Lagrange's equations.
  b. For small x and θ, formulate the modal problem. Obtain the characteristic equation for natural frequencies of vibration of the system.
  c. If $k = 0$ (i.e., in the absence of a flexible coupler), what are the natural frequencies and mode shapes of the system?
5.13 A simplified model that can be used for studying the pitch (or roll) and heave motions of a vehicle is shown in Figure P5.13. Let y be the vertical displacement (heave) of the vehicle body and θ be the associated angle of rotation of the body, as measured from

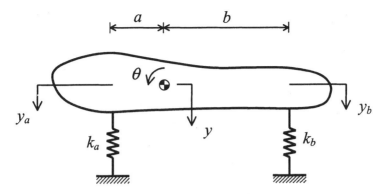

FIGURE P5.13 A simplified vehicle model for pitch/roll and heave motions.

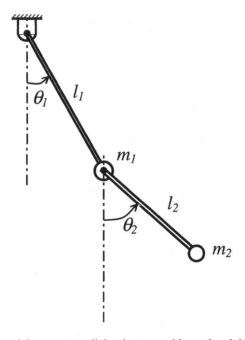

FIGURE P5.14 A double pendulum or a two-link robot arm with revolute joints.

the static equilibrium configuration. Suspensions of stiffness k_a and k_b are located at horizontal distances a and b, respectively, from the centroid. The mass of the vehicle is m and the moment of inertia about the centroid is J.

Obtain equations of motion for this system using:
a. direct application of Newton's second law;
b. Lagrange's equations.

Next, using the vertical displacements y_a and y_b at the suspensions k_a and k_b, with respect to the static equilibrium position as the motion variables, obtain a new set of equations of motion.

In each of the two generalized coordinate systems, outline the procedure of modal analysis. Comment on the expected results (natural frequencies and mode shapes) in each case. Neglect damping and external excitation forces throughout the problem.

5.14 Consider the double pendulum (or a two-link robot with revolute joints) having arm lengths l_1 and l_2, and the end masses m_1 and m_2, as shown in Figure P5.14.

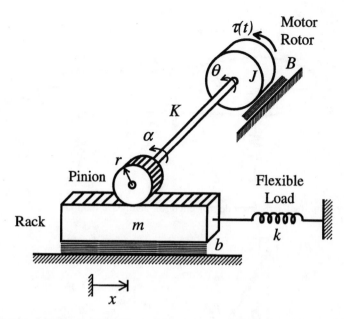

FIGURE P5.15 Motor-driven rack-and-pinion mechanism with a flexible load.

a. Use Lagrange's equations to obtain the equations of motion for the system in terms of the absolute angles of swing θ_1 and θ_2 about the vertical equilibrium configuration. Linearize the equations for small motions θ_1, $\dot{\theta}_1$, θ_2, and $\dot{\theta}_2$.

b. For the special case of $m_1 = m_2 = m$ and $l_1 = l_2 = l$, solve the modal problem of this system. Normalize the mode shape vectors so that the first element (corresponding to θ_1) of each vector is unity. What are the corresponding modal masses and modal stiffnesses? Verify that the natural frequencies can be obtained from these modal parameters. Using the modal solution, express the free response of the system to an initial condition excitation of $\theta(0)$ and $\dot{\theta}(0)$.

c. Express the free response as obtained in part (b) for the case $l = 9.81$ m with
$$\theta(0) = \begin{bmatrix} 1 & -\dfrac{1}{\sqrt{2}} \end{bmatrix}^T \text{ and } \dot{\theta}(0) = \mathbf{0}.$$
Sketch this response for a time period of 20 seconds.

5.15 Consider a rack-and-pinion system driven by a DC motor and pushing against a purely elastic load. A representation of this system is shown in Figure P5.15. The following parameters are defined:

J = moment of inertia of the motor rotor about its axis of rotation
B = equivalent damping constant at the motor rotor
K = torsional stiffness of the drive shaft of the motor
r = radius of the pinion at the end of drive shaft
m = mass of the rack
b = equivalent damping constant at the rack
k = stiffness of the elastic load.

Neglect the inertia of the pinion. Motor torque (magnetic torque) is $\tau(t)$. Angle of rotation of the motor rotor is θ, and the corresponding displacement of the rack is x, as measured from the relaxed configuration of the load spring.

a. Derive the equations of motion of the system in terms of the motion variables θ and x. Express these in the vector-matrix form. What is the characteristic equation of the system?
b. Derive a purely rotational system that is equivalent.
c. Derive a purely translatory system that is equivalent.
d. If $k = 0$, what are the natural frequencies of motion of the system?

5.16 a. Define the following terms:
  i. Modal matrix
  ii. Degrees of freedom
  iii. Rigid body modes
  iv. Static modes
  v. Proportional damping.
  b. Explain the significance of these terms in modal analysis of a vibrating system.
  c. Suppose that a mechanical system has two identical natural frequencies $\omega_1 = \omega_2$. If and ψ_1 are ψ_2 mode shapes corresponding to these natural frequencies, show that

$$\psi = \alpha\psi_1 + \beta\psi_2$$

for any arbitrary α and β, will also serve as a mode shape for either of these repeated frequencies.
  d. Do damped systems possess real modes? Explain your answer, clearly justifying the arguments. Sketch a system by adding viscous damping elements to the undamped system shown in Figure P5.5(b) so that it has proportional damping.
  e. In modal analysis, what common assumptions are made with regard to the system? Are these assumptions justified in practice? Explain.

5.17 a. Sketch a two-degree-of-freedom mechanical system that has two identical natural frequencies. What are its mode shapes? Comment about the modal motions of a system of this type.
  b. Consider a mechanical system given by the equations of motion

$$\begin{bmatrix} 1 & 0 \\ 0 & 2 \end{bmatrix} \begin{bmatrix} \ddot{y}_1 \\ \ddot{y}_2 \end{bmatrix} + \begin{bmatrix} 2 & -1 \\ -1 & 4 \end{bmatrix} \begin{bmatrix} y_1 \\ y_2 \end{bmatrix} = \begin{bmatrix} f_1(t) \\ f_2(t) \end{bmatrix}$$

Obtain the natural frequencies and mode shapes of the system. Express the modal matrix of the system, using M-normal mode shapes.
Suppose that $f_1(t) = 2\sin 3t$ and $f_2(t) = 0$. The system starts from rest (i.e., zero velocities) with initial conditions $y_1(0) = 1.0$ and $y_2(0) = 1.0$. Using the solution for the response of a second-order undamped system subjected to a harmonic excitation, obtain the complete response (y_1 and y_2) of the system.
  c. Obtain the mode shapes of the *damped* system

$$\begin{bmatrix} 1 & 0 \\ 0 & 2 \end{bmatrix} \begin{bmatrix} \ddot{y}_1 \\ \ddot{y}_2 \end{bmatrix} + \begin{bmatrix} 2 & 1 \\ 1 & 4 \end{bmatrix} \begin{bmatrix} \dot{y}_1 \\ \dot{y}_2 \end{bmatrix} + \begin{bmatrix} 2 & -1 \\ -1 & 4 \end{bmatrix} \begin{bmatrix} y_1 \\ y_2 \end{bmatrix} = \begin{bmatrix} f_1(t) \\ f_2(t) \end{bmatrix}$$

5.18 a. Explain the terms:
  i. Mode shapes
  ii. Natural frequencies
  iii. Modal analysis

iv. Modal testing
v. Experimental modal analysis.
b. Briefly describe the process of experimental modal analysis of a mechanical system.
c. Consider the mechanical system expressed in the vector-matrix form

$$M\ddot{y} + B\dot{y} + Ky = f(t)$$

Its frequency transfer function matrix $G(j\omega)$ is given by $Y(j\omega) = G(j\omega)F(j\omega)$ where Y and F are the frequency-response (Fourier spectra) vectors of y and f, respectively. Express G in terms of M, B, and K.

5.19 The reverse problem of modal analysis (which is useful in experimental modeling of vibratory systems) is to determine the mass matrix M and the stiffness matrix K (and perhaps the damping matrix C) with the knowledge of the modal information such as mode shapes and natural frequencies.

Consider a two-degree-of-freedom system. The following information is given. The M-normal mode shape vectors are:

$$\psi_1 = \begin{bmatrix} 1/2 \\ 0 \end{bmatrix} \quad \text{and} \quad \psi_2 = \begin{bmatrix} \sqrt{3}/2 \\ 2/\sqrt{3} \end{bmatrix}$$

The modal stiffness parameters are:

$$K_1 = 0 \quad \text{and} \quad K_2 = 4\frac{g}{l}$$

where g is the acceleration due to gravity and l is some length parameter.

a. What are the natural frequencies of the system? From this information, comment about the nature of this system.
b. What is the modal matrix Ψ? Determine its inverse Ψ^{-1}.
c. What is the modal mass matrix $\overline{M}$ and what is the modal stiffness matrix $\overline{K}$ corresponding to the given mode shape vectors?
d. Determine the mass matrix M and the stiffness matrix K of the system.
e. If the system has proportional damping and the damping ratios of the two modes are $\zeta_1 = 0$ and $\zeta_2 = 0.1$, what is the modal damping matrix $\overline{C}$ and what is the damping matrix C?
f. Sketch a mechanical system that has the M, C, and K matrices as obtained in this problem.

5.20 Suppose the viscous damping matrix of a vibrating system is given by

$$C = a\left(MK^{-1}\right)^r M + b\left(KM^{-1}\right)^m K$$

in which M and K are the mass matrix and the stiffness matrix respectively, and r and m are positive integers. Show that the resulting damped system possesses real modes of vibration, which are identical to those of the undamped system.

5.21 Show that the mode shapes and natural frequencies of the system

$$M\ddot{y} + Ky = f(t)$$

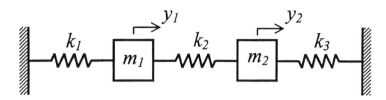

FIGURE P5.21 A two-degree-of-freedom vibrating system.

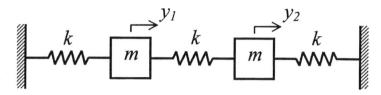

FIGURE P5.22 A two-degree-of-freedom model.

are given by the solutions for ψ and ω in the equation

$$(\omega^2 M - K)\psi = 0$$

Give a matrix whose eigenvalues are ω^2 and the eigenvectors are ψ. Is this matrix symmetric? Are the mode shapes orthogonal in general?

Suppose Ψ denotes the modal matrix (the matrix formed by using modal vectors ψ as columns) of the system. Show that it diagonalizes M and K by the *congruence transformations*

$$\Psi^T M \Psi \quad \text{and} \quad \Psi^T K \Psi$$

and that it diagonalizes the matrix by the *similarity transformation*

$$\Psi^{-1} M^{-1} K \Psi$$

at least in the case of distinct (unequal) natural frequencies. Note, however, that this result is true in general, even with repeated (equal) natural frequencies.

Consider the system shown in Figure P5.21. Determine the mass matrix M and the stiffness matrix K. Show that M and K do not commute in general. What does this tell us?

5.22 Consider the two-degree-of-freedom system shown in Figure P5.22. Using one-degree-of-freedom results and possibly using the concepts of "node" and "symmetry," determine the natural frequencies and the corresponding mode shapes of the system. Then, verify your results using a complete two-degree-of-freedom analysis.

Sketch how viscous dampers should be connected to the system in Figure P5.22 so that proportional damping of the type

  a. Momentum
  b. Strain rate

is introduced into the system. Give the corresponding damping matrices.

5.23 Consider the simplified model of a vehicle shown in Figure P5.23 that can be used to study the heave (verticals up and down) and pitch (front-back rotation) motions due to the road profile. For our purposes, assume that the road profiles that excite the front and back suspensions are independent. They are the displacement inputs $u_1(t)$ and $u_2(t)$. The

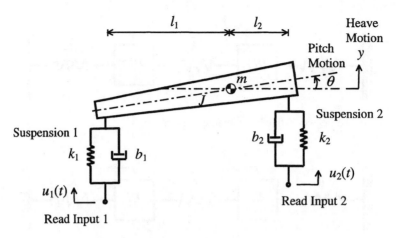

FIGURE P5.23 A simplified model of a vehicle.

mass and the pitch moment of inertia of the vehicle body are denoted by m and J, respectively. The suspension inertia is neglected and the stiffness and the damping constant of the suspension systems are denoted by k and b, respectively, with appropriate subscripts, as shown.

a. Write the differential equations for the pitch angle θ and the vertical (heave) displacement y of the centroid of the vehicle body using $u_1(t)$ and $u_2(t)$ as the inputs. Assume small motions so that linear approximations hold.
b. What is the order of the system?
c. Determine the transfer function relation for this system.
d. Identify a mass matrix M, a stiffness matrix K, and a damping matrix C for the system.

5.24 A manufacturer of rubber parts uses a conventional process of steam-cured moulding of natural latex. The moulded rubber parts are first cooled and buffed (polished), and then sent for inspection and packing.

A simple version of a rubber buffing machine is shown in Figure P5.24(a). It consists of a large hexagonal drum whose inside surfaces are all coated with a layer of bonded emery. The shaft of the drum is supported horizontally on two heavy-duty, self-aligning bearings at the two ends, and is rotated using a three-phase induction motor. The drive shaft of the drum is connected to the motor shaft through a flexible coupling.

The buffing process consists of filling the drum with rubber parts, steadily rotating the drum for a specified period of time, and finally vacuum cleaning the drum and its contents. Dynamics of the machine affect the loading on various components such as the motor, coupling, bearings, shafts, and the support structure. In order to study the dynamics and vibration behavior, particularly at the startup stage and under disturbances during steady-state operation, an engineer develops a simplified model of the buffing machine. This model is shown in Figure P5.24(b).

The motor is modeled as a torque source T_m that is applied on the rotor that has moment of inertia J_m, and resisted by a viscous damping torque of damping constant b_m. The connecting shafts and the coupling unit are represented by an equivalent torsional spring of stiffness k_L. The drum and its contents are represented by an equivalent constant moment of inertia J_L. There is a resisting torque on the drum even at steady operating speed, due to misalignments and the eccentricity of the contents of the drum. This load is represented by a constant torque T_r. Furthermore, energy dissipation due to the buffing action (between the rubber parts and the emery surfaces of the drum) is represented by a nonlinear damping torque T_{NL}, which is approximated as:

Modal Analysis

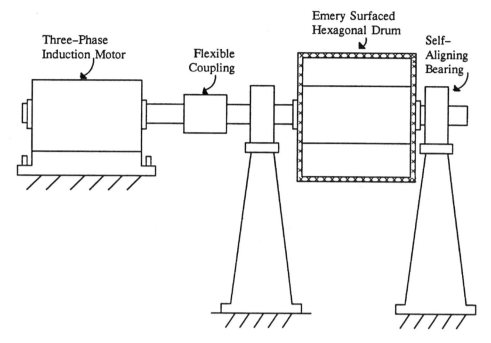

FIGURE P5.24(a) A rubber buffing machine.

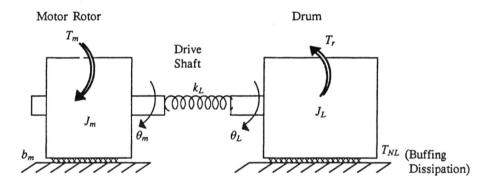

FIGURE P5.24(b) A dynamic model of the buffing machine.

$$T_{Nl} = c|\dot{\theta}_L|\dot{\theta}_L \qquad \text{with } c > 0$$

Note that θ_m and θ_L are the angles of rotation of the motor rotor and the drum, respectively, and these are measured from inertial reference lines that correspond to a relaxed configuration of spring k_L.

a. Comment on the assumptions made in the modeling process of this problem, and briefly discuss the validity (or accuracy) of the model.
b. Show that the model equations are:

$$J_m \ddot{\theta}_m = T_m - k_l(\theta_m - \theta_L) - b_m \dot{\theta}_m$$

$$J_L \ddot{\theta}_L = k_l(\theta_m - \theta_L) - c|\dot{\theta}_L|\dot{\theta}_L - T_r$$

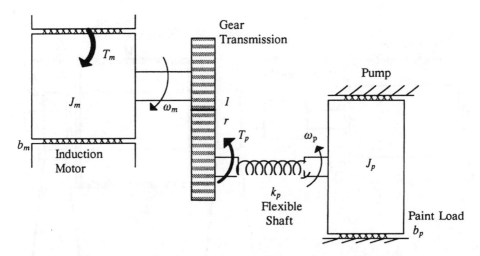

FIGURE P5.25 A model for a paint pumping system in an automobile assembly plant.

What are the excitation inputs of this system?
c. If the buffing dissipation is represented by the linear viscous damping term $b_L \dot\theta_L$, obtain the mass, damping, and stiffness matrices of the system.
d. Using the speeds $\dot\theta_m$ and $\dot\theta_L$, and the spring torque T_k as the state variables, and the twist of the spring as the output, obtain a complete state model for the *nonlinear* system.

What is the order of the state model?

5.25 The robotic spray-painting system of an automobile assembly plant employs an induction motor and pump combination to supply paint at an overall peak rate of 15 gal/min to clusters of spray-paint heads in several painting booths. The painting booths are an integral part of the assembly lines in the plant. The pumping and filtering stations are on the ground level of the building and the painting booths are on an upper level. Not all booths or painting heads operate at a given time. The pressure in the paint supply lines is maintained at a desired level (approximately 275 psi) by controlling the speed of the pump, which is achieved through a combination of voltage control and frequency control of the induction motor.

An approximate model for the paint pumping system is shown in Figure P5.25.

The induction motor is linked to the pump through a gear transmission of efficiency η, speed ratio 1:r, and a flexible shaft of torsional stiffness k_p. The moments of inertia of the motor rotor and the pump impeller are denoted by J_m and J_p, respectively. The gear inertia is neglected (or lumped with J_m). The mechanical dissipation in the motor and its bearings is modeled as linear viscous damping of damping constant b_m. The load on the pump (the paint load plus any mechanical dissipation) is also modeled as viscous damping, and the equivalent damping constant is b_p. The magnetic torque T_m generated by the induction motor is given by

$$T_m = \frac{T_0 q \omega_0 (\omega_0 - \omega_m)}{(q\omega_0^2 - \omega_m^2)}$$

in which ω_m is the motor speed. The parameter T_0 depends directly (quadratically) on the phase voltage supplied to the motor. The second parameter ω_0 is directly proportional

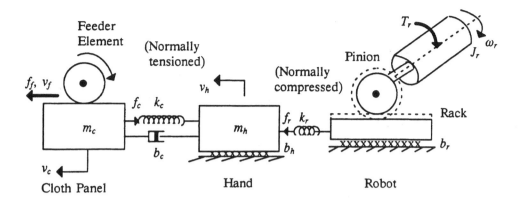

FIGURE P5.26 A robotic sewing system.

to the line frequency of the AC supply. The third parameter q is positive and greater than unity, and this parameter is assumed constant in the control system.

a. Comment about the accuracy of the model shown in Figure P5.25.
b. For vibration analysis of the system, develop its equations and identify the mass (M), stiffness (K), and damping (C) matrices. Comment on the nature of K.

5.26 A robotic sewing system consists of a conventional sewing head. During operation, a panel of garment is fed by a robotic hand into the sewing head. The sensing and control system of the robotic hand ensures that the seam is accurate and the cloth tension is correct in order to guarantee the quality of the stitch. The sewing head has a frictional feed mechanism that pulls the fabric in an intermittent cyclic manner, away from the robotic hand, using a toothed feeding element.

When there is slip between the feeding element and the garmet, the feeder functions as a *force source*. The applied force is assumed cyclic with a constant amplitude. When there is no slip, however, the feeder functions as a *velocity source*, which is the case during normal operation. The robot hand has inertia. There is some flexibility at the mounting location of the hand on the robot. The links of the robot are assumed rigid, and some of its joints can be locked to reduce the number of degrees of freedom when desired.

Consider the simplified case of a single-degree-of-freedom robot. The corresponding robotic sewing system is modeled as in Figure P5.26. Note that the robot is modeled as a single moment of inertia J_r that is linked to the hand with a light rack-and-pinion device of speed transmission given by:

$$\frac{\text{Rack translatory movement}}{\text{Pinion rotatory movement}} = r$$

Assume that this transmission is 100% efficient (no loss).

The drive torque of the robot is T_r and the associated rotatory speed is ω_r. Under conditions of slip, the feeder input to the cloth panel is force f_f, and with no slip the input is the velocity v_f. Various energy dissipation mechanisms are modeled as linear viscous damping of damping constant b (with appropriate subscripts). The flexibility of various system elements is modeled by linear springs with stiffness k. The inertia effects of the cloth panel and the robotic hand are denoted by the lumped masses m_c and m_h, respectively, having velocities v_c and v_h as shown in Figure P5.26. Note that the cloth panel is normally in tension, with tensile force f_c, and in order to push the panel, the robotic wrist is normally in compression, with compressive force f_r.

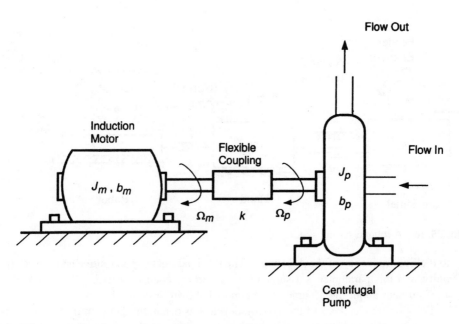

FIGURE P5.27 A centrifugal pump driven by an induction motor.

a. First consider the case of the feeding element with slip. How many degrees of freedom does the system have? Formulate the system equations and identify the *M*, *K*, and *C* matrices.

b. Now consider the case where there is no slip at the feeder element. How many degrees of freedom does the system have now? Formulate the system equations and identify the *M*, *K*, and *C* matrices for this case.

5.27 a. Linearized models of nonlinear systems are commonly used in the vibration analysis and control of dynamic systems. What is the main assumption that is made in using a linearized model to represent a nonlinear system?

b. A three-phase induction motor is used to drive a centrifugal pump for incompressible fluids. To reduce misalignment and associated problems such as vibration, noise, and wear, a flexible coupling is used for connecting the motor shaft to the pump shaft. A schematic representation of the system is shown in Figure P5.27.

Assume that the motor is a "torque source" of torque T_m that is being applied to the motor rotor of inertia J_m. Also, the following variables and parameters are defined:

J_p = moment of inertia of the pump impeller assembly
Ω_m = angular speed of the motor rotor/shaft
Ω_p = angular speed of the pump impeller/shaft
k = torsional stiffness (linear) of the flexible coupling
T_f = torque transmitted through the flexible coupling
Q = volume flow rate of the pump
b_m = equivalent viscous damping constant at the motor rotor including bearings.

Also, assume that the net torque required at the pump shaft, to pump fluid steadily at a volume flow rate of Q is given by $b_p \Omega_p$, where

$$Q = V_p \Omega_p$$

Modal Analysis

and V_p = volumetric parameter of the pump (assumed constant).

How many degrees of freedom does the system have? Using angular displacements θ_m and θ_p as the motion variables, where $\dot\theta_m = \Omega_m$ and $\dot\theta_p = \Omega_p$, develop a linear analytical model for the dynamic system in terms of an inertia matrix (M), a damping matrix (B), and a stiffness matrix (K). What is the order of the system? Comment on the modes of motion of the system.

c. Using T_m as the input, Q as the output of the system, and Ω_m, Ω_p, and the torque T_f of the flexible coupler as the state variables, develop a complete state-space model for the system. Identify the matrices A, B, and C in the usual notation in this model. What is the order of the system? Compare/contrast this result with the answer to part (b).

d. Suppose that the motor torque is given by

$$T_m = \frac{aSV_f^2}{\left[1+(S/S_b)^2\right]}$$

where motor slip S is defined as

$$S = 1 - \frac{\Omega_m}{\Omega_s}$$

Note that a and S_b are constant parameters of the motor.
Also,

Ω_s = no-load (i.e., synchronous) speed of the motor
V_f = amplitude of voltage applied to each phase winding (field) of the motor

In voltage control, V_f is used as the input and, in frequency control, Ω_s is used as the input. For combined voltage and frequency control, derive a linearized state-space model using the incremental variables $\hat{V}_f$ and $\hat{\Omega}_s$ about operating values $\overline{V}_f$ and $\overline{\Omega}_s$, as the inputs to the system and the incremental flow $\hat{Q}$ as the output.

5.28 Consider an automobile traveling at a constant speed on a rough road, as shown in Figure P5.28(a). The disturbance input due to road irregularities can be considered either as a displacement source $u(t)$ or as a velocity $\dot{u}(t)$ at the tires, in the vertical direction. An approximate, one-dimensional model is shown in Figure P5.28(b), and this can be used to study the "heave" (up and down) motion of the automobile. Note that v_1 and v_2 are the velocities of the lumped masses m_1 and m_2, respectively, and x_1 and x_2 are the corresponding displacements.

a. Briefly state what physical components of the automobile are represented by the model parameters k_1, m_1, k_2, m_2, and b_2. Also, discuss the validity of the assumptions that are made in arriving at this model.

b. Using x_1 and x_2 as the response variables, obtain an analytical model for the dynamic system in terms of a mass matrix (M), stiffness matrix (K), and a damping matrix (B). Does this system possess rigid-body modes?

c. Using v_1, v_2, f_1, and f_2 as the state variables, $\dot{u}(t)$ as the input variable, and v_1 and v_2 as the output variables, obtain a state-space model for the system. The compressive forces in springs k_1 and k_2 are denoted by f_1 and f_2, respectively. What is the order of the model?

d. If, instead of a motion source $u(t)$, a force source $f(t)$ that is applied at the same location, is considered as the system input, determine an analytical model similar to

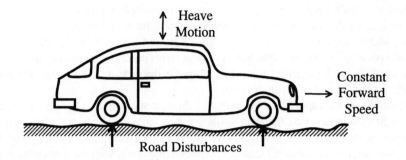

FIGURE P5.28(a) An automobile traveling at constant speed.

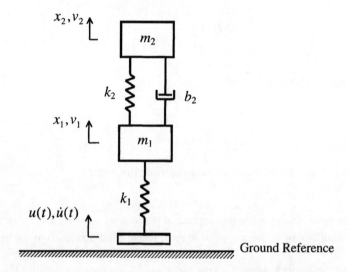

FIGURE P5.28(b) A crude model of an automobile for the heave-motion analysis.

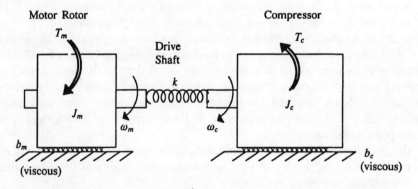

FIGURE P5.29 A model of a motor-compressor unit.

that in part (b), for the resulting system. Does this system possess rigid-body modes? Does this damped system possess real modes of vibration? Explain your answer.

e. For the case of a force source $f(t)$, derive a state-space model similar to that obtained in part (c). What is the order of this model? Explain your answer.

Note: In this problem, you may assume that gravitational effects are completely balanced by the initial compression of the springs with reference to which all motions are defined.

Modal Analysis

5.29 An approximate model for a motor-compressor combination that is used in a process application is shown in Figure P5.29.

Note that T, J, k, b, and ω denote torque, moment of inertia, torsional stiffness, angular viscous damping constant, and angular speed, respectively, and the subscripts m and c denote the motor rotor and the compressor impeller, respectively.

   a. Sketch a translatory mechanical model that is analogous to this rotary mechanical model.

   b. Formulate an analytical model for the systems in terms of a mass matrix (M), a stiffness matrix (K), and a damping matrix (B), with θ_m and θ_c as the response variables, and the motor magnetic torque T_m and the compressor load torque T_c as the input variables.

Comment on the modes of motion of the system.

   c. Obtain a state-space representation of the given model. The outputs of the system are compressor speed ω_c and the torque T transmitted through the drive shaft. What is the order of this model? Comment.

5.30 A model for a single joint of a robotic manipulator is shown in Figure P5.30.

The usual notation is used. The gear inertia is neglected and the gear reduction ratio is taken as $1:r$.

   a. Obtain an analytical model in terms of a mass matrix, a stiffness matrix, and a damping matrix, assuming that no external (load) torque is present at the robot arm. Use the motor rotation θ_m and the robot arm rotation θ_r as the response variables.

   b. Derive a state model for this system. The input is the motor magnetic torque T_m, and the output is the angular speed ω_r of the robot arm. What is the order of the system? Comment.

   c. Discuss the validity of various assumptions that were made in arriving at this simplified model for a commercial robotic manipulator.

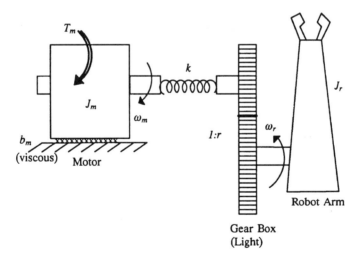

FIGURE P5.30 A model of a single-degree-of-freedom robot.

6 Distributed-Parameter Systems

Most of the vibration analysis examples encountered in the previous chapters assumed that the inertial (mass), flexibility (spring), and dissipative (damping) characteristics could be "lumped" as a finite number of "discrete" elements. Such models are termed *lumped-parameter* or *discrete-parameter systems*. If all the mass elements move in the same direction, one has a one-dimensional system (i.e., "rectilinear" motion), with each mass having a single degree of freedom. If the masses can move independently of each other, the number of degrees of freedom of such a one-dimensional system will be equal to the number of lumped masses, and will be finite. In a "planar" system, each lumped mass will be able to move in two orthogonal directions and hence, will have two degrees of freedom; and similarly in a "spatial" system, each mass will have three degrees of freedom. As long as the number of lumped inertia elements is finite, then one has a lumped-parameter (or, discrete) system with a finite degree of freedom. Note that time is a system variable and not a system parameter. Discrete-time models are used in computer analysis and simulation, regardless of whether the system is a discrete-parameter or continuous-parameter one.

Generally, in practical vibrating systems, inertial, elastic, and dissipative effects are found continuously distributed in one, two, or three dimensions. Correspondingly, there are line structures, surface/planar structures, or spatial structures. They will possess an infinite number of mass elements, continuously distributed in the structure, and integrated with some connecting flexibility (elasticity) and energy dissipation. In view of the connecting flexibility, each small element of mass will be able to move out of phase (or somewhat independently) with the remaining mass elements. It follows that a continuous system (or a distributed-parameter system) will have an infinite number of degrees of freedom and will require an infinite number of coordinates to represent its motion. In other words, when extending the concept of a finite-degree-of-freedom system as analyzed previously, an infinite-dimensional vector is needed to represent the general motion of a continuous system. Equivalently, a one-dimensional continuous system (a line structure) will need one independent spatial variable, in addition to time, to represent its response. In view of the need for two independent variables in this case — one for time and the other for space — the representation of system dynamics will require partial differential equations (PDEs) rather than ordinary differential equations (ODEs). Furthermore, the system will depend on the boundary conditions as well as the initial conditions.

The present chapter concerns vibration analysis of continuous systems. Strings, cables, rods, shafts, beams, membranes, plates, and shells are example of continuous members. In special cases, closed-form analytical solutions can be obtained for the vibration of these members. A general structure may consist of more than one such member and, furthermore, boundary conditions may be various, individual members may be nonuniform, and the material characteristics may be inhomogeneous and anistropic. Closed-form analytical solutions would not be generally possible in such cases. Nevertheless, the insight gained by analyzing the vibration of standard members will be quite beneficial in studying the vibration behavior of more complex structures. The vibration analysis of a few representative continuous members is discussed in this chapter.

The concepts of modal analysis can be extended from lumped-parameter systems to continuous systems. In particular, because the number of principal modes is equal to the number of degrees of freedom of the system, a distributed-parameter system will have an infinite number of natural modes of vibration. A particular mode can be excited by deflecting the member so that its *elastic curve* assumes the shape of the particular mode, and then releasing from this initial condition.

When damping is significant and nonproportional, however, there is no guarantee that such an initial condition could accurately excite the required mode. A general excitation consisting of a force or an initial condition will excite more than one mode of motion. But, as in the case of discrete-parameter systems, the general motion can be analyzed and expressed in terms of modal motions, through modal analysis. As discussed in Chapter 5, in a modal motion, the mass elements will move at a specific frequency (the natural frequency), and bearing a constant proportion in displacement (i.e., maintaining the mode shape), and passing the static equilibrium of the system simultaneously. In view of this behavior, it is possible to separate the time response and spatial response of a vibrating system in a modal motion. This separability is fundamental to modal analysis of a continuous system. Furthermore, in practice, all infinite number of natural frequencies and mode shapes are not significant and typically the very high modes can be neglected. Such a modal-truncation procedure, although carried out by continuous-system analysis, is equivalent to approximating the original infinite-degree-of-freedom system by a finite-degree-of-freedom system. Vibration analysis of continuous systems can be applied in modeling, analysis, design, and evaluation of such practical systems as cables; musical instruments; transmission belts and chains; containers of fluid; animals; structures including buildings, bridges, guideways, and space stations; and transit vehicles, including automobiles, ships, aircraft, and spacecraft.

6.1 TRANSVERSE VIBRATION OF CABLES

The first continuous member to be studied in this chapter is a string or cable in tension. This is a line structure for which its geometric configuration can be completely defined by the position of its axial line, with reference to a fixed coordinate line. We will study the transverse (lateral) vibration problem; that is, the vibration in a direction perpendicular to its axis and in a single plane. Applications will include stringed musical instruments, overhead transmission lines (of electric power or telephone signals), drive systems (belt drives, chain drives, pulley ropes, etc.), suspension bridges, and structural cables carrying cars (e.g., ski lifts, elevators, overhead sightseeing systems, and cable cars).

As usual, some simplifying assumptions will be made for analytical convenience; but the results and insight obtained in this manner will be useful in understanding the behavior of more complex systems containing cable-like structures. The main assumptions are:

1. The system is a line structure. The lateral dimensions are much smaller compared to the longitudinal dimension (normally in the x direction).
2. The structure stays in a single plane, and the motion of every element of the structure will be in a fixed transverse direction (y).
3. The cable tension (T) remains constant during motion. In other words, the initial tension is sufficiently large that the variations during motion are negligible.
4. Variations in slope (θ) along the structure are small. Hence, for example,

$$\theta \cong \sin\theta \cong \tan\theta = \frac{\partial v}{\partial x}.$$

A general configuration of a cable (or string) is shown in Figure 6.1(a). Consider a small element of length dx of the cable at location x, as shown in Figure 6.1(b). The equation (Newton's second law) of motion (transverse) of this element is given by

$$f(x,t)dx - T\sin\theta + T\sin(\theta + d\theta) = m(x)\cdot dx \frac{\partial^2 v(x,t)}{\partial t^2} \tag{6.1}$$

Distributed-Parameter Systems

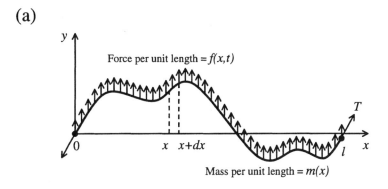

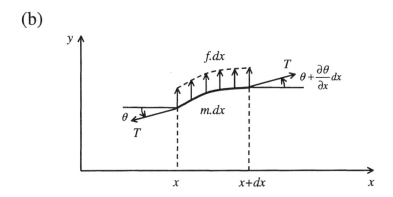

FIGURE 6.1 (a) Transverse vibration of a cable in tension, and (b) motion of a general element.

where

$v(x,t)$ = transverse displacement of the cable
$f(x,t)$ = lateral force per unit length of the cable
$m(x)$ = mass per unit length of the cable
T = cable tension
θ = cable slope at location x.

Note that the dynamic loading $f(x,t)$ may arise due to such causes as aerodynamic forces, fluid drag, and electromagnetic forces, depending on the specific application.

Using the small slope assumption, one obtains $\sin\theta \cong \theta$ and $\sin(\theta + d\theta) \cong \theta + d\theta$ with $\theta = \dfrac{\partial v}{\partial x}$ and $d\theta = \dfrac{\partial^2 v}{\partial x^2} dx$ as $dx \to 0$

On substitution of these approximations into equation (6.1) and canceling out dx, one obtains

$$m(x)\frac{\partial^2 v(x,t)}{\partial t^2} = T\frac{\partial^2 v(x,t)}{\partial x^2} + f(x,t) \tag{6.2}$$

Now consider the case of free vibration, where $f(x,t) = 0$. Then,

$$\frac{\partial^2 v(x,t)}{\partial t^2} = c^2 \frac{\partial^2 v(x,t)}{\partial x^2} \tag{6.3}$$

with

$$c = \sqrt{T/m} \qquad (6.4)$$

Also assume that the cable is uniform so that m is constant.

6.1.1 Wave Equation

The solution to any equation of the form (6.3) will appear as a wave, traveling either in the forward (+ve x) or in the backward (–ve x) direction at speed c. Hence, (6.3) is called the wave equation and c is the *wave speed*. To prove this fact, first show that a solution to equation (6.3) can take the form

$$v(x,t) = v_1(x - ct) \qquad (6.5)$$

To this end, let $x - ct = z$. Hence, $v_1(x - ct) = v_1(z)$
Then,

$$\frac{\partial v_1}{\partial x} = \frac{dv_1}{dz} \cdot \frac{\partial z}{\partial x} \quad \text{and} \quad \frac{\partial v_1}{\partial t} = \frac{dv_1}{dz} \cdot \frac{\partial z}{\partial t}$$

with

$$\frac{\partial z}{\partial x} = 1 \quad \text{and} \quad \frac{\partial z}{\partial t} = -c$$

It follows that

$$\frac{\partial^2 v_1}{\partial x^2} = v_1'' \quad \text{and} \quad \frac{\partial^2 v_1}{\partial t^2} = c^2 v_1''$$

where

$$v_1'' = \frac{d^2 v_1}{dz^2}$$

Clearly then v_1 satisfies equation (6.3).

Now examine the nature of the solution $v_1(x - ct)$. It is clear that v_1 will be constant when $x - ct$ = constant. But, the equation $x - ct$ = constant corresponds to a point moving along the x-axis in the positive direction at speed c. What this means is that the shape of the cable at $t = 0$ will "appear" to travel along the cable at speed c. This is analogous to the waves one observes in a pond when excited by dropping a stone. Note that the particles of the cable do not travel along x, and it is the deformation "shape" (the wave) that travels.

Similarly, it can be shown that

$$v(x,t) = v_2(x + ct) \qquad (6.6)$$

Distributed-Parameter Systems

is also a solution to equation (6.3), and this corresponds to a wave that travels backward (–ve x direction) at speed c. The general solution, of course, will be of the form

$$v(x,t) = v_1(x - ct) + v_2(x + ct) \tag{6.7}$$

which represents two waves, one traveling forward and the other backward.

6.1.2 General (Modal) Solution

As usual in modal analysis, one looks for a separable solution of the form

$$v(x,t) = Y(x) \cdot q(t) \tag{6.8}$$

for the cable/string vibration problem given by the wave equation (6.3). If a solution of the form of equation (6.8) is obtained, it will be essentially a modal solution. This should be clear from the separability of the solution. Specifically, at any given time t, the time function $q(t)$ will be fixed, and the structure will have a shape give by $Y(x)$. Hence, at all times, the structure will maintain a particular "shape" $Y(x)$ and this will be a *mode shape*. Also, at a given point x of the structure, the value of the space function $Y(x)$ will be fixed, and the structure will vibrate according to the time response $q(t)$. It will be shown that $q(t)$ will obey the simple harmonic motion of a specific frequency. This is the *natural frequency* of vibration corresponding to that particular mode. Note that, for a continuous system, there will be an infinite number of solutions of the form (6.8), with different natural frequencies. The corresponding functions $Y(x)$ will be "orthogonal" in some sense. Hence, they are called *normal modes* (normal meaning "perpendicular"). The systems will be able to move independently in each mode, and this collection of solutions of the form (6.8) will be a complete set. With this qualitative understanding, one can now seek a solution of the form of equation (6.8) for the system equation (6.3).

Substitute equation (6.8) in (6.3) to obtain

$$Y(x) \frac{d^2 q(t)}{dt^2} = c^2 \frac{d^2 Y(\alpha)}{dx^2} q(t)$$

or

$$\frac{1}{Y(x)} \frac{d^2 Y(x)}{dx^2} = \frac{1}{c^2 q(t)} \frac{d^2 q(t)}{dt^2} = -\lambda^2 \tag{6.9}$$

In equation (6.9), because the left-hand terms are a function of x only and the right-hand terms are a function of t only, for the two sides to be equal in general, each function should be a constant (that is, independent of both x and t). This constant is denoted by $-\lambda^2$, which is called the *separation constant* and is designated to be negative. There are two good reasons for that. If this common constant were positive, the function $q(t)$ would be nonoscillatory and transient, which is contrary to the nature of undamped vibration. Furthermore, it can be shown that a nontrivial solution for $Y(x)$ would not be possible if the common constant were positive.

The unknown constant λ is determined by solving the space equation (mode shape equation) of (6.9); specifically,

$$\frac{d^2Y(x)}{dx^2} + \lambda^2 Y(x) = 0 \tag{6.10}$$

and then applying the boundary conditions of the problem. There will be an infinite number of solutions for λ, with corresponding natural frequencies ω and mode shapes $Y(x)$.

The characteristic equation of (6.10) is

$$p^2 + \lambda^2 = 0 \tag{6.11}$$

which has the *characteristic roots* (or eigenvalues)

$$p = \pm j\lambda \tag{6.12}$$

The general solution is

$$Y(x) = A_1 e^{j\lambda x} + A_2 e^{-j\lambda x}$$
$$= C_1 \cos \lambda x + C_2 \sin \lambda x \tag{6.13}$$

Note that, since $Y(x)$ is a real function representing a geometric shape, the constants A_1 and A_2 have to be complex conjugates and C_1 and C_2 have to be real. Specifically, in view of the fact that $\cos \lambda x = \dfrac{e^{j\lambda x} + e^{-j\lambda x}}{2}$ and $\sin \lambda x = \dfrac{e^{j\lambda x} - e^{-j\lambda x}}{2j}$, one can show that

$$A_1 = \frac{1}{2}(C_1 - jC_2) \quad \text{and} \quad A_2 = \frac{1}{2}(C_1 + jC_2)$$

For analytical convenience, use the real-parameter form of equation (6.13).

Note that one cannot determine both constants C_1 and C_2 using boundary conditions. Only their ratio is determined, and the constant multiplier is absorbed into $q(t)$ in equation (6.8) and then determined using the appropriate initial conditions (at $t = 0$). It follows that the ratio of C_1 and C_2 and the value of λ are determined using the boundary conditions. Two boundary conditions will be needed. Some useful situations and appropriate relations are given in Table 6.1.

6.1.3 CABLE WITH FIXED ENDS

One can obtain the complete solution for free vibration of a taut cable that is fixed at both ends. The applicable boundary conditions are

$$Y(0) = Y(l) = 0 \tag{6.14}$$

where l is the length of the cable. Substitution into equation (6.13) gives

$$C_1 \times 1 + C_2 \times 0 = 0$$

$$C_1 \cos \lambda l + C_2 \sin \lambda l = 0$$

Hence,

TABLE 6.1
Some Useful Boundary Conditions for the Cable Vibration Problem

Type of End Condition	Nature of End $x = x_o$	Boundary Condition	Modal Boundary Condition
Fixed		$v(x_o, t) = 0$	$Y_i(x_o) = 0$
Free		$T \dfrac{\partial v(x_o, t)}{\partial x} = 0$	$\dfrac{dY_i(x_o)}{dx} = 0$
Flexible		$T \dfrac{\partial v(x_o, t)}{\partial x} - kv(x_o, t) = 0$	$T \dfrac{dY_i(x_o)}{dx} - kY_i(x_o) = 0$
Flexible and inertial		$T \dfrac{\partial v(x_o, t)}{\partial x} - kv(x_o, t) = M \dfrac{\partial^2 v(x_o, t)}{\partial t^2}$	$T \dfrac{dY_i(x_o)}{dx} - (k - \omega_i^2 M) Y_i(x_o) = 0$

$$C_1 = 0 \quad \text{and}$$

$$C_2 \sin \lambda l = 0 \tag{6.15}$$

A possible solution for equation (6.15) is $C_2 = 0$. But this is the *trivial solution*, which corresponds to $Y(x) = 0$; that is, a stationary cable with no vibration. It follows that the applicable, nontrivial solution is

$$\sin \lambda l = 0$$

which produces an *infinite* number of solutions for λ given by

$$\lambda_i = \frac{i\pi}{l} \quad \text{with, } i = 1, 2, \ldots, \infty \tag{6.16}$$

As mentioned earlier, the corresponding infinite number of mode shapes is given by

$$Y_i(x) = C_i \sin\frac{i\pi x}{l} \tag{6.17}$$

for $i = 1, 2, \ldots, \infty$.

Note: If one had used a positive constant λ^2 instead of $-\lambda^2$ in equation (6.9), only a trivial solution (with $C_1 = 0$ and $C_2 = 0$) would be possible for $Y(x)$. This further justifies the decision to use $-\lambda^2$. Substitute equation (6.16) into (6.9) to determine the corresponding time response (generalized coordinates) $q_i(t)$; thus,

$$\frac{d^2 q_i(t)}{dt^2} + \omega_i^2 q_i(t) = 0 \tag{6.18}$$

in which

$$\omega_i = \lambda_i c = \frac{i\pi}{l}\sqrt{\frac{T}{m}} \quad \text{for } i = 1, 2, \ldots, \infty \tag{6.19}$$

Equation (6.18) represents a simple harmonic motion with the *modal natural frequencies* ω_i given by equation (6.19). It follows that there are an infinite number of natural frequencies, as mentioned earlier. The general solution of equation (6.18) is given by

$$q_i(t) = c_i \sin(\omega_i t + \phi_i) \tag{6.20}$$

where the amplitude parameter c_i and the phase parameter ϕ_i are determined using two initial conditions of the system. It should be clear that it is redundant to use a separate constant C_i for $Y_i(x)$ in equation (6.17) as it can be absorbed into the amplitude constant in equation (6.20), to express the general free response of the cable as

$$v(x,t) = \sum c_i \sin\frac{i\pi x}{l} \sin(\omega_i t + \phi_i) \tag{6.21}$$

In this manner, the complete solution has been expressed as a summation of the modal solutions. This is known as the *modal series expansion*. Such a solution is quite justified because of the fact that the mode shapes are *orthogonal* in some sense, and what was obtained above is a complete set of *normal modes* (normal in the sense of perpendicular or orthogonal). The system is able to move independently in each mode, with a unique spatial shape, at the corresponding natural frequency, because each modal solution is separable into a space function $Y_i(x)$ and a time function (generalized coordinate) $q_i(t)$. Of course, the system will be able to simultaneously move in a linear combination of two modes (say, $C_1 Y_1(x) q_1(t) + C_2 Y_2(x) q_2(t)$) because this combination satisfies the original system equation (6.3), in view of its linearity, and because each modal component satisfies the equation. But, clearly, this solution (with two modes) is not separable into a product of a space

Distributed-Parameter Systems

function and a time function. Hence, it is not a modal solution. In this manner, it can be argued that the infinite sum of modal solutions $\sum c_i Y_i(x) q_i(t)$ is the most general solution to system (6.3). Orthogonality of mode shapes plays a key role in this argument and, furthermore, it is useful in the analysis of the system. In particular, in equation (6.21), the unknown constants c_i and ϕ_i are determined using the system initial conditions, and the orthogonality property of modes is useful in that procedure. This will be addressed next.

6.1.4 Orthogonality of Natural Modes

A cable can vibrate at frequency ω_i called natural frequency while maintaining a unique natural shape $Y_i(x)$ called mode shape of the cable. It has been shown that for the fixed-ended cable, the natural mode shapes are given by $\sin\frac{i\pi x}{l}$, with the corresponding natural frequencies ω_i given by equation (6.19). It can be easily verified that

$$\int_0^l \sin\frac{i\pi x}{l}\sin\frac{j\pi x}{l}dx = 0 \quad \text{for } i \neq j \tag{6.22}$$

$$= \frac{l}{2} \quad \text{for } i = j$$

In other words, the natural modes are orthogonal. Equation (6.22) represents the principle of orthogonality of natural modes in this case.

Orthogonality makes the modal solutions independent and the corresponding mode shapes "normal." It also makes the infinite set of modal solutions a complete set or a *basis* so that any arbitrary response can be formed as a linear combination of these normal mode solutions.

The orthogonality holds for other types of boundary conditions as well. To show this, one sees from equation (6.9) that

$$\frac{d^2 Y_i(x)}{dx^2} + \lambda_i^2 Y_i(x) = 0 \quad \text{for mode } i \tag{6.23}$$

$$\frac{d^2 Y_j(x)}{dx^2} + \lambda_j^2 Y_j(x) = 0 \quad \text{for mode } j \tag{6.24}$$

Multiply equation (6.23) by $Y_j(x)$, equation (6.24) by $Y_i(x)$, subtract that second result from the first and integrate with respect to x along the cable length from $x = 0$ to l. One obtains

$$\int_0^l \left[Y_j \frac{d^2 Y_i}{dx^2} - Y_i \frac{d^2 Y_j}{dx^2} \right] dx + \left(\lambda_i^2 - \lambda_j^2\right) \int_0^l Y_i Y_j dx = 0 \tag{6.25}$$

Integrating by parts, one obtains the results

$$\int_0^l Y_j \frac{d^2 Y_i}{dx^2} dx = Y_j \frac{dY_i}{dx}\bigg|_0^l - \int_0^l \frac{dY_i}{dx}\frac{dY_j}{dx} dx$$

$$\int_0^l Y_j \frac{d^2 Y_j}{dx^2} dx = Y_i \frac{dY_j}{dx}\bigg|_0^l - \int_0^l \frac{dY_i}{dx} \frac{dY_j}{dx} dx$$

Hence, the first term of equation (6.25) becomes

$$\left[Y_j \frac{dY_i}{dx} - Y_i \frac{dY_j}{dx} \right]_0^l$$

which will vanish for typical boundary conditions. Then, since $\lambda_i \neq \lambda_j$ for $i \neq j$, one has

$$\int_0^l Y_i(x) Y_j(x) dx = 0 \quad \text{for } i \neq j$$

One can pick the value of the multiplication constant in the general solution for $Y(x)$, given by equation (6.13), so as to *normalize* the mode shapes such that

$$\int_0^l Y_i^2(x) dx = \frac{l}{2}$$

which is consistent with the result (6.22). Hence, the general condition of orthogonality of natural modes, may be expressed as

$$\int_0^l Y_i(x) \cdot Y_j(x) dx = 0 \quad \text{for } i \neq j$$

$$= \frac{l}{2} \quad \text{for } i = j$$

(6.26)

Nodes: When vibrating in a particular mode, one or more points of the system (cable) that are not physically fixed, can remain stationary at all times. These points are called *nodes* of that mode. For example, in the second mode of a cable with its ends fixed, there will be a node at the midspan. This should be clear from the fact that the mode shape of the second mode is $\sin 2\pi x/l$, which becomes zero at $x = l/2$. Similarly, in the third mode, with mode shape $\sin 3\pi x/l$, there will be nodes at $x = l/3$ and $2l/3$.

EXAMPLE 6.1

If the cable tension varies along the length x, what is the corresponding equation of free lateral vibration?

A hoist mechanism has a rope of freely hanging length l in a particular equilibrium configuration and carrying a load of mass M, as shown in Figure 6.2(a). Determine the equation of lateral vibration and the applicable boundary conditions for the rope segment.

SOLUTION

With reference to Figure 6.1(b), equation (6.1) can be modified for the case of variable T as

Distributed-Parameter Systems

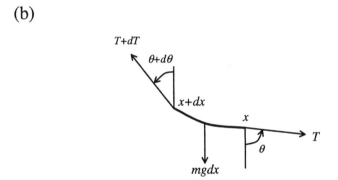

FIGURE 6.2 (a) Free segment of a stationary hoist, and (b) a small element of the rope.

$$-T\sin\theta + (T+dT)\sin(\theta+d\theta) = mdx\frac{\partial^2 v}{\partial t^2}dx \qquad (6.27)$$

where $f(x,t) = 0$ for free vibration. Now, with the assumption of small θ and by neglecting the second-order product term $dT\,d\theta$, one obtains

$$Td\theta + \theta dT = m\frac{\partial^2 v}{\partial x^2}dx$$

Next, using

$\theta = \dfrac{\partial v}{\partial x}$, $d\theta = \dfrac{\partial^2 v}{\partial x^2}$, and $dT = \dfrac{\partial T}{\partial x}dx$, and canceling dx, one gets the equation of lateral vibration of a cable as

$$m\frac{\partial^2 v}{\partial t^2} = T\frac{\partial^2 v}{\partial x^2} + \frac{\partial T}{\partial x}\frac{\partial v}{\partial x} \qquad (6.28)$$

Longitudinal (axial) dynamics of the rope are negligible for the case of a stationary hoist. Then, longitudinal equilibrium (in the x direction) of the small element of rope shown in Figure 6.2(b) gives

$$(T+dT)\cos(\theta+d\theta) - T\cos\theta - mgdx = 0$$

For small θ, $\cos\theta \cong 1$ and $\cos(\theta + d\theta) \cong 1$ up to the first-order term in the Taylor series expansion. Hence,

$$dT = mgdx \tag{6.29}$$

Integration gives

$$T = T_o + mgx \tag{6.30}$$

with the end condition

$$T = Mg \quad \text{at } x = 0$$

Hence,

$$T = Mg + mgx \tag{6.31}$$

Note from equation (6.29) that $\dfrac{\partial T}{\partial x} = \dfrac{dT}{dx} = mg$ for this problem. Substitute in (6.28) this fact and equation (6.31) to obtain

$$m\frac{\partial^2 v}{\partial t^2} = (M + mx)g\frac{\partial^2 v}{\partial x^2} + mg\frac{\partial v}{\partial x}$$

or

$$\frac{\partial^2 v}{\partial t^2} = \left(\frac{M}{m} + x\right)g\frac{\partial^2 v}{\partial x^2} + g\frac{\partial v}{\partial x} \tag{6.32}$$

The boundary condition at $x = 0$ is obtained by applying Newton's second law to the end mass in the lateral (y) direction. This gives

$$T_o \frac{\partial v(0,t)}{\partial x} = M\frac{\partial^2 v(0,t)}{\partial t^2}$$

Now, using the fact that $T_o = Mg$, one has the boundary condition:

$$g\frac{\partial v(0,t)}{\partial x} = \frac{\partial^2 v(0,t)}{\partial t^2}$$

For mode i

$$\frac{\partial v(0,t)}{\partial x} = \frac{dY_i(0)}{dx}q_i(t)$$

and

$$\frac{\partial^2 v(0,t)}{\partial t^2} = Y_i(0)\frac{d^2 q_i(t)}{dt^2} = -\omega_i^2 Y_i(0)q_i(t)$$

which holds for all t, and where ω_i is the ith natural frequency of vibration. Hence, the modal boundary condition at $x = 0$ is

$$g\frac{dY_i(0)}{dx} + \omega_i^2 Y_i(0) = 0 \qquad \text{for } i = 1, 2, \ldots \tag{6.33}$$

The boundary condition at $x = l$ is

$$v(l,t) = 0 \tag{6.34}$$

which hold for all t. Hence, the corresponding modal boundary condition is

$$Y_i(l) = 0 \qquad \text{for } i = 1, 2, \ldots \tag{6.35}$$

$\square$

6.1.5 Application of Initial Conditions

The general solution to the cable vibration problem is given by

$$v(x,t) = \sum c_i Y_i(x)\sin(\omega_i t + \phi_i) \tag{6.36}$$

where $Y_i(x)$ are the *normalized* mode shapes that satisfy the orthogonality property (6.26). The unknown constants c_i and ϕ_i are determined using the initial conditions

$$v(x,0) = d(x) \tag{6.37}$$

$$\frac{\partial v(x,0)}{\partial t} = s(x) \tag{6.38}$$

By substituting equation (6.36) into (6.37) and (6.38), one obtains

$$d(x) = \sum c_i Y_i(x)\sin\phi_i \tag{6.39}$$

$$s(x) = \sum c_i \omega_i Y_i(x)\cos\phi_i \tag{6.40}$$

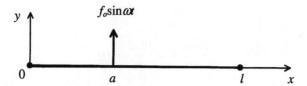

FIGURE 6.3 A cable excited by a point harmonic force.

Multiply equations (6.39) and (6.40) by $Y_j(x)$ and integrate with respect to x from 0 to l, making use of the orthogonality condition (6.26). Thus,

$$\int_0^l d(x)Y_j(x)dx = c_j \frac{l}{2}\sin\phi_j$$

$$\int_0^l s(x)Y_j(x)dx = c_j \omega_j \frac{l}{2}\cos\phi_j$$

Solving these two equations, one obtains

$$\tan\phi_j = \omega_j \frac{\int_0^l d(x)Y_j(x)dx}{\int_0^l s(x)Y_j(x)dx} \qquad \text{for } j = 1, 2, 3, \ldots \qquad (6.41)$$

Once ϕ_j is determined in this manner, one can obtain c_j using

$$c_j = \frac{2}{l\sin\phi_j}\int_0^l d(x)Y_j(x)dx \qquad \text{for } j = 1, 2, 3, \ldots \qquad (6.42)$$

EXAMPLE 6.2

Consider a taut horizontal cable of length l and mass m per unit length, as shown in Figure 6.3, excited by a transverse point force $f_0\sin\omega t$ at location $x = a$, where ω is the frequency of (harmonic) excitation and f_0 is the forcing amplitude. Determine the resulting response of the cable under general end conditions and initial conditions. For the special case of fixed ends, what is the steady-state response of the cable?

SOLUTION

It has been shown that the forced transverse response of a cable is given by equation (6.2):

$$\frac{\partial^2 v(x,t)}{\partial t^2} = c^2 \frac{\partial^2 v(x,t)}{\partial x^2} + \frac{f(x,t)}{m} \qquad (6.2)$$

where $v(x,t)$ is the transverse displacement and $f(x,t)$ is the external force per unit length of the cable.

Distributed-Parameter Systems

For the point force F at $x = a$, the analytical representation of the equivalent distributed force per unit length is

$$f(x,t) = F\delta(x-a) \tag{6.43}$$

where the Dirac delta function (unit impulse function) $\delta(x)$ is such that

$$\int_{a_1}^{a_2} g(x)\delta(x-a)dx = g(a) \tag{i}$$

for an arbitrary function $g(x)$, provided that the point a is within the interval of integration $[a_1, a_2]$. Seek a "modal superposition" solution of the form

$$v(x,t) = \sum Y_i(x)\bar{q}_i(t) \tag{6.44}$$

where $\bar{q}_i(t)$ are the generalized coordinates of the forced response solution (which are generally different from those for the free solution, i.e., $q_i(t)$).

Substitute the solution (6.44) into the system equation (6.2) and make use of the governing equation of the mode shapes (see equation (6.10)).

$$\frac{d^2 Y_i(x)}{dx^2} = -\lambda_i^2 Y_i(x) \tag{6.45}$$

we get

$$m \sum Y_i(x)\ddot{\bar{q}}_i(t) = -T \sum \lambda_i^2 Y_i(x)\bar{q}_i(t) + f_0 \sin\omega t \, \delta(x-a) \tag{ii}$$

Multiply equation (ii) by $Y_j(x)$, and integrate from $x = 0$ to l using the orthogonality property (6.26) and also (i). One obtains

$$m\frac{l}{2}\ddot{\bar{q}}_j(t) = -T\frac{l}{2}\lambda_j^2 \bar{q}_j(t) + f_0 Y_j(a)\sin\omega t$$

Because $\omega_j = \lambda_j \sqrt{T/m}$ (see equation (6.19)), one obtains

$$\ddot{\bar{q}}_j(t) + \omega_j^2 \bar{q}_j(t) = \frac{2f_0}{lm} Y_j(a)\sin\omega t \qquad \text{for } j = 1, 2, 3, \ldots \tag{6.46}$$

This has the familiar form of a simple oscillator excited by a harmonic force and its solution is well known. The initial conditions $\bar{q}_j(0)$ and $\dot{\bar{q}}_j(0)$ are needed. Suppose that the initial transverse displacement and the speed of the cable are

$$v(x,0) = d(x) \quad \text{and} \quad \dot{v}(x,0) = s(x)$$

Then, in view of equation (6.44), one can write

$$\sum_i Y_i(x)\bar{q}_i(0) = d(x) \qquad (6.47)$$

$$\sum_i Y_i(x)\dot{\bar{q}}_i(0) = s(x) \qquad (6.48)$$

Multiply equations (6.47) and (6.48) by $Y_j(x)$ and integrate from $x = 0$ to l using the orthogonality property (6.26). The necessary initial conditions are obtained:

$$\bar{q}_j(0) = \frac{2}{l}\int_0^l d(x)Y_j(x)dx \qquad (6.49)$$

$$\dot{\bar{q}}_j(0) = \frac{2}{l}\int_0^l s(x)Y_j(x)dx \qquad (6.50)$$

which will provide the complete solution for equation (6.46) and, hence, will completely determine (6.44).

For a fixed ended cable,

$$Y_i(x) = \sin\frac{i\pi x}{l} \qquad \text{(iii)}$$

and at steady state, the time response $\bar{q}_j(t)$ will be harmonic at the same frequency as the excitation frequency ω. Hence,

$$\bar{q}_j(t) = q_{0j}\sin(\omega t + \phi_j) \qquad (6.51)$$

For equation (6.51) to satisfy (6.46) in this undamped problem, one must have $\phi_j = 0$. Direct substitution gives

$$\left[-\omega^2 + \omega_j^2\right]q_{0j} = \frac{2f_0}{lm}Y_j(a)$$

which determines q_{0j}. Hence, from equation (6.45), the complete solution for the fixed-ended problem, at steady state, is

$$v(x,t) = \frac{2f_0}{lm}\sin\omega t \sum \frac{\sin i\pi a/l}{(\omega_i^2 - \omega^2)}\sin\frac{i\pi x}{l} \qquad (6.52)$$

□

Some important results for transverse vibration of strings and cables are summarized in Box 6.1.

Distributed-Parameter Systems

BOX 6.1 Transverse Vibration of Strings and Cables

Equation of motion:

$$m(x)\frac{\partial^2 v(x,t)}{\partial t^2} = T\frac{\partial^2 v(x,t)}{\partial x^2} + f(x,t)$$

Separable (modal) solution for free vibration:

$$v(x,t) = \sum Y_i(x)q_i(t)$$

with

$$\frac{d^2 Y_i(x)}{dx^2} + \lambda_i^2 Y_i(x) = 0$$

(Needs two boundary conditions)

and

$$\frac{d^2 q_i(t)}{dt^2} + \omega_i^2 q_i(t) = 0$$

(Needs two initial conditions)

Natural frequency: $\omega_i = \lambda_i c$

Wave speed: $c = \sqrt{\dfrac{T}{m}}$

Traveling-wave solution (long cable, independent of end conditions):

$$v(x,t) = v_1(x - ct) + v_2(x + ct)$$

Orthogonality:

$$\int_0^l Y_i(x)Y_j(x)dx = 0 \quad \text{for } i \neq j$$

$$= \frac{l}{2} \quad \text{for } i = j$$

Initial conditions:
(for initial displacement $d(x)$ and speed $s(x)$)

$$q_i(0) = \frac{2}{l}\int_0^l d(x)Y_i(x)dx$$

$$\dot{q}_i(0) = \frac{2}{l}\int_0^l s(x)Y_i(x)dx$$

Variable-tension problem:

$$m\frac{\partial^2 v}{\partial t^2} = T\frac{\partial^2 v}{\partial x^2} + \frac{\partial T}{\partial x}\frac{\partial v}{\partial x}$$

6.2 LONGITUDINAL VIBRATION OF RODS

It can be shown that the governing equation of longitudinal vibration of line structures such as rods and bars is identical to that of the transverse vibration of cables and strings. Hence, it is not necessary to repeat the complete analysis here. This section first develops the equation of motion, then considers boundary conditions, next identifies the similarity with the cable vibration problem, and concludes with an illustrative example.

6.2.1 EQUATION OF MOTION

Consider a rod that is mounted horizontally (so that the gravitational effects can be neglected) as shown in Figure 6.4(a). A small element of length dx (the limiting case of δx) at position x is shown in Figure 6.4(b). The longitudinal strain at x is given by

$$\varepsilon = \frac{\partial u}{\partial x} \tag{6.53}$$

where

$u(x,t)$ = longitudinal displacement of the rod at distance x from a fixed reference.

Note that the fixed reference can be chosen arbitrarily. However, if the assumption of small u is needed, the reference may be chosen as the relaxed (unstrained) position of the element. The longitudinal stress at the cross section at x is $\sigma = E\varepsilon$ and, hence, the longitudinal force is

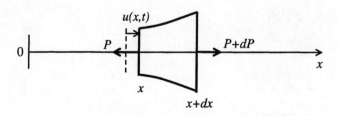

FIGURE 6.4 (a) A rod with distributed loading and in longitudinal vibration; and (b) a small element of the rod.

Distributed-Parameter Systems

$$P = EA \frac{\partial u}{\partial x} \tag{6.54}$$

where

E = Young's modulus of the rod
A = area of cross section.

It is not necessary at this point to assume a uniform rod. Hence, A may depend on x. The equation of motion for the small element shown in Figure 6.4(b) is

$$\rho A dx \frac{\partial^2 u(x,t)}{\partial t^2} = P + dP - P + f(x,t)dx$$

or

$$\rho A \frac{\partial^2 u}{\partial t^2} dx = dP + f(x,t)dx \tag{6.55}$$

From equation (6.54), one has

$$dP = \frac{\partial}{\partial x} EA(x) \frac{\partial u}{\partial x} dx \tag{6.56}$$

which when substituted into (6.55) gives

$$\rho A \frac{\partial^2 u(x,t)}{\partial t^2} = \frac{\partial}{\partial x} EA(x) \frac{\partial u(x,t)}{\partial x} + f(x,t) \tag{6.57}$$

For the case of a uniform rod (constant A) in free vibration ($f(x,t) = 0$),

$$\frac{\partial^2 u}{\partial t^2} = c^2 \frac{\partial^2 u(x,t)}{\partial x^2} \tag{6.58}$$

which is identical to the cable vibration equation (6.3), but with the wave speed parameter given by

$$c = \sqrt{\frac{E}{\rho}} \tag{6.59}$$

which should be compared with equation (6.4). The analysis of the present problem may be done exactly as for the cable vibration. In particular, the traveling wave solution will hold. Mode shape orthogonality will hold as well. Even the boundary conditions are similar to that for the cable vibration problem.

6.2.2 Boundary Conditions

As for the cable vibration problem, two boundary conditions will be needed, along with two initial conditions, in order to obtain the complete solution to the longitudinal vibration of a rod. Both free and forced vibration can be analyzed as before. For a fixed end at $x = x_0$, there will be no deflection. Hence,

$$u(x_0, t) = 0 \qquad (6.60)$$

with the corresponding modal end condition

$$X_i(x_0) = 0 \qquad \text{for } i = 1, 2, 3, \ldots \qquad (6.61)$$

For a free end at $x = x_0$, there will not be an end force. Hence, in view of equation (6.54), the applicable boundary condition will be

$$\frac{\partial u(x_0, t)}{\partial x} = 0 \qquad (6.62)$$

with the corresponding modal boundary condition

$$\frac{dX_i(x_0)}{dx} = 0 \qquad \text{for } i = 1, 2, 3, \ldots \qquad (6.63)$$

The mode shapes $X_i(x)$ will satisfy the orthogonality property

$$\int_0^l X_i(x) X_j(x) dx = 0 \qquad \text{for } i \neq j$$
$$= l_j \qquad \text{for } i = j \qquad (6.64)$$

as before. It can be easily verified, for example, that for a rod with both ends fixed,

$$X_i(x) = \sin\frac{i\pi x}{l} \qquad (6.65)$$

EXAMPLE 6.3

A uniform structural column of length l, mass M, and area of cross section A hangs from a rigid platform and is supported on a flexible base of stiffness k. A model is shown in Figure 6.5. Initially, the system remains stationary, in static equilibrium. Suddenly, an axial (vertical) speed of u_0 is imparted uniformly on the entire column, due to a seismic jolt. Determine the subsequent vibration motion of the column from its initial equilibrium configuration.

SOLUTION

The gravitational force corresponds to a force per unit length

$$f(x,t) = \frac{Mg}{l}$$

and equation (6.57) becomes

$$\frac{\partial^2 u(x,t)}{\partial t^2} = c^2 \frac{\partial^2 u(x,t)}{\partial x^2} + \frac{Mg}{\rho Al}$$

But $M = \rho Al$. Hence,

$$\frac{\partial^2 u(x,t)}{\partial t^2} = c^2 \frac{\partial^2 u(x,t)}{\partial x^2} + g \tag{6.66}$$

Boundary conditions are

$$u(0,t) = 0 \tag{6.67}$$

$$EA \frac{\partial u(l,t)}{\partial x} + k u(l,t) = 0 \tag{6.68}$$

Initial conditions are

$$u(x,0) = 0 \tag{6.69}$$

$$\frac{\partial u(x,0)}{\partial t} = u_0 \tag{6.70}$$

We seek the modal summation solution

$$u(x,t) = \sum X_i(x) q_i(t) \tag{6.71}$$

where the mode shapes $X_i(x)$ satisfy

$$\frac{d^2 X_i(x)}{dx^2} + \lambda_i^2 X_i(x) = 0 \tag{6.72}$$

whose solution is

$$X_i(x) = C_1 \sin \lambda_i x + C_2 \cos \lambda_i x \tag{6.73}$$

According to equations (6.67) and (6.68), the modal boundary conditions are

$$X_i(0) = 0 \tag{6.74}$$

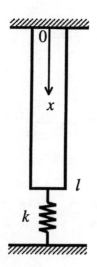

FIGURE 6.5 A column suspended from a fixed platform and supported on a flexible base.

$$EA\frac{dX_i(l)}{dx} + kX_i(l) = 0 \tag{6.75}$$

Substitute equation (6.74) into (6.73). We have $C_2 = 0$. Next, use equation (6.75) and obtain

$$EA\lambda_i C_1 \cos\lambda_i l + kC_1 \sin\lambda_i l = 0$$

Since $C_1 \neq 0$ for a nontrivial solution, the required condition is

$$EA\lambda_i \cos\lambda_i l + k \sin\lambda_i l = 0$$

which can be expressed as

$$\tan\lambda_i l + \frac{EA}{k}\lambda_i = 0 \tag{6.76}$$

This *transcendental equation* has an infinite number of solutions λ_i that correspond to the modes of vibration. The solution can be made computationally and the corresponding natural frequencies are obtained using

$$\omega_i = \lambda_i c = \lambda_i \sqrt{\frac{E}{\rho}} = \lambda_i \sqrt{\frac{EAl}{M}} \tag{6.77}$$

Substitute equation (6.71) in (6.66) and use (6.72) to get

$$\sum X_i(x)\ddot{q}_i(t) = -c^2 \sum \lambda_i^2 X_i(x)q_i(t) + g \tag{6.78}$$

Multiply equation (6.78) by $X_j(x)$ and integrate from $x = 0$ to l, using the orthogonality property (6.64), to get

$$l_j \ddot{q}_j(t) + c^2 \lambda_j^2 l_j q_j(t) = g \int_0^l X_j(x) dx \qquad (6.79)$$

One can normalize the mode shapes as

$$X_i(x) = \sin \lambda_i x \qquad (6.80)$$

where the constant multiplier (C_1) has been absorbed into $q_i(t)$ in equation (6.71). Then,

$$l_j = \int_0^l \sin^2 \lambda_j x \, dx = \int_0^l \frac{1}{2}[1 - \cos 2\lambda_j x] dx$$

$$= \frac{1}{2}\left[x - \frac{1}{2\lambda_j}\sin 2\lambda_j x\right]_0^l = \frac{1}{2}\left[l - \frac{1}{2\lambda_j}\sin 2\lambda_j l\right] \qquad (6.81)$$

and

$$\int_0^l \sin \lambda_j x \, dx = \frac{1}{\lambda_j}[1 - \cos \lambda_j l]$$

Accordingly, equation (6.79) becomes

$$\ddot{q}_j(t) + \omega_j^2 q_j(t) = \frac{g}{l_j \lambda_j}[1 - \cos \lambda_j l] \qquad (6.82)$$

where the RHS is a constant and is completely known from equations (6.81) and (6.76), and ω_j is given by equation (6.77). Now equation (6.82), which corresponds to a simple oscillator with a constant force input, can be solved using any convenient approach. For example, the particular solution is (see Chapter 2)

$$q_{jp} = \frac{g}{\omega_j^2 l_j \lambda_j}[1 - \cos \lambda_j l] \qquad (6.83)$$

and the overall solution is

$$q_j(t) = A_j \sin \omega_j t + B_j \cos \omega_j t + q_{jp} \qquad (6.84)$$

The constants A_j and B_j are determined using the initial conditions $q_j(0)$ and $\dot{q}_j(0)$. These are obtained by substituting equation (6.71) into (6.69) and (6.70), multiplying by $X_j(x)$, and integrating from $x = 0$ to l, making use of the orthogonality property (6.64). Specifically, one obtains

$$q_j(0) = 0 \tag{6.85}$$

and

$$\dot{q}_j(0) = \frac{u_0}{l_j} \int_0^l \sin\lambda_j x\, dx = \frac{u_0}{l_j \lambda_j}\left[1 - \cos\lambda_j l\right] \tag{6.86}$$

□

6.3 TORSIONAL VIBRATION OF SHAFTS

Torsional vibrations are oscillating angular motions of a device about some axis of rotation. Examples are vibration in shafts, rotors, vanes, and propellers. The governing partial differential equation of torsional vibration of a shaft is quite similar to that encountered previously for transverse vibration of a cable in tension and longitudinal vibration of a rod. But, in the present case, the vibrations are rotating (angular) motions with resulting shear strains, shear stresses, and torques in the torsional member. Furthermore, the parameters of the equation of motion will take different meanings. When bending and torsional motions occur simultaneously, there can be some interaction, thereby making the analysis more difficult. Here, one neglects such interactions by assuming that only the torsional effects are present or the motions are quite small.

Because the form of the torsional vibration equation is similar to what was studied before, the same procedures of analysis can be employed and, in particular, the concepts of modal analysis will be similar. But, the torsional parameters will be rather complex for members with noncircular cross sections. For this reason, and also because a vast majority of torsional devices have circular cross sections, this case will be considered first.

6.3.1 SHAFT WITH CIRCULAR CROSS SECTION

Here, one can formulate the problem of torsional vibration of a shaft having circular cross section. The general case of non-uniform cross section, along the shaft, is considered. However, the usual assumptions such as homogeneous, isotropic, and elastic material are made.

First, obtain a relationship between torque (T) and angular deformation or twist (θ) for a circular shaft. Consider a small element of length dx along the shaft axis and observe the cylindrical surface at a general radius r (in the interior of the shaft segment), as shown in Figure 6.6(a). During vibration, this element will deform (twist) through a small angle $d\theta$.

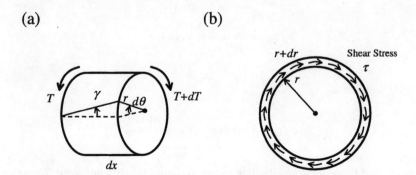

FIGURE 6.6 (a) Small element of a circular shaft in torsion, and (b) shear stresses in a small annular cross section carrying torque.

A point on the circumference will deform through $rd\theta$ as a result, and a longitudinal line on the cylindrical surface will deform through angle γ as shown in Figure 6.6(a). From solid mechanics (strength of materials or theory of elasticity), one knows that γ is the shear strain. Hence,

$$\text{Shear strain } \gamma = \frac{rd\theta}{dx}$$

But, allowing for the fact that the angular shift θ is a function of t as well as x in the general case of dynamics, one uses partial derivatives, and writes

$$\gamma = r\frac{\partial \theta}{\partial x} \tag{6.87}$$

The corresponding shear stress at the deformed point at radius r is

$$\tau = G\gamma = Gr\frac{\partial \theta}{\partial x} \tag{6.88}$$

where

G = shear modulus.

This shear stress acts tangentially. Consider a small annular cross section of width dr at radius r of the shaft, as shown in Figure 6.6(b). By symmetry, the shear stress will be the same throughout this region, and will form a torque of $r \times \tau \times 2\pi r dr = 2\pi r^2 \tau dr$. Hence, the overall torque at the shaft cross section is

$$T = \int 2\pi r^2 \tau dr$$

which, in view of equation (6.88), is written as

$$T = G\frac{\partial \theta}{\partial x} \int 2\pi r^3 dr \tag{6.89}$$

It is clear that the integral term is the *polar moment of area* of the shaft cross section:

$$J = \int 2\pi r^3 dr \tag{6.90}$$

In particular, for a solid shaft of radius r,

$$J = \frac{\pi r^4}{2} \tag{6.91}$$

and for a hollow shaft of inner radius r_1 and the outer radius r_2,

$$J = \frac{\pi}{2}\left(r_2^4 - r_1^4\right) \tag{6.92}$$

Thus, one can write equation (6.89) as

$$T = GJ(x)\frac{\partial \theta}{\partial x} \tag{6.93}$$

The combined parameter GJ is termed the *torsional rigidity* of the shaft. It has been emphasized that the shaft can be non-uniform and, hence, J is a function of x. Consider a uniform shaft segment of length l, with associated overall angular deformation θ. Equation (6.93) can be written as

$$\text{Torsional stiffness } K = \frac{T}{\theta} = \frac{GJ}{l} \tag{6.94}$$

Note: For a shaft with noncircular cross section, replace J by J_t in this equation. It follows that the larger the torsional rigidity GJ, the higher the torsional stiffness K, as expected. Furthermore, longer members have a lower torsional stiffness (and smaller natural frequencies).

Now apply Newton's second law for rotatory motion of the small element dx shown in Figure 6.6(a). The *polar moment of inertia* of the element is $\int r^2 dm = \int r^2 \rho dx dA = \rho dx \int r^2 dA = \rho J dx$, where J is the polar moment of area, as discussed before. Also, suppose that a distributed external torque of $\tau(x,t)$ per unit length is applied along the shaft. Hence, the equation of motion is

$$\rho J dx \frac{\partial^2 \theta}{\partial t^2} = T + dT - T + \tau(x,t)dx = \frac{\partial T}{\partial x}dx + \tau(x,t)dx$$

Substitute equation (6.93) and cancel dx to get the equation of torsional vibration of a circular shaft as

$$\rho J \frac{\partial^2 \theta(x,t)}{\partial t^2} = \frac{\partial}{\partial x} GJ(x) \frac{\partial \theta(x,t)}{\partial x} + \tau(x,t) \tag{6.95}$$

For the case of a uniform shaft (constant J) in free vibration ($\tau(x,t) = 0$),

$$\frac{\partial^2 \theta(x,t)}{\partial t^2} = c^2 \frac{\partial^2 \theta(x,t)}{\partial x^2} \tag{6.96}$$

with

$$c = \sqrt{\frac{G}{\rho}} \tag{6.97}$$

Note that equation (6.96) is quite similar to that for transverse vibration of a cable in tension and longitudinal vibration of a rod. Hence, the same concepts and procedures of analysis can be used. In particular, two boundary conditions will be needed in the solution; for example,

$$\text{Fixed end at } x = x_o: \quad \theta(x_0, t) = 0 \tag{6.98}$$

Distributed-Parameter Systems

$$\text{Free end at } x = x_o: \quad \frac{\partial \theta(x_0, t)}{\partial x} = 0 \tag{6.99}$$

Orthogonality property of mode shapes $\Theta_i(x)$:

$$\int_0^l \Theta_i(x) \Theta_j(x) dx = 0 \quad \text{for } i \neq j \tag{6.100}$$

$$= l_j \quad \text{for } i = j$$

6.3.2 TORSIONAL VIBRATION OF NONCIRCULAR SHAFTS

Unlike for longitudinal and transverse vibrations of rods and beams, in the case of torsional vibration of shafts, the same equation of motion for circular shafts (equations (6.95) and (6.96)) cannot be used for shafts with noncircular cross sections. The reason is that the shear stress distributions in the two cases can be quite different, and equation (6.88) does not hold for noncircular sections. Hence, the parameter J in the torque-deflection relations (e.g., equations (6.93) and (6.94)) is not the polar moment of area in the case of noncircular sections. In this case, one writes

$$T = GJ_t \frac{\partial \theta}{\partial x} \tag{6.101}$$

where

J_t = torsional parameter.

The Saint Venant theory of torsion and related membrane analogy, developed by Prandtl, have provided equations for J_t in special cases. For example, for a thin hollow section,

$$J_t = \frac{4tA_s^2}{p} \tag{6.102}$$

where

A_s = enclosed (contained) area of the hollow section
p = perimeter of the section
t = wall thickness of the section.

For a thin, solid section, one has

$$J_t = \frac{t^3 a}{3} \tag{6.103}$$

where

a = length of the narrow section
t = thickness of the narrow section.

Torsional parameters for some useful sections are given in Table 6.2.

TABLE 6.2
Torsional Parameters for Several Sections

Section	Shape	Torsional Parameter J_t
Solid circular	(circle of radius r)	$\dfrac{\pi}{2} r^4$
Hollow circular	(annulus, inner radius r_1, outer r_2)	$\dfrac{\pi}{2}\left(r_2^4 - r_1^4\right)$
Thin closed	(closed thin-walled section, enclosed area A_s, perimeter p, thickness t)	$\dfrac{4 t A_s^2}{p}$
Thin open	(thin open strip, thickness t, length a)	$\dfrac{t^3 a}{3}$
Solid square	(square of side a)	$0.1406\, a^4$
Hollow square	(hollow square, outer a_2, inner a_1)	$0.1406\left(a_2^4 - a_1^4\right)$

Example 6.4

Consider a thin rectangular hollow section of thickness t, height a, and width $a/2$, as shown in Figure 6.7(a). Suppose that the section is opened by making a small slit as in Figure 6.7(b). Study the effect on the torsional parameter J_t and torsional stiffness K of the member due to the opening.

Distributed-Parameter Systems

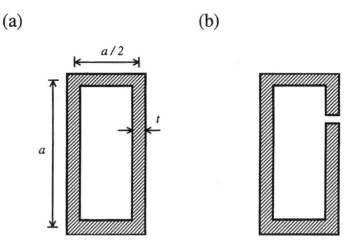

FIGURE 6.7 (a) A thin closed section, and (b) a thin open section.

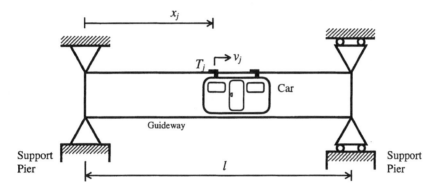

FIGURE 6.8 A torsional guideway transit system.

Solution

a. Closed section:

The contained area of the section $A_s = \dfrac{a^2}{2}$

The perimeter of the section $p = 3a$

Using equation (6.102), the torsional parameter

$$J_{tc} = \frac{ta^3}{3}$$

b. Open section:
Solid length of the section $= 3a$
Using equation (6.103), the torsional parameter

$$J_{to} = at^3$$

Ratio of the torsional parameters

$$\frac{J_{to}}{J_{tC}} = \frac{3t^2}{a^2}$$

For members of equal length, their torsional stiffness satisfy the same ratio as given by this expression. Since t is small compared to a, there will be a significant drop in torsional stiffness due to the opening (cutout).

□

EXAMPLE 6.5

An innovative automated transit system uses an elevated guideway with cars whose suspensions are attached to (and slide on) the side of the guideway. Due to this eccentric loading on the guideway, there is a significant component of torsional dynamics in addition to bending. Assume that the torque T_j acting on the guideway due to the jth suspension of the vehicle is constant, acting at a point x_j as measured from one support pier and moving at speed v_j. A schematic representation is given in Figure 6.8. The guideway span shown has a length l and a cross section that is a thin-walled rectangular box of height a, width b, and thickness t. The ends of the guideway span are restrained for angular motion (i.e., fixed).

  a. Formulate and analyze the torsional (angular) motion of the guideway.
  b. For a single point vehicle entering a guideway that is at rest, what is the resulting dynamic response of the guideway? What is the critical speed that should be avoided?
  c. Given the parameter values:

$$l = 60 \text{ ft } (18.3 \text{ m})$$

$$a \times b \times t = 5 \text{ ft} \times 2.2 \text{ ft} \times \frac{1}{2} \text{ ft } (1.52 \text{ m} \times 0.67 \text{ m} \times 0.15 \text{ m})$$

$$\rho = 4.66 \text{ slugs ft}^{-3} \ (2.4 \times 10^3 \text{ kg m}^{-3})$$

$$G = 1.55 \times 10^6 \text{ psi } (1.07 \times 10^{10} \text{ N m}^{-2})$$

and vehicle speed

$$v = 60 \text{ mph } (26.8 \text{ m s}^{-1})$$

compute the crossing frequency ratio given by

$$v_c = \frac{\text{Rate of span crossing}}{\text{Fundamental natural frequency of guideway}}$$

and discuss its implications.

SOLUTION

  a. For a uniform guideway with distributed torque load $\tau(x,t)$, and noncircular cross section having torsional parameter J_t, the governing equation is

$$\rho J \frac{\partial^2 \theta(x,t)}{\partial t^2} = GJ_t \frac{\partial^2 \theta}{\partial x^2} + \tau(x,t) \qquad (6.104)$$

As usual, the mode shapes are obtained by solving

$$\frac{d^2 \Theta}{dx^2} + \lambda^2 \Theta = 0 \qquad (6.105)$$

and the corresponding natural frequencies are given by

$$\omega_i = \lambda_i \sqrt{GJ/\rho J_t} \qquad (6.106)$$

The general solution of equation (6.105) is

$$\Theta(x) = A_1 \sin \lambda x + A_2 \cos \lambda x$$

where $A_i (i = 1, 2)$ are the constants of integration. The torsional boundary conditions corresponding to the fixed ends (no twist) are

$$\Theta(0) = \Theta(l) = 0,$$

where l = guideway span length. For a nontrivial solution, one needs

$$A_2 = 0$$

and

$$\lambda_i = \frac{i\pi}{l} \qquad i = 1, 2, \ldots \qquad (6.107)$$

The solution corresponds to an infinite set of eigenfunctions $\Theta_i(x)$ satisfying equation (6.105). Each of these represents a *natural mode* in which the beam can undergo free torsional vibrations. The actual motion consists of a linear combination of the normal modes, depending on the beam initial conditions and the forcing term $\tau(x,t)$. The integration constant A_1 can be incorporated (partially) into the generalized coordinate q, which is still unknown and is determined through initial conditions. Here, the normalized eigenfunctions are used:

$$\Theta_i(x) = \sqrt{2} \sin \frac{i\pi x}{l} \qquad i = 1, 2, \ldots \qquad (6.108)$$

The orthogonality condition given by

$$\frac{1}{l}\int_0^l \Theta_i \Theta_j \, dx = 0 \quad \text{for } i \neq j \qquad (6.109)$$

$$= 1 \quad \text{for } i = j$$

is satisfied. In view of the relations (6.106) and (6.107), the natural frequencies corresponding to different eigenfunctions (natural mode shapes) are

$$\omega_i = \frac{i\pi}{l}\sqrt{GJ/\rho J_t} \qquad i = 1, 2, \ldots \qquad (6.110)$$

For n number of vehicle suspensions located on the analyzed span,

$$\tau(x,t) = \sum_{j=1}^{n} T_j \delta(x - x_{0j} - v_j t) \qquad (6.111)$$

where T_j = torque exerted on guideway by the jth suspension
v_j = speed of the jth suspension
x_{0j} = initial position of the jth suspension (at $t = 0$) along the guideway
$\delta(\cdot)$ = Dirac delta function.

The forced motion can be represented in terms of the normalized eigenfunctions as

$$\theta(x,t) = \sum_{i=1}^{\infty} q_i(t) \Theta_i(x) \qquad (6.112)$$

where $q_i(t)$ = generalized coordinate for forced motion in the ith mode. On substituting relations (6.111) and (6.112) into (6.104) and integrating the result over the span length, after multiplying by a general eigenfunction while making use of the orthogonality relation (6.109), one obtains

$$\frac{d^2 q_i}{dt^2} + \omega_i^2 q_i = \frac{1}{\rho J_t l} \sum_{j=1}^{n} T_j \Theta_i(x_{0j} + v_j t) \qquad i = 1, 2, \ldots \qquad (6.113)$$

b. For a single suspension entering the guideway at $t = 0$, with the guideway initially at rest $[q_i(0) = \dot{q}_i(0) = 0]$, one has $n = 1$ and $x_{01} = 0$. Then the complete solution of equation (6.113) is

$$q_i(t) = \frac{\sqrt{2lT}\left[\sin\frac{i\pi v t}{l} - 2v_c \sin \omega_i t\right]}{GJ_t \pi^2 i^2 (1 - v_c^2)} \quad \text{for } i = 1, 2, \ldots \qquad (6.114)$$

where the crossing frequency ratio v_c is given by

$$v_C = \frac{v}{l\omega_1} \tag{6.115}$$

Note from equation (6.114) that the critical speed corresponds to $v_C = 1$ and should be avoided. In typical transit systems, v_C is considerably less than 1.

c. For the given numerical values, by straightforward computation, it can be shown that:

$$J_t = 3.485 \times 10^5 \text{ in}^4 \ (1.451 \times 10^7 \text{ cm}^4)$$

$$J = 6.574 \times 10^5 \text{ in}^4 \ (2.736 \times 10^7 \text{ cm}^4)$$

$$\omega_1 = 263.8 \text{ rad s}^{-1}$$

$$= 42.0 \text{ Hz}$$

Note: The expression for thin hollow section given in Table 6.2 was used to compute J_t. Finally, the crossing frequency ratio is computed to be $v_C = 0.017$, which is much less than 1.0, as expected.

□

6.4 FLEXURAL VIBRATION OF BEAMS

This section will discuss yet another continuous member in vibration. Specifically, a beam (or rod or shaft) in flexural vibration is considered. The vibration is in the "transverse" or "lateral" direction, which is accompanied by bending (or flexure) of the member. Hence, the vibrations are perpendicular to the main axis of the member, as in the case of the cable or string that was studied in Section 6.1. But a beam — unlike a string — can support shear forces and bending moments at its cross section. In the initial analysis of bending vibration, assume that there is no axial force at the ends of the beam. Further simplifying assumptions will be made, which will be clear in the development of the governing equation of motion. The analysis procedure will be quite similar to that followed in the previous sections for other continuous members.

The study of bending vibration (or lateral or transverse vibration) of beams is very important in a variety of practical situations. Noteworthy, here, are the vibration analysis of structures like bridges, vehicle guideways, tall buildings, and space stations; ride quality and structural integrity analysis of buses, trains, ships, aircraft and spacecraft; dynamics and control of rockets, missiles, machine tools, and robots; and vibration testing, evaluation, and qualification of products with continuous members.

6.4.1 Governing Equation for Thin Beams

To develop the *Bernoulli-Euler equation*, which governs transverse vibration of thin beams, consider a beam in bending, in the x-y plane, with x as the longitudinal axis and y as the transverse axis of bending deflection, as shown in Figure 6.9. The required equation is developed by considering the bending moment-deflection relation, rotational equilibrium, and transverse dynamics of a beam element.

Moment-Deflection Relation

A small beam element of length δx subjected to bending moment M is shown in Figure 6.9. Neglect any transverse deflections due to shear stresses. Consider a strip-like area element δA in the cross

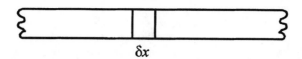

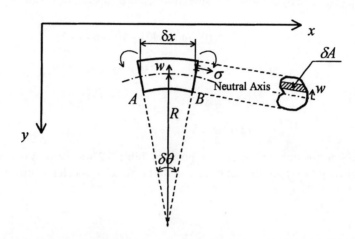

FIGURE 6.9 A thin beam in bending.

section A of the beam element, at a distance w (measure parallel to y) from the neutral axis of bending.

$$\text{Normal strain (at } \delta A\text{)} \quad \varepsilon = \frac{(R+w)\delta\theta - R\delta\theta}{R\delta\theta}$$

Note that the neutral axis joins the points along the beam where the normal strain and stress are zero. Hence,

$$\varepsilon = \frac{w}{R} \tag{6.116}$$

where R = radius of curvature of the bent element. Normal stress in the axial direction is

$$\sigma = E\varepsilon = E\frac{w}{R} \tag{6.117}$$

where E = Young's modulus (of elasticity). Then, the bending moment is

$$M = \int_A w\sigma \cdot dA = \int w^2 \frac{E}{R} dA = \frac{E}{R} \int w^2 dA = \frac{EI}{R}$$

where I = second moment of area of the beam cross section, about the neutral axis. Thus,

$$M = \frac{EI}{R} \tag{6.118}$$

Distributed-Parameter Systems

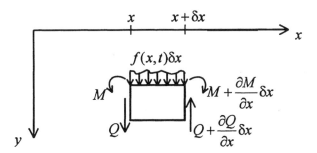

FIGURE 6.10 Dynamics of a beam element in bending.

Slope at $A = \dfrac{\partial v}{\partial x}$; Slope at $B = \dfrac{\partial v}{\partial x} + \dfrac{\partial^2 v}{\partial x^2}\delta x$, where, v = lateral deflection of the beam at element δx. Hence, change in slope $= \dfrac{\partial^2 v}{\partial x^2}\delta x = \delta\theta$, where $\delta\theta$ is the arc angle of bending for the beam element δx, as shown in Figure 6.9.

Also, $\delta x = R\delta\theta$. Hence, $\dfrac{\partial^2 v}{\partial x^2} R\delta\theta = \delta\theta$. Cancel $\delta\theta$, and obtain

$$\frac{1}{R} = \frac{\partial^2 v}{\partial x^2} \qquad (6.119)$$

Substitute equation (6.119) in (6.118) and obtain

$$M = EI\frac{\partial^2 v}{\partial x^2} \qquad (6.120)$$

Rotatory Dynamics (Equilibrium)

Again consider the beam element δx, as shown in Figure 6.10, where forces and moments acting on the element are indicated. Here, $f(x,t)$ = excitation force per unit length acting on the beam, in the transverse direction, at location x. Neglect *rotatory inertia* of the beam element.

The equation of angular motion is given by the equilibrium condition of moments:

$$M + Q\delta x - \left(M + \frac{\partial M}{\partial x}\partial x\right) = 0 \quad \text{or}$$

$$Q = \frac{\partial M}{\partial x} = \frac{\partial}{\partial x}\left(EI\frac{\partial^2 v}{\partial x^2}\right) \qquad (6.121)$$

where the previously obtained result (6.120) for M has been substituted. Note that a uniform beam is not assumed and, hence, $I = I(x)$ will be variable along the beam length.

Transverse Dynamics

The equation of transverse motion (Newton's second law) for element δx is

$$(\rho A \delta x) \frac{\partial^2 v}{\partial t^2} = f(x,t)\delta x + Q - \left(Q + \frac{\delta Q}{\partial x}\delta x\right)$$

Here, ρ = mass density of the beam material. Thus,

$$\rho A \frac{\partial^2 v}{\partial t^2} + \frac{\partial Q}{\partial x} = f(x,t)$$

or, in view of equation (6.121), one obtains the governing equation of forced transverse motion of the beam as

$$\rho A \frac{\partial^2 v}{\partial t^2} + \frac{\partial^2}{\partial x^2}\left(EI \frac{\partial^2 v}{\partial x^2}\right) = f(x,t) \tag{6.122}$$

This is the well-known Bernoulli-Euler beam equation.

6.4.2 Modal Analysis

The solution to the flexural vibration problem given by equation (6.122) can be obtained exactly as followed previously for other continuous members. Specifically, first obtain the natural frequencies and mode shapes, and express the general solution as a summation of the modal responses. The approach is similar for both free and forced problems, but the associated generalized coordinates will be different. This approach is followed here.

For modal (natural) vibration, consider the free motion described by

$$\frac{\partial^2}{\partial x^2}\left(EI \frac{\partial^2 v}{\partial x^2}\right) + \rho A \frac{\partial^2 v}{\partial t^2} = 0 \tag{6.123}$$

For a uniform beam, EI will be constant and equation (6.123) can be expressed as

$$\frac{\partial^2 v(x,t)}{\partial t^2} = -c^2 \frac{\partial^4 v(x,t)}{\partial t^4} \tag{6.124}$$

where

$$c = \sqrt{\frac{EI}{\rho A}} \tag{6.125}$$

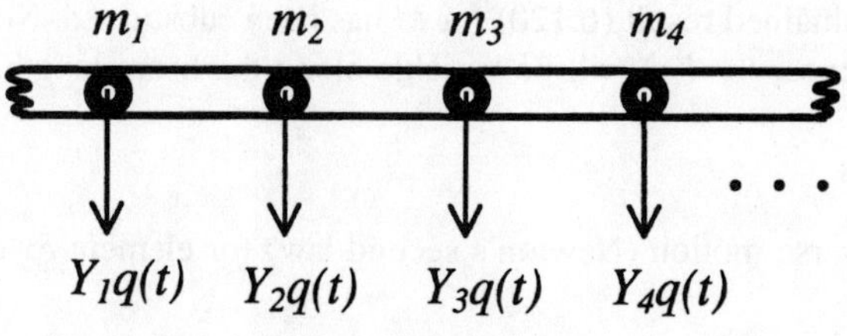

FIGURE 6.11 Modal motions of a lumped-parameter system.

Distributed-Parameter Systems

Observe from equation (6.124) that it is fourth order in x and second order in t, whereas the governing equations for transverse vibration of a cable, longitudinal vibration of a road, and torsional vibration of a shaft are all identical in form and second order in x. So, the behavior of transverse vibration of beams will not be exactly identical to that of the other three types of continuous system. In particular, the traveling wave solution (6.7) will not be satisfied. However, there are many similarities.

In each mode, the system will vibrate in a fixed shape ratio. Hence, the time and space functions will be separable for a modal motion; seek a solution of the form

$$v(x,t) = Y(x)q(t) \qquad (6.126)$$

This separable solution for a modal response is justified as usual. Note that even in the lumped parameter case (Chapter 5), the same assumption was made — except in that case, one had a modal vector:

$$Y = \begin{bmatrix} Y_1 \\ Y_2 \\ \vdots \\ Y_n \end{bmatrix}$$

instead of a mode shape function $Y(x)$. For a given mode of a lumped parameter system, Y_i values denote the "relative" displacements of various inertia elements m_i, as shown in Figure 6.11. Hence, the vector Y corresponds to the *mode shape*. Note that Y_i can be either positive or negative. Also, $q(t)$ is the harmonic function corresponding to the natural frequency.

It should be clear that Y and $q(t)$ are separable in this lumped-parameter case of modal motion. Then, in the limit, $Y(x)$ and $q(t)$ also should be separable for the distributed-parameter case.

Substitute equation (6.126) in (6.123), and bring terms containing x to the LHS and terms containing t to the RHS. Then,

$$\frac{1}{\rho AY} \frac{d^2}{dx^2}\left(EI \frac{d^2Y}{dx^2}\right) = -\frac{1}{q(t)} \frac{d^2q}{dt^2} = \omega^2 \qquad (6.127)$$

Because a function of x cannot be equal to a function of t in general (unless each function is equal to the same constant), ω^2 is defined as a constant.

It has not been shown that this *separation constant* (ω^2) should be positive. This requirement can be verified due to the nature of the particular vibration problem; that is, $q(t)$ should have an oscillatory solution in general. It is also clear that the physical interpretation of ω is a natural frequency of the system. Equation (6.127) corresponds to the two ordinary differential equations — one in t and the other in x — as

$$\frac{d^2q(t)}{dt^2} + \omega^2 q(t) = 0 \qquad (6.128)$$

$$\frac{d^2}{dx^2} EI \frac{d^2Y(x)}{dx^2} - \omega^2 \rho A Y(x) = 0 \qquad (6.129)$$

The solution of these two equations will provide the natural frequencies ω and the corresponding mode shapes $Y(x)$ of the beam.

For further analysis of the modal behavior, assume a uniform beam. Then, *EI* will be constant and equation (6.129) can be expressed as

$$\frac{d^4 Y(x)}{dx^4} - \lambda^4 Y(x) = 0 \tag{6.130}$$

where

$$\omega = \lambda^2 c = \lambda^2 \sqrt{\frac{EI}{\rho A}} \tag{6.131}$$

The positive parameter λ is yet to be determined, and will come from the mode shape analysis. The characteristic equation corresponding to equation (6.130) is

$$p^4 - \lambda^4 = 0; \quad \text{or} \quad (p^2 - \lambda^2)(p^2 + \lambda^2) = 0 \tag{6.132}$$

The roots are

$$p = \pm \lambda, \ \pm j\lambda \tag{6.133}$$

Hence, the general solution for a mode shape (eigenfunction) is given by

$$\begin{aligned} Y(x) &= A_1 e^{\lambda x} + A_2 e^{-\lambda x} + A_3 e^{+j\lambda x} + A_4 e^{-j\lambda x} \\ &= C_1 \cosh \lambda x + C_2 \sinh \lambda x + C_3 \cos \lambda x + C_4 \sin \lambda x \end{aligned} \tag{6.134}$$

There are five unknowns (C_1, C_2, C_3, C_4, and λ) here. The mode shapes can be normalized and one of the first four unknowns can be incorporated into $q(t)$ as usual. The remaining four unknowns are determined by the end conditions of the beam. Thus, four boundary conditions will be needed. *Note*:

$$\cosh \lambda x = \frac{e^{\lambda x} + e^{-\lambda x}}{2}; \quad \sinh \lambda x = \frac{e^{\lambda x} - e^{-\lambda x}}{2}$$

$$\cos \lambda x = \frac{e^{j\lambda x} + e^{-j\lambda x}}{2}; \quad \sin \lambda x = \frac{e^{j\lambda x} - e^{-j\lambda x}}{2j}$$

$$\frac{d}{dx} \cosh \lambda x = \lambda \sinh \lambda x; \quad \frac{d}{dx} \sinh \lambda x = \lambda \cosh \lambda x$$

6.4.3 BOUNDARY CONDITIONS

The four modal boundary conditions that are needed can be derived in the usual manner, depending on the conditions at the two ends of the beam. The procedure is to apply the separable (modal) solution (6.126) to the end relation with the understanding that this relation has to be true for all possible values of $q(t)$. The relation (6.128) can be substituted as well, if needed.

For example, consider an end $x = x_0$ that is completely free. Then, both bending moment and shear force have to be zero at this end. From equations (6.120) and (6.121), one obtains

Distributed-Parameter Systems

$$EI\frac{\partial^2 v(x_0,t)}{\partial x^2} = 0 \qquad (6.135)$$

$$\frac{\partial}{\partial x}\left(EI\frac{\partial^2 v(x_0,t)}{\partial x^2}\right) = 0 \qquad (6.136)$$

Substitute equation (6.126) into (6.135) and (6.136)

$$EI\frac{d^2 Y(x_0)}{dx^2} q(t) = 0$$

$$\frac{d}{dx} EI\frac{d^2 Y(x_0)}{dx^2} q(t) = 0$$

which are true for all $q(t)$. Hence, the following modal boundary conditions result for a free end:

$$\frac{d^2 Y(x_0)}{dx^2} = 0 \qquad (6.137)$$

$$\frac{d}{dx} EI\frac{d^2 Y(x_0)}{dx^2} = 0 \qquad (6.138)$$

For a uniform beam, equation (6.138) becomes

$$\frac{d^3 Y(x_0)}{dx^3} = 0 \qquad (6.138b)$$

Some common conditions and the corresponding modal boundary condition equations for bending vibration of a beam are listed in Box 6.2.

6.4.4 FREE VIBRATION OF A SIMPLY SUPPORTED BEAM

To illustrate the approach, consider a uniform beam of length l that is pinned (simply supported) at both ends. In this case, both displacement and the bending moment will be zero at each end. Accordingly, one has the following modal boundary conditions (BCs):

$$Y(0) = 0 = Y(l) \qquad (6.139)$$

$$\frac{d^2 Y(0)}{dx^2} = 0 = \frac{d^2 Y(l)}{dx^2} \qquad (6.140)$$

where, l = length of the beam. Substitute equation (6.134) in (6.139).

$$C_1 + C_3 = 0 \qquad (6.141)$$

BOX 6.2 Boundary Conditions for Transverse Vibration of Beams

1. Simply supported: Deflection $= 0 \Rightarrow Y = 0$
   (pinned)

   Bending moment $= 0 \Rightarrow \dfrac{d^2Y}{dx^2} = 0$

2. Clamped: Deflection $= 0 \Rightarrow Y = 0$
   (fixed)

   Slope $= 0 \Rightarrow \dfrac{dY}{dx} = 0$

3. Free: Bending moment $= 0 \Rightarrow \dfrac{d^2Y}{dx^2} = 0$

   Shear force $= 0 \Rightarrow \dfrac{d}{dx} EI \dfrac{d^2Y}{dx^2} = 0$

4. Sliding: Slope $= 0 \Rightarrow \dfrac{dY}{dx} = 0$

   Shear force $= 0 \Rightarrow \dfrac{d}{dx} EI \dfrac{d^2Y}{dx^2} = 0$

5. Dynamic: Transverse equation of motion (or Force balance).
   (flexible, inertial, etc.)   Substitute equation (6.128), if needed.
       Rotatory equation of motion (or Moment balance).

$$C_1 \cosh \lambda l + C_2 \sinh \lambda l + C_3 \cos \lambda l + C_4 \sin \lambda l = 0 \tag{6.142}$$

To apply the bending moment BCs, first differentiate equation (6.134) to get

$$\frac{dY}{dx} = \lambda C_1 \sinh \lambda x + \lambda C_2 \cosh \lambda x - \lambda C_3 \sin \lambda x + \lambda C_4 \cos \lambda x$$

$$\frac{d^2Y}{dx^2} = \lambda^2 C_1 \cosh \lambda x + \lambda^2 C_2 \sinh \lambda x - \lambda^2 C_3 \cos \lambda x - \lambda^2 C_4 \sin \lambda x$$

and then substitute these in the bending moment BCs (6.140). One obtains

$$C_1 - C_3 = 0 \tag{6.143}$$

$$C_1 \cosh \lambda l + C_2 \sinh \lambda l - C_3 \cos \lambda l - C_4 \sin \lambda l = 0 \tag{6.144}$$

as $\lambda \neq 0$ in general, due to the oscillatory nature of most modes.
Equations (6.141) and (6.143) give $C_1 = 0 = C_3$. Then,

Equation (6.142): $C_2 \sinh \lambda l + C_4 \sin \lambda l = 0$

Equation (6.144): $C_2 \sinh \lambda l - C_4 \sin \lambda l = 0$

Distributed-Parameter Systems

Add:
$$C_2 \sinh \lambda l = 0$$

But $\sinh \lambda l = 0$ if and only if $\lambda = 0$. This corresponds to zero-frequency conditions (no oscillations), and is rejected as it is not true in general. Hence, $C_2 = 0$. Accordingly, one is left with the remaining equation:

$$C_4 \sin \lambda l = 0 \qquad (6.145)$$

However, if $C_4 = 0$, then $Y(x) = 0$, which corresponds to a stationary beam with no oscillations, and is rejected as the trivial solution. Hence, the valid solution is given by $\sin \lambda l = 0$, which gives the infinite set of solutions:

$$\lambda_i l = i\pi \qquad \text{for } i = 1, 2, 3, \ldots \qquad (6.146)$$

Note that one must have $i > 0$ because λ has to be non-zero, thereby giving non-zero natural frequencies according to equation (6.131) as required for the given problem.

Normalization of Mode Shape Functions

For absorbing the yet unknown constant C_4 into $q(t)$, one can normalize the mode shape functions. The commonly used normalization condition is

$$\int_0^l Y_i^2 dx = \frac{l}{2} \qquad (6.147)$$

Hence,

$$\frac{l}{2} = \int_0^l C_4^2 \sin^2 \frac{i\pi x}{l} dx = C_4^2 \int_0^l \sin^2 \frac{i\pi x}{l} dx = \frac{C_4^2}{2} l$$

Note that $\cos 2\theta = 1 - 2\sin^2\theta$ was used prior to integration. Then, for normalized mode shape functions, $C_4 = 1$. Hence, the normalized eigenfunctions (mode shape functions) for various modes are given by

$$Y_i(x) = \sin \frac{i\pi x}{l} \qquad \text{for } i = 1, 2, 3, \ldots \qquad (6.148)$$

Using the result (6.146) in (6.131), the natural frequencies of the ith mode are

$$\omega_i = \frac{i^2 \pi^2}{l^2} \sqrt{\frac{EI}{\rho A}} \qquad \text{for } i = 1, 2, 3, \ldots \qquad (6.149)$$

In this manner, one obtains an infinite set of mode shape functions $Y_i(x)$ for a simply supported beam. Hence, according to the solution (6.126), there is a corresponding infinite set of generalized coordinates, $q_i(t)$, $i = 1, 2, 3, \ldots$, that satisfy equation (6.128). It follows that the overall response of the beam is

$$v(x,t) = \sum Y_i(x)q_i(t) \qquad (6.150)$$

Initial Conditions

One has yet to solve equation (6.128) for determining $q_i(t)$. For this, one needs to know the initial conditions $q_i(0)$ and $\dot{q}_i(0)$. These are determined from the beam initial conditions of displacement and speed, which must be known:

$$v(x,0) = d(x) \qquad (6.151)$$

$$\frac{\partial v}{\partial t}(x,0) = s(x) \qquad (6.152)$$

Substitute equation (6.150) in (6.151) and (6.152) to get

$$\sum Y_i(x)q_i(0) = d(x) \qquad (6.153)$$

$$\sum Y_i(x)\dot{q}_i(0) = s(x) \qquad (6.154)$$

Multiply by $Y_j(x)$ and integrate from $x = 0$ to l, using the *orthogonality property* of $Y_i(x) = \sin\frac{i\pi x}{l}$; namely,

$$\int_0^l \sin\frac{i\pi x}{l} \sin\frac{j\pi x}{l} dx = 0 \quad \text{for } i \neq j$$
$$= \frac{l}{2} \quad \text{for } i = j \qquad (6.155)$$

One obtains the necessary initial conditions

$$q_j(0) = \frac{2}{l}\int_0^l d(x)Y_j(x)dx \qquad (6.156)$$

$$\dot{q}_j(0) = \frac{2}{l}\int_0^l s(x)Y_j(x)dx \qquad (6.157)$$

In this manner, $q_i(t)$ is completely determined for each ω_i by solving equation (6.128) using the initial conditions (6.156) and (6.157). Hence, the complete solution (6.150) is determined for the free bending vibration of a simply supported beam.

6.4.5 Orthogonality of Mode Shapes

It has been seen that the mode shapes of a simply supported beam in bending vibration are orthogonal [see equation (6.155)]. This property is not limited to simply supported beams, but holds for all typical boundary conditions, as will now be shown. First, from *integration by parts*, twice, one has

$$\int_0^l Y_i \frac{d^4 Y_j}{dx^4} dx = Y_i \frac{d^3 Y_j}{dx^3} \bigg|_0^l - \int_0^l \frac{dY_i}{dx} \frac{d^3 Y_j}{dx^3} dx$$

$$= \left[Y_i \frac{d^3 Y_j}{dx^3} - \frac{dY_i}{dx} \frac{d^2 Y_j}{dx^2} \right]_0^l + \int_0^l \frac{d^2 Y_i}{dx^2} \cdot \frac{d^2 Y_j}{dx^2} dx \qquad (6.158)$$

Now consider two separate modes i and j, which have the modal equations

$$\text{Mode } i: \quad \frac{d^4 Y_i}{dx^4} = \lambda_i^4 Y_i \qquad (a)$$

$$\text{Mode } j: \quad \frac{d^4 Y_j}{dx^4} = \lambda_j^4 Y_j \qquad (b)$$

Multiply (a) by Y_j; (b) by Y_i; integrate both with respect to x from 0 to l; make use of equation (6.158) and subtract the second result from the first to obtain

$$(\lambda_i^4 - \lambda_j^4) \int_0^l Y_i Y_j dx = \int_0^l \left(Y_j \frac{d^4 Y_i}{dx^4} - Y_i \frac{d^4 Y_j}{dx^4} \right) dx$$

$$= \left[Y_j \frac{d^3 Y_i}{dx^3} - \frac{dY_j}{dx} \frac{d^2 Y_i}{dx^2} \right]_0^l - \left[Y_i \frac{d^3 Y_j}{dx^3} - \frac{dY_i}{dx} \frac{d^2 Y_j}{dx^2} \right]_0^l \qquad (6.159)$$

Clearly, the two right-hand-side terms are zero for typical boundary conditions, such as pinned, fixed, free, and sliding. Now, because $\lambda_i \neq \lambda_j$ for $i \neq j$ (unequal modes), one has

$$\int_0^l Y_i Y_j dx = 0 \qquad \text{for } i \neq j$$

$$= l_j \qquad \text{for } i = j \qquad (6.160)$$

Note that normalized mode shape functions can be used here to make the constant $l_j = \frac{l}{2}$.

Case of Variable Cross Section

Orthogonality of mode shapes holds for non-uniform beams as well. Here, EI is not constant. Then one must use integration by parts in the following form:

$$\int_0^l Y_j \frac{d^2}{dx^2} EI \frac{d^2 Y_i}{dx^2} dx = \left[Y_j \underbrace{\frac{d}{dx} EI \frac{d^2 Y_i}{dx^2}}_{Q} \right]_0^l - \left[\frac{dY_j}{dx} \underbrace{EI \frac{d^2 Y_i}{dx^2}}_{M} \right]_0^l + \int_0^l EI \frac{d^2 Y_j}{dx^2} \frac{d^2 Y_i}{dx^2} dx \qquad (6.161)$$

Again, the first two terms on the RHS are zero for typical BCs. Then, as before, use the modal equations (6.129) for two different modes i and j:

$$\frac{d^2}{dx^2} EI \frac{d^2 Y_i}{dx^2} = \omega_i^2 \rho A Y_i$$

$$\frac{d^2}{dx^2} EI \frac{d^2 Y_j}{dx^2} = \omega_j^2 \rho A Y_j$$

Multiply the first equation by Y_j, and the second equation by Y_i; subtract the second result from the first, integrate the result form $x = 0$ to l, and finally use equation (6.161) to cancel the equal terms. One then obtains

$$\left(\omega_i^2 - \omega_j^2\right) \int_0^l \rho A Y_i Y_j dx - \left[Y_j \frac{d}{dx} EI \frac{d^2 Y_i}{dx} - Y_i \frac{d}{dx} EI \frac{d^2 Y_j}{dx}\right]_0^l + \left[\frac{dY_j}{dx} EI \frac{d^2 Y_i}{dx^2} - \frac{dY_i}{dx} EI \frac{d^2 Y_j}{dx^2}\right]_0^l = 0 \quad (6.162)$$

Now, as before, for common boundary conditions, the second and the third boundary terms in equation (6.162) will vanish. Hence, after canceling the term $\omega_i^2 - \omega_j^2$, which is $\neq 0$ for $i \neq j$, the orthogonality condition for non-uniform beams is given by

$$\int_0^l \rho A Y_i Y_j dx = 0 \qquad \text{for } i \neq j \qquad (6.163)$$

$$= \alpha_j \qquad \text{for } i = j$$

The general steps for the modal analysis of a distributed-parameter vibrating system are summarized in Box 6.3.

6.4.6 Forced Bending Vibration

The equation of motion is

$$\frac{\partial^2}{\partial x^2} EI \frac{\partial^2 v}{\partial x^2} + \rho A \frac{\partial^2 v}{\partial t^2} = f(x,t) \qquad (6.164)$$

Assume a separable, forced response:

$$v(x,t) = \sum_i \bar{q}_i(t) Y_i(x) \qquad (6.165)$$

where $\bar{q}_i(t)$ are the generalized coordinates in the forced case. Substitute equation (6.165) into the beam equation (6.164):

$$\sum_i \bar{q}_i(t) \frac{d^2}{dx^2} EI \frac{d^2 Y_i(x)}{dx^2} + \rho A \sum_i \ddot{\bar{q}}_i(t) Y_i(x) = f(x,t)$$

Distributed-Parameter Systems

BOX 6.3 Modal Analysis of Continuous Systems

Equation of free (unforced) motion:

$$L(x,t)v(x,t) = 0 \qquad (i)$$

where $v(x,t)$ = system response
$L(x,t)$ = partial differential operator in space (x) and time (t).

Modal Solution:
Assume a separable solution

$$v(x,t) = Y(x)q(t) \qquad (ii)$$

because a modal response is separable in time and space.

Note 1: $Y(x)$ = mode shape
$q(t)$ = generalized coordinate for free response.

Note 2: For two- and three-dimensional space systems, time and space will still be separable for a modal response; but the space function itself may not be separable along various coordinate directions.

Steps:
1. Substitute (*ii*) in (*i*) and separate the space function (of x) and the time function (of t), each of which should be equal to the same constant.
2. Solve the resulting ordinary differential equation (ODE) for $Y(x)$ using system boundary conditions. One obtains an infinite set of mode shapes $Y_i(x)$ up to one unknown (removed by normalization) and natural frequencies ω_i.
3. Solve the ODE for $q(t)$ using system initial conditions to determine $q_i(t)$ for mode i. (Orthogonality of $Y_i(x)$ will be needed to establish the initial conditions for $q_i(t)$.)
4. Overall response

$$v(x,t) = \sum_i Y_i(x) q_i(t)$$

The first term on the LHS, on using the mode-shape equation (6.129), becomes $\bar{q}_i(t)\rho A \omega_i^2 Y_i(x)$. Multiply the result by $Y_j(x)$ and integrate with respect to $x[0, l]$ to obtain

$$\omega_j^2 q_j(t) \int_0^l \rho A Y_j^2(x) dx + \ddot{\bar{q}}_j(t) \int_0^l \rho A Y_j^2(x) dx = \underbrace{\int_0^l Y_j(x) f(x,t) dx}_{f_j(t)}$$

Each of the two integrals on the LHS evaluates to α_j according to equation (6.163). Hence,

$$\ddot{\bar{q}}_j(t) + \omega_j^2 \bar{q}_j(t) = \frac{1}{\alpha_j} f_j(t) \qquad \text{for } j = 1, 2, 3, \ldots \qquad (6.166)$$

BOX 6.4 Forced Response of Continuous Systems

Equation of forced motion:

$$L(x,t)v(x,t) = L_1(x,t)f(x,t) \qquad (i)$$

where $v(x,t)$ = forced response of the system
$f(x,t)$ = distributed force per unit length (space)
L, L_1 = partial differential operators in space and time.

Steps:
1. Substitute the modal expansion

$$v(x,t) = \sum Y_i(x)\bar{q}_i(t) \qquad (ii)$$

in (*i*), where $Y_i(x)$ = mode shapes
$\bar{q}_i(t)$ = generalized coordinates for forced motion.
2. Multiply by $Y_j(x)$ and integrate with respect to space (x) using orthogonality

$$\int_0^l m(x)Y_i(x)Y_j(x)dx = 0 \qquad \text{for } i \neq j \qquad (iii)$$

Note: Additional boundary terms enter into (*iii*) when there are lumped elements at the system boundary.
3. Determine initial conditions for $\bar{q}_i(t)$. One will need (*iii*) for this.
4. Solve the ordinary differential equation for $\bar{q}_i(t)$ using initial conditions.
5. Substitute the results in (*ii*).

One can then solve this equation to determine the generalized coordinates $\bar{q}_j(t)$ using the knowledge of the forcing function $f_j(t)$ and the initial conditions $\bar{q}_j(0)$ and $\dfrac{d\bar{q}_j}{dt}(0)$. Specifically, if the initial displacement and speed of the beam are given by equations (6.151) and (6.152), respectively, then by following the procedure that was adopted to obtain the results (6.156) and (6.157), one determines that

$$\bar{q}_j(0) = \frac{1}{\alpha_j}\int_0^l \rho A d(x) Y_j(x) dx \qquad (6.167)$$

$$\frac{d\bar{q}_j(0)}{dt} = \frac{1}{\alpha_j}\int_0^l \rho A s(x) Y_j(x) dx \qquad (6.168)$$

Finally, one obtains the overall response of the forced system as

$$v(x,t) = \sum_i \bar{q}_j(t) Y_j(x) \qquad (6.169)$$

The main steps in the forced response analysis are summarized in Box 6.4.

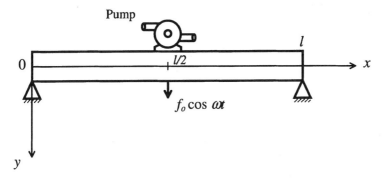

FIGURE 6.12 A pump mounted on a simply supported beam.

EXAMPLE 6.6

A pump is mounted at the mid-span of a simply supported thin beam of uniform cross section and length l. The pump rotation generates a transverse force $f_0\cos\omega t$ as schematically shown in Figure 6.12. Initially, the system starts from rest, from the static equilibrium position of the beam, such that

$$v(x,t) = 0 \quad \text{and} \quad \frac{\partial v(x,t)}{\partial t} = 0 \quad \text{at } t = 0.$$

It is required to obtain the transverse response $v(x, t)$ of the beam in the form of a modal summation during operation of the pump.

First determine $q_j(t)$ in terms of f_0, ω, ω_j and the beam parameters ρ, A, and l, assuming that the beam is completely undamped. Are all modes of the beam excited by the pump? If the beam is lightly damped, what would be its steady-state response?

In particular, what is the steady-state response of the beam at the pump location? Sketch its amplitude as a function of the excitation frequency ω.

SOLUTION

Using the Dirac delta function $\delta(x)$, one can express the equation of forced motion of the beam as

$$EI\frac{\partial^4 v}{\partial x^4} + \rho A\frac{\partial^2 v}{\partial t^2} = f_0 \cos\omega t\, \delta\!\left(x - \frac{l}{2}\right) \tag{6.170}$$

Substitute $v(x,t) = \sum_i Y_i(x)q_i(t)$, where normalized mode shapes for the simply supported beam are $Y_i(x) = \sin i\pi x/l$; multiply by $Y_j(x)$; integrate over $x = [0, l]$ and use the orthogonality of mode shapes to obtain

$$EI\left(\frac{i\pi}{l}\right)^4 \frac{l}{2}q_j(t) + \rho A\frac{l}{2}\ddot{q}_j = f_0 \cos\omega t \sin\frac{j\pi}{2}$$

Note:

$$\int_{a^-}^{a^+} p(x)\delta(x-a)\,dx = p(a)$$

for some function $p(x)$.
Hence, from equation (6.149), one obtains

$$\ddot{q}_j + \omega_j^2 q_j = \alpha_j \cos \omega t \tag{6.171}$$

where

$$\alpha_j = \frac{2f_0}{\rho Al} \sin j\frac{\pi}{2}$$

The given initial conditions are satisfied if and only if $q_j(0) = 0$ and $\dot{q}_j(0) = 0$ for all j. Hence, the complete solution is

$$q_j(t) = \frac{\alpha_j}{\left(\omega_j^2 - \omega^2\right)}\left[\cos \omega t - \cos \omega_j t\right] \tag{6.172}$$

It follows that the total response is

$$v(x,t) = \frac{2f_0}{\rho Al} \sum \frac{\sin j\pi/2}{\left(\omega_j^2 - \omega^2\right)} \sin j\frac{\pi x}{l}\left[\cos \omega t - \cos \omega_j t\right] \tag{6.173}$$

Clearly, $\sin \frac{j\pi}{2} = 0$ for even values of j. One notes that even modes of the beam are not excited by the pump; this is to be expected because, for even modes, the mid-span is a node point and has no motion.

If the beam is lightly damped, its natural response terms ($\cos \omega_j t$) will decay to zero with time. Hence, the steady-state response will be

$$v_{ss}(x,t) = \frac{2f_0}{\rho Al} \cos \omega t \sum \frac{\sin j\pi/2 \, \sin j\pi x/l}{\left(\omega_j^2 - \omega^2\right)} \tag{6.174}$$

At the pump location ($x = l/2$), the steady state response is

$$v_{ss}(l/2, t) = \frac{2f_0}{\rho Al} \cos \omega t \sum \frac{\sin^2 j\pi/2}{\left(\omega_j^2 - \omega^2\right)}$$

$$= \frac{2f_0}{\rho Al} \cos \omega t \left[\frac{1}{\left(\omega_1^2 - \omega^2\right)} + \frac{1}{\left(\omega_3^2 - \omega^2\right)} + \frac{1}{\left(\omega_5^2 - \omega^2\right)} + \cdots\right]$$

Note again that only the odd modes contribute to the response. Furthermore, for a simply supported beam,

Distributed-Parameter Systems

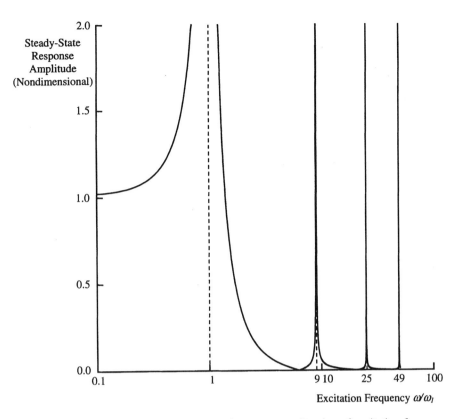

FIGURE 6.13 Amplitude of the steady response at the pump as a function of excitation frequency.

$$\omega_j = j^2 \omega_1 \qquad \text{for } j = 1, 2, 3, \ldots$$

where

$$\omega_1 = \left(\frac{\pi}{l}\right)^2 \sqrt{\frac{EI}{\rho A}}$$

Non-dimensionalize the mid-span response at steady state as

$$\bar{v} = \frac{\rho A l \omega_1^2}{2 f_0} v_{ss}(l/2, t) = \left[\frac{1}{(1^2 - \bar{\omega}^2)} + \frac{1}{(3^2 - \bar{\omega}^2)} + \frac{1}{(5^2 - \bar{\omega}^2)} + \cdots\right] \cos \omega t$$

where the nondimensional excitation frequency $\bar{\omega} = \omega / \omega_1$. The amplitude is

$$v_0(\bar{\omega}) = \left|\frac{1}{(1^4 - \bar{\omega}^2)} + \frac{1}{(3^4 - \bar{\omega}^2)} + \frac{1}{(5^4 - \bar{\omega}^2)} + \cdots\right|$$

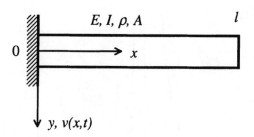

FIGURE 6.14 A cantilever in bending vibration.

The characteristic of the amplitude as a function of the nondimensional excitation frequency is sketched in Figure 6.13.

□

EXAMPLE 6.7

Perform a modal analysis to determine natural frequencies and mode shapes of transverse vibration of a thin cantilever (i.e., a beam with "fixed-free" or "clamped-free" end conditions). The coordinate system and the beam parameters are as shown in Figure 6.14.

SOLUTION

As usual, the mode shapes are given by

$$Y(x) = C_1 \cosh \lambda x + C_2 \sinh \lambda x + C_3 \cos \lambda x + C_4 \sin \lambda x \tag{6.134}$$

Its first three derivatives are

$$\frac{dY(x)}{dx} = C_1 \lambda \sinh x + C_2 \lambda \cosh \lambda x - C_3 \lambda \sin \lambda x + C_4 \lambda \cos \lambda x \tag{6.175}$$

$$\frac{d^2Y(x)}{dx^2} = C_1 \lambda^2 \cosh \lambda x + C_2 \lambda^2 \sinh \lambda x - C_3 \lambda^2 \cos \lambda x - C_4 \lambda^2 \sin \lambda x \tag{6.176}$$

$$\frac{d^3Y(x)}{dx^3} = C_1 \lambda^3 \sinh \lambda x + C_2 \lambda^3 \cosh \lambda x + C_3 \lambda^3 \sin \lambda x - C_4 \lambda^3 \cos \lambda x \tag{6.177}$$

The boundary conditions of the beam are:

$$\text{At } x = 0: \quad v(0,t) = 0 \quad \text{and} \quad \frac{\partial v(0,t)}{\partial x} = 0$$

$$\text{At } x = l: \quad EI\frac{\partial^2 v(l,t)}{\partial x^2} = 0 \quad \text{and} \quad EI\frac{\partial^3 v(l,t)}{\partial x^3} = 0$$

The corresponding modal boundary conditions are:

Distributed-Parameter Systems

$$Y(0) = 0; \quad \frac{dY(0)}{dx} = 0; \quad \frac{d^2Y(l)}{dx^2} = 0; \quad \frac{d^3Y(l)}{dx^3} = 0 \qquad (6.178)$$

Substitute equations (6.134), (6.175)–(6.177) into (6.178) to obtain

$$C_1 + C_3 = 0 \qquad (i)$$

$$C_2 + C_4 = 0 \qquad (ii)$$

$$C_1 \cosh \lambda l + C_2 \sinh \lambda l - C_3 \cos \lambda l - C_4 \sin \lambda l = 0 \qquad (iii)$$

$$C_1 \sinh \lambda l + C_2 \cosh \lambda l + C_3 \sin \lambda l - C_4 \cos \lambda l = 0 \qquad (iv)$$

Eliminate C_1 and C_2 in equations (iii) and (iv) by substituting (i) and (ii). We get

$$[\cosh \lambda l + \cos \lambda l] C_3 + [\sinh \lambda l + \sin \lambda l] C_4 = 0 \qquad (v)$$

$$[\sinh \lambda l - \sin \lambda l] C_3 + [\cosh \lambda l + \cos \lambda l] C_4 = 0 \qquad (vi)$$

or, in the vector-matrix form,

$$\begin{bmatrix} \cosh \lambda l + \cos \lambda l & \sinh \lambda l + \sin \lambda l \\ \sinh \lambda l - \sin \lambda l & \cosh \lambda l + \cos \lambda l \end{bmatrix} \begin{bmatrix} C_3 \\ C_4 \end{bmatrix} = \begin{bmatrix} 0 \\ 0 \end{bmatrix} \qquad (6.179)$$

The trivial solution of equation (6.179) is $C_3 = 0 = C_4$. Then, from equations (i) and (ii), $C_1 = 0 = C_2$. This solution corresponds to $Y(x) = 0$ and is not acceptable in general for a vibrating system. Hence, one must have the matrix in equation (6.179) non-invertible (i.e, singular). For this, the determinant of the matrix must vanish (see Appendix C); thus,

$$(\cosh \lambda l + \cos \lambda l)^2 - (\sinh \lambda l + \sin \lambda l)(\sinh \lambda l - \sin \lambda l) = 0$$

or

$$\cosh^2 \lambda l + 2 \cosh \lambda l \cos \lambda l + \cos^2 \lambda l - \sinh^2 \lambda l + \sin^2 \lambda l = 0$$

But, it is well known that

$$\cosh^2 \lambda l - \sinh^2 \lambda l = 1 \quad \text{and} \quad \cos^2 \lambda l + \sin^2 \lambda l = 1$$

Hence,

$$\cos \lambda l \cosh \lambda l = -1 \qquad (6.180)$$

This transcendental equation has an infinite number of solutions λ_i for $i = 1, 2, 3, \ldots$, correspondingly giving an infinite number of natural frequencies

$$\omega_i = \lambda_i^2 \sqrt{\frac{EI}{\rho A}} \qquad (6.181)$$

The corresponding mode shapes are given by equation (6.134) subject to (*i*), (*ii*), and (*v*) or (*iv*). This gives $Y_i(x) = C_3(\cos\lambda_i x - \cosh\lambda_i x) + C_4(\sin\lambda_i x - \sinh\lambda_i x)$ with

$$C_3 = -\frac{[\sinh\lambda_i l + \sin\lambda_i l]}{[\cosh\lambda_i l + \cos\lambda_i l]} C_4$$

It follows that

$$Y_i(x) = C_4[\sin\lambda_i x - \sinh\lambda_i x] + C_4\left[\frac{\sinh\lambda_i l + \sin\lambda_i l}{\cosh\lambda_i l + \cos\lambda_i l}\right][-\cos\lambda_i x + \cosh\lambda_i x]$$

The unknown multiplier C_4 simply scales the mode shape and is absorbed into the generalized coordinate $q_i(t)$ as usual. In fact, this is a process of normalization of mode shapes, where $C_4 = 1$ is used. So, one has the normalized mode shapes:

$$Y_i(x) = a\sin\lambda_i x + b\sinh\lambda_i x + \alpha_i[c\cos\lambda_i x + d\cosh\lambda_i x] \qquad (6.182)$$

with

$$a = 1,\ b = -1,\ c = -1,\ d = 1, \quad \text{and} \quad \alpha_i = \frac{\sinh\lambda_i l + \sin\lambda_i l}{\cosh\lambda_i l + \cos\lambda_i l} \qquad (6.183)$$

The first three roots of equation (6.180) are

$$\lambda_1 l = 1.875104; \quad \lambda_2 l = 4.694091; \quad \lambda_3 l = 7.854757$$

The corresponding three mode shapes are sketched in Figure 6.15. Note, in particular, the *node points*, which are not physically fixed but remain stationary during a modal motion. This completes the solution.

□

The modal information corresponds to the infinite set of natural frequencies obtained by solving equation (6.180) subject to (6.181); together with the mode shapes given by (6.182) subject to (6.183). Modal information corresponding to other common boundary conditions can also be put in this form. Table 6.3 summarizes such data.

Table 6.4 provides numerical values corresponding to this modal information for the first three modes.

6.4.7 BENDING VIBRATION OF BEAMS WITH AXIAL LOADS

Previous sections have considered the problems of longitudinal vibration of beams with axial loads and also transverse vibration of cables in tension. In practice, beam-type members that undergo flexural (transverse) vibrations also can carry axial forces. Examples are structural members such as columns, struts, and towers. Generally, a tension will increase the natural frequencies of bending,

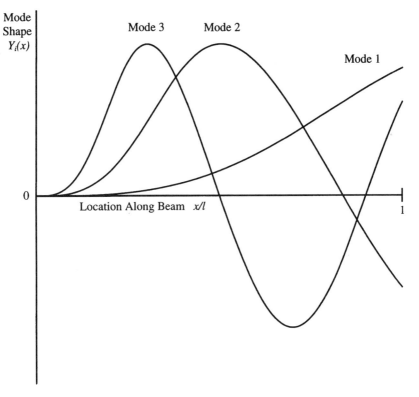

FIGURE 6.15 First three modes of a cantilever (fixed-free beam) in transverse vibration.

and compression will decrease them. Hence, one way to avoid the excitation of a particular natural frequency (and mode) of bending vibration is to use a suitable tension or compression in the axial direction.

The equation of motion for transverse motion of a thin beam subjected to an axial tension P can be easily derived by following the same procedure that led to equation (6.122), but now including P. For simplicity, assume a thin beam subjected to a constant tensile force P. A small element δx of the beam is shown in Figure 6.16. The vertical component (in the +ve y direction) of the axial force is

$$-P\sin\theta + P\sin(\theta+\delta\theta) \cong -P\theta + P(\theta+\delta\theta) = P\delta\theta$$

because the slope $\theta = \dfrac{\partial v}{\partial x}$ is small. Also, the change in slope is

$$\delta\theta = \frac{\partial^2 v}{\partial x^2}\delta x$$

It is clear that the previous equation of transverse dynamics of element δx must be modified simply by adding the term $P\dfrac{\partial^2 v}{\partial x^2}\delta x$ to the $f(x,t)\delta x$ side of the equation. Then, the resulting equation of transverse vibration will be

$$\rho A \frac{\partial^2 v}{\partial t^2} + \frac{\partial^2}{\partial x^2}\left(EI\frac{\partial^2 v}{\partial x^2}\right) - P\frac{\partial^2 v}{\partial x^2} = f(x,t) \tag{6.184}$$

TABLE 6.3
Modal Information for Bending Vibration of Beams

End Conditions	Natural Frequencies $\omega_i = \lambda_i^2 \sqrt{EI/\rho A}$ Where λ_i are Roots of:	\multicolumn{4}{c}{Mode Shapes $Y_i(x) = a\sin\lambda_i x + b\sinh\lambda_i x + \alpha_i[c\cos\lambda_i x + d\cosh\lambda_i x]$}	
		a	b	c	d	α_i
Pinned-pinned	$\sin\lambda_i l = 0$ $i = 1, 2, 3, \ldots$	1	0	0	0	0
Fixed-fixed	$\cosh\lambda_i l \cos\lambda_i l = 1$ $i = 1, 2, 3, \ldots$	1	-1	-1	1	$\dfrac{\sinh\lambda_i l - \sin\lambda_i l}{\cosh\lambda_i l - \cos\lambda_i l}$
Free-free	$\cosh\lambda_i l \cos\lambda_i l = 1$ $i = 0, 1, 2, \ldots$	1	1	-1	-1	Same
Fixed-pinned	$\tanh\lambda_i l = \tan\lambda_i l$ $i = 1, 2, 3, \ldots$	1	-1	-1	1	Same
Fixed-free	$\cosh\lambda_i l \cos\lambda_i l = -1$	1	-1	-1	1	$\dfrac{\sinh\lambda_i l + \sin\lambda_i l}{\cosh\lambda_i l + \cos\lambda_i l}$
Fixed-sliding	$\tanh\lambda_i l = -\tan\lambda_i l$	1	-1	-1	1	$\dfrac{\cosh\lambda_i l + \cos\lambda_i l}{\sinh\lambda_i l - \sin\lambda_i l}$
Pinned-free	$\tanh\lambda_i l = \tan\lambda_i l$	1	α_i	0	0	$\dfrac{\sin\lambda_i l}{\sinh\lambda_i l}$

For modal analysis of a uniform beam, then, use the equation of free motion

$$\rho A \frac{\partial^2 v}{\partial t^2} + EI \frac{\partial^4 v}{\partial x^4} - P \frac{\partial^2 v}{\partial x^2} = 0 \qquad (6.185)$$

With a separable (modal) solution of the form

$$v(x,t) = Y(x)q(t) \qquad (6.186)$$

one has

$$\frac{\ddot{q}(t)}{q(t)} = -\frac{EI\, d^4Y/dx^4 - P\, d^2Y/dx^2}{\rho A Y} = -\omega^2 \qquad (6.187)$$

which gives, as before, the time response equation of the generalized coordinates:

$$\ddot{q}(t) + \omega^2 q(t) = 0 \qquad (6.188)$$

and the mode shape equation

TABLE 6.4
Roots of the Frequency Equation for Bending Vibration of Beams

End Conditions	First Three Roots $\lambda_i l$
Pinned-pinned	π
	2π
	3π
Fixed-fixed	4.730041
	7.853205
	10.995608
Free-free	0
	4.730041
	7.853205
	10.995608
Fixed-pinned	3.926602
	7.068583
	10.210176
Fixed-free	1.875104
	4.694091
	7.854757
Fixed-sliding	2.365020
	5.497804
	8.639380
Pinned-free	0
	3.926602
	7.068583
	10.210176

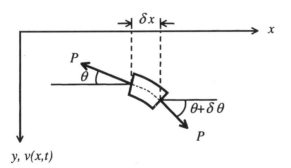

FIGURE 6.16 Beam element in transverse vibration and subjected to axial tension.

$$EI\frac{d^4Y}{dx^4} - P\frac{d^2Y}{dx^2} - \rho A\omega^2 Y = 0 \qquad (6.189)$$

Note that the mode shape equation is still fourth order, but is different. The analysis, however, can be done as before, by using four boundary conditions at the two ends of the beam, to determine the natural frequencies (an infinite set) and the corresponding normalized mode shapes.

6.4.8 BENDING VIBRATION OF THICK BEAMS

In the derivation of the governing equation for lateral vibration of thin beams (known as the Bernoulli-Euler beam equation), the following effects, in particular, were neglected:

1. Deformation and associated lateral motion due to shear stresses
2. Moment of inertia of beam elements in rotatory motion.

Note, however, that use was made of the fact that shear forces (Q) are present in a beam cross-section, although the resulting deformations were not taken into account. Also, in writing the equation for rotational motion of a beam element δx, the moments were simply summed to zero, without including the inertial moment. These assumptions are valid for a beam whose cross-sectional dimensions are small compared to its length. But, for a thick beam, the effects of *shear deformation* and *rotatory inertia* have to be included in deriving the governing equation. The resulting equation is known as the *Timoshenko beam equation*. Important steps in the derivation of the equation of motion for forced transverse vibration of beams, including the effects of shear deformation and rotatory inertia, are given below.

Consider a small element δx of a beam. Figure 6.17(a) illustrates the contribution of the bending of an element and the deformation due to transverse shear stresses toward the total slope of the beam neutral axis. Let

θ = angle of rotation of the beam element due to bending
ϕ = increase in slope of the element due to shear deformation in transverse shear (this is equal to shear strain).

Then, the total slope of the beam element is

$$\frac{\partial v}{\partial x} = \theta + \phi \tag{6.190}$$

Here, v and x take the usual meanings — as for a thin beam.

Figure 6.17(b) shows an element δx of the beam with the forces, moments, and the linear and angular accelerations marked. With the sign convention shown in Figure 6.17(b), the linear shear-stress-shear-strain relation can be stated as

$$Q = -kGA\phi \tag{6.191}$$

where the Timoshenko shear coefficient

$$k = \frac{\text{Average shear stress on beam cross-section at } x}{\text{Shear stress at the neutral axis at } x}$$

and

$$G = \text{shear modulus.}$$

The equation for translatory motion is

$$f(x,t)\delta x + Q - \left(Q + \frac{\partial Q}{\partial x}\partial x\right) = (\rho A \delta x)\frac{\partial^2 v}{\partial t^2}$$

Distributed-Parameter Systems

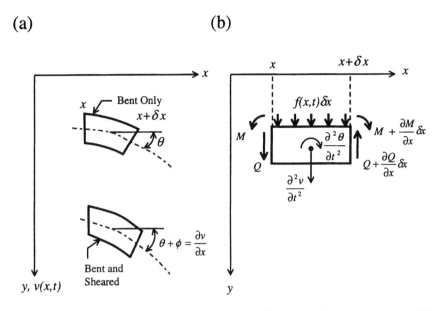

FIGURE 6.17 A Timoshenko beam element: (a) combined effect of bending and shear; and (b) dynamic effect, including rotatory inertia.

Hence,

$$f(x,t) - \frac{\partial Q}{\partial x} = \rho A \frac{\partial^2 v}{\partial t^2} \quad (6.192)$$

The equation for rotatory motion of the element, taking into account the rotatory inertia, is

$$M + \frac{\partial M}{\partial x}\delta x - M - Q\delta x = (I\rho\delta x)\frac{\partial^2 \theta}{\partial t^2}$$

which becomes

$$\frac{\partial M}{\partial x} - Q = I\rho \frac{\partial^2 \theta}{\partial t^2} \quad (6.193)$$

From the elementary theory of bending, as before,

$$M = EI\frac{\partial \theta}{\partial x} \quad (6.194)$$

The relationship between the shear modulus and the modulus of elasticity is known to be

$$E = 2(1+\upsilon)G \quad (6.195)$$

where υ = Poisson's ratio. This relation can be substituted if desired. Manipulation of these equations yields

$$EI\frac{\partial^4 v}{\partial x^4}+\rho A\frac{\partial^2 v}{\partial t^2}-\rho I\left(1+\frac{E}{kG}\right)\frac{\partial^4 v}{\partial x^2 \partial t^2}+\frac{\rho^2 I}{kG}\frac{\partial^4 v}{\partial t^4}=f(x,t)-\frac{EI}{kGA}\frac{\partial^2 f(x,t)}{\partial x^2}+\frac{\rho I}{kGA}\frac{\partial^2 f(x,t)}{\partial t^2} \quad (6.196)$$

This is the Timoshenko beam equation for forced transverse motion. Note that this equation is fourth order in time, whereas the thin beam equation is second order in time. The modal analysis may proceed as before, by using the free ($f = 0$) equation and a separable solution. However, the resulting differential equation for the generalized coordinates will be fourth order in time and, as a result, additional natural frequency bands will be created. The reason is the independent presence of shear and bending motions. The differential equation of mode shapes will be fourth order in x, and the solution procedure will be the same as before, through the use of four boundary conditions at the two ends of the beam.

6.4.9 Use of the Energy Approach

Thus far, only the direct, Newtonian approach has been used in deriving the governing equations for continuous members in vibration. Of course, the same results can be obtained using the Lagrangian (energy) approach (see Appendix B). The general approach here is to first express the Lagrangian L of the system as

$$L = T^* - V \quad (6.197)$$

where

T^* = total kinetic co-energy (equal to kinetic energy T for typical systems)
V = total potential energy.

Then, for a virtual increment (variation) of the system through incrementing the system response variable, the following condition will hold:

$$\int_{t_1}^{t_2}[\delta L + \delta W]dt = 0 \quad (6.198)$$

where δL is the increment in the Lagrangian and δW is the work done by the external forces on the system due to the increment (virtual work). Finally, using the arbitrariness of the variation, the equation of motion, along with the boundary conditions, can be obtained. This approach is illustrated now.

First consider the free motion. One can easily extend the result to the case of forced motion. The kinetic energy is given by

$$T = \frac{1}{2}\int_0^l \rho A dx \left(\frac{\partial v}{\partial t}\right)^2 = \frac{1}{2}\int_0^l \rho A \left(\frac{\partial v}{\partial t}\right)^2 dx \quad (6.199)$$

The potential energy due to bending results from the work needed to bend the beam (angle denoted by $\theta = \frac{\partial v}{\partial x}$ as usual) under bending moment M; thus

Distributed-Parameter Systems

$$V = \frac{1}{2} \int_{\text{End 1}}^{\text{End 2}} M d\phi = \frac{1}{2} \int_0^l EI\left(\frac{\partial^2 v}{\partial x^2}\right) dx \qquad (6.200)$$

Before proceeding further, note the following steps of variation, and integration by parts with respect to t:

$$\delta \int_{t_1}^{t_2} \frac{1}{2}\left(\frac{\partial v}{\partial t}\right)^2 dt = \int_{t_1}^{t_2} \frac{\partial v}{\partial t} \delta\left(\frac{\partial v}{\partial t}\right) dt = \int_{t_1}^{t_2} \frac{\partial v}{\partial t} \frac{\partial \delta v}{\partial t} dt = \left[\frac{\partial v}{\partial t} \delta v\right]_{t_1}^{t_2} - \int_{t_1}^{t_2} \frac{\partial^2 v}{\partial t^2} \delta v \, dt \qquad (6.201)$$

Note the interchange of the operations δ and $\frac{\delta}{\delta t}$ prior to integrating by parts. Also, by convention, assume that no variations are performed at the starting and ending times (t_1 and t_2) of integration. Hence,

$$\delta v(t_1) = 0 \quad \text{and} \quad \delta v(t_2) = 0 \qquad (6.202)$$

Similarly, variation, and integration by parts with respect to x, are done as follows:

$$\delta \int_0^l \frac{1}{2} EI\left(\frac{\partial^2 v}{\partial x^2}\right)^2 dx = \int_0^l EI \frac{\partial^2 v}{\partial x^2} \delta\left(\frac{\partial^2 v}{\partial x^2}\right) dx = \int_0^l EI \frac{\partial^2 v}{\partial x^2} \frac{\partial}{\partial x} \delta\left(\frac{\partial v}{\partial x}\right) dx$$

$$= \left[EI \frac{\partial^2 v}{\partial x^2} \delta\left(\frac{\partial v}{\partial x}\right)\right]_0^l - \int_0^l \frac{\partial}{\partial x} EI \frac{\partial^2 v}{\partial x^2} \delta\left(\frac{\partial v}{\partial x}\right) dx$$

$$= \left[EI \frac{\partial^2 v}{\partial x^2} \delta\left(\frac{\partial v}{\partial x}\right)\right]_0^l - \int_0^l \frac{\partial}{\partial x} EI \frac{\partial^2 v}{\partial x^2} \frac{\partial \delta v}{\partial x} dx \qquad (6.201)^*$$

$$= \left[EI \frac{\partial^2 v}{\partial x^2} \delta\left(\frac{\partial v}{\partial x}\right)\right]_0^l - \left[\frac{\partial}{\partial x} EI \frac{\partial^2 v}{\partial x^2} \delta v\right]_0^l + \int_0^l \frac{\partial^2}{\partial x^2} EI \frac{\partial^2 v}{\partial x^2} \delta v \, dx$$

Now, for the case of free vibration ($\delta W = 0$), substitute equations (6.201) and (6.201)* in δT and δV from equations (6.199) and (6.200) to obtain

$$\int_{t_1}^{t_2} \delta L \, dt = 0 = -\int_{t_1}^{t_2} dt \int_0^l dx \left[\rho A \frac{\partial^2 v}{\partial t^2} + \frac{\partial^2}{\partial x^2} EI \frac{\partial^2 v}{\partial x^2}\right] \delta v$$

$$+ \int_0^l dx \rho A \left[\frac{\partial v}{\partial t} \delta v\right]_{t_1}^{t_2} - \int_{t_1}^{t_2} dt \left[EI \frac{\partial^2 v}{\partial x^2} \delta\left(\frac{\partial v}{\partial x}\right)\right]_0^l + \int_{t_1}^{t_2} dt \left[\frac{\partial}{\partial x} EI \frac{\partial^2 v}{\partial x^2} \delta v\right]_0^l \qquad (6.203)$$

Because equation (6.203) holds for all arbitrary variations $\delta v(t)$, its coefficient should vanish. Hence,

$$\rho A \frac{\partial^2 v}{\partial t^2} + \frac{\partial^2}{\partial x^2} EI \frac{\partial^2 v}{\partial x^2} = 0 \qquad (6.204)$$

which is the same Bernoulli-Euler beam equation for free motion as previously derived.

The second integral term on the RHS of equation (6.203) has no consequence. One can conventionally pick $\delta v(t_1) = 0$ and $\delta v(t_2) = 0$ at the time points t_1 and t_2 as given by equation (6.202).

The third integral term on the RHS of (6.203) gives some boundary conditions. Specifically, if the slope boundary condition $\frac{\partial v}{\partial x}$ is zero (i.e., fixed end), then the corresponding bending moment at the end is arbitrary, as expected. But if the slope at the boundary is arbitrary, then the bending moment $EI \frac{\partial^2 v}{\partial x^2}$ at the end should be zero (i.e., pinned or free end).

The last integral term on the RHS of (6.203) gives some other boundary conditions. Specifically, if the displacement boundary condition v is zero (i.e., pinned or fixed end), then the corresponding shear force at the end is arbitrary. But if the displacement at the boundary is arbitrary, then the shear force $\frac{\partial}{\partial x} EI \frac{\partial^2 v}{\partial x^2}$ at that end should be zero (i.e., free or sliding end).

Next consider a forced beam with force per unit length given by $f(x,t)$. Then, the work done by the $f(x,t)dx$ in a small element dx of the beam, when moved through a displacement of δv, is

$$f(x,t)dx\delta v \qquad (6.205)$$

Then, by combining equation (6.205) with (6.203), for arbitrary variation δv, one obtains the forced vibration equation:

$$\rho A \frac{\partial^2 v}{\partial t^2} + \frac{\partial^2}{\partial x^2} EI \frac{\partial^2 v}{\partial x^2} = f(x,t) \qquad (6.206)$$

Note that external forces and moments applied at the ends of the beam can be incorporated into the boundary conditions in the same manner.

6.4.10 Orthogonality with Inertial Boundary Conditions

It can be verified that the conventional orthogonality condition (6.163) holds for beams in transverse vibration under common non-inertial boundary conditions. When an inertia element (rectilinear or rotatory) is present at an end of the beam, this condition is violated. A modified and more general orthogonality condition can be derived for application to beams with inertial boundary conditions.

To illustrate the procedure, consider a beam with a mass m attached at the end $x = l$, as shown in Figure 6.18(a). A free-body diagram giving the sign convention for shear force Q acting on m is shown in Figure 6.18(b).

The boundary conditions at $x = l$ are:

1. Bending moment vanishes, because there is no rotatory inertia at the end that is free. Hence,

$$EI \frac{\partial^2 v(l,t)}{\partial x^2} = 0 \qquad (6.207)$$

2. Equation of rectilinear motion of the end mass:

Distributed-Parameter Systems

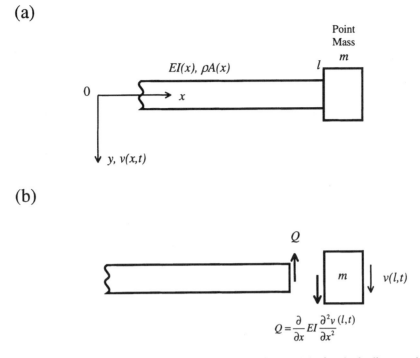

FIGURE 6.18 (a) A beam with an end mass in transverse vibration, and (b) free-body diagram showing the shear force acting on the end mass.

$$m\frac{\partial^2 v(l,t)}{\partial t^2} = \frac{\partial}{\partial x} EI \frac{\partial^2 v(l,t)}{\partial x^2} \qquad (6.208)$$

In the usual manner, by substituting $v(x,t) = Y_i(x)q_i(t)$ for mode i, along with $\ddot{q}_i(t) = -\omega_i^2 q_i(t)$, into equations (6.207) and (6.208) with the understanding that the result should hold for any $q_i(t)$, one obtains the corresponding modal boundary conditions:

$$\frac{d^2 Y_i(l)}{dx^2} = 0 \qquad (6.209)$$

$$\frac{d}{dx} EI \frac{d^2 Y_i(l)}{dx^2} + \omega_i^2 m Y_i(l) = 0 \qquad \text{for mode } i. \qquad (6.210)$$

Now return to equation (6.162):

$$\left(\omega_i^2 - \omega_j^2\right)\int_0^l \rho A Y_i Y_j dx - \left[Y_j \frac{d}{dx} EI \frac{d^2 Y_i}{dx^2} - Y_i \frac{d}{dx} EI \frac{d^2 Y_j}{dx^2}\right]_0^l + \left[\frac{dY_j}{dx} EI \frac{d^2 Y_i}{dx^2} - \frac{dY_i}{dx} EI \frac{d^2 Y_j}{dx^2}\right]_0^l = 0 \quad (6.162)$$

The second and the third terms of equation (6.162) will vanish at $x = 0$ for non-inertial boundary conditions, as usual. At $x = l$, the third term will vanish in view of equation (6.209). So, one is left with the second term at $x = l$. Substitute equation (6.210):

$$\left(Y_j \frac{d}{dx} EI \frac{d^2 Y_i}{dx^2} - Y_i \frac{d}{dx} EI \frac{d^2 Y_j}{dx^2}\right)_l = -\left(\omega_i^2 - \omega_j^2\right) m Y_i(l) Y_j(l) \qquad (6.211)$$

Now substitute equation (6.211) into (6.162) and cancel $\omega_i^2 - \omega_j^2 \neq 0$ for $i \neq j$; one obtains

$$\int_0^l \rho A Y_i Y_j \, dx + m Y_i(l) Y_j(l) = 0 \qquad \text{for } i \neq j$$

$$= \alpha_j \qquad \text{for } i = j \qquad (6.212)$$

This is the modified and more general orthogonality property. If the mass is at $x = 0$, the direction of Q that acts on m will reverse and, hence, the second term in equation (6.212) will become $-mY_i(0)Y_j(0)$.

Rotatory Inertia

If there is a free rotatory inertia at $x = l$, without an associated rectilinear inertia, then the shear force will vanish, giving

$$\frac{d}{dx} EI \frac{d^2 Y_i(l)}{dx^2} = 0 \qquad (6.213)$$

The equation of rotational motion of J will give

$$EI \frac{d^2 Y_i(l)}{dx^2} - \omega_i^2 J \frac{dY_i(l)}{dx} = 0 \qquad (6.214)$$

Here, the second term in equation (6.162) will vanish in view of (6.213). Then, by substituting equation (6.214) into the third term of (6.162), one gets the modified orthogonality relation

$$\int_0^l \rho A Y_i Y_j \, dx + J \frac{dY_i(l)}{dx} \frac{dY_j(l)}{dx} = 0 \qquad \text{for } i \neq j$$

$$= \alpha_j \qquad \text{for } i = j \qquad (6.215)$$

6.5 DAMPED CONTINUOUS SYSTEMS

All practical mechanical systems have some form of energy dissipation (damping). When the level of dissipation is small, damping is neglected, as has been done thus far in this chapter. Yet, some effects of damping — for example, the fact that at steady state, the natural (modal) vibration components decay to zero leaving only the steady forcing component — are tacitly assumed even in undamped analysis. The subject of damping is treated in Chapter 7.

The natural behavior of a system is expected to change due to the presence of damping. In particular, the system natural frequencies will decrease (and be called *damped natural frequencies*) as a result of damping. Furthermore, it is quite possible that a damped system would not possess "real" modes in which it could independently vibrate. Mathematically, in that case, the modes will become complex (as opposed to real), and physically, all points of the system will not move, maintaining a constant phase at a given damped natural frequency. In other words, a real solution that is separable in space (x) and time (t) may not be possible for the free vibration problem of a damped system. Also, node points of an undamped system can vary with time as a result of damping. With light damping, of course, such effects of damping will be negligible.

Because there are damped systems that do not possess real natural modes of vibration, care should be exercised when extending the results of modal analysis from an undamped system to a damped one. But, in some cases, the mode shapes will remain the same after including damping (although the natural frequencies will change). This is analogous to the case of *proportional damping*, which was discussed under lumped-parameter (multi-dof) vibrating systems. The modal analysis of a damped system will become significantly easier if one assumes that the mode shapes will remain the same as those for the undamped system. Even when the actual type of damping in the system results in complex modes, for analytical convenience, an equivalent damping model that gives real modes is used in simplified analysis. This is analogous to the use of linear viscous damping in lumped-parameter systems.

6.5.1 Modal Analysis of Damped Beams

Consider the problem of free damped transverse vibration of a thin beam, given by

$$\frac{\partial^2}{\partial x^2} EI \frac{\partial^2 v}{\partial x^2} + L\left(\frac{\partial v}{\partial t}\right) + \rho A \frac{\partial^2 v}{\partial t^2} = 0 \qquad (6.216)$$

where L is a spatial differential operator (in x). Consider the following two possible models of damping:

1. $L = \dfrac{\partial^2}{dx^2} E^* I \dfrac{\partial^2}{\partial x^2}$ \hfill (6.217)

2. $L = c$ \hfill (6.218)

Model 1 corresponds to the *Kelvin-Voigt model* of material (internal) damping given by the stress-strain relation

$$\sigma = E\varepsilon + E^* \frac{\partial \varepsilon}{\partial t} \qquad (6.219)$$

where E^* is the damping parameter of the beam material. Hence, one obtains the damped beam equation simply by replacing E in the undamped beam equation by $E + E^* \dfrac{\partial}{\partial t}$. Also, E^* is independent of the frequency of vibration for the *viscoelastic damping* model, but will be frequency dependent for the *hysteretic damping* model. Modal analysis is done regardless of any frequency dependence of E^* and, in the final modal result for a particular modal frequency ω_i, the appropriate frequency function for $E^*(\omega)$ is used with $\omega = \omega_i$, if the damping is of the hysteretic type. It can be easily verified that the mode shapes of the damped system with model (6.217) are identical to those of the undamped system, regardless of whether the beam cross section is uniform or not.

In model 2 [equation (6.218)], the operator is a constant c. This corresponds to external damping of the linear viscous type, distributed along the beam length. For example, imagine a beam resting on a foundation of viscous damping material. For model 2, it can be shown that the damped mode shapes are identical to the undamped ones — assuming that the beam cross section is uniform. If the beam is non-uniform, the damped and the undamped mode shapes are identical if one assumes that the damping constant c varies along the beam in proportion to the area of cross section $A(x)$ of the beam. This is shown in Example 6.8.

EXAMPLE 6.8

Perform the modal analysis for transverse vibration of a thin non-uniform beam with linear viscous damping distributed along its length and satisfying the beam equation

$$\frac{\partial^2}{\partial x^2} EI(x)\frac{\partial^2 v}{\partial x^2} + \rho A(x) b \frac{\partial v}{\partial t} + \rho A(x)\frac{\partial^2 v}{\partial t^2} = 0 \qquad (6.220)$$

Determine damped natural frequencies, modal damping ratios, and the response $v(x,t)$ as a modal series expansion, given $v(x,0) = d(x)$ and $\dot{v}(x,0) = s(x)$.

SOLUTION

Substitute the separable solution

$$v(x,t) = Y(x)q(t) \qquad (6.221)$$

in equation (6.220) to obtain

$$\frac{d^2}{dx^2} EI \frac{d^2 Y(x)}{dx^2} q(t) + \rho A(x) b Y(x) \dot{q}(t) + \rho A(x) Y(x) \ddot{q}(t) = 0$$

Group the functions of x and t separately, and equate to the same constant ω^2, as usual:

$$\frac{\dfrac{d^2}{dx^2} EI \dfrac{d^2 Y(x)}{dx^2}}{\rho A(x) Y(x)} = -\frac{\ddot{q}(t) + b\dot{q}(t)}{q(t)} = \omega^2 \qquad (6.222)$$

Thus,

$$\frac{d^2}{dx^2} EI \frac{d^2 Y(x)}{dx^2} - \omega^2 \rho A Y(x) = 0 \qquad (6.223)$$

and

$$\ddot{q}(t) + b\dot{q}(t) + \omega^2 q(t) = 0 \qquad (6.224)$$

Note that equation (6.223) is identical to that for the undamped beam. Hence, with known boundary conditions, one will obtain the same mode shapes $Y_i(x)$ and the same *undamped* natural frequencies ω_i in the usual manner. But, the equation of modal generalized coordinates $q(t)$ given by equation (6.224) is different from that for the undamped case ($b = 0$). For mode i,

Distributed-Parameter Systems

$$\ddot{q}_i(t) + 2\zeta_i\omega_i\dot{q}_i(t) + \omega_i^2 q_i(t) = 0 \tag{6.225}$$

where

$$\zeta_i = \frac{b}{2\omega_i} = \text{modal damping ratio for mode } i. \tag{6.226a}$$

Damped natural frequencies are

$$\omega_{di} = \sqrt{1-\zeta_i^2}\ \omega_i \tag{6.226b}$$

Equation (6.225) can be solved in the usual manner, with initial conditions $q_i(0)$ and $\dot{q}_i(0)$ determined *a priori*, using known $v(x,0)$ and $\dot{v}(x,0)$.

The modal series solution is

$$v(x,t) = \sum Y_i(x) q_i(t) \tag{6.227}$$

The initial conditions are

$$\sum Y_i(x) q_i(0) = d(x) \tag{6.228}$$

$$\sum Y_i(x) \dot{q}_i(0) = s(x) \tag{6.229}$$

Multiply equations (6.228) and (6.229) by $\rho A(x) Y_j(x)$ and integrate form $x = 0$ to l using the orthogonality condition

$$\int_0^l \rho A(x) Y_i(x) Y_j(x) dx = 0 \quad \text{for } i \neq j$$
$$= \alpha_j \quad \text{for } i = j \tag{6.230}$$

Thus,

$$q_j(0) = \frac{1}{\alpha_j} \int_0^l d(x) \rho A(x) dx \tag{6.231}$$

$$\dot{q}_j(0) = \frac{1}{\alpha_j} \int_0^l s(x) \rho A(x) dx \tag{6.232}$$

This completes the solution for the free damped beam. The forced damped case can be analyzed in the same manner as for the forced undamped case because the mode shapes are the same.

□

6.6 VIBRATION OF MEMBRANES AND PLATES

The cables, rods, shafts, and beams for which the vibration has been studied thus far in this chapter, are *one-dimensional members* or *line structures*. These continuous members need one spatial variable (x), in addition to the time variable (t), as an independent variable to represent their governing equation of motion. Membranes and plates are *two-dimensional members* or *planar structures*. They need two independent spatial variables (x and y) in addition to time (t), to represent their dynamics.

A membrane can be interpreted as a two-dimensional extension of a string or cable. In particular, it has to be in tension and cannot support any bending moment. A plate is a two-dimensional extension of a beam. It can support a bending moment. The governing equations of these two-dimensional members, hence, will resemble two-dimensional versions of their respective one-dimensional counterparts. Modal analysis will also follow the familiar steps, after accounting for the extra dimension. This section provides an introduction to the modal analysis of membranes and plates. For simplicity, only special cases of rectangular members with relatively simple boundary conditions will be considered. Analysis of more complicated boundary geometries and conditions will follow analogous procedures, although requiring greater effort and producing more complicated results.

6.6.1 Transverse Vibration of Membranes

Consider a stretched membrane (in tension) that lies on the x–y plane, as shown in Figure 6.19. Transverse vibration $v(x,y,t)$ in the z-direction is of interest. By following a procedure that is somewhat analogous to the derivation of the cable equation, one can obtain the governing equation as

$$\frac{\partial^2 v(x,y,t)}{\partial t^2} = c^2 \left[\frac{\partial^2}{\partial x^2} + \frac{\partial^2}{\partial y^2} \right] v(x,y,t) \tag{6.233}$$

with

$$c = \sqrt{\frac{T'}{m'}} \tag{6.234}$$

where

T' = tension per unit length of membrane section (assumed constant)
m' = mass per unit surface area of membrane.

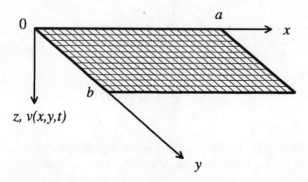

FIGURE 6.19 A membrane or a plate in Cartesian coordinates.

Distributed-Parameter Systems

For modal analysis, seek a separable solution of the form

$$v(x,y,t) = Y(x)Z(y)q(t) \tag{6.235}$$

and substitute equation (6.235) into (6.233) and divide throughout by $Y(x)Z(y)q(t)$ to get

$$\frac{\ddot{q}(t)}{q(t)} = c^2\left[\frac{1}{Y(x)}\frac{d^2Y(x)}{dx^2} + \frac{1}{Z(y)}\frac{d^2Z(y)}{dy^2}\right] \tag{6.236}$$

Because equation (6.236) is true for all possible values of t, x, and y, which are independent, the three function groups should separately equal to constants; thus,

$$\frac{1}{Y(x)}\frac{d^2Y}{dx^2} = -\alpha^2 \quad \text{or} \quad \frac{d^2Y(x)}{dx^2} + \alpha^2 Y(x) = 0 \tag{6.237}$$

$$\frac{1}{Z(y)}\frac{d^2Z(y)}{dy^2} = -\beta^2 \quad \text{or} \quad \frac{d^2Z(y)}{dy^2} + \beta^2 Z(y) = 0 \tag{6.238}$$

$$\frac{\ddot{q}(t)}{q(t)} = -\omega^2 \quad \text{or} \quad \ddot{q}(t) + \omega^2 q(t) = 0 \tag{6.239}$$

with

$$\omega^2 = c^2(\alpha^2 + \beta^2) \tag{6.240}$$

The argument for using positive constants α^2, β^2, and ω^2 is similar to that given for the one-dimensional case. Next, equations (6.237) and (6.238) must be solved using two end conditions for each direction, as usual. This will provide an infinite number of solutions α_i and β_j and the corresponding natural frequencies:

$$\omega_{ij}^2 = c(\alpha_i^2 + \beta_j^2)^{1/2} \quad \text{for } i = 1, 2, 3, \ldots \text{ and } j = 1, 2, 3, \ldots \tag{6.241}$$

along with the mode shape components $Y_i(x)$ and $Z_j(y)$ for the two dimensions.

6.6.2 Rectangular Membrane with Fixed Edges

Consider a rectangular membrane of length a and width b as shown in Figure 6.19 and with the four edges fixed. The boundary conditions are

$$v(0,y,t) = 0; \quad v(a,y,t) = 0; \quad v(x,0,t) = 0; \quad v(x,b,t) = 0$$

Using these in solving equations (6.237) and (6.238) as usual, one obtains

$$Y_i(x) = \sin\alpha_i x \quad \text{with } \alpha_i = \frac{i\pi}{a} \tag{6.242}$$

$$Z_j(y) = \sin\beta_j y \qquad \text{with } \beta_j = \frac{j\pi}{b} \qquad (6.243)$$

$$\omega_{ij} = c\left[\alpha_i^2 + \beta_j^2\right]^{1/2} = \pi c\left[\frac{i^2}{a^2} + \frac{j^2}{b^2}\right]^{1/2} \qquad \text{for } i = 1, 2, 3, \dots \text{ and } j = 1, 2, 3, \dots \qquad (6.244)$$

Note that the spatial mode shapes are given by

$$Y_j(x)Z_j(y) = \sin\frac{i\pi x}{a}\sin\frac{j\pi y}{b} \qquad (6.245)$$

6.6.3 Transverse Vibration of Thin Plates

Consider a thin plate of thickness h in a Cartesian coordinate system as shown in Figure 6.19. The usual assumptions as for the derivation of the Bernoulli-Euler beam equation are used. In particular, h is assumed small compared to the surface dimensions (a and b for a rectangular plate). Then, shear deformation and rotatory inertia can be neglected, and normal stresses in the transverse direction (z) can also be neglected. Furthermore, any end forces in the planar directions (x and y) are neglected. The governing equation is

$$\frac{\partial^2 v(x,y,t)}{\partial t^2} + c^2\left[\frac{\partial^2}{\partial x^2} + \frac{\partial^2}{\partial y^2}\right]^2 v(x,y,t) = 0 \qquad (6.246)$$

with

$$c^2 = \frac{E'I'}{\rho A'} = \frac{Eh^2}{12(1-\upsilon^2)\rho} \qquad (6.247)$$

where

$$E' = \frac{E}{(1-\upsilon^2)} \qquad (6.248)$$

$$I' = \frac{h^3}{12} = \text{second moment of area per unit length of section} \qquad (6.249)$$

A' = area per unit length of section
ρ = mass density of material
E = Young's modulus of elasticity of the plate material
υ = Poisson's ratio of the plate material.

If one attempts modal analysis by assuming a completely separable solution of the form $v(x,y,t) = Y(x)Z(y)q(t)$ in equation (6.246), a separable grouping of functions of x and y will not be achieved in general. But, the space and the time will be separable in modal motions. Hence, seek a solution of the form

$$v(x,y,t) = Y(x,y)q(t) \qquad (6.250)$$

Distributed-Parameter Systems

One will get

$$\ddot{q}(t) + \omega^2 q(t) = 0 \tag{6.251}$$

and

$$\nabla^2 \nabla^2 Y(x,y) - \lambda^4 Y(x,y) = 0 \tag{6.252}$$

with the natural frequencies ω given by

$$\omega = \lambda^2 c \tag{6.253}$$

and ∇^2 is the Laplace operator given by

$$\nabla^2 = \frac{\partial^2}{\partial x^2} + \frac{\partial^2}{\partial y^2} \tag{6.254}$$

and, hence, $\nabla^2\nabla^2$ is the biharmonic operator ∇^4 given by

$$\nabla^4 = \frac{\partial^4}{\partial x^4} + 2\frac{\partial^4}{\partial x^2 \partial y^2} + \frac{\partial^4}{\partial y^4} \tag{6.255}$$

Solution of equation (6.252) will require two sets of boundary conditions for each edge of the plate (as for a beam), but will be mathematically involved. Instead of a direct solution, a logical trial solution that satisfies equation (6.252) and the boundary conditions is employed next for a simply supported rectangular plate. The solution tried is in fact the correct solution for the particular problem.

6.6.4 Rectangular Plate with Simply Supported Edges

As a special case, now consider a thin rectangular plate of length a, width b, and thickness h, as shown in Figure 6.19, whose edges are simply supported. For each edge, the boundary conditions are that the displacement is zero and the bending moment about the edge is zero. Specifically,

$$v(x,y,t) = 0 \quad \text{and} \quad M_x = E'I'\left(\frac{\partial^2 v}{\partial x^2} + v\frac{\partial^2 v}{\partial y^2}\right) = 0 \quad \text{for } x = 0 \text{ and } a;\ 0 \leq y \leq b$$

$$\tag{6.256}$$

$$v(x,y,t) = 0 \quad \text{and} \quad M_y = E'I'\left(\frac{\partial^2 v}{\partial y^2} + v\frac{\partial^2 v}{\partial x^2}\right) = 0 \quad \text{for } y = 0 \text{ and } b;\ 0 \leq x \leq a$$

where E' and I' are given by equations (6.248) and (6.249), respectively. In this case, the mode shapes are found to be

$$Y_{ij}(x,y) = \sin\frac{i\pi x}{a}\sin\frac{j\pi y}{b} \quad \text{for } i = 1,\ 2,\ \ldots \text{ and } j = 1,\ 2,\ \ldots \tag{6.257}$$

which clearly satisfy the boundary conditions (6.256) and the governing model equation (6.252). There exists an infinite set of solutions for λ given by

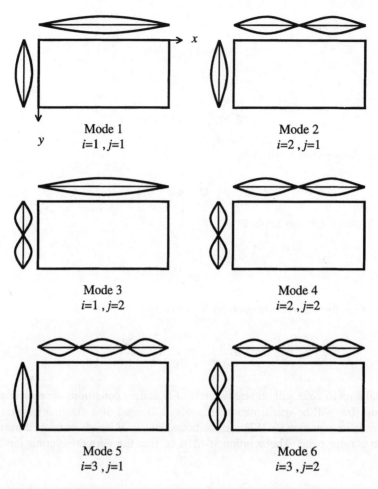

FIGURE 6.20 Mode shapes of transverse vibration of a simply supported rectangular plate.

$$\lambda_{ij}^2 = \pi^2 \left(\frac{i^2}{a^2} + \frac{j^2}{b^2} \right) \tag{6.258}$$

and, hence, from equation (6.253), the natural frequencies are

$$\omega_{ij} = \pi^2 c \left(\frac{i^2}{a^2} + \frac{j^2}{b^2} \right) \tag{6.259}$$

where c is given by equation (6.247). The overall response, then, is given by

$$v(x,y,t) = \sum_{i=1}^{\infty} \sum_{j=1}^{\infty} \left[A_{ij} \sin \omega_{ij} t + B_{ij} \cos \omega_{ij} t \right] \sin \frac{i\pi x}{a} \sin \frac{i\pi y}{b} \tag{6.260}$$

The unknown constants A_{ij} and B_{ij} are determined by the system initial conditions $v(x,y,0)$ and $\dot{v}(x,y,0)$. The first six mode shapes of transverse vibration of a rectangular plate are sketched in Figure 6.20.

PROBLEMS

6.1 Consider the traveling wave solution given by

$$v(x,t) = v_1(x-ct) + v_2(x+ct)$$

to the problem of free transverse vibration of a taut cable (wave equation):

$$\frac{\partial^2 v}{\partial t^2} = c^2 \frac{\partial^2 v}{\partial x^2}$$

Suppose that the system is excited by an initial displacement $d(x)$ and an initial speed $s(x)$. Show that the functions $v_1(x-ct)$ and $v_2(x+ct)$ can be completely determined by these two initial conditions. This also means that the traveling wave solution does not depend on the boundary conditions of the system. Under what conditions would this assumption be satisfied and, hence, the solution be valid? Provide some justification for the solution.

6.2 a. List several differences between lumped-parameter (discrete) and distributed-parameter (continuous) vibratory systems.

b. Consider a cable of length l that is stretched across a flexible pole and a rigid pole, modeled as shown in Figure P6.2. The flexible pole is represented by a spring of stiffness k. The cable is assumed to have a uniform cross section, with a mass of m per unit length. The cable tension T is assumed constant. Analyze the problem of free transverse vibration of the cable, giving appropriate boundary actions.

6.3 Show that the orthogonality property is satisfied by the mode shapes of transverse vibration of a taut cable under a flexibly supported end with stiffness k.

6.4 The cord of a musical instrument is mounted horizontally with fixed ends and maintained in tension T. The length of the cord is l and mass per unit length is m. An impulsive speed of $a\delta\left(x - \frac{1}{2}\right)$ is applied to the stationary cord in the transverse direction; for

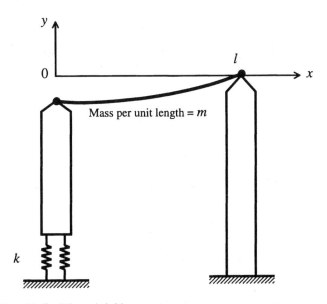

FIGURE P6.2 Cable with flexible and rigid supports.

example, by gently hitting the mid-span with a hammer. Determine the resulting vibration of the cord.

Note: $\delta(x - x_0)$ is the Dirac delta function such that

$$\int_a^b f(x)\delta(x - x_0)dx = f(x_0)$$

where x_0 is within the integration interval $[a,b]$ and $f(x)$ is an arbitrary function.

6.5 a. Discuss why boundary conditions have to be known explicitly in solving a continuous vibrating system, but not for solving a discrete vibrating system.
   b. Determine the modal boundary condition at $x = l$ for a rod carrying a mass M and restrained by a fixture of stiffness k, as shown in Figure P6.5. The area of cross section of the rod is A, and the Young's modulus is E. Show that the orthogonality property, in the conventional form, is not satisfied by the mode shapes of the rod, and that the following modified orthogonality condition is satisfied:

$$\rho l \int_0^l X_i X_j dx + MX_i(l)X_j(l) = 0 \quad \text{for } i \neq j$$

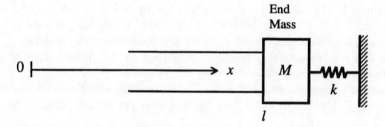

FIGURE P6.5 A rod with a dynamic restraint.

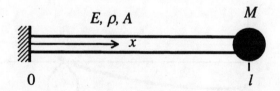

FIGURE P6.6 Longitudinal vibration of a rod with an end mass.

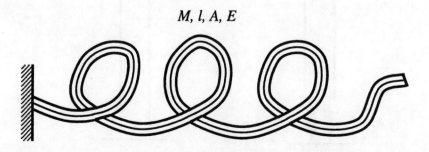

FIGURE P6.7 A helical spring with dominant axial deformation in the coil wire.

How would the result change if $I = I(x)$ and if M is located at $x = 0$?

6.6 a. What factors govern the use of the discrete-parameter assumption for a continuous-parameter vibrating system?

b. A uniform rod of length l, cross-sectional area A, mass density ρ, and Young's modulus E is fixed horizontally to a rigid wall at one end and carries a mass M at the other end, as shown in Figure P6.6. Analyze the system to determine the natural frequencies of longitudinal vibration.

6.7 When a helical (coil) spring is stretched, the coil wire will be subjected to bending and torsion at each cross section, as well as tension along the axis of the wire. The former effects can dominate unless the spring is almost stretched or loosely wound so that the coil pitch (distance between adjacent coils turns) is large compared to the coil diameter and that the coil diameter is not large compared to the wire diameter.

Assume that for a (somewhat unusual) helical spring, the bending and torsion effects are negligible compared to the longitudinal deflection of the wire. For such a spring with one end fixed and the other end free, as shown in Figure P6.7, determine the stiffness k and the natural frequencies of vibration. The following parameters are given:

M = mass of the spring
l = nominal, relaxed length of the coil wire
A = area of cross section of the coil wire
E = Young's modulus of the coil material.

6.8 Consider a uniform structural column of height l, mass density ρ, and Young's modulus E, and mounted on a rigid base, as schematically shown in Figure P6.8. Two types of initial conditions are considered:

a. An impulsive impact is made at the top end of the column so as to impart an instantaneous velocity of $v_0 \delta(x - l)$ at that point.

b. The column is pressed down at the top through a displacement u_0 and released suddenly from rest.

In each case, determine the subsequent longitudinal vibration of the column.

6.9 A uniform metal post of height l, mass density ρ, and Young's modulus E is fixed vertically on a rigid floor as schematically shown in Figure P6.9. The top end of the post is harmonically excited using a shaker device in the following two ways:

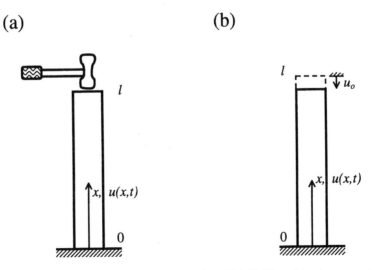

FIGURE P6.8 A structural column in longitudinal vibration: (a) initial impulsive impact at the top end, and (b) releasing from rest with initial elastic displacement.

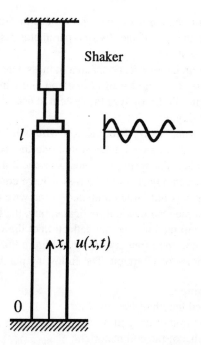

FIGURE P6.9 A post mounted on a rigid floor and excited at the top by a shaker.

a. The shaker head is displacement-feedback controlled so that a sinusoidal displacement of $u_0 \sin\omega t$ is generated.
b. The shaker head is force-feedback controlled so that a sinusoidal force of $f_0 \sin\omega t$ is generated.

In each case, determine the longitudinal vibratory displacement of the post, under steady-state conditions.

6.10 Consider a uniform shaft, not necessarily of circular cross section, in torsional vibration. The following shaft parameters are known:

J = polar moment of area of the shaft cross section (about the axis of rotation)
J_t = Saint Venant torsional parameter (equal to J for a circular cross section)
ρ = mass density
G = shear modulus.

Six sets of boundary conditions, as shown in Figure P6.10, are studied. In cases (e) and (f), there is an element of moment of inertia I_e, about the common axis of rotation, attached to one end of the shaft.

For each case of boundary conditions, determine the natural frequencies and mode shapes of torsional vibration. In cases (e) and (f), show that the orthogonality property, in the conventional form, is not satisfied, but the following modified orthogonality condition governs:

$$\int_0^l \rho J \Theta_i(x)\Theta_j(x)dx + I_e\Theta_i(l)\Theta_j(l) = 0 \quad \text{for } i \neq j$$

6.11 A mounted drill can slide along its guidepost so as to engage the drill bit with a workpiece. A schematic diagram is given in Figure P6.11(a). Suppose that the drill bit rotates at a constant angular speed ω_0 prior to engagement. At the instant of engagement, the power

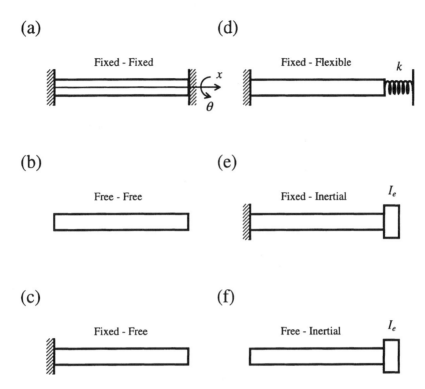

FIGURE P6.10 Some boundary conditions for torsional vibration of shafts.

is lost and the motor drive torque become zero. Assuming a large workpiece, the drill bit can be assumed fixed at the engagement end, under these conditions, as represented in Figure P6.11(b). Analyze the resulting torsional vibrations in the drill bit. The following parameters are known:

Length of the drill bit = l
Moment of inertia of the drive rotor = I_r
The usual parameters G, ρ, J, and J_t of the drill bit.

6.12 A vibration engineer proceeds to estimate the shear modulus G of an unknown material. She prepares a uniform shaft of circular cross section and length 50.0 cm of the material, rigidly mounts one end on a heavy fixture, and leaves the other end free. She excites the shaft and measures the fundamental natural frequency of torsion. It is found to be 1.6 kHz. Also, the density of the material is measured to be 7.8×10^3 kg m^{-3}.
  a. Indicate a possible method of exciting the shaft and measuring the frequency in this experiment.
  b. Estimate the value of G.
  c. Guess the material type.

6.13 A grinding tool is modeled as in Figure P6.13. The drive torque that is applied at one end of the tool, through electromagnetic means, is given by $T_0 + T_a \sin\omega t$, where T_0 is a constant denoting the steady torque. There is a torque ripple of amplitude T_a and frequency ω. The grinding process is represented by an energy dissipation as in a viscous torsional damper with damping constant b. The tool cross section is circular and has a polar moment of area J about its axis of rotation. The tool length is l, the mass density is ρ, and the shear modulus is G. Determine the rotational motion $\theta(x,t)$

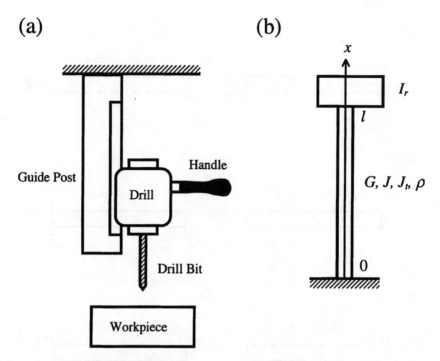

FIGURE P6.11 The problem of a failed drill.

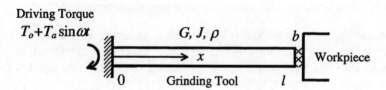

FIGURE P6.13 Torsional vibration of a grinding tool.

of the tool at steady state, where x is measured from the driven end, along the tool axis, as shown.

6.14 Consider the torsional guideway transit problem shown in Figure 6.8 and analyzed in Example 6.5.

a. If the crossing frequency ratio v_c is small compared to 1, give an expression for the angle of twist $\theta(t)$ of the guideway at the kth vehicle suspension, when there are N_v vehicle suspensions moving along the particular guideway span. Use only the first N_t modes in the modal summation.

b. Neglecting the dynamics of the guideway, what is the angle of twist at a vehicle suspension, when only that suspension is on the guideway span? What is its maximum value θ_{max}?

c. For a single suspension, determine an expression for $e = \dfrac{(\theta_s - \theta)}{\theta_{max}} \times 100\%$, where θ is the angle of twist obtained in (a), with N_t modes, when there is only one suspension on the guideway span. Plot e versus the fractional vehicle location $\dfrac{vt}{l}$ for the five cases of $N_t = 1, 2, 3, 4,$ and 5. How many modes would be adequate for a "good" approximation in the present application?

Distributed-Parameter Systems

6.15 An AC induction motor drives a pump through a shaft. The free torsional vibration of the system is to be analyzed. A schematic representation of this free system (i.e., with zero motor torque) is shown in Figure P6.15. It is not necessary to assume a circular cross section for the shaft. The usual parameters G, J_t, ρ, J, and l are known for the shaft. Also, given are:

I_m = moment of inertia of the motor rotor
I_p = moment of inertia of the pump impeller.

a. Neglecting bearing friction, formulate the problem and identify the boundary conditions.
b. What are the modal boundary conditions?
c. Giving all the steps and the necessary equations, describe how the natural frequencies and mode shapes of torsional vibration can be determined for this system.
d. If the shaft is massless, what are the natural frequencies of the system? Show how these frequencies could be derived from the general solution given in (c), as the shaft inertia approaches zero.

6.16 a. A beam in transverse vibration is represented by the equation

$$\frac{\partial^2}{\partial x^2} EI \frac{\partial^2 v(x,t)}{\partial x^2} + \rho A \frac{\partial^2 v(x,t)}{\partial t^2} = f(x,t)$$

i. Define the parameters E, I, ρ, A, and the variables t, x, f, and v in this equation.
ii. What are the assumptions made in deriving this equation?
b. Prove that the mode shape functions $Y_i(x)$ of the beam in part (a) are orthogonal.
c. For a simply supported beam with uniform cross section, it can be shown that the mode shape functions are given by:

$$Y_i(x) = \sin \frac{i\pi x}{l} \quad i = 1, 2, \ldots$$

Express the natural frequencies ω_i corresponding to these modes, in terms of E, I, ρ, A, l, and the mode number i.

6.17 a. Compare and contrast linear, lumped-parameter systems and distributed-parameter systems, considering for example the nature of their
i. equations of motion
ii. natural frequencies
iii. mode shapes.

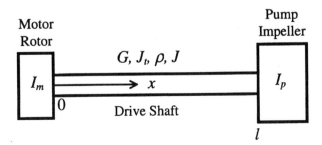

FIGURE P6.15 A pump driven by an induction motor.

What are possible practical problems that might arise when a distributed-parameter system is approximated by a lumped-parameter system?

b. Consider the Bernoulli-Euler beam equation given by

$$\frac{\partial^2}{\partial x^2} EI \frac{\partial^2 v}{\partial x^2} + \rho A \frac{\partial^2 v}{\partial t^2} = f(x,t)$$

i. Define all the parameters and variables in this equation.
ii. What are the assumptions made in deriving this equation?
iii. What are the analytical steps involved in solving for natural frequencies and mode shapes of this system?
iv. What analytical steps could be followed in obtaining the response of the beam to a specified forcing function $f(x,t)$?
v. Suppose that the beam is simply supported and a constant force F_0 moves along the beam at constant speed u_0 from one end to the other. What is $f(x,t)$ in this case?

6.18 What are the modal boundary conditions of a thin beam in transverse vibration, with the end $x = 0$ fixed (clamped) and the other end $x = l$ sliding?

Giving all the necessary steps, derive an equation for which the solutions provide the natural frequencies of transverse vibration of this cantilever. What are the corresponding mode shape functions? Give three different forms of these eigenfunctions.

6.19 For mode shapes of a thin and non-uniform beam in bending vibration, show that the orthogonality condition

$$\int_0^l EI(x) \frac{d^2 Y_i(x)}{dx^2} \frac{d^2 Y_j(x)}{dx^2} dx = 0 \quad \text{for } i \neq j$$

holds under common boundary conditions. The parameter EI represents flexural stiffness of the beam.

6.20 One way of normalizing the mode shapes $Y_i(x)$ of a beam in transverse vibration is according to

$$\int_0^l \rho A Y_i^2 dx = \int_0^l \rho A dx \quad \text{for all } i$$

Note that the right-hand-side integral is equal to the beam mass. Determine the normalized mode shape functions according to this approach for a simply supported beam of uniform cross section.

6.21 A non-uniform beam is excited in the transverse direction by a distributed harmonic force of

$$f(x,t) = f(x) \sin \omega t$$

as sketched in Figure P6.21.

Determine the resulting transverse response $v(x,t)$ of the beam for initial conditions

Distributed-Parameter Systems

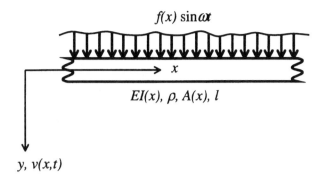

FIGURE P6.21 A beam excited by a transverse, distributed, harmonic force.

$$v(x,0) = d(x) \quad \text{and} \quad \frac{\partial v}{\partial t}(x,0) = s(x)$$

What is the steady-state response, assuming very light damping?
Specialize your results to the case of a uniform simply supported beam.
Assume that the usual parameters E, I, ρ, A, and l of the beam are known.

6.22 Beams on elastic foundations are useful in many applications such as railroad tracks, engine baseblocks and mounts, and seismic motions of structures. For a beam resting on an elastic foundation, the following parameters are defined:

k_f = foundation elastic modulus
   = force per unit length of the beam that causes a unit deflection in the foundation
b_f = foundation damping modulus
   = force per unit length of the beam that causes a unit velocity in the foundation.

Indicate how the transverse vibration of a beam can be modified to include k_f and b_f. A schematic diagram of the system is shown in Figure P6.22. For the undamped case ($b_f = 0$), explain how modal analysis could be performed for this system. In particular, solve the case of a uniform, simply supported beam.

6.23 Consider a thin beam with a constant axial force P. Perform a modal analysis for transverse vibration. For the special case of a uniform, simply supported beam (with the usual parameters EI, ρ, A, l), obtain the complete solution giving all the natural frequencies and mode shapes. Show that the natural frequencies increase due to the tensile force.

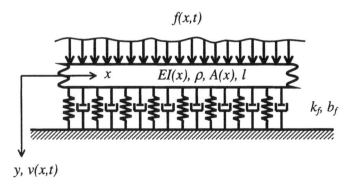

FIGURE P6.22 A beam on an elastic and dissipative foundation.

a. Show that each natural frequency has two contributions: one for a beam without an axial force and the other for a cable that cannot support a bending moment. In particular, show that when $P = 0$, one gets the former component; and when the beam cannot support a bending moment (i.e., $EI = 0$), one gets the latter component.

b. Show that the compressive force P_{cr} at which the fundamental natural frequency of transverse vibration becomes zero is the first Euler buckling load for the beam.

6.24 Consider a non-uniform and thin beam in bending vibration. The left end of the beam has a translatory-dynamic boundary condition consisting of a mass m, a spring of stiffness k, and a linear viscous damper of damping constant b. The right end of the beam has a rotatory-dynamic boundary condition consisting of a moment of inertia J, a spring of torsional stiffness K, and a linear viscous damper with rotatory damping constant B. A schematic representation of the system is given in Figure P6.24.

Assume that there are no translatory-dynamic effects (mass, translatory stiffness, translatory damping) due to the rotatory dynamic element at the right-hand end (which is *not* usually the case).

a. Express the boundary conditions of the beam.

b. For the undamped case ($b = 0$, $B = 0$), determine the modal boundary conditions.

6.25 Consider a circular shaft of length l, mass density ρ, shear modulus G, and polar moment of area $J(x)$ carrying a circular disk of moment of inertia I at the far end. Using the energy variational approach, determine the equation of torsional vibration of the system, including the boundary conditions for the case where the near end of the shaft is

a. fixed

b. free.

Now consider the modal analysis of a uniform circular shaft with its near end fixed and the far end carrying a disk. The analysis was done in Problem 6.10(e). For the special case of $I = l\rho J$, determine the fundamental natural frequency (ω_1) and the corresponding normalized mode shape ($\Theta_1(x)$).

6.26 Using the energy variational approach, derive the terms that should be added to the governing equation and boundary conditions of a circular shaft with an end disk, in torsional vibration, as in Problem 6.25, for the following situations:

a. A distributed torque $\tau(x,t)$ per unit length along the shaft length

b. A point torque T_e at $x = l$.

Consider a uniform circular shaft (ρ, l, G, J) with one end ($x = 0$) fixed and the other end ($x = l$) carrying a circular disk of moment of inertia I. A harmonic point torque $T_0 \sin\omega t$ is applied to the disk at $x = l$. Determine the torsional response of the shaft at steady state. A schematic diagram of the system is shown in Figure P6.26.

i. What are the critical excitation frequencies at which the steady-state response of the system would become very high?

ii. What is the amplitude of the angular response of the end disk at steady state?

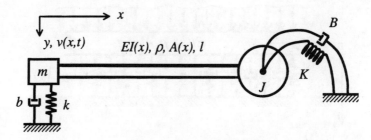

FIGURE P6.24 A beam with translatory-dynamic and rotatory-dynamic boundary conditions.

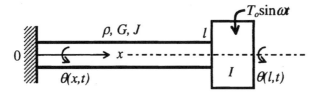

FIGURE P6.26 A shaft with an end disk excited by a harmonic torque.

6.27 In thin beam theory of transverse vibration (Bernoulli-Euler beam), does the rotation of a beam element vary in phase with the displacement?
In thick beam theory (Timoshenko beam), show that if one assumes that displacement and rotation of a beam element are in phase, it will lead to erroneous results.

6.28 Perform a modal analysis for transverse vibration of a thin damped beam given by

$$\frac{\partial^2}{\partial x^2} EI(x)\frac{\partial^2 v}{\partial x^2} + L\frac{\partial v}{\partial t} + \rho A(x)\frac{\partial^2 v}{\partial t^2} = 0$$

for the two damping models $L\frac{\partial v}{\partial t}$ given by:

a. $L = \dfrac{\partial^2}{\partial x^2} E^* I(x) \dfrac{\partial^2}{\partial x^2}$

b. $L = c\dfrac{\partial}{\partial x}$

In (a), determine the damped natural frequencies and modal damping ratios if the energy dissipation in the beam material is represented by the hysteretic damping model:

$$E^* = \frac{g_1}{\omega} + g_2$$

6.29 Consider a single vehicle moving at constant speed p along a span of an elevated guideway. Assume that the guideway is uniform with parameters EI and ρA; the span is simply supported at the support piers; and the vehicle is a point suspension with constant load W, inclusive of the vehicle weight. A schematic diagram is given in Figure P6.29.

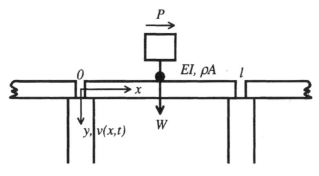

FIGURE P6.29 A vehicle on an elevated guideway.

a. Determine the transverse motion of the guideway as the vehicle travels along the span.
b. What are the critical speeds that should be avoided by the vehicle?
c. If initially the span is at rest, what is the deflection of the guideway just underneath the vehicle, measured from the equilibrium configuration of the guideway?

6.30 An elevated guideway of a transit system consists of two-span single beam segments on three support piers, as shown in Figure P6.30. The span lengths are equal at l, and the ends of each two-span beam are simply supported. In order to determine the guideway response to vehicles moving on it, first the natural frequencies and mode shapes of each guideway beam must be determined. Clearly state the steps that need to be carried out in accomplishing this, giving the equations that need to be solved, along with appropriate boundary conditions. Assume that the piers do not receive bending moments from the guideway and that the guideway is always attached to the piers.

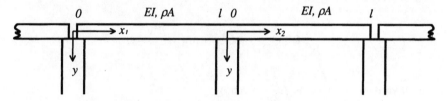

FIGURE P6.30 A single-beam, two-span elevated guideway segment.

7 Damping

Damping is the phenomenon by which mechanical energy is dissipated (usually converted into internal thermal energy) in dynamic systems. A knowledge of the level of damping in a dynamic system is important in utilization, analysis, and testing of the system. For example, a device having natural frequencies within the seismic range (i.e., less than 33 Hz) and having relatively low damping, could produce damaging motions under resonant conditions when subjected to a seismic disturbance. Also, the device motions could be further magnified by low-frequency support structures and panels having low damping. This illustrates that a knowledge of damping in constituent devices, components, and support structures is particularly useful in the design and operation of a complex mechanical system. The nature and the level of component damping should be known in order to develop a dynamic model of the system and its peripherals. A knowledge of damping in a system is also important in imposing dynamic environmental limitations on the system (i.e., the maximum dynamic excitation the system could withstand) under in-service conditions. Furthermore, a knowledge of its damping could be useful in order to make design modifications in a system that has failed the acceptance test. The significance of knowledge of damping level in a test object, for the development of test excitation (input), is often overemphasized, however. Specifically, if the response-spectrum method is used to represent the required excitation in a vibration test, it is not necessary that the damping value used in the development of the required response spectrum specification be equal to actual damping in the test object. It is only necessary that the damping used in the specified response spectrum be equal to that used in the test-response spectrum (see Chapter 10). The degree of dynamic interaction between test object and shaker table, however, will depend on the actual level of damping in these systems. Furthermore, when testing near the resonant frequency of a test object, it is desirable to have a knowledge of damping in the test object, because it is in this neighborhood that the object response is most sensitive to damping.

In characterizing damping in a dynamic system, it is important, first, to understand the major mechanisms associated with mechanical-energy dissipation in the system. Then, a suitable damping model should be chosen to represent the associated energy dissipation. Finally, damping values (model parameters) are determined, for example, by testing the system or a representative physical model, by monitoring system response under transient conditions during normal operation, or by employing already available data.

7.1 TYPES OF DAMPING

There is some form of mechanical-energy dissipation in any dynamic system. In the modeling of systems, damping can be neglected if the mechanical energy that is dissipated during the time duration of interest is small in comparison to the initial total mechanical energy of excitation in the system. Even for highly damped systems, it is useful to perform an analysis with the damping terms neglected, in order to study several crucial dynamic characteristics; for example, modal characteristics (undamped natural frequencies and mode shapes).

Several types of damping are inherently present in a mechanical system. If the level of damping that is available in this manner is not adequate for proper functioning of the system, external damping devices can be added either during the original design or in a subsequent stage of design modification of the system. Three primary mechanisms of damping are important in the study of mechanical systems. They are:

1. Internal damping (of material)
2. Structural damping (at joints and interfaces)
3. Fluid damping (through fluid-structure interactions).

Internal (material) damping results from mechanical-energy dissipation within the material due to various microscopic and macroscopic processes. Structural damping is caused by mechanical-energy dissipation resulting from relative motions between components in a mechanical structure that has common points of contact, joints, or supports. Fluid damping arises from the mechanical-energy dissipation resulting from drag forces and associated dynamic interactions when a mechanical system or its components move in a fluid.

Two general types of external dampers can be added to a mechanical system in order to improve its energy dissipation characteristics. They are

1. passive dampers
2. active dampers.

A passive damper is a device that dissipates energy through some motion, without needing an external power source or actuator. Active dampers have actuators that need external sources of power. They operate by actively controlling the motion of the system that needs damping. Dampers may be considered as vibration controllers (see Chapter 12). The present chapter emphasizes damping that is inherently present in a mechanical system.

7.1.1 MATERIAL (INTERNAL) DAMPING

Internal damping of materials originates from the energy dissipation associated with microstructure defects, such as grain boundaries and impurities; thermoelastic effects caused by local temperature gradients resulting from non-uniform stresses, as in vibrating beams; eddy-current effects in ferromagnetic materials; dislocation motion in metals; and chain motion in polymers. Several models have been employed to represent energy dissipation caused by internal damping. This variability is primarily a result of the vast range of engineering materials; no single model can satisfactorily represent the internal damping characteristics of all materials. Nevertheless, two general types of internal damping can be identified: viscoelastic damping and hysteretic damping. The latter term is actually a misnomer, because all types of internal damping are associated with hysteresis-loop effects. The stress (σ) and strain (ϵ) relations at a point in a vibrating continuum possess a hysteresis loop, such as the one shown in Figure 7.1. The area of the hysteresis loop gives the energy dissipation per unit volume of the material, per stress cycle. This is termed per-unit-volume damping capacity, and is denoted by d. It is clear that d is given by the cyclic integral

$$d = \oint \sigma d\epsilon \qquad (7.1)$$

In fact, for any damped device, there is a corresponding hysteresis loop in the displacement-force plane as well. In this case, the cyclic integral of force with respect to the displacement, which is the area of the hysteresis loop, is equal to the work done against the damping force. It follows that this integral (loop area) is the energy dissipated per cycle of motion. This is the *damping capacity*, which, when divided by the material volume, gives the per-unit-volume damping capacity as before.

It should be clear that, unlike a pure elastic force (e.g., spring force), a damping force cannot be a function of displacement (q) alone. The reason is straightforward. Consider a force $f(q)$ that depends on q alone. Then, for a particular displacement point q of the component, the force will be the same regardless of the magnitude and direction of motion (i.e., the value and sign of $\dot{q}$).

Damping

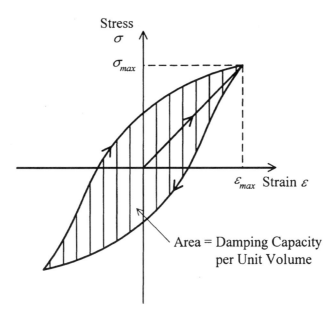

FIGURE 7.1 A typical hysteresis loop for mechanical damping.

It follows that, in a loading and unloading cycle, the same path will be followed in both directions of motion. Hence, a hysteresis loop will not be formed. In other words, the net work done in a complete cycle of motion will be zero. Next consider a force $f(q, \dot{q})$ that depends on both q and $\dot{q}$. Then, at a given displacement point q, the force will depend on $\dot{q}$, as well. Hence, even at low speeds, force in one direction of motion can be significantly different from that in the opposite direction. As a result, a hysteresis loop will be formed, which corresponds to work done against the damping force (i.e., energy dissipation). One can conclude then that damping force has to depend on a relative velocity $\dot{q}$ in some manner. In particular, Coulomb friction, which does not depend on the magnitude of $\dot{q}$, does depend on the sign (direction) of $\dot{q}$.

Viscoelastic Damping

For a linear viscoelastic material, the stress-strain relationship is given by a linear differential equation with respect to time, having constant coefficients. A commonly employed relationship is

$$\sigma = E\varepsilon + E^* \frac{d\varepsilon}{dt} \tag{7.2}$$

which is known as the *Kelvin-Voigt model*. In equation (7.2), E is Young's modulus and E^* is a viscoelastic parameter that is assumed to be time independent. The elastic term $E\varepsilon$ does not contribute to damping, and, as noted before, mathematically, its cyclic integral vanishes. Consequently, for the Kelvin-Voigt model, damping capacity per unit volume is

$$d_v = E^* \oint \frac{d\varepsilon}{dt} d\varepsilon \tag{7.3}$$

For a material that is subjected to a harmonic (sinusoidal) excitation, at steady state, one obtains

$$\varepsilon = \varepsilon_{max} \cos \omega t \tag{7.4}$$

When equation (7.4) is substituted in equation (7.3), one obtains

$$d_v = \pi\omega E^* \varepsilon_{max}^2 \tag{7.5}$$

Now, $\varepsilon = \varepsilon_{max}$ when $t = 0$ in equation (7.4), or when $\dfrac{d\varepsilon}{dt} = 0$. The corresponding stress, according to equation (7.2), is $\sigma_{max} = E\varepsilon_{max}$. It follows that

$$d_v = \frac{\pi\omega E^* \sigma_{max}^2}{E^2} \tag{7.6}$$

These expressions for d_v depend on the frequency of excitation, ω.

Apart from the Kelvin-Voigt model, two other models of viscoelastic damping are also commonly used. They are, the *Maxwell model* given by

$$\sigma + c_s \frac{d\sigma}{dt} = E^* \frac{d\varepsilon}{dt} \tag{7.7}$$

and the *standard linear solid model* given by

$$\sigma + c_s \frac{d\sigma}{dt} = E\varepsilon + E^* \frac{d\varepsilon}{dt} \tag{7.8}$$

It is clear that the standard linear solid model represents a combination of the Kelvin-Voigt model and the Maxwell model, and is the most accurate of the three. But, for most practical purposes, the Kelvin-Voigt model is adequate.

Hysteretic Damping

It was noted that the stress, and hence the internal damping force, of a viscoelastic damping material depend on the frequency of variation of the strain (and consequently on the frequency of motion). For some types of material, it has been observed that the damping force does not significantly depend on the frequency of oscillation of strain (or frequency of harmonic motion). This type of internal damping is known as *hysteretic damping*.

Damping capacity per unit volume (d_h) for hysteretic damping is also independent of the frequency of motion and can be represented by

$$d_h = J\sigma_{max}^n \tag{7.9}$$

As clear from equation (7.6), a simple model that satisfies equation (7.9), for the case of $n = 2$, is given by

$$\sigma = E\varepsilon + \frac{\tilde{E}}{\omega} \frac{d\varepsilon}{dt} \tag{7.10}$$

which is equivalent to using a viscoelastic parameter E^* that depends on the frequency of motion in equation (7.2) according to $E^* = \tilde{E}/\omega$. Consider the case of harmonic motion at frequency ω, with the material strain given by

Damping

$$\varepsilon = \varepsilon_0 \cos \omega t \qquad (7.11)$$

Then, equation (7.10) becomes

$$\sigma = E\varepsilon_0 \cos \omega t - \tilde{E}\varepsilon_0 \sin \omega t = E\varepsilon \cos \omega t + \tilde{E}\varepsilon_0 \cos\left(\omega t + \frac{\pi}{2}\right) \qquad (7.12)$$

Note that the material stress consists of two components, as given by the right-hand side of equation (7.12). The first component corresponds to the linear elastic behavior of a material and is in phase with the strain. The second component of stress, which corresponds to hysteretic damping, is 90° out of phase (this stress component leads the strain by 90°). A convenient mathematical representation would be possible, by using the usual complex form of the response according to

$$\varepsilon = \varepsilon_0 e^{j\omega t} \qquad (7.13)$$

Then, equation (7.10) becomes

$$\sigma = \left(E + j\tilde{E}\right)\varepsilon \qquad (7.14)$$

It follows that this form of simplified hysteretic damping can be represented using a complex modulus of elasticity, consisting of a real part that corresponds to the usual linear elastic (energy storage) modulus (or Young's modulus) and an imaginary part that corresponds to the hysteretic loss (energy dissipation) modulus.

By combining equations (7.2) and (7.10), a simple model for combined viscoelastic and hysteretic damping can be given by

$$\sigma = E\varepsilon + \left(E^* + \frac{\tilde{E}}{\omega}\right)\frac{d\varepsilon}{dt} \qquad (7.15)$$

in which the parameters E, E^* and $\tilde{E}$ are independent of the frequency ω.

The equation of motion for a system for which the damping is represented by equation (7.15) can be deduced from the pure elastic equation of motion by simply substituting E with the operator

$$E + \left(E^* + \frac{\tilde{E}}{\omega}\right)\frac{\partial}{\partial t}$$

in the time domain.

Example 7.1

Determine the equation of flexural motion of a non-uniform slender beam whose material has both viscoelastic and hysteretic damping.

Solution

The Bernoulli-Euler equation of bending motion of an undamped beam subjected to a dynamic load of $f(x,t)$ per unit length is given by (see Chapter 6):

$$\frac{\partial^2}{\partial x^2} EI \frac{\partial^2 q}{\partial x^2} + \rho A \frac{\partial^2 q}{\partial t^2} = f(x,t) \tag{7.16}$$

Here, q is the transverse motion at a distance x along the beam. Then, for a beam with material damping (both viscoelastic and hysteretic), one can write,

$$\frac{\partial^2}{\partial x^2} EI \frac{\partial^2 q}{\partial x^2} + \frac{\partial^2}{\partial x^2}\left(E^* + \frac{\tilde{E}}{\omega}\right) I \frac{\partial^3 q}{\partial t \partial x^2} + \rho A \frac{\partial^2 q}{\partial t^2} = f(x,t) \tag{7.17}$$

in which ω is the frequency of the external excitation $f(x,t)$ in the case of steady forced vibrations. In the case of free vibration, however, ω represents the frequency of free-vibration decay. Consequently, when analyzing the modal decay of free vibrations, ω in equation (7.17) should be replaced by the appropriate frequency (ω_i) of modal vibration in each modal equation. Here, the resulting damped vibratory system possesses the same normal mode shapes as the undamped system. The analysis of the damped case is very similar to that for the undamped system, as noted in Chapter 6.

□

7.1.2 Structural Damping

Structural damping is a result of the mechanical-energy dissipation caused by rubbing friction resulting from relative motion between components and by impacting or intermittent contact at the joints in a mechanical system or structure. Energy-dissipation behavior depends on the details of the particular mechanical system in this case. Consequently, it is extremely difficult to develop a generalized analytical model that would satisfactorily describe structural damping. Energy dissipation caused by rubbing is usually represented by a Coulomb-friction model. Energy dissipation caused by impacting, however, should be determined from the coefficient of restitution of the two members that are in contact.

The common method of estimating structural damping is by measurement. The measured values, however, represent the overall damping in the mechanical system. The structural damping component is obtained by subtracting the values corresponding to other types of damping, such as material damping present in the system (estimated by environment-controlled experiments, previous data, etc.), from the overall damping value.

Usually, internal damping is negligible compared to structural damping. A large portion of mechanical-energy dissipation in tall buildings, bridges, vehicle guideways, and many other civil engineering structures, and in machinery such as robots and vehicles takes place through the structural-damping mechanism. A major form of structural damping is the slip damping that results from the energy dissipation by interface shear at a structural joint. The degree of slip damping that is directly caused by Coulomb (dry) friction depends on such factors as joint forces (e.g., bolt tensions), surface properties, and the nature of the materials of the mating surfaces. This is associated with wear, corrosion, and general deterioration of the structural joint. In this sense, slip damping is time dependent. It is common practice to place damping layers at joints, to reduce undesirable deterioration of the joints. Sliding will cause shear distortions in the damping layers, causing energy dissipation by material damping and also through Coulomb friction. In this way, a high level of equivalent structural damping can be maintained without causing excessive joint deterioration. These damping layers should have a high stiffness (as well as a high specific-damping capacity) in order to take the structural loads at the joint.

For structural damping at a joint, the damping force varies as slip occurs at the joint. This is primarily caused by local deformations at the joint, which occur with slipping. A typical hysteresis loop for this case is shown in Figure 7.2(a). The arrows on the hysteresis loop indicate the direction

Damping

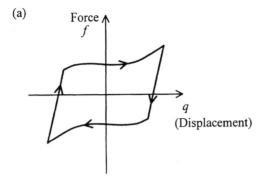
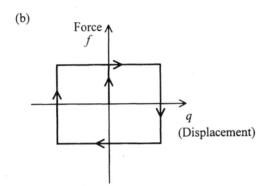
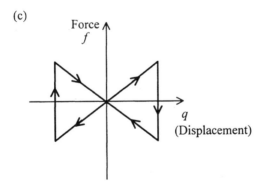

FIGURE 7.2 Some representative hysteresis loops: (a) typical structural damping; (b) Coulomb friction model; and (c) simplified structural damping model.

of relative velocity. For idealized Coulomb friction, the frictional force (F) remains constant in each direction of relative motion. An idealized hysteresis loop for structural Coulomb damping is shown in Figure 7.2(b). The corresponding constitutive relation is

$$f = c\,\mathrm{sgn}(\dot{q}) \tag{7.18}$$

in which f is the damping force, q is the relative displacement at the joint, and c is a friction parameter. A simplified model for structural damping caused by local deformation can be given by

$$f = c|q|\,\mathrm{sgn}(\dot{q}) \tag{7.19}$$

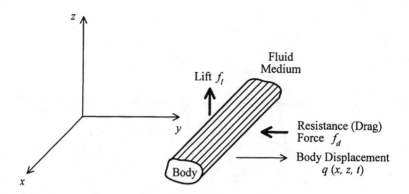

FIGURE 7.3 A body moving in a fluid medium.

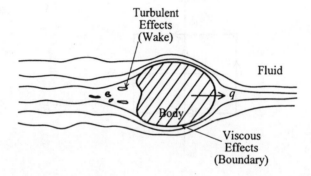

FIGURE 7.4 Mechanics of fluid damping.

The corresponding hysteresis loop is shown in Figure 7.2(c). Note that the signum function is defined by

$$\text{sgn}(v) = 1 \quad \text{for } v \geq 0$$
$$\phantom{\text{sgn}(v)} = -1 \quad \text{for } v < 0 \quad (7.20)$$

7.1.3 Fluid Damping

Consider a mechanical component moving in a fluid medium. The direction of relative motion is shown parallel to the y-axis in Figure 7.3. Local displacement of the element relative to the surrounding fluid is denoted by $q(x,z,t)$. The resulting drag force per unit area of projection on the x-z plane is denoted by f_d. This resistance is the cause of mechanical-energy dissipation in fluid damping. It is usually expressed as

$$f_d = \frac{1}{2} c_d \rho \dot{q}^2 \text{sgn}(\dot{q}) \quad (7.21)$$

in which $\dot{q} = \partial q(x,z,t)/\partial t$ is the relative velocity. The drag coefficient c_d is a function of the Reynold's number and the geometry of the structural cross section. Net damping effect is generated by viscous drag produced by the boundary-layer effects at the fluid–structure interface, and by pressure drag produced by the turbulent effects resulting from flow separation at the wake. The two effects are illustrated in Figure 7.4. Fluid density is ρ. For fluid damping, the damping capacity per unit volume associated with the configuration shown in Figure 7.3 is given by

Damping

$$d_f = \frac{\oint \int_0^{L_x} \int_0^{L_z} f_d \, dz \, dx \, dq(x,z,t)}{L_x L_z q_0} \quad (7.22)$$

in which, L_x and L_z are cross-sectional dimensions of the element in the x and y directions, respectively, and q_0 is a normalizing amplitude parameter for relative displacement.

Example 7.2

Consider a beam of length L and uniform rectangular cross section, that is undergoing transverse vibration in a stationary fluid. Determine an expression for the damping capacity per unit volume for this fluid–structure interaction.

Solution

Suppose that the beam axis is along the x-direction and the transverse motion is in the y-direction. There is no variation in the z-direction, and hence, the length parameters in this direction cancel out. Thus,

$$d_f = \frac{\oint \int_0^{L} f_d \, dx \, dq(x,t)}{L q_0}$$

or

$$d_f = \frac{\int_0^T \int_0^L f_d \dot{q}(x,t) \, dx \, dt}{L q_0} \quad (7.23)$$

in which T is the period of the oscillations. Assuming constant c_d, substitute equation (7.21) into equation (7.23):

$$d_f = \frac{1}{2} \frac{c_d \rho}{L q_0} \int_0^L \int_0^T |\dot{q}|^3 \, dt \, dx \quad (7.24)$$

For steady-excited harmonic vibration of the beam at frequency ω and shape function $Q(x)$ (or for free-modal vibration at natural frequency ω and mode shape $Q(x)$), one has

$$q(x,t) = q_{max} Q(x) \sin \omega t \quad (7.25)$$

In this case, with the change of variable $\theta = \omega t$, equation (7.24) becomes

$$d_f = 2 c_d \rho \frac{q_{max}^3}{L q_0} \int_0^L |Q(x)|^3 \, dx \omega^2 \int_0^{\pi/2} \cos^3 \theta \, d\theta$$

BOX 7.1 Damping Classification

Type of Damping	Origin	Typical Constitutive Relation
Internal damping	Material properties	Viscoelastic: $\sigma = E\varepsilon + E^* \dfrac{d\varepsilon}{dt}$ Hysteretic: $\sigma = E\varepsilon + \dfrac{\tilde{E}}{\omega}\dfrac{d\varepsilon}{dt}$
Structural damping	Structural joints and interfaces	Structural deformation: $f = c\|q\|\mathrm{sgn}(\dot{q})$ Coulomb: $f = c\,\mathrm{sgn}(\dot{q})$ General interface: $f = f_s$ for $v = 0$ $ = f_{sb}(v)\mathrm{sgn}(v)$ for $v \neq 0$
Fluid damping	Fluid-structure interactions	$f_d = \dfrac{1}{2}c_d \rho \dot{q}^2 \mathrm{sgn}(\dot{q})$

or

$$d_f = \frac{4}{3} c_d \rho q_{\max}^3 \omega^2 \frac{\int_0^L |Q(x)|^3\, dx}{L q_0}$$

Note: The integration interval of $t = 0$ to T becomes $\theta = 0$ to 2π or four times that from $\theta = 0$ to $\pi/2$.

If the normalizing parameter is defined as

$$q_0 = \frac{1}{L} q_{\max} \int_0^L |Q(x)|^3\, dx$$

then one obtains

$$d_f = \frac{4}{3} c_d \rho q_{\max}^2 \omega^2 \qquad (7.26)$$

□

A useful classification of damping is given in Box 7.1.

7.2 REPRESENTATION OF DAMPING IN VIBRATION ANALYSIS

It is not practical to incorporate detailed microscopic representations of damping in the dynamic analysis of systems. Instead, simplified models of damping that are representative of various types of energy dissipation are typically used. Consider a general n-degree-of-freedom mechanical system. Its motion can be represented by the vector x of n generalized coordinates x_i, representing the independent motions of the inertia elements. For small displacements, linear spring elements can be assumed. As seen in Chapter 5, the corresponding equations of motion can be expressed in the vector-matrix form:

$$M\ddot{x} + d + Kx = f(t) \qquad (7.27)$$

in which M is the mass (inertia) matrix and K is the stiffness matrix. The forcing-function vector is $f(t)$. The damping-force vector $d(x,\dot{x})$ is generally a nonlinear function of x and $\dot{x}$. The type of damping used in the system model can be represented by the nature of d that is employed in the system equations. Several possibilities of damping models that can be used, as discussed in the previous section, are listed in Table 7.1. Only the linear viscous damping term given in Table 7.1 is amenable to simplified mathematical analysis. In simplified dynamic models, other types of damping terms are usually replaced by an equivalent viscous damping term. Equivalent viscous damping is chosen so that its energy dissipation per cycle of oscillation is equal to that for the original damping. The resulting equations of motion are expressed by

$$M\ddot{x} + C\dot{x} + Kx = f(t) \qquad (7.28)$$

It was seen in Chapter 5 that in modal analysis of vibratory systems it is the proportional damping model, where the damping matrix satisfies

$$C = c_m M + c_k K \qquad (7.29)$$

TABLE 7.1
Some Common Damping Models Used in Dynamic System Equations

Damping Type	Simplified Model d_i
Viscous	$\sum_j c_{ij} \dot{x}_j$
Hysteretic	$\sum_j \frac{1}{\omega} c_{ij} \dot{x}_j$
Structural	$\sum_j c_{ij} |x_j| \text{sgn}(\dot{x}_j)$
Structural Coulomb	$\sum_j c_{ij} \text{sgn}(\dot{x}_j)$
Fluid	$\sum_j c_{ij} |\dot{x}_j| \dot{x}_j$

that is commonly used. The first term on the right-hand side of equation (7.29) is known as the *inertial damping matrix*. The corresponding damping force on each concentrated mass is proportional to its momentum. It represents the energy loss associated with change in momentum (e.g., during an impact). The second term is known as the *stiffness damping matrix*. The corresponding damping force is proportional to the rate of change of the local deformation forces at joints near the concentrated mass elements. Consequently, it represents a simplified form of linear structural damping. If damping is of the proportional type, it follows that the damped motion can be uncoupled into individual modes. This means that, if the damping model is of the proportional type, the damped system (as well as the undamped system) will possess real modes.

7.2.1 Equivalent Viscous Damping

Consider a linear, single-degree-of-freedom system with viscous damping, subjected to an external excitation. The equation of motion, for a unit mass, is given by

$$\ddot{x} + 2\zeta\omega_n \dot{x} + \omega_n^2 x = \omega_n^2 u(t) \tag{7.30}$$

If the excitation force is harmonic, with frequency ω, one has

$$u(t) = u_0 \cos \omega t \tag{7.31}$$

Then, as discussed in Chapter 3, the response of the system at steady state is given by

$$x = x_0 \cos(\omega t + \phi) \tag{7.32}$$

in which the response amplitude is

$$x_0 = u_0 \frac{\omega_n^2}{\left[\left(\omega_n^2 - \omega^2\right)^2 + 4\zeta^2 \omega_n^2 \omega^2\right]^{1/2}} \tag{7.33}$$

and the response phase lead is

$$\phi = -\tan^{-1} \frac{2\zeta\omega_n \omega}{\left(\omega_n^2 - \omega^2\right)} \tag{7.34}$$

The energy dissipation (i.e., damping capacity) ΔU per unit mass, in one cycle, is given by the net work done by the damping force f_d; thus,

$$\Delta U = \oint f_d dx = \int_{-\phi/\omega}^{(2\pi-\phi)/\omega} f_d \dot{x} dt \tag{7.35}$$

Since the viscous damping force, normalized with respect to mass (see equation (7.30)), is given by

$$f_d = 2\zeta\omega_n \dot{x} \tag{7.36}$$

Damping

the damping capacity ΔU_v, for viscous damping, can be obtained as

$$\Delta U_v = 2\zeta\omega_n \int_0^{2\pi/\omega} \dot{x}^2 dt \tag{7.37}$$

Finally, by using equation (7.32) in (7.37), one obtains

$$\Delta U_v = 2\pi x_0^2 \omega_n \omega \zeta \tag{7.38}$$

For any general type of damping (see Table 7.1), the equation of motion becomes

$$\ddot{x} + d(x, \dot{x}) + \omega_n^2 x = \omega_n^2 u(t) \tag{7.39}$$

The energy dissipation per unit mass in one cycle [equation (7.35)] is given by

$$\Delta U = \int_{-\phi/\omega}^{(2\pi-\phi)/\omega} d(x, \dot{x}) \dot{x} dt \tag{7.40}$$

Various damping force expressions $d(x, \dot{x})$, normalized with respect to mass, are given in Table 7.2. For fluid damping, for example, the damping capacity is

$$\Delta U_f = \int_{-\phi/\omega}^{(2\pi-\phi)/\omega} c|\dot{x}|\dot{x}^2 dt \tag{7.41}$$

By substituting equation (7.32) in equation (7.41), for steady, harmonic motion, one obtains

TABLE 7.2
Equivalent Damping-Ratio Expressions for Some Common Types of Damping

Damping Type	Damping Force $d(x,\dot{x})$ per Unit Mass	Equivalent Damping Ratio ζ_{eq}
Viscous	$2\zeta\omega_n\dot{x}$	ζ
Hysteretic	$\dfrac{c}{\omega}\dot{x}$	$\dfrac{c}{2\omega_n\omega}$
Structural	$c|x|\mathrm{sgn}(\dot{x})$	$\dfrac{c}{\pi\omega_n\omega}$
Structural Coulomb	$c\,\mathrm{sgn}(\dot{x})$	$\dfrac{2c}{\pi x_0 \omega_n \omega}$
Fluid	$c|\dot{x}|\dot{x}$	$\dfrac{4}{3\pi}\left(\dfrac{\omega}{\omega_n}\right) x_0 c$

$$\Delta U_f = \frac{8}{3} c x_0^3 \omega^2 \tag{7.42}$$

By comparing equation (7.42) with equation (7.38), the equivalent damping ratio for fluid damping is obtained as

$$\zeta_f = \frac{4}{3\pi}\left(\frac{\omega}{\omega_n}\right) x_0 c \tag{7.43}$$

in which x_0 is the amplitude of steady-state vibrations, as given by equation (7.33). For other types of damping that are listed in Table 7.1, expressions for the equivalent damping ratio can be obtained in a similar manner. The corresponding equivalent damping-ratio expressions are give in Table 7.2. It should be noted that, for non-viscous damping types, ζ is generally a function of the frequency of oscillation ω and the amplitude of excitation u_0. Also note that the expressions given in Table 7.2 are derived assuming harmonic excitation. Engineering judgment should be exercised when employing these expressions for non-harmonic excitations.

For multi-degree-of-freedom systems that incorporate proportional damping, the equations of motion can be transformed into a set of one-degree-of-freedom equations (modal equations) of the type given by equation (7.30). In this case, damping ratio and natural frequency correspond to the respective modal values, and in particular, $\omega = \omega_n$.

7.2.2 Complex Stiffness

Consider a linear spring of stiffness k connected in parallel with a linear viscous damper of damping constant c, as shown in Figure 7.5(a). Suppose that a force f is applied to the system, moving it through distance x from the relaxed position of the spring. Then,

$$f = kx + c\dot{x} \tag{7.44}$$

Also suppose that the motion is harmonic, given by

$$x = x_0 \cos \omega t \tag{7.45}$$

It is clear that the spring force kx is in phase with the displacement, but the damping force $c\dot{x}$ has a 90° phase lead with respect to the displacement. This is because the velocity $\dot{x} = -x_0 \omega \sin\omega t = x_0 \omega \cos\left(\omega t + \frac{\pi}{2}\right)$ has a 90° phase lead with respect to x. Specifically,

$$f = kx_0 \cos \omega t + cx_0 \omega \cos\left(\omega t + \frac{\pi}{2}\right) \tag{7.46}$$

This same fact can be represented using complex numbers where the in-phase component is considered as the real part and the 90° phase-lead component is considered as the imaginary part, each component oscillating at the same frequency ω. Then, one can write equation (7.46) in the equivalent form

$$f = kx + j\omega cx \tag{7.47}$$

This is exactly what is obtained by starting with the complex representation of the displacement

$$x = x_0 e^{j\omega t} \tag{7.48}$$

and substituting it in equation (7.44). Note that equation (7.47) can be written as

$$f = k^* x \tag{7.49}$$

where k^* is a "complex" stiffness, given by

$$k^* = k + j\omega c \tag{7.50}$$

Clearly, the system itself and its two components (spring and damper) are real. Their individual forces are also real. The complex stiffness is simply a mathematical representation of the two force components (spring force and damping force), which are 90° out of phase, when subjected to harmonic motion. It follows that a linear damping element can be "mathematically" represented by an "imaginary" stiffness. In the case of viscous damping, this imaginary stiffness (and, hence, the damping force magnitude) increases linearly with the frequency ω of the harmonic motion. The concept of complex stiffness that is used when dealing with discrete dampers is analogous to the use of complex elastic modulus in material damping, as discussed earlier in this chapter.

It has been noted that, for hysteretic damping, the damping force (or damping stress) is independent of the frequency in harmonic motion. It follows that a hysteretic damper can be represented by an equivalent damping constant of

$$c = \frac{h}{\omega} \tag{7.51}$$

which is valid for a harmonic motion (e.g., modal motion or forced motion) of frequency ω. This situation is shown in Figure 7.5(b). It is seen that the corresponding complex stiffness is

$$k^* = k + jh \tag{7.52}$$

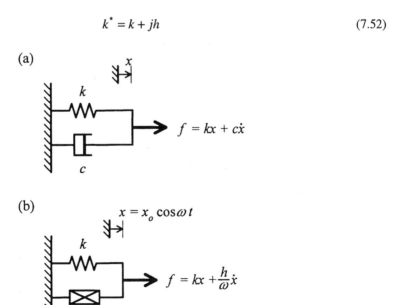

FIGURE 7.5 Spring element in parallel with: (a) viscous damper, and (b) hysteretic damper.

Example 7.3

A flexible system consists of a mass m attached to the hysteretic-damper-and-spring combination shown in Figure 7.5(b). What is the frequency response function of the system, relating an excitation force f applied to the mass and the resulting displacement response x? Obtain the resonant frequency of the system. Compare the results with the case of viscous damping.

Solution

For a harmonic motion of frequency ω, the equation of motion of the system is

$$m\ddot{x} + \frac{h}{\omega}\dot{x} + kx = f \qquad (7.53)$$

with a forcing excitation of $f = f_0 e^{j\omega t}$ and the resulting steady-state response $x = x_0 e^{j\omega t}$, where x_0 has a phase difference (i.e., it is a complex function) with respect to f_0. Then, in the frequency domain, substituting the harmonic response $x = x_0 e^{j\omega t}$ into equation (7.53), one obtains

$$\left[-\omega^2 m + \frac{h}{\omega}j\omega + k\right]x = f$$

resulting in the frequency transfer function

$$\frac{x}{f} = \frac{1}{\left[k - \omega^2 m + jh\right]} \qquad (7.54)$$

Note that, as usual, this result is obtained simply by substituting $j\omega$ for $\frac{d}{dt}$. The magnitude of transfer function is maximum at resonance. This corresponds to the minimum value of

$$p(\omega) = \left(k - \omega^2 m\right)^2 + h^2$$

Set $\frac{dp}{d\omega} = 0$. One then obtains

$$2\left(k - \omega^2 m\right)(-2\omega) = 0$$

Hence, the resonant frequency corresponds to the root of

$$k - \omega^2 m = 0$$

This gives the resonant frequency

$$\omega_r = \sqrt{\frac{k}{m}} \qquad (7.55)$$

Note that, in the case of hysteretic damping, the resonant frequency is equal to the undamped natural frequency ω_n and, unlike in the case of viscous damping (see Chapter 3), does not depend on the level of damping itself.

For convenience, consider the system response as the spring force

Damping

$$f_s = kx \tag{7.56}$$

rather than the displacement (x) itself. Then, a normalized transfer function is obtained, as given by

$$\frac{f_s}{f} = G(j\omega) = \frac{1}{\left[1 - \omega^2 \frac{m}{k} + j\frac{h}{k}\right]} \tag{7.57}$$

or

$$\frac{f_s}{f} = \frac{1}{\left[1 - r^2 + j\alpha\right]} \tag{7.58}$$

where

$$r = \frac{\omega}{\omega_n} \quad \text{and} \quad \alpha = \frac{h}{k} \tag{7.59}$$

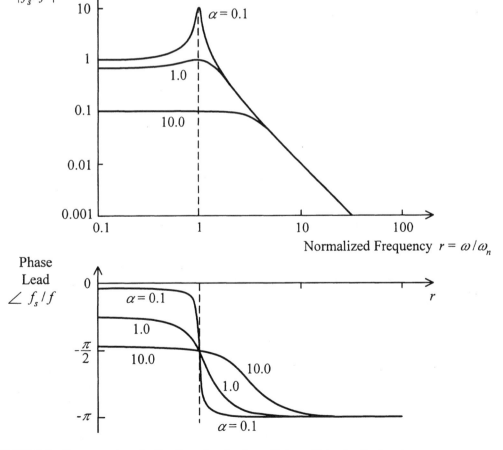

FIGURE 7.6 Frequency transfer function of a simple oscillator with hysteretic damping.

which are the normalized frequency and the normalized hysteretic damping coefficient, respectively.
The magnitude of the transfer function is

$$\left|\frac{f_s}{f}\right| = \frac{1}{\sqrt{(1-r^2)+\alpha^2}} \tag{7.60}$$

and the phase angle (phase lead) is

$$\angle f_s/f = -\tan^{-1}\frac{\alpha}{(1-r^2)} \tag{7.61}$$

These results are sketched in Figure 7.6.

□

7.2.3 Loss Factor

We define *damping capacity* of a device (damper) as the energy dissipated in a complete cycle of motion; specifically,

$$\Delta U = \oint f_d dx \tag{7.62}$$

This is given by the area of the hysteresis loop in the displacement-force plane. If the initial (total) energy of the system is denoted by U_{max}, the *specific damping capacity D* is given by the ratio

$$D = \frac{\Delta U}{U_{max}} \tag{7.63}$$

The *loss factor* η is the specific damping capacity per radian of the damping cycle. Hence,

$$\eta = \frac{\Delta U}{2\pi U_{max}} \tag{7.64}$$

Note that U_{max} is approximately equal to the maximum kinetic energy and also to the maximum potential energy of the device, when the damping is low.

Equation (7.38) gives the damping capacity per unit mass of a device with viscous damping as

$$\Delta U = 2\pi x_0^2 \omega_n \omega \zeta \tag{7.65}$$

Here, x_0 is the amplitude and ω is the frequency of harmonic motion of the device, ω_n is the undamped natural frequency and ζ is the damping ratio. Also, the maximum potential energy per unit mass of the system is

$$U_{max} = \frac{1}{2}\frac{k}{m}x_0^2 = \frac{1}{2}\omega_n^2 x_0^2 \tag{7.66}$$

Hence, from equation (7.64), the loss factor for a viscous-damped simple oscillator is given by

$$\eta = \frac{2\pi x_0^2 \omega_n \omega \zeta}{2\pi \times \frac{1}{2}\omega_n^2 x_0^2} = \frac{2\omega\zeta}{\omega_n} \qquad (7.67)$$

For free decay of the system, $\omega = \omega_d \cong \omega_n$, where the latter approximation holds for low damping. For forced oscillation, the worst response conditions occur when $\omega = \omega_r \cong \omega_n$, which is what one must consider with regard to energy dissipation. In either case, the loss factor is approximately given by

$$\eta = 2\zeta \qquad (7.68)$$

For other types of damping, equation (7.68) will still hold when the equivalent damping ratio ζ_{eq} (see Table 7.2) is used in place of ζ.

The loss factors of some common materials are given in Table 7.3. Definitions of useful damping parameters, as defined here, are summarized in Table 7.4.

TABLE 7.3
Loss Factors of Some Useful Materials

Material	Loss Factor $\eta \cong 2\zeta$
Aluminum	2×10^{-5} to 2×10^{-3}
Concrete	0.02 to 0.06
Glass	0.001 to 0.002
Rubber	0.1 to 1.0
Steel	0.002 to 0.01
Wood	0.005 to 0.01

TABLE 7.4
Definitions of Damping Parameters

Parameter	Definition	Mathematical Formula
Damping capacity (ΔU)	Energy dissipated per cycle of motion (area of displacement-force hysteresis loop)	$\oint f_d dx$
Damping capacity per volume (d)	Energy dissipated per cycle per unit material volume (area of strain-stress hysteresis loop)	$\oint \sigma d\varepsilon$
Specific damping capacity (D)	Ratio of energy dissipated per cycle (ΔU) to the initial maximum energy (U_{max}) *Note*: For low damping, U_{max} = max. potential energy = max. kinetic energy	$\dfrac{\Delta U}{U_{max}}$
Loss factor (η)	Specific damping capacity per unit angle of cycle. *Note*: For low damping, $\eta = 2\times$ damping ratio.	$\dfrac{\Delta U}{2\pi U_{max}}$

7.3 MEASUREMENT OF DAMPING

Damping can be represented by various parameters (such as specific damping capacity, loss factor, Q-factor, and damping ratio) and models (such as viscous, hysteretic, structural, and fluid). Before attempting to measure damping in a system, one should decide on a representation (model) that will adequately characterize the nature of mechanical-energy dissipation in the system. Next, one should decide on the parameter (or parameters) of the model that need to be measured.

It is extremely difficult to develop a realistic yet tractable model for damping in a complex piece of equipment operating under various conditions of mechanical interaction. Even if a satisfactory damping model is developed, experimental determination of its parameters could be tedious. A major difficulty arises because it usually is not possible to isolate various types of damping (e.g., material, structural, and fluid) from an overall measurement. Furthermore, damping measurements must be conducted under actual operating conditions for them to be realistic.

If one type of damping (e.g., fluid damping) is eliminated during the actual measurement, it would not represent true operating conditions. This would also eliminate possible interacting effects of the eliminated damping type with the other types. In particular, overall damping in a system is not generally equal to the sum of individual damping values when they are acting independently. Another limitation of computing equivalent damping values using experimental data arises because it is assumed, for analytical simplicity, that the dynamic system behavior is linear. If the system is highly nonlinear, a significant error could be introduced into the damping estimate. Nevertheless, it is customary to assume linear viscous behavior when estimating damping parameters using experimental data.

There are two general ways by which damping measurements can be made: time-response methods and frequency-response methods. The basic difference between the two types of measurements is that the first type uses a time-response record of the system to estimate damping, whereas the second type uses a frequency-response record (see Chapters 2 and 3).

7.3.1 LOGARITHMIC DECREMENT METHOD

This is perhaps the most popular time-response method used to measure damping. When a single-degree-of-freedom oscillatory system with viscous damping [see equation (7.30)] is excited by an impulse input (or an initial condition excitation), its response takes the form of a time decay (see Figure 7.7), given by

$$y(t) = y_0 \exp(-\zeta \omega_n t) \sin \omega_d t \qquad (7.69)$$

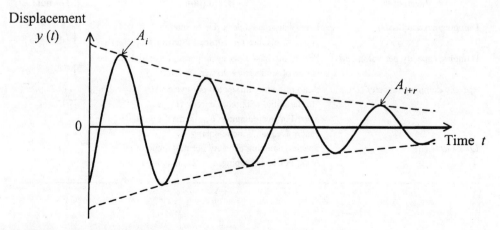

FIGURE 7.7 Impulse response of a simple oscillator.

Damping

in which the damped natural frequency is given by

$$\omega_d = \sqrt{1-\zeta^2}\,\omega_n \qquad (7.70)$$

If the response at $t = t_i$ is denoted by y_i, and the response at $t = t_i + 2\pi r/\omega_d$ is denoted by y_{i+r}, then, from equation (7.69):

$$\frac{y_{i+r}}{y_i} = \exp\left(-\zeta\frac{\omega_n}{\omega_d}2\pi r\right) \qquad i = 1, 2, \ldots, n \qquad (7.71a)$$

In particular, suppose that y_i corresponds to a peak point in the time decay, having magnitude A_i, and y_{i+r} corresponds to the peak-point r cycles later in the time history, and its magnitude is denoted by A_{i+r} (see Figure 7.7). Although the above equation holds for any pair of points that are r periods apart in the time history, the peak points seem to be the appropriate choice for measurement in the present procedure, as these values would be more prominent than any arbitrary points in a response time history. Then,

$$\frac{A_{i+r}}{A_i} = \exp\left(-\zeta\frac{\omega_n}{\omega_d}2\pi r\right) = \exp\left[-\frac{\zeta}{\sqrt{1-\zeta^2}}2\pi r\right] \qquad (7.71b)$$

where equation (7.70) has been used. Then, the logarithmic decrement δ (per unit cycle) is given by

$$\delta = \frac{1}{r}\ln\left(\frac{A_i}{A_{i+r}}\right) = \frac{2\pi\zeta}{\sqrt{1-\zeta^2}} \qquad (7.71)$$

or, the damping ratio can be expressed as

$$\zeta = \frac{1}{\sqrt{1+(2\pi/\delta)^2}} \qquad (7.72)$$

For low damping (typically, $\zeta < 0.1$), $\omega_d \cong \omega_n$, and equation (7.71b) becomes

$$\frac{A_{i+r}}{A_i} \cong \exp(-\zeta 2\pi r) \qquad (7.73)$$

or

$$\zeta = \frac{1}{2\pi r}\ln\left(\frac{A_i}{A_{i+r}}\right) = \frac{\delta}{2\pi} \qquad \text{for } \zeta < 0.1 \qquad (7.74)$$

This is, in fact, the "per-radian" logarithmic decrement.

The damping ratio can be estimated from a free-decay record using equation (7.74). Specifically, the ratio of the extreme amplitudes in prominent r cycles of decay is determined and substituted into equation (7.74) to get the equivalent damping ratio. Alternatively, if n cycles of damped oscillation are needed for the amplitude to decay by a factor of two, for example, then, from equation (7.74), one obtains

$$\zeta = \frac{1}{2\pi n}\ln(2) = \frac{0.11}{n} \quad \text{for } \zeta < 0.1 \tag{7.75}$$

For slow decays (low damping) the logarithmic decay in one cycle may be approximated by:

$$\ln\left(\frac{A_i}{A_{i+1}}\right) \cong \frac{2(A_i - A_{i+1})}{(A_i + A_{i+1})} \tag{7.76}$$

Then, from equation (7.74), one obtains

$$\zeta = \frac{A_i - A_{i+1}}{\pi(A_i + A_{i+1})} \quad \text{for } \zeta < 0.1 \tag{7.77}$$

Any one of the equations (7.72), (7.74), (7.75), and (7.77) can be employed in computing ζ from test data. It should be cautioned that the results assume single-degree-of-freedom system behavior. For multi-degree-of-freedom systems, the modal damping ratio for each mode can be determined using this method if the initial excitation is such that the decay takes place primarily in one mode of vibration. In other words, substantial modal separation and the presence of "real" modes (not "complex" modes with non-proportional damping) are assumed.

7.3.2 Step-Response Method

This is also a time-response method. If a unit-step excitation is applied to the single-degree-of-freedom oscillatory system given by equation (7.30), its time response is given by

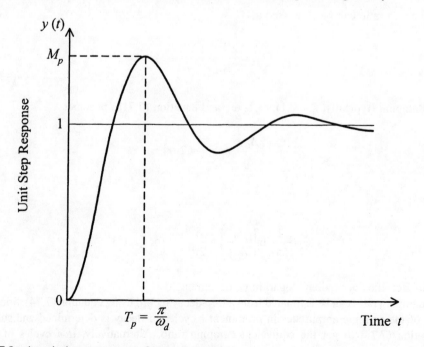

FIGURE 7.8 A typical step response of a simple oscillator.

Damping

$$y(t) = 1 - \frac{1}{\sqrt{1-\zeta^2}} \exp(-\zeta\omega_n t)\sin(\omega_d t + \phi) \tag{7.78}$$

in which $\phi = \cos\zeta$. A typical step-response curve is shown in Figure 7.8. The time at the first peak (peak time), T_p, is given by

$$T_p = \frac{\pi}{\omega_d} = \frac{\pi}{\sqrt{1-\zeta^2}\,\omega_n} \tag{7.79}$$

The response at peak time (peak value), M_p, is given by

$$M_p = 1 + \exp(-\zeta\omega_n T_p) = 1 + \exp\left(\frac{-\pi\zeta}{\sqrt{1-\zeta^2}}\right) \tag{7.80}$$

The percentage overshoot, PO, is given by,

$$PO = (M_p - 1) \times 100\% = 100\exp\left(\frac{-\pi\zeta}{\sqrt{1-\zeta^2}}\right) \tag{7.81}$$

It follows that, if any one parameter of T_p, M_p, or PO is known from a step-response record, the corresponding damping ratio ζ can be computed using the appropriate relationship from the following:

$$\zeta = \sqrt{1 - \left(\frac{\pi}{T_p \omega_n}\right)^2} \tag{7.82}$$

$$\zeta = \frac{1}{\sqrt{1 + \dfrac{1}{\left[\dfrac{\ln(M_p - 1)}{\pi}\right]^2}}} \tag{7.83}$$

$$\zeta = \frac{1}{\sqrt{1 + \dfrac{1}{\left[\dfrac{\ln(PO/100)}{\pi}\right]^2}}} \tag{7.84}$$

It should be noted that, when determining M_p, the response curve should be normalized to unit steady-state value. Furthermore, the results are valid only for single-degree-of freedom systems and modal excitations in multi-degree-of-freedom systems.

7.3.3 Hysteresis Loop Method

For a damped system, the force versus displacement cycle produces a hysteresis loop. Depending on the inertial and elastic characteristics and other conservative loading conditions (e.g., gravity) in the system, the shape of the hysteresis loop will change; but the work done by conservative forces (e.g., intertial, elastic, and gravitational) in a complete cycle of motion will be zero. Consequently, the net work done will be equal to the energy dissipated due to damping only. Accordingly, the area of the displacement-force hysteresis loop will give the damping capacity ΔU (see equation (7.62)). Also, the maximum energy in the system can be determined from the displacement-force curve. Then, the loss factor η can be computed using equation (7.64), and the damping ratio from equation (7.68). This approach of damping measurement can also be considered basically as a time domain method.

Note that equation (7.65) is the work done against (i.e., energy dissipation in) a single loading–unloading cycle, per unit mass. It should be recalled that $2\zeta\omega_n = c/m$, where, c = viscous damping constant, and m = mass. Accordingly, from equation (7.65), the energy dissipation per unit mass, and per hystereris loop, is $\Delta U = \pi x_0^2 \omega c/m$. Hence, without normalizing with respect to mass, the energy dissipation per hysteresis loop of viscous damping is

$$\Delta U_v = \pi x_0^2 \omega c \tag{7.85}$$

Equation (7.85) can be derived directly by performing the cyclic integration indicated in equation (7.62), with the damping force $f_d = c\dot{x}$, harmonic motion $x = x_0 e^{j\omega t}$, and the integration interval $t = 0$ to 2π.

Similarly, in view of equation (7.51), the energy dissipation per hysteresis loop of hysteretic damping is

$$\Delta U_h = \pi x_0^2 h \tag{7.86}$$

Now, since the initial maximum energy can be represented by the initial maximum potential energy, one obtains

$$U_{max} = \frac{1}{2} k x_0^2 \tag{7.87}$$

Note that the stiffness k can be measured as the average slope of the displacement-force hysteresis loop measured at low speed. Hence, in view of equation (7.64), the loss factor for hysteretic damping is given by

$$\eta = \frac{h}{k} \tag{7.88}$$

Then, from equation (7.68), the equivalent damping ratio for hysteretic damping is

$$\zeta = \frac{h}{2k} \tag{7.89}$$

Example 7.4

A damping material was tested by applying a low-speed loading cycle of –900 N to +900 N and back to –900 N, on a thin bar made of the material, and measuring the corresponding deflection. The smoothed load vs. deflection curve that was obtained in this experiment is shown in Figure 7.9. Assuming that the damping is predominantly of the hysteretic type, estimate

   a. the hysteretic damping constant
   b. the equivalent damping ratio.

Solution

Approximating the top and the bottom segments of the hysteresis loop by triangles, one can estimate the area of the loop as

$$\Delta U_h = 2 \times \frac{1}{2} \times 2.5 \times 900 \text{ N} \cdot \text{mm}$$

Alternatively, one can obtain this result by counting the squares within the hysteresis loop. The deflection amplitude is,

$$x_0 = 9.0 \text{ mm}$$

Hence, from equation (7.86),

$$h = \frac{2 \times \frac{1}{2} \times 2.5 \times 900}{\pi \times 9.0^2} \text{ N mm}^{-1} = 8.8 \text{ N mm}^{-1}$$

The stiffness of the damping element is estimated as the average slope of the hysteresis loop; thus,

$$k = \frac{600}{4.5} \text{ N mm}^{-1} = 133.3 \text{ N mm}^{-1}$$

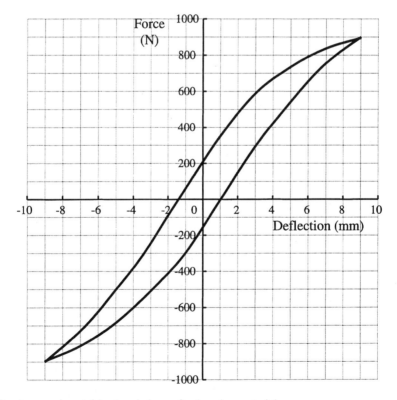

FIGURE 7.9 An experimental hysteresis loop of a damping material.

Hence, from equation (7.89), the equivalent damping ratio is

$$\zeta = \frac{8.8}{2 \times 133.3} \approx 0.03$$

□

7.3.4 Magnification-Factor Method

This is a frequency-response method. Consider the single-degree-of-freedom oscillatory system with viscous damping. The magnitude of its frequency-response function is

$$|H(\omega)| = \frac{\omega_n^2}{\left[(\omega_n^2 - \omega^2)^2 + 4\zeta^2 \omega_n^2 \omega^2\right]^{1/2}} \quad (7.90)$$

A plot of this expression with respect to ω, the frequency of excitation, is given in Figure 7.10. The peak value of magnitude occurs when the denominator of the expression is minimum. This corresponds to

$$\frac{d}{d\omega}\left[(\omega_n^2 - \omega^2)^2 + 4\zeta^2 \omega_n^2 \omega^2\right] = 0 \quad (7.91)$$

The resulting solution for ω is termed the resonant frequency ω_r (see Chapter 3):

$$\omega_r = \sqrt{1 - 2\zeta^2}\, \omega_n \quad (7.92)$$

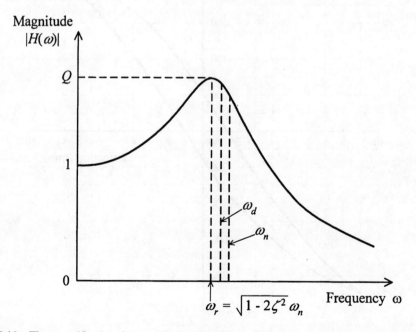

FIGURE 7.10 The magnification factor method of damping measurement applied to a single-degree-of-freedom system.

Damping

It is noted that $\omega_r < \omega_d$ (see equation (7.70)), but for low damping ($\zeta < 0.1$), the values of ω_n, ω_d, and ω_r are nearly equal. The amplification factor Q, which is the magnitude of the frequency-response function at resonant frequency, is obtained by substituting equation (7.92) in equation (7.90):

$$Q = \frac{1}{2\zeta\sqrt{1-\zeta^2}} \tag{7.93}$$

For low damping ($\zeta < 0.1$),

$$Q = \frac{1}{2\zeta} \tag{7.94}$$

In fact, equation (7.94) corresponds to the magnitude of the frequency-response function at $\omega = \omega_n$.

It follows that, if the magnitude curve of the frequency-response function (or a Bode plot) is available, then the system damping ratio ζ can be estimated using equation (7.94). In using this method, it should be remembered to normalize the frequency-response curve so that its magnitude at zero frequency (termed *static gain*) is unity.

For a multi-degree-of-freedom system, modal damping values can be estimated from the magnitude Bode plot of its frequency-response function, provided that the modal frequencies are not too closely spaced and the system is lightly damped. Consider the logarithmic (base 10) magnitude plot shown in Figure 7.11. The magnitude is expressed in decibels (dB), which is done by multiplying the $\log_{10}$ (magnitude) by the factor 20. At the ith resonant frequency ω_i, the amplification factor q_i (in dB) is obtained by drawing an asymptote to the preceding segment of the curve and measuring the peak value from the asymptote. Then,

$$Q_i = (10)^{q_i/20} \tag{7.95}$$

and the modal damping ratio

$$\zeta = \frac{1}{2Q_i} \quad i = 1, 2, \ldots, n \tag{7.96}$$

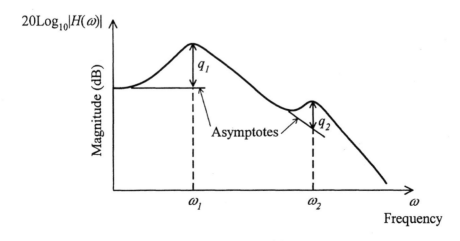

FIGURE 7.11 Magnification factor method applied to a multi-degree-of-freedom system.

If the significant resonances are closely spaced, curve-fitting to a suitable function might be necessary in order to determine the corresponding modal damping values. The Nyquist plot can also be used in computing damping using frequency domain data. This will be discussed under experimental modal analysis in Chapter 11.

7.3.5 Bandwidth Method

The bandwidth method of damping measurement is also based on frequency response. Consider the frequency-response-function magnitude given by equation (7.90) for a single-degree-of-freedom oscillatory system with viscous damping. The peak magnitude is given by equation (7.94) for low damping. Bandwidth (half-power) is defined as the width of the frequency-response magnitude curve when the magnitude is $\frac{1}{\sqrt{2}}$ times the peak value. This is denoted by $\Delta\omega$ (see Figure 7.12). An expression for $\Delta\omega = \omega_2 - \omega_1$ is obtained below using equation (7.90). By definition, ω_1 and ω_2 are the roots of the equation:

$$\frac{\omega_n^2}{\left[\left(\omega_n^2 - \omega^2\right)^2 + 4\zeta^2\omega_n^2\omega^2\right]^{1/2}} = \frac{1}{\sqrt{2}\,2\zeta} \tag{7.97}$$

for ω. Equation (7.97) can be expressed in the form

$$\omega^4 - 2\left(1 - 2\zeta^2\right)\omega_n^2\omega^2 + \left(1 - 8\zeta^2\right)\omega_n^4 = 0 \tag{7.98}$$

This is a quadratic equation in ω^2, having roots ω_1^2 and ω_2^2, which satisfy

$$\left(\omega^2 - \omega_1^2\right)\left(\omega^2 - \omega_2^2\right) = \omega^4 - \left(\omega_1^2 + \omega_2^2\right)\omega^2 + \omega_1^2\omega_2^2 = 0$$

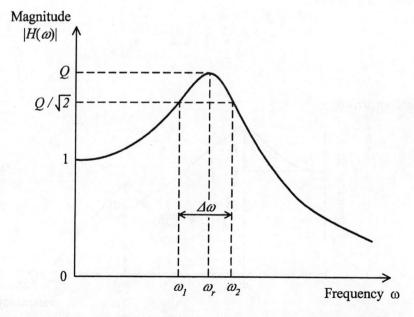

FIGURE 7.12 Bandwidth method of damping measurement in a single-degree-of-freedom system.

Damping

Consequently,

$$\omega_1^2 + \omega_2^2 = 2(1 - 2\zeta^2)\omega_n^2 \tag{7.99}$$

and

$$\omega_1^2 \omega_2^2 = (1 - 8\zeta^2)\omega_n^4 \tag{7.100}$$

It follows that

$$(\omega_2 - \omega_1)^2 = \omega_1^2 + \omega_2^2 - 2\omega_1\omega_2$$
$$= 2(1 - 2\zeta^2)\omega_n^2 - 2\sqrt{1 - 8\zeta^2}\;\omega_n^2$$

For small ζ (in comparison to 1):

$$\sqrt{1 - 8\zeta^2} \cong 1 - 4\zeta^2$$

Hence,

$$(\omega_2 - \omega_1)^2 \cong 4\zeta^2 \omega_n^2$$

or; for low damping,

$$\Delta\omega = 2\zeta\omega_n = 2\zeta\omega_r \tag{7.101}$$

From equation (7.101), it follows that the damping ratio can be estimated from bandwidth using the relation

$$\zeta = \frac{1}{2}\frac{\Delta\omega}{\omega_r} \tag{7.102}$$

For a multi-degree-of-freedom system having widely spaced resonances, the foregoing method can be extended to estimate modal damping. Consider the frequency-response magnitude plot (in dB) shown in Figure 7.13. Since a factor of $\sqrt{2}$ corresponds to 3 dB, the bandwidth corresponding to a resonance is given by the width of the magnitude plot at 3 dB below that resonant peak. For the ith mode, the damping ratio is given by

$$\zeta_i = \frac{1}{2}\frac{\Delta\omega_i}{\omega_i} \tag{7.103}$$

The bandwidth method of damping measurement indicates that the bandwidth at a resonance is a measure of the energy dissipation in the system in the neighborhood of that resonance. The simplified relationship given by equation (7.103) is valid for low damping, however, and is based on linear system analysis. Several methods of damping measurement are summarized in Box 7.2.

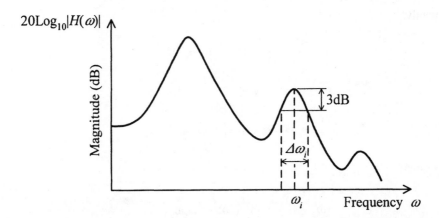

FIGURE 7.13 Bandwidth method of damping measurement in a multi-degree-of-freedom system.

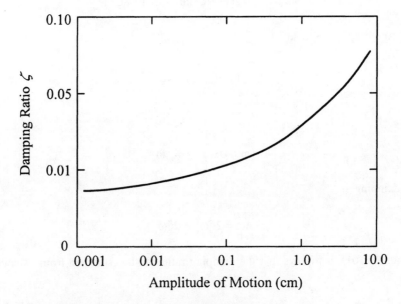

FIGURE 7.14 Effect of vibration amplitude on damping in structures.

7.3.6 GENERAL REMARKS

There are limitations to the use of damping values that are experimentally determined. For example, consider time-response methods of determining the modal damping of a device for higher modes. The customary procedure is to first excite the system at the desired resonant frequency, using a harmonic exciter, and then to release the excitation mechanism. In the resulting transient vibration, however, there invariably will be modal interactions, except in the case of proportional damping. In this type of test, it is tacitly assumed that the device can be excited in the particular mode. In essence, proportional damping is assumed in modal damping measurements. This introduces a certain amount of error into the measured damping values.

Expressions used in computing damping parameters from test measurements are usually based on linear system theory. All practical devices exhibit some nonlinear behavior, however. If the degree of nonlinearity is high, the measured damping values will not be representative of the actual system behavior. Furthermore, testing to determine damping is usually done at low amplitudes of vibration. The corresponding responses could be an order of magnitude lower than, for instance, the amplitudes exhibited under extreme operating conditions. Damping in practical devices

Damping

BOX 7.2 Damping Measurement Methods

Method	Measurements	Formulas
Logarithmic decrement method	A_i = first significant amplitude A_{i+r} = amplitude after r cycles.	Logarithmic decrement $\delta = \frac{1}{r}\ln\frac{A_i}{A_{i+r}}$; $\frac{\delta}{2\pi} = \frac{\zeta}{\sqrt{1-\zeta^2}}$ (per cycle) (per radian) or $\zeta = \frac{1}{\sqrt{1+(2\pi/\delta)^2}}$ For Low Damping: $\zeta = \frac{\delta}{2\pi}$ $\zeta = \frac{A_i - A_{i+1}}{\pi(A_i + A_{i+1})}$
Step-response method	M_p = first peak value normalized w.r.t. steady-state value PO = percentage overshoot (over steady-state value).	$M_p = 1 + \exp\left[\frac{-\pi\zeta}{\sqrt{1-\zeta^2}}\right]$ $PO = 100\exp\left[\frac{-\pi\zeta}{\sqrt{1-\zeta^2}}\right]$
Hysteresis loop method	ΔU = area of displacement-force hysteresis loop x_0 = maximum displacement of the hysteresis loop k = average slope of the hysteresis loop.	Hysteretic damping constant: $h = \frac{\Delta U}{\pi x^2}$ Loss factor: $\eta = \frac{h}{k}$ Equivalent damping ratio: $\zeta = \frac{h}{2k}$
Magnification-factor method	Q = amplification at resonance w.r.t. zero-frequency value.	$Q = \frac{1}{2\zeta\sqrt{1-\zeta^2}}$ For low damping: $\zeta = \frac{1}{2Q}$
Bandwidth method	$\Delta\omega$ = bandwidth at $1/\sqrt{2}$ of resonant peak (i.e., half-power bandwidth) ω_r = resonant frequency.	$\zeta = \frac{\Delta\omega}{2\omega_r}$

increases with the amplitude of motion, except for relatively low amplitudes. A typical nonlinear behavior is illustrated in Figure 7.14. Consequently, the damping values determined from experiments should be extrapolated when they are used to study the system behavior under various operating conditions. Alternatively, damping could be associated with a stress level in the device. Different components in a device are subjected to varying levels of stress, however, and it might be difficult to obtain a representative stress value for the entire device. A recommended method for estimating damping in structures under seismic disturbances, for example, is by analyzing earthquake-response records for structures that are similar to the one being considered. Some typical

TABLE 7.5
Typical Damping Values for Seismic Applications

System	Damping Ratio ($\zeta\%$)	
	OBE	SSE
Equipment and large-diameter piping systems[a] (>12 in. diameter)	2	3
Small-diameter piping systems (≤ 12 in. diameter)	1	2
Welded-steel structures	2	4
Bolted-steel structures	4	7
Prestressed-concrete structures	2	5
Reinforced-concrete structures	4	7

[a] Includes both material and structural damping. If the piping system consists of only one or two spans, with little structural damping, use values for small-diameter piping.

Reprinted from ASME BPVC, Section III-Division 1, Appendices, by permission of The American Society of Mechanical Engineers. All rights reserved.

damping ratios that are applicable under operating basis earthquake (OBE) and safe-shutdown earthquake (SSE) conditions for a range of items are given in Table 7.5.

When damping values are estimated using frequency-response magnitude curves, accuracy becomes poor at very low damping ratios (<1%). The main reason for this is the difficulty in obtaining a sufficient number of points in the magnitude curve near a poorly damped resonance when the frequency-response function is determined experimentally. As a result, the magnitude curve is poorly defined in the neighborhood of a weakly damped resonance. For low damping (<2%) time-response methods are particularly useful. At high damping values, the decay could be so fast that the measurements would contain large errors. Modal interference in closely spaced modes could also affect measured damping results.

7.4 INTERFACE DAMPING

In many practical applications, damping is generated at the interface of two sliding surfaces. This is the case, for example, in bearings, gears, screws, and guideways. Although this type is commonly treated under structural damping, due to its significance it will be considered here again in more detail, as the category of interface damping.

Interface damping was formally considered by DaVinci in the early 1500s and again by Coulomb in the 1700s. The simplified model used is the well-known Coulomb friction model as given by

$$f = \mu R \, \mathrm{sgn}(v) \qquad (7.104)$$

where

f = frictional force that opposes the motion
R = normal reaction force between the sliding surfaces
v = relative velocity between the sliding surfaces
μ = coefficient of friction.

Note that the signum function "sgn" is used to emphasize that f is in the opposite direction of v. This simple model is not expected to provide accurate results in all situations of interface damping.

Damping

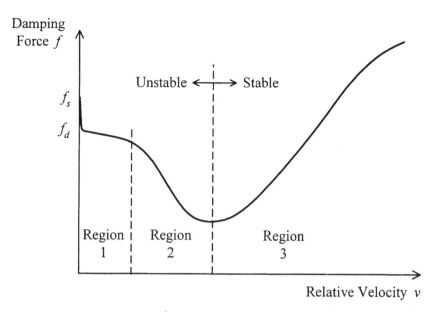

FIGURE 7.15 Main characteristics of interface damping.

It is known that, apart from the loading conditions, interface damping depends on a variety of factors such as material properties, surface characteristics, nature of lubrication, geometry of the moving parts, and the magnitude of the relative velocity.

A somewhat more complete model for interface damping, which incorporates the following characteristics, is shown in Figure 7.15:

1. Static and dynamic friction, with stiction and stick-slip behavior
2. Conventional Coulomb friction (Region 1)
3. Drop in dynamic friction, with a negative slope, before increasing again (this is known as the "Stribeck effect" (Region 2))
4. Conventional viscous damping (Region 3).

These characteristics agree with the behavior of interface damping that is commonly observed in practice. In particular, suppose that a force is exerted to generate a relative motion between two surfaces. For small values of the force, there will not be a relative motion, in view of friction. The minimum force f_s needed for the motion to start is the *static frictional force*. The force needed to maintain the motion will drop instantaneously to f_d as the motion begins. It is as though initially the two surfaces were "stuck," and f_s is the necessary breakaway force. Hence, this characteristic is known as *stiction*. The minimum force f_d needed to maintain the relative motion between the two surfaces is called *dynamic friction*. In fact, under dynamic conditions, it is possible for "stick-slip" to occur where repeated sticking and breaking away cycles of intermittent motion take place. Clearly, such "chattering" motion corresponds to instability (e.g., in machine tools). It is an undesirable effect and should be avoided.

After the relative motion begins, conventional Coulomb type damping behavior may dominate for small relative velocities, as represented in Region 1. For lubricated surfaces at low relative velocities, there will be some solid-to-solid contact that generates a Coulomb-type damping force. As the relative speed increases, the degree of this solid-to-solid contact will decrease and the damping force will drop, as in Region 2 of Figure 7.15. This characteristic is known as the *Stribeck effect*. Because the slope of the friction curve is negative in Regions 1 and 2, this corresponds to the unstable region. As the relative velocity is further increased, in fully lubricated surfaces, viscous-type damping will dominate as shown in Region 3 of Figure 7.15. This is the stable region. It follows that a combined model of interface damping can be expressed as

$$f = f_s \qquad \text{for } v = 0$$
$$= f_{sb}(v)\text{sgn}(v) + bv \qquad \text{for } v \neq 0 \tag{7.105}$$

Note that $f_{sb}(v)$ is a nonlinear function of velocity that will represent both dynamic friction (for $v > 0$) and the Stribeck effect. Models that have been used to represent this effect include the following:

$$f_{sb} = \frac{f_d}{1 + (v/v_c)^2} \tag{7.106}$$

$$f_{sb} = f_d e^{-(v/v_c)^2} \tag{7.107}$$

and

$$f_{sb} = \left(f_d + \alpha|v|^{1/2}\right)\text{sgn}(v) \tag{7.108}$$

Here, f_d represents dynamic Coulomb friction and v_c and α are modal parameters.

Example 7.5

An object of mass m rests on a horizontal surface and is attached to a spring of stiffness k, as shown in Figure 7.16. the mass is pulled so that the extension of the spring is x_0, and is released from rest from that position. Determine the subsequent sliding motion of the object. The coefficient of friction between the object and the horizontal surface is μ.

Solution

Note that when the object moves to the left, the frictional force μmg acts to the right, and vice versa. Consider the first cycle of motion, stating from rest with $x = x_0$, moving to the left, coming to rest with the spring compressed, and then moving to the right.

First Half Cycle (Moving to Left)

The equation of motion is:

$$m\ddot{x} = -kx + \mu mg \tag{i}$$

or

$$\ddot{x} + \omega_n^2 x = \mu g \tag{ii}$$

with $\omega_n = \sqrt{k/m}$ is the undamped material frequency. Equation (ii) has a homogeneous solution of

$$x_h = A_1 \sin(\omega_n t) + A_2 \cos(\omega_n t) \tag{iii}$$

Damping

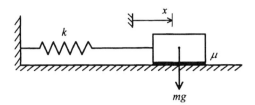

FIGURE 7.16 An object sliding against Coulomb friction

and a particular solution of

$$x_p = \frac{\mu g}{\omega_n^2} \tag{iv}$$

Hence, the total solution is

$$x = A_1 \sin(\omega_n t) + A_2 \cos(\omega_n t) + \frac{\mu g}{\omega_n^2} \tag{v}$$

Use the initial conditions $x = x_0$ and $\dot{x} = 0$ at $t = 0$; then, $A_1 = 0$ and $A_2 = x_0 - \frac{\mu g}{\omega_n^2}$.
Hence, equation (v) becomes

$$x = \left(x_0 - \frac{\mu g}{\omega_n^2}\right)\cos(\omega_n t) + \frac{\mu g}{\omega_n^2} \tag{vi}$$

At the end of this half cycle, $\dot{x} = 0$, or $\sin\omega_n t = 0$. Hence, the corresponding time is $t = \pi/\omega_n$. Substituting in equation (vi), the corresponding position of the object is (note: $\cos\pi = -1$)

$$x_{l1} = -\left(x_0 - \frac{2\mu g}{\omega_n^2}\right) \tag{vii}$$

Second Half Cycle (Moving to Right)

The equation of motion is

$$m\ddot{x} = -kx - \mu mg \tag{viii}$$

or

$$\ddot{x} + \omega_n^2 x = -\mu g \tag{ix}$$

The corresponding response is given by

$$x = B_1 \sin(\omega_n t) + B_2 \cos(\omega_n t) - \frac{\mu g}{\omega_n^2} \tag{x}$$

Use the initial conditions $x = -\left(x_0 - \dfrac{2\mu g}{\omega_n^2}\right)$ and $\dot{x} = 0$ at $t = \pi/\omega_n$; then, $B_1 = 0$ and $B_2 = x_0 - \dfrac{3\mu g}{\omega_n^2}$. Hence, equation (x) becomes

$$x = \left(x_0 - \dfrac{3\mu g}{\omega_n^2}\right)\cos(\omega_n t) - \dfrac{\mu g}{\omega_n^2} \qquad (xi)$$

The object will come to rest ($\dot{x} = 0$) next at $t = 2\pi/\omega_n$. Hence, the position of the object at the end of the present half cycle would be

$$x_{r1} = x_0 - \dfrac{4\mu g}{\omega_n^2} \qquad (xii)$$

The response for the next cycle is determined by substituting x_{r1}, as given by equation (xii) which is the initial condition, into equation (v) for the left motion; determining the subsequent end point x_{l2}, and using it as the initial condition for equation (x) for the right motion; and soon. Then, one can express the general response as:

Left motion in cycle i: $\qquad x = [x_0 - (4i - 3)\Delta]\cos\omega_n t + \Delta \qquad (xiii)$

Right motion in cycle i: $\qquad x = [x_0 - (4i - 1)\Delta]\cos\omega_n t - \Delta \qquad (xiv)$

where

$$\Delta = \dfrac{\mu g}{\omega_n^2} \qquad (xv)$$

Note that the amplitude of the harmonic part of the response should be positive for that half cycle of motion to be possible. Hence, one must have

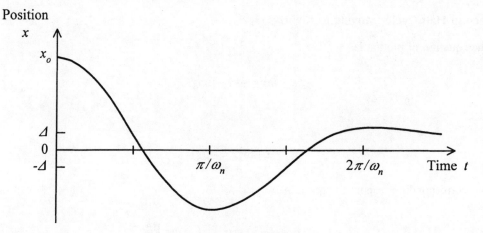

FIGURE 7.17 A typical cyclic response under Coulomb friction.

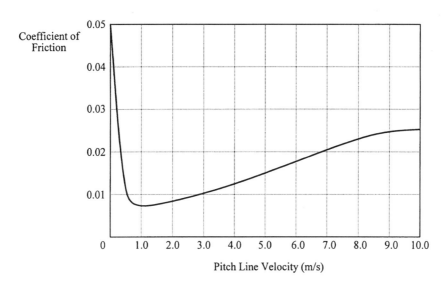

FIGURE 7.18 Frictional characteristics of a pair of spur gears.

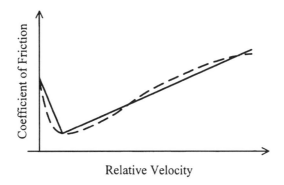

FIGURE 7.19 A friction model for rotatory devices.

$$x_0 > (4i-3)\Delta \quad \text{for left motion in cycle } i$$

$$x_0 > (4i-1)\Delta \quad \text{for right motion in cycle } i$$

Also note from equations (*xiii*) and (*xiv*) that the equilibrium (central) position for the left motion is $+\Delta$, and for the right motion it is $-\Delta$. A typical response curve is sketched in Figure 7.17.

□

7.4.1 Friction in Rotational Interfaces

Friction in gear transmissions, rotary bearings, and other rotary joints has somewhat similar behavior. Of course, the friction characteristics will depend on the nature of the device and also the loading conditions. However, experiments have shown that the frictional behavior of these devices can be represented by the interface damping model given here. Typically, experimental results are presented as curves of coefficient of friction (frictional force/normal force) versus relative velocity of the two sliding surfaces. While in the case of rotary bearings the rotational speed of the shaft is used as the relative velocity, it is the pitch line velocity that is used for gears. An experimental result for a pair of spur gears is shown in Figure 7.18.

What is interesting to notice from the result is the fact that, for this type of rotational device, the damping behavior can be approximated by two straight-line segments in the velocity-friction plane — the first segment having a sharp negative slope and the second segment having a moderate positive slope, which represents the equivalent viscous damping constant, as shown in Figure 7.19.

7.4.2 INSTABILITY

Unstable behavior or self-excited vibrations such as stick-slip and chatter that is exhibited by interacting devices such as metal removing tools (e.g., lathes, drills, and milling machines) can be easily explained using the interface damping model. In particular, it is noted that the model has a region of negative slope (or negative damping constant) that corresponds to low relative velocities, and a region of positive slope that corresponds to high relative velocities. Consider the single-degree-of-freedom model.

$$m\ddot{x} + b\dot{x} + kx = 0 \tag{7.109}$$

without an external excitation force. Initially, the velocity is $\dot{x} = 0$. But, in this region, the damping constant b will be negative and hence the system will be unstable. Thus, a slight disturbance will result in a steadily increasing response. Subsequently, $\dot{x}$ will increase above the critical velocity where b will be positive and the system will be stable. As a result, the response will steadily decrease. This growing and decaying cycle will be repeated at a frequency that primarily depends on the inertia and stiffness parameters (m and k) of the system. Chatter is caused in this manner in interfaced devices.

PROBLEMS

7.1 a. Give three desirable effects and three undesirable effects of damping.
  b. The moment of inertia of a door about its hinges is J kg·m². An automatic door closer of torsional stiffness K N·m·rad^{-1} is attached to it. What is the damping constant C needed for critical damping with this door closer? Give the units of C.

7.2 a. Compare and contrast viscoelastic (material) damping and hysteretic (material) damping.
  b. The stress-strain relations for the Kelvin-Voigt, Maxwell, and standard linear solid models of material damping are:

$$\sigma = E\varepsilon + E^* \frac{d\varepsilon}{dt}$$

$$\sigma + c_s \frac{d\sigma}{dt} = E^* \frac{d\varepsilon}{dt}$$

$$\sigma + c_s \frac{d\sigma}{dt} = E\varepsilon + E^* \frac{d\varepsilon}{dt}$$

Sketch spring and dashpot lumped-parameter systems that represent these three damping models.

7.3 a. The Kelvin-Voigt model of material damping is represented by the stress-strain model

Damping

$$\sigma = E\varepsilon + E^* \frac{d\varepsilon}{dt}$$

and the standard linear solid model of material damping is represented by

$$\sigma + c_s \frac{d\sigma}{dt} = E^* \frac{d\varepsilon}{dt}$$

Under what condition could the latter model be approximated by the former?

b. Damping capacity per unit volume of a material is given by

$$d = \oint \sigma \, d\varepsilon$$

Also, the maximum elastic potential energy per unit volume is given by $u_{max} = \frac{1}{2}\sigma_{max}\varepsilon_{max}$, or, in view of the relation $\sigma_{max} = E\varepsilon_{max}$ for an elastic material, by $u_{max} = \frac{1}{2}E\varepsilon_{max}^2$, where ε_{max} is the maximum strain in a load cycle. Show that the loss factor for a Kelvin-Voigt viscoelastic material is given by

$$\eta = \frac{\omega E^*}{E}$$

7.4 Verify that the loss factors for the material damping models given in Table P7.4 are as given in the last column of the table. What would be the corresponding damping ratio expressions?

7.5 a. A damping material is represented by the frequency-dependent standard linear solid model:

$$\sigma + c_s \frac{d\sigma}{dt} = E\varepsilon + \left(\frac{g_1}{\omega} + g_2\right)\frac{d\varepsilon}{dt}$$

Obtain an approximate expression for the damping ratio of this material.

TABLE P7.4
Loss Factors for Several Material Damping Models

Material Damping Model	Stress-Strain Constitutive Relation	Loss Factor (η)
Viscoelastic Kelvin-Voigt	$\sigma = E\varepsilon + E^*\dfrac{d\varepsilon}{dt}$	$\dfrac{\omega E^*}{E}$
Hysteretic Kelvin-Voigt	$\sigma = E\varepsilon + \dfrac{\tilde{E}}{\omega}\dfrac{d\varepsilon}{dt}$	$\dfrac{\tilde{E}}{E}$
Viscoelastic standard linear solid	$\sigma + c_s\dfrac{d\sigma}{dt} = E\varepsilon + E^*\dfrac{d\varepsilon}{dt}$	$\dfrac{\omega E^*}{E}\dfrac{(1 - c_s E/E^*)}{(1 + \omega^2 c_s^2)}$
Hysteretic standard linear solid	$\sigma + c_s\dfrac{d\sigma}{dt} = E\varepsilon + \dfrac{\tilde{E}}{\omega}\dfrac{d\varepsilon}{dt}$	$\dfrac{\tilde{E}}{E}\dfrac{(1 - \omega c_s E/\tilde{E})}{(1 + \omega^2 c_s^2)}$

TABLE P7.6
Lumped-Parameter Models of Damping

Damping Type	Damping Force $d(x,\dot{x})$ per Unit Mass
Hysteretic	$\dfrac{c}{\omega}\dot{x}$
Structural	$c|x|\mathrm{sgn}(\dot{x})$
Structural Coulomb	$c\,\mathrm{sgn}(\dot{x})$

b. A thin cantilever beam that is made of this material has a mode of transverse vibration with natural frequency 15.0 Hz. Also, the following parameter values are known: $E = 1.9 \times 10^{11}$ Pa, $g_1 = 6.2 \times 10^9$ Pa, $g_2 = 8.6 \times 10^7$ Pa·s, and $c_s = 1.6 \times 10^{-4}$ s. Estimate the modal damping ratio for this mode of vibration. (*Note*: 1 Pa = 1 N·m^{-2})

7.6 Consider the single-degree-of-freedom damped system given by the dynamic equation

$$\ddot{x} + d(x,\dot{x}) + \omega_n^2 x = \omega_n^2 u(t)$$

where

x = response of the lumped mass
$u(t)$ = normalized excitation
ω_n = undamped natural frequency
d = damping force per unit mass.

Three possible cases of damping are given in Table P7.6.
Determine an expression for the equivalent damping ratio in each of these three cases of lumped-parameter models.

7.7 a. A load cycle is applied at low speed during axial testing of a test specimen. The applied force and the corresponding deflection are measured and the area A_f of the resulting hysteresis loop is determined. The longitudinal stiffness of the specimen is k and the amplitude of the deflection during the loading cycle is x_0. Obtain an expression for the loss factor η of the material. Comment on the accuracy of this expression.
A hysteresis loop that was obtained from a cyclic tensile test on a specimen is shown in Figure P7.7. Estimate the damping ratio of the material.

7.8 a. A torque cycle is applied during torsional testing of a shaft. The applied torque and the corresponding angle of twist are measured and the area A_t of the resulting hysteresis loop is determined. The torsional stiffness of the shaft is K and the amplitude of the angle of twist is θ_0. Show that the loss factor η of the shaft material is given by

$$\eta = \frac{A_t}{\pi \theta_0^2 K}$$

b. A hysteresis loop obtained from a low-speed, cyclic torsional test on a shaft is shown in Figure P7.8. Estimate the damping ratio of the material.

7.9 a. A cyclic tensile test was carried out at low speed on a specimen of metal and the stress versus strain hysteresis loop was obtained. The area of the hysteresis loop was found to be A_s. The Young's modulus of the specimen was E and the amplitude of the axial strain was ε_0. Show that the loss factor η of the material can be expressed as

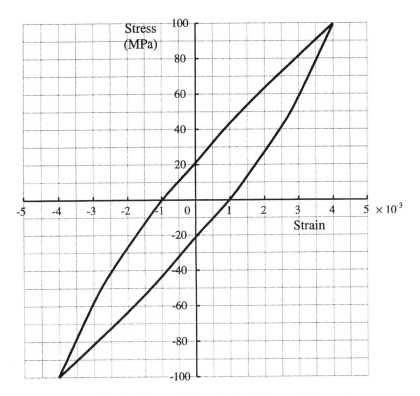

FIGURE P7.7 A force vs. deflection hysteresis loop obtained from a cyclic tensile test.

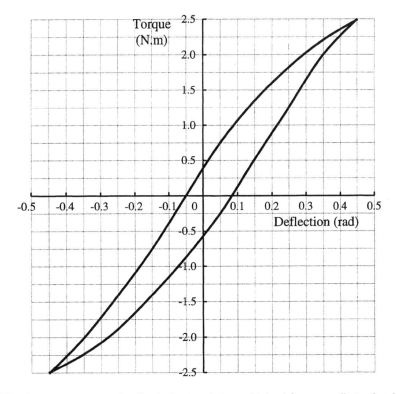

FIGURE P7.8 A torque versus angle of twist hysteresis loop obtained from a cyclic torsional test.

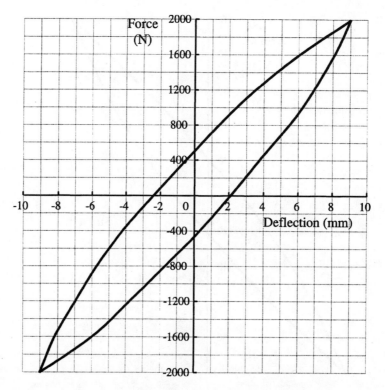

FIGURE P7.9 A stress-strain hysteresis loop obtained from a cyclic tensile test.

$$\eta = \frac{A_s}{\pi \varepsilon_0^2 E}$$

b. A hysteresis loop that was obtained from a low-speed, cyclic stress-strain test is shown in Figure P7.9. Estimate the damping ratio of the material.

7.10 a. Consider a single-degree-of-freedom system with the displacement coordinate x. If the damping in the system is of hysteretic type, the damping force can be given by

$$f_d = \frac{h}{\omega} \dot{x}$$

where h is the hysteretic damping constant and ω is the frequency of motion. Now, for a harmonic motion given by $x = x_0 \sin(\omega t)$, show that the energy dissipation per cycle is

$$\Delta U_h = \pi x_0^2 h$$

Also, if the stiffness of the system is k, show that the loss factor is given by

$$\eta = \frac{h}{k}$$

b. Consider a uniform cylindrical rod of length l, area of cross section A, and Young's modulus E that is used as a specimen of tensile testing. What is its longitudinal stiffness k? Suppose that for a single cycle of loading, the area of the stress-strain

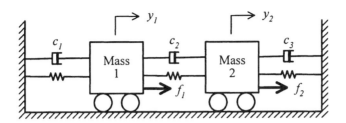

FIGURE P7.11 A damped lumped-parameter system.

hysteresis loop is A_s and the amplitude of the corresponding strain is ε_0. Show that, for this rod,

$$h = \frac{A_s A}{\pi \varepsilon_0^2 l}$$

What is the damping ratio of the rod in axial motion?

7.11 What is proportional damping? What is its main advantage?
A two-degree-of-freedom lumped-parameter system is shown in Figure P7.11. Using the influence coefficient approach, determine the damping matrix of this system.

7.12 An automated wood-cutting system contains a cutting unit that consists of a DC motor and a cutting blade, which are linked by a flexible shaft and coupling. The purpose of the flexible shaft is to locate the blade unit at any desirable configuration, away from the motor itself. A simple, lumped-parameter dynamic model of the cutting unit is shown in Figure P7.12.
The following parameters and variables are shown in the figure:

J_m = axial moment of inertia of the motor rotor
b_m = equivalent viscous damping constant of the motor bearings
k = torsional stiffness of the flexible shaft
J_c = axial moment of inertia of the cutter blade
b_c = equivalent viscous damping constant of the cutter bearings
T_m = magnetic torque of the motor
θ_m = motor angle of rotation
ω_m = motor speed
T_k = torque transmitted through the flexible shaft
θ_c = cutter angle of rotation
ω_c = cutter speed
T_L = load torque on the cutter from the workpiece (wood).

In comparison with the flexible shaft, the coupling unit is assumed rigid, and is also assumed light. The cutting load is given by

$$T_L = c|\omega_c|\omega_c$$

The parameter c, which depends on factors such as the depth of cut and the material properties of the workpiece, is assumed to be constant in the present problem.
a. Comment on the suitability of the damping models used in this problem.

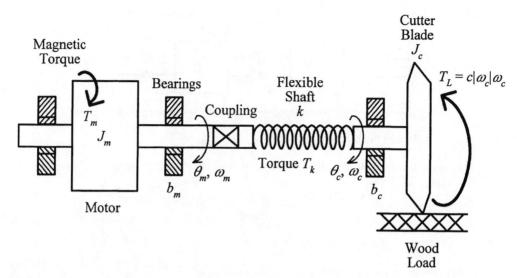

FIGURE P7.12 A wood-cutting machine.

b. Using T_m as the input, T_L as the output, and $[\omega_m\ T_k\ \omega_c]^T$ as the state vector, develop a complete (nonlinear) state model for the system shown in Figure P7.12. What is the order of the system?

c. Using the state model derived in part (a), obtain a single input-output differential equation for the system, with T_m as the input and ω_c as the output.

d. Consider the steady operating conditions, where $T_m = \overline{T}_m$, $\omega_m = \overline{\omega}_m$, $T_k = \overline{T}_k$, $\omega_c = \overline{\omega}_c$, and $T_L = \overline{T}_L$ are all constants. Express the operating point values $\overline{\omega}_m$, $\overline{T}_k$, $\overline{\omega}_c$, and $\overline{T}_L$ in terms of $\overline{T}_m$ and model parameters only. You must consider both cases, $\overline{T}_m > 0$ and $\overline{T}_m < 0$.

e. Now consider an incremental change $\hat{T}_m$ in the motor torque and the corresponding changes $\hat{\omega}_m$, $\hat{T}_k$, $\hat{\omega}_c$, and $\hat{T}_L$ in the system variables. Determine a linear state model (A, B, C, D) for the incremental dynamics of the system in this case, using $x = [\hat{\omega}_m, \hat{T}_k, \hat{\omega}_c]^T$ as the state vector, $u = [\hat{T}_m]^T$ as the input and $y = [\hat{T}_L]^T$ as the output.

f. In the nonlinear model (see part (b)), if the twist angle of the flexible shaft (i.e., $\theta_m - \theta_c$) is used as the output, what would be a suitable state model? What is the system order then?

g. In the nonlinear model, if the angular position θ_c of the cutter blade is used as the output variable, explain how the state model obtained in part (b) should be modified. What is the system order in this case?

h. For vibration analysis of the wood-cutting machine, the damped natural frequencies and the associated damping ratios are required. How many natural frequencies and damping ratios would you expect for this problem? How would you determine them?

Hint for Part (e):

$$\frac{d}{dt}\left(|\omega_c|\omega_c\right) = 2|\omega_c|\dot{\omega}_c$$

$$\frac{d^2}{dt^2}\left(|\omega_c|\omega_c\right) = 2|\omega_c|\ddot{\omega}_c + 2\dot{\omega}_c^2\,\text{sgn}(\omega_c)$$

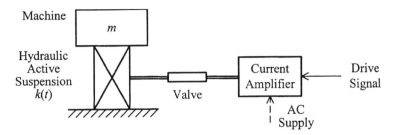

FIGURE P7.13(a) A machine with an active suspension.

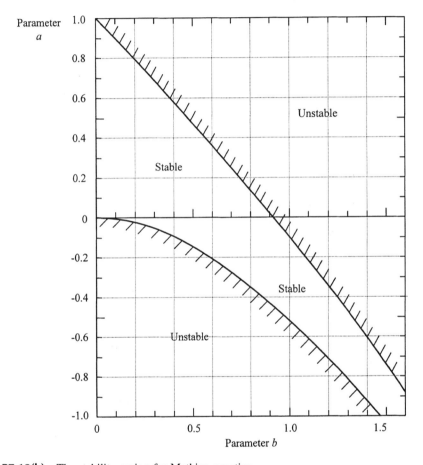

FIGURE P7.13(b) The stability region for Mathieu equation.

7.13 A machine with an active suspension system is schematically shown in Figure P7.13(a). The mass of the machine is m. The active suspension system provides a variable stiffness $k(t)$ by means of a hydraulic actuator. The vertical displacement of the machine is denoted by y. Under steady conditions, it was found that the suspension stiffness fluctuates about an average value k_0 according to the relation

$$k(t) = k_0 - k_1 \cos \omega t.$$

It is suspected that this is due to an error in the current amplifier that provides the drive signal to the actuator. The frequency of the fluctuation is in fact the line frequency (of the AC supply) and is 60 Hz. Also, $k_1 = 2.84 \times 10^7$ N·m^{-1} and $m = 1000$ kg. Determine the range of k_0 for which the system will be stable.

Hint: For a system given by the Mathieu equation:

$$\frac{d^2y}{dt^2} + (a - 2b\cos 2t)y = 0$$

the stability depends on the values of a and b, as given by the stability curves of Figure P7.13(b).

7.14 a. Prepare a table to compare and contrast the following methods of damping measurement:
1. Logarithmic decrement method
2. Step-response method
3. Hysteresis loop method
4. Magnification-factor method
5. Bandwidth method

with regard to the following considerations:
1. Domain of analysis (time or frequency?)
2. Whether it can measure several modes simultaneously
3. Accuracy restrictions
4. Cost
5. Speed
6. Model limitations.

b. A machine with its suspension system weighs 500 kg. The logarithmic decrement of its free decay under an initial-condition excitation was measured to be 0.63, and the corresponding frequency was 8.0 Hz.
   i. Compute the undamped natural frequency and the damping ratio of the system.
   ii. Suppose that the machine, under normal operating conditions, generates an unbalance force $f = f_0 \cos\omega t$ with the force amplitude $f_0 = 4.8 \times 10^4$ N and the frequency $\omega = 15.0 \times 2\pi$ rad·s^{-1}. What is the amplitude of the steady-state vibration of the machine under this excitation force?
   iii. Estimate the resonant frequency and the half-power bandwidth of the system.

7.15 A commercial fish processing machine (known as the "Iron Butcher") has a conveyor belt with holding pockets. The fish are placed in the holding pockets and are held from the top using a stationary belt, as schematically shown in Figure P7.15(a). It was found that, under some conditions, the fish undergo stick-slip type vibratory motion during conveying. A model that can be used to analyze this unstable behavior is shown in Figure P7.15(b). The model parameters are:

m = mass of a fish
k = stiffness of a fish
b = equivalent damping constant of dissipation between the stationary holding belt and a fish
v = velocity of the conveyor
x = absolute displacement of a fish.

Using this model, explain the stick-slip motion of a fish.

7.16 a. Compare the free decay response of a system under linear viscous damping, with that under Coulomb friction.

b. An object of mass m is restrained by a spring of stiffness k and slides on a surface against a constant Coulomb frictional force F. Obtain expressions for the peak motion

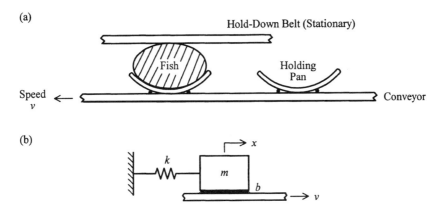

FIGURE P7.15 (a) Conveying of fish in a fish processing machine, and (b) dynamic model for analyzing the stick-slip response of a fish.

in the $i+r$th cycle in terms of that of the ith cycle, separately, for the two directions of motion.

7.17 a. Consider an object in cross flow of fluid with velocity v as shown in Figure P7.17. Suppose that the object vibrates in the direction transverse to the fluid flow, at cyclic frequency f. The representative transverse dimension of the object is d. A nondimensional velocity $\frac{v}{df}$ is known to determine the nature of vibration of the object. In particular,

   i. For small $\frac{v}{df}$ (in the range of 1.0 to 10.0), vortex shedding predominates (e.g., for large d and f or for stationary fluid)

   ii. For intermediate $\frac{v}{df}$ (in the range of 10 to 100), galloping predominates (e.g., for cylindrical objects at reasonably high v, as in transmission lines).

   iii. For large $\frac{v}{df}$ (in the range of 100 to 1000), flutter predominates (e.g., thin objects such as aerofoils at high fluid flow speeds).

   What are other factors that determine the nature of vibration of the object?

b. Suppose that the object is stationary in a fluid flowing at speed v. Then there will be a *drag force* f_d acting on the body in the direction of v, and a *lift force* f_l acting on the body in the transverse direction. Thus,

$$f_d = \frac{1}{2}c_d v^2$$

$$f_l = \frac{1}{2}c_l v^2$$

where

c_d = drag coefficient
c_l = lift coefficient

are nonlinear parameters and will vary with the direction of the flow.

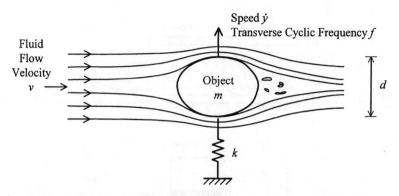

FIGURE P7.17 A vibrating object in a cross flow of fluid.

Now consider the system shown in Figure P7.17 where the object has a transverse speed $\dot{y}$, and the fluid flows at a steady speed v. Show that the equivalent linear viscous damping constant for the fluid-structure interaction is given by

$$b = \frac{1}{2}\left[c_d(0) - c_l(0) - \frac{\partial c_l(0)}{\partial \theta}\right]v$$

where θ = angle of attack = $\tan^{-1} \dot{y}/v$ and $c_d(0)$, $c_l(0)$, and $\dfrac{\partial c_l(0)}{\partial \theta}$ are the values of c_d, c_l, and $\dfrac{\partial c_l}{\partial \theta}$ when the object is stationary (i.e., $\dot{y} = 0$ or $\theta = 0$).

7.18 The Bernoulli-Euler beam equation is given by

$$\frac{\partial^2}{\partial x^2} EI \frac{\partial^2 v}{\partial x^2} + \rho A \frac{\partial^2 v}{\partial t^2} = f(x,t)$$

where

$v(x,t)$ = beam response
$f(x,t)$ = applied force per unit length of the beam
E = Young's modulus
I = second moment of area of beam cross-section, about the neutral axis bending
ρ = mass density
A = area of cross section.

Derive the corresponding beam equation with material damping represented by the
i. Kelvin-Voigt model

$$\sigma = E\varepsilon + E^* \frac{d\varepsilon}{dt}$$

ii. Standard linear solid model

$$\sigma + c_s \frac{d\sigma}{dt} = E\varepsilon + E^* \frac{d\varepsilon}{dt}$$

where

Damping

E^* = viscoelastic damping parameter
c_s = standard linear solid model parameter.

7.19 Consider a nonuniform rod of length l, area of cross section A, mass density ρ, and Young's modulus E. Assume that the ends are free and the rod executes longitudinal vibrations of displacement $u(x,t)$ at location x (see Figure P7.19). The equation of free motion is known to be

$$-\rho A \frac{\partial^2 u}{\partial t^2} + \frac{\partial}{\partial x} EI \frac{\partial u}{\partial x} = 0$$

with boundary conditions

$$\frac{\partial u}{\partial x}(0,t) = 0$$

$$\frac{\partial u}{\partial x}(l,t) = 0$$

The corresponding modal motions are, for the ith mode,

$$u(x,t) = Y_i(x)\sin\omega_i t$$

with the mode shapes

$$Y_i(x) = \cos\frac{i\pi x}{x}$$

and the natural frequencies

$$\omega_i = \frac{i\pi}{l}\sqrt{\frac{E}{\rho}}$$

Consider the following two cases of damping:
i. External damping of linear viscous type given by a damping force per unit length along the beam:

$$-b\frac{\partial u}{\partial t}$$

ii. Material damping of the Kelvin-Voigt type, given by the stress-strain equation

$$\sigma = E\left(1 + c\frac{\partial}{\partial t}\right)\varepsilon$$

For each case, determine the modal loss factor and modal damping ratio. Compare/contrast these results for the two cases of damping.

7.20 Consider a vibrating system with damping, given by the normalized equation

$$\ddot{y} + 2\zeta\omega_n\dot{y} + \omega_n^2 y = u(t)$$

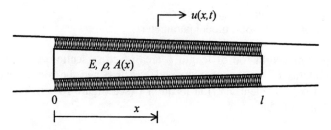

FIGURE P7.19 Damped longitudinal vibration of a rod.

where

- $u(t)$ = forcing function
- y = response
- ω_n = undamped natural frequency
- ζ = damping ratio.

Suppose that the system is excited by a harmonic force so that at steady state, the response is given by

$$y = y_0 \sin \omega t$$

a. Derive the shape of the u vs. y curve for this motion.
b. What is the energy dissipation per cycle of motion?

8 Vibration Instrumentation

Measurement and associated experimental techniques play a significant role in the practice of vibration. The objective of this chapter is to introduce instrumentation that is important in vibration applications. Chapter 9 will provide complementary material on signal conditioning associated with vibration instrumentation.

Academic exposure to vibration instrumentation usually arises in relation to learning, training, and research. In vibration practice, perhaps the most important task of instrumentation is the measurement or sensing of vibration. Vibration sensing is useful in the following applications:

1. Design and development of a product
2. Testing (screening) of a finished product for quality assurance
3. Qualification of a good-quality product to determine its suitability for a specific application
4. Mechanical aging of a product prior to carrying out a test program
5. Exploratory testing of a product to determine its dynamic characteristic such as resonances, mode shapes, and even a complete dynamic model
6. Vibration monitoring for performance evaluation
7. Control and suppression of vibration.

Figure 8.1 indicates a typical procedure of experimental vibration, highlighting the essential instrumentation. Vibrations are generated in a device (test object) in response to some excitation. In some experimental procedures (primarily in vibration testing, see Figure 8.1), the excitation signal must be generated in a signal generator, in accordance with some requirement (specification), and applied to the object through an exciter after amplification and conditioning. In some other situations (primarily in performance monitoring and vibration control), the excitations are generated as an integral part of the operating environment of the vibrating object and can originate either within the object (e.g., engine excitations in an automobile) or in the environment with which the object interacts during operation (e.g., road disturbances on an automobile). Sensors are needed to measure vibrations in the test object. In particular, a control sensor is used to check whether the specified excitation is applied to the object, and one or more response sensors can be used to measure the resulting vibrations at key locations of the object.

The sensor signals must be properly conditioned (e.g., by filtering and amplification) and modified (e.g., through modulation, demodulation, and analog-to-digital conversion) prior to recording, analyzing, and display. These considerations will be discussed in Chapter 9. The purpose of the controller is to guarantee that the excitation is correctly applied to the test object. If the signal from the control sensor deviates from the required excitation, the controller modifies the signal to the exciter so as to reduce this deviation. Furthermore, the controller will stabilize or limit (compress) the vibrations in the object. It follows that instrumentation in experimental vibration can be generally classified into the following categories:

1. Signal-generating devices
2. Vibration exciters
3. Sensors and transducers
4. Signal conditioning/modifying devices
5. Signal analysis devices
6. Control devices
7. Vibration recording and display devices.

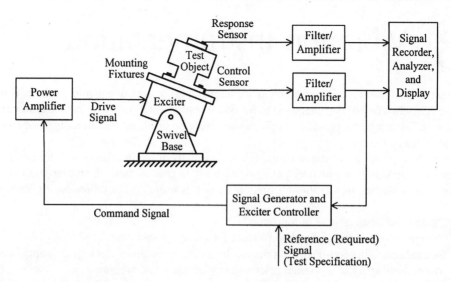

FIGURE 8.1 Typical instrumentation in experimental vibration.

Note that one instrument can perform the tasks of more than one category listed above. Also, more than one instrument may be needed to carry out tasks in a single category. The following sections will give some representative types of vibration instrumentation, along with characteristics, operating principles, and important practical considerations. Signal conditioning and modification techniques are described in Chapter 9.

An experimental vibration system generally consists of four main subsystems:

1. Test object
2. Excitation system
3. Control system
4. Signal acquisition and modification system

as schematically shown in Figure 8.2. Note that various components shown in Figure 8.1 can be incorporated into one of these subsystems. In particular, component matching hardware and object mounting fixtures can be considered interfacing devices that are introduced through the interaction between the main subsystems shown in Figure 8.2. Some important issues of vibration testing and instrumentation are summarized in Box 8.1.

8.1 VIBRATION EXCITERS

Vibration experimentation may require an external exciter to generate the necessary vibration. This is the case in controlled experiments such as product testing where a specified level of vibration is applied to the test object and the resulting response is monitored. A variety of vibration exciters are available, with different capabilities and principles of operation.

Three basic types of vibration exciters (shakers) are widely used: hydraulic shakers, inertial shakers, and electromagnetic shakers. The operation-capability ranges of typical exciters in these three categories are summarized in Table 8.1. Stroke, or maximum displacement, is the largest displacement the exciter is capable of imparting onto a test object whose weight is assumed to be within its design load limit. Maximum velocity and acceleration are similarly defined. Maximum force is the largest force that could be applied by the shaker to a test object of acceptable weight (within the design load). The values given in Table 8.1 should be interpreted with caution. Maximum displacement is achieved only at very low frequencies. Maximum velocity corresponds to interme-

Vibration Instrumentation

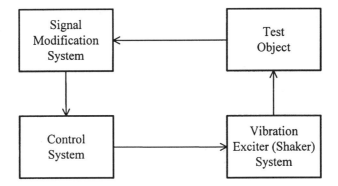

FIGURE 8.2 Interactions between major subsystems of an experimental vibration system.

BOX 8.1 Vibration Instrumentation

Vibration Testing Applications for Products:
- Design and development
- Production screening and quality assessment
- Utilization and qualification for special applications.

Testing Instrumentation:
- Exciter (excites the test object)
- Controller (controls the exciter for accurate excitation)
- Sensors and transducers (measure excitations and responses and provide excitation error signals to controller)
- Signal conditioning (converts signals to appropriate form)
- Recording and display (for processing, storage, and documentation).

Exciters:
- Shakers
  - Electrodynamic (high bandwidth, moderate power, complex and multifrequency excitations)
  - Hydraulic (moderate to high bandwidth, high power, complex and multifrequency excitations)
  - Inertial (low bandwidth, low power, single-frequency harmonic excitations).
- Transient/initial-condition
  - Hammers (impulsive, bump tests)
  - Cable release (step excitations)
  - Drop (impulsive).

Signal Conditioning:
- Filters
- Modulators/demodulators
- Amplifiers
- ADC/DAC.

Sensors:
- Motion (displacement, velocity, acceleration)
- Force (strain, torque).

TABLE 8.1
Typical Operation-Capability Ranges for Various Shaker Types

	Typical Operational Capabilities					
Shaker Type	Frequency	Maximum Displacement (Stroke)	Maximum Velocity	Maximum Acceleration	Maximum Force	Excitation Waveform
Hydraulic (electrohydraulic)	Intermediate 0.1–500 Hz	High 20 in. 50 cm	Intermediate 50 in·s⁻¹ 125 cm·s⁻¹	Intermediate 20 g	High 100,000 lbf 450,000 N	Average flexibility (simple to complex and random)
Inertial (counter-rotating mass)	Low 2–50 Hz	Low 1 in. 2.5 cm	Intermediate 50 in·s⁻¹ 125 cm·s⁻¹	Intermediate 20 g	Intermediate 1000 lbf 4500 N	Sinusoidal only
Electromagnetic (electrodynamic)	High 2–10,000 Hz	Low 1 in. 2.5 cm	Intermediate 50 in·s⁻¹ 125 cm·s⁻¹	High 100 g	Low to intermediate 450 lbf 2000 N	High flexibility and accuracy (simple to complex and random)

diate frequencies in the operating-frequency range of the shaker. Maximum acceleration and force ratings are usually achieved at high frequencies. It is not feasible, for example, to operate a vibration exciter at its maximum displacement and its maximum acceleration simultaneously.

Consider a loaded exciter that is executing harmonic motion. Its displacement is given by

$$x = s \sin \omega t \quad (8.1)$$

in which s is the displacement amplitude (or stroke). The corresponding velocity and acceleration are

$$\dot{x} = s\omega \cos \omega t \quad (8.2)$$

$$\ddot{x} = -s\omega^2 \sin \omega t \quad (8.3)$$

If the velocity amplitude is denoted by v and the acceleration amplitude by a, it follows from equations (8.2) and (8.3) that

$$v = \omega s \quad (8.4)$$

and

$$a = \omega v \quad (8.5)$$

An idealized performance curve of a shaker has a constant displacement-amplitude region, a constant velocity-amplitude region, and a constant acceleration-amplitude region for low, intermediate, and high frequencies, respectively, in the operating frequency range. Such an ideal performance curve is shown in Figure 8.3(a) on a frequency–velocity plane. Logarithmic axes are used. In practice, typical shaker-performance curves would be rather smooth yet nonlinear curves, similar to those shown in Figure 8.3(b). As the mass increases, the performance curve compresses. Note that the acceleration limit of a shaker depends on the mass of the test object (load). Full load corresponds to the heaviest object that could be tested. No load condition corresponds to a shaker without a test object. To standardize the performance curves, they usually are defined at the rated load of the shaker. A performance curve in the frequency–velocity plane can be converted to a

Vibration Instrumentation

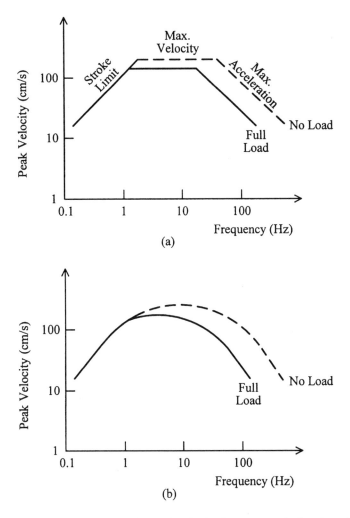

FIGURE 8.3 Performance curve of a vibration exciter in the frequency–velocity plane (log): (a) ideal and (b) typical.

curve in the frequency–acceleration plane simply by increasing the slope of the curve by a unit magnitude (i.e., 20 dB·decade^{-1}).

Several general observations can be made from equations (8.4) and (8.5). In the constant-peak displacement region of the performance curve, the peak velocity increases proportionally with the excitation frequency, and the peak acceleration increases with the square of the excitation frequency. In the constant-peak velocity region, the peak displacement varies inversely with the excitation frequency, and the peak acceleration increases proportionately. In the constant-peak acceleration region, the peak displacement varies inversely with the square of the excitation frequency, and the peak velocity varies inversely with the excitation frequency. This further explains why rated stroke, maximum velocity, and maximum acceleration values are not simultaneously realized in general.

8.1.1 Shaker Selection

Vibration testing is accomplished by applying a specified excitation to a test package, using a shaker apparatus, and monitoring the response of the test object. Test excitation can be represented by its response spectrum (see Chapter 10). The test requires that the response spectrum of the actual excitation, known as the *test response spectrum* (TRS), envelop the response spectrum specified for the particular test, known as the *required response spectrum* (RRS).

A major step in the planning of any vibration testing program is the selection of a proper shaker (exciter) system for a given test package. The three specifications that are of primary importance in selecting a shaker are the force rating, the power rating, and the stroke (maximum displacement) rating. Force and power ratings are particularly useful in moderate to high frequency excitations and the stroke rating is the determining factor for low frequency excitations. In this section, a procedure is given to determine conservative estimates for these parameters in a specified test for a given test package. Frequency domain considerations (see Chapters 3 and 4) are used here.

Force Rating

In the frequency domain, the (complex) force at the exciter (shaker) head is given by

$$F = mH(\omega)a_s(\omega) \tag{8.6}$$

in which ω is the excitation frequency variable, m is the total mass of the test package including mounting fixture and attachments, $a_s(\omega)$ is the Fourier spectrum of the support-location (exciter head) acceleration, and $H(\omega)$ is the frequency-response function that takes into account flexibility and damping effects (dynamics) of the test package, per unit mass. In the simplified case where the test package can be represented by a simple oscillator of natural frequency ω_n and damping ratio ζ_t, this function becomes

$$H(\omega) = \left\{1 + 2j\zeta_t\omega/\omega_n\right\} / \left\{1 - (\omega/\omega_n)^2 + 2j\zeta_t\omega/\omega_n\right\} \tag{8.7}$$

in which $j = \sqrt{-1}$. This approximation is adequate for most practical purposes. The static weight of the test object is not included in equation (8.6). Most heavy-duty shakers, which are typically hydraulic, have static load support systems such as pneumatic cushion arrangements that can exactly balance the deadload. The exciter provides only the dynamic force. In cases where the shaker directly supports the gravity load, in the vertical test configuration, equation (8.6) should be modified by adding a term to represent this weight.

A common practice in vibration test applications is to specify the excitation signal by its response spectrum (see Chapter 10). This is simply the peak response of a simple oscillator, expressed as a function of its natural frequency when its support location is excited by the specified signal. Clearly, damping of the simple oscillator is an added parameter in a response spectrum specification. Typical damping ratios (ζ_r) used in response spectra specifications are less than 0.1 (or 10%). It follows that an approximate relationship between the Fourier spectrum of the support acceleration and its response spectrum is

$$a_s = 2j\zeta_r a_r(\omega) \tag{8.8}$$

Here we have used the fact that for low damping ζ_r the transfer function of a simple oscillator may be approximated by $1/(2j\zeta_r)$ near its peak response. The magnitude $|a_r(\omega)|$ is the response spectrum as discussed in Chapter 10.

Equation (8.8) substituted into equation (8.6) gives

$$F = mH(\omega)2j\zeta_r a_r(\omega) \tag{8.9}$$

Vibration Instrumentation

In view of equation (8.7), for test packages having low damping, the peak value of $H(\omega)$ is approximately $1/(2j\zeta_t)$, which should be used in computing the force rating if the test package has a resonance within the frequency range of testing. On the other hand, if the test package is assumed rigid, $H(\omega) \cong 1$. A conservative estimate for the force rating is

$$F_{max} = m(\zeta_r/\zeta_t)|a_r(\omega)|_{max} \qquad (8.10)$$

It should be noted that $|a_r(\omega)|_{max}$ is the peak value of the specified (required) response spectrum (RRS) for acceleration (see Chapter 10). It follows from equation (8.10) that the peak value of the acceleration RRS curve will correspond to the force rating.

Power Rating

The exciter head does not develop its maximum force when driven at maximum velocity. Output power is determined using

$$p = \text{Re}[Fv_s(\omega)] \qquad (8.11)$$

in which $v_s(\omega)$ is the Fourier spectrum of the exciter velocity, and Re [ ] denotes the real part of a complex function. Note that $a_s = j\omega v_s$. Substituting equations (8.8) and (8.9) into equation (8.11) yields

$$p = (4m\zeta_r^2/\omega)\text{Re}[jH(\omega)a_r^2(\omega)] \qquad (8.12)$$

It follows that a conservative estimate for the power rating is

$$P_{max} = 2m(\zeta_r^2/\zeta_t)[|a_r(\omega)|^2/\omega]_{max} \qquad (8.13)$$

Representative segments of typical acceleration RRS curves have slope n, as given by

$$a = k_1\omega^n \qquad (8.14)$$

It should be clear from equation (8.13) that the maximum output power is given by

$$P_{max} = k_2\omega^{2n-1} \qquad (8.15)$$

This is an increasing function of ω for $n > \frac{1}{2}$ and a decreasing function of ω for $n < \frac{1}{2}$. It follows that the power rating corresponds to the highest point of contact between the acceleration RRS curve and a line of slope equal to $\frac{1}{2}$. A similar relationship can be derived if velocity RRS curves (having slopes $n - 1$) are used.

Stroke Rating

From equation (8.8), it should be clear that the Fourier spectrum x_s of the exciter displacement

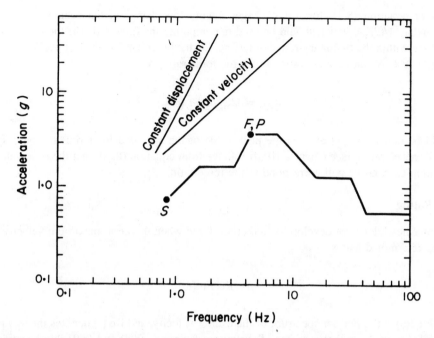

FIGURE 8.4 Test excitation specified by an acceleration RRS (5% damping).

time history can be expressed as

$$x_s = 2\zeta_r a_r(\omega)/j\omega^2 \qquad (8.16)$$

An estimate for stroke rating is

$$x_{max} = 2\zeta_r \left[|a_r(\omega)|/\omega^2\right]_{max} \qquad (8.17)$$

This is of the form

$$x_{max} = k\omega^{n-2} \qquad (8.18)$$

It follows that the stroke rating corresponds to the highest point of contact between the acceleration RRS curve and a line of slope equal to 2.

EXAMPLE 8.1

A test package of overall mass 100 kg is to be subjected to dynamic excitation represented by the acceleration RRS (at 5% damping) shown in Figure 8.4. The estimated damping of the test package is 7%. The test package is known to have a resonance within the frequency range of the specified test. Determine the exciter specifications for the test.

SOLUTION

From the development presented in the previous section, it is clear that point F (or P) in Figure 8.4 corresponds to the force and output power ratings, and point S corresponds to the stroke rating. The coordinates of these critical points are $F, P = (4.2 \text{ Hz}, 4.0 \text{ g})$, and $S = (0.8 \text{ Hz}, 0.75 \text{ g})$. Equation (8.10) gives the force rating as $F_{max} = 100 \times (0.05/0.07) \times 4.0 \times 9.81$ N = 2803 N. Equation (8.13) gives the power rating as

$$p_{max} = 2 \times 100 \times (0.05^2/0.07)\left[(4.0 \times 9.81)^2/(4.2 \times 2\pi)\right] \text{watts} = 417 \text{ W}$$

Equation (8.17) gives the stroke rating as

$$x_{max} = 2 \times 0.05 \times \left[(0.75 \times 9.8)/(0.8 \times 2\pi)^2\right] \text{m} = 3 \text{ cm}$$

Hydraulic Shakers

A typical hydraulic shaker consists of a piston-cylinder arrangement (also called a ram), a servo-valve, a fluid pump, and a driving electric motor. Hydraulic fluid (oil) is pressurized (typical operating pressure, 4000 psi) and pumped into the cylinder through a servo-valve by means of a pump that is driven by an electric motor (typical power, 150 hp). The flow (typical rate, 100 gal·min^{-1}) that enters the cylinder is controlled (modulated) by the servo-valve, which, in effect, controls the resulting piston (ram) motion. A typical servo-valve consists of a two-stage spool valve that provides a pressure difference and a controlled (modulated) flow to the piston, which sets it in motion.

The servo-valve itself is moved by means of a linear torque motor, which is driven by the excitation-input signal (electrical). A primary function of the servo-valve is to provide stabilizing feedback to the ram. In this respect, the servo-valve complements the main control system of the test setup. The ram is coupled to the shaker table by means of a link with some flexibility. The cylinder frame is mounted on the support foundation with swivel joints. This allows for some angular and lateral misalignment, which might primarily be caused by test-object dynamics as the table moves.

Two-degree-of-freedom testing requires two independent sets of actuators, and three-degree-of-freedom testing requires three independent actuator sets (see Chapter 10). Each independent actuator set can consist of several actuators operating in parallel, using the same pump and the same excitation-input signal to the torque motors.

If the test table is directly supported on the vertical actuators, they must withstand the total dead weight (i.e., the weight of the test table, the test object, the mounting fixtures, and the instrumentation). This is usually prevented by providing a pressurized air cushion in the gap between the test table and the foundation walls. Air should be pressurized so as to balance the total dead weight exactly (typical required gage pressure, 3 psi).

Figure 8.5(a) shows the basic components of a typical hydraulic shaker. The corresponding operational block diagram is shown in Figure 8.5(b). It is desirable to locate the actuators in a pit in the test laboratory so that the test tabletop is flush with the test laboratory floor under no-load conditions. This minimizes the effort required to place the test object on the test table. Otherwise, the test object will have to be lifted onto the test table with a forklift. Also, installation of an air cushion to support the system dead weight would be difficult under these circumstances of elevated mounting.

Hydraulic actuators are most suitable for heavy load testing and are widely used in industrial and civil engineering applications. They can be operated at very low frequencies (almost DC), as well as at intermediate frequencies (see Table 8.1). Large displacements (strokes) are possible at low frequencies.

Hydraulic shakers have the advantage of providing high flexibility of operation during the test, including the capabilities of variable-force and constant-force testing and wide-band random-input testing. Velocity and acceleration capabilities of hydraulic shakers are intermediate. Although any general excitation-input motion (e.g., sine wave, sine beat, wide-band random) can be used in hydraulic shakers, faithful reproduction of these signals is virtually impossible at high frequencies because of distortion and higher-order harmonics introduced by the high noise levels that are

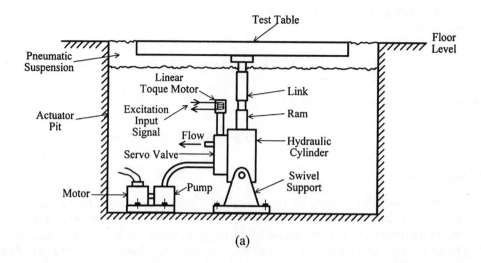

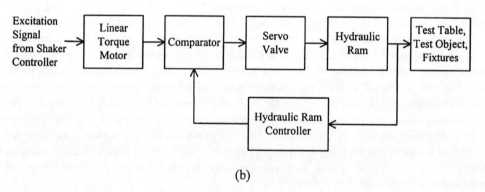

FIGURE 8.5 A typical hydraulic shaker arrangement: (a) schematic diagram, and (b) operational block diagram.

common in hydraulic systems. This is only a minor drawback in heavy-duty, intermediate-frequency applications. Dynamic interactions are reduced through feedback control.

Inertial Shakers

In inertial shakers or "mechanical exciters," the force that causes the shaker-table motion is generated by inertia forces (accelerating masses). Counterrotating-mass inertial shakers are typical in this category. To explain their principle of operation, consider two equal masses rotating in opposite directions at the same angular speed ω and in the same circle of radius r (see Figure 8.6). This produces a resultant force equal to $2m\omega^2 r\cos\omega t$ in a fixed direction (the direction of symmetry of the two rotating arms). Consequently, a sinusoidal force with a frequency of ω and an amplitude proportional to ω^2 are generated. This reaction force is applied to the shaker table.

Figure 8.7 shows a sketch of a typical counterrotating-mass inertial shaker. It consists of two identical rods rotating at the same speed in opposite directions. Each rod has a series of slots to place weights. In this manner, the magnitude of the eccentric mass can be varied to achieve various force capabilities. The rods are driven by a variable-speed electric motor through a gear mechanism that usually provides several speed ratios. A speed ratio is selected, depending on the required test-frequency range. The whole system is symmetrically supported on a carriage that is directly connected to the test table. The test object is mounted on the test table. The preferred mounting

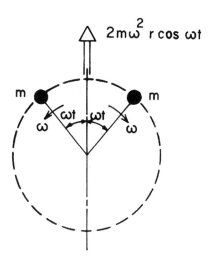

FIGURE 8.6 Principle of operation of a counter-rotating-mass inertial shaker.

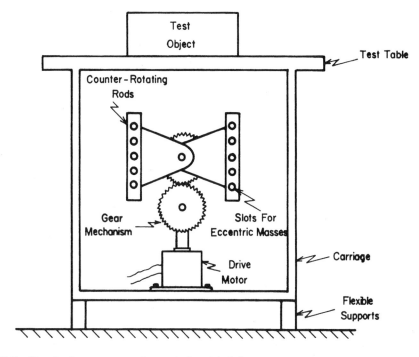

FIGURE 8.7 Sketch of a counterrotating-mass inertial shaker.

configuration is horizontal so that the excitation force is applied to the test object in a horizontal direction. In this configuration, there are no variable gravity moments (weight × distance to center of gravity) acting on the drive mechanism. Figure 8.7 shows the vertical configuration. In dynamic testing of large structures, the carriage can be mounted directly on the structure at a location where the excitation force should be applied. By incorporating two pairs of counterrotating masses, it is possible to generate test moments as well as test forces.

Inertially driven reaction-type shakers are widely used for prototype testing of civil engineering structures. Their first application dates back to 1935. Inertial shakers are capable of producing intermediate excitation forces. The force generated is limited by the strength of the carriage frame.

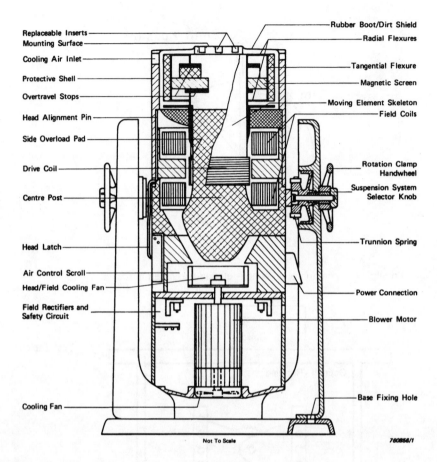

FIGURE 8.8 Schematic sectional view of a typical electromagnetic shaker. (Courtesy of Bruel and Kjaer. With permission.)

The frequency range of operation and the maximum velocity and acceleration capabilities are low to intermediate for inertial shakers, whereas the maximum displacement capability is typically low. A major limitation of inertial shakers is that their excitation force is exclusively sinusoidal and the force amplitude is directly proportional to the square of the excitation frequency. As a result, complex and random excitation testing, constant-force testing (e.g., transmissibility tests and constant-force sine-sweep tests), and flexibility to vary the force amplitude or the displacement amplitude during a test are not generally possible with this type of shaker. Excitation frequency and amplitude can be varied during testing, however, by incorporating a variable-speed drive for the motor. The sinusoidal excitation generated by inertial shakers is virtually undistorted, which is an advantage over the other types of shakers when used in sine-dwell and sine-sweep tests. Small portable shakers with low-force capability are available for use in on-site testing.

Electromagnetic Shakers

In electromagnetic shakers or "electrodynamic exciters," the motion is generated using the principle of operation of an electric motor. Specifically, the excitation force is produced when a variable excitation signal (electrical) is passed through a moving coil placed in a magnetic field.

The components of a commercial electromagnetic shaker are shown in Figure 8.8. A steady magnetic field is generated by a stationary electromagnet that consists of field coils wound on a ferromagnetic base that is rigidly attached to a protective shell structure. The shaker head has a

Vibration Instrumentation

coil wound on it. When the excitation electrical signal is passed through this drive coil, the shaker head, which is supported on flexure mounts, will be set in motion. The shaker head consists of the test table on which the test object is mounted. Shakers with interchangeable heads are available. The choice of appropriate shaker head is based on the geometry and mounting features of the test object. The shaker head can be turned to different angles by means of a swivel joint. In this manner, different directions of excitation (in biaxial and triaxial testing) can be obtained.

8.1.2 Dynamics of Electromagnetic Shakers

Consider a single-axis electromagnetic shaker (Figure 8.8) with a test object having a single natural frequency of importance within the test frequency range. The dynamic interactions between the shaker and the test object give rise to two significant natural frequencies (and, correspondingly, two significant resonances). These appear as peaks in the frequency-response curve of the test setup. Furthermore, the natural frequency (resonance) of the test package alone causes a "trough" or depression (anti-resonance) in the frequency-response curve of the overall test setup. To explain this characteristic, consider the dynamic model shown in Figure 8.9. The following mechanical parameters are defined in Figure 8.9(a): m, k, and b are the mass, stiffness, and equivalent viscous damping constant, respectively, of the test package, and m_e, k_e, and b_e are the corresponding parameters of the exciter (shaker). Also, in the equivalent electrical circuit of the shaker head, as shown in Figure 8.9(b), the following electrical parameters are defined: R_e and L_e are the resistance and (leakage) inductance, and k_b is the back electromotive force (back emf) of the linear motor. Assuming that the gravitational forces are supported by the static deflection of the flexible elements, and that the displacements are measured from the static equilibrium position, one obtains the system equations:

$$\text{Test object}: \quad m\ddot{y} = -k(y - y_e) - b(\dot{y} - \dot{y}_e) \quad (8.19)$$

$$\text{Shaker head}: \quad m_e\ddot{y}_e = f_e + k(y - y_e) + b(\dot{y} - \dot{y}_e) - k_e y - b_e \dot{y}_e \quad (8.20)$$

$$\text{Electrical}: \quad L_e \frac{di_e}{dt} + R_e i_e + k_b \dot{y}_e = v(t) \quad (8.21)$$

The electromagnetic force f_e generated in the shaker head is a result of the interaction of the magnetic field generated by the current i_e with coil of the moving shaker head and the constant magnetic field (stator) in which the head coil is located. Thus,

$$f_e = k_b i_e \quad (8.22)$$

Note that $v(t)$ is the voltage signal applied by the amplifier to the shaker coil, y_e is the displacement of the shaker head, and y is the displacement response of the test package. It is assumed that k_b has consistent electrical and mechanical units (V·m^{-1}·s^{-1} and N·A^{-1}). Usually, the electrical time constant of the shaker is quite small compared to the primarily mechanical time constants (of the shaker and the test package). Then, $L_e \frac{di_e}{dt}$ term in equation (8.21) can be neglected. Consequently,

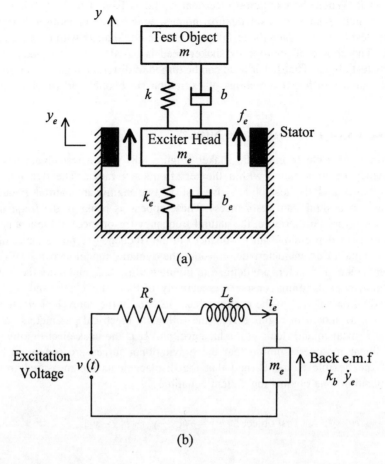

FIGURE 8.9 Dynamic model of an electromagnetic shaker and a flexible test package: (a) mechanical model and (b) electrical model.

equations (8.19) through (8.22) can be expressed in the Laplace (frequency) domain, with the Laplace variable s taking the place of the derivative $\frac{d}{dt}$, as

$$(ms^2 + bs + k)y = (bs + k)y_e \qquad (8.23)$$

$$\left[m_e s^2 + (b + b_e)s + (k + k_e)\right]y_e = (bs + k)y + \frac{k_b}{R_e}v - \frac{k_b^2 s}{R_e}y_e \qquad (8.24)$$

It follows that the transfer function of the shaker head motion with respect to the excitation voltage is given by

$$\frac{y_e}{v} = \frac{k_b}{R_e}\frac{\Delta(s)}{\Delta_d(s)} \qquad (8.25)$$

Vibration Instrumentation

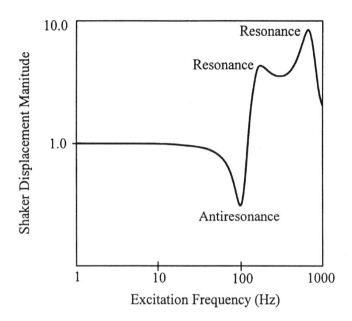

FIGURE 8.10 Frequency-response curve of a typical electromagnetic shaker with a test object.

where $\Delta(s)$ = characteristic function of the primary dynamics of the test object

$$\Delta(s) = ms^2 + bs + k \tag{8.26}$$

$\Delta_d(s)$ = characteristic function of the primary dynamic interactions between the shaker and the test object.

$$\Delta_c(s) = mm_e s^4 + \left[m(b_e + b + b_0) + m_e b\right]s^3 + \left[m(k_e + k) + m_e k + b(b_e + b_0)\right]s^2 \\ + \left[bk_e + (b_e + b_0)k\right]s + kk_e \tag{8.27}$$

where

$$b_0 = \frac{k_b^2}{R_e} \tag{8.28}$$

It is clear that under low damping conditions, $\Delta_d(s)$ will produce two resonances as it is fourth order in s and, similarly, $\Delta(s)$ will produce one antiresonance (trough) corresponding to the resonance of the test object. Note that in the frequency domain, $s = j\omega$ and, hence, the frequency-response function given by equation (8.25) is in fact

$$\frac{y_e}{v} = \frac{k_b}{R_b} \frac{\Delta(j\omega)}{\Delta_d(j\omega)} \tag{8.29}$$

The magnitude of this frequency response function, for a typical test system, is sketched in Figure 8.10. Note that this curve is for the "open-loop" case where there is no feedback from the shaker

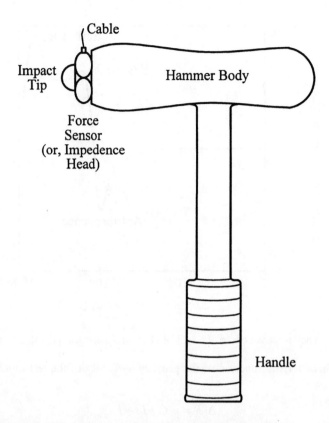

FIGURE 8.11 An instrumented hammer used in bump tests or hammer tests.

controller. In practice, the shaker controller will be able to compensate for the resonances and antiresonances to some degree, depending on its effectiveness.

The main advantages of electromagnetic shakers are their high frequency range of operation, their high degree of operating flexibility, and the high level of accuracy of the generated shaker motion. Faithful reproduction of complex excitations is possible because of the advanced electronics and control systems used in this type of shaker. Electromagnetic shakers are not suitable for heavy-duty applications (large test objects), however. High test-input accelerations are possible at high frequencies, when electromagnetic shakers are used, but displacement and velocity capabilities are limited to low or intermediate values (see Table 8.1).

Transient Exciters

Other varieties of exciters are commonly used in transient-type vibration testing. In these tests, either an impulsive force or an initial excitation is applied to the test object and the resulting response is monitored (see Chapter 10). The excitations and the responses are "transient" in this case. Hammer test, drop tests, and pluck tests, which are described in Chapter 10, fall into this category. For example, a hammer test can be conducted by hitting the object with an instrumented hammer and then measuring the response of the object. The hammer has a force sensor at its tip, as sketched in Figure 8.11. A piezoelectric or strain-gage type force sensor can be used. More sophisticated hammers have impedance heads in place of force sensors. An impedance head measures force and acceleration simultaneously. The results of a hammer test will depend on many factors; for example, dynamics of the hammer body, how firmly the hammer is held during the impact, how quickly the impact was applied, and whether there were multiple impacts.

8.2 CONTROL SYSTEM

The two primary functions of the shaker control system in vibration testing are to (1) guarantee that the specified excitation is applied to the test object, and (2) ensure that the dynamic stability (motion constraints) of the test setup is preserved. An operational block diagram illustrating these control functions is given in Figure 8.12. The reference input to the control system represents the desired excitation force that should be applied to the test object. In the absence of any control, however, the force reaching the test object will be distorted, primarily because of (1) dynamic interactions and nonlinearities of the shaker, the test table, the mounting fixtures, the auxiliary instruments, and the test object itself; (2) noise and errors in the signal generator, amplifiers, filters, and other equipment; and (3) external loads and disturbances (e.g., external restraints, aerodynamic forces, friction) acting on the test object and other components. To compensate for these distorting factors, response measurements (displacements, velocities, acceleration, etc.) are made at various locations in the test setup and are used to control the system dynamics. In particular, responses of the shaker, the test table, and the test object are measured. These responses are used to compare the actual excitation felt by the test object at the shaker interface, with the desired (specified) input. The drive signal to the shaker is modified, depending on the error present.

Two types of control are commonly employed in shaker apparatus: simple manual control and complex automatic control. Manual control normally consists of simple, open-loop, trial-and-error methods of manual adjustments (or calibration) of the control equipment to obtain a desired dynamic response. The actual response is usually monitored (on an oscilloscope or frequency analyzer screen, for example) during manual-control operations. The pretest adjustments in manual control can be very time-consuming; as a result, the test object might be subjected to overtesting (which could produce cumulative damage), which is undesirable and could defeat the test purpose. Furthermore, the calibration procedure for the experimental setup must be repeated for each new test object.

The disadvantages of manual control suggest that automatic control is desirable in complex test schemes in which high accuracy of testing is desired. The first step of automatic control involves automatic measurement of the system response, using control sensors and transducers. The measurement is then fed back into the control system, which instantaneously determines the best drive signal to actuate the shaker in order to get the desired excitation. This can be done by either analog means or digital methods.

Some control systems require an accurate mathematical description of the test object. This dependency of the control system on the knowledge of test-object dynamics is clearly a disadvantage. Performance of a good control system should not be considerably affected by the dynamic interactions and nonlinearities of the test object or by the nature of the excitation. Proper selection of feedback signals and control-system components can reduce such effects and will make the system robust.

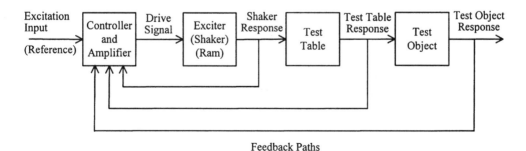

FIGURE 8.12 Operational block diagram illustrating a general shaker control system.

In the response-spectrum method of vibration testing, it is customary to use displacement control at low frequencies, velocity control at intermediate frequencies, and acceleration control at high frequencies. This necessitates feedback of displacement, velocity, and acceleration responses. Generally, however, the most important feedback is the *velocity feedback*. In sine-sweep tests, the shaker velocity must change steadily over the frequency band of interest. In particular, the velocity control must be precise near the resonances of the test object. Velocity (speed) feedback has a stabilizing effect on the dynamics, which is desirable. This effect is particularly useful in ensuring stability in motion when testing is done near resonances of lightly damped test objects. On the contrary, displacement (position) feedback can have a destabilizing effect on some systems, particularly when high feedback gains are used.

The controller usually consists of various instruments, equipment, and computation hardware and software. Often, the functions of the data-acquisition and processing system overlap with those of the controller to some extent. An example might be the digital controller of vibration testing apparatus. First, the responses are measured through sensors (and transducers), filtered, and amplified (conditioned). These data channels can be passed through a multiplexer, the purpose of which is to select one data channel at a time for processing. Most modern data-acquisition hardware do not need a separate multiplexer to handle multiple signals. The analog data are converted into digital data using analog-to-digital converters (ADCs), as described in Chapter 9. The resulting sampled data are stored on a disk or as block data in the computer memory. The reference input signal (typically a signal recorded on an FM tape) is also sampled (if it is not already in the digital form), using an ADC, and fed into the computer. Digital processing is done on the reference signal and the response data, with the objective of computing the command signal to drive the shaker. The digital command signal is converted into an analog signal, using a digital-to-analog converter (DAC), and amplified (conditioned) before it is used to drive the exciter.

The nature of the control components depends to a large extent on the nature and objectives of the particular test to be conducted. Some of the basic components in a shaker controller are described in the following subsections.

8.2.1 COMPONENTS OF A SHAKER CONTROLLER

Compressor

A compressor circuit is incorporated in automatic excitation control devices to control the excitation-input level automatically. The level of control depends on the feedback signal from a control sensor and the specified (reference) excitation signal. Usually, the compressor circuit is included in the excitation-signal generator (e.g., a sine generator). The control by this means can be done on the basis of a single-frequency component (e.g., the fundamental frequency).

Equalizer (Spectrum Shaper)

Random-signal equalizers are used to shape the spectrum of a random signal in a desired manner. In essence, and equalizer consists of a bank of narrow-band filters (e.g., 80 filters) in parallel over the operating frequency range. By passing the signal through each filter, the spectral density (or the mean square value) of the signal in that narrow frequency band (e.g., each one-third-octave band) is determined. This is compared with the desired spectral level, and automatic adjustment is made in that filter in case there is an error. In some systems, response-spectrum analysis is made in place of power spectral density analysis (see Chapters 4 and 10). In that case, the equalizer consists of a bank of simple oscillators, in which the resonant frequencies are distributed over the operating frequency range of the equalizer. The feedback signal is passed through each oscillator, and the peak value of its output is determined. This value is compared with the desired response spectrum value at that frequency. If there is an error, automatic gain adjustment is made in the appropriate excitation signal components.

Random-noise equalizers are used in conjunction with random signal generators. They receive feedback signals from the control sensors. In some digital control systems, there are algorithms (software) that are used to iteratively converge the spectrum of the excitation signal felt by the test object into the desired spectrum.

Tracking filter

Many vibration tests are based on single-frequency excitations. In such cases, the control functions should be performed on the basis of amplitudes of the fundamental-frequency component of the signal. A tracking filter is simply a frequency-tuned bandpass filter. It automatically tunes the center frequency of its very narrow bandpass filter to the frequency of a carrier signal. Then, when a noisy signal is passed through the tuned filter, the output of the filter will be the required fundamental frequency component in the signal. Tracking filters are also useful in obtaining amplitude–frequency plots using an X-Y plotter. In such cases, the frequency value comes from the signal generator (sweep oscillator), which produces the carrier signal to the tracking filter. The tracking filter then determines the corresponding amplitude of a signal that is fed into it. Most tracking filters have dual channels so that two signals can be handled (tracked) simultaneously.

Excitation Controller (Amplitude Servo-Monitor)

An excitation controller is typically an integral part of the signal generator. It can be set so that automatic sweep between two frequency limits can be performed at a selected sweep rate (linear or logarithmic). More advanced excitation controllers have the capability of automatic switch-over between constant-displacement, constant-velocity, and constant-acceleration excitation-input control at specified frequencies over the sweep frequency interval. Consequently, integrator circuits, to determine velocities and displacements from acceleration signals, should be present within the excitation controller unit. Sometimes, integration is performed by a separate unit called a *vibration meter*. This unit also offers the operator the capability of selecting the desired level of each signal (acceleration, velocity, or displacement). There is an automatic cutoff level for large displacement values that could result from noise in acceleration signals. A compressor is also a subcomponent of the excitation controller. The complete unit is sometimes known as an *amplitude servo-monitor*.

8.2.2 Signal-Generating Equipment

Shakers are force-generating devices that are operated using drive (excitation) signals generated from a source. The excitation-signal source is known as the *signal generator*. Three major types of signal generators are used in vibration testing applications: (1) oscillators or sine-signal generators, (2) random-signal generators, and (3) storage devices. In some units, oscillators and random-signal generators are combined (sine-random generators). These two generators are discussed separately, however, because of their difference in functions. It should also be noted that almost any digital signal (deterministic or random) can be generated by a digital computer using a suitable computer program; it eventually can be passed through a DAC to obtain the corresponding analog signal. These "digital" signal generators, along with analog sources such as magnetic tape players (FM), are classified into the category of storage devices.

The dynamic range of any equipment is the ratio of the maximum and minimum output levels (expressed in decibels) within which it is capable of operating without significant error. This is an important specification for many types of equipment, and particularly signal-generating devices. The output level of the signal generator should be set to a value within its dynamic range.

Oscillators

Oscillators are essentially single-frequency generators. Typically, sine signals are generated, but other waveforms (such as rectangular and triangular pulses) are also available in many oscillators. Normally, an oscillator has two modes of operation: (1) up-and-down sweep between two frequency limits, and (2) dwell at a specified frequency. In the sweep operation, the sweep rate should be specified. This can be done either on a linear scale (Hz·min^{-1}) or on a logarithmic scale (octaves·min^{-1}). In the dwell operation, the frequency points (or intervals) should be specified. In either case, a desired signal level can be chosen using the gain-control knob. An oscillator that is operated exclusively in the sweep mode is called a *sweep oscillator*.

The early generation of oscillators employed variable inductor-capacitor types of electronic circuits to generate signals oscillating at a desired frequency. The oscillator is tuned to the required frequency by varying the capacitance or inductance parameters. A DC voltage is applied to energize the capacitor and to obtain the desired oscillating voltage signal, which is subsequently amplified and conditioned. Modern oscillators use operational amplifier circuits along with resistor, capacitor, and semiconductor elements. Also commonly used are crystal (quartz) parallel-resonance oscillators to generate voltage signals accurately at a fixed frequency. The circuit is activated using a DC voltage source. Other frequencies of interest are obtained by passing this high-frequency signal through a frequency converter. The signal is then conditioned (amplified and filtered). Required shaping (e.g., rectangular pulse) is obtained using a shape circuit. Finally, the required signal level is obtained by passing the resulting signal through a variable-gain amplifier. A block diagram of an oscillator, illustrating various stages in the generation of a periodic signal, is given in Figure 8.13.

A typical oscillator offers a choice of several (typically six) linear and logarithmic frequency ranges and a sizable level of control capability (e.g., 80 dB). Upper and lower frequency limits in a sweep can be preset on the front panel to any of the available frequency ranges. Sweep-rate settings are continuously variable (typically, 0 to 10 octaves·min^{-1} in the logarithmic range, and 0 to 60 kHz·min^{-1} in the linear range), but one value must be selected for a given test or part of a test. Most oscillators have a repetitive-sweep capability, which allows the execution of more than one sweep continuously (e.g., for mechanical aging and in product-qualification single-frequency tests). Some oscillators have the capability of also varying the signal level (amplitude) during each test cycle (sweep or dwell). This is known as *level programming*. Also, automatic switching between acceleration, velocity, and displacement excitations at specified frequency points in each test cycle can be implemented with some oscillators. A frequency counter, which is capable of recording the fundamental frequency of the output signal, is usually an integral component of the oscillator.

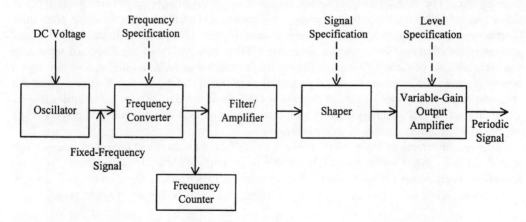

FIGURE 8.13 Block diagram of an oscillator-type signal generator.

Vibration Instrumentation

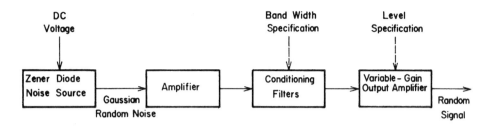

FIGURE 8.14 Block diagram of a random signal generator.

Random Signal Generators

In modern random signal generators, semiconductor devices (e.g., Zener diodes) are used to generate a random signal that has a required (e.g., Gaussian) distribution. This is accomplished by applying a suitable DC voltage to a semiconductor circuit. The resulting signal is then amplified and passed through a bank of conditioning filters, which effectively acts as a spectrum shaper. In this manner, the bandwidth of the signal can be adjusted in a desired manner. Extremely wide-band signals (white noise), for example, can be generated for random excitation vibration testing in this manner. The block diagram in Figure 8.14 shows the essential steps in a random signal generation process. A typical random signal generator has several (typically eight) bandwidth selections over a wide frequency range (e.g., 1 Hz to 100 kHz). A level-control capability (typically 80 dB) is also available.

Tape Players

Vibration testing for product qualification can be performed using a tape player as the signal source. A tape player is essentially a signal reproducer. The test input signal that has a certain specified response spectrum is obtained by playing a magnetic tape and mixing the contents in the several tracks of the tape in a desirable ratio. Typically, each track contains a sine-beat signal (with a particular beat frequency, amplitude, and number of cycles per beat) or a random signal component (with a desired spectral characteristic).

In frequency modulation (FM) tapes, the signal amplitude is proportional to the frequency of a carrier signal. The carrier signal is the one that is recorded on the tape. When played back, the actual signal is reproduced, based on detecting the frequency content of the carrier signal in different time points. The FM method is usually satisfactory, particularly for low-frequency testing (below 100 Hz).

Performance of a tape player is determined by several factors, including tape type and quality, signal reproduction (and recording) circuitry, characteristics of the magnetic heads, and the tape-transport mechanism. Some important specifications for tape players are (1) the number of tracks per tape (e.g., 14 or 28); (2) the available tape speeds (e.g., 3.75, 7.5, 15, or 30 in·s^{-1}); (3) reproduction filter-amplifier capabilities (e.g., 0.5% third-harmonic distortion in a 1-kHz signal recorded at 15 in·s^{-1} tape speed, peak-to-peak output voltage of 5 V at 100-ohm load, signal-to-noise ratio of 45 dB, output impedance of 50 ohms); and (4) the available control options and their capabilities (e.g., stop, play, reverse, fast-forward, record, speed selection, channel selection). Tape player specifications for vibration testing are governed by an appropriate regulatory agency, according to a specified standard (e.g., the Communication and Telemetry Standard of the Intermediate Range Instrumentation Group (IRIG Standard 106-66).

A common practice in vibration testing is to generate the test input signal by repetitively playing a closed tape loop. In this manner, the input signal becomes periodic but has the desired frequency content. Frequency modulation players can be fitted with special loop adaptors for playing tape loops. In spectral (Fourier) analysis of such signals, the analyzing filter bandwidth should be an order of magnitude higher than the repetition frequency (tape speed per loop length). Extraneous noise is caused by discontinuities at the tape joint. This can be suppressed using suitable filters or gating circuits.

A technique that can be employed to generate low-frequency signals with high accuracy is to record the signal first at a very low tape speed and then play it back at a high tape speed (e.g., r times higher). This has the effect of multiplying all frequency components in the signal by the speed ratio (r). Consequently, the filter circuits in the tape player will allow some low-frequency components in the signal that would normally be cut off, and will cut off some high-frequency components that would normally be allowed. Hence, this process is a way of emphasizing the low-frequency components in a signal.

Data Processing

A controller generally has some data processing functions as well. A data-acquisition and processing system usually consists of response sensors (and transducers), signal conditioners, an input-output (I/O) board including a multiplexer, ADCs, etc., and a digital computer with associated software. The functions of a digital data-acquisition and processing system can be quite general, as listed below.

1. Measuring, conditioning, sampling, and storing the response signals and operational data of test object (using input commands through a user interface, as necessary)
2. Digital processing of the measured data according to the test objectives (and using input commands, as necessary)
3. Generation of drive signals for the control system
4. Generation and recording of test results (responses) in the required format.

The capacity and the capabilities of a data acquisition and processing system are determined by such factors as:

1. The number of response data channels that can be handled simultaneously
2. The data-sampling rate (samples per second) for each data channel
3. Computer memory size
4. Computer processing speed
5. External storage capability (hard disks, floppy disks, etc.)
6. The nature of the input and output devices
7. Software capabilities and features.

Commercial data-acquisition and processing systems with a wide range of processing capabilities are available for use in vibration testing. Some of the standard processing capabilities are the following (also see Chapters 4 and 10):

1. Response-spectrum analysis
2. FFT analysis (spectral densities, correlations, coherence, Fourier spectra, etc.)
3. Frequency-response function, transmissibility, and mechanical-impedance analysis
4. Natural frequency and mode-shape analysis
5. System parameter identification (e.g., damping parameters).

Most processing is done in realtime, which means that the signals are analyzed as they are being measured. The advantage of this is that outputs and command signals are available simultaneously as the monitoring is done, so that any changes can be detected as they occur (e.g., degradation in the test object or deviations in the excitation signal from the desired form) and automatic feedback control can be effected. For realtime processing to be feasible, the data acquisition rate (sampling rate) and the processing speed of the computer should be sufficiently fast. In realtime frequency analysis, the entire frequency range (not narrow bands separately) is analyzed at a given instant. Results are presented as Fourier spectra, power spectral densities, cross-spectral densities, coherence functions, correlation functions, and response spectra curves. Averaging of frequency plots can be done over small frequency bands (e.g., one-third-octave analysis), or the running average of each instantaneous plot can be determined.

8.3 PERFORMANCE SPECIFICATION

Proper selection and integration of sensors and transducers are crucial in instrumenting a vibrating system. The response variable that is being measured (e.g., acceleration) is termed the *measurand*. A measuring device passes through two stages in making a measurement. First, the measurand is sensed; then, the measured signal is transduced (converted) into a form that is particularly suitable for signal conditioning, processing, or recording. Often, the output from the transducer stage is an electrical signal. It is common practice to identify the combined sensor-transducer unit as either a sensor or a transducer.

The measuring device itself might contain some of the signal-conditioning circuitry and recording (or display) devices or meters. These are components of an overall measuring system. For the purposes here, these components are considered separately.

In most applications, the following four variables are particularly useful in determining the response and structural integrity of a vibrating system:

1. Displacement (potentiometer or LVDT)
2. Velocity (tachometer)
3. Acceleration (accelerometer)
4. Stress and strain (strain gage).

In each case, the usual measuring devices are indicated in parentheses. It is somewhat common in vibration practice to measure acceleration first and then determine velocity and displacement by direct integration. Any noise and DC components in the measurement, however, could give rise to erroneous results in such cases. Consequently, it is good practice to measure displacement, velocity, and acceleration using separate sensors, particularly when the measurements are employed in feedback control of the vibratory system. It is not recommended to differentiate a displacement (or velocity) signal to obtain velocity (or acceleration) because this process would amplify any noise present in the measured signal. Consider, for example, a sinusoidal signal given by $A\sin\omega t$. Since $d/dt(A\sin\omega t) = A\omega\cos\omega t$, it follows that any high-frequency noise would be amplified by a factor proportional to its frequency. Also, any discontinuities in noise components would produce large deviations in the results. Using the same argument, it can be concluded that acceleration measurements are desirable for high-frequency signals and displacement measurements are desirable for low-frequency signals. It follows that the selection of a particular measurement transducer should depend on the frequency content of the useful portion of the measured signal.

Transducers are divided into two broad categories: active transducers and passive transducers. *Passive transducers* do not require an external electric source for activation. Some examples are electromagnetic, piezoelectric, and photovoltaic transducers. *Active transducers* do not possess self-contained energy sources and thus need external activation. A good example is a resistive transducer, such as potentiometer.

In selecting a particular transducer (measuring device) for a specific vibration application, special attention should be given to its ratings, which are usually provided by the manufacturer, and the required performance specifications as provided by the customer (or developed by the system designer).

8.3.1 Parameters for Performance Specification

A *perfect measuring device* can be defined as one that possesses the following characteristics:

1. Output instantly reaches the measured value (fast response).
2. Transducer output is sufficiently large (high gain, low output impedance, high sensitivity).
3. Output remains at the measured value (without drifting or being affected by environmental effects and other undesirable disturbances and noise) unless the measurand itself changes (stability and robustness).
4. The output signal level of the transducer varies in proportion to the signal level of the measurand (static linearity).
5. Connection of a measuring device does not distort the measurand itself (loading effects are absent and impedances are matched).
6. Power consumption is small (high input impedance).

All of these properties are based on dynamic characteristics and therefore can be explained in terms of dynamic behavior of the measuring device. In particular, items 1 through 4 can be specified in terms of the device (response), either in the *time domain* or in the *frequency domain*. Items 2, 5, and 6 can be specified using the *impedance* characteristics of a device. First, response characteristics that are important in performance specification of a sensor/transducer unit are discussed.

Time-Domain Specifications

Several parameters that are useful for the time-domain performance specification of a device are as follows:

1. Rise time (T_r): This is the time taken to pass the steady-state value of the response for the first time. In overdamped systems, the response is nonoscillatory; consequently, there is no overshoot. So that the definition would be valid for all systems, rise time is often defined as the time taken to pass 90% of the steady-state value for the first time. Rise time is often measured from 10% of the steady-state value in order to leave out start-up irregularities and time lags that might be present in a system. Rise time represents the speed of response of a device — a small rise time indicates a fast response.
2. Delay time (T_d): This is usually defined as the time taken to reach 50% of the steady-state value for the first time. This parameter is also a measure of the speed of response.
3. Peak time (T_p): This is the time at the first peak. This parameter also represents the speed of response of the device.
4. Settling time (T_s): This is the time taken for the device response to settle down within a certain percentage (e.g., ±2%) of the steady-state value. This parameter is related to the degree of damping present in the device as well as the degree of stability.
5. Percentage overshoot (P.O.): This is defined as

$$\text{P.O.} = 100(M_p - 1)\% \qquad (8.30)$$

using the normalized-to-unity step response curve, where M_p is the peak value. Percentage overshoot is a measure of damping or relative stability in the device.
6. Steady-state error: This is the deviation of the actual steady-state value from the desired value. Steady-state error can be expressed as a percentage with respect to the (desired) steady-state value. In a measuring device, steady-state error manifests itself as an offset.

This is a systematic (deterministic) error that normally can be corrected by recalibration. In servo-controlled devices, steady-state error can be reduced by increasing the loop gain or by introducing a lag compensation. Steady-state error can be completely eliminated using the integral control (reset) action.

For the best performance of a measuring device, it is desirable to have the values of all the foregoing parameters as small as possible. In actual practice, however, it might be difficult to meet all specifications, particularly under conflicting requirements. For example, T_r can be decreased by increasing the dominant natural frequency ω_n of the device. This, however, increases the P.O. and sometimes the T_s. On the other hand, the P.O. and T_s can be decreased by increasing device damping, but it has the undesirable effect of increasing T_r.

Frequency-Domain Specifications

Because any time signal can be decomposed into sinusoidal components through Fourier transform, it is clear that the response of a system to an arbitrary input excitation can also be determined using transfer-function (frequency response-function) information for that system. For this reason, one could argue that it is redundant to use both time-domain specifications and frequency-domain specifications, as they carry the same information. Often, however, both specifications are used simultaneously because this can provide a better picture of the system performance. Frequency-domain parameters are more suitable in representing some characteristics of a system under some types of excitation.

Consider a device with the frequency-response function (transfer function) $G(j\omega)$ Some useful parameters for performance specification of the device, in the frequency domain, are:

1. Useful frequency range (operating interval): This is given by the flat region of the frequency response magnitude $|G(j\omega)|$ of the device.
2. Bandwidth (speed of response): This can be represented by the primary natural frequency (or resonant frequency) of the device.
3. Static gain (steady-state performance): Because static conditions correspond to zero frequencies, this is given by $G(0)$.
4. Resonant frequency (speed and critical frequency region) ω_r: This corresponds to the lowest frequency at which $|G(j\omega)|$ peaks.
5. Magnitude at resonance (stability): This is given by $|G(j\omega_r)|$.
6. Input impedance (loading, efficiency, interconnectability): This represents the dynamic resistance as felt at the input terminals of the device. This parameter will be discussed in more detail under component interconnection and matching (Section 8.6).
7. Output impedance (loading, efficiency, interconnectability): This represents the dynamic resistance as felt at the output terminals of the device.
8. Gain margin (stability): This is the amount by which the device gain could be increased before the system becomes unstable.
9. Phase margin (stability): This is the amount by which the device phase lead could be decreased (i.e., phase lag increased) before the system becomes unstable.

8.3.2 Linearity

A device is considered *linear* if it can be modeled by linear differential equations, with time t as the independent variable. Nonlinear devices are often analyzed using linear techniques by considering small excursions about an operating point. This linearization is accomplished by introducing incremental variables for the excitations (inputs) and responses (outputs). If one increment can cover the entire operating range of a device with sufficient accuracy, it is an indication that the device is linear. If the input/output relations are nonlinear algebraic equations, that represents a *static nonlinearity*. Such a situation can be handled simply by using nonlinear calibration curves, which linearize the device without introducing nonlinearity errors. If, on the other hand, the input/output relations are nonlinear differential equations, analysis usually becomes quite complex. This situation represents a *dynamic nonlinearity*.

Transfer-function representation is a "linear" model of an instrument. Hence, it implicitly assumes linearity. According to industrial terminology, a linear measuring instrument provides a measured value that varies linearly with the value of the measurand. This is consistent with the definition of static linearity. All physical devices are *nonlinear* to some degree. This stems from any deviation from the ideal behavior, due to causes such as saturation, deviation from Hooke's law in elastic elements, Coulomb friction, creep at joints, aerodynamic damping, backlash in gears and other loose components, and component wearout. Nonlinearities in devices are often manifested as some peculiar characteristics. In particular, the following properties are important in detecting nonlinear behavior in dynamic systems:

1. *Saturation*: The response does not increase when the excitation is increased beyond some level. This may result from such causes as magnetic saturation, which is common in transformer devices such as differential transformers, plasticity in mechanical components, or nonlinear deformation in springs.
2. *Hysteresis*: In this case, the input/output curve changes, depending on the direction of motion, resulting in a hysteresis loop. This is common in loose components such as gears, which have backlash; in components with nonlinear damping, such as Coulomb friction; and in magnetic devices with ferromagnetic media and various dissipative mechanisms (e.g., eddy current dissipation).
3. *The jump phenomenon*: Some nonlinear devices exhibit an instability known as the *jump phenomenon* (or *fold catastrophe*). Here, the frequency-response (transfer) function curve suddenly jumps in magnitude at a particular frequency, while the excitation frequency is increased or decreased. A device with this nonlinearity will exhibit a characteristic "tilt" of its resonant peak either to the left (softening nonlinearity) or to the right (hardening nonlinearity). Furthermore, the transfer function itself may change with the level of input excitation in the case of nonlinear devices.
4. *Limit cycles*: A limit cycle is a closed trajectory in the state space that corresponds to sustained oscillations without decay or growth. The amplitude of these oscillations is independent of the initial location from which the response started. In the case of a stable limit cycle, the response will return to the limit cycle irrespective of the location in the neighborhood of the limit cycle from which the response was initiated. In the case of an unstable limit cycle, the response will steadily move away from it with the slightest disturbance.
5. *Frequency creation*: At steady state, nonlinear devices can create frequencies that are not present in the excitation signals. These frequencies might be harmonics (integer multiples of the excitation frequency), subharmonics (integer fractions of the excitation frequency), or nonharmonics (usually rational fractions of the excitation frequency).

Several methods are available to reduce or eliminate nonlinear behavior in vibrating systems. They include calibration (in the static case), the use of linearizing elements such as resistors and amplifiers to neutralize the nonlinear effects, and the use of nonlinear feedback. It is also a good practice to take the following precautions.

1. Avoid operating the device over a wide range of signal levels.
2. Avoid operation over a wide frequency band.
3. Use devices that do not generate large mechanical motions.

4. Minimize Coulomb friction.
5. Avoid loose joints and gear coupling (i.e., use *direct-drive* mechanisms).

8.3.3 INSTRUMENT RATINGS

Instrument manufacturers do not usually provide complete dynamic information for their products. In most cases, it is unrealistic to expect complete dynamic models (in the time domain or the frequency domain) and associated parameter values for complex instruments. Performance characteristics provided by manufacturers and vendors are primarily static parameters. Known as instrument ratings, these are available as parameter values, tables, charts, calibration curves, and empirical equations. Dynamic characteristics such as transfer functions (e.g., transmissibility curves expressed with respect to excitation frequency) might also be provided for more sophisticated instruments, but the available dynamic information is never complete. Furthermore, definitions of rating parameters used by manufacturers and vendors of instruments are in some cases not the same as analytical definitions used in textbooks. This is particularly true in relation to the term *linearity*. Nevertheless, instrument ratings provided by manufacturers and vendors are very useful in the selection, installation, operation, and maintenance of instruments. Some of these performance parameters are indicated below.

Rating Parameters

Typical rating parameters supplied by instrument manufacturers are:

1. Sensitivity
2. Dynamic range
3. Resolution
4. Linearity
5. Zero drift and full-scale drift
6. Useful frequency range
7. Bandwidth
8. Input and output impedances.

The conventional definitions given by instrument manufacturers and vendors are summarized below.

Sensitivity of a transducer is measured by the magnitude (peak, rms value, etc.) of the output signal corresponding to a unit input of the measurand. This can be expressed as the ratio of (incremental output)/(incremental input) or, analytically, as the corresponding partial derivative. In the case of vectorial or tensorial signals (e.g., displacement, velocity, acceleration, strain, force), the direction of sensitivity should be specified.

Cross-sensitivity is the sensitivity along directions that are orthogonal to the direction of primary sensitivity; it is expressed as a percentage of the direct sensitivity. High sensitivity and low cross-sensitivity are desirable for measuring instruments. Sensitivity to parameter changes, disturbances, and noise must be small in any device, however, and this is an indication of its *robustness*. Often, sensitivity and robustness are conflicting requirements.

Dynamic range of an instrument is determined by the allowed lower and upper limits of its input or output (response) so as to maintain a required level of measurement accuracy. This range is usually expressed as a ratio, in *decibels*. In many situations, the lower limit of the dynamic range is equal to the resolution of the device. Hence, the dynamic range is usually expressed as the ratio (range of operation)/(resolution), in decibels.

Resolution is the smallest change in a signal that can be detected and accurately indicated by a transducer, a display unit, or any pertinent instrument. It is usually expressed as a percentage of the maximum range of the instrument, or as the inverse of the dynamic range ratio, as defined above. It follows that dynamic range and resolution are closely related.

Linearity is determined by the calibration curve of an instrument. The curve of output amplitude (peak or rms value) versus input amplitude under static conditions within the dynamic range of an instrument is known as the *static calibration curve*. Its closeness to a straight line measures the degree of linearity. Manufacturers provide this information either as the maximum deviation of the calibration curve from the least-squares straight-line fit of the calibration curve or from some other reference straight line. If the least-squares fit is used as the reference straight line, the maximum deviation is called *independent linearity* (more correctly, independent nonlinearity, because the larger the deviation, the greater the nonlinearity). Nonlinearity can be expressed as a percentage of either the actual reading at an operating point or the full-scale reading.

Zero drift is defined as the drift from the null reading of the instrument when the measurand is maintained steady for a long period. Note that in this case, the measurand is kept at zero or any other level that corresponds to null reading of the instrument. Similarly, *full-scale drift* is defined with respect to the full-scale reading (the measurand is maintained at the full-scale value). Usual causes of drift include instrument instability (e.g., instability in amplifiers), ambient changes (e.g., changes in temperature, pressure, humidity, and vibration level), changes in power supply (e.g., changes in reference DC voltage or AC line voltage), and parameter changes in an instrument (due to aging, wearout, nonlinearities, etc.). Drift due to parameter changes that are caused by instrument nonlinearities is known as *parametric drift, sensitivity drift*, or *scale-factor drift*. For example, a change in spring stiffness or electrical resistance due to changes in ambient temperature results in a parametric drift. Note that the parametric drift depends on the measurand level. Zero drift, however, is assumed to be the same at any measurand level if the other conditions are kept constant. For example, a change in reading caused by thermal expansion of the readout mechanism due to changes in the ambient temperature is considered a zero drift. In electronic devices, drift can be reduced using alternating current (AC) circuitry rather than direct current (DC) circuitry. For example, AC-coupled amplifiers have fewer drift problems than DC amplifiers. Intermittent checking for the instrument response level for zero input is a popular way to calibrate for zero drift. In digital devices, for example, this can be done automatically from time to time between sample points, when the input signal can be bypassed without affecting the system operation.

Useful frequency range corresponds to the interval of both flat gain and zero phase in the frequency-response characteristics of an instrument. The maximum frequency in this band is typically less than half (say, one fifth of) the dominant resonant frequency of the instrument. This is a measure of instrument bandwidth.

Bandwidth of an instrument determines the maximum speed or frequency at which the instrument is capable of operating. High bandwidth implies faster speed of response. Bandwidth is determined by the dominant natural frequency ω_n or the dominant resonant frequency ω_r of the transducer. (Note: For low damping, ω_r is approximately equal to ω_n). It is inversely proportional to the rise time and the dominant time constant. Half-power bandwidth is also a useful parameter. Instrument bandwidth must be sufficiently greater than the maximum frequency of interest in the measured signal. The bandwidth of a measuring device is important, particularly when measuring transient signals. Note that the bandwidth is directly related to the useful frequency range.

8.3.4 Accuracy and Precision

The instrument ratings mentioned above affect the overall *accuracy* of an instrument. Accuracy can be assigned either to a particular reading or to an instrument. Note that instrument accuracy depends not only on the physical hardware of the instrument, but also on the operating conditions (e.g., design conditions that are the normal, steady operating conditions or extreme transient conditions, such as emergency start-up and shutdown). *Measurement accuracy* determines the closeness of the measured value to the true value. *Instrument accuracy* is related to the worst accuracy obtainable within the dynamic range of the instrument in a specific operating environment. *Measurement error* is defined as

$$\text{Error} = (\text{Measured value}) - (\text{True value}) \tag{8.31}$$

Correction, which is the negative of error, is defined as

$$\text{Correction} = (\text{True value}) - (\text{Measured value}) \tag{8.32}$$

Each of these can also be expressed as a percentage of the true value. The accuracy of an instrument can be determined by measuring a parameter whose true value is known, near the extremes of the dynamic range of the instrument, under certain operating conditions. For this purpose, standard parameters or signals that can be generated at very high levels of accuracy would be needed. The National Institute for Standards and Testing (NIST) is usually responsible for the generation of these standards. Nevertheless, accuracy and error values cannot be determined to 100% exactness in typical applications because the true value is not known to begin with. In a given situation, one can only make estimates for accuracy by using ratings provided by the instrument manufacturer or by analyzing data from previous measurements and models.

Causes of error include instrument instability, external noise (disturbances), poor calibration, inaccurate information (e.g., poor analytical models, inaccurate control), parameter changes (e.g., due to environmental changes, aging, and wearout), unknown nonlinearities, and improper use of the instrument.

Errors can be classified as *deterministic* (or *systematic*) and *random* (or *stochastic*). Deterministic errors are those caused by well-defined factors, including nonlinearities and offsets in readings. These usually can be accounted for by proper calibration and analytical practices. Error ratings and calibration charts are used to remove systematic errors from instrument readings. Random errors are caused by uncertain factors entering into the instrument response. These include device noise, line noise, and effects of unknown random variations in the operating environment. A statistical analysis using sufficiently large amounts of data is necessary to estimate random errors. The results are usually expressed as a mean error, which is the systematic part of random error, and a standard deviation or confidence interval for instrument response.

Precision is not synonymous with accuracy. Reproducibility (or repeatability) of an instrument reading determines the precision of an instrument. Two or more identical instruments that have the same high offset error might be able to generate responses at high precision, although these readings are clearly inaccurate. For example, consider a timing device (clock) that very accurately indicates time increments (say, up to the nearest microsecond). If the reference time (starting time) is set incorrectly, the time readings will be in error, although the clock has a very high precision.

Instrument error can be represented by a random variable that has a mean value μ_e and a standard deviation σ_e. If the standard deviation is zero, the variable is considered deterministic. In that case, the error is said to be deterministic or repeatable. Otherwise, the error is said to be random. The precision of an instrument is determined by the standard deviation of error in the instrument response. Readings of an instrument may have a large mean value of error (e.g., large offset); but if the standard deviation is small, the instrument has a high precision. Hence, a quantitative definition for precision would be

$$\text{Precision} = (\text{Measured range})/\sigma_e \tag{8.33}$$

Lack of precision originates from random causes and poor construction practices. It cannot be compensated for by recalibration, just as the precision of a clock cannot be improved by resetting the time. On the other hand, accuracy can be improved by recalibration. Repeatable (deterministic) accuracy is inversely proportional to the magnitude of the mean error μ_e.

In selecting instruments for a particular application, in addition to matching instrument ratings with specifications, several additional considerations should be looked into. These include geometric limitations (size, shape, etc.), environmental conditions (e.g., chemical reactions including corrosion, extreme temperatures, light, dirt accumulation, electromagnetic fields, radioactive environments, shock, and vibration), power requirements, operational simplicity, availability, past record and reputation of the manufacturer and of the particular instrument, and cost-related economic aspects (initial cost, maintenance cost, cost of supplementary components such as signal-conditioning and processing devices, design life and associated frequency of replacement, and cost of disposal and replacement). Often, these considerations become the ultimate deciding factors in the selection process.

8.4 MOTION SENSORS AND TRANSDUCERS

Motion sensing is considered the most important measurement in vibration applications. Other variables such as force, torque, stress, strain, and material properties are also important, either directly or indirectly, in the practice of vibration. This section will describe some useful measuring devices for motion in the field of mechanical vibration.

8.4.1 POTENTIOMETER

The potentiometer, or *pot*, is a displacement transducer. This active transducer consists of a uniform coil of wire or a film of high-resistive material — such as carbon, platinum, or conductive plastic — whose resistance is proportional to its length. A fixed voltage v_{ref} is applied across the coil (or film) using an external, constant DC voltage supply. The transducer output signal v_o is the DC voltage between the movable contact (wiper arm) sliding on the coil and one terminal of the coil, as shown schematically in Figure 8.15(a). Slider displacement x is proportional to the output voltage:

$$v_o = kx \qquad (8.34)$$

This relationship assumes that the output terminals are open-circuit; that is, infinite-impedance load (or resistance in the present DC case) is present at the output terminal, so that the output current is zero. In actual practice, however, the load (the circuitry into which the pot signal is fed; e.g., conditioning or processing circuitry) has a finite impedance. Consequently, the output current (the current through the load) is non-zero, as shown in Figure 8.15(b). The output voltage thus drops to $\bar{v}_o$, even if the reference voltage v_{ref} is assumed to remain constant under load variations (i.e., the voltage source has zero output impedance); this consequence is known as the *loading effect* of the transducer. Under these conditions, the linear relationship given by equation (8.34) would no longer be valid. This causes an error in the displacement reading. Loading can affect the transducer reading in two ways: by changing the reference voltage (i.e., loading the voltage source) and by loading the transducer. To reduce these effects, a voltage source that is not seriously affected by load variations (e.g., a regulated or stabilized power supply that has low output impedance) and data acquisition circuitry (including signal-conditioning circuitry) that has high input impedance should be used.

The resistance of a potentiometer should be chosen with care. On the one hand, an element with high resistance is preferred because this results in reduced power dissipation for a given voltage, which has the added benefit of reduced thermal effects. On the other hand, increased resistance increases the output impedance of the potentiometer and results in loading nonlinearity error unless the load resistance is also increased proportionately. Low-resistance pots have resistances less than $10\,\Omega$. High-resistance pots can have resistances on the order of $100\,k\Omega$. Conductive

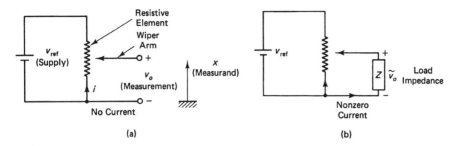

FIGURE 8.15 (a) Schematic diagram of a potentiometer, and (b) potentiometer loading.

plastics can provide high resistances — typically about 100 Ω·m/m — and are increasingly used in potentiometers. Reduced friction (low mechanical loading), reduced wear, reduced weight, and increased resolution are advantages of using conductive plastics in potentiometers.

Potentiometer Resolution

The force required to move the slider arm comes from the motion source, and the resulting energy is dissipated through friction. This energy conversion, unlike pure mechanical-to-electrical conversions, involves relatively high forces, and the energy is wasted rather than being converted into the output signal of the transducer. Furthermore, the electrical energy from the reference source is also dissipated through the resistor coil (or film), resulting in an undesirable temperature rise. These are two obvious disadvantages of this *resistively coupled transducer*. Another disadvantage is the finite *resolution* in coil-type pots.

Coils, instead of straight wire, are used to increase the resistance per unit travel of the slider arm. But the slider contact jumps from one turn to the next in this case. Accordingly, the resolution of a coil-type potentiometer is determined by the number of turns in the coil. For a coil that has N turns, the resolution r, expressed as a percentage of the output range, is given by

$$r = \frac{100}{N}\% \tag{8.35}$$

Resolutions better (smaller) than 0.1% (i.e., 1000 turns) are available with coil potentiometers. Infinitesimal (incorrectly termed infinite) resolutions are now possible with high-quality resistive film potentiometers that use conductive plastics, for example. In this case, the resolution is limited by other factors, such as mechanical limitations and signal-to-noise ratio. Nevertheless, resolutions on the order of 0.01 mm are possible with good rectilinear potentiometers.

Some limitations and disadvantages of potentiometers as displacement measuring devices are as follows:

1. The force needed to move the slider (against friction and arm inertia) is provided by the vibration source. This mechanical loading distorts the measured signal itself.
2. High-frequency (or highly transient) measurements are not feasible because of such factors as slider bounce, friction and inertia resistance, and induced voltages in the wiper arm and primary coil.
3. Variations in the supply voltage cause error.
4. Electrical loading error can be significant when the load resistance is low.
5. Resolution is limited by the number of turns in the coil and by the coil uniformity. This will limit small-displacement measurements such as fine vibrations.
6. Wearout and heating up (with associated oxidation) in the coil (film) and slider contact cause accelerated degradation.

There are several advantages associated with potentiometer devices, however, including the following:

1. They are relatively less costly.
2. Potentiometers provide high-voltage (low-impedance) output signals, requiring no amplification in most applications. Transducer impedance can be varied simply by changing the coil resistance and supply voltage.

Optical Potentiometer

The optical potentiometer, shown schematically in Figure 8.16(a), is a displacement sensor. A layer of photoresistive material is sandwiched between a layer of regular resistive material and a layer of conductive material. The layer of resistive material has a total resistance of R_c, and it is uniform (i.e., it has a constant resistance per unit length). The photoresistive layer is practically an electrical insulator when no light is projected on it. The displacement of the moving object (whose displacement is being measured) causes a moving light beam to be projected onto a rectangular area of the photoresistive layer. This light-projected area attains a resistance of R_p, which links the resistive layer that is above the photoresistive layer and the conductive layer that is below it. The supply voltage to the potentiometer is v_{ref}, and the length of the resistive layer is L. The light spot is projected at a distance x from one end of the resistive element, as shown in the figure.

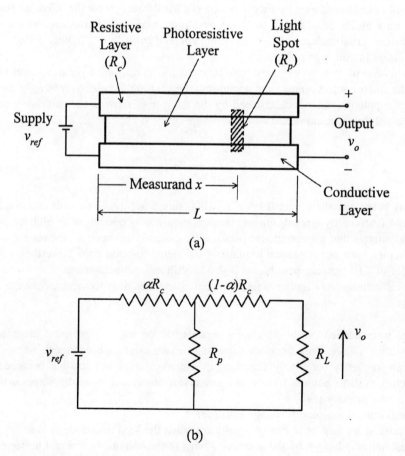

FIGURE 8.16 (a) An optical potentiometer, and (b) equivalent circuit ($\alpha = x/L$).

An equivalent circuit for the optical potentiometer is shown in Figure 8.16(b). Here it is assumed that a load of resistance R_L is present at the output of the potentiometer, voltage across which being v_o. Current through the load is v_o/R_L. Hence, the voltage drop across $(1-\alpha)R_c + R_L$, which is also the voltage across R_p, is given by $[(1-\alpha)R_c + R_L]v_o/R_L$. Note that $\alpha = x/L$, is the fractional position of the light spot. The current balance at the junction of the three resistors in Figure 8.16(b) is

$$\frac{v_{ref} - [(1-\alpha)R_c + R_L]v_o/R_L}{\alpha R_c} = \frac{v_o}{R_L} + \frac{[(1-\alpha)R_c + R_L]v_o/R_L}{R_p}$$

which can be written as

$$\frac{v_o}{v_{ref}}\left\{\frac{R_c}{R_L} + 1 + \frac{x}{L}\frac{R_c}{R_p}\left[\left(1-\frac{x}{L}\right)\frac{R_c}{R_L} + 1\right]\right\} = 1 \qquad (8.36)$$

When the load resistance R_L is quite large in comparison to the element resistance R_c, then $R_c/R_L \simeq 0$. Hence, equation (8.36) becomes

$$\frac{v_o}{v_{ref}} = \frac{1}{\left[\dfrac{x}{L}\dfrac{R_c}{R_p} + 1\right]} \qquad (8.37)$$

This relationship is still nonlinear in v_o/v_{ref} vs. x/L. The nonlinearity decreases, however, with decreasing R_c/R_p.

8.4.2 Variable-Inductance Transducers

Motion transducers that employ the principle of electromagnetic induction are termed *variable-inductance transducers*. When the flux linkage (defined as magnetic flux density times the number of turns in the conductor) through an electrical conductor changes, a voltage is induced in the conductor. This, in turn, generates a magnetic field that opposes the primary field. Hence, a mechanical force is necessary to sustain the change of flux linkage. If the change in flux linkage is brought about by a relative motion, the mechanical energy is directly converted (induced) into electrical energy. This is the basis of electromagnetic induction, and it is the principle of operation of electrical generators and variable-inductance transducers. Note that in these devices, the change of flux linkage is caused by a mechanical motion, and mechanical-to-electrical energy transfer takes place under near-ideal conditions. The induced voltage or change in inductance can be used as a measure of the motion. Variable-inductance transducers are generally electromechanical devices coupled by a magnetic field.

There are many different types of variable-inductance transducers. Three primary types can be identified:

1. Mutual-induction transducers
2. Self-induction transducers
3. Permanent-magnet transducers.

Variable-inductance transducers that use a nonmagnetized ferromagnetic medium to alter the reluctance (magnetic resistance) of the flux path are known as *variable-reluctance transducers*. Some of the mutual-induction transducers and most of the self-induction transducers are of this type. Permanent-magnet transducers do not fall into the category of variable-reluctance transducers.

Mutual-Induction Transducers

The basic arrangement of a mutual-induction transducer constitutes two coils: the *primary windings* and the *secondary windings*. One of the coils (primary windings) carries an AC excitation that induces a steady AC voltage in the other coil (secondary windings). The level (amplitude, rms value, etc.) of the induced voltage depends on the flux linkage between the coils. In mutual-induction transducers, a change in the flux linkage is effected by one of two common techniques. One technique is to move an object made of ferromagnetic material within the flux path. This changes the reluctance of the flux path, with an associated change of the flux linkage in the secondary coil. This is the operating principle of the linear-variable differential transformer (LVDT), the rotatory-variable differential transformer (RVDT), and the mutual-induction proximity probe. All of these are, in fact, variable-reluctance transducers. The other common way to change the flux linkage is to move one coil with respect to the other. This is the operating principle of the resolver, the synchro-transformer, and some types of AC tachometers. These are not variable-reluctance transducers, however.

The motion can be measured using the secondary signal in several ways. For example, the AC signal in the secondary windings can be demodulated by rejecting the *carrier frequency* (primary-winding excitation frequency) and directly measuring the resulting signal, which represents the motion. This method is particularly suitable for measuring transient motions. Alternatively, the amplitude or the rms (root-mean-square) value of the secondary (induced) voltage can be measured. Yet another method is to measure the change of inductance in the secondary circuit directly, using a device such as an inductance bridge circuit.

Linear-Variable Differential Transformer (LVDT)

The LVDT is a displacement (vibration) measuring device that can overcome most of the shortcomings of the potentiometer. It is considered a passive transducer because the measured displacement provides energy for "changing" the induced voltage, although an external power supply is used to energize the primary coil, which in turn induces a steady carrier voltage in the secondary coil. The LVDT is a variable-reluctance transducer of the mutual-induction type. In its simplest form, the LVDT consists of a cylindrical, insulating, nonmagnetic form that has a primary coil in the midsegment and a secondary coil symmetrically wound in the two end segments, as depicted schematically in Figure 8.17(a). The primary coil is energized by an AC supply voltage v_{ref}. This will generate, by mutual induction, an AC of the same frequency in the secondary winding. A core made of ferromagnetic material is inserted coaxially into the cylindrical form without actually touching it, as shown. As the core moves, the reluctance of the flux path changes.

Hence, the degree of flux linkage depends on the axial position of the core. Because the two secondary coils are connected in series opposition, so that the potentials induced in these two coil segments oppose each other, the net induced voltage is zero when the core is centered between the two secondary winding segments. This is known as the *null position*. When the core is displaced from this position, a non-zero induced voltage will be generated. At steady state, the amplitude v_o of this induced voltage is proportional, in the linear (operating) region, to the core displacement x. Consequently, v_o can be used as a measure of the displacement.

Note: Because of opposed secondary windings, the LVDT provides the direction as well as the magnitude of displacement. If the output signal is not demodulated, the direction is determined by the phase angle between the primary (reference) voltage and the secondary (output) voltage, which include the carrier signal as well.

For an LVDT to measure transient motions accurately, the frequency of the reference voltage (the carrier frequency) must be about ten times larger than the largest significant frequency component in the measured motion. For quasi-dynamic displacements and slow transients on the order of a few hertz, a standard AC supply (at 60-Hz line frequency) is adequate. The performance (particularly sensitivity and accuracy) is known to improve with the excitation frequency, however.

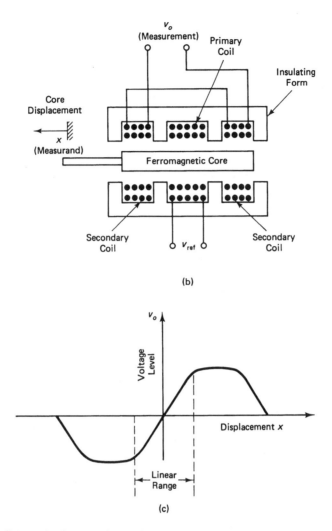

FIGURE 8.17 (a) Schematic diagram of an LVDT, and (b) a typical operating curve.

Because the amplitude of the output signal is proportional to the amplitude of the primary signal, the reference voltage should be regulated to get accurate results. In particular, the power source should have a low output impedance.

The output signal from a differential transformer is normally not in phase with the reference voltage. Inductance in the primary windings and the leakage inductance in the secondary windings are mainly responsible for this phase shift. Because demodulation involves extraction of the modulating signal by rejecting the carrier frequency component from the secondary signal (see Chapter 9), it is important to understand the size of this phase shift. An error known as *null voltage* is present in some differential transformers. This manifests itself as a non-zero reading at the null position (i.e., at zero displacement). This is usually 90° out of phase from the main output signal and, hence, is known as *quadrature error*. Nonuniformities in the windings (unequal impedances in the two segments of the secondary windings) are a major reason for this error. The null voltage can also result from harmonic noise components in the primary signal and nonlinearities in the device. Null voltage is usually negligible (typically about 0.1% of full scale). This error can be eliminated from the measurements by employing appropriate signal-conditioning and calibration practices.

Signal Conditioning

Signal conditioning associated with differential transformers includes filtering and amplification. Filtering is needed to improve the signal-to-noise ratio of the output signal. Amplification is necessary to increase the signal strength for data acquisition and processing. Because the reference frequency (carrier frequency) is embedded in the output signal, it is also necessary to interpret the output signal properly, particularly for transient motions. Two methods are commonly used to interpret the amplitude-modulated output signal from a differential transformer: (1) rectification and (2) demodulation.

In the first method (*rectification*), the AC output from the differential transformer is rectified to obtain a DC signal. This signal is amplified and then low-pass filtered to eliminate any high-frequency noise components. The amplitude of the resulting signal provides the transducer reading. In this method, the phase shift in the LVDT output must be checked separately to determine the direction of motion. In the second method (*demodulation*), the carrier frequency component is rejected from the output signal by comparing it with a phase-shifted and amplitude-adjusted version of the primary (reference) signal. Note that phase shifting is necessary because the output signal is not in phase with the reference signal. The modulating signal that is extracted in this manner is subsequently amplified and filtered. As a result of advances in miniature integrated circuit (LSI and VLSI) technology, differential transformers with built-in microelectronics for signal conditioning are commonly available today. DC differential transformers have built-in oscillator circuits to generate the carrier signal powered by a DC supply. The supply voltage is usually on the order of 25 V, and the output voltage is about 5 V. The demodulation approach of signal conditioning for an LVDT is now illustrated, using an example.

EXAMPLE 8.2

Figure 8.18 shows a schematic diagram of a simplified signal conditioning system for an LVDT. The system variables and parameters are as indicated in Figure 8.18. In particular,

$u(t)$ = displacement of the LVDT core (to be measured)

ω_c = frequency of the carrier voltage

v_o = output signal of the system (measurement).

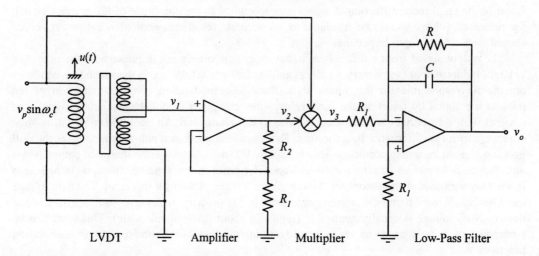

FIGURE 8.18 Signal conditioning system for an LVDT.

Vibration Instrumentation

The resistances R_1, R_2, and R, and the capacitance C are as marked. In addition, one can introduce a transformer parameter r for the LVDT, as required.

1. Explain the functions of the various components of the system shown in Figure 8.18.
2. Write equations for the amplifier and filter circuits and, using them, give expressions for the voltage signals v_1, v_2, v_3, and v_o marked in Figure 8.18. Note that the excitation in the primary coil is $v_p \sin\omega_c t$.
3. Suppose that the carrier frequency is $\omega_c = 500$ rad·s^{-1} and the filter resistance $R = 100$ kΩ. If no more than 5% of the carrier component should pass through the filter, estimate the required value of the filter capacitance C. Also, what is the useful frequency range (measurement bandwidth) of the system, in radians per second, with these parameter values?

SOLUTION

1. The LVDT has a primary coil that is excited by an AC voltage of $v_p \sin\omega_c t$. The ferromagnetic core is attached to the moving object whose displacement $x(t)$ is to be measured. The two secondary coils are connected in *series opposition* so that the LVDT output is zero at the null position, and that the direction of motion can be detected as well. The amplifier is a non-inverting type (see Chapter 9). It amplifies the output of the LVDT, which is an AC (carrier) signal of frequency ω_c that is modulated by the core displacement $x(t)$.

    The multiplier circuit determines the product of the primary (carrier) signal and the secondary (LVDT output) signal. This is an important step in demodulating the LVDT output.

    The product signal from the multiplier has a high-frequency ($2\omega_c$) carrier component, added to the modulating component ($x(t)$). The low-pass filter removes this unnecessary high-frequency component, to obtain the demodulated signal, which is proportional to the core displacement $x(t)$.

2. Non-Inverting Amplifier: Note that the potentials at the + and − terminals of the op-amp are nearly equal. Also, currents through these leads are nearly zero. (These are the two common assumptions used for an op-amp, as discussed in Chapter 9). Then, the current balance at node A gives

$$\frac{v_2 - v_1}{R_2} = \frac{v_1}{R_1}$$

or

$$v_2 = \frac{R_1 + R_2}{R_1} v_1$$

Then,

$$v_2 = k v_1 \qquad (i)$$

where

$$k = \frac{R_1 + R_2}{R_1} = \text{amplifier gain}.$$

Low-Pass Filter: Since the + lead of the op-amp has approximately a zero potential (ground), the voltage at point B is also approximately zero. The current balance for node B gives

$$\frac{v_3}{R_1} + \frac{v_o}{R} + C\dot{v}_o = 0$$

Hence,

$$\tau \frac{dv_o}{dt} + v_o = -\frac{R}{R_1} v_3 \qquad (ii)$$

where $\tau = RC$ = filter time constant. The transfer function of the filter is

$$\frac{v_o}{v_3} = -\frac{k_o}{(1+\tau s)} \qquad (iii)$$

with the filter gain $k_o = R/R_1$. In the frequency domain,

$$\frac{v_o}{v_3} = -\frac{k_o}{(1+\tau j\omega)} \qquad (8.38)$$

Finally, neglecting the phase shift in the LVDT, one obtains

$$v_1 = v_p r\, u(t) \sin \omega_c t$$
$$v_2 = v_p rk\, u(t) \sin \omega_c t$$
$$v_3 = v_p^2 rk\, u(t) \sin^2 \omega_c t$$

or

$$v_3 = \frac{v_p^2 rk}{2} u(t)[1 - \cos 2\omega_c t] \qquad (iv)$$

Due to the low-pass filter, with an appropriate cutoff frequency, the carrier signal will be filtered out. Then,

$$v_o = \frac{v_p^2 rk_o}{2} u(t) \qquad (8.39)$$

3. Filter magnitude $= \dfrac{k_o}{\sqrt{1+\tau^2 \omega^2}}$. For no more than 5% of the carrier ($2\omega_c$) component to pass through, one must have

$$\frac{5}{100}k_o \geq \frac{k}{\sqrt{1+\tau^2\omega^2}} \quad \text{for } \omega = 2\omega_c \qquad (v)$$

or $400 \leq 1 + 4\tau^2\omega_c^2$; or $\tau^2\omega_c^2 \geq \frac{399}{4}$; or $\tau\omega_c \geq 10$ (approximately). Pick $\tau\omega_c = 10$. With $R = 100$ kΩ and $\omega_c = 500$ rad·s^{-1}, then $C \times 100 \times 10^3 \times 500 = 10$; hence, $C = 0.2$ μF.

According to the carrier frequency (500 rad·s^{-1}), one should be able to measure displacements $u(t)$ up to about 50 rad·s^{-1}. But the flat region of the filter is up to about $\omega\tau = 0.1$, which with the present value of $\tau = 0.02$ s, gives a bandwidth of only 5 rad·s^{-1}.

□

Advantages of the LVDT include the following:

1. It is essentially a noncontacting device with no frictional resistance. Near-ideal electromechanical energy conversion and lightweight core will result in very small resistive mechanical forces. Hysteresis (both magnetic hysteresis and mechanical backlash) is negligible.
2. It has low output impedance, typically on the order of 100 Ω. (Signal amplification is usually not needed.)
3. Directional measurements (positive/negative) are obtained.
4. It is available in small sizes (e.g., 1 cm long with maximum travel of 2 mm).
5. It has a simple and robust construction (less expensive and durable).
6. Fine resolutions are possible (theoretically, infinitesimal resolution; practically, much better than that of a coil potentiometer).

The *rotatory-variable differential transformer* (RVDT) operates using the same principle as the LVDT, except that in an RVDT, a rotating ferromagnetic core is used. The RVDT is used for measuring angular displacements. The rotating core is shaped such that a reasonably wide linear operating region is obtained. Advantages of the RVDT are essentially the same as those cited for the LVDT. The linear range is typically ±40° with a nonlinearity error less than 1%.

In variable-inductance devices, the induced voltage is generated through the rate of change of the magnetic flux linkage. Therefore, displacement readings are distorted by velocity and, similarly, velocity readings are affected by acceleration. For the same displacement value, the transducer reading will depend on the velocity at that displacement. This error is known to increase with the ratio: (cyclic velocity of the core)/(carrier frequency). Hence, these *rate errors* can be reduced by increasing the carrier frequency. The reason for this follows.

At high carrier frequencies, the induced voltage due to the transformer effect (frequencies of the primary signal) is greater than the induced voltage due to the rate (velocity) effect of the moving member. Hence, the error will be small. To estimate a lower limit for the carrier frequency in order to reduce rate effects, one can proceed as follows:

1. For LVDT: $\dfrac{\text{Max. speed of operation}}{\text{Stroke of LVDT}} = \omega_o$

   Then the excitation frequency of the primary coil should be chosen as at least $5\omega_o$.
2. For RVDT: for ω_o, use the maximum angular frequency of operation (of the rotor).

8.4.3 Mutual-Induction Proximity Sensor

This displacement transducer operates on the mutual-induction principle. A simplified schematic diagram of such a device is shown in Figure 8.19(a). The insulated "E core" carries the primary windings in its middle limb. The two end limbs carry secondary windings that are connected in series. Unlike the LVDT and the RVDT, the two voltages induced in the secondary winding segments are additive in this case. The region of the moving surface (target object) that faces the coils must be made of ferromagnetic material so that as it moves, the magnetic reluctance and the flux linkage will change. This, in turn, changes the induced voltage in the secondary windings, and this change is a measure of the displacement. Note that, unlike the LVDT, which has an "axial" displacement configuration, the proximity probe has a "transverse" displacement configuration. Hence, it is particularly suitable for measuring transverse displacements or proximities of moving objects (e.g., transverse vibrations of a beam or whirling of a rotating shaft). One can see from the operating curve shown in Figure 8.19(b) that the displacement–voltage relation of a proximity probe is nonlinear. Hence, these proximity sensors should be used only for measuring small displacements, such as linear vibrations (e.g., a linear range of 5.0 mm or 0.2 in.), unless accurate nonlinear calibration curves are available. Because the proximity sensor is a noncontacting device, mechanical loading is small and the product life is high. Because a ferromagnetic object is used to alter the reluctance of the flux path, the mutual-induction proximity sensor is a variable-reluctance device. The operating frequency limit is about one tenth the excitation frequency of the primary coil (carrier frequency). As for an LVDT, demodulation of the induced voltage (secondary) would be required to obtain direct (DC) output readings.

8.4.4 Self-Induction Transducers

These transducers are based on the principle of self-induction. Unlike mutual-induction transducers, only a single coil is employed. This coil is activated by an AC supply voltage v_{ref}. The current produces a magnetic flux, which is linked with the coil. The level of flux linkage (or self-inductance) can be varied by moving a ferromagnetic object within the magnetic field.

This changes the reluctance of the flux path and the inductance in the coil. This change is a measure of the displacement of the ferromagnetic object. The change in inductance is measured using an inductance measuring circuit (e.g., an inductance bridge). Note that self-induction transducers are usually variable-reluctance devices.

A typical self-induction transducer is a self-induction proximity sensor. A schematic diagram of this device is shown in Figure 8.20. This device can be used as a displacement or vibration sensor for transverse displacements. For example, the distance between the sensor tip and ferromagnetic surface of a moving object, such as a beam or shaft, can be measured. Applications are essentially the same as those for mutual-induction proximity sensors. High-speed displacement (vibration) measurements can result in velocity error (rate error) when variable-inductance displacement sensors (including self-induction transducers) are used. This effect can be reduced, as in other AC-powered variable-inductance sensors, by increasing the carrier frequency.

8.4.5 Permanent-Magnet Transducers

In discussing this third type of variable-inductance transducer, first consider the permanent-magnet DC velocity sensors (DC tachometers). A distinctive feature of permanent-magnet transducers is that they have a permanent magnet to generate a uniform and steady magnetic field. A relative motion between the magnetic field and an electrical conductor induces a voltage that is proportional to the speed at which the conductor crosses the magnetic field. In some designs, a unidirectional magnetic field generated by a DC supply (i.e., an electromagnet) is used in place of a permanent magnet. Nevertheless, this is generally termed a *permanent-magnet transducer*.

Vibration Instrumentation

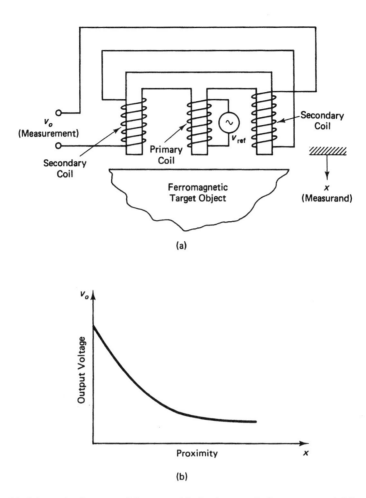

FIGURE 8.19 (a) Schematic diagram of the mutual-induction proximity sensor, and (b) operating curve.

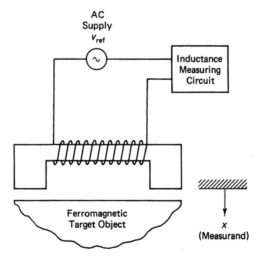

FIGURE 8.20 Schematic diagram of a self-induction proximity sensor.

The principle of electromagnetic induction between a permanent magnet and a conducting coil is used in speed measurement by permanent-magnet transducers. Depending on the configuration, either rectilinear speeds or angular speeds can be measured. Schematic diagrams of the two configurations are shown in Figure 8.21. Note that these are passive transducers because the energy for the output signal v_o is derived from the motion (measured signal) itself. The entire device is usually enclosed in a steel casing to isolate it from ambient magnetic fields.

In the rectilinear velocity transducer [Figure 8.21(a)], the conductor coil is wrapped on a core and placed centrally between two magnetic poles that produce a cross-magnetic field. The core is attached to the moving object whose velocity must be measured. The velocity v is proportional to the induced voltage v_o. An alternative design — a moving-magnet and fixed-coil arrangement — can be used as well, thus eliminating the need for any sliding contacts (slip rings and brushes) for the output leads, and thereby reducing mechanical loading error, wearout, and related problems. The tachogenerator (or tachometer) is a very common permanent-magnet device. The principle of operation of a DC tachogenerator is shown in Figure 8.21(b). The rotor is directly connected to the rotating object. The output signal that is induced in the rotating coil is picked up as DC voltage v_o using a suitable *commutator* device — typically consisting of a pair of low-resistance carbon brushes — that is stationary but makes contact with the rotating coil through split slip rings so as to maintain the positive direction of induced voltage throughout each revolution. The induced voltage is given by

$$v_o = (2nhr\beta)\omega_c \qquad (8.40)$$

for a coil of height h and width $2r$ that has n turns, moving at an angular speed ω_c in a uniform magnetic field of flux density β. This proportionality between v_o and ω_c is used to measure the angular speed ω_c.

When tachometers are used to measure transient velocities, some error will result from the rate (acceleration) effect. This error generally increases with the maximum significant frequency that must be retained in the transient velocity signal. Output distortion can also result because of reactive (inductive and capacitive) loading of the tachometer. Both types of error can be reduced by increasing the load impedance.

For illustration, consider the equivalent circuit of a tachometer with an impedance load, as shown in Figure 8.22. The induced voltage $k\omega_c$ is represented by a voltage source. Note that the constant k depends on the coil geometry, the number of turns, and the magnetic flux density [see equation (8.40)]. Coil resistance is denoted by R, and leakage inductance is denoted by L_ℓ. The load impedance is Z_L. From straightforward circuit analysis in the frequency domain, the output voltage at the load is given by

$$v_o = \left[\frac{Z_L}{R + j\omega L_\ell + Z_L}\right] k\omega_c \qquad (8.41)$$

It can be seen that because of the leakage inductance, the output signal attenuates more at higher frequencies ω of the velocity transient. In addition, loading error is present. If Z_L is much larger than the coil impedance, however, the ideal proportionality, as given by

$$v_o = k\omega_c \qquad (8.42)$$

is achieved.

Some tachometers operate in a different manner. For example, *digital tachometers* generate voltage pulses at a frequency proportional to the angular speed. These are considered as digital transducers.

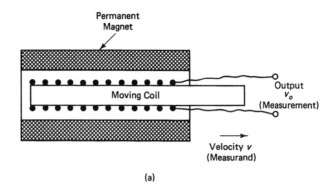

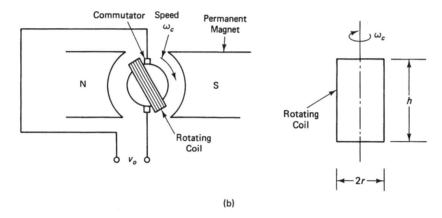

FIGURE 8.21 Permanent-magnet transducers: (a) rectilinear velocity transducer, and (b) DC tachometer generator.

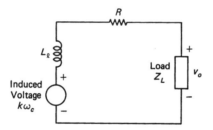

FIGURE 8.22 Equivalent circuit for a tachometer with an impedance load.

8.4.6 AC Permanent-Magnet Tachometer

This device has a permanent magnet rotor and two separate sets of stator windings, as schematically shown in Figure 8.23(a). One set of windings is energized using an AC reference voltage. Induced voltage in the other set of windings is the tachometer output. When the rotor is stationary or moving in a quasi-static manner, the output voltage is a constant-amplitude signal much like the reference voltage. As the rotor moves at a finite speed, an additional induced voltage that is proportional to the rotor speed is generated in the secondary windings. This is due to the rate of change of flux linkage from the magnet in the secondary coil. The net output is an amplitude-modulated signal whose amplitude is proportional to the rotor speed. For transient velocities, it will be necessary to demodulate this signal in order to extract the transient velocity signal (i.e., the modulating signal)

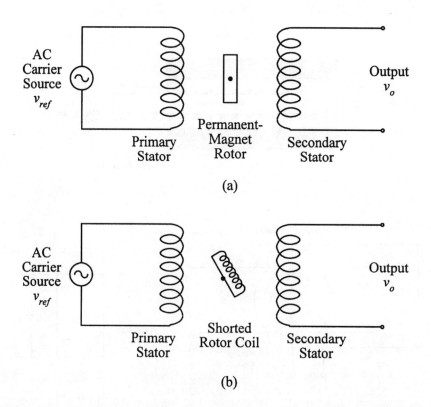

FIGURE 8.23 (a) AC Permanent-magnet tachometer, and (b) AC induction tachometer.

from the modulated output, as described in Chapter 9. The direction of velocity is determined from the phase angle of the modulated signal with respect to the carrier signal. Note that in an LVDT, the amplitude of the AC magnetic flux is altered by the position of the ferromagnetic core; but in an AC permanent-magnet tachometer, the DC magnetic flux generated by the magnetic rotor is linked with the stator windings, and the associated induced voltage is caused by the speed of rotation of the rotor.

For low-frequency applications (5 Hz or less), a standard AC supply (60 Hz) can be used to power an AC tachometer. For moderate-frequency applications, a 400-Hz supply is widely used. Typical sensitivity of an AC permanent-magnet tachometer is on the order of 50 to 100 mV·rad^{-1}·s^{-1}.

8.4.7 AC Induction Tachometer

These tachometers are similar in construction to the two-phase induction motors. The stator arrangement is identical to that of the AC permanent-magnet tachometer. The rotor, however, has windings that are shorted and not energized by an external source, as shown in Figure 8.23(b). One set of stator windings is energized with an AC supply. This induces a voltage in the rotor windings, and it has two components. One component is due to the direct transformer action of the supply AC. The other component is induced by the speed of rotation of the rotor, and its magnitude is proportional to the speed of rotation. The nonenergized stator windings provide the output of the tachometer. Voltage induced in the output stator windings is due to both the primary stator windings and the rotor windings. As a result, the tachometer output has a carrier AC component and a modulating component that is proportional to the speed of rotation. Demodulation would be needed to extract the output component that is proportional to the angular speed of the rotor.

The main advantage of AC tachometers over their DC counterparts is the absence of slip-ring and brush devices. In particular, the signal from a DC tachometer usually has a voltage ripple,

Vibration Instrumentation

known as commutator ripple, which is generated as the split segments of the slip ring pass over the brushes. The frequency of the commutator ripple depends on the speed of operation; consequently, filtering it out using a notch filter is difficult (a tunable notch filter would be necessary). Also, there are problems with frictional loading and contact bounce in DC tachometers, and these problems are absent in AC tachometers. It is known, however, that the output from an AC tachometer is somewhat nonlinear (saturation effect) at high speeds. Furthermore, for measuring transient speeds, signal demodulation would be necessary. Another disadvantage of AC tachometers is that the output signal level depends on the supply voltage; hence, a stabilized voltage source that has a very small output impedance is necessary for accurate measurements.

8.4.8 Eddy Current Transducers

If a conducting (i.e., low-resistivity) medium is subjected to a fluctuating magnetic field, eddy currents are generated in the medium. The strength of eddy currents increases with the strength of the magnetic field and the frequency of the magnetic flux. This principle is used in eddy current proximity sensors. Eddy current sensors can be used as either dimensional gaging devices or high-frequency vibration sensors.

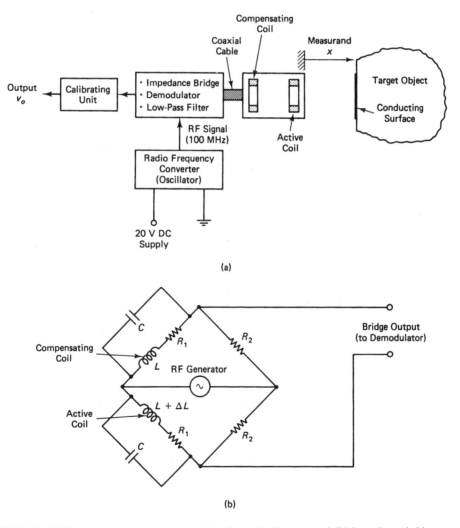

FIGURE 8.24 Eddy current proximity sensor: (a) schematic diagram, and (b) impedance bridge.

A schematic diagram of an eddy current proximity sensor is shown in Figure 8.24(a). Unlike variable-inductance proximity sensors, the target object of the eddy current sensor does not have to be made of ferromagnetic material. A conducting target object is needed, but a thin film conducting material — such as household aluminum foil glued onto a nonconducting target object — would be adequate. The probe head has two identical coils, which will form two arms of an impedance bridge. The coil closer to the probe face is the *active coil*. The other coil is the *compensating coil*. It compensates for ambient changes, particularly thermal effects. The other two arms of the bridge will consist of purely resistive elements [see Figure 8.24(b)]. The bridge is excited by a radiofrequency voltage supply. The frequency can range from 1 MHz to 100 MHz. This signal is generated from a radiofrequency converter (an oscillator) that is typically powered by a 20-VDC supply. In the absence of the target object, the output of the impedance bridge is zero, which corresponds to the balanced condition. When the target object is moved close to the sensor, eddy currents are generated in the conducting medium because of the radiofrequency magnetic flux from the active coil. The magnetic field of the eddy currents opposes the primary field that generates these currents. Hence, the inductance of the active coil increases, creating an imbalance in the bridge. The resulting output from the bridge is an amplitude-modulated signal containing the radiofrequency carrier. This signal is demodulated by removing the carrier. The resulting signal (modulating signal) measures the transient displacement (vibration) of the target object. Low-pass filtering is used to remove the high-frequency leftover noise in the output signal once the carrier is removed. For large displacements, the output is not linearly related to the displacement. Furthermore, the sensitivity of the eddy current probe depends nonlinearly on the nature of the conducting medium, particularly the resistivity. For example, for low resistivities, sensitivity increases with resistivity; for high resistivities, sensitivity decreases with resistivity. A calibrating unit is usually available with commercial eddy current sensors to accommodate various target objects and nonlinearities. The gage factor is usually expressed in volts per millimeter. Note that eddy current probes can also be used to measure resistivity and surface hardness (which affects resistivity) in metals.

The facial area of the conducting medium on the target object has to be slightly larger than the frontal area of the eddy current probe head. If the target object has a curved surface, its radius of curvature has to be at least four times the diameter of the probe. These are not serious restrictions because the typical diameter of the probe head is about 2 mm. Eddy current sensors are medium-impedance devices; 1000 Ω output impedance is typical. Sensitivity is on the order of 5 V·m/m. Since the carrier frequency is very high, eddy current devices are suitable for highly transient vibration measurements — for example, bandwidths up to 100 kHz. Another advantage of an eddy current sensor is that it is a noncontacting device; there is no mechanical loading on the moving (target) object.

8.4.9 Variable-Capacitance Transducers

Capacitive or *reactive* transducers are commonly used to measure small transverse displacements such as vibrations, large rotations, and fluid level oscillations. They can also be employed to measure angular velocities. In addition to analog capacitive sensors, digital (pulse-generating) capacitive tachometers are also available.

Capacitance C of a two-plate capacitor is given by

$$C = \frac{kA}{x} \tag{8.43}$$

where A is the common (overlapping) area of the two plates, x is the gap width between the two plates, and k is the *dielectric constant*, which depends on dielectric properties of the medium between the two plates. A change in any one of these three parameters can be used in the sensing process. Schematic diagrams for measuring devices that use this feature are shown in Figure 8.25.

In Figure 8.25(a), angular displacement of one of the plates causes a change in A. In Figure 8.25(b), a transverse displacement of one of the plates changes x. Finally, in Figure 8.25(c), a change in k is produced as the fluid level between the capacitor plates changes. Liquid oscillations can be sensed in this manner. In all three cases, the associated change in capacitance is measured directly or indirectly, and is used to estimate the measurand. A popular method is to use a capacitance bridge circuit to measure the change in capacitance, in a manner similar to how an inductance bridge is used to measure changes in inductance. Other methods include measuring a change in such quantities as charge (using a charge amplifier), voltage (using a high input-impedance device in parallel), and current (using a very low impedance device in series) that will result from the change in capacitance in a suitable circuit. An alternative method is to make the capacitor a part of an inductance-capacitance (L-C) oscillator circuit; the natural frequency of the oscillator $\left(1/\sqrt{LC}\right)$ measures the capacitance. (Incidentally, this method can also be used to measure inductance.)

Capacitive Displacement Sensors

For the arrangement shown in Figure 8.25(a), since the common area A is proportional to the angle of rotation θ, equation (8.43) can be written as

$$C = K\theta \qquad (8.44)$$

where K is a sensor constant. This is a linear relationship between C and θ. The capacitance can be measured by any convenient method. The sensor is linearly calibrated to give the angle of rotation.

For the arrangement shown in Figure 8.25(b), the sensor relationship is

$$C = \frac{K}{x} \qquad (8.45)$$

The constant K has a different meaning here. Note that equation (8.45) is a nonlinear relationship. A simple way to linearize this transverse displacement sensor is to use an inverting amplifier, as shown in Figure 8.26. Note that C_{ref} is a fixed reference capacitance. Because the gain of the operational amplifier is very high, the voltage at the negative lead (point A) is zero for most practical purposes (because the positive lead is grounded). Furthermore, because the input impedance of the op-amp is also very high, the current through the input leads is negligible. These are the two common assumptions used in op-amp analysis. Accordingly, the charge balance equation for node point A is

$$v_{ref} C_{ref} + v_o C = 0$$

Now, in view of equation (8.45), one obtains the following linear relationship for the output voltage v_o in terms of the displacement x:

$$v_o = -\frac{v_{ref} C_{ref}}{K} x \qquad (8.46)$$

Hence, measurement of v_o gives the displacement through linear calibration. The sensitivity of the device can be increased by increasing v_{ref} and C_{ref}. The reference voltage could be DC as well as AC. With an AC reference voltage, the output voltage is a modulated signal that has to be demodulated to measure transient displacements.

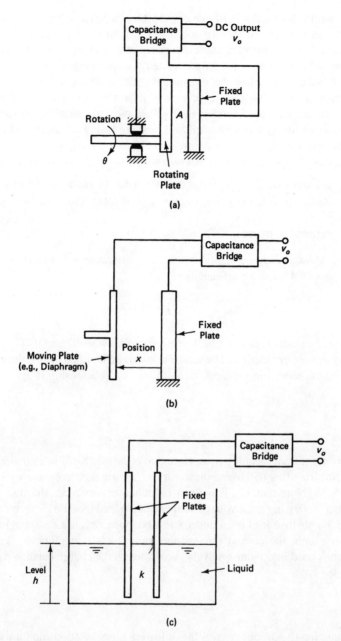

FIGURE 8.25 Schematic diagrams of capacitive sensors: (a) capacitive rotation sensor; (b) capacitive displacement sensor; and (c) capacitive liquid oscillation sensor.

Capacitive Angular Velocity Sensor

The schematic diagram for an angular velocity sensor that uses a rotating-plate capacitor is shown in Figure 8.27. Because the current sensor has negligible resistance, the voltage across the capacitor is almost equal to the supply voltage v_{ref}, which is constant. It follows that the current in the circuit is given by

$$i = \frac{d}{dt}\left(Cv_{ref}\right) = v_{ref}\frac{dC}{dt}$$

Vibration Instrumentation

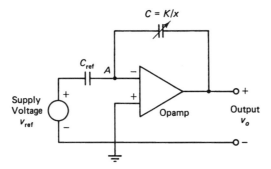

FIGURE 8.26 Inverting amplifier circuit used to linearize the capacitive transverse displacement sensor.

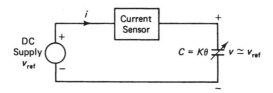

FIGURE 8.27 Rotating-plate capacitive angular velocity sensor.

which, in view of equation (8.44), can be expressed as

$$\frac{d\theta}{dt} = \frac{i}{Kv_{ref}} \qquad (8.47)$$

This is a linear relationship for the angular velocity in terms of the measured current i. Care must be exercised to guarantee that the current-measuring device does not interfere with the basic circuit.

An advantage of capacitance transducers is that because they are noncontacting devices, mechanical loading effects are negligible. There is some loading due to inertial and frictional resistance in the moving plate. This can be eliminated by using the moving object itself to function as the moving plate. Variations in the dielectric properties due to humidity, temperature, pressure, and impurities introduce errors. A capacitance bridge circuit can compensate for these effects. Extraneous capacitances, such as cable capacitance, can produce erroneous readings in capacitive sensors. This problem can be reduced using a charge amplifier to condition the sensor signal. Another drawback of capacitance displacement sensors is low sensitivity. For a transverse displacement transducer, the sensitivity is typically less than one picofarad (pF) per millimeter (1 pF = 10^{-12} F). This problem is not serious because high supply voltages and amplifier circuitry can be used to increase the sensor sensitivity.

Capacitance Bridge Circuit

Sensors that are based on the change in capacitance (reactance) will require some means of measuring that change. Furthermore, changes in capacitance that are not caused by a change in the measurand — for example, due to change in humidity, temperature, etc. — will cause errors and should be compensated for. Both these goals are accomplished using a capacitance bridge circuit. An example is shown in Figure 8.28. In this circuit,

Z_2 = reactance (i.e., capacitive impedance) of the capacitive sensor (of capacitance C_2)

$$= \frac{1}{j\omega C_2}$$

Z_1 = reactance of the compensating capacitor C_1

$$= \frac{1}{j\omega C_1}$$

Z_4, Z_3 = bridge completing impedances (typically reactances)
v_{ref} = excitation AC voltage
$= v_a \sin \omega t$
v_o = bridge output
$= v_b \sin(\omega t - \phi)$
ϕ = phase lag of the output with respect to the excitation.

Using the two assumptions for the op-amp (potentials at the negative and positive leads are equal and the current through these leads is zero), as discussed in Chapter 9, one can write the current balance equations:

$$\frac{v_{ref} - v}{Z_1} + \frac{v_o - v}{Z_2} = 0 \tag{i}$$

$$\frac{v_{ref} - v}{Z_3} + \frac{0 - v}{Z_4} = 0 \tag{ii}$$

where v is the common voltage at the op-amp leads. Next, eliminate v in equations (i) and (ii) to obtain

$$v_o = \frac{(Z_4/Z_3 - Z_2/Z_1)}{1 + Z_4/Z_3} v_{ref} \tag{8.48}$$

It is noted that when

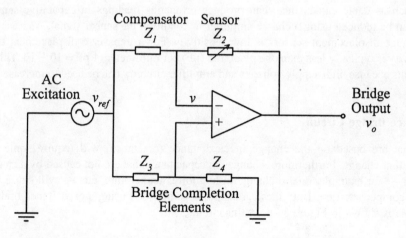

FIGURE 8.28 A bridge circuit for capacitive sensors.

$$\frac{Z_2}{Z_1} = \frac{Z_4}{Z_3} \tag{8.49}$$

the bridge output $v_o = 0$, and the bridge is said to be balanced. Because all capacitors in the bridge are similarly affected by ambient changes, a balanced bridge will maintain that condition even under ambient changes, unless the sensor reactance Z_2 is changed due to the measurand itself. It follows that the ambient effects are compensated for (at least up to the first order) by a bridge circuit. From equation (8.48), it is clear that the bridge output due to a sensor change of δZ, starting from a balanced state, is given by

$$\delta v_o = -\frac{v_{ref}}{Z_1\left(1 + Z_4/Z_3\right)} \delta Z \tag{8.50}$$

The amplitude and phase angle of δv_o with respect to v_{ref} will determine δZ, assuming that Z_1 and Z_4/Z_3 are known.

8.4.10 Piezoelectric Transducers

Some substances, such as barium titanate and single-crystal quartz, can generate an electrical charge and an associated potential difference when subjected to mechanical stress or strain. This *piezoelectric effect* is used in piezoelectric transducers. Direct application of the piezoelectric effect is found in pressure and strain measuring devices, and many indirect applications also exist. They include piezoelectric accelerometers and velocity sensors and piezoelectric torque sensors and force sensors. It is also interesting to note that piezoelectric materials deform when subjected to a potential difference (or charge). Some delicate test equipment (e.g., in vibration testing) use piezoelectric actuating elements (reverse piezoelectric action) to create fine motions. Also, piezoelectric valves (e.g., flapper valves), directly actuated using voltage signals, are used in pneumatic and hydraulic control applications and in ink-jet printers. Miniature stepper motors based on the reverse piezoelectric action are available.

Consider a piezoelectric crystal in the form of a disc with two electrodes plated on the two opposite faces. Because the crystal is a dielectric medium, this device is essentially a capacitor that can be modeled by a capacitance C, as in equation (8.43). Accordingly, a piezoelectric sensor can be represented as a charge source with a series capacitive impedance (Figure 8.29) in an equivalent circuit. The impedance from the capacitor is given by

$$Z = \frac{1}{j\omega C} \tag{8.51}$$

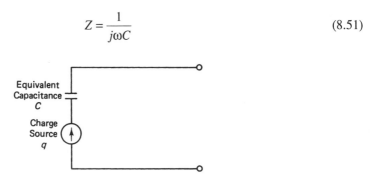

FIGURE 8.29 Equivalent circuit representation of a piezoelectric sensor.

As is clear from equation (8.51), the output impedance of piezoelectric sensors is very high, particularly at low frequencies. For example, a quartz crystal may present an impedance of several megohms at 100 Hz, increasing hyperbolically with decreasing frequencies. This is one reason why piezoelectric sensors have a limitation on the useful lower frequency. The other reason is the charge leakage.

Sensitivity

The sensitivity of a piezoelectric crystal can be represented either by its *charge sensitivity* or by its *voltage sensitivity*. Charge sensitivity is defined as

$$S_q = \frac{\partial q}{\partial F} \tag{8.52}$$

where q denotes the generated charge and F denotes the applied force. For a crystal with surface area A, equation (8.52) can be expressed as

$$S_q = \frac{1}{A}\frac{\partial q}{\partial p} \tag{8.53}$$

where p is the stress (normal or shear) or pressure applied to the crystal surface. Voltage sensitivity S_v is given by the change in voltage due to a unit increment in pressure (or stress) per unit thickness of the crystal. Thus, in the limit, one obtains

$$S_v = \frac{1}{d}\frac{\partial v}{\partial p} \tag{8.54}$$

where d denotes the crystal thickness. Now, because

$$\delta q = C\delta v \tag{8.55}$$

by using equation (8.43) for a capacitor element, the following relationship between charge sensitivity and voltage sensitivity is obtained:

$$S_q = kS_v \tag{8.56}$$

Note that k is the dielectric constant of the crystal capacitor, as defined by equation (8.43).

EXAMPLE 8.3

A barium titanate crystal has a charge sensitivity of 150.0 picocoulombs per newton (pC·N^{-1}). (*Note*: 1 pC = 1 × 10^{-12} coulombs; coulombs = farads × volts). The dielectric constant for the crystal is 1.25 × 10^{-8} farads per meter (F·m^{-1}). What is the voltage sensitivity of the crystal?

SOLUTION

The voltage sensitivity of the crystal is given by

$$S_v = \frac{150.0 \text{ pC} \cdot \text{N}^{-1}}{1.25 \times 10^{-8} \text{ F} \cdot \text{m}^{-1}} = \frac{150.0 \times 10^{-12} \text{ C} \cdot \text{N}^{-1}}{1.25 \times 10^{-8} \text{ F} \cdot \text{m}^{-1}}$$

or

$$S_v = 12.0 \times 10^{-3} \text{ V} \cdot \text{m} \cdot \text{N}^{-1} = 12.0 \text{ mV} \cdot \text{m} \cdot \text{N}^{-1}$$

□

The sensitivity of a piezolectric element is dependent on the direction of loading. This is because the sensitivity depends on the crystal axis. Sensitivities of several piezoelectric materials along their most sensitive crystal axis are listed in Table 8.2.

Piezoelectric Accelerometer

A more detailed discussion of a piezoelectric motion transducer or vibration sensor — the piezoelectric accelerometer — in more detail is now provided. A piezoelectric velocity transducer is simply a piezoelectric accelerometer with a built-in integrating amplifier in the form of a miniature integrated circuit.

Accelerometers are acceleration-measuring devices. It is known from Newton's second law that a force (*f*) is necessary to accelerate a mass (or inertia element), and its magnitude is given by the product of mass (*m*) and acceleration (*a*). This product (*ma*) is commonly termed *inertia force*. The rationale for this terminology is that if a force of magnitude *ma* were applied to the accelerating mass in the direction opposing the acceleration, then the system could be analyzed using static equilibrium considerations. This is known as *D'Alembert's principle*. The force that causes acceleration is itself a measure of the acceleration (mass is kept constant). Accordingly, mass can serve as a front-end element to convert acceleration into a force. This is the principle of operation of common accelerometers. There are many different types of accelerometers, ranging from strain gage devices to those that use electromagnetic induction. For example, force that causes acceleration can be converted into a proportional displacement using a spring element, and this displacement can be measured using a convenient displacement sensor. Examples of this type are differential-transformer accelerometers, potentiometer accelerometers, and variable-capacitance accelerometers. Alternatively, the strain, at a suitable location of a member that was deflected due to inertia force, can be determined using a strain-gage. This method is used in strain gage accelerometers. Vibrating-wire accelerometers use the accelerating force to tension a wire. The force is measured by detecting the natural frequency of vibration of the wire (which is proportional to the square root of tension). In servo force-balance (or null-balance) accelerometers, the inertia element is restrained from accelerating by detecting its motion and feeding back a force (or torque) to exactly cancel out the accelerating force (torque). This feedback force is determined, for example, by knowing the motor current, and it is a measure of the acceleration.

TABLE 8.2
Sensitivities of Several Piezoelectric Materials

Material	Charge Sensitivity S_q (pC·N^{-1})	Voltage Sensitivity S_v (mV·m·N^{-1})
Lead zirconate titanate (PZT)	110	10
Barium titanate	140	6
Quartz	2.5	50
Rochelle salt	275	90

The advantages of piezoelectric accelerometers (also known as *crystal accelerometers*) over other types of accelerometers are their light weight and high-frequency response (up to about 1 MHz). However, piezoelectric transducers are inherently high-output impedance devices that generate small voltages (on the order of 1 mV). For this reason, special impedance-transforming amplifiers (e.g., charge amplifiers) must be employed to condition the output signal and to reduce loading error.

A schematic diagram for a compression-type piezoelectric accelerometer is shown in Figure 8.30. The crystal and the inertia mass are restrained by a spring of very high stiffness. Consequently, the fundamental natural frequency or resonant frequency of the device becomes high (typically 20 kHz). This gives a reasonably wide, useful range (typically up to 5 kHz). The lower limit of the useful range (typically 1 Hz) is set by factors such as the limitations of the signal-conditioning systems, the mounting methods, the charge leakage in the piezoelectric element, the time constant of the charge-generating dynamics, and the signal-to-noise ratio. A typical frequency response curve for a piezoelectric accelerometer is shown in Figure 8.31.

In compression-type crystal accelerometers, the inertia force is sensed as a compressive normal stress in the piezoelectric element. There are also piezoelectric accelerometers that sense the inertia force as a shear strain or tensile strain. For an accelerometer, acceleration is the signal that is being measured (the measurand). Hence, accelerometer sensitivity is commonly expressed in terms of electrical charge per unit acceleration or voltage per unit acceleration (compare this with equations (8.53) and (8.54)). Acceleration is measured in units of acceleration due to gravity (g), and charge is measured in picocoulombs (pC), which are units of 10^{-12} coulombs (C). Typical accelerometer sensitivities are 10 pC·g^{-1} and 5 mV·g^{-1}. Sensitivity depends on the piezoelectric properties and on the mass of the inertia element. If a large mass is used, the reaction inertia force on the crystal will be large for a given acceleration, thus generating a relatively large output signal. A large accelerometer mass results in several disadvantages, however. In particular:

1. The accelerometer mass distorts the measured motion variable (mechanical loading effect).
2. A heavier accelerometer has a lower resonant frequency and, hence, a lower useful frequency range (Figure 8.31).

For a given accelerometer size, improved sensitivity can be obtained by using the shear-strain configuration. In this configuration, several shear layers can be used (e.g., in a *delta arrangement*) within the accelerometer housing, thereby increasing the effective shear area and, hence, the sensitivity in proportion to the shear area. Another factor that should be considered in selecting an accelerometer is its *cross-sensitivity* or transverse sensitivity. Cross-sensitivity primarily results from manufacturing irregularities of the piezoelectric element, such as material unevenness and incorrect orientation of the sensing element. Cross-sensitivity should be less than the maximum error (percentage) that is allowed for the device (typically 1%).

The technique employed to mount the accelerometer to an object can significantly affect the useful frequency range of the accelerometer. Some common mounting techniques include:

1. Screw-in base
2. Glue, cement, or wax
3. Magnetic base
4. Spring-base mount
5. Hand-held probe.

Drilling holes in the object can be avoided by using the second through fifth methods, but the useful range can decrease significantly when spring-base mounts or hand-held probes are used (typical upper limit of 500 Hz). The first two methods usually maintain the full useful range,

Vibration Instrumentation

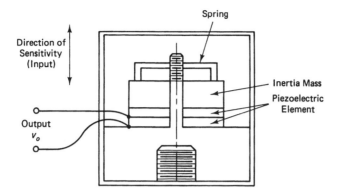

FIGURE 8.30 A compression-type piezoelectric accelerometer.

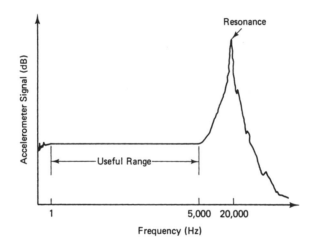

FIGURE 8.31 A typical frequency-response curve for a piezoelectric accelerometer.

whereas the magnetic attachment method reduces the upper frequency limit to some extent (typically 1.5 kHz).

Piezoelectric signals cannot be read using low-impedance devices. The two primary reasons for this are:

1. High output impedance in the sensor results in small output signal levels and large loading errors.
2. The charge can quickly leak out through the load.

Charge Amplifier

The charge amplifier, which has a very high input impedance and a very low output impedance, is the commonly used signal-conditioning device for piezoelectric sensors. Clearly, the impedance at the charge amplifier output is much smaller than the output impedance of the piezoelectric sensor. These impedance characteristics of a charge amplifier virtually eliminate loading error. Also, by using a charge amplifier circuit with a large time constant, charge leakage speed can be decreased. For example, consider a piezoelectric sensor and charge amplifier combination, as represented by the circuit in Figure 8.32. One can examine how the charge leakage rate is slowed down by using this arrangement. Sensor capacitance, feedback capacitance of the charge amplifier, and feedback

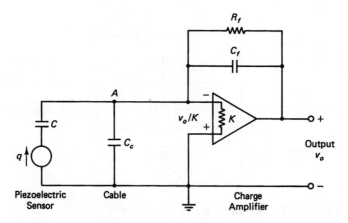

FIGURE 8.32 A peizoelectric sensor and charge amplifier combination.

resistance of the charge amplifier are denoted by C, C_f, and R_f, respectively. The capacitance of the cable that connects the sensor to the charge amplifier is denoted by C_c.

For an op-amp of gain K, the voltage at its negative (inverting) input is $-v_o/K$, where v_o is the voltage at the amplifier output. Note that the positive input of the op-amp is grounded (zero potential). Current balance at point A gives

$$\dot{q} + C_c \frac{\dot{v}_o}{K} + C_f \left(\dot{v}_o + \frac{\dot{v}_o}{K} \right) + \frac{v_o + v_o/K}{R_f} = 0 \tag{8.57}$$

Since gain K is very large (typically 10^5 to 10^9) compared to unity, this differential equation can be approximated to

$$R_f C_f \frac{dv_o}{dt} + v_o = -R_f \frac{dq}{dt} \tag{8.58}$$

Alternatively, instead of using equation (8.57), it is possible to directly obtain equation (8.58) from the two common assumptions (equal inverting and noninverting lead potentials and zero lead currents) for an op-amp. Then the potential at the negative (inverting) lead would be zero, as the positive lead is grounded. Also, as a result, the voltage across C_c would be zero. Hence, the current balance at point A gives

$$\dot{q} + \frac{v_o}{R_f} + C_f \dot{v}_o = 0$$

This is identical to equation (8.58). The corresponding transfer function is

$$\frac{v_o(s)}{q(s)} = -\frac{R_f s}{[R_f C_f s + 1]} \tag{8.59}$$

where s is the Laplace variable. Now, in the frequency domain ($s = j\omega$),

Vibration Instrumentation

$$\frac{v_o(j\omega)}{q(j\omega)} = -\frac{R_f j\omega}{\left[R_f C_f j\omega + 1\right]} \tag{8.60}$$

Note that the output is zero at zero frequency ($\omega = 0$). Hence, a piezoelectric sensor cannot be used for measuring constant (DC) signals. At very high frequencies, on the other hand, the transfer function approaches the constant value $-1/C_f$, which is the calibration constant for the device.

From equation (8.58) or (8.60), which represent a first-order system, it is clear that the time constant τ_c of the sensor-amplifier unit is

$$\tau_c = R_f C_f \tag{8.61}$$

Suppose that the charge amplifier is properly calibrated (by the factor $-1/C_f$) so that the frequency transfer function (equation 8.60) can be written as

$$G(j\omega) = \frac{j\tau_c \omega}{\left[j\tau_c \omega + 1\right]} \tag{8.62}$$

Magnitude M of this transfer function is given by

$$M = \frac{\tau_c \omega}{\sqrt{\tau_c^2 \omega^2 + 1}} \tag{8.63}$$

As $\omega \to \infty$, note that $M \to 1$. Hence, at infinite frequency, there is no error. Measurement accuracy depends on the closeness of M to 1. Suppose that one wants the accuracy to be better than a specified value M_o. Accordingly, one must have

$$\frac{\tau_c \omega}{\sqrt{\tau_c^2 \omega^2 + 1}} > M_0 \tag{8.64}$$

or

$$\tau_c \omega > \frac{M_0}{\sqrt{1 - M_0^2}} \tag{8.65}$$

If the required lower frequency limit is $\omega_{\min}$, the time constant requirement is

$$\tau_c > \frac{M_0}{\omega_{\min}\sqrt{1 - M_0^2}} \tag{8.66}$$

or

$$R_f C_f > \frac{M_0}{\omega_{\min}\sqrt{1 - M_0^2}} \tag{8.67}$$

It follows that a specified lower limit on frequency of operation, for a specified level of accuracy, can be achieved by increasing the charge-amplifier time constant (i.e., by increasing R_f, C_f, or both). For example, an accuracy better than 99% is obtained if

$$\frac{\tau_c \omega}{\sqrt{\tau_c^2 \omega^2 + 1}} > 0.99, \text{ or } \tau_c \omega > 7.0.$$

The minimum frequency of a transient signal that can tolerate this level of accuracy is $\omega_{min} = \dfrac{7.0}{\tau_c}$. Now, ω_{min} can be set by adjusting the time constant.

8.5 TORQUE, FORCE, AND OTHER SENSORS

The forced vibrations in a mechanical system depend on the forces and torques (excitations) applied to the system. Also, the performance of the system can be specified in terms of forces and torques that are generated, as in machine-tool operations such as grinding, cutting, forging, extrusion, and rolling. Performance monitoring and evaluation, failure detection and diagnosis, and vibration testing may depend considerably on accurate measurement of associated forces and torques. In mechanical applications such as parts assembly, slight errors in motion can generate large forces and torques. These observations highlight the importance of measuring forces and torques. The strain gage is a sensor that is commonly used in this context. There are numerous other types of sensors and transducers that are useful in the practice of mechanical vibration. This section outlines several of these sensors.

8.5.1 STRAIN-GAGE SENSORS

Many types of force and torque sensors and also motion sensors such as accelerometers are based on strain-gage measurements. Hence, strain gages are very useful in vibration instrumentation. Although strain gages measure strain, the measurements can be directly related to stress and force. Note, however, that strain gages can be used in a somewhat indirect manner (using auxiliary front-end elements) to measure other types of variables, including displacement and acceleration.

Equations for Strain-Gage Measurements

The change of electrical resistance in material when mechanically deformed is the property used in resistance-type strain gages. The resistance R of a conductor that has length ℓ and area of cross section A is given by

$$R = \rho \frac{\ell}{A} \tag{8.68}$$

where ρ denotes the *resistivity* of the material. Taking the logarithm of equation (8.68), then $\log R = \log \rho + \log(\ell/A)$. Now, taking the differential, one obtains

$$\frac{dR}{R} = \frac{d\rho}{\rho} + \frac{d(\ell/A)}{\ell/A} \tag{8.69}$$

The first term on the right-hand side of equation (8.69) depends on the change in resistivity, and the second term represents deformation. It follows that the change in resistance comes from the change in shape as well as from the change in resistivity of the material. For linear deformations, the two terms on the right-hand side of equation (8.69) are linear functions of strain ε; the proportionality constant of the second term, in particular, depends on Poisson's ratio of the material. Hence, the following relationship can be written for a strain-gage element:

$$\frac{\delta R}{R} = S_s \varepsilon \qquad (8.70)$$

The constant S_s is known as the *sensitivity* or *gage factor* of the strain-gage element. The numerical value of this constant ranges from 2 to 6 for most *metallic strain-gage* elements and from 40 to 200 for *semiconductor strain gages*. These two types of strain gages are discussed later. The change in resistance of a strain-gage element, which determines the associated strain [equation (8.70)], is measured using a suitable electrical circuit.

Resistance strain gages are based on resistance change due to strain, or the *piezoresistive* property of materials. Early strain gages were fine metal filaments. Modern strain gages are manufactured primarily as metallic foil (e.g., using the copper-nickel alloy known as constantan) or semiconductor elements (e.g., silicon with trace impurity boron). They are manufactured by first forming a thin film (foil) of metal or a single crystal of semiconductor material and then cutting it into a suitable grid pattern, either mechanically or by using photoetching (chemical) techniques. This process is much more economical and is more precise than making strain gages with metal filaments. The strain-gage element is formed on a backing film of electrically insulated material (e.g., plastic). This element is cemented onto the member whose strain is to be measured. Alternatively, a thin film of insulating ceramic substrate is melted onto the measurement surface, on which the strain gage is mounted directly. The direction of sensitivity is the major direction of elongation of the strain-gage element [Figure 8.33(a)]. To measure strains in more than one direction, multiple strain gages (e.g., various rosette configurations) are available as single units. These units have more than one direction of sensitivity. Principal strains in a given plane (the surface of the object on which the strain gage is mounted) can be determined using these multiple strain-gage units. Typical foil-type strain gages have relatively large output signals.

A direct way to obtain strain-gage measurement is to apply a constant DC voltage across a series-connected strain-gage element and a suitable resistor and to measure the output voltage v_o across the strain gage under open-circuit conditions (using a voltmeter with high-input impedance). It is known as a *potentiometer circuit* or *ballast circuit* (see Figure 8.34(a)). This arrangement has several weaknesses. Any ambient temperature variation will directly introduce some error because of associated change in the strain-gage resistance and the resistance of the connecting circuitry. Also, measurement accuracy will be affected by possible variations in the supply voltage v_{ref}. Furthermore, the electrical loading error will be significant unless the load impedance is very high. Perhaps the most serious disadvantage of this circuit is that the change in signal due to strain is usually a very small percentage of the total signal level in the circuit output.

A more favorable circuit for use in strain-gage measurements is the *Wheatstone bridge*, shown in Figure 8.34(b). One or more of the four resistors R_1, R_2, R_3, and R_4 in the circuit can represent strain gages. To obtain the output relationship for the Wheatstone bridge circuit, assume that the load impedance R_L is very high. Hence, the load current i is negligibly small. Then, the potentials at nodes A and B are

$$v_A = \frac{R_1}{(R_1 + R_2)} v_{ref} \quad \text{and} \quad v_B = \frac{R_3}{(R_3 + R_4)} v_{ref}$$

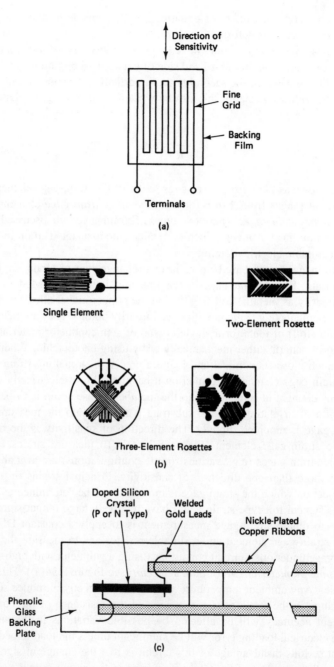

FIGURE 8.33 (a) Strain-gage nomenclature; (b) typical foil-type strain gages; and (c) a semiconductor strain gage.

and the output voltage $v_o = v_A - v_B$ is given by

$$v_o = \left[\frac{R_1}{(R_1 + R_2)} - \frac{R_3}{(R_3 + R_4)} \right] v_{ref} \qquad (8.71)$$

Now, by using straightforward algebra, one obtains

Vibration Instrumentation

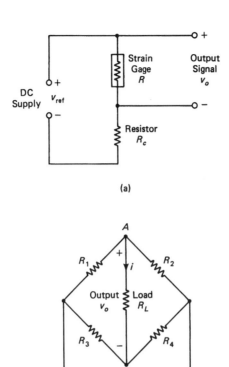

FIGURE 8.34 (a) A potentiometer circuit (ballast circuit) for strain-gage measurements, and (b) a Wheatstone bridge circuit for strain-gage measurements.

$$v_o = \frac{(R_1 R_4 - R_2 R_3)}{(R_1 + R_2)(R_3 + R_4)} v_{ref} \qquad (8.72)$$

When this output voltage is zero, the bridge is said to be "balanced." It follows from equation (8.72) that for a balanced bridge,

$$\frac{R_1}{R_2} = \frac{R_3}{R_4} \qquad (8.73)$$

Note that equation (8.73) is valid for any value of R_L, not just for large R_L, because when the bridge is balanced, current i will be zero, even for small R_L.

Bridge Sensitivity

Strain-gage measurements are calibrated with respect to a balanced bridge. When the strain gages in the bridge deform, the balance is upset. If one of the arms of the bridge has a variable resistor, it can be changed to restore the balance. The amount of this change measures the amount by which the resistance of the strain gages changed, thereby measuring the applied strain. This is known as the *null-balance method* of strain measurement. This method is inherently slow because of the time required to balance the bridge each time a reading is taken. Hence, the null-balance method is generally not suitable for dynamic (time-varying) measurements. This approach to strain measurement can be speeded up using servo balancing, whereby the output error signal is fed back into an actuator that automatically adjusts the variable resistance so as to restore the balance.

A more common method, which is particularly suitable for making dynamic readings from a strain-gage bridge, is to measure the output voltage resulting from the imbalance caused by the deformation of active strain gages in the bridge. To determine the *calibration constant* of a strain-gage bridge, the sensitivity of the bridge output to changes in the four resistors in the bridge should be known. For small changes in resistance, this can be determined using the differential relation (or, equivalently, the first-order approximation for the Taylor series expansion):

$$\delta v_o = \sum_{i=1}^{4} \frac{\partial v_o}{\partial R_i} \delta R_i \tag{8.74}$$

The partial derivatives are obtained directly from equation (8.71). Specifically,

$$\frac{\partial v_o}{\partial R_1} = \frac{R_2}{(R_1 + R_2)^2} v_{ref} \tag{8.75}$$

$$\frac{\partial v_o}{\partial R_2} = -\frac{R_1}{(R_1 + R_2)^2} v_{ref} \tag{8.76}$$

$$\frac{\partial v_o}{\partial R_3} = -\frac{R_4}{(R_3 + R_4)^2} v_{ref} \tag{8.77}$$

$$\frac{\partial v_o}{\partial R_4} = \frac{R_3}{(R_3 + R_4)^2} v_{ref} \tag{8.78}$$

The required relationship is obtained by substituting equations (8.75) through (8.78) into (8.74); thus,

$$\frac{\delta v_o}{v_{ref}} = \frac{(R_2 \delta R_1 - R_1 \delta R_2)}{(R_1 + R_2)^2} - \frac{(R_4 \delta R_3 - R_3 \delta R_4)}{(R_3 + R_4)^2} \tag{8.79}$$

This result is subject to equation (8.73) because changes are measured from the balanced condition. Note from equation (8.79) that if all four resistors are identical (in value and material), resistance changes due to ambient effects cancel out among the first-order terms (δR_1, δR_2, δR_3, δR_4), producing no net effect on the output voltage from the bridge. Closer examination of equation (8.79) will reveal that only the adjacent pairs of resistors (e.g., R_1 with R_2 and R_3 with R_4) must be identical in order to achieve this environmental compensation. Even this requirement can be relaxed. As a matter of fact, compensation is achieved if R_1 and R_2 have the same temperature coefficient and if R_3 and R_4 have the same temperature coefficient.

The Bridge Constant

Numerous activation combinations of strain gages are possible in a bridge circuit; for example, tension in R_1 and compression in R_2, as in the case of two strain gages mounted symmetrically at

Vibration Instrumentation

45° about the axis of a shaft in torsion. In this manner, the overall sensitivity of a strain-gage bridge can be increased. It is clear from equation (8.79) that if all four resistors in the bridge are active, the best sensitivity is obtained if, for example, R_1 and R_4 are in tension and R_2 and R_3 are in compression, so that all four differential terms have the same sign. If more than one strain gage is active, the bridge output can be expressed as

$$\frac{\delta v_o}{v_{ref}} = k \frac{\delta R}{4R} \quad (8.80)$$

where

$$k = \frac{\text{Bridge output in the general case}}{\text{Bridge output if only one strain gage is active}}$$

This constant is known as the *bridge constant*. The larger the bridge constant, the better the sensitivity of the bridge.

EXAMPLE 8.4

A strain-gage load cell (force sensor) consists of four identical strain gages, forming a Wheatstone bridge, which are mounted on a rod that has square cross section. One opposite pair of strain gages is mounted axially, and the other pair is mounted in the transverse direction, as shown in Figure 8.35(a). To maximize the bridge sensitivity, the strain gages are connected to the bridge as shown in Figure 8.35(b). Determine the bridge constant k in terms of *Poisson's ratio* υ of the rod material.

SOLUTION

Suppose that $\delta R_1 = \delta R$. Then, for the given configuration,

$$\delta R_2 = -\upsilon \delta R$$

$$\delta R_3 = -\upsilon \delta R$$

$$\delta R_4 = \delta R$$

Note that from the definition of Poisson's ratio, the transverse strain = $(-\upsilon) \times$ longitudinal strain. Now, it follows from equation (8.79) that

$$\frac{\delta v_o}{v_{ref}} = 2(1+\upsilon)\frac{\delta R}{4R} \quad (8.81)$$

according to which the bridge constant is given by

$$k = 2(1+\upsilon)$$

□

The Calibration Constant

The calibration constant C of a strain-gage bridge relates the strain that is measured to the output of the bridge. Specifically,

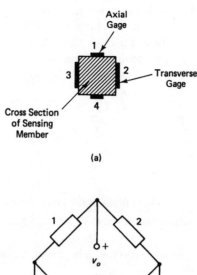

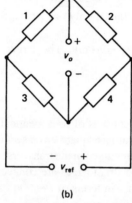

FIGURE 8.35 A strain-gage force sensor: (a) mounting configuration, and (b) bridge circuit.

$$\frac{\delta v_o}{v_{ref}} = C\varepsilon \qquad (8.82)$$

Now, in view of equations (8.70) and (8.80), the calibration constant can be expressed as

$$C = \frac{k}{4} S_s \qquad (8.83)$$

where k is the *bridge constant* and S_s is the *sensitivity* or *gage factor* of the strain gage. Ideally, the calibration constant should remain constant over the measurement range of the bridge (i.e., independent of strain ε and time t) and should be stable with respect to ambient conditions. In particular, there should not be any creep, nonlinearities such as hysteresis, or thermal effects.

EXAMPLE 8.5

A schematic diagram of a strain-gage accelerometer is shown in Figure 8.36(a). A point mass of weight W is used as the acceleration-sensing element, and a light cantilever with rectangular cross section, mounted inside the accelerometer casing, converts the inertia force of the mass into a strain. The maximum bending strain at the root of the cantilever is measured using four identical active semiconductor strain gages. Two of the strain gages (A and B) are mounted axially on the top

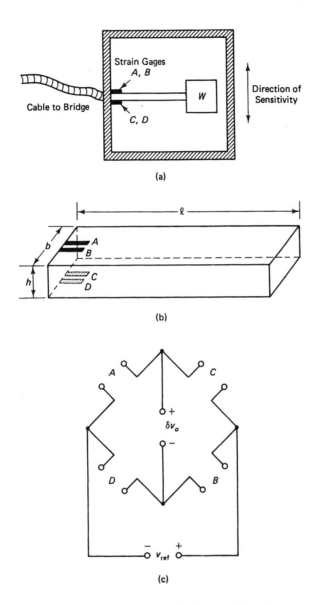

FIGURE 8.36 A strain-gage accelerometer: (a) schematic diagram; (b) strain-gage mounting configuration; and (c) bridge connections.

surface of the cantilever, and the remaining two gages (C and D) are mounted on the bottom surface, as shown in Figure 8.36(b). In order to maximize the sensitivity of the accelerometer, indicate the manner in which the four strain gages — A, B, C, and D — should be connected to a Wheatstone bridge circuit. What is the bridge constant of the resulting circuit?

Obtain an expression relating the applied acceleration a (in units of g, which denotes acceleration due to gravity) to the bridge output δv_o (measured using a bridge balanced at zero acceleration) in terms of the following parameters:

W = weight of the seismic mass at the free end of the cantilever element
E = Young's modulus of the cantilever
ℓ = length of the cantilever

b = cross-section width of the cantilever
h = cross-section height of the cantilever
S_s = sensitivity (gage factor) of each strain gage
v_{ref} = supply voltage to the bridge.

If $W = 0.02$ lb, $E = 10 \times 10^6$ lbf·in^{-2}, $\ell = 1$ in, $b = 0.1$ in, $h = 0.05$ in, $S_s = 200$, and $v_{ref} = 20$ V, determine the sensitivity of the accelerometer in mV·g^{-1}.

If the yield strength of the cantilever element is 10×10^3 lbf·in^{-2}, what is the maximum acceleration that could be measured using the accelerometer?

Is the cross-sensitivity [i.e., the sensitivity in the two directions orthogonal to the direction of sensitivity shown in Figure 8.36(a)] small with your arrangement of the strain-gage bridge? Explain.

Note: For a cantilever subjected to force F at the free end, the maximum stress at the root is given by

$$\sigma = \frac{6F\ell}{bh^2} \tag{8.84}$$

with the present notation.

SOLUTION

Clearly, the bridge sensitivity is maximized by connecting the strain gages A, B, C, and D to the bridge as shown in Figure 8.36(c). This follows from equation (8.79), noting that the contributions from all four strain gages are positive when δR_1 and δR_4 are positive, and δR_2 and δR_3 are negative. The bridge constant for the resulting arrangement is $k = 4$. Hence, from equation (8.80), one obtains

$$\frac{\delta v_o}{v_{ref}} = \frac{\delta R}{R}$$

or, from equations (8.82) and (8.83),

$$\frac{\delta v_o}{v_{ref}} = S_s \varepsilon$$

Also,

$$\varepsilon = \frac{\sigma}{E} = \frac{6F\ell}{Ebh^2}$$

where F denotes the inertia force:

$$F = \frac{W}{g}\ddot{x} = Wa$$

Note that $\ddot{x}$ is the acceleration in the direction of sensitivity, and $\ddot{x}/g = a$ is the acceleration in units of g. Thus,

$$\varepsilon = \frac{6W\ell}{Ebh^2}a \tag{8.85}$$

or

$$\delta v_o = \frac{6W\ell}{Ebh^2} S_s v_{ref} a \qquad (8.86)$$

Now, with the given values,

$$\frac{\delta v_o}{a} = \frac{6 \times 0.02 \times 1 \times 200 \times 20}{10 \times 10^6 \times 0.1 \times (0.05)^2} \text{ V/g}$$

$$= 0.192 \text{ V/g} = 192 \text{ mV/g}$$

$$\frac{\varepsilon}{a} = \frac{1}{S_s v_{ref}} \frac{\delta v_o}{a} = \frac{0.192}{200 \times 20} \text{ strain/g}$$

$$= 48 \times 10^{-6} \text{ }\varepsilon/\text{g} = 48\,\mu\varepsilon/\text{g}$$

$$\text{Yield strain} = \frac{\text{Yield strength}}{E} = \frac{10 \times 10^3}{10 \times 10^6} = 1 \times 10^{-3} \text{ strain}$$

Hence,

$$\text{Number of } gs \text{ to yielding} = \frac{1 \times 10^{-3}}{48 \times 10^{-6}} \text{ g} = 20.8 \text{ g}$$

Cross-sensitivity comes from accelerations in the two directions (y and z) orthogonal to the direction of sensitivity (x). In the lateral (y) direction, the inertia force causes lateral bending. This will produce equal tensile (or compressive) strains in B and D, and equal compressive (or tensile) strains in A and C. According to the bridge circuit, one sees that these contributions cancel each other. In the axial (z) direction, the inertia force causes equal tensile (or compressive) stresses in all four strain gages. These also will cancel out, as is clear from the following relationship for the bridge:

$$\frac{\delta v_o}{v_{ref}} = \frac{(R_C \delta R_A - R_A \delta R_C)}{(R_A + R_C)^2} - \frac{(R_B \delta R_D - R_D \delta R_B)}{(R_D + R_B)^2} \qquad (8.87)$$

with

$$R_A = R_B = R_C = R_D = R$$

which gives

$$\frac{\delta v_o}{v_{ref}} = \frac{(\delta R_A - \delta R_C - \delta R_D + \delta R_B)}{4R} \qquad (8.88)$$

It follows that this arrangement is good with respect to cross-sensitivity problems.

□

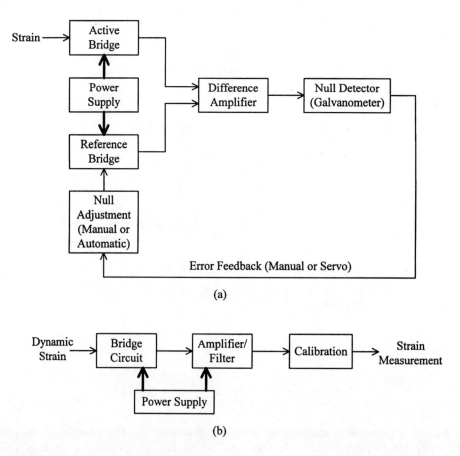

FIGURE 8.37 Strain-gage bridge measurement: (a) null-balance method, and (b) imbalance output method.

Data Acquisition

As noted earlier, the two common methods of measuring strains using a Wheatstone bridge circuit are the (1) null-balance method and (2) imbalance output method. One possible scheme for using the first method is shown in Figure 8.37(a). In this particular arrangement, two bridge circuits are used. The *active bridge* contains the active strain gages, dummy gages, and bridge-completion resistors. The *reference bridge* has four resistors, one of which is micro-adjustable, either manually or automatically. The outputs from the two bridges are fed into a difference amplifier, which provides an amplified difference of the two signals. This error signal is indicated on a null detector, such as a galvanometer. Initially, both bridges are balanced. When the measurement system is in use, the active gages are subjected to the strain that is being measured. This upsets the balance, giving a net output, which is indicated on the null detector. In manual operation of the null-balance mechanism, the resistance knob in the reference bridge is adjusted carefully until the galvanometer indicates a null reading. The knob can be calibrated to indicate the measured strain directly. In servo operation, which is much faster than the manual method, the error signal is fed into an actuator that automatically adjusts the variable resistor in the reference bridge until the null balance is achieved. Actuator movement measures the strain.

For measuring dynamic strains in vibrating systems, either the servo null-balance method or the imbalance output method should be employed. A schematic diagram for the imbalance output method is shown in Figure 8.37(b). In this method, the output from the active bridge is directly measured as a voltage signal and calibrated to provide the measured strain. An AC bridge can be used, where the bridge is powered by an AC voltage. The supply frequency should be about ten times the maximum frequency of interest in the dynamic strain signal (bandwidth). A supply frequency on the order of 1 kHz is typical. This signal is generated by an oscillator and then fed into the bridge. The transient component of the output from the bridge is very small (typically less than 1 mV and possibly a few microvolts). This signal must be amplified, demodulated (especially if the signals are transient), and filtered to provide the strain reading. The calibration constant of the bridge should be known in order to convert the output voltage to strain.

Strain-gage bridges powered by DC voltages are very common. They have the advantages of simplicity with regard to necessary circuitry and portability. The advantages of AC bridges include improved stability (reduced drift) and accuracy, and reduced power consumption.

Accuracy Considerations

Foil gages are available with resistances as low as 50 Ω and as high as several kilohms. The power consumption of the bridge decreases with increased resistance. This has the added advantage of decreased heat generation. Bridges with a high range of measurement (e.g., a maximum strain of 0.01 m·m^{-1}) are available. The accuracy depends on the linearity of the bridge, environmental (particularly temperature) effects, and mounting techniques. For example, zero shift, due to the strains produced when the cement that is used to mount the strain gage dries, will result in calibration error. Creep will introduce errors during static and low-frequency measurements. Flexibility and hysteresis of the bonding cement will bring about errors during high-frequency strain measurements. Resolutions on the order of 1 μm·m^{-1} (i.e., 1 *microstrain*) are common. The cross-sensitivity should be small (say, less than 1% of the direct sensitivity). Manufacturers usually provide the values of the cross-sensitivity factors for their strain gages. This factor, when multiplied by the cross strain present in a given application, gives the error in the strain reading due to cross-sensitivity.

Often, measurements of strains in moving members are needed, for example, in real-time monitoring and failure detection in machine tools. If the motion is small or the device has a limited stroke, strain gages mounted on the moving member can be connected to the signal-conditioning circuitry and the power source using coiled flexible cables. For large motions, particularly in rotating shafts, some form of commutating arrangement must be used. Slip rings and brushes are commonly used for this purpose. When AC bridges are used, a mutual-induction device (rotary transformer) can be used, with one coil located on the moving member and the other coil stationary. To accommodate and compensate for errors (e.g., losses and glitches in the output signal) caused by commutation, it is desirable to place all four arms of the bridge, rather than just the active arms, on the moving member.

Semiconductor Strain Gages

In some low-strain applications (e.g., dynamic torque measurement), the sensitivity of foil gages is not adequate to produce an acceptable strain gage signal. Semiconductor (SC) strain gages are particularly useful in such situations. The strain element of an SC strain gage is made of a single crystal of *piezoresistive* material such as silicon, doped with a trace impurity such as boron. A typical construction is shown in Figure 8.38. The sensitivity (gage factor) of an SC strain gage is about two orders of magnitude higher than that of a metallic foil gage (typically, 40 to 200). The resistivity is also higher, providing reduced power consumption and heat generation. Another advantage of SC strain gages is that they deform elastically to fracture. In particular, mechanical hysteresis is negligible. Furthermore, they are smaller and lighter, providing less cross-sensitivity, reduced distribution error (i.e., improved spatial resolution), and negligible error due to mechanical loading. The maximum strain that is measurable using a semiconductor strain gage is typically 0.003 m/m (i.e., 3000 με). Strain-gage resistance can be several hundred ohms (typically, 120 Ω or 350 Ω).

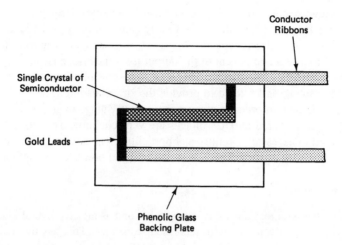

FIGURE 8.38 Details of a semiconductor strain gage.

There are several disadvantages associated with semiconductor strain gages, however, that can be interpreted as advantages of foil gages. Undesirable characteristics of SC gages include:

1. The strain-resistance relationship is more nonlinear.
2. They are brittle and difficult to mount on curved surfaces.
3. The maximum strain that can be measured is an order of magnitude smaller (typically, less than 0.01 m·m^{-1}).
4. They are more costly.
5. They have a much higher temperature sensitivity.

The first disadvantage is illustrated in Figure 8.39. There are two types of semiconductor strain gages: the P-type and the N-type. In P-type strain gages, the direction of sensitivity is along the (1, 1, 1) crystal axis, and the element produces a "positive" (P) change in resistance in response to a positive strain. In N-type strain gages, the direction of sensitivity is along the (1, 0, 0) crystal axis, and the element responds with a "negative" (N) change in resistance to a positive strain. In both types, the response is nonlinear and can be approximated by the quadratic relationship

$$\frac{\delta R}{R} = S_1 \varepsilon + S_2 \varepsilon^2 \tag{8.89}$$

The parameter S_1 represents the *linear sensitivity*, which is positive for P-type gages and negative for N-type gages. Its magnitude is usually somewhat larger for P-type gages, thereby providing better sensitivity. The parameter S_2 represents the degree of nonlinearity, which is usually positive for both types of gages. Its magnitude, however, is typically a little smaller for P-type gages. It follows that P-type gages are less nonlinear and have higher strain sensitivities. The nonlinear relationship given by equation (8.89) or the nonlinear characteristic curve (Figure 8.39) should be used when measuring moderate to large strains with semiconductor strain gages. Otherwise, the nonlinearity error would be excessive.

Force and Torque Sensors

Torque and force sensing is useful in vibration applications, including the following:

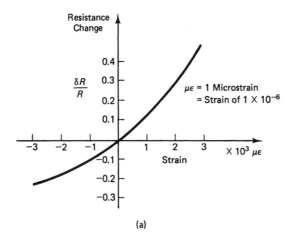

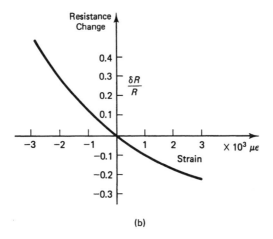

FIGURE 8.39 Nonlinear behavior of a semiconductor (silicon-boron) strain gage: (a) a P-type gage, and (b) an N-type gage.

1. In vibration control of machinery where a small motion error can cause large damaging forces or performance degradation
2. In high-speed vibration control when motion feedback alone is not fast enough; here, force feedback and feedforward force control can be used to improve the accuracy and bandwidth
3. In vibration testing, monitoring, and diagnostic applications, where torque and force sensing can detect, predict, and identify abnormal operation, malfunction, component failure, or excessive wear (e.g., in monitoring of machine tools such as milling machines and drills)
4. In experimental modal analysis where both excitation forces and response motioning may be needed to experimentally determine the system model.

In most applications, torque (or force) is sensed by detecting either an effect or the cause of torque (or force). There are also methods for measuring torque (or force) directly. Common methods of torque sensing include:

1. measuring the strain in a sensing member between the drive element and the driven load, using a strain-gage bridge
2. measuring the displacement in a sensing member (as in the first method) — either directly, using a displacement sensor, or indirectly, by measuring a variable, such as magnetic inductance or capacitance, that varies with displacement
3. measuring the reaction in the support structure or housing (by measuring a force) and the associated lever arm length
4. in electric motors, measuring the field or armature current that produces motor torque; in hydraulic or pneumatic actuators, measuring the actuator pressure
5. measuring the torque directly, using piezoelectric sensors, for example
6. employing the servo method to balance the unknown torque with a feedback torque generated by an active device (say, a servomotor) whose torque characteristics are known precisely
7. measuring the angular acceleration in a known inertia element when the unknown torque is applied.

Note that force sensing can be accomplished by essentially the same techniques. Some types of force sensors (e.g., strain-gage force sensor) have been introduced before. The discussion here is primarily limited to torque sensing. The extension of torque-sensing techniques to force sensing is somewhat straightforward.

Strain–Gage Torque Sensors

The most straightforward method of torque sensing is to connect a torsion member between the drive unit and the load in series, and then to measure the torque in the torsion member. If a circular shaft (solid or hollow) is used as the torsion member, the torque–strain relationship becomes relatively simple. A complete development of the relationship is found in standard textbooks on elasticity, solid mechanics, or strength of materials (also, see Chapter 6). With reference to Figure 8.40, it can be shown that the torque T can be expressed in terms of the direct strain ε on the shaft surface along a principal stress direction (i.e., at 45° to the shaft axis) as

$$T = \frac{2GJ}{r}\varepsilon \qquad (8.90)$$

where G = shear modulus of the shaft material; J = polar moment of area of the shaft; and r = shaft radius (outer). This is the basis of torque sensing using strain measurements. Using the general bridge equation (8.82) along with (8.83) in equation (8.90), one can obtain torque T from bridge output δv_o:

$$T = \frac{8GJ}{kS_s r}\frac{\delta v_o}{v_{ref}} \qquad (8.91)$$

where S_s is the gage factor (or sensitivity) of the strain gages. The bridge constant k depends on the number of active strain gages used. Strain gages are assumed to be mounted along a principal direction. Three possible configurations are shown in Figure 8.41. In configurations (a) and (b), only two strain gages are used, and the bridge constant $k = 2$. Note that both axial loads and bending are compensated with the given configurations because resistance in both gages will be changed by the same amount (same sign and same magnitude) that cancels out up to first order, for the bridge circuit connection shown in Figure 8.41. Configuration (c) has two pairs of gages, mounted

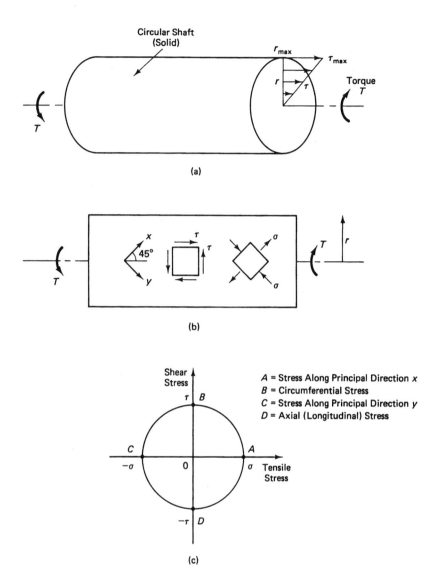

FIGURE 8.40 (a) Linear distribution of shear stress in a circular shaft under pure torsion; (b) pure shear state of stress and principal directions x and y; and (c) Mohr's circle.

on the two opposite surfaces of the shaft. The bridge constant is doubled in this configuration and, here again, the sensor clearly self-compensates for axial and bending loads up to first order $[O(\delta R)]$.

For a circular-shaft torque sensor that uses semiconductor strain gages, design criteria for obtaining a suitable value for the polar moment of area (J) are listed in Table 8.3. Note that ϕ is a safety factor.

Although the manner in which strain gages are configured on a torque sensor can be exploited to compensate for cross-sensitivity effects arising from factors such as tensile and bending loads, it is advisable to use a torque-sensing element that inherently possesses low sensitivity to those factors that cause error in a torque measurement. A tubular torsion element is convenient for analytical purposes because of the simplicity of the associated expressions for design parameters. Unfortunately, such an element is not very rigid to bending and tensile loading. Alternative shapes and structural arrangements must be considered if inherent rigidity (insensitivity) to cross-loads is needed. Furthermore, a tubular element has the same strain at all locations on the element surface.

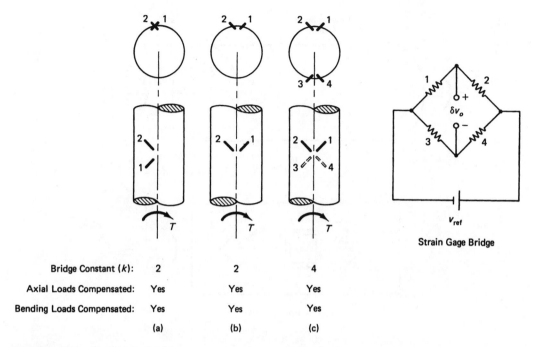

FIGURE 8.41 Strain-gage configurations for a circular shaft torque sensor.

TABLE 8.3
Design Criteria for a Strain-Gage Torque-Sensing Element

Criterion	Specification	Governing Design Formula for the Polar Moment of Area (J)
Strain capacity of strain-gage element	ε_{max}	$\geq \dfrac{\phi r}{2G} \cdot \dfrac{T_{max}}{\varepsilon_{max}}$
Strain-gage nonlinearity	$N_p = \dfrac{\text{Max. strain error}}{\text{Strain range}} \times 100\%$	$\geq \dfrac{25\phi r S_2}{GS_1} \cdot \dfrac{T_{max}}{N_p}$
Sensor sensitivity (output voltage)	$v_o = K_a \delta v_o$, where K_a = transducer gain.	$\leq \dfrac{K_a k S_s r v_{ref}}{8G} \cdot \dfrac{T_{max}}{v_o}$
Sensor stiffness (system bandwidth and gain)	$K = \dfrac{\text{Torque}}{\text{Twist angle}}$	$\geq \dfrac{L}{G} \cdot K$

This does not give us a choice with respect to mounting locations of strain gages in order to maximize the torque sensor sensitivity. Another disadvantage of the basic tubular element is that the surface is curved; therefore, much care is needed in mounting fragile semiconductor gages, which could be easily damaged even with slight bending. Hence, a sensor element that has flat surfaces to mount the strain gages would be desirable. A torque-sensing element having the foregoing desirable characteristics (i.e., good strength, inherent insensitivity to cross-loading, non-uniform strain distribution on the surface, and availability of flat surfaces to mount strain gages) is shown in Figure 8.42. Note that two sensing elements are connected radially between the drive unit and the driven member. The sensing elements undergo bending while transmitting a torque between the driver and the driven member. Bending strains are measured at locations of high sensitivity and are taken to be proportional to the transmitted torque. Analytical determination of

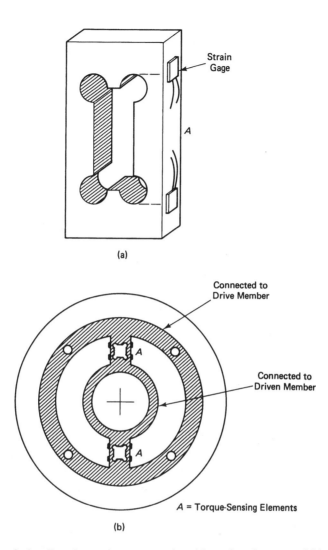

FIGURE 8.42 Use of a bending element in torque sensing: (a) sensing element, and (b) element connection.

the calibration constant is not easy for such complex sensing elements, but experimental determination is straightforward. Note that the strain-gage torque sensors measure the direction as well as the magnitude of the torque transmitted through it.

Deflection Torque Sensors

Instead of measuring strain in the sensor element, the actual deflection (twisting or bending) can be measured and used to determine torque, through a suitable calibration constant. For a circular-shaft (solid or hollow) torsion element, the governing relationship is given by:

$$T = \frac{GJ}{L}\theta \tag{8.92}$$

The calibration constant GJ/L must be small in order to achieve high sensitivity. This means that the element stiffness should be low. This will limit the bandwidth (which measures speed of response) and gain (which determines steady-state error) of the overall system. The twist angle θ

is very small (e.g., a fraction of a degree) in systems with high bandwidth. This requires very accurate measurement of θ in order to determine the torque T. A type of displacement sensor that can be used is as follows: two ferromagnetic gear wheels are splined at two axial locations of the torsion element. Two stationary proximity probes of the magnetic induction type (self-induction or mutual induction) are placed radially, facing the gear teeth, at the two locations. As the shaft rotates, the gear teeth change the flux linkage of the proximity sensor coils. The resulting output signals of the two probes are pulse sequences, shaped somewhat like sine waves. The phase shift of one signal with respect to the other determines the relative angular deflection of one gear wheel with respect to the other, assuming that the two probes are synchronized under no-torque conditions. Both the magnitude and the direction of the transmitted torque are determined using this method. A 360° phase shift corresponds to a relative deflection by an integer multiple of the gear pitch. It follows that deflections less than half the pitch can be measured without ambiguity. Assuming that the output signals of the two probes are sine waves (narrow-band filtering can be used to achieve this), the phase shift will be proportional to the angle of twist θ.

Variable–Reluctance Torque Sensor

A torque sensor that is based on the sensor element deformation and does not require a contacting commutator is a variable-reluctance device that operates like a differential transformer (RVDT or LVDT). The torque-sensing element is a ferromagnetic tube that has two sets of slits, typically oriented along the two principal stress directions of the tube (45°) under torsion. When a torque is applied to the torsion element, one set of gaps closes and the other set opens as a result of the principal stresses normal to the slit axes. Primary and secondary coils are placed around the slitted tube, and they remain stationary. One segment of the secondary coil is placed around one set of slits, and the second segment is placed around the other (perpendicular) set. The primary coil is excited by an AC supply, and the induced voltage v_o in the secondary coil is measured. As the tube deforms, it changes the magnetic reluctance in the flux linkage path, thus changing the induced voltage. The two segments of the secondary coil should be connected so that the induced voltages are absolutely additive (algebraically subtractive) — because one voltage increases and the other decreases — to obtain the best sensitivity. The output signal should be demodulated (by removing the carrier frequency component) to measure transient torques effectively. Note that the direction of torque is given by the sign of the demodulated signal.

Reaction Torque Sensors

The foregoing methods of torque sensing use a sensing element that is connected between the drive member and the driven member. A major drawback of such an arrangement is that the sensing element modifies the original system in an undesirable manner, particularly by decreasing the system stiffness and adding inertia. Not only will the overall bandwidth of the system decrease, but the original torque will also be changed (mechanical loading) because of the inclusion of an auxiliary sensing element. Furthermore, under dynamic conditions, the sensing element will be in motion, thereby making the torque measurement more difficult. The reaction method of torque sensing eliminates these problems to a large degree. This method can be used to measure torque in a rotating machine. The supporting structure (or housing) of the rotating machine (e.g., motor, pump, compressor, turbine, generator) is cradled by releasing its fixtures, and the effort necessary to keep the structure from moving is measured. A schematic representation of the method is shown in Figure 8.43(a). Ideally, a lever arm is mounted on the cradled housing, and the force required to fix the housing is measured using a force sensor (load cell). The reaction torque on the housing is given by

$$T_R = F_R \cdot L \tag{8.93}$$

Vibration Instrumentation

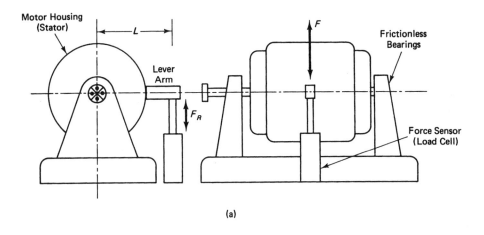

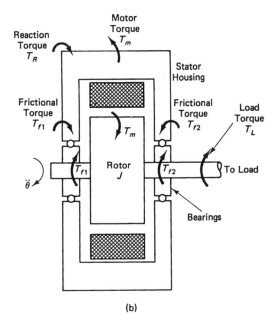

FIGURE 8.43 (a) Schematic representation of a reaction torque sensor setup (reaction dynamometer), and (b) various torque components.

where F_R = reation force measured using load cell and L = lever arm length.

Alternatively, strain gages or other types of force sensors can be mounted directly at the fixture locations (e.g., at the mounting bolts) of the housing to measure the reaction forces without cradling the housing. Then, the reaction torque is determined with a knowledge of the distance of the fixture locations from the shaft axis.

The reaction torque method of torque sensing is widely used in dynamometers (reaction dynamometers) that determine the transmitted power in rotating machinery through torque and shaft speed measurements. A drawback of reaction-type torque sensors can be explained using Figure 8.43(b). A motor with rotor inertia J, which rotates at angular acceleration $\ddot{\theta}$, is shown. By Newton's third law (action = reaction), the electromagnetic torque generated at the rotor of the motor T_m and the frictional torques T_{f1} and T_{f2} will be reacted back onto the stator and housing. By applying Newton's second law to the motor rotor and the housing combination, one obtains

$$T_L = T_R - J\ddot{\theta} \qquad (8.94)$$

Note that T_L is what must be measured. Under accelerating or decelerating conditions, the reaction torque T_R is not equal to the actual torque T_L that is transmitted. One method of compensating for this error is to measure the shaft acceleration, compute the inertia torque, and adjust the measured reaction torque using this inertia torque. Note that the frictional torque in the bearings does not enter the final equation (8.94). This is an advantage of this method.

8.5.2 Miscellaneous Sensors

Motion and force/torque sensors of the types described thus far are widely used in vibration instrumentation. Several other types of sensors are also useful. A few of them are indicated now.

Stroboscope

Consider an object that executes periodic motions, such as vibrations or rotations, in a rather dark environment. Suppose that a light is flashed at the object at the same frequency as the moving object. Because the object completes a full cycle of motion during the time period between two adjacent flashes, the object will appear to be stationary. This is the principle of operation of a stroboscope. The main components of a stroboscope are a high-intensity "strobe" lamp and circuitry to vary the frequency of the electrical pulse signal that energizes the lamp. The flashing frequency can be varied either manually using a knob or according to the frequency of an external periodic signal (trigger signal) that is applied to the stroboscope.

It is clear that by synchronizing the stroboscope with a moving (vibrating, rotating) object so that the object appears stationary, and then noting the flashing (strobe) frequency, the frequency of vibration or speed of rotation of the object can be measured. In this sense, stroboscope is a non-contacting vibration frequency sensor or a tachometer (rotating speed sensor). Note that the object appears stationary for any integer multiple of the synchronous flashing frequency. Hence, once the strobe is synchronized with the moving object, it is a good practice to check whether the strobe synchronizes also at an integer fraction of that flashing frequency (typically, trying 1/2, 1/3, 1/4, and 1/5 the original synchronous frequency would be adequate). The lowest synchronous frequency thus obtained is the correct speed (frequency) of the object. Because the frequency of visual persistence of a human is about 15 Hz, the stationary appearance will not be possible using a stroboscope below this frequency. Hence, the low-frequency limit for a stroboscope is about 15 Hz.

In addition to serving as a sensor for vibration frequency and rotating speed, a stroboscope has many other applications. For example, by maintaining the strobe (flashing) frequency close (but not equal) to the object frequency, the object will appear to move very slowly. In this manner, visual inspection of objects that execute periodic motions at high speed are possible. Also, stroboscopes are widely used in dynamic balancing of rotating machinery (see Chapter 12). In this case, it is important to measure the phase angle of the resultant imbalance force with respect to a coordinate axis (direction) that is fixed to the rotor. Suppose that a radial line is marked on the rotor. If a stroboscope is synchronized with the rotor such that the marked line appears not only stationary but also oriented in a fixed direction (e.g., horizontal or vertical), then, in effect, the strobe signal is in phase with the rotation of the rotor. Then, by comparing the imbalance force signal of the rotor (obtained, for example, by an accelerometer or a force sensor at the bearings of the rotor) with the synchronized strobe signal (with a fixed reference), by means of an oscilloscope or a phase meter, it is possible to determine the orientation of the imbalance force with respect to a fixed body reference of the rotating machine.

Vibration Instrumentation

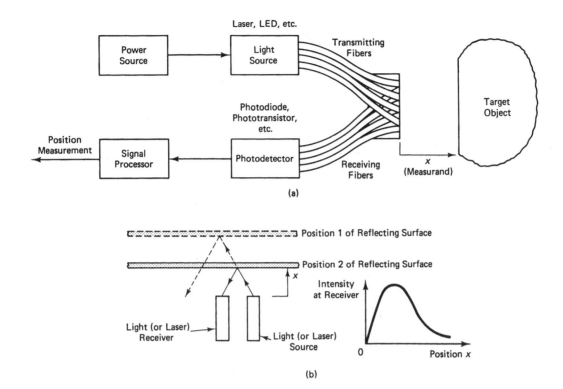

FIGURE 8.44 (a) A fiber-optic proximity sensor, and (b) a nonlinear characteristic curve.

Fiber-Optic Sensors and Lasers

The characteristic component in a fiber-optic sensor is a bundle of glass fibers (typically a few hundred) that can carry light. Each optical fiber may have a diameter on the order of 0.01 mm. There are two basic types of fiber-optic sensors. In one type — the "indirect" or the *extrinsic* type — the optical fiber acts only as the medium in which the sensed light is transmitted. In this type, the sensing element itself does not consist of optical fibers. In the second type — the "direct" or the *intrinsic* type — the optical fiber bundle itself acts as the sensing element. Then, when the conditions of the sensed medium change, the light-propagation properties of the optical fibers change, providing a measurement of the change in the conditions. Examples of the first (extrinsic) type of sensor include fiber-optic position sensors and tactile (distributed touch) sensors. The second (intrinsic) type of sensor is found, for example, in fiber-optic gyroscopes, fiber-optic hydrophones, and some types of micro-displacement or force sensors.

A schematic representation of a fiber-optic position sensor (or proximity sensor or displacement sensor) is shown in Figure 8.44(a). The optical fiber bundle is divided into two groups: transmitting fibers and receiving fibers. Light from the light source is transmitted along the first bundle of fibers to the target object whose position is being measured. Light reflected onto the receiving fibers by the surface of the target object is carried to a photodetector. The intensity of the light received by the photodetector will depend on the position x of the target object. In particular, if $x = 0$, the transmitting bundle will be completely blocked off, and the light intensity at the receiver will be zero. As x is increased, the received light intensity will increase because more and more light will be reflected onto the receiving bundle tip. This will reach a peak at some value of x. When x is increased beyond that value, more and more light will be reflected outside the receiving bundle; thus, the intensity of the received light will decrease. Hence, in general, the proximity–intensity curve for an optical proximity sensor will be nonlinear and will have the shape shown in Figure 8.44(b). Using this (calibration) curve, one can determine the position (x) once the intensity of the light received at the photosensor is known. The light source could be a *laser* (*l*ight *a*mplification by *s*timulated *e*mission of *r*adiation, structured light), an infrared light source, or some other type, such as a light-emitting diode (LED). The light sensor (photodetector) could be some light-sensitive discrete semiconductor element such as a photodiode or a photo field effect transistor (photo FET). Very fine resolutions better than 1×10^{-6} cm can be obtained using a fiber-optic position sensor. An *optical encoder* is a digital (or pulse-generating) motion transducer. Here, a light beam is intercepted by a moving disk that has a pattern of transparent windows. The light that passes through, as detected by a photosensor, provides the transducer output. These sensors can also be considered in the extrinsic category.

The advantages of fiber optics include insensitivity to electrical and magnetic noise (due to optical coupling), safe operation in explosive, high temperature, and hazardous environments, and high sensitivity. Furthermore, mechanical loading and wear problems do not exist because fiber-optic position sensors are non-contacting devices with stationary sensor heads. The disadvantages include direct sensitivity to variations in the intensity of the light source and dependence on ambient conditions (ambient light, dirt, moisture, smoke, etc.).

As an *intrinsic* application of fiber optics in sensing, consider a straight optical fiber element that is supported at the two ends. In this configuration, almost 100% of the light at the source end will transmit through the optical fiber and will reach the detector (receiver) end. Then, suppose that a slight load is applied to the optical fiber segment at its mid-span. The fiber will deflect slightly due to the load and, as a result, the amount of light received at the detector can significantly drop. For example, a deflection of just 50 μm can result in a drop in intensity at the detector by a factor of 25. Such an arrangement can be used in deflection, force, and tactile sensing. Another intrinsic application is the fiber-optic gyroscope, as described below.

Fiber-Optic Gyroscope

This is an angular speed sensor that uses fiber optics. Contrary to the implication of its name, however, it is not a gyroscope in the conventional sense. Two loops of optical fibers wrapped around a cylinder are used in this sensor. One loop carries a monochromatic light (or laser) beam in the clockwise direction, and the other loop carries a beam from the same light (laser) source in the counterclockwise direction. Because the laser beam traveling in the direction of rotation of the cylinder has a higher frequency than that of the other beam, the difference in frequencies of the two laser beams received at a common location will measure the angular speed of the cylinder. This can be accomplished through interferometry, as the light and dark patterns of the detected light will measure the frequency difference. Note that the length of the optical fiber in each loop can exceed 100 m. Angular displacements can be measured with the same sensor, simply by counting the number of cycles and clocking the fractions of cycles. Acceleration can be determined by digitally determining the rate of change of speed.

Laser Doppler Interferometer

The laser (*l*ight *a*mplification by *s*timulated *e*mission of *r*adiation) produces electromagnetic radiation in the ultraviolet, visible, or infrared bands of the spectrum. A laser can provide a single-frequency (*monochromatic*) light source. Furthermore, the electromagnetic radiation in a laser is *coherent* in the sense that all waves generated have constant phase angles. The laser uses oscillations of atoms or molecules of various elements. The helium–neon (HeNe) laser and the semiconductor laser are commonly used in industrial applications.

As noted earlier, the laser is useful in fiber optics; but it can also be used directly in sensing and gaging applications. The laser Doppler interferometer is one such sensor. It is useful in the accurate measurement of small displacements; for example, in strain measurements. To understand the operation of this device, two phenomena should be explained: the Doppler effect and light wave

Vibration Instrumentation

interference. Consider a wave source (e.g., a light source or sound source) that is moving with respect to a receiver (observer). If the source moves toward the receiver, the frequency of the received wave appears to have increased; if the source moves away from the receiver, the frequency of the received wave appears to have decreased. The change in frequency is proportional to the velocity of the source relative to the receiver. This phenomenon is known as the *Doppler effect*. Now consider a monochromatic (single-frequency) light wave of frequency f (say, 5×10^{14} Hz) emitted by a laser source. If this ray is reflected by a target object and received by a light detector, the frequency of the received wave would be

$$f_2 = f + \Delta f \tag{8.95}$$

The frequency increase Δf will be proportional to the velocity v of the target object, which is assumed positive when moving toward the light source. Hence,

$$\Delta f = cv \tag{8.96}$$

Now, by comparing the frequency f_2 of the reflected wave with the frequency

$$f_1 = f \tag{8.97}$$

of the original wave, one can determine Δf and, hence, the velocity v of the target object.

The change in frequency Δf due to the Doppler effect can be determined by observing the fringe pattern due to light wave interference. To understand this, consider the two waves

$$v_1 = a \sin 2\pi f_1 t \tag{8.98}$$

and

$$v_2 = a \sin 2\pi f_2 t \tag{8.99}$$

If one adds these two waves, the resulting wave would be

$$v = v_1 + v_2 = a(\sin 2\pi f_1 t + \sin 2\pi f_2 t)$$

which can be expressed as

$$v = 2a \sin \pi (f_2 + f_1) t \cos \pi (f_2 - f_1) t \tag{8.100}$$

It follows that the combined signal will beat at the beat frequency $\Delta f/2$. When f_2 is very close to f_1 (i.e., when Δf is small compared to f), these beats will appear as dark and light lines (fringes) in the resulting light wave. This is known as *wave interference*. Note that Δf can be determined by two methods:

1. by measuring the spacing of the fringes
2. by counting the beats in a given time interval, or by timing successive beats using a high-frequency clock signal.

The velocity of the target object is determined in this manner. Displacement can be obtained simply by digital integration (or by accumulating the count). A schematic diagram for the laser Doppler

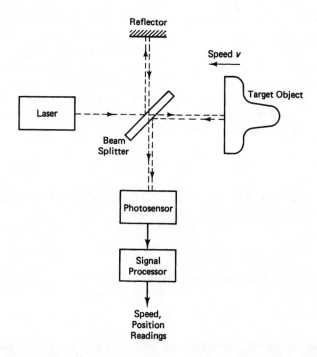

FIGURE 8.45 A laser Doppler interferometer for measuring velocity and displacement.

interferometer is shown in Figure 8.45. Industrial interferometers usually employ a helium–neon laser that has waves of two frequencies close together. In that case, the arrangement shown in Figure 8.45 must be modified to take into account the two frequency components.

Note that there are laser interferometers that directly measure *displacement* rather than speed. They are based on measuring *phase difference* between the direct and the returning laser, not the Doppler effect (frequency difference). In this case, integration is not needed to obtain displacement from a measured velocity.

Ultrasonic Sensors

Audible sound waves have frequencies in the range of 20 Hz to 20 kHz. Ultrasound waves are pressure waves, just like sound waves, but their frequencies are higher than audible frequencies. Ultrasonic sensors are used in many applications, including displacement and vibration sensing, medical imaging, ranging systems for cameras with autofocusing capabilities, level sensing, machine monitoring, and speed sensing. For example, in medical applications, ultrasound probes with frequencies of 40 kHz, 75 kHz, 7.5 MHz, and 10 MHz are commonly used. Ultrasound can be generated according to several principles. For example, high-frequency (gigahertz) oscillations in piezoelectric crystals subjected to electrical potentials are used to generate very high-frequency ultrasound. Another method is to use the magnetostrictive property of ferromagnetic materials. Ferromagnetic materials deform when subjected to magnetic fields. Resonant oscillations generated by this principle can produce ultrasonic waves. Another method of generating ultrasound is to apply a high-frequency voltage to a metal-film capacitor. A microphone can serve as an ultrasound detector (receiver).

Analogous to fiber-optic sensing, there are two common ways of employing ultrasound in a sensor. In one approach — the *intrinsic* method — the ultrasound signal undergoes changes as it passes through an object, due to acoustic impedance and absorption characteristics of the object. The resulting signal (image) can be interpreted to determine properties of the object, such as texture, firmness, and deformation. This approach can be utilized, for example, in machine monitoring and

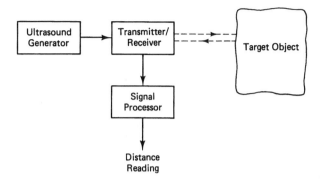

FIGURE 8.46 An ultrasonic position sensor.

object firmness sensing. In the other approach — the *extrinsic* method — the time of flight of an ultrasound burst from its source to an object and then back to a receiver is measured. This approach is used in distance, position, and vibration measurement and in dimensional gaging.

In distance (vibration, proximity, displacement) measurement using ultrasound, a burst of ultrasound is projected at the target object, and the time taken for the echo to be received is clocked. A signal processor computes the position of the target object, possibly compensating for environmental conditions. This configuration is shown in Figure 8.46. Alternatively, the velocity of the target object can be measured, using the Doppler effect, by measuring (clocking) the change in frequency between the transmitted wave and the received wave. The "beat" phenomenon can be employed here. Position measurements with fine resolution (e.g., a fraction of a millimeter) can be achieved using the ultrasonic method. Because the speed of ultrasonic wave propagation depends on the temperature of the medium (typically air), errors will enter into the ultrasonic readings unless the sensor is compensated for temperature variations.

Gyroscopic Sensors

Consider a rigid body spinning about an axis at angular speed ω. If the moment of inertia of the body about that axis is J, the angular momentum H about the same axis is given by:

$$H = J\omega \qquad (8.101)$$

Newton's second law (torque = rate of change of angular momentum) implies that to rotate (precess) the spinning axis slightly, a torque must be applied, because precession causes a change in the spinning angular momentum vector (the magnitude remains constant but the direction changes), as shown in Figure 8.47(a). This is the principle of the operation of a gyroscope. Gyroscopic sensors are commonly used in control systems for stabilizing vehicle systems.

Consider the gyroscope shown in Figure 8.47(b). The disk is spun about frictionless bearings using a torque motor. Because the gimbal (the framework on which the disk is supported) is free to turn about frictionless bearings on the vertical axis, it will remain fixed with respect to an inertial frame, even if the bearing housing (the main structure in which the gyroscope is located) rotates. Hence, the relative angle between the gimbal and the bearing housing (angle θ in the figure) can be measured, and this gives the angle of rotation of the main structure. In this manner, angular displacements in systems such as aircraft, space vehicles, ships, and land vehicles can be measured and stabilized with respect to an inertial frame. Note that bearing friction introduces an error that must be compensated for, perhaps, by recalibration before a reading is taken.

The *rate gyro* — which has the same arrangement as shown in Figure 8.47(b), with a slight modification — can be used to measure angular speeds. In this case, the gimbal is not free; it is

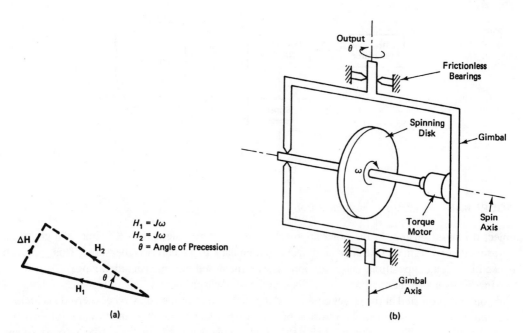

FIGURE 8.47 (a) Illustration of the gyroscopic torque needed to change the direction of an angular momentum vector, and (b) a simple single-axis gyroscope for sensing angular displacements.

restrained by a torsional spring. A viscous damper is provided to suppress any oscillations. By analyzing this gyro as a mechanical tachometer, note that the relative angle of rotation θ gives the angular speed of the structure about the gimbal axis.

Several areas can be identified where new developments and innovations are being made in sensor technology, including:

1. Microminiature sensors (IC-based, with built-in signal processing)
2. Intelligent sensors (built-in reasoning or information preprocessing to provide high-level knowledge)
3. Integrated and distributed sensors (sensors are integral with the components and agents of the overall multi-agent system that communicate with each other)
4. Hierarchical sensory architectures (low-level sensory information is preprocessed to match higher-level requirements).

These four areas of activity are also representative of future trends in sensor technology development. To summarize, rating parameters of a selected set of sensors/transducers are listed in Table 8.4.

8.6 COMPONENT INTERCONNECTION

When two or more components are interconnected, the behavior of the individual components in the overall system can deviate significantly from their behavior when each component operates independently. Matching of components in a multicomponent system, particularly with respect to their impedance characteristics, should be done carefully in order to improve system performance and accuracy. This is particularly true in vibration instrumentation.

8.6.1 IMPEDANCE CHARACTERISTICS

When components such as measuring instruments, digital processing boards, process (plant) hardware, and signal-conditioning equipment are interconnected, it is necessary to *match* impedances

TABLE 8.4
Rating Parameters of Several Sensors and Transducers

Transducer	Measurand	Measurand Frequency Max/Min	Output Impedance	Typical Resolution	Accuracy	Sensitivity
Potentiometer	Displacement	10 Hz/DC	Low	0.1 mm	0.1%	200 mV·m/m
LVDT	Displacement	2500 Hz/DC	Moderate	0.001 mm or less	0.3%	50 mV·m/m
Resolver	Angular displacement	500 Hz/DC (limited by excitation frequency)	Low	2 min.	0.2%	10 mV·deg^{-1}
Tachometer	Velocity	700 Hz/DC	Moderate (50 Ω)	0.2 mm·s^{-1}	0.5%	5 mV·m/m·s^{-1} 75 mV·rad^{-1}·s^{-1}
Eddy current proximity sensor	Displacement	100 kHz/DC	Moderate	0.001 mm 0.05% full scale	0.5%	5 V·m/m
Piezoelectric accelerometer	Acceleration (and velocity, etc.)	25 kHz/1Hz	High	1 mm·s^{-2}	1%	0.5 mV·m^{-1}·s^{-2}
Semiconductor strain gage	Strain (displacement, acceleration, etc.)	1 kHz/DC (limited by fatigue)	200 Ω	1 to 10$^{\mu\epsilon}$ (1$^{\mu\epsilon}$ = 10^{-6} unity strain)	1%	1 V/ε, 2000 με max
Load cell	Force (10–1000 N)	500 Hz/DC	Moderate	0.01 N	0.05%	1 mV·N^{-1}
Laser	Displacement/shape	1 kHz/DC	100 Ω	1.0 μm	0.5%	1 V·m/m
Optical encoder	Motion	100 kHz/DC	500 Ω	10 bit	±1/2 bit	10^4/rev

properly at each interface in order to realize their rated performance level. One adverse effect of improper impedance matching is the *loading effect*. For example, in a measuring system, the measuring instrument can distort the signal that is being measured. The resulting error can far exceed other types of measurement error. Loading errors will result from connecting a measuring device with low input impedance to a signal source.

Impedance can be interpreted either in the traditional electrical sense or in the mechanical sense, depending on the signal that is being measured. For example, a heavy accelerometer can introduce an additional dynamic load that will modify the actual acceleration at the monitoring location. Similarly, a voltmeter can modify the currents (and voltages) in a circuit. In mechanical and electrical systems, loading errors can appear as phase distortions as well. Digital hardware can also produce loading errors. For example, an analog-to-digital conversion (ADC) board can load the amplifier output from a strain-gage bridge circuit, thereby significantly affecting digitized data.

Another adverse effect of improper impedance consideration is inadequate output signal levels, which can make signal processing and transmission very difficult. Many types of transducers (e.g., piezoelectric accelerometers, impedance heads, and microphones) have high output impedances on the order of a thousand megohms. These devices generate low output signals, and they require conditioning to step up the signal level. *Impedance-matching amplifiers*, which have high input impedances (megohms) and low output impedances (a few ohms), are used for this purpose (e.g., charge amplifiers are used in conjunction with piezoelectric sensors). A device with a high input impedance has the further advantage in that it usually consumes less power (v^2/R is low) for a given input voltage. The fact that a low input impedance device extracts a high level of power from the preceding output device can be interpreted as the reason for loading error.

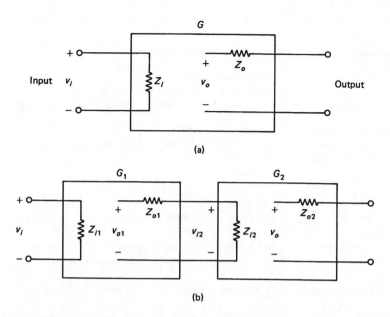

FIGURE 8.48 (a) Schematic representation of input impedance and output impedance, and (b) the influence of cascade connection of devices on the overall impedance characteristics.

Cascade Connection of Devices

Consider a standard two-port electrical device. The output impedance Z_o of such a device is defined as the ratio of the open-circuit (i.e., no-load) voltage at the output port to the short-circuit current at the output port.

Open-circuit voltage at the output is the output voltage present when there is no current flowing at the output port. This is the case if the output port is not connected to a load (impedance). As soon as a load is connected at the output of the device, a current will flow through it, and the output voltage will drop to a value less than that of the open-circuit voltage. To measure open-circuit voltage, the rated input voltage is applied at the input port and maintained constant, and the output voltage is measured using a voltmeter that has a very high (input) impedance. To measure short-circuit current, a very low-impedance ammeter is connected at the output port.

The input impedance Z_i is defined as the ratio of the rated input voltage to the corresponding current through the input terminals, while the output terminals are maintained as an open circuit.

Note that these definitions are associated with electrical devices. A generalization is possible to include both electrical and mechanical devices, by interpreting voltage and velocity as *across variables*, and current and force as *through variables*. Then, *mechanical mobility* should be used in place of electrical impedance in the associated analysis.

Using these definitions, input impedance Z_i and output impedance Z_o can be represented schematically as in Figure 8.48(a). Note that v_o is the open-circuit output voltage. When a load is connected at the output port, the voltage across the load will be different from v_o. This is caused by the presence of a current through Z_o. In the frequency domain, v_i and v_o are represented by their respective *Fourier spectra* (see Chapters 3 and 4). The corresponding transfer relation can be expressed in terms of the complex frequency response (transfer) function $G(j\omega)$ under open-circuit (no-load) conditions:

$$v_o = Gv_i \qquad (8.102)$$

Now consider two devices connected in cascade, as shown in Figure 8.48(b). It can be easily verified that the following relations apply.

$$v_{o1} = G_1 v_i \qquad (8.103)$$

$$v_{i2} = \frac{Z_{i2}}{Z_{o1} + Z_{i2}} v_{o1} \qquad (8.104)$$

$$v_o = G_2 v_{i2} \qquad (8.105)$$

These relations can be combined to give the overall input/output relation:

$$v_o = \frac{Z_{i2}}{Z_{o1} + Z_{i2}} G_2 G_1 v_i \qquad (8.106)$$

From equation (8.106), one sees that the overall frequency transfer function differs from the ideally expected product $(G_2 G_1)$ by the factor

$$\frac{Z_{i2}}{Z_{o1} + Z_{i2}} = \frac{1}{Z_{o1}/Z_{i2} + 1} \qquad (8.107)$$

Note that cascading has "distorted" the frequency-response characteristics of the two devices. If $Z_{o1}/Z_{i2} \ll 1$, this deviation becomes insignificant. From this observation, it can be concluded that when frequency response characteristics (i.e., dynamic characteristics) are important in a cascaded device, cascading should be done such that the output impedance of the first device is much smaller than the input impedance of the second device.

Impedance-Matching Amplifiers

From the analysis given in the preceding subsection, it is clear that the signal-conditioning circuitry should have a considerably large input impedance in comparison to the output impedance of the sensor-transducer unit in order to reduce loading errors. The problem is quite serious in measuring devices such as piezoelectric sensors, which have very high output impedances. In such cases, the input impedance of the signal-conditioning unit might be inadequate to reduce loading effects; also, the output signal level of these high-impedance sensors can be quite low for signal transmission, processing, and recording. The solution for this problem is to introduce several stages of amplifier circuitry between the sensor output and the data acquisition unit input. The first stage is typically an *impedance-matching amplifier* that has very high input impedance, very low output impedance, and almost unity gain. The last stage is typically a stable high-gain amplifier stage to step up the signal level. Impedance-matching amplifiers are, in fact, *operational amplifiers* with feedback.

Operational Amplifiers

Operational amplifiers (op-amps) are voltage amplifiers with very high gain K (typically 10^5 to 10^9), high input impedance Z_i (typically greater than $1 M\Omega$), and low output impedance Z_o (typically smaller than $100\ \Omega$). Thanks to the advances in integrated circuit technology, op-amps — originally made with discrete elements such as conventional transistors, diodes, and resistors — are now available as miniature units with monolithic integrated circuit elements. Because of their small size, the recent

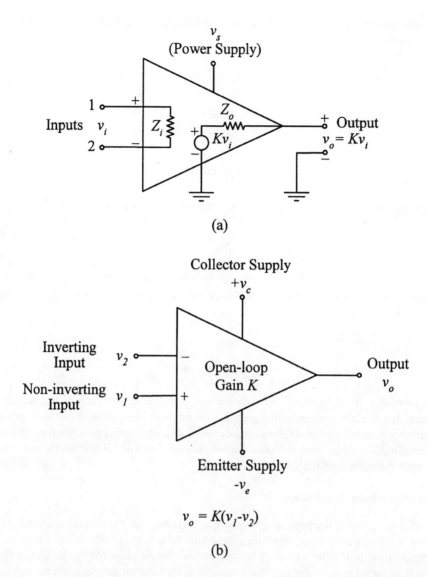

FIGURE 8.49 Operational amplifier: (a) schematic representation, and (b) symbolic representation in circuits.

trend has been to make signal-conditioning hardware an integral part of the sensor transducer unit. The name "operational amplifier" originated due to the fact that, historically, op-amps were used in analog computers for carrying out mathematical "operations" such as addition and integration.

A schematic diagram for an op-amp is shown in Figure 8.49(a). Supply voltage v_s is essential to power the op-amp. Actually, both a positive voltage v_c (collector supply) and a negative voltage v_e (emitter supply) are provided to an op-amp, with respect to ground. The associated terminals (leads) can be omitted, however, in schematic diagrams and equivalent circuits within the scope of present considerations. In the standard design of op-amps, there are two input leads (terminals), denoted by 1 and 2 in Figure 8.49(a). The lead denoted by a plus (+) sign corresponds to the *non-inverting input* and that denoted by a minus (−) sign corresponds to the *inverting input* of the op-amp. If one of the two input leads is grounded, it is a *single-ended amplifier*. If neither lead is grounded, it is a *differential amplifier* that requires two input signals. The latter arrangement rejects noise common to the two inputs (e.g., line noise, thermal noise, magnetic noise) because signal 2

Vibration Instrumentation

(at the negative terminal of the op-amp) is subtracted from signal 1 (at the positive terminal) and amplified to give the output signal. Specifically,

$$v_o = K(v_1 - v_2) \tag{8.108}$$

where K is the open-loop gain of the op-amp. The conventional circuit symbol of an op-amp is shown in Figure 8.49(b). Note that all voltages are given with respect to ground (zero). The power supply terminals $+v_c$ and $-v_e$ are often not shown in the symbolic representation.

In view of the very large values of K and Z_i, one can make the following approximations, with a high degree of accuracy, for an op-amp:

1. The voltage at the inverting input (v_2) is equal to the voltage at the non-inverting input v_1. This is clear from equation (8.108) because the output voltage v_o is a typical circuit voltage that is not large, and K is very large.
2. The current through the input terminals, both inverting and non-inverting, (its DC component is called *input-bias current*) is zero; this follows from the fact that Z_i is very large.

Note that, according to equation (8.108), when $v_1 = v_2$ (this is called the *common-mode voltage*), the output should be zero. But, in practice, there will be a very small signal, called common-mode output, under these conditions. Some compensation can be made to reject this unwanted noise. This is known as *common-mode rejection* (CMR).

Strictly speaking, K is a transfer function, and it depends on the frequency variable ω of the input signal. Typically, however, the bandwidth of an op-amp is on the order of 10 kHz; consequently, K can be assumed frequency independent in the operating frequency range. This assumption is satisfactory for most practical applications. Nevertheless, an operational amplifier in its basic open-loop form has poor stability characteristics; hence, the amplifier output can drift while the input is maintained steady. Furthermore, its gain is too high and also not very steady for direct voltage amplification of practical signals. For these reasons, additional passive elements, such as feedback resistors, are used in conjunction with op-amps in practical applications. The open-loop gain K can be eliminated in the circuit equations, retaining only the accurately known parameter values of the externally connected circuit elements. The topic of operational amplifiers is revisited in Chapter 9.

Voltage Followers

Voltage followers are impedance-matching amplifiers (or *impedance transformers*) with very high input impedance, very low output impedance, and almost-unity gain. For these reasons, they are suitable for use with high output impedance sensors such as piezoelectric devices. A schematic diagram for a voltage follower is shown in Figure 8.50(a). It consists of a standard (differential) op-amp with a feedback resistor R_f connected between the output lead and the negative (inverting) input lead. The sensor output, which is the amplifier input v_i, is connected to the positive (non-inverting) input lead of the op-amp with a series resistor R_s. The amplifier output is v_o, as shown. An equivalent circuit for a voltage follower can be drawn by combining Figures 8.49(a) and 8.50(a). Because the input impedance Z_i of the op-amp is much larger than the other impedances (Z_o, R_s, and R_f) in the circuit, the simplified equivalent circuit shown in Figure 8.50(b) is obtained. Note that v_i' is the voltage drop across Z_i. It can be shown that gain $\tilde{K}$ of the voltage follower is given by

$$\tilde{K} = \frac{K}{1+K} \tag{8.109}$$

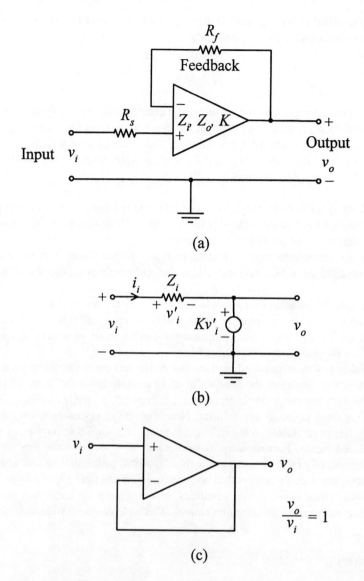

FIGURE 8.50 Voltage follower: (a) schematic representation; (b) simplified equivalent circuit; and (c) circuit symbol.

which is almost unity for large K. The input impedance $\tilde{Z}_i$ of the voltage follower is given by

$$\tilde{Z}_i = (1+K)Z_i \tag{8.110}$$

Because both Z_i and K are very large, it follows that a voltage follower clearly provides a high input impedance. Accordingly, it is able to reduce loading effects of sensors that have high output impedances. The output impedance $\tilde{Z}_o$ of a voltage follower is given by

$$\tilde{Z}_o = \frac{Z_o}{1+K} \tag{8.111}$$

Vibration Instrumentation

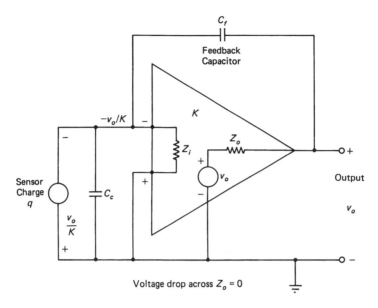

FIGURE 8.51 Charge amplifier.

Because Z_o is small to begin with and K is very large, it is clear that the output impedance of a voltage follower is very small, as desired. It follows that the voltage follower has a practically unity gain, a very high input impedance, and a very low output impedance; it can be used as an impedance transformer. This fact is emphasized in the practical circuit symbol of a voltage follower, as shown in Figure 8.50(c). The fact that $v_o = v_i$ for this circuit directly follows from a property of an op-amp, as stated before; the voltage at the inverting lead is equal to that at the non-inverting lead. Note also that in Figure 8.50(c), the feedback resistance R_f is not present (or, it is assumed that $R_f = 1$). Again, this assumption is valid due to another property of an op-amp: namely, the current through the input leads (terminals) is zero, so that the voltage at the inverting (−) terminal becomes equal to v_i, regardless of the size of the feedback resistance. By connecting a voltage follower to a high-impedance measuring device (sensor-transducer), a low-impedance output signal is obtained. Signal amplification might be necessary before this signal is transmitted or processed, however.

In many data acquisition systems, output impedance of the output amplifier is made equal to the transmission line impedance. When maximum power amplification is desired, *conjugate matching* is recommended. In this case, input impedance and output impedance of the matching amplifier are made equal to the complex conjugates of the source impedance and the load impedance, respectively.

Charge Amplifiers

The principle of capacitance feedback is utilized in charge amplifiers. These amplifiers are commonly used for conditioning the output signals from piezoelectric transducers. A schematic diagram for this device is shown in Figure 8.51. The feedback capacitance is denoted by C_f and the connecting cable capacitance by C_c. The charge amplifier views the sensor as a charge source (q), although there is an associated voltage. Using the fact that Charge = Voltage × Capacitance, a charge balance equation can be written:

$$q + \frac{v_o}{K} C_c + \left(v_o + \frac{v_o}{K}\right) C_f = 0$$

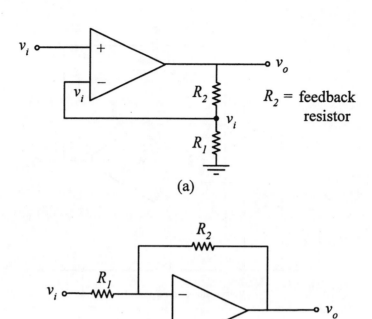

FIGURE 8.52 (a) Non-inverting amplifier, and (b) inverting amplifier.

From this, one obtains

$$v_o = -\frac{K}{[(K+1)C_f + C_c]} q \qquad (8.112)$$

If the feedback capacitance is large in comparison with the cable capacitance, the latter can be neglected. This is desirable in practice. In any event, for large values of gain K, one obtains the approximate relationship

$$v_o = -\frac{q}{C_f} \qquad (8.113)$$

Note that the output voltage is proportional to the charge generated at the sensor and depends only on the feedback parameter C_f. This parameter can be appropriately chosen in order to obtain the required output impedance characteristics. Practical charge amplifiers also have a feedback resistor R_f in parallel with the feedback capacitor C_f. Then the relationship corresponding to equation (8.112) becomes a first-order ordinary differential equation, which in turn determines the time constant of the charge amplifier. This time constant should be high. If it is low, the charge generated by the piezoelectric sensor will leak out quickly, giving erroneous results at low frequencies.

Common op-amp circuits can be modeled and analyzed without having to deal with their parameters such as the open-loop gain K and input impedance Z_i. Basically, one uses the two assumptions that were mentioned before: specifically, the voltages at the two input leads are equal and the currents through the input leads are zero. For example, consider the *non-inverting amplifier* shown in Figure 8.52(a). This is similar to a voltage follower, but with resistors R_1 and R_2 connected

Vibration Instrumentation

in a specific manner. From the assumption of equal voltages at the input leads, it is seen that the input voltage v_i is directly transmitted to the junction of the two resistors. Also, from the assumption of zero current through the input leads, it follows that there is no current in the feedback path from the junction of the two resistors. Accordingly, straightforward application of Ohms' law to the two-resistor circuit element gives (because one end is grounded):

$$\text{Current through } R_1 \text{ is } i = \frac{v_i}{R_1}$$

$$\text{Voltage drop across } R_1 + R_2 \text{ is } v_o = (R_1 + R_2)i$$

Hence,

$$v_o = (R_1 + R_2)\frac{v_i}{R_1}$$

or

$$\frac{v_o}{v_i} = 1 + \frac{R_2}{R_1} \quad (8.114)$$

This is a simple voltage amplifier of gain $1 + \frac{R_2}{R_1}$ and the sign of the output voltage is the same as that of the input voltage (hence, non-inverting). Note that the op-amp gain does not enter into the picture, and the amplifier gain is determined by the values of R_1 and R_2, which are accurately known.

Similarly, consider the *inverting amplifier* shown in Figure 8.52(b). Here, using the two assumptions for an op-amp as given before, the current summation at the common junction of the resistors R_1 and R_2 gives

$$\frac{v_i}{R_1} + \frac{v_o}{R_2} = 0$$

Note that because the non-inverting terminal of the op-amp is grounded (zero voltage), the inverting terminal, and hence the junction of the two resistors, is also at zero voltage. Rearranging the equation, one obtains

$$\frac{v_o}{v_i} = -\frac{R_2}{R_1} \quad (8.115)$$

This is a voltage amplifier of gain $\frac{R_2}{R_1}$ that is accurately defined in terms of R_1 and R_2 only. Also, the sign of the output voltage is opposite to that of the input voltage (hence, inverting).

8.6.2 Instrumentation Amplifier

In instrumentation practice, it is often required to obtain the difference between two signals (e.g., between the input and the output, giving the error signal) and then amplify this difference by a

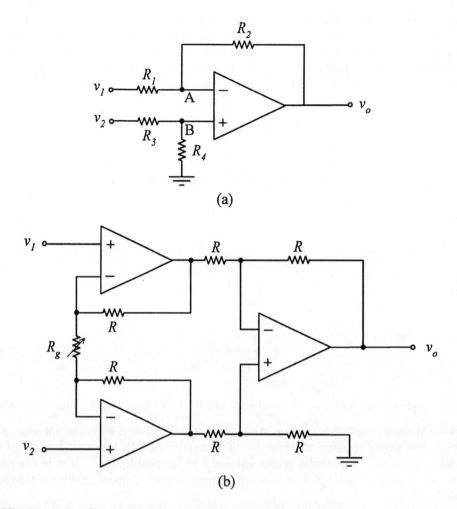

FIGURE 8.53 (a) A difference amplifier, and (b) an instrumentation amplifier.

gain parameter. A *difference amplifier* (or *differential amplifier*) can be used for this purpose. A simple op-amp circuit for a difference amplifier is shown in Figure 8.53(a). To obtain its governing equation, use the two well-known assumptions for an op-amp. It is easily seen that the voltage at the non-inverting terminal B is

$$v_B = \frac{R_4}{(R_3 + R_4)} v_2 \qquad (i)$$

Similarly, voltage v_A at the inverting terminal A is determined by the current summation

$$\frac{v_1 - v_A}{R_1} + \frac{v_o - v_A}{R_2} = 0 \qquad (ii)$$

Now, using the fact that $v_A = v_B$, from the op-amp assumption, one obtains

$$v_o = \frac{R_4}{R_3} \frac{(1 + R_2/R_1)}{(1 + R_3/R_4)} v_2 - \frac{R_2}{R_1} v_1 \qquad (8.116)$$

Vibration Instrumentation

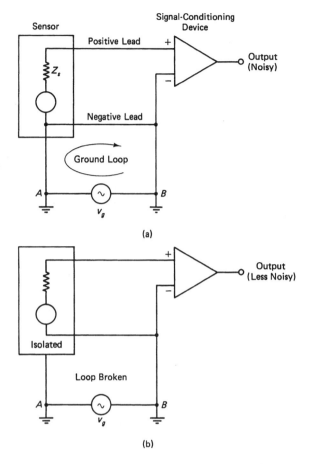

FIGURE 8.54 (a) Illustration of a ground loop, and (b) device isolation to eliminate ground loops (an example of internal isolation).

If one sets $\dfrac{R_4}{R_3} = \dfrac{R_2}{R_1}$, one obtains

$$v_o = \frac{R_2}{R_1}(v_2 - v_1) \qquad (8.117)$$

Thus, one has a difference amplifier with an accurately definable gain of R_2/R_1. One disadvantage of this arrangement is the requirement of $R_4/R_3 = R_2/R_1$. It is not convenient to maintain this relation because, in order to change the amplifier gain (while maintaining the relation), at least two parameters must be changed. For example, if one changes R_2, then one must change R_4 in proportion in order to maintain the governing equation. This problem has been overcome using the arrangement known as the *instrumentation amplifier* as shown in Figure 8.53(b). In this circuit, only the resistor R_g is varied to obtain a desired gain. It can be shown that the governing equation of the instrumentation amplifier is

$$v_o = \left(1 + \frac{2R}{R_g}\right)(v_2 - v_1) \qquad (8.118)$$

where R is a fixed resistor. The topic of instrumentation amplifier is further explored in Chapter 9.

Ground Loop Noise

In devices that handle low-level signals (e.g., accelerometers and strain-gage bridge circuitry), electrical noise can create excessive error. One form of noise is caused by fluctuating magnetic fields due to nearby AC lines. This can be avoided either by taking precautions not to have strong magnetic fields and fluctuating currents near delicate instruments or by using *fiber-optic* (optically coupled) signal transmission. Furthermore, if the two signal leads (positive and negative) are twisted or if shielded cables are used, the induced noise voltages become equal in the two leads, which cancel each other.

Another cause of electrical noise is ground loops. If two interconnected devices are grounded at two separate locations, ground loop noise can enter the signal leads because of the possible potential difference between the two ground points. The reason is that the ground itself is not generally a uniform-potential medium, and a non-zero (and finite) impedance may exist from point to point within the ground medium. This is, in fact, the case with typical ground media, such as instrument housings and common ground wire. An example is shown schematically in Figure 8.54(a). In this example, the two leads of a sensor are directly connected to a signal-conditioning device such as an amplifier. Because of nonuniform ground potentials, the two ground points A and B are subjected to a potential difference v_g. This will create a ground loop with the common negative lead of the two interconnected devices. The solution to this problem is to isolate (i.e., provide an infinite impedance to) either one of the two devices. Figure 8.54(a) shows internal isolation of the sensor. External isolation, by insulating the casing, is also acceptable. Floating off the power supply ground will also help eliminate ground loops.

PROBLEMS

8.1 What do you consider a perfect measuring device? Suppose that you are asked to develop an analog device for measuring angular position in an application related to a kinematic linkage system (a robotic manipulator, for example). What instrument ratings (or specifications) would you consider crucial in this application? Discuss their significance.

8.2 Discuss and then contrast the following terms:
   a. Measurement accuracy
   b. Instrument accuracy
   c. Measurement error
   d. Precision.

   Also, for an analog sensor-transducer unit of your choice, identify and discuss various sources of error and ways to minimize or account for their influence.

8.3 Four sets of measurements were taken on the same response variable of a machine using four different sensors. The true value of the response was known to be constant. Suppose that the four sets of data are as shown in Figure P8.3(a)–(d). Classify these data sets, and hence the corresponding sensors, with respect to precision and deterministic (repeatable) accuracy.

8.4 a. Explain why mechanical loading error due to tachometer inertia can be significantly higher when measuring transient speeds than when measuring constant speeds.
   b. A DC tachometer has an equivalent resistance $R_a = 20\ \Omega$ in its rotor windings. In a position plus velocity servo system of a mechanical positioning device, the tachometer signal is connected to a feedback control circuit with equivalent resistance 2 kΩ. Estimate the percentage error due to electrical loading of the tachometer at steady state.
   c. If the conditions were not steady, how would the electrical loading be affected in this application?

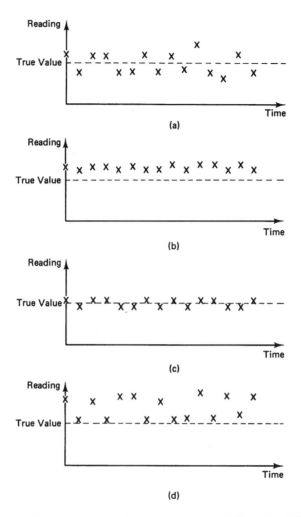

FIGURE P8.3 Four sets of measurements on the same response variable using different sensors.

8.5 Active vibration isolators known as electronic mounts have been proposed for automobile engines. Their purpose is to actively filter out the cyclic excitation forces generated by the internal-combustion engines before they would adversely vibrate components such as seats, floor, and steering column that come into contact with the vehicle occupants (see Chapter 12). Consider a four-stroke, four-cylinder engine. It is known that the excitation frequency on the engine mounts is twice the crankshaft speed, as a result of the firing cycles of the cylinders. A schematic representation of an active engine mount is shown in Figure P8.5(a). The crankshaft speed is measured and supplied to the controller of a valve actuator. The servo–valve of a hydraulic cylinder is operated on the basis of this measurement. The hydraulic cylinder functions as an active suspension with a variable (active) spring and a damper. A simplified model of the mechanical interactions is shown in Figure P8.5(b).

   a. Neglecting gravity forces (which cancel out due to the static spring force), show that a linear model for system dynamics can be expressed as

$$m\ddot{y} + b\dot{y} + ky = f_i$$

$$b\dot{y} + ky + f_o$$

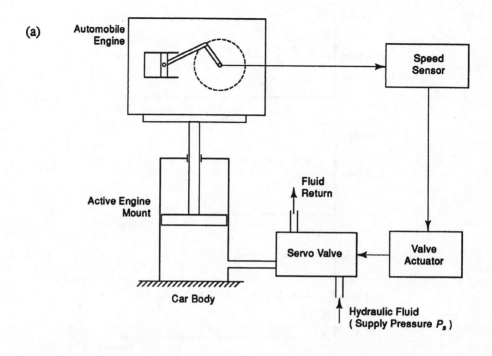

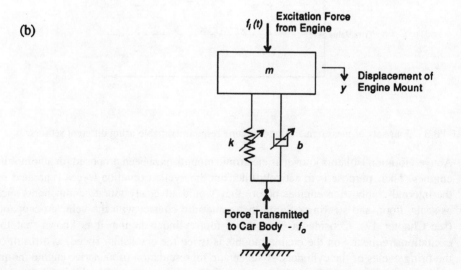

FIGURE P8.5 An active engine mount for an automobile: (a) schematic diagram, and (b) approximate model.

where

f_i = excitation force from the engine
f_o = force transmitted to the passenger compartment (car body)
y = displacement of the engine mount with respect to a frame fixed to the passenger compartment
m = mass of the engine unit
k = equivalent stiffness of the active mount
b = equivalent viscous damping constant of the active mount.

b. Determine the transfer function (with the Laplace variable s) f_o/f_i for the system.
c. Sketch the magnitude versus frequency curve of the transfer function obtained in part (b), and show a suitable operating range for the active mount.
d. For a damping ratio $\zeta = 0.2$, what is the magnitude of the transfer function when the excitation frequency ω is 5 times the natural frequency ω_n of the suspension (engine mount) system?
e. Suppose that the magnitude estimated in part (d) is satisfactory for the purpose of vibration isolation. If the engine speed varies from 600 rpm to 1200 rpm, what is the range in which the spring stiffness k (N·m^{-1}) should be varied by the control system in order to maintain this level of vibration isolation? Assume that the engine mass $m = 100$ kg, and the damping ratio is approximately constant at $\zeta = 0.2$.

8.6 Giving examples, discuss situations in which measurement of more than one type of kinematic variable using the same measuring device is:
a. An advantage
b. A disadvantage.

8.7 Giving examples for suitable auxiliary front-end elements, discuss the use of a force sensor to measure:
a. Displacement
b. Velocity
c. Acceleration.

8.8 Write the expression for loading nonlinearity error (percentage) in a rotatory potentiometer in terms of the angular displacement, maximum displacement (stroke), potentiometer element resistance, and load resistance. Plot the percentage error as a function of the fractional displacement for the three cases $R_L/R_c = 0.1$, 1.0, and 10.0.

8.9 A vibrating system has an effective mass M, an effective stiffness K, and an effective damping constant B in its primary mode of vibration at point A with respect to coordinate y. Write expressions for the undamped natural frequency, the damped natural frequency, and the damping ratio for this first mode of vibration of the system. A displacement trandsducer is used to measure the fundamental undamped natural frequency and the damping ratio of the system by subjecting the system to an initial excitation and recording the displacement trace at a suitable location (point A along y in the Figure P8.9) in the system. This trace will provide the period of damped oscillations and the logarithmic decrement of the exponential decay from which the required parameters can be computed using well-known relations (see Chapter 7). It was found, however, that the mass m of the moving part of the displacement sensor and the associated equivalent viscous damping constant b are not negligible. Using the model shown in Figure P8.9, derive expressions for the measured undamped natural frequency and damping ratio. Suppose that $M = 10$ kg, $K = 10$ N·m^{-1}, and $B = 2$ N·m^{-1}·s. Consider an LVDT whose core weighs 5 g and has negligible damping, and a potentionmeter whose slider arm weighs 5 g and has an equivalent viscous damping constant of 0.05 N·m^{-1}·s. Estimate the percentage error of the results for the undamped natural frequency and damping ratio measured using each of these two displacement sensors.

8.10 It is known that some of the factors that should be considered in selecting an LVDT for a particular application are linearity, sensitivity, response time, size and mass of core, size of the housing, primary excitation frequency, output impedance, phase change between primary and secondary voltages, null voltage, stroke, and environmental effects (temperature compensation, magnetic shielding, etc.). Explain why and how each of these factors is an important consideration.

8.11 A high-performance LVDT has a linearity rating of 0.01% in its output range of 0.1 to 1.0 V AC. The response time of the LVDT is known to be 10 ms. What should be the frequency of the primary excitation?

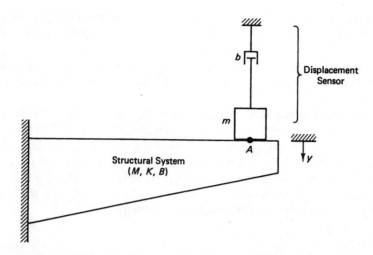

FIGURE P8.9 The use of a displacement sensor to measure the natural frequency and damping ratio of a structure.

8.12 For directional sensing using an LVDT, it is necessary to determine the phase angle of the induced signal. In other words, *phase-sensitive demodulation* is needed.
   a. First, consider a linear core displacement starting from a positive value, moving to zero, and then returning to the same position in an equal time period. Sketch the output of the LVDT for this "triangular" core displacement.
   b. Next, sketch the output if the core continued to move to the negative side at the same speed.

   By comparing the two outputs, show that phase-sensitive demodulation would be needed to distinguish between the two cases of displacement.

8.13 Compare and contrast the principles of operation of a DC tachometer and an AC tachometer (both permanent-magnet and induction types). What are the advantages and disadvantages of these two types of tachometers?

8.14 Discuss the relationships among displacement or vibration sensing, distance sensing, position sensing, and proximity sensing. Explain why the following characteristics are important in using some types of motion sensors:
   a. Material of the moving (or target) object
   b. Shape of the moving object
   c. Size (including mass) of the moving object
   d. Distance (large or small) of the target object
   e. Nature of motion (transient or not, what speed, etc.) of the moving object
   f. Environmental conditions (humidity, temperature, magnetic fields, dirt, lighting conditions, shock and vibration, etc.).

8.15 Compression molding is used in making parts of complex shapes and varying sizes. Typically, the mold consists of two platens, the bottom platen fixtured to the press table and the top platen operated by a hydraulic press. Metal or plastic sheets — for example, for the automotive industry — can be compression-molded in this manner. The main requirement in controlling the press is to accurately position the top platen with respect to the bottom platen (e.g., with a 0.001 in or 0.025 mm tolerance), and it has to be done quickly (e.g., in a few seconds) without residual vibrations. How many degrees of freedom have to be sensed (how many position sensors are needed) in controlling the mold? Suggest typical displacement measurements that would be made in this application and the types of sensors that could be employed. Indicate sources of error that cannot be perfectly compensated for in this application.

Vibration Instrumentation

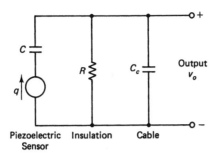

FIGURE P8.18 Equivalent circuit for a quartz crystal (piezoelectric) accelerometer.

8.16 Discuss factors that limit the lower and upper frequency limits of the output from the following measuring devices:
   a. Potentiometer
   b. LVDT
   c. Eddy current proximity sensor
   d. DC tachometer
   e. Piezoelectric transducer.

8.17 An active suspension system is proposed for a high-speed ground transit vehicle in order to achieve improved ride quality. The system senses jerk (rate of change of acceleration) due to road disturbances and adjusts system parameters accordingly.
   a. Draw a suitable schematic diagram for the proposed control system and describe appropriate measuring devices.
   b. Suggest a way to specify the "desired" ride quality for a given type of vehicle. (Would you specify one value of jerk, a jerk range, or a curve with respect to time or frequency?)
   c. Discuss the drawbacks and limitations of the proposed control system with respect to such facts as reliability, cost, feasibility, and accuracy.

8.18 A design objective in most control system applications is to achieve small time constants. An exception is the time constant requirement for a piezoelectric sensor. Explain why a large time constant, on the order of 10 s, is desirable for a piezoelectric sensor in combination with its signal conditioning system. An equivalent circuit for a piezoelectric accelerometer that uses a quartz crystal as the sensing element is shown in Figure P8.18. The charge generated is denoted by q, and the voltage output at the end of the accelerometer cable is v_o. The sensor capacitance is modeled by C, and the overall capacitance experienced at the sensor output, whose primary contribution is due to cable capacitance, is denoted by C_c. The resistance of the electric insulation in the accelerometer is denoted by R. Write a differential equation relating v_o to q. What is the corresponding transfer function? Using this result, show that the accuracy of the accelerometer improves when the sensor time constant is large and when the frequency of the measured acceleration is high. For a quartz crystal sensor with $R = 1 \times 10^{11}$ Ω and $C_c = 1000$ pF, compute the time constant.

8.19 Applications of accelerometers are found in the following areas:
   a. Transit vehicles (automobiles, aircraft, ships, etc.)
   b. Power cable monitoring
   c. Robotic manipulator control
   d. Building structures
   e. Shock and vibration testing
   f. Position and velocity sensing.
   Describe one direct use of acceleration measurement in each application area.

8.20 a. A standard accelerometer that weighs 100 g is mounted on a test object that has an equivalent mass of 3 kg. Estimate the accuracy in the first natural frequency of the

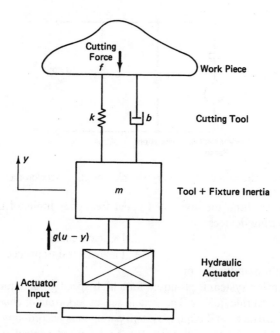

FIGURE P8.21 A model of a machining operation.

object measured using this arrangement, considering mechanical loading due to the accelerometer mass alone. If a miniature accelerometer that weighs 0.5 g is used instead, what is the resulting accuracy?

b. A strain-gage accelerometer uses a semiconductor strain gage mounted at the root of a cantilever element, with the seismic mass mounted at the free end of the cantilever. Suppose that the cantilever element has a square cross section with dimensions 1.5×1.5 mm². The equivalent length of the cantilever element is 25 mm, and the equivalent seismic mass is 0.2 g. If the cantilever is made of an aluminum alloy with Young's modulus $E = 69 \times 10^9$ N·m⁻², estimate the useful frequency range of the accelerometer in hertz. *Hint*: When a force F is applied to the free end of a cantilever, the deflection y at that location can be approximated by the formula

$$y = \frac{F\ell^3}{3EI}$$

where ℓ = cantilever length
I = second moment area of the cantilever cross section about the bending neutral axis = $bh^3/12$
b = cross-section width
h = cross-section height.

8.21 A model for a machining operation is shown in Figure P8.21. The cutting force is denoted by f, and the cutting tool with its fixtures is modeled by a spring (stiffness k), a viscous damper (damping constant b), and a mass m. The actuator (hydraulic) with its controller is represented by an active stiffness g. Obtain a transfer relation between the actuator input u and the cutting force f. Discuss a control strategy for counteracting effects due to random variations in the cutting force. Note that this is important for controlling the product quality.

Vibration Instrumentation

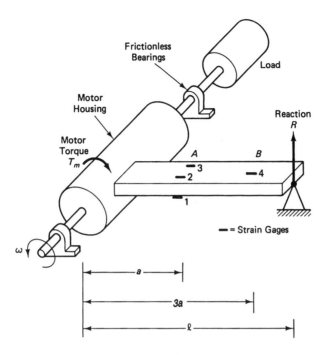

FIGURE P8.22 Use of a strain-gage sensor for measuring motor torque.

8.22 The use of strain-gage sensors to measure the torque T_m generated by a motor is shown schematically in Figure P8.22. The motor is floated on frictionless bearings. A uniform rectangular lever arm is rigidly attached to the motor housing, and its projected end is restrained by a pin joint. Four identical strain gages are mounted on the lever arm, as shown. Three of the strain gages are at point A, which is located at a distance a from the motor shaft; and the fourth strain gage is at point B, which is located at a distance $3a$ from the motor shaft. The pin joint is at a distance ℓ from the motor shaft. Strain gages 2, 3, and 4 are on the top surface of the lever arm, and gage 1 is on the bottom surface. Obtain an expression for T_m in terms of the bridge output δv_o and the following additional parameters:

S_s = gage factor (strain-gage sensitivity)
v_{ref} = supply voltage to the bridge
b = width of the lever arm cross section
h = height of the lever arm cross section
E = Young's modulus of the lever arm.

Verify that the bridge sensitivity does not depend on ℓ. Describe means to improve the bridge sensitivity. Explain why the sensor reading is only an approximation to the torque transmitted to the load. Give a relation to determine the net normal reaction force at the bearings, using the bridge output.

8.23 A bridge with two active strain gages is being used to measure bending moment M [Figure P8.23(a)] and torque T [Figure P8.23(b)] in a machine part. Using sketches, suggest the orientations of the two gages mounted on the machine part and the corresponding bridge connections in each case in order to obtain the best sensitivity from the bridge. What is the value of the bridge constant in each case?

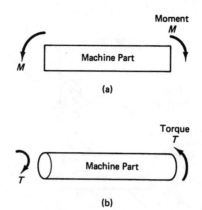

FIGURE P8.23 Sensing elements: (a) bending member, and (b) torsion member.

8.24 Compare the potentiometer (ballast) circuit with the Wheatstone bridge circuit for strain-gage measurements with respect to the following considerations:
   a. Sensitivity to the measured strain
   b. Error due to ambient effects (e.g., temperature change)
   c. Signal-to-noise ratio of the output voltage
   d. Circuit complexity and cost
   e. Linearity.

8.25 Discuss the advantages and disadvantages of the following techniques in the context of measuring transient signals:
   a. DC bridge circuits versus AC bridge circuits
   b. Slip ring and brush commutators versus AC transformer commutators
   c. Strain-gage torque sensors versus variable-inductance torque sensors
   d. Piezoelectric accelerometers versus strain-gage accelerometers
   e. Tachometer velocity transducers versus piezoelectric velocity transducers

8.26 For a semiconductor strain gage characterized by the quadratic strain-resistance relationship

$$\frac{\delta R}{R} = S_1 \varepsilon + S_2 \varepsilon^2$$

obtain an expression for the equivalent gage factor (sensitivity) S_s using the least-squares error linear approximation. Assume that only positive strains up to ε_{max} are measured with the gage. Derive an expression for the percentage nonlinearity. Taking $S_1 = 117$, $S_2 = 3600$, and $\varepsilon_{max} = 0.01$ strain, compute S_s and the percentage nonlinearity.

8.27 Briefly describe how strain gages can be used to measure:
   a. Force
   b. Displacement
   c. Acceleration.

Show that if a compensating resistance R_c is connected in series with the supply voltage v_{ref} to a strain-gage bridge that has four identical members, each with resistance R, the output equation is given by

$$\frac{\delta v_o}{v_{ref}} = \frac{R}{(R+R_c)} \frac{kS_s}{4} \varepsilon$$

in the usual rotation.

Vibration Instrumentation

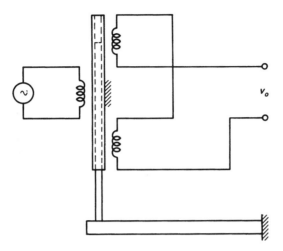

FIGURE P8.28 An analog sensor.

A foil-gage load cell uses a simple (one-dimensional) tensile member to measure force. Suppose that k and S_s are insensitive to temperature change. If the temperature coefficient of R is α_1, that of the series compensating resistance R_c is α_2, and that of the Young's modulus of the tensile member is $(-\beta)$, determine an expression for R_c that would result in automatic (self-) compensation for temperature effects. Under what conditions is this arrangement realizable?

8.28 Figure P8.28 shows a schematic diagram of a measuring device.
   a. Identify the various components in this device.
   b. Describe the operation of the device, explaining the function of each component and identifying the nature of the measurand and the output of the device.
   c. List the advantages and disadvantages of the device.
   d. Describe a possible application of this device.

8.29 Discuss factors that limit the lower and upper frequency limits of measurements obtained from the following devices:
   a. Strain gage
   b. Rotating shaft torque sensor
   c. Reaction torque sensor.

8.30 Briefly describe a situation in which tension in a moving belt or cable has to be measured under transient conditions. What are some of the difficulties associated with measuring tension in a moving member? A strain-gage tension sensor for a belt-drive system is shown in Figure P8.30. Two identical active strain gages, G_1 and G_2, are mounted at the root of a cantilever element with rectangular cross section, as shown. A light, frictionless pulley is mounted at the free end of the cantilever element. The belt makes a 90° turn when passing over this idler pulley.
   a. Using a circuit diagram, show the Wheatstone bridge connections necessary for the strain gages G_1 and G_2 so that the strains due to the axial forces in the cantilever member have no effect on the bridge output (i.e., effects of axial loads are compensated) and the sensitivity to the bending loads is maximized.
   b. Obtain an equation relating the belt tension T and the bridge output δv_o in terms of the following additional parameters:

   S_s = gage factor (sensitivity) of each strain gage
   E = Young's modulus of the cantilever element

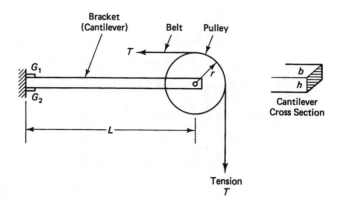

FIGURE P8.30 A strain-gage tension sensor.

L = length of the Cantilever element
b = width of the cantilever cross section
h = height of the cantilever cross section.

Note that the radius of the pulley does not enter into this equation.

8.31 Show that in a Wheatstone bridge circuit if the resistance elements R_1 and R_2 have the same temperature coefficient of resistance and if R_3 and R_4 have the same temperature coefficient of resistance, the temperature effects are compensated up to first order. A strain-gage accelerometer uses two semiconductor strain gages, one integral with the cantilever element near the fixed end (root) and the other mounted at an unstrained location in the accelerometer housing. Describe the operation of the accelerometer. What is the purpose of the second strain gage?

8.32 Consider the following types of sensors, and briefly explain whether they can be used in measuring liquid oscillations. Also, what are the limitations of each type?
  a. Capacitive sensors
  b. Inductive sensors
  c. Ultrasonic sensors.

8.33 Consider the following types of vibration sensors: inductive, capacitive, eddy current, fiber optic, and ultrasonic. For the following conditions, indicate which of these types are not suitable and explain why.
  a. Environment with variable humidity
  b. Target object made of aluminum
  c. Target object made of steel
  d. Target object made of plastic
  e. Target object several feet away from the sensor location
  f. Environment with significant temperature fluctuations
  g. Smoke-filled environment.

8.34 Discuss advantages and disadvantages of fiber-optic sensors. Consider a fiber-optic vibration sensor. In which region of the light intensity curve would you prefer to operate the sensor, and what are the corresponding limitations?

8.35 Analyze a single-axis rate gyro. Obtain a relationship between the gimbal angle θ and the angular velocity Ω of the mounting structure (e.g., a missile) about the gimbal axis. Use the following parameters:

J = moment of inertia of the gyroscopic disk about the spinning axis
ω = angular speed of spin

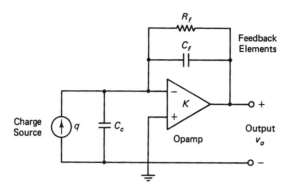

FIGURE P8.37 Schematic diagram for a charge amplifier.

k = torsional stiffness of the gimbal restraint.

Assume that Ω is constant and the conditions are steady. How would you improve the sensitivity of this device? Discuss any problems associated with the suggested methods of sensitivity improvement and ways to reduce them.

8.36 Define electrical impedance and mechanical impedance. Identify a defect in these definitions in relation to the force–current analogy. What improvements would you suggest? What roles do input impedance and output impedance play in relation to the accuracy of a measuring device?

8.37 A schematic diagram for a charge amplifier (with resistive feedback) is shown in Figure P8.37. Obtain the differential equation governing the response of the charge amplifier. Identify the time constant of the device and discuss its significance. Would you prefer a charge amplifier to a voltage follower for conditioning signals from a piezoelectric accelerometer? Explain.

8.38 What is meant by "loading error" in a signal measurement? Also, suppose that a piezoelectric sensor of output impedance Z_s is connected to a voltage-follower amplifier of input impedance Z_i. The sensor signal is v_i volts and the amplifier output is v_o volts. The amplifier output is connected to a device with very high input impedance. Plot to scale the signal ratio v_o/v_i against the impedance ratio Z_i/Z_s for values of the impedance ratio in the range 0.1 to 10.

8.39 Thevenin's theorem states that with respect to the characteristics at an output port, an unknown subsystem consisting of linear passive elements and ideal source elements can be represented by a single across-variable (voltage) source v_{eq} connected in series with a single impedance Z_{eq}. This is illustrated in Figure P8.39(a) and P8.39(b). Note that v_{eq} is equal to the open-circuit across variable v_{oc} at the output port because the current through Z_{eq} is zero. Consider the network shown in Figure P8.39(c). Determine the equivalent voltage source v_{eq} and the equivalent series impedance Z_{eq}, in the frequency domain, for this circuit.

8.40 Using suitable impedance circuits, explain why a voltmeter should have a high resistance and an ammeter should have a very low resistance. What are some of the design implications of these general requirements for the two types of measuring instruments, particularly with respect to instrument sensitivity, speed of response, and robustness? Use a classical moving-coil meter as the model for your discussion.

8.41 Define the following terms:
a. Mechanical loading
b. Electrical loading
in the context of motion sensing, and explain how these loading effects can be reduced.

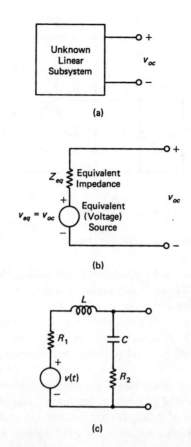

FIGURE P8.39 Illustration of Thevenin's theorem: (a) unknown linear subsystem; (b) equivalent representation; and (c) example.

The following table gives ideal values for some parameters of an operational amplifier. Give typical, practical values for these parameters (e.g., output impedance of 50 Ω).

Parameter	Ideal Value	Typical Value
Input impedance	Infinity	?
Output impedance	Zero	50 Ω
Gain	Infinity	?
Bandwidth	Infinity	?

Also, ideally, inverting-lead voltage is equal to the noninverting-lead voltage (i.e., offset voltage is zero).

8.42 A light-emitting diode (LED) and a photodetector (phototransistor or photodiode) in a single package can be used to measure tip vibrations of a cantilever beam, as schematically shown in Figure P8.42. Alternatively, a strain gage mounted at the root of the cantilever can be used. Identify several advantages and disadvantages of each of these two approaches to vibration sensing. Indicate a practical application to which these concepts of vibration sensing can be extended.

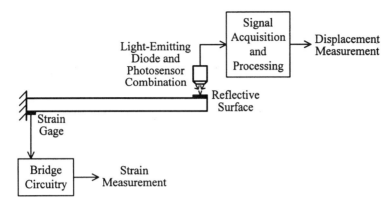

FIGURE P8.42 Optical and strain gage methods of vibration sensing.

9 Signal Conditioning and Modification

Signal modification is an important function in many applications of vibration. The tasks of signal modification can include *signal conditioning* (e.g., amplification, and analog and digital filtering), *signal conversion* (e.g., analog-to-digital conversion, digital-to-analog conversion, voltage-to-frequency conversion, and frequency-to-voltage conversion), *modulation* (e.g., amplitude modulation, frequency modulation, phase modulation, pulse-width modulation, pulse-frequency modulation, and pulse-code modulation), and *demodulation* (the reverse process of modulation). In addition, many other types of useful signal modification operations can be identified. For example, *sample-and-hold circuits* are used in digital data acquisition systems. Devices such as *analog and digital multiplexers* and *comparators* are needed in many applications of data acquisition and processing. Phase shifting, curve shaping, offsetting, and linearization can also be classified as signal modification. This chapter describes signal conditioning and modification operations that are useful in vibration applications. Signal modification plays a crucial role in component interfacing. When two devices are interfaced, it is essential to guarantee that a signal leaving one device and entering the other will do so at proper signal levels (voltage, current, power), in the proper form (analog, digital), and without distortion (loading and impedance considerations). For transmission, a signal should be properly modified (by amplification, modulation, digitizing, etc.) so that the signal-to-noise ratio of the transmitted signal is sufficiently large at the receiver. The significance of signal modification is clear from these observations. The material covered in this chapter is intimately related to what has been discussed in chapters on signal analysis and instrumentation (see Chapters 4 and 8).

9.1 AMPLIFIERS

The level of an electrical signal can be represented by variables such as voltage, current, and power. Across variables, through variables, and power variables that are analogous can be defined for other types of signals (e.g., mechanical) as well. Signal levels at various interface locations of components in a vibratory system must be properly adjusted for correct performance of these components and the overall system. For example, input to an actuator should possess adequate power to drive the actuator. A signal should maintain its signal level above some threshold during transmission so that errors due to signal weakening are not excessive. Signals applied to digital devices must remain within the specified logic levels. Many types of sensors produce weak signals that must be upgraded before they can be fed into a monitoring system, data processor, controller, or data logger.

Signal amplification concerns proper adjustment of the signal level for performing a specific task. Amplifiers are used to accomplish signal amplification. An amplifier is an active device that needs an external power source to operate. Although active circuits — amplifiers in particular — can be developed in the monolithic form using an original integrated-circuit (IC) layout so as to accomplish a particular amplification task, it is convenient to study their performance using the *operational amplifier* (op-amp) as the basic building block. Of course, operational amplifiers are widely used not only for modeling and analyzing other types of amplifiers, but also as basic building blocks in building these various kinds of amplifiers. For these reasons, the present discussion on amplifiers will focus on the operational amplifier. An introduction to this topic was presented in Chapter 8.

9.1.1 OPERATIONAL AMPLIFIER

The origin of the operational amplifier dates back to the 1940s when the vacuum tube operational amplifier was introduced. Operational amplifier or *op-amp* got its name due to the fact that it was originally used almost exclusively to perform mathematical operations; for example, in analog computers. Subsequently, in the 1950s, the transistorized op-amp was developed. It used discrete elements such as *bipolar junction transistors* and *resistors*. Still, it was too large in size, consumed too much power, and was too expensive for widespread use in general applications. This situation changed in the late 1960s when the *integrated-circuit* (IC) op-amp was developed in the monolithic form, as a single IC chip. Today, the IC op-amp, which consists of a large number of circuit elements on a *substrate* of typically a single *silicon crystal* (the monolithic form), is a valuable component in almost any signal modification device.

An op-amp can be manufactured in the discrete-element form using, say, ten bipolar junction transistors and as many discrete resistors or alternatively (and preferably) in the modern monolithic form as an IC chip that may be equivalent to over 100 discrete elements. In any form, the device has an *input impedance* Z_i, an *output impedance* Z_o, and a *gain K*. Hence, a schematic model for an op-amp can be given as in Figure 9.1(a). The conventional symbol of an op-amp is shown in Figure 9.1(b). Typically, there are about six terminals (lead connections) to an op-amp. For example, there are two input leads (a *positive lead* with voltage v_{ip} and a *negative load* with voltage v_{in}), an output lead (voltage v_o), two bipolar power supply leads ($+v_s$ and $-v_s$), and a ground lead.

Note from Figure 9.1(a) that under open-loop (no feedback) conditions,

$$v_o = K v_i \tag{9.1}$$

in which the input voltage v_i is the differential input voltage defined as the algebraic difference between the voltages at the positive and negative lead; thus,

$$v_i = v_{ip} - v_{in} \tag{9.2}$$

The *open-loop voltage gain K* is very high (10^5 to 10^9) for a typical op-amp. Furthermore, the input impedance Z_i could be as high as 1 MΩ and the output impedance is low, on the order of 10 Ω. Because v_o is typically 1 to 10 V, from equation (9.1), it follows that $v_i \cong 0$ since K is very large. Hence, from equation (9.2), $v_{ip} \cong v_{in}$. In other words, the voltages at the two input leads are nearly equal. Now, if one applies a large voltage differential v_i (say, 1 V) at the input, then according to equation (9.1), the output voltage should be extremely high. This never happens in practice, however, because the device saturates quickly beyond moderate output voltages (of the order of 15 V).

From equations (9.1) and (9.2), it is clear that if the negative input lead is grounded (i.e., $v_{in} = 0$), then

$$v_o = K v_{ip} \tag{9.3}$$

and if the positive input lead is grounded (i.e., $v_{ip} = 0$), then

$$v_o = -K v_{in} \tag{9.4}$$

Accordingly, v_{ip} is termed *noninverting input* and v_{in} is termed *inverting input*.

Signal Conditioning and Modification

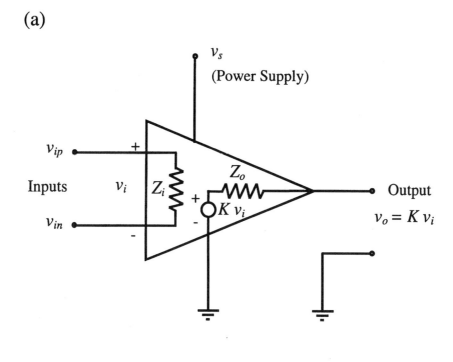

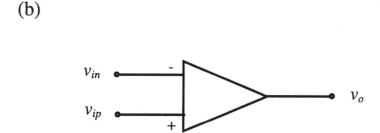

FIGURE 9.1 Operational amplifier: (a) a schematic model, and (b) conventional symbol.

EXAMPLE 9.1

Consider an op-amp having an open loop gain of 1×10^5. If the saturation voltage is 15 V, determine the output voltage in the following cases:

(a) 5 μV at the positive lead and 2 μV at the negative lead
(b) −5 μV at the positive lead and 2 μV at the negative lead
(c) 5 μV at the positive lead and −2 μV at the negative lead
(d) −5 μV at the positive lead and −2 μV at the negative lead
(e) 1 V at the positive lead and negative lead grounded
(f) 1 V at the negative lead and positive lead grounded.

SOLUTION

This problem can be solved using equations (9.1) and (9.2). The results are given in Table 9.1. Note that in the last two cases, the output will saturate and equation (9.1) will no longer hold.

TABLE 9.1
Solution to Example 9.1

v_{ip}	v_{in}	v_i	v_o
5 μV	2 μV	3 μV	0.3 V
−5 μV	2 μV	−7 μV	−0.7 V
5 μV	−2 μV	7 μV	0.7 V
−5 μV	−2 μV	−3 μV	−0.3 V
1 V	0	1 V	15 V
0	1 V	−1 V	−15 V

□

Field effect transistors (FETs) — for example, metal oxide semiconductor field effect transistors (MOSFETs) — could be used in the IC form of an op-amp. The MOSFET type has advantages over many other types; for example, higher input impedance and more stable output (almost equal to the power supply voltage) at saturation, making the MOSFET op-amps preferable over bipolar junction transistor op-amps in many applications.

In analyzing operational amplifier circuits under unsaturated conditions, one can use the following two characteristics of an op-amp:

1. Voltages of the two input leads should be (almost) equal.
2. Currents through each of the two input leads should be (almost) zero.

As explained earlier, the first property is credited to high open-loop gain, and the second property to high input impedance in an operational amplifier. These two properties are repeatedly used to obtain input-output equations for amplifier circuits and systems.

9.1.2 USE OF FEEDBACK IN OP-AMPS

An operational amplifier is a very versatile device, primarily due to its very high input impedance, low output impedance, and very high gain. However, it cannot be used without modification as an amplifier because it is not very stable in the form shown in Figure 9.1. Two factors that contribute to this problem are:

1. Frequency response
2. Drift.

Stated another way, op-amp gain K does not remain constant; it can vary with the frequency of the input signal (i.e., frequency response function is not flat in the operating range); and also, it can vary with time (i.e., *drift*). Frequency response problems arise due to circuit dynamics of an operational amplifier. This problem is usually not severe unless the device is operated at very high frequencies. Drift problems arise due to the sensitivity of gain K to environmental factors such as temperature, light, humidity, and vibration, and as a result of the variation of K due to aging. Drift in an op-amp can be significant, and steps should be taken to remove this problem.

It is virtually impossible to avoid gain drift and frequency-response error in an operational amplifier. But an ingenious way has been found to remove the effect of these two problems at the amplifier output. Because gain K is very large, by using feedback, one can virtually eliminate its effect at the amplifier output. This *closed-loop* form of an op-amp is preferred in almost every application. In particular, *voltage followers* and *charge amplifiers* are devices that use the properties of high Z_i, low Z_o, and high K of the op-amp, along with feedback through a precision resistor, to eliminate errors due to non-constant K. In summary, an operational amplifier is not very useful in its *open-loop* form, particularly because gain K is not steady. But because K is very large, the problem can be removed using feedback. It is this *closed-loop* form that is commonly used in practical applications of an op-amp.

In addition to the nonsteady nature of gain, there are other sources of error that contribute to the less-than-ideal performance of an operational amplifier circuit. Noteworthy are:

1. The *offset current* present at the input leads due to bias currents that are needed to operate the solid-state circuitry
2. The *offset voltage* that might be present at the output even when the input leads are open
3. The unequal gains corresponding to the two input leads (i.e., the *inverting gain* not equal to the *noninverting gain*).

Such problems can produce nonlinear behavior in op-amp circuits, and they can be reduced by proper circuit design and through the use of compensating circuit elements.

9.1.3 Voltage, Current, and Power Amplifiers

Any type of amplifier can be constructed from scratch in the monolithic form as an IC chip, or in the discrete form as a circuit containing several discrete elements such as discrete bipolar junction transistors or discrete field effect transistors, discrete diodes, and discrete resistors. However, almost all types of amplifiers can also be built using operational amplifier as the basic building block. Because one is already familiar with op-amps, and because op-amps are extensively used in general amplifier circuitry, the latter approach — which uses discrete op-amps for the modeling of general amplifiers — is preferred.

If an electronic amplifier performs a voltage amplification function, it is termed a *voltage amplifier*. These amplifiers are so common that the term *amplifier* is often used to denote a voltage amplifier. A voltage amplifier can be modeled as

$$v_o = K_v v_i \tag{9.5}$$

where

v_o = output voltage
v_i = input voltage
K_v = voltage gain.

Voltage amplifiers are used to achieve voltage compatibility (or level shifting) in circuits.

Current amplifiers are used to achieve current compatibility in electronic circuits. A *current amplifier* can be modeled by

$$i_o = K_i i_i \tag{9.6}$$

where

i_o = output current
i_i = input current
K_i = current gain.

Note that the voltage follower has $K_v = 1$ and, hence, it can be considered as a current amplifier. Also, it provides impedance compatibility and acts as a buffer between a low-current (high-impedance) output device (the device that provides the signal) and a high-current (low-impedance) input device (device that receives the signal) that are interconnected. Hence, the name *buffer amplifier* or *impedance transformer* is sometimes used for a current amplifier with unity voltage gain.

If the objective of signal amplification is to upgrade the associated power level, then a *power amplifier* should be used for that purpose. A simple model for a power amplifier is

$$p_o = K_p p_i \tag{9.7}$$

where

p_o = output power
p_i = input power
K_p = power gain.

It is easy to see from equations (9.5) through (9.7) that

$$K_p = K_v K_i \tag{9.8}$$

Note that all three types of amplification can be achieved simultaneously from the same amplifier. Furthermore, a current amplifier with unity voltage gain (e.g., a voltage follower) is a power amplifier as well. Usually, voltage amplifiers and current amplifiers are used in the first stages of a signal path (e.g., sensing, data acquisition, and signal generation) where signal levels and power levels are relatively low. Power amplifiers are typically used in the final stages (e.g., actuation, recording, display) where high signal levels and power levels are usually required.

Figure 9.2(a) shows an op-amp-based voltage amplifier. Note the feedback resistor R_f that serves the purposes of stabilizing the op-amp and providing an accurate voltage gain. The negative lead is grounded through an accurately known resistor R. To determine the voltage gain, recall that the voltages at the two input leads of an op-amp should be virtually equal. The input voltage v_i is applied to the positive lead of the op-amp. Then the voltage at point A should also be equal to v_i. Next, recall that the current through the input lead of an op-amp is virtually 0. Hence, by writing the current balance equation for the node point A, one obtains

$$\frac{v_o - v_i}{R_f} = \frac{v_i}{R}$$

This gives the amplifier equation

$$v_o = \left(1 + \frac{R_f}{R}\right) v_i \tag{9.9}$$

Hence, the voltage gain is given by

$$K_v = 1 + \frac{R_f}{R} \tag{9.10}$$

Note the K_v depends on R and R_f, and not on the op-amp gain. Hence, the voltage gain can be accurately determined by selecting the two resistors R and R_f precisely. Also note that the output

Signal Conditioning and Modification

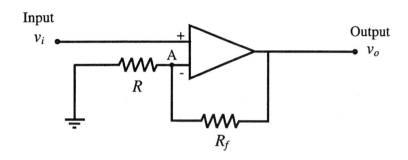

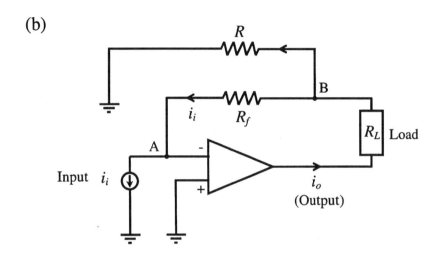

FIGURE 9.2 (a) A voltage amplifier, and (b) a current amplifier.

voltage has the same sign as the input voltage. Hence, this is a *noninverting amplifier*. If the voltages are of the opposite sign, it will be an *inverting amplifier*.

A current amplifier is shown in Figure 9.2(b). The input current i_i is applied to the negative lead of the op-amp as shown, and the positive lead is grounded. There is a feedback resistor R_f connected to the negative lead through the load R_L. The resistor R_f provides a path for the input current because the op-amp takes in virtually zero current. There is a second resistor R through which the output is grounded. This resistor is needed for current amplification. To analyze the amplifier, note that the voltage at point A (i.e., at the negative lead) should be 0 because the positive lead of the op-amp is grounded (zero voltage). Furthermore, the entire input current i_i passes through resistor R_f as shown. Hence, the voltage at point B is $R_f i_i$. Consequently, current through resistor R is $R_f i_i / R$, which is positive in the direction shown. It follows that the output current i_o is given by

$$i_o = i_i + \frac{R_f}{R} i_i$$

or

$$i_o = \left(1 + \frac{R_f}{R}\right) i_i \qquad (9.11)$$

The current gain of the amplifier is

$$K_i = 1 + \frac{R_f}{R} \qquad (9.12)$$

This gain can be accurately set using high-precision resistors R and R_f.

9.1.4 Instrumentation Amplifiers

An instrumentation amplifier is typically a special-purpose voltage amplifier dedicated to a particular instrumentation application. Examples include amplifiers used for producing the output from a bridge circuit (bridge amplifier) and amplifiers used with various sensors and transducers. An important characteristic of an instrumentation amplifier is the adjustable-gain capability. The gain value can be adjusted manually in most instrumentation amplifiers. In more sophisticated instrumentation amplifiers, gain is *programmable* and can be set by means of digital logic. Instrumentation amplifiers are normally used with low-voltage signals.

Differential Amplifier

Usually, an instrumentation amplifier is also a *differential amplifier* (sometimes termed a *difference amplifier*). Note that in a differential amplifier, both input leads are used for signal input, whereas in a single-ended amplifier, one of the leads is grounded and only one lead is used for signal input. *Ground-loop noise* can be a serious problem in single-ended amplifiers. Ground-loop noise can be effectively eliminated using a differential amplifier because noise loops are formed with both inputs of the amplifier and, hence, these noise signals are subtracted at the amplifier output. Because the noise level is almost the same for both inputs, it is canceled. Note that any other noise (e.g., 60-Hz line noise) that might enter both inputs with the same intensity will also be canceled out at the output of a differential amplifier.

A basic differential amplifier that uses a single op-amp is shown in Figure 9.3(a). The input-output equation for this amplifier can be obtained in the usual manner. For example, because current through the op-amp is negligible, current balance at point B gives

$$\frac{v_{i2} - v_B}{R} = \frac{v_B}{R_f} \qquad (i)$$

in which v_B is the voltage at B. Similarly, current balance at point A gives

$$\frac{v_o - v_A}{R_f} = \frac{v_A - v_{i1}}{R} \qquad (ii)$$

Now we use the property

$$v_A = v_B \qquad (iii)$$

for an operational amplifier, to eliminate v_A and v_B from equations (i) and (ii). This gives

Signal Conditioning and Modification

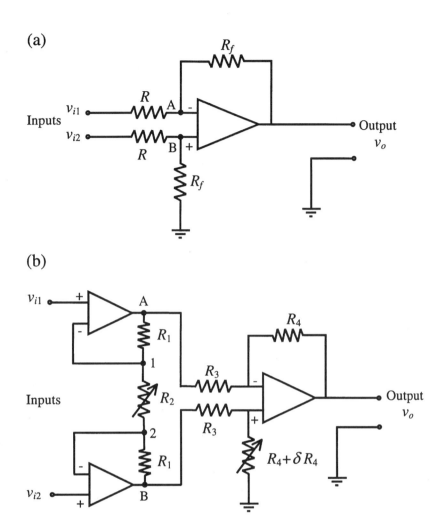

FIGURE 9.3 (a) A basic differential amplifier, and (b) a basic instrumentation amplifier.

$$\frac{v_{i2}}{\left(1+R/R_f\right)} = \frac{\left(v_o R/R_f + v_{i1}\right)}{\left(1+R/R_f\right)}$$

or

$$v_o = \frac{R_f}{R}\left(v_{i2} - v_{i1}\right) \tag{9.13}$$

Two things are clear from equation (9.13). First, the amplifier output is proportional to the "difference" and not the absolute value of the two inputs v_{i1} and v_{i2}. Second, the voltage gain of the amplifier is R_f/R. This is known as the *differential gain*. Note that the differential gain can be accurately set using high-precision resistors R and R_f.

The basic differential amplifier, shown in Figure 9.3(a) and discussed above, is an important component of an instrumentation amplifier. In addition, an instrumentation amplifier should possess the adjustable gain capability. Furthermore, it is desirable to have a very high input impedance and very low output impedance at each input lead. An instrumentation amplifier that possesses these basic requirements is shown in Figure 9.3(b). The amplifier gain can be adjusted using the precisely variable resistor R_2. Impedance requirements are provided by two voltage-follower type amplifiers, one for each input, as shown. The variable resistance δR_4 is necessary to compensate for errors due to unequal *common-mode gain*. First consider this last aspect, and then obtain an equation for the instrumentation amplifier.

Common Mode

The voltage that is "common" to both input leads of a differential amplifier is known as the *common-mode voltage*. This is equal to the smaller of the two input voltages. If the two inputs are equal, then the common-mode voltage is obviously equal to each one of the two inputs. When $v_{i1} = v_{i2}$, ideally, the output voltage v_o should be 0. In other words, ideally, *common-mode signals* are rejected by a differential amplifier. But because operational amplifiers are not ideal and because they usually do not have exactly identical gains with respect to the two input leads, the output voltage v_o will not be 0 when the two inputs are identical. This *common-mode error* can be compensated for by providing a variable resistor with fine resolution at one of the two input leads of the differential amplifier. Hence, in Figure 9.3(b), to compensate for the common-mode error (i.e., to achieve a satisfactory level of common-mode rejection), first the two inputs are made equal and then δR_4 is carefully varied until the output voltage level is sufficiently small (minimum). Usually, δR_4 that is required to achieve this compensation is small compared to the nominal feedback resistance R_4.

Since ideally $\delta R_4 = 0$, one can neglect δR_4 in the derivation of the instrumentation amplifier equation. Now, note from the basic characteristics of an op-amp with no saturation (voltages at the two input leads have to be almost identical), that in Figure 9.3(b), the voltage at point 2 should be v_{i2} and the voltage at point 1 should be v_{i1}. Furthermore, current through each input lead of an op-amp is negligible. Hence, current through the circuit path $B \to 2 \to 1 \to A$ must be the same. This gives the current continuity equations

$$\frac{v_B - v_{i2}}{R_1} = \frac{v_{i2} - v_{i1}}{R_2} = \frac{v_{i1} - v_A}{R_1}$$

in which V_A and V_B are the voltages at points A and B, respectively. Hence, one obtains

$$v_B = v_{i2} + \frac{R_1}{R_2}(v_{i2} - v_{i1})$$

$$v_A = v_{i1} - \frac{R_1}{R_2}(v_{i2} - v_{i1})$$

Now, by subtracting the second equation from the first, one obtains the equation for the first stage of the amplifier; thus,

$$v_B - v_A = \left(1 + \frac{2R_1}{R_2}\right)(v_{i2} - v_{i1}) \qquad (i)$$

Next, from the previous result [see equation (9.13)] for a differential amplifier, one obtains (with $\delta R_4 = 0$)

Signal Conditioning and Modification

$$v_o = \frac{R_4}{R_3}(v_B - v_A) \qquad \text{(ii)}$$

Note that only the resistor R_2 is varied to adjust the gain (differential gain) of the amplifier. In Figure 9.3(b), the two input op-amps (the voltage-follower op-amps) do not have to be exactly identical as long as the resistors R_1 and R_2 are chosen to be accurate. This is so because the op-amp parameters such as open-loop gain and input impedance do not enter the amplifier equations, provided their values are sufficiently high, as noted earlier.

9.1.5 Amplifier Performance Ratings

The main factors that affect the performance of an amplifier are:

1. Stability
2. Speed of response (bandwidth, slew rate)
3. Unmodeled signals.

The significance of some of these factors has already been discussed.

The level of stability of an amplifier, in the conventional sense, is governed by the dynamics of the amplifier circuitry, and can be represented by a *time constant*. But more important consideration for an amplifier is the "parameter variation" due to aging, temperature, and other environmental factors. Parameter variation is also classified as a stability issue, in the context of devices such as amplifiers, because it pertains to the steadiness of the response when the input is maintained steady. Of particular importance is the *temperature drift*. This can be specified as a drift in the output signal per unity change in temperature (e.g., mV·°C^{-1}).

The speed of response of an amplifier dictates the ability of the amplifier to faithfully respond to transient inputs. Conventional time-domain parameters such as *rise time* can be used to represent this. Alternatively, in the frequency domain, speed of response can be represented by a *bandwidth* parameter. For example, the frequency range over which the frequency response function is considered constant (flat) can be taken as a measure of bandwidth. Because there is some nonlinearity in any amplifier, bandwidth can depend on the signal level itself. Specifically, *small-signal bandwidth* refers to the bandwidth that is determined using small input signal amplitudes.

Another measure of the speed of response is the *slew rate*. Slew rate is defined as the largest possible rate of change of the amplifier output for a particular frequency of operation. Since for a given input amplitude, the output amplitude depends on the amplifier gain, slew rate is usually defined for unity gain.

Ideally, for a linear device, the frequency response function (transfer function) does not depend on the output amplitude (i.e., the product of the DC gain and the input amplitude). But for a device that has a limited slew rate, the bandwidth (or the maximum operating frequency at which output distortions can be neglected) will depend on the output amplitude. The larger the output amplitude, the smaller the bandwidth for a given slew rate limit.

Example 9.2

Obtain a relationship between the slew rate and the bandwidth for a slew rate-limited device. An amplifier has a slew rate of 1 V·µs^{-1}. Determine the bandwidth of this amplifier when operating at an output amplitude of 5 V.

Solution

Clearly, the amplitude of the rate of change of output signal, divided by the amplitude of the output signal, yields an estimate of output frequency. Consider a sinusoidal output voltage given by

$$v_o = a \sin 2\pi f t \tag{9.14}$$

The rate of change of output is

$$\frac{dv_o}{dt} = 2\pi f a \cos 2\pi f t$$

Hence, the maximum rate of change of output is $2\pi f a$. Since this corresponds to the slew rate when f is the maximum allowable frequency, one obtains

$$s = 2\pi f_b a \tag{9.15}$$

where

s = slew rate
f_b = bandwidth
a = output amplitude.

Now, with $s = 1$ V·μs^{-1} and $a = 5$ V, one obtains

$$f_b = \frac{1}{2\pi} \times \frac{1}{1 \times 10^{-8}} \times \frac{1}{5} \text{ Hz}$$

$$= 31.8 \text{ kHz}$$

☐

It has been noted that stability problems and frequency response errors are prevalent in the open-loop form of an operational amplifier. These problems can be eliminated using feedback because the effect of the open-loop transfer function on the closed-loop transfer function is negligible if the open-loop gain is very large — which is the case for an operational amplifier.

Unmodeled signals can be a major source of amplifier error. Unmodeled signals include:

1. Bias currents
2. Offset signals
3. Common-mode output voltage
4. Internal noise.

In analyzing operational amplifiers, it is assumed that the current through the input leads is 0. This is not strictly true because bias currents for the transistors within the amplifier circuit have to flow through these leads. As a result, the output signal of the amplifier will deviate slightly from the ideal value.

Another assumption made in analyzing op-amps is that the voltage is equal at the two input leads. But, in practice, offset currents and offset voltages are present at the input leads, due to minute discrepancies inherent to the internal circuits within an op-amp.

Signal Conditioning and Modification

Common-Mode Rejection Ratio (CMRR)

Common-mode error in a differential amplifier was discussed earlier. Note that, ideally, the *common mode input voltage* (the voltage common to both input leads) should have no effect on the output voltage of a differential amplifier. But because a practical amplifier has unbalances in the internal circuitry (e.g., gain with respect to one input lead is not equal to the gain with respect to the other input lead and, furthermore, bias signals are needed for operation of the internal circuitry), there will be an error voltage at the output that depends on the common-mode input. The *common-mode rejection ratio* (CMRR) of a differential amplifier is defined as

$$\text{CMRR} = \frac{K v_{cm}}{v_{ocm}} \quad (9.16)$$

where

K = gain of the differential amplifier (i.e., differential gain)
v_{cm} = common-mode input voltage (i.e., voltage common to both input leads)
v_{ocm} = common-mode output voltage (i.e., output voltage due to common-mode input voltage).

Note that, ideally, $v_{ocm} = 0$ and CMRR should be infinity. It follows that the larger the CMRR, the better the differential amplifier performance.

The three types of unmodeled signals mentioned above can be considered as noise. In addition, there are other types of noise signals that degrade the performance of an amplifier. For example, ground-loop noise can enter the output signal. Furthermore, stray capacitances and other types of unmodeled circuit effects can generate internal noise. Usually in amplifier analysis, unmodeled signals (including noise) can be represented by a noise voltage source at one of the input leads. Effects of unmodeled signals can be reduced using suitably connected compensating circuitry, including variable resistors that can be adjusted to eliminate the effect of unmodeled signals at the amplifier output [e.g., see δR_4 in Figure 9.3(b)]. Some useful information about operational amplifiers is summarized in Box 9.1.

AC-Coupled Amplifiers

The DC component of a signal can be blocked off by connecting the signal through a capacitor. (Note that the impedance of a capacitor is $1/(j\omega C)$ and hence, at zero frequency, there will be an infinite impedance.) If the input lead of a device has a series capacitor, the input is AC-coupled; and if the output lead has a series capacitor, then the output is AC-coupled. Typically, an AC-coupled amplifier has a series capacitor both at the input lead and the output lead. Hence, its frequency response function will have a high-pass characteristic; in particular, the DC components will be filtered out. Errors due to bias currents and offset signals are negligible for an AC-coupled amplifier. Furthermore, in an AC-coupled amplifier, stability problems are not very serious.

9.2 ANALOG FILTERS

Unwanted signals can seriously degrade the performance of a vibration monitoring and analysis system. External disturbances, error components in excitations, and noise generated internally within system components and instrumentation are such spurious signals. A filter is a device that allows through only the desirable part of a signal, rejecting the unwanted part.

BOX 9.1 Operational Amplifiers

Ideal op-amp properties:
- Infinite open-loop differential gain
- Infinite input impedance
- Zero output impedance
- Infinite bandwidth
- Zero output for zero differential input.

Ideal analysis assumptions:
- Voltages at the two input leads are equal
- Current through either input lead is zero.

Definitions:

- Open-loop gain = $\left| \dfrac{\text{Output voltage}}{\text{Voltage difference at input leads}} \right|$ with no feedback

- Input impedance = $\dfrac{\text{Voltage between an input lead and ground}}{\text{Current through that lead}}$

  (with the other input lead grounded and the output in open circuit)

- Output impedance = $\dfrac{\text{Voltage between output lead and ground in open circuit}}{\text{Current through that lead with normal input conditions}}$

- Bandwidth = Frequency range in which the frequency response is flat (gain is constant)
- Input bias current = Average (DC) current through one input lead
- Input offset current = Difference in the two input bias currents
- Differential input voltage = Voltage at one input lead with the other grounded when the output voltage is zero

- Common-mode gain = $\dfrac{\text{Output voltage when input leads are at the same voltage}}{\text{Common input voltage}}$

- Common-mode rejection ratio (CMRR) = $\dfrac{\text{Open-loop differential gain}}{\text{Common-mode gain}}$

- Slew rate = Speed at which steady output is reached for a step input.

In typical applications of acquisition and processing of a vibration signal, the filtering task would require allowing through certain frequency components and filtering out certain other frequency components in the signal. In this context, one can identify four broad categories of filters:

1. Low-pass filters
2. High-pass filters
3. Bandpass filters
4. Band-reject (or notch) filter.

The ideal frequency-response characteristic of each of these four types of filters is shown in Figure 9.4. Note that only the magnitude of the frequency response function is shown. It is understood, however, that the phase distortion of the input signal should also be small within the *pass band* (the allowed frequency range). Practical filters are less than ideal. Their frequency-response functions do not exhibit sharp cutoffs as in Figure 9.4 and, furthermore, some phase distortion will be unavoidable.

A special type of bandpass filter widely used in acquisition and monitoring of vibration signals (e.g., in vibration testing) is the *tracking filter*. This is simply a bandpass filter with a narrow pass

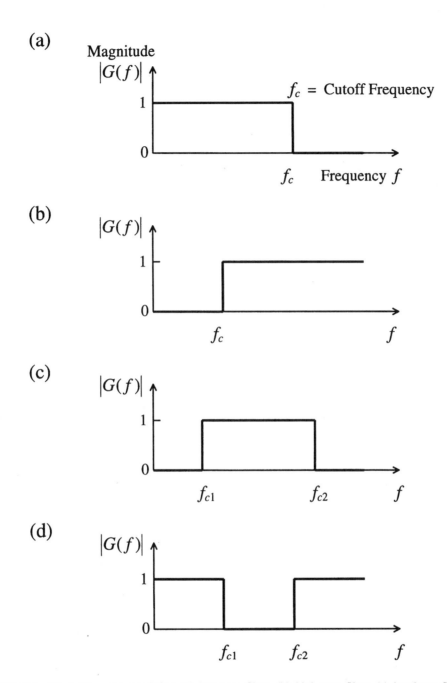

FIGURE 9.4 Ideal filter characteristics: (a) low-pass filter; (b) high-pass filter; (c) bandpass filter; and (d) band-reject (notch) filter.

band that is frequency-tunable. The center frequency (mid-value) of the pass band is variable, usually by coupling it to the frequency of a carrier signal. In this manner, signals whose frequency varies with some basic variable in the system (e.g., rotor speed, frequency of a harmonic excitation signal, frequency of a sweep oscillator) can be accurately tracked in the presence of noise. The inputs to a tracking filter are the signal that is being tracked and the variable *tracking frequency*

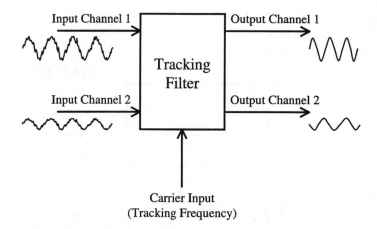

FIGURE 9.5 Schematic representation of a two-channel tracking filter.

(*carrier input*). A typical tracking filter that can simultaneously track two signals is schematically shown in Figure 9.5.

Filtering can be achieved using *digital filters* as well as *analog filters*. Before digital signal processing became efficient and economical, analog filters were used exclusively for signal filtering, and are still widely used. In an analog filter, the signal is passed through an analog circuit. Dynamics of the circuit will be such that the desired signal components will be passed through, and the unwanted signal components will be rejected. Earlier versions of analog filters employed discrete circuit elements such as discrete transistors, capacitors, resistors, and even discrete inductors. Because inductors have several shortcomings, such as susceptibility to electromagnetic noise, unknown resistance effects, and large size, they are rarely used today in filter circuits. Furthermore, due to well-known advantages of integrated-circuit (IC) devices, analog filters in the form of monolithic IC chips are extensively used today in modem applications and are preferred over discrete-element filters. Digital filters that employ digital signal processing to achieve filtering are also widely used today.

9.2.1 Passive Filters and Active Filters

Passive analog filters employ analog circuits containing passive elements such as resistors and capacitors (and sometimes inductors) only. An external power supply is not needed in a passive filter. Active analog filters employ active elements and components such as transistors and operational amplifiers in addition to passive elements. Because external power is needed for the operation of the active elements and components, an active filter is characterized by the need for an external power supply. Active filters are widely available in monolithic integrated-circuit (IC) form and are usually preferred over passive filters.

Advantages of active filters include:

1. Loading effects are negligible because active filters can provide a very high input impedance and very low output impedance.
2. They can be used with low-level signals because signal amplification and filtering can be provided by the same active circuit.
3. They are widely available in a low-cost and compact integrated-circuit form.
4. They can be easily integrated with digital devices.
5. They are less susceptible to noise from electromagnetic interference.

Commonly mentioned disadvantages of active filters include:

1. They need an external power supply.
2. They are susceptible to "saturation"-type nonlinearity at high signal levels.
3. They can introduce many types of internal noise and unmodeled signal errors (offset, bias signals, etc.).

Note that the advantages and disadvantages of passive filters can be directly inferred from the disadvantages and advantages of active filters, as given above.

Number of Poles

Analog filters are dynamic systems and can be represented by transfer functions, assuming linear dynamics. The number of poles of a filter is the number of poles in the associated transfer function. This is also equal to the order of the characteristic polynomial of the filter transfer function (i.e., *order of the filter*). Note that poles (or eigenvalues) are the roots of the characteristic equation.

The following discussion will show simplified versions of filters, typically consisting of a single filter stage. The performance of such a basic filter can be improved at the expense of circuit complexity (and increased pole count). Only simple discrete-element circuits are shown for passive filters. Simple operational-amplifier circuits are given for active filters. Even here, much more complex devices are commercially available, but the purpose is to illustrate underlying principles rather than to provide descriptions and data sheets for commercial filters.

9.2.2 Low-Pass Filters

The purpose of a low-pass filter is to allow through all signal components below a certain (cutoff) frequency and block off all signal components above that cutoff. Analog low-pass filters are widely used as *anti-aliasing filters* in digital signal processing (see Chapter 4). An error known as *aliasing* will enter the digitally processed results of a signal if the original signal has frequency components above half the *sampling frequency* (half the sampling frequency is called the *Nyquist frequency*). Hence, aliasing distortion can be eliminated if the signal is filtered using a low-pass filter with its cutoff set at the Nyquist frequency, prior to sampling and digital processing. This is one of numerous applications of analog low-pass filters. Another typical application would be to eliminate high-frequency noise in a measured vibration response.

A single-pole, passive low-pass filter circuit is shown in Figure 9.6(a). An active filter corresponding to the same low-pass filter is shown in Figure 9.6(b). It can be shown that the two circuits have identical transfer functions. Hence, it might seem that the op-amp in Figure 9.6(b) is redundant. This is not true, however. If two passive filter stages, each similar to Figure 9.6(a), are connected together, then the overall transfer function is not equal to the product of the transfer functions of the individual stages. The reason for this apparent ambiguity is the circuit loading that arises due to the fact that the input impedance of the second stage is not sufficiently larger than the output impedance of the first stage. But, if two active filter stages similar to Figure 9.6(b) are connected together, such loading errors will be negligible because the op-amp with feedback (i.e., a voltage follower) introduces a very high input impedance and very low output impedance, while maintaining the voltage gain at unity.

To obtain the filter equation for Figure 9.6(a), note that since the output is open circuit (zero load current), the current through capacitor C is equal to the current through resistor R. Hence,

$$C\frac{dv_o}{dt} = \frac{v_i - v_o}{R}$$

or

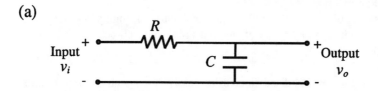

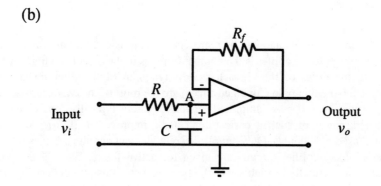

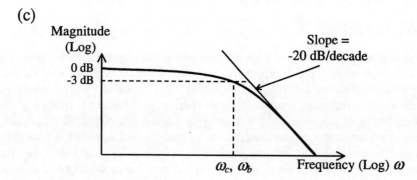

FIGURE 9.6 A single-pole low-pass filter: (a) a passive filter stage; (b) an active filter stage; and (c) the frequency response characteristic.

$$\tau \frac{dv_o}{dt} + v_o = v_i \qquad (9.17)$$

where the filter time constant is

$$\tau = RC \qquad (9.18)$$

Now, from equation (9.17), it follows that the filter transfer function is

$$\frac{v_o}{v_i} = G(s) = \frac{1}{(\tau s + 1)} \qquad (9.19)$$

From this transfer function, it is clear that an analog low-pass filter is essentially a *lag circuit* (i.e., it provides a phase lag).

Signal Conditioning and Modification

It can be shown that the active filter stage in Figure 9.6(b) has the same input/output equation. First, because the current through an op-amp lead is almost 0, one obtains from the previous analysis of the passive circuit stage

$$\frac{v_A}{v_i} = \frac{1}{(\tau s + 1)} \quad (i)$$

in which v_A is the voltage at the node point A. Now, because the op-amp with feedback resistor is in fact a voltage follower, one has

$$\frac{v_o}{v_A} = 1 \quad (ii)$$

Next, by combining equations (i) and (ii), one obtains equation (9.19), as required. Repeating, a major advantage of the active filter version is that the resulting loading error is negligible.

The frequency-response function corresponding to equation (9.19) is obtained by setting $s = j\omega$; thus,

$$G(j\omega) = \frac{1}{(\tau j\omega + 1)} \quad (9.20)$$

This gives the response of the filter when a sinusoidal signal of frequency ω is applied. The magnitude $|G(j\omega)|$ of the frequency transfer function gives the signal amplification, and the phase angle $\angle G(j\omega)$ gives the phase lead of the output signal with respect to the input. The magnitude curve (*Bode magnitude curve*) is shown in Figure 9.6(c). Note from equation (9.20) that for small frequencies (i.e., $\omega \ll 1/\tau$), the magnitude is approximately unity. Hence, $1/\tau$ can be considered the cutoff frequency ω_c:

$$\omega_c = \frac{1}{\tau} \quad (9.21)$$

EXAMPLE 9.3

Show that the cutoff frequency given by equation (9.21) is also the *half-power bandwidth* for the low-pass filter. Show that for frequencies much larger than this, the filter transfer function on the Bode magnitude plane (i.e., log magnitude vs. log frequency) can be approximated by a straight line with slope –20 dB per decade. This slope is known as the *roll-off rate*.

SOLUTION

The frequency corresponding to half power (or $1/\sqrt{2}$ magnitude) is given by

$$\frac{1}{|\tau j\omega + 1|} = \frac{1}{\sqrt{2}}$$

or

$$\frac{1}{\tau^2\omega^2+1} = \frac{1}{2}$$

or

$$\tau^2\omega^2 + 1 = 2$$

or

$$\tau^2\omega^2 = 1$$

Hence, the half-power bandwidth is

$$\omega_b = \frac{1}{\tau} \tag{9.22}$$

This is identical to the cutoff frequency given by equation (9.11).

Now, for $\omega \gg 1/\tau$ (i.e., $\tau\omega \gg 1$), equation (9.20) can be approximated by

$$G(j\omega) = \frac{1}{\tau j\omega}$$

This has the magnitude

$$|G(j\omega)| = \frac{1}{\tau\omega}$$

On the log scale,

$$\log_{10}|G(j\omega)| = -\log_{10}\omega - \log_{10}\tau$$

It follows that the $\log_{10}$ (magnitude) vs. $\log_{10}$ (frequency) curve is a straight line with a slope of -1. In other words, when frequency increases by a factor of ten (i.e., a decade), the $\log_{10}$ (magnitude) decreases by unity (i.e., by 20 dB). Hence, the roll-off rate is -20 dB per decade. These observations are shown in Figure 9.6(c). Note that an amplitude change by a factor of $\sqrt{2}$ (or power by a factor of 2) corresponds to 3 dB. Hence, when the DC (zero-frequency) magnitude is unity (0 dB), the half-power magnitude is -3dB.

□

The cutoff frequency and the roll-off rate are the two main design specifications for a low-pass filter. Ideally, one would like a low-pass filter magnitude curve to be flat up to the required pass-band limit (cutoff frequency) and then roll off very rapidly. The low-pass filter shown in Figure 9.6 only approximately meets these requirements. In particular, the roll-off rate is not large enough. One would like a roll-off rate of at least -40 dB per decade and, preferably, -60 dB per decade in practical filters. This can be realized using a higher-order filter (i.e., a filter having many poles). The low-pass Butterworth filter is a widely used filter of this type.

Signal Conditioning and Modification

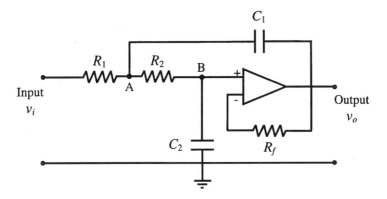

FIGURE 9.7 A two-pole, low-pass Butterworth filter.

Low-Pass Butterworth Filter

A low-pass Butterworth filter having two poles can provide a roll-off rate of –40 dB per decade, and one having three poles can provide a roll-off rate of –60 dB per decade. Furthermore, the steeper the roll-off slope, the flatter the filter magnitude curve within the pass band. A two-pole, low-pass Butterworth filter is shown in Figure 9.7. One can construct a two-pole filter simply by connecting together two single-pole stages of the type shown in Figure 9.6(b). One would then require two op-amps; whereas the circuit shown in Figure 9.7 achieves the same objective using only one op-amp (i.e., at a lower cost).

EXAMPLE 9.4

Show that the opamp circuit in Figure 9.7 is a low-pass filter having two poles. What is the transfer function of the filter? Estimate the cutoff frequency under suitable conditions. Show that the roll-off rate is –40 dB per decade.

SOLUTION

To obtain the filter equation, one writes the current balance equations. Specifically, the sum of currents through R_1 and C_1 passes through R_2. The same current passes through C_2 because current through the op-amp lead must be 0. Hence,

$$\frac{v_i - v_A}{R_1} + C_1 \frac{d}{dt}(v_o - v_A) = \frac{v_A - v_B}{R_2} = C_2 \frac{dv_B}{dt} \quad (i)$$

Also, because the op-amp with a feedback resistor R_f is a voltage follower (with unity gain), one obtains

$$v_B = v_o \quad (ii)$$

From equations (*i*) and (*ii*),

$$\frac{v_i - v_A}{R_1} + C_1 \frac{dv_o}{dt} - C_1 \frac{dv_A}{dt} = C_2 \frac{dv_o}{dt} \quad (iii)$$

$$\frac{v_A - v_o}{R_2} = C_2 \frac{dv_o}{dt} \qquad (iv)$$

Now, defining the constants

$$\tau_1 = R_1 C_1 \qquad (9.23)$$

$$\tau_2 = R_2 C_2 \qquad (9.24)$$

$$\tau_3 = R_1 C_2 \qquad (9.25)$$

and introducing the Laplace variable s, one can eliminate v_A by substituting equation (iv) into (iii); thus,

$$\frac{v_o}{v_i} = \frac{1}{\left[\tau_1 \tau_2 s^2 + (\tau_2 + \tau_3)s + 1\right]} = \frac{\omega_n^2}{\left[s^2 + 2\zeta\omega_n^2 + \omega_n^2\right]} \qquad (9.26)$$

This second-order transfer function becomes oscillatory if $(\tau_2 + \tau_3)^2 < 4\tau_1\tau_2$. Ideally, one would like to have a zero resonant frequency, which corresponds to a damping ratio value $\zeta = 1/\sqrt{2}$. Since the undamped natural frequency is

$$\omega_n = \frac{1}{\sqrt{\tau_1 \tau_2}} \qquad (9.27)$$

the damping ratio is

$$\zeta = \frac{\tau_2 + \tau_3}{\sqrt{4\tau_1 \tau_2}} \qquad (9.28)$$

and the resonant frequency is

$$\omega_r = \sqrt{1 - 2\zeta^2}\,\omega_n \qquad (9.29)$$

then, under ideal conditions (i.e., for $\omega_r = 0$), one obtains

$$(\tau_2 + \tau_3)^2 = 2\tau_1 \tau_2 \qquad (9.30)$$

The frequency response function of the filter is (see equation (9.26))

$$G(j\omega) = \frac{\omega_n^2}{\left[\omega_n^2 - \omega^2 + 2j\zeta\omega_n\omega\right]} \qquad (9.31)$$

Now, for $\omega \ll \omega_n$, the filter frequency response is flat with a unity gain. For $\omega \gg \omega_n$, the filter frequency response can be approximated by

Signal Conditioning and Modification

$$G(j\omega) = -\frac{\omega_n^2}{\omega^2}$$

On a log (magnitude) vs. log (frequency) scale, this function is a straight line with slope = –2. Hence, when the frequency increases by a factor of 10 (i.e., one decade), the $\log_{10}$ (magnitude) drops by 2 units (i.e., 40 dB). In other words, the roll-off rate is –40 dB per decade. Also, ω_n can be taken as the filter cutoff frequency. Hence,

$$\omega_c = \frac{1}{\sqrt{\tau_1 \tau_2}} \qquad (9.32)$$

It can be easily verified that when $\zeta = 1/\sqrt{2}$, this frequency is identical to the half-power bandwidth (i.e., the frequency at which the transfer function magnitude becomes $1/\sqrt{2}$).

□

Note that if two single-pole stages (of the type shown in Figure 9.6(b)) are cascaded, the resulting two-pole filter has an overdamped (nonoscillatory) transfer function, and it is not possible to achieve $\zeta = 1/\sqrt{2}$ as in the present case. Also, note that a three-pole, low-pass Butterworth filter can be obtained by cascading a two-pole unit as shown in Figure 9.7 with a single-pole unit shown in Figure 9.6(b). Higher order low-pass Butterworth filters can be obtained in a similar manner by cascading an appropriate selection of basic units.

9.2.3 High-Pass Filters

Ideally, a high-pass filter allows through it all signal components above a certain (cutoff) frequency, and blocks off all signal components below that frequency. A single-pole, high-pass filter is shown in Figure 9.8. As for the low-pass filter discussed earlier, the passive filter stage [Figure 9.8(a)] and the active filter stage [Figure 9.8(b)] have identical transfer functions. The active filter is desired, however, because of its many advantages, including negligible loading error due to high input impedance and low output impedance of the op-amp voltage follower that is present in this circuit.

The filter equation is obtained by considering the current balance in Figure 9.8(a), noting that the output is in open circuit (zero load current). Accordingly,

$$C \frac{d}{dt}(v_i - v_o) = \frac{v_o}{R}$$

or

$$\tau \frac{dv_o}{dt} + v_o = \tau \frac{dv_i}{dt} \qquad (9.33)$$

in which the filter time constant

$$\tau = RC \qquad (9.34)$$

Introducing the Laplace variable s, the filter transfer function is obtained as

$$\frac{v_o}{v_i} = G(s) = \frac{\tau s}{(\tau s + 1)} \qquad (9.35)$$

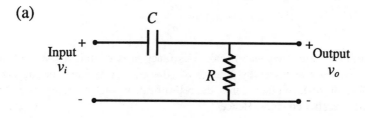

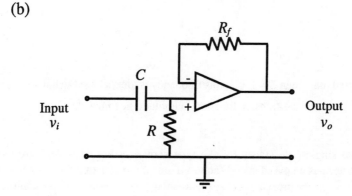

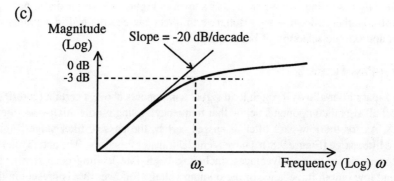

FIGURE 9.8 A single-pole, high-pass filter: (a) a passive filter stage; (b) an active filter stage; and the (c) frequency response characteristic.

Note that this corresponds to a *lead circuit* (i.e., an overall phase lead is provided by this transfer function). The frequency response function is

$$G(j\omega) = \frac{\tau j\omega}{(\tau j\omega + 1)} \tag{9.36}$$

Since its magnitude is 0 for $\omega \ll 1/\tau$ and it is unity for $\omega \gg 1/\tau$, the cutoff frequency becomes

$$\omega_c = \frac{1}{\tau} \tag{9.37}$$

Signals above this cutoff frequency should be allowed undistorted by an ideal high-pass filter, and signals below the cutoff should be completely blocked. The actual behavior of the basic high-pass filter discussed above is not that perfect, as observed from the frequency-response characteristic shown in Figure 9.8(c). It can be easily verified that the half-power bandwidth of the basic high-

Signal Conditioning and Modification

pass filter is equal to the cutoff frequency given by equation (9.37), as in the case of the basic low-pass filter. The roll-up slope of the single-pole, high-pass filter is 20 dB per decade. Steeper slopes are desirable. Multiple-pole, high-pass Butterworth filters can be constructed to give steeper roll-up slopes and reasonably flat pass-band magnitude characteristics.

9.2.4 Bandpass Filters

An ideal bandpass filter passes all signal components within a finite frequency band and blocks off all signal components outside that band. The lower frequency limit of the pass band is called the *lower cutoff frequency* (ω_{c1}), and the upper frequency limit of the band is called the *upper cutoff frequency* (ω_{c2}).

The most straightforward way to form a bandpass filter is to cascade a high-pass filter of cutoff frequency ω_{c1} with a low-pass filter of cutoff frequency ω_{c2}. Such an arrangement is shown in Figure 9.9. The passive circuit shown in Figure 9.9(a) is obtained by connecting together the circuits shown in Figures 9.6(a) and 9.8(a). The passive circuit shown in Figure 9.9(b) is obtained by connecting a voltage follower op-amp circuit to the original passive circuit. Passive and active filters have the same transfer function, assuming that loading problems are not present in the passive filter. Because loading errors can be serious in practice, however, the active version is preferred.

To obtain the filter equation, first consider the high-pass portion of the circuit shown in Figure 9.9(a). Since the output is open-circuit (zero current), from equation (9.35), one obtains

$$\frac{v_o}{v_A} = \frac{\tau_2 s}{(\tau_2 s + 1)} \qquad (i)$$

where

$$\tau_2 = R_2 C_2 \qquad (9.38)$$

Next, on writing the current balance at node A of the circuit, one has

$$\frac{v_i - v_A}{R_1} = C_1 \frac{dv_A}{dt} + C_2 \frac{d}{dt}(v_A - v_o) \qquad (ii)$$

Introducing the Laplace variable s, one obtains

$$v_i = (\tau_1 s + \tau_3 s + 1) v_A - \tau_3 s v_o \qquad (iii)$$

in which

$$\tau_1 = R_1 C_1 \qquad (9.39)$$

and

$$\tau_3 = R_1 C_2 \qquad (9.40)$$

Now, on eliminating v_A by substituting equation (i) into (iii), one obtains the bandpass filter transfer function:

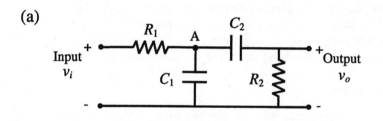

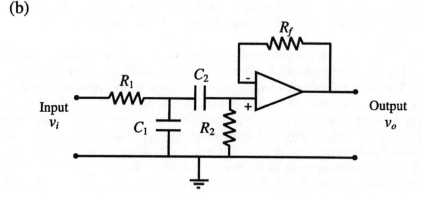

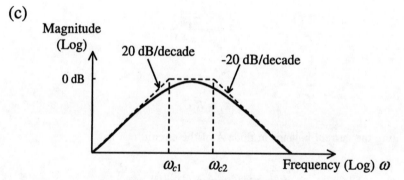

FIGURE 9.9 Bandpass filter: (a) a basic passive filter stage; (b) a basic active filter stage; and the (c) frequency response characteristic.

$$\frac{v_o}{v_i} = G(s) = \frac{\tau_2 s}{\left[\tau_1 \tau_2 s^2 + (\tau_1 + \tau_2 + \tau_3)s + 1\right]} \quad (9.41)$$

One can show that the roots of the characteristic equation

$$\tau_1 \tau_2 s^2 + (\tau_1 + \tau_2 + \tau_3)s + 1 = 0 \quad (9.42)$$

are real and negatives. The two roots are denoted by $-\omega_{c1}$ and $-\omega_{c2}$, and they provide the two cutoff frequencies shown in Figure 9.9(c). It can be verified that, for this basic bandpass filter, the roll-up slope is +20 dB per decade, and the roll-down slope is −20 dB per decade. These slopes are not sufficient in many applications. Furthermore, the flatness of the frequency response within the pass band of the basic filter is also not adequate. More complex (higher-order) bandpass filters with sharper cutoffs and flatter pass bands are commercially available.

Signal Conditioning and Modification

Resonance-Type Bandpass Filters

There are many applications where a filter with a very narrow pass band is required. The tracking filter mentioned in the beginning of this section on analog filters is one such application. A filter circuit with a sharp resonance can serve as a narrow-band filter. Note that the cascaded RC circuit shown in Figure 9.9 does not provide an oscillatory response (filter poles are all real) and, hence, it does not form a resonance-type filter. A slight modification to this circuit using an additional resistor R_1 as shown in Figure 9.10(a) will produce the desired effect.

To obtain the filter equation, note that for the voltage follower unit,

$$v_A = v_o \quad (i)$$

Next, since the current through an op-amp lead is 0, for the high-pass circuit unit [see equation (9.35)], one has

$$\frac{v_A}{v_B} = \frac{\tau_2 s}{(\tau_2 s + 1)} \quad (ii)$$

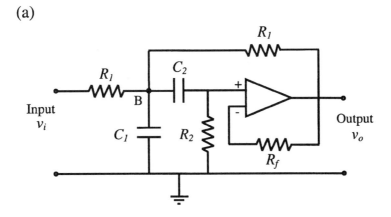

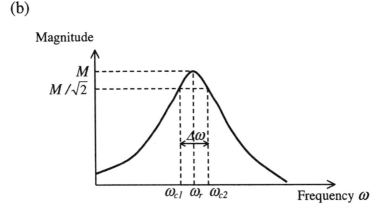

FIGURE 9.10 A resonance-type narrow bandpass filter: (a) an active filter stage, and the (b) frequency response characteristic.

where

$$\tau_2 = R_2 C_2$$

Finally, current balance at node B gives

$$\frac{v_i - v_B}{R_1} = C_1 \frac{dv_B}{dt} + C_2 \frac{d}{dt}(v_B - v_A) + \frac{v_B - v_o}{R_1}$$

or, using the Laplace variable, one obtains

$$v_i = (\tau_1 s + \tau_3 s + 2) v_B - \tau_3 s v_A - v_o \qquad (iii)$$

Now, by eliminating v_A and v_B in equations (i) through (iii), the filter transfer function is obtained as

$$\frac{v_o}{v_i} = G(s) = \frac{\tau_2 s}{\left[\tau_1 \tau_2 s^2 + (\tau_1 + \tau_2 + \tau_3) s + 2\right]} \qquad (9.43)$$

It can be shown that, unlike equation (9.41), the present characteristic equation

$$\tau_1 \tau_2 s^2 + (\tau_1 + \tau_2 + \tau_3) s + 2 = 0 \qquad (9.44)$$

can possess complex roots.

EXAMPLE 9.5

Verify that the bandpass filter shown in Figure 9.10(a) can have a frequency response with a resonant peak as shown in Figure 9.10(b). Verify that the half-power bandwidth $\Delta\omega$ of the filter is given by $2\zeta\omega_r$ at low damping values. (*Note*: ζ = damping ratio and ω_r = resonant frequency.)

SOLUTION

One can verify that the transfer function given by equation (9.43) can have a resonant peak by showing that the characteristic equation (9.44) can have complex roots. For example, for parameter values $C_1 = 2$, $C_2 = 1$, $R_1 = 1$, and $R_2 = 2$, one has $\tau_1 = 2$, $\tau_2 = 2$, and $\tau_3 = 1$. The corresponding characteristic equation is

$$4s^2 + 5s + 2 = 0$$

which has the roots

$$s = -\frac{5}{8} \pm j \frac{\sqrt{7}}{8}$$

which are obviously complex.

To obtain an expression for the half-power bandwidth of the filter, note that the filter transfer function can be written as

Signal Conditioning and Modification

$$G(s) = \frac{ks}{\left(s^2 + 2\zeta\omega_n s + \omega_n^2\right)} \tag{9.45}$$

where

ω_n = undamped natural frequency
ζ = damping ratio
k = a gain parameter.

The frequency response function is given by

$$G(j\omega) = \frac{kj\omega}{\left[\omega_n^2 - \omega^2 + 2j\zeta\omega_n\omega\right]} \tag{9.46}$$

For low damping, resonant frequency $\omega_r \cong \omega_n$. The corresponding peak magnitude M is obtained by substituting $\omega = \omega_n$ in equation (9.46) and taking the transfer function magnitude; thus,

$$M = \frac{k}{2\zeta\omega_n} \tag{9.47}$$

At half-power frequencies,

$$|G(j\omega)| = \frac{M}{\sqrt{2}}$$

or

$$\frac{k\omega}{\sqrt{\left(\omega_n^2 - \omega^2\right)^2 + 4\zeta^2\omega_n^2\omega^2}} = \frac{k}{2\sqrt{2}\zeta\omega_n}$$

This gives

$$\left(\omega_n^2 - \omega^2\right)^2 = 4\zeta^2\omega_n^2\omega^2 \tag{9.48}$$

the positive roots of which provide the pass-band frequencies ω_{c1} and ω_{c2}. Note that the roots are given by

$$\omega_n^2 - \omega^2 = \pm 2\zeta\omega_n\omega$$

Hence, the two roots ω_{c1} and ω_{c2} satisfy the following two equations:

$$\omega_{c1}^2 + 2\zeta\omega_n\omega_{c1} - \omega_n^2 = 0$$

$$\omega_{c2}^2 - 2\zeta\omega_n\omega_{c2} - \omega_n^2 = 0$$

Accordingly, by solving these two quadratic equations and selecting the appropriate sign, one obtains

$$\omega_{c1} = -\zeta\omega_n + \sqrt{\omega_n^2 + \zeta^2\omega_n^2} \qquad (9.49)$$

$$\omega_{c2} = \zeta\omega_n + \sqrt{\omega_n^2 + \zeta^2\omega_n^2} \qquad (9.50)$$

The half-power bandwidth is

$$\Delta\omega = \omega_{c2} - \omega_{c1} = 2\zeta\omega_n \qquad (9.51)$$

Now, since $\omega_n \cong \omega_r$ for low ζ, one has

$$\Delta\omega = 2\zeta\omega_r \qquad (9.52)$$

$\square$

A notable shortcoming of a resonance-type filter is that the frequency response within the bandwidth (pass band) is not flat. Hence, quite nonuniform signal attenuation takes place inside the pass band.

9.2.5 BAND-REJECT FILTERS

Band-reject filters, or *notch filters*, are commonly used to filter out a narrow band of noise components from a signal. For example, 60-Hz line noise in signals can be eliminated by using a notch filter with a notch frequency of 60 Hz.

An active circuit that could serve as a notch filter is shown in Figure 9.11(a). This is known as the Twin T circuit because its geometric configuration resembles two T-shaped circuits connected together. To obtain the filter equation, note that the voltage at point P is v_o because of unity gain of the voltage follower. Now, one can write the current balance at nodes A and B; thus,

$$\frac{v_i - v_B}{R} = 2C\frac{dv_B}{dt} + \frac{v_B - v_o}{R}$$

$$C\frac{d}{dt}(v_i - v_A) = \frac{v_A}{R/2} + C\frac{d}{dt}(v_A - v_o)$$

Next, since the current through the + lead of the op-amp (voltage follower) is 0, the current continuity through point P is given by

$$\frac{v_B - v_o}{R} = C\frac{d}{dt}(v_o - v_A)$$

These three equations are written in the Laplace form as

$$v_i = 2(\tau s + 1)v_B - v_o \qquad (i)$$

$$\tau s v_i = 2(\tau s + 1)v_A - \tau s v_o \qquad (ii)$$

Signal Conditioning and Modification

$$v_B = (\tau s + 1)v_o - \tau s v_A \quad (iii)$$

where

$$\tau = RC \quad (9.53)$$

Finally, eliminating v_A and v_B in equations (i) through (iii), one obtains

$$\frac{v_o}{v_i} = G(s) = \frac{(\tau^2 s^2 + 1)}{(\tau^2 s^2 + 4\tau s + 1)} \quad (9.54)$$

The frequency response function of the filter (with $s = j\omega$) is

$$G(j\omega) = \frac{(1 - \tau^2 \omega^2)}{(1 - \tau^2 \omega^2 + 4j\tau\omega)} \quad (9.55)$$

Note that the magnitude of this function becomes 0 at frequency

$$\omega_o = \frac{1}{\tau} \quad (9.56)$$

This is known as the *notch frequency*. The magnitude of the frequency-response function of the notch filter is sketched in Figure 9.11(b). One notices that any signal component at frequency ω_o will be completely eliminated by the notch filter. Sharp roll-down and roll-up are needed to allow the other (desirable) signal components through without too much attenuation.

Whereas the previous three types of filters achieve their frequency-response characteristics through the poles of the filter transfer function, a notch filter achieves its frequency-response characteristic through its zeros (roots of the numerator polynomial equation). Some useful information about filters is summarized in Box 9.2.

9.3 MODULATORS AND DEMODULATORS

Sometimes, signals are deliberately modified to maintain the accuracy during signal transmission, conditioning, and processing. In signal *modulation*, the data signal, known as the *modulating signal*, is used to vary a property (such as amplitude or frequency) of a *carrier signal*. Then, one can say that the carrier signal is modulated by the data signal. After transmitting or conditioning the modulated signal, the data signal is usually recovered by removing the carrier signal. This is known as *demodulation* or *discrimination*.

Many modulation techniques exist, and several other types of signal modification (e.g., digitizing) can be classified as signal modulation although they might not be commonly termed as such. Four types of modulation are illustrated in Figure 9.12. In *amplitude modulation* (AM), the amplitude of a periodic carrier signal is varied according to the amplitude of the data signal (modulating signal), the frequency of the carrier signal (*carrier frequency*) being kept constant. Suppose that the transient signal shown in Figure 9.12(a) is used as the modulating signal. A high-frequency sinusoidal signal is used as the carrier signal. The resulting amplitude-modulated signal is shown in Figure 9.12(b). Amplitude modulation is used in telecommunications, radio and TV signal transmission, instrumen-

(a)

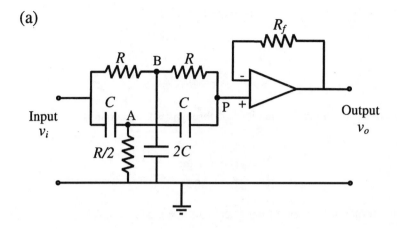

(b)

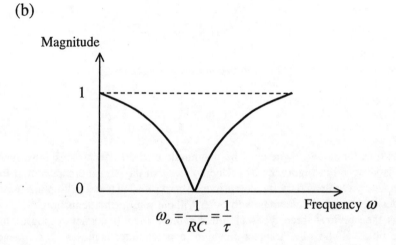

FIGURE 9.11 A notch filter: (a) an active twin T filter circuit, and the (b) frequency response characteristic.

tation, and signal conditioning. The underlying principle is useful in other applications such as fault detection and diagnosis in rotating machinery.

In frequency modulation (FM), the frequency of the carrier signal is varied in proportion to the amplitude of the data signal (modulating signal), while keeping the amplitude of the carrier signal constant. If the data signal shown in Figure 9.12(a) is used to frequency-modulate a sinusoidal carrier signal, then the result will appear as in Figure 9.12(c). Because information is carried as frequency rather than amplitude, any noise that might alter the signal amplitude will have virtually no effect on the transmitted data. Hence, FM is less susceptible to noise than AM. Furthermore, since in FM the carrier amplitude is kept constant, signal weakening and noise effects that are unavoidable in long-distance data communication will have less effect than in the case of AM, particularly if the data signal level is low in the beginning. But more sophisticated techniques and hardware are needed for signal recovery (demodulation) in FM transmission because FM demodulation involves frequency discrimination rather than amplitude detection. Frequency modulation is also widely used in radio transmission and in data recording and replay.

In *pulse-width modulation* (PWM), the carrier signal is a pulse sequence. The pulse width is changed in proportion to the amplitude of the data signal, while keeping the pulse spacing constant. This is illustrated in Figure 9.12(d). Pulse-width-modulated signals are extensively used in controlling electric motors and other mechanical devices such as valves (hydraulic, pneumatic) and machine

BOX 9.2 Filters

Active Filters (Need External Power)
Advantages:
- Smaller loading errors (have high input impedance and low output impedance, and hence do not affect the input circuit conditions and output signals)
- Lower cost
- Better accuracy.

Passive Filters (No External Power, Use Passive Elements)
Advantages:
- Usable at very high frequencies (e.g., radio frequency)
- No need for a power supply.

Filter Types
- *Low pass*: Allows frequency components up to the cutoff and rejects the higher-frequency components.
- *High pass*: Rejects frequency components up to the cutoff and allows the higher-frequency components.
- *Bandpass*: Allows frequency components within an interval and rejects the rest.
- *Notch (or band reject)*: Rejects frequency components within an interval (usually, narrow) and allows the rest.

Definitions
- *Filter order*: Number of poles in the filter circuit transfer function.
- *Anti-aliasing filter*: Low-pass filter with cutoff at less than half the sampling rate (i.e., Nyquist frequency), for digital processing.
- *Butterworth filter*: A high-order filter with a very flat pass band.
- *Chebyshev filter*: An optimal filter with uniform ripples in the pass band.
- *Sallen-Key filter*: An active filter whose output is in phase with input.

tools. Note that in a given (short) time interval, the average value of the pulse-width-modulated signal is an estimate of the average value of the data signal in that period. Hence, PWM signals can be used directly in controlling a process, without having to demodulate it. Advantages of pulse-width modulation include better energy efficiency (less dissipation) and better performance with nonlinear devices. For example, a device may stick at low speeds due to Coulomb friction. This can be avoided by using a PWM signal that provides the signal amplitude necessary to overcome friction, while maintaining the required average control signal, which might be very small.

In *pulse-frequency modulation* (PFM) as well, the carrier signal is a pulse sequence. In this method, the frequency of the pulses is changed in proportion to the data signal level, while keeping the pulse width constant. Pulse-frequency modulation has the advantages of ordinary frequency modulation. Additional advantages result due to the fact that electronic circuits (digital circuits in particular) can handle pulses very efficiently. Furthermore, pulse detection is not susceptible to noise because it involves distinguishing between presence and absence of a pulse rather than accurate determination of the pulse amplitude (or width). Pulse-frequency modulation can be used in place of pulse-width modulation in most applications, with better results.

Another type of modulation is *phase modulation* (PM). In this method, the phase angle of the carrier signal is varied in proportion to the amplitude of the data signal.

Conversion of discrete (sampled) data into the digital (binary code) form is also considered modulation. In fact, this is termed *pulse-code modulation* (PCM). In this case, each discrete data sample is represented by a binary number containing a fixed number of binary digits (bits). Since

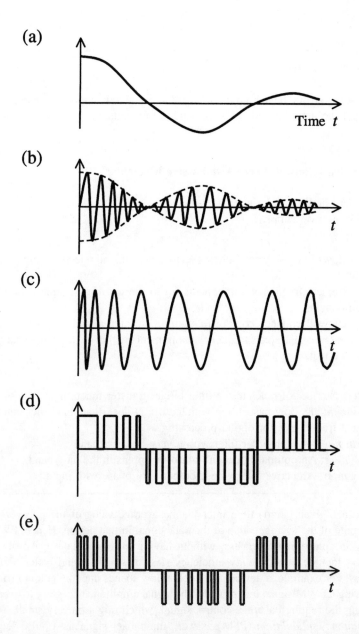

FIGURE 9.12 (a) Modulating signal (data signal); (b) amplitude-modulated (AM) signal; (c) frequency-modulated (FM) signal; (d) pulse-width-modulated (PWM) signal; and (e) pulse-frequency-modulated (PFM) signal.

each digit in the binary number can take only two values, 0 or 1, it can be represented by the absence or presence of a voltage pulse. Hence, each data sample can be transmitted using a set of pulses. This is known as *encoding*. At the receiver, the pulses have to be interpreted (or decoded) in order to determine the data value. As with any other pulse technique, PCM is quite immune to noise because decoding involves detection of the presence or absence of a pulse rather than determination of the exact magnitude of the pulse signal level. Also, because pulse amplitude is constant, long-distance signal transmission (of this digital data) can be accomplished without the danger of signal weakening and associated distortion. Of course, there will be some error introduced by the digitization process itself, which is governed by the finite word size (or dynamic range) of

Signal Conditioning and Modification

the binary data element. This is known as *quantization error* and is unavoidable in signal digitization.

In any type of signal modulation, it is essential to preserve the algebraic sign of the modulating signal (data). Different types of modulators handle this in different ways. For example, in pulse-code modulation (PCM), an extra *sign bit* is added to represent the sign of the transmitted data sample. In amplitude modulation and frequency modulation, a *phase-sensitive demodulator* is used to extract the original (modulating) signal with the correct algebraic sign. Note that in these two modulation techniques, a sign change in the modulating signal can be represented by a 180° phase change in the modulated signal. This is not quite noticeable in Figures 9.12(b) and (c). In pulse-width modulation and pulse-frequency modulation, a sign change in the modulating signal can be represented by changing the sign of the pulses, as shown in figures 9.12(d) and (e). In phase modulation, a positive range of phase angles (say 0 to π) could be assigned for the positive values of the data signal, and a negative range of phase angles (say $-\pi$ to 0) could be assigned for the negative values of the signal.

9.3.1 Amplitude Modulation

Amplitude modulation can naturally enter into many physical phenomena. More important, perhaps, is the deliberate (artificial) use of amplitude modulation to facilitate data transmission and signal conditioning. First, examine the related mathematics.

Amplitude modulation is achieved by multiplying the data signal (modulating signal) $x(t)$ by a high-frequency (periodic) carrier signal $x_c(t)$. Hence, the amplitude-modulated signal $x_a(t)$ is given by

$$x_a(t) = x(t) x_c(t) \tag{9.57}$$

Note that the carrier can be any periodic signal, such as harmonic (sinusoidal), square wave, or triangular. The main requirement is that the fundamental frequency of the carrier signal (*carrier frequency*) f_c be significantly larger (say, by a factor of 5 or 10) than the highest frequency of interest (*bandwidth*) of the data signal. Analysis can be simplified by assuming a sinusoidal carrier frequency; thus,

$$x_c(t) = a_c \cos 2\pi f_c t \tag{9.58}$$

Modulation Theorem

This is also known as the *frequency-shifting theorem* and relates the fact that if a signal is multiplied by a sinusoidal signal, the Fourier spectrum of the product signal is simply the Fourier spectrum of the original signal shifted through the frequency of the sinusoidal signal. In other words, the Fourier spectrum $X_a(f)$ of the amplitude-modulated signal $x_a(t)$ can be obtained from the Fourier spectrum $X(f)$ of the data signal $x(t)$ simply by shifting through the carrier frequency f_c.

To mathematically explain the modulation theorem, one can use the definition of the Fourier integral transform (see Chapter 4) to obtain

$$X_a(f) = a_c \int_{-\infty}^{\infty} x(t) \cos 2\pi f_c t \exp(-j2\pi ft) dt$$

But, since

$$\cos 2\pi f_c t = \frac{1}{2}\left[\exp(j2\pi f_c t) + \exp(-j2\pi f_c t)\right]$$

one has

$$X_a(f) = \frac{1}{2}a_c \int_{-\infty}^{\infty} x(t)\exp\left[-j2\pi(f-f_c)t\right]dt + \frac{1}{2}a_c \int_{-\infty}^{\infty} x(t)\exp\left[-j2\pi(f+f_c)t\right]dt$$

$$X_a(f) = \frac{1}{2}a_c\left[X(f-f_c) + X(f+f_c)\right] \tag{9.59}$$

Equation (9.59) is the mathematical statement of the modulation theorem. It is illustrated by an example in Figure 9.13. Consider a transient signal $x(t)$ with a (continuous) Fourier spectrum $X(f)$ whose magnitude $|X(f)|$ is as shown in Figure 9.13(a). If this signal is used to amplitude-modulate a high-frequency sinusoidal signal, the resulting modulated signal $x_a(t)$ and the magnitude of its Fourier spectrum are as shown in Figure 9.13(b). It should be kept in mind that the magnitude has been multiplied by $a_c/2$. Note that the data signal is assumed to be *band limited*, with bandwidth f_b. Of course, the theorem is not limited to band-limited signals; but for practical reasons, there needs to be some upper limit on the useful frequency of the data signal. Also, for practical reasons (not for the theorem itself), the carrier frequency f_c should be several times larger than f_b so that there is a reasonably wide frequency band from 0 to $(f_c - f_b)$ within which the magnitude of the modulated signal is virtually zero. The significance of this should be clear when the applications of amplitude modulation are discussed.

Figure 9.13 shows only the magnitude of the frequency spectra. It should be remembered, however, that every Fourier spectrum has a phase-angle spectrum as well. This is not shown for conciseness; but, clearly the phase-angle spectrum is also similarly affected (frequency shifted) by amplitude modulation.

Side Frequencies and Side Bands

The modulation theorem, as described above, assumed transient data signals with associated continuous Fourier spectra. The same ideas are applicable to periodic signals (with discrete spectra) as well. The case of periodic signals is merely a special case of what was discussed above. This case can be analyzed using Fourier integral transform itself, from the start. In that case, however, one must cope with impulsive spectral lines. Alternatively, Fourier series expansion could be employed to avoid the introduction of impulsive discrete spectra into the analysis. But, as shown in Figure 9.13(c) and (d), no analysis is actually needed for the periodic signal case because the final answer can be deduced from the transient signal results. Specifically, each frequency component f_o with amplitude $a/2$ in the Fourier series expansion of the data signal will be shifted by $\pm f_c$ to the two new frequency locations $f_c + f_o$ and $-f_c + f_o$ with an associated amplitude $aa_c/4$. The negative frequency component $-f_o$ should also be considered in the same way, as illustrated in Figure 9.13(d). Note that the modulated signal does not have a spectral component at carrier frequency f_c but, rather, on each side of it, at $f_c \pm f_o$. Hence, these spectral components are termed *side frequencies*. When a band of side frequencies is present, one has a *side band*. Side frequencies are very useful in fault detection and diagnosis of rotating machinery.

Signal Conditioning and Modification

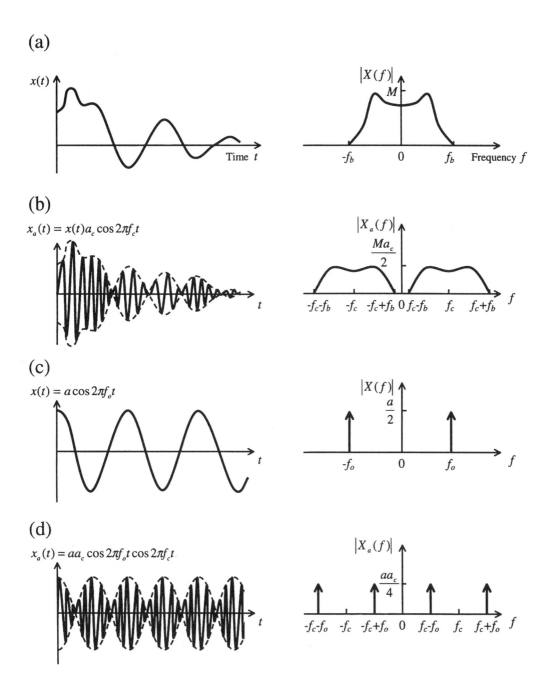

FIGURE 9.13 Illustration of the modulation theorem: (a) a transient data signal and its Fourier spectrum magnitude; (b) amplitude-modulated signal and its Fourier spectrum magnitude; (c) a sinusoidal data signal; and (d) amplitude modulation by a sinusoidal signal.

9.3.2 Application of Amplitude Modulation

The main hardware component of an amplitude modulator is an *analog multiplier*. It is commercially available in the monolithic IC form, or one can be assembled using integrated-circuit op-amps and other discrete circuit elements. A schematic representation of an amplitude modulator is shown in

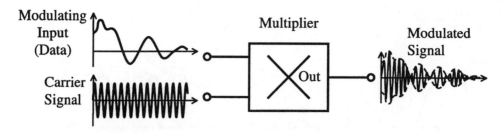

FIGURE 9.14 Representation of an amplitude modulator.

Figure 9.14. Note that, in practice, to achieve satisfactory modulation, other components such as signal preamplifiers and filters will be needed.

There are many applications of amplitude modulation. In some applications, modulation is performed intentionally. In others, modulation occurs naturally as a consequence of the physical process, and the resulting signal is used to meet a practical objective. Typical applications of amplitude modulation include the following:

1. Conditioning of general signals (including DC, transient, and low-frequency) by exploiting the advantages of AC signal conditioning hardware
2. Improvement of the immunity of low-frequency signals to low-frequency noise
3. Transmission of general signals (DC, low-frequency, etc.) by exploiting the advantages of AC signals
4. Transmission of low-level signals under noisy conditions
5. Transmission of several signals simultaneously through the same medium (e.g., same telephone line, same transmission antenna, etc.)
6. Fault detection and diagnosis of rotating machinery.

The role of amplitude modulation in many of these applications should be obvious if one understands the frequency-shifting property of amplitude modulation. Several other types of applications are also feasible due to the fact that the power of the carrier signal can be increased somewhat arbitrarily, irrespective of the power level of the data (modulating) signal. The six categories of applications mentioned above will be discussed — one by one.

AC signal-conditioning devices such as AC amplifiers are known to be more "stable" than their DC counterparts. In particular, *drift* problems are not as severe and nonlinearity effects are lower in AC signal-conditioning devices. Hence, instead of conditioning a DC signal using DC hardware, one can first use the signal to modulate a high-frequency carrier signal. Then, the resulting high-frequency modulated signal can be conditioned more effectively using AC hardware.

The frequency-shifting property of amplitude modulation can be exploited in making low-frequency signals immune to low-frequency noise. Note from Figure 9.13 that by amplitude modulation, the low-frequency spectrum of the modulating signal can be shifted out into a very high-frequency region by choosing the carrier frequency f_c sufficiently large. Then, any low-frequency noise (within the band 0 to $f_c - f_b$) would not distort the spectrum of the modulated signal. Hence, this noise can be removed by a high-pass filter (with cutoff at $f_c - f_b$) without affecting the data. Finally, the original data signal can be recovered by demodulation. Note that the frequency of a noise component can very well be within the bandwidth f_b of the data signal and, hence, if amplitude modulation is not employed, noise can directly distort the data signal.

Transmission of AC signals is more efficient than that of DC signals. Advantages of AC transmission include lower energy dissipation problems. Hence, a modulated signal can be transmitted over long distances more effectively than can the original data signal alone. Furthermore, transmission of low-frequency (large wavelength) signals require large antennas. Hence, when amplitude modulation is employed (with an associated reduction in signal wavelength), the size of broadcast antenna can be effectively reduced.

Transmission of weak signals over long distances is not desirable because further signal weakening and corruption by noise could produce disastrous results. By increasing the power of the carrier signal to a sufficiently high level, the strength of the modulated signal can be elevated to an adequate level for long-distance transmission.

It is impossible to simultaneously transmit two or more signals in the same frequency range using a single telephone line. This problem can be resolved using carrier signals with significantly different carrier frequencies to amplitude-modulate the data signals. By picking the carrier frequencies sufficiently farther apart, the spectra of the modulated signals can be made non-overlapping, thereby making simultaneous transmission possible. Similarly, with amplitude modulation, simultaneous broadcasting by several radio (AM) broadcast stations in the same broadcast area has become possible.

Fault Detection and Diagnosis

One use of the amplitude modulation principle that is particularly important in the practice of mechanical vibration is in the fault detection and diagnosis of rotating machinery. In this method, modulation is not deliberately introduced, but rather it results from the dynamics of the machine. Flaws and faults in a rotating machine are known to produce periodic forcing signals at frequencies higher than, and typically at an integer multiple of, the rotating speed of the machine. For example, backlash in a gear pair will generate forces at the tooth-meshing frequency (equal to the product: number of teeth × gear rotating speed). Flaws in roller bearings can generate forcing signals at frequencies proportional to the rotating speed times the number of rollers in the bearing race. Similarly, blade passing in turbines and compressors, and eccentricity and unbalance in a rotor, can produce forcing components at frequencies that are integer multiples of the rotating speed. Now, the resulting vibration response will be an amplitude modulated signal, where the rotating response of the machine modulates the high frequency forcing response. This can be confirmed experimentally by Fourier analysis (fast Fourier transform or FFT) of the resulting vibration signals. For a gear box, for example, it will be noticed that, instead of getting a spectral peak at the gear tooth-meshing frequency, two side bands are produced around that frequency. Faults can be detected by monitoring the evolution of these side bands. Furthermore, because side bands are the result of modulation of a specific forcing phenomenon (e.g., gear-tooth meshing, bearing-roller hammer, turbine-blade passing, unbalance, eccentricity, misalignment, etc.), one can trace the source of a particular fault (i.e., diagnose the fault) by studying the Fourier spectrum of the measured vibrations.

Amplitude modulation is an integral part of many types of sensors. In these sensors a high-frequency carrier signal (typically the AC excitation in a primary winding) is modulated by the motion. Actual motion can be detected by demodulation of the output. Examples of sensors that generate modulated outputs are differential transformers (LVDT, RVDT), magnetic-induction proximity sensors, eddy current proximity sensors, AC tachometers, and strain-gage devices that use AC bridge circuits. Signal conditioning and transmission will be facilitated by amplitude modulation in these cases. However, the signal has to be demodulated at the end, for most practical purposes such as analysis and recording.

9.3.3 DEMODULATION

Demodulation (or *discrimination* or *detection*) is the process of extracting the original data signal from a modulated signal. In general, demodulation must be phase sensitive in the sense that the algebraic sign of the data signal should be preserved and determined by the demodulation process. In *full-wave demodulation*, an output is generated continuously. In *half-wave demodulation*, no output is generated for every alternate half-period of the carrier signal.

A simple and straightforward method of demodulation is by detection of the envelope of the modulated signal. For this method to be feasible, the carrier signal must be quite powerful (i.e., the signal level must be high) and the carrier frequency should also be very high. An alternative method of demodulation that generally provides more reliable results involves a further step of modulation performed on the already-modulated signal, followed by low-pass filtering. This method can be explained by referring to Figure 9.13.

Consider the amplitude-modulated signal $x_a(t)$ shown in Figure 9.13(b). If this signal is multiplied by the sinusoidal carrier signal $2/a_c \cos 2\pi f_c t$, one obtains

$$\tilde{x}(t) = \frac{2}{a_c} x_a(t) \cos 2\pi f_c t \tag{9.60}$$

Now, by applying the modulation theorem [equation (9.59)] to equation (9.60), one obtains the Fourier spectrum of $\tilde{x}(t)$ as

$$\tilde{X}(f) = \frac{1}{2}\frac{2}{a_c}\left[\frac{1}{2}a_c\{X(f-2f_c)+X(f)\}+\frac{1}{2}a_c\{X(f)+X(f+2f_c)\}\right]$$

or

$$\tilde{X}(f) = X(f) + \frac{1}{2}X(f-2f_c) + \frac{1}{2}X(f+2f_c) \tag{9.61}$$

The magnitude of this spectrum is shown in Figure 9.15(a). Note that the spectrum $X(f)$ of the original data signal has been recovered, except for the two side bands that are present at locations far removed (centered at $\pm 2f_c$) from the bandwidth of the original signal. Hence, one can easily low-pass filter this signal $\tilde{x}(t)$ using a filter with cutoff at f_b to recover the original data signal. A schematic representation of this method of amplitude demodulation is shown in Figure 9.15(b).

9.4 ANALOG–DIGITAL CONVERSION

Data acquisition systems in machine condition monitoring, fault detection and diagnosis, and vibration testing employ digital computers for various tasks, including signal processing, data analysis and reduction, parameter identification, and decision-making. Typically, the measured response (output) of a dynamic system is available in the analog form as a continuous signal (function of continuous time). Furthermore, typically, the excitation signals (inputs) for a dynamic system must be provided in the analog form.

Inputs to a digital device (e.g., a digital computer) and outputs from a digital device are necessarily present in the digital form. Hence, when a digital device is interfaced with an analog device, the interface hardware and associated driver software have to perform several important functions. Two of the most important interface functions are *digital-to-analog conversion* (DAC) and *analog-to-digital conversion* (ADC). The digital output from a digital device must be converted into the analog form for feeding into an analog device such as an actuator or analog recording or display unit. Also, an analog signal must be converted into the digital form, according to an appropriate code, before being read by a digital processor or computer. DACs are simpler and lower in cost than ADCs. Furthermore, some types of ADCs employ a DAC to perform their function. For these reasons, the DAC will be discussed first.

Signal Conditioning and Modification

(a)

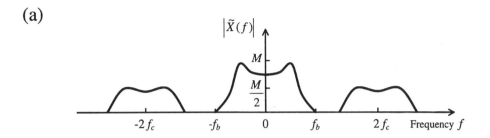

(b)

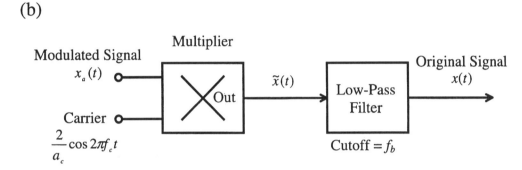

FIGURE 9.15 Amplitude demodulation: (a) spectrum of the signal after the second modulation, and (b) demodulation schematic diagram (modulation + filtering).

9.4.1 DIGITAL-TO-ANALOG CONVERSION (DAC)

The function of a digital-to-analog converter (DAC) is to convert a sequence of digital words stored in a *data register* (called *DAC register*), typically in straight binary form, into an analog signal. The data in the DAC register may be arriving from a data bus of a computer. Each binary digit (bit) of information in the register can be present as a state of a *bistable* (two-stage) logic device that can generate a voltage pulse or a voltage level to represent that bit. For example, the *off* state of a bistable logic element, or absence of a voltage *pulse*, or *low level* of a voltage signal, or *no change* in a voltage level can represent binary 0. Then, the *on* state of a bistable device, or *presence* of a voltage pulse, or *high level* of a voltage signal, or *change* in a voltage level will represent binary 1. The combination of these bits forming the digital word in the DAC register will correspond to some numerical value for the output signal. Then, the purpose of DAC is to generate an output voltage (signal level) that has this numerical value, and maintain the value until the next digital word is converted Because a voltage output cannot be arbitrarily large or small for practical reasons, some form of scaling will have to be employed in the DAC process. This scale will depend on the reference voltage v_{ref} used in the particular DAC circuit.

A typical DAC unit is an active circuit in the integrated-circuit form and may consist of a data register (digital circuits), solid-state switching circuits, resistors, and operational amplifiers powered by an external power supply that can provide a reference voltage. The reference voltage will determine the maximum value of the output (*full-scale voltage*). An integrated-circuit (IC) chip that represents the DAC is usually one of many components mounted on a *printed circuit* (PC) board. This PC board can be identified by several names, including input/output (I/O) board, I/O card, interface board, or data acquisition and control board. Typically, the same board will provide both DAC and ADC capabilities for many output and input channels.

There are many types and forms of DAC circuits. The form will depend mainly on the manufacturer, and requirements of the user or of the particular application. Most types of DAC are variations of two basic types: the *weighted type* (or *summer type* or *adder type*) and the *ladder type*. The latter type of DAC is more desirable although the former type could be somewhat simpler and less expensive.

Weighted-Resistor DAC

A schematic representation of a weighted-resistor DAC (or *summer DAC* or *adder DAC*) is shown in Figure 9.16. Note that this is a general n-bit DAC, where n is the number of bits in the output register. The binary word in the register is

$$w = [b_{n-1} b_{n-1} b_{n-3} \ldots b_1 b_0] \tag{9.62}$$

in which b_i is the bit in the ith position, and it can take the value 0 or 1, depending on the value of the digital output. The decimal value (D) of this binary word is given by

$$D = 2^{n-1} b_{n-1} + 2^{n-2} b_{n-2} + \ldots + 2^0 b_0 \tag{9.63}$$

Note that the least significant bit (LSB) is b_0, and the most significant bit (MSB) is b_{n-1}. The analog output voltage v of the DAC must be proportional to D.

Each bit b_i in the digital word w will activate a solid-state microswitch in the switching circuit, typically by sending a switching voltage pulse. If $b_i = 1$, the circuit lead will be connected to the $-v_{ref}$ supply, providing an input voltage $v_i = -v_{ref}$ to the corresponding weighting resistor $2^{n-i-1}R$. If, on the other hand, $b_i = 0$, then the circuit lead will be connected to ground, thereby providing an input voltage $v_i = 0$ to the same resistor. Note that the MSB is connected to the smallest resistor (R), and the LSB is connected to the largest resistor $2^{n-1}R$. By writing the current summation at node A of the output op-amp, one obtains

$$\frac{v_{n-1}}{R} + \frac{v_{n-2}}{2R} + \ldots + \frac{v_0}{2^{n-1}R} + \frac{v}{R/2} = 0$$

In writing this equation, we have used the two principal facts for an op-amp: the voltage is the same at both input leads, and the current through each lead is 0. Note that the + lead is grounded and, hence, node A should have zero voltage. Now, since

$$v_i = -b_i v_{ref}$$

where $b_i = 0$ or 1, depending on the bit value (state of the corresponding switch), one has

$$v = \left[b_{n-1} + \frac{b_{n-2}}{2} + \ldots + \frac{b_0}{2^{n-1}} \right] \frac{v_{ref}}{2} \tag{9.64}$$

Clearly, the output voltage v is proportional to the value D of the digital word w, as required.

The *full-scale value* (FSV) of the analog output occurs when all b_i are equal to 1. Hence,

$$\text{FSV} = \left[1 + \frac{1}{2} + \ldots + \frac{1}{2^{n-1}} \right] \frac{v_{ref}}{2}$$

Signal Conditioning and Modification

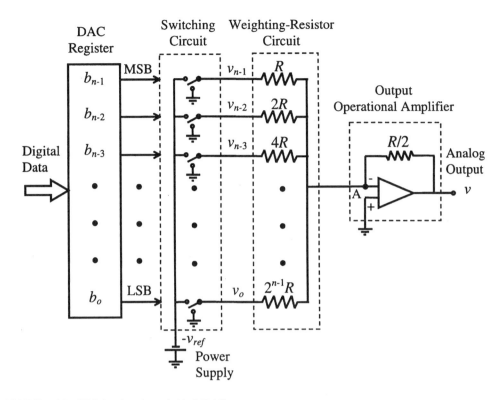

FIGURE 9.16 Weighted-resistor (adder) DAC.

Using the commonly known formula for the sum of a geometric series

$$1 + r + r^2 + \ldots + r^{n-1} = \frac{(1-r^n)}{(1-r)} \tag{9.65}$$

one obtains

$$\text{FSV} = \left(1 - \frac{1}{2^n}\right) v_{ref} \tag{9.66}$$

Note that this value is slightly smaller than the reference voltage v_{ref}.

A major drawback of the weighted-resistor DAC is that the range of the resistance value in the weighting circuit is very wide. This presents a practical difficulty, particularly when the size (number of bits n) of the DAC is large. Use of resistors having widely different magnitudes in the same circuit can create accuracy problems. For example, because the MSB corresponds to the smallest weighting resistor, it follows that the resistors must have a very high precision.

Ladder DAC

A DAC that uses an $R - 2R$ ladder circuit is known as a ladder DAC. This circuit uses only two types of resistors — one having resistance R and the other having $2R$. Hence, the tight tolerance requirements in resistance accuracy for the weighted-resistor DAC can be relieved using a ladder DAC. Schematic representation of an $R - 2R$ ladder DAC is shown in Figure 9.17. Microswitching in the switching circuit in this case can operate just like in the previous case of a weighted-resistor

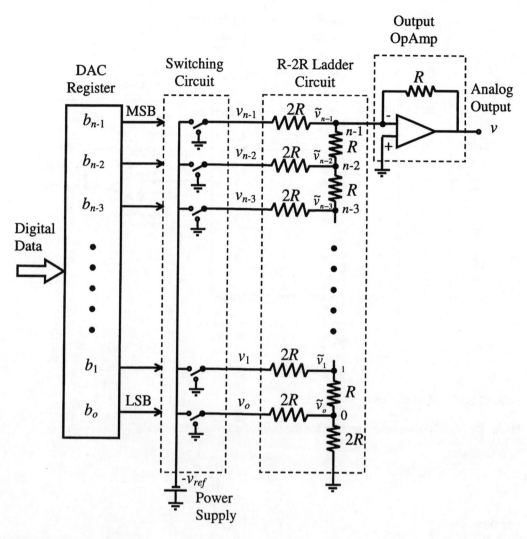

FIGURE 9.17 Ladder DAC.

DAC. To obtain the input-output equation for the ladder DAC, suppose that, as before, the voltage output from the solid-state switch associated with b_i of the digital word is v_i. Furthermore, suppose that $\tilde{v}_i$ is the voltage at node i of the ladder circuit, as shown in Figure 9.17. Now, writing the current summation at node i, one obtains

$$\frac{v_i - \tilde{v}_i}{2R} + \frac{\tilde{v}_{i+1} - \tilde{v}_i}{R} + \frac{\tilde{v}_{i-1} - \tilde{v}_i}{R} = 0$$

or

$$\frac{1}{2}v_i = \frac{5}{2}\tilde{v}_i - \tilde{v}_{i-1} - \tilde{v}_{i+1} \quad \text{for } i = 1, 2, \ldots, n-2 \qquad (i)$$

Note that equation (*i*) is valid for all nodes except node 0 and node $n - 1$. It is seen that the current summation for node 0 gives

Signal Conditioning and Modification

$$\frac{v_0 - \tilde{v}_0}{2R} + \frac{\tilde{v}_1 - \tilde{v}_0}{R} + \frac{0 - \tilde{v}_0}{2R} = 0$$

or

$$\frac{1}{2}v_0 = 2\tilde{v}_0 - \tilde{v}_1 \qquad (ii)$$

and the current summation for node $n-1$ gives

$$\frac{v_{n-1} - \tilde{v}_{n-1}}{2R} + \frac{v - \tilde{v}_{n-1}}{R} + \frac{\tilde{v}_{n-2} - \tilde{v}_{n-1}}{R} = 0$$

But, since the + lead of the op-amp is grounded, then $\tilde{v}_{n-1} = 0$. Hence,

$$\frac{1}{2}v_{n-1} = -\tilde{v}_{n-2} - v \qquad (iii)$$

Next, using equations (i) through (iii) and remembering that $\tilde{v}_{n-1} = 0$, one can write the following series of equations:

$$\begin{aligned}
\frac{1}{2}v_{n-1} &= & -\tilde{v}_{n-2} & & -v \\
\frac{1}{2^2}v_{n-2} &= \frac{1}{2}\frac{5}{2}\tilde{v}_{n-2} & -\frac{1}{2}\tilde{v}_{n-3} \\
\frac{1}{2^3}v_{n-3} &= \frac{1}{2^2}\frac{5}{2}\tilde{v}_{n-3} & -\frac{1}{2^2}\tilde{v}_{n-4} & -\frac{1}{2^2}\tilde{v}_{n-2} \\
&\vdots \\
\frac{1}{2^{n-1}}v_1 &= \frac{1}{2^{n-2}}\frac{5}{2}\tilde{v}_1 & -\frac{1}{2^{n-2}}\tilde{v}_0 & -\frac{1}{2^{n-2}}\tilde{v}_2 \\
\frac{1}{2^n}v_0 &= \frac{1}{2^{n-1}}2\tilde{v}_0 & & -\frac{1}{2^{n-1}}\tilde{v}_1
\end{aligned} \qquad (iv)$$

If one sums these n equations, first denoting that

$$S = \frac{1}{2^2}\tilde{v}_{n-2} + \frac{1}{2^3}\tilde{v}_{n-3} + \ldots + \frac{1}{2^{n-1}}\tilde{v}_1$$

then one obtains

$$\frac{1}{2}v_{n-1} + \frac{1}{2^2}v_{n-2} + \ldots + \frac{1}{2^n}v_0 = 5S - 4S - S + \frac{1}{2^{n-1}}2\tilde{v}_0 - \frac{1}{2^{n-2}}\tilde{v}_0 - v = -v$$

Finally, since $v_i = -b_i v_{ref}$, the analog output is given as

$$v = \left[\frac{1}{2}b_{n-1} + \frac{1}{2^2}b_{n-2} + \ldots + \frac{1}{2^n}b_0\right]v_{ref} \qquad (9.67)$$

which is identical to equation (9.64) obtained for the weighted-resistor DAC. Hence, as before, the analog output is proportional to the value D of the digital word; furthermore, the *full-scale value* of the ladder DAC is also given by the previous equation (9.66).

DAC Error Sources

For a given digital word, the analog output voltage from a DAC will not be exactly equal to what is given by the analytical formulas (e.g., equation (9.64)) derived earlier. The difference between the actual output and the ideal output is the error. The DAC error can be normalized with respect to the full-scale value.

There are many causes of DAC error. Typical error sources include parametric uncertainties and variations, circuit time constants, switching errors, and variations and noise in the reference voltage. Several types of error sources and representations are discussed below.

1. *Code ambiguity*: In many digital codes (e.g., in the straight binary code), incrementing a number by an LSB will involve more than one bit switching. If the speed of switching from 0 to 1 is different from that for 1 to 0, and if switching pulses are not applied to the switching circuit simultaneously, the bit switchings will not take place simultaneously. For example, in a 4-bit DAC, incrementing from decimal 2 to decimal 4 will involve changing the digital word from 0011 to 0100. This requires two bit switchings from 1 to 0 and one bit switching from 0 to 1. If 1 to 0 switching is faster than the 0 to 1 switching, then an intermediate value given by 0000 (decimal zero) will be generated, with a corresponding analog output. Hence, there will be a momentary code ambiguity and associated error in the DAC signal. This problem can be reduced (and eliminated in single-bit increments) if a *gray code* is used to represent the digital data. Improved switching circuitry will also help reduce this error.
2. *Settling time*: The circuit hardware in a DAC unit will have some dynamics, with associated time constants and perhaps oscillations (underdamped response). Hence, the output voltage cannot instantaneously settle to its ideal value upon switching. The time required for the analog output to settle within a certain band (say, ±2% of the final value or $\pm\frac{1}{2}$ resolution), following the application of the digital data, is termed the *settling time*. Naturally, settling time should be smaller for better (faster and more accurate) performance. As a rule of thumb, the settling time should be approximately half the data arrival time. Note that the data arrival time is the time interval between the arrival of two successive data values, and is given by the inverse of the data arrival rate.
3. *Glitches*: Switching of a circuit will involve sudden changes in magnetic flux due to current changes. This will induce voltages that produce unwanted signal components. In a DAC circuit, these induced voltages due to rapid switching can cause signal spikes that will appear at the output. The error due to these noise signals is not significant at low conversion rates.
4. *Parametric errors*: As discussed before, resistor elements in a DAC might not be precise, particularly when resistors within a wide range of magnitudes are employed, as in the case of a weighted-resistor DAC. These errors appear at the analog output. Furthermore, aging and environmental changes (primarily changes in temperature) will change the values of circuit parameters — resistance in particular. This also will result in DAC error. These types of errors due to imprecision of circuit parameters and variations of parameter values are termed *parametric errors*. Effects of such errors can be reduced in several ways, including the use of compensation hardware (and perhaps software) and directly by using precise and robust circuit components and employing good manufacturing practices.

5. *Reference voltage variations*: Because the analog output of a DAC is proportional to the reference voltage v_{ref}, any variations in the voltage supply will directly appear as an error. This problem can be overcome using stabilized voltage sources with sufficiently low output impedance.
6. *Monotonicity*: Clearly, the output of a DAC should change by its resolution ($\delta y = v_{ref}/2^n$) for each step of one LSB (least-significant bit) increment in the digital value. This ideal behavior might not exist in some practical DACs due to errors such as those mentioned above. At least the analog output should not decrease as the value of the digital input increases. This is known as the *monotonicity* requirement that should be met by a practical digital-to-analog converter.
7. *Nonlinearity*: Suppose that the digital input to a DAC is varied from [0 0 ... 0] to [1 1 ... 1] in steps of one LSB. As mentioned above, ideally the analog output should increase in constant jumps of $\delta y = v_{ref}/2^n$, giving a staircase-shaped analog output. If one draws the best linear fit for this ideally montonic staircase response, it will have a slope equal to the resolution/step. This slope is known as the *ideal scale factor*. Nonlinearity of a DAC is measured by the largest deviation of the DAC output from this best linear fit. Note that, in the ideal case, the nonlinearity is limited to half the resolution $\left(\frac{1}{2}\delta y\right)$.

One cause of nonlinearity is clearly the faulty bit transitions. Another cause is circuit nonlinearity in the conventional sense. Specifically, due to nonlinearities in circuit elements such as op-amps and resistors, the analog output will not be proportional to the value of the digital word dictated by the bit switchings (faulty or not). This latter type of nonlinearity can be accounted for using calibration.

9.4.2 Analog-to-Digital Conversion (ADC)

Analog signals, which are continuously defined with respect to time, must be sampled at discrete time points, and the sample values must be represented in the digital form (according to a suitable code) to be read into a digital system such as a microcomputer. An analog-to-digital converter (ADC) is used to accomplish this. For example, because response measurements of dynamic systems are usually available as analog signals, these signals must be converted into the digital form before passing on to a signal analysis computer. Hence, the computer interface for the measurement channels should contain one or more ADCs.

DACs and ADCs are usually situated on the same digital interface board. But, the analog to digital conversion process is more complex and time consuming than the digital to analog conversion process. Furthermore, many types of ADCs use DACs to accomplish the analog to digital conversation. Hence, ADCs are usually more costly and their conversion rate is usually slower in comparison to DACs.

Several types of analog to digital converters are commercially available. The principle of operation may vary depending on the type. A few commonly-known types are discussed here.

Successive-Approximation ADC

This type of analog-to-digital converter is very fast, and is suitable for high-speed applications. The speed of conversion depends on the number of bits in the output register of the ADC, but is virtually independent of the nature of the analog input signal. A schematic diagram for a successive-approximation ADC is shown in Figure 9.18. Note that a DAC is an integral component of this ADC. The sampled analog signal (from a *sample and hold circuit*) is applied to a *comparator* (typically a *differential amplifier*). Simultaneously, a "start conversion" (SC) control pulse is sent into the *control logic unit* by the external device (perhaps a microcomputer) that controls the operation of the ADC. Then, until the "conversion complete" (CC) pulse is sent out by the control

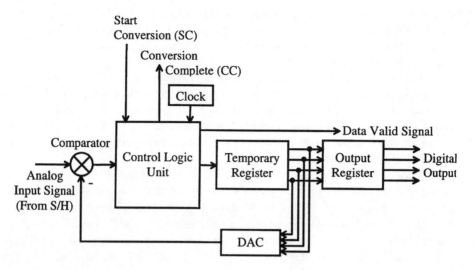

FIGURE 9.18 Successive-approximation ADC.

logic unit, no new data will be accepted by the ADC. Initially, the registers are cleared so that they contain all zero bits. Now, the ADC is ready for its first conversion approximation.

The first approximation begins with a clock pulse. Then, the control logic unit will set the most significant bit (MSB) of the temporary register (DAC control register) to 1, all the other bits in that register being 0. This digital word in the temporary register is supplied to the DAC. Note that the analog output of the DAC is now half the full-scale value. This analog signal is subtracted from the analog input by the comparator. If the output of the comparator is positive, the control logic unit will leave the MSB of the temporary register at binary 1 and proceed to the next approximation. If the comparator output is negative, the control logic unit will change the MSB to binary 0 and proceed to the next approximation.

The second approximation will start with another clock pulse. This approximation will consider the second most significant bit of the temporary register. As before, this bit is set to 1 and the comparison is made. If the comparator output is positive, this bit is left at value 1 and the third most significant bit is considered. If the comparator output is negative, the bit value will be changed to 0 before proceeding to the third most significant bit.

In this manner, all bits in the temporary register are set successively, starting from the MSB and ending with the LSB. The contents of the temporary register are then transferred to the output register and the data valid signal is set by the control logic unit, signaling the interfaced device (e.g., a microcomputer) to read the contents of the output register. The interfaced device will not read the register if the data valid signal is not present. Next, a "conversion complete" (CC) pulse is sent out by the control logic unit and the temporary register is cleared. The ADC is now ready to accept another data sample for digital conversion.

Note that the conversion process is essentially the same for every bit in the temporary register. Hence, the total conversion time is approximately n times the conversion time for one bit. Typically, one bit conversion can be completed within one clock period.

It should be clear that if the maximum value of the analog input signal exceeds the full-scale value of the DAC, then the excess signal amount cannot be converted by the ADC. The excess amount will directly contribute to error in the digital output of the ADC. Hence, this situation should be avoided by properly scaling the analog input, or by properly selecting the reference voltage for the internal DAC unit.

In the foregoing discussion, it was assumed that the analog input signal is positive. Otherwise, the sign of the signal should be accounted for by some means. For example, the sign of the signal

Signal Conditioning and Modification

can be detected from the sign of the comparator output initially, when all bits are 0. Then, the same conversion process can be used after switching the polarity of the comparator. Finally, the sign must be represented in the digital output (e.g., by the two's complement representation for negative quantities). Another approach to account for signed (bipolar) input signals is to offset the signal by a sufficiently large constant voltage such that the analog input is always positive. After the conversion, the digital number corresponding to this offset must be subtracted from the converted data in the output register in order to obtain the correct digital output. In what follows, it will be assumed that the analog input signal is positive.

Dual-Slope ADC

This analog-to-digital converter uses an RC integrating circuit. Hence, it is also known as an *integrating ADC*. This ADC is simple and inexpensive. In particular, an internal DAC is not utilized and, hence, DAC errors mentioned previously will not enter the ADC output. Furthermore, the parameters R and C in the integrating circuit do not enter the ADC output. Hence, the device is self-compensating in terms of circuit-parameter variations due to temperature, aging, etc. A shortcoming of this ADC is its slow conversion rate because, for accurate results, the signal integration must proceed for a longer time in comparison to the conversion time for a successive-approximation ADC.

Analog-to-digital conversion in a dual-slope ADC is based on timing (i.e., counting the number of clock pulses during) a capacitor-charging process. The principle of operation can be explained with reference to the integrating circuit shown in Figure 9.19(a). Note that v_i is a constant input voltage to the circuit, and v is the output voltage. Since the + lead of the op-amp is grounded, the − lead (and node A) will also have zero voltage. Also, the current through the op-amp leads is negligible. Hence, the current balance at node A gives

$$\frac{v_i}{R} + C\frac{dv}{dt} = 0$$

Integrating this equation for constant v_i one obtains

$$v(t) = v(0) - \frac{v_i t}{RC} \tag{9.68}$$

Equation (9.68) will be utilized in obtaining a principal result for the dual-slope ADC.

A schematic diagram for a dual-slope ADC is shown in Figure 9.19(b). Initially, the capacitor C in the integrating circuit is discharged (zero voltage). Then, the analog signal v_s is supplied to the switching element and held constant by the sample and hold circuit (S/H). Simultaneously, a "conversion start" (CS) control signal is sent to the control logic unit. This will clear the timer and the output register (i.e., all bits are set to 0) and will send a pulse to the switching element to connect the input v_s to the integrating circuit. Also, a signal is sent to the timer to initiate timing (counting). The capacitor C will begin to charge. Equation (9.68) is now applicable with input $v_i = v_s$ and the initial state $v(0) = 0$. Suppose that the integrator output v becomes $-v_c$ at time $t = t_1$. Hence, from equation (9.68), one has

$$v_c = \frac{v_s t_1}{RC} \tag{i}$$

The timer will keep track of the capacitor charging time (as a clock pulse count n) and will inform the control logic unit when the elapsed time is t_1 (i.e., when the count is n_1). Note that t_1 and n_1 are fixed (and known), but voltage v_c depends on the value of v_s and is unknown.

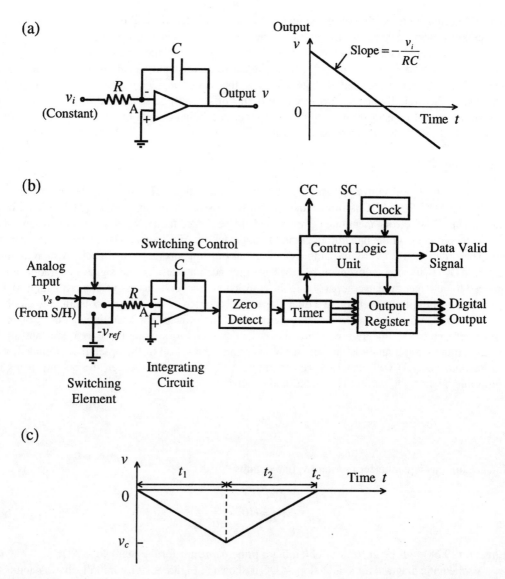

FIGURE 9.19 (a) RC integrating circuit; (b) dual-slope ADC; and (c) dual-slope charging-discharging curve.

At this point, the control logic unit will send a signal to the switching unit that will connect the input lead of the integrator to a negative supply voltage $-v_{ref}$. Simultaneously, a signal is sent to the timer to clear its contents and start timing (counting) again. The capacitor begins to discharge now. The output of the integrating circuit is monitored by the *zero-detect unit*. When this output becomes zero, the zero-detect unit sends a signal to the timer to stop counting. The zero-detect unit could be a comparator (differential amplifier) having one of the two input leads set at zero potential. Now suppose that the elapsed time is t_2 (with a corresponding count of n_2).

One can use equation (9.68) for the capacitor discharging process as well. Note that $v_i = -v_{ref}$ and $v(0) = -v_c$ in this case. Also, $v(t) = 0$ at $t = t_2$. Hence, from equation (9.68), one obtains

$$0 = -v_c + \frac{v_{ref} t_2}{RC}$$

or

$$v_c = \frac{v_{ref} t_2}{RC} \quad (ii)$$

On dividing equation (*i*) by (*ii*), one obtains

$$v_s = v_{ref} \frac{t_2}{t_1}$$

However, the timer pulse count is proportional to the elapsed time; hence,

$$\frac{t_2}{t_1} = \frac{n_2}{n_1}$$

Then,

$$v_s = \frac{v_{ref}}{n_1} n_2 \quad (9.69)$$

Since v_{ref} and n_1 are fixed quantities, v_{ref}/n_1 can be interpreted as a scaling factor for the analog input. It follows from equation (9.69) then that the second count n_2 is proportional to the analog signal sample v_s. Note that the timer output is available in the digital form. Accordingly, the count n_2 is used as the digital output of the ADC.

At the end of the capacitor discharge period, the count n_2 in the timer is transferred to the output register of the ADC and the "data valid" signal is set. The contents of the output register are now ready to be read by the interfaced digital system, and the ADC is ready to convert a new sample.

The *charging-discharging curve* for the capacitor during the conversion process is shown in Figure 9.19(c). The slope of the curve during charging is $-\frac{v_s}{RC}$, and the slope during discharging is $+\frac{v_{ref}}{RC}$. The reason for the use of the term "dual-slope" to denote this ADC is therefore clear.

As mentioned before, any variations in R and C do not affect the accuracy of the output, but it is clear from the above discussion that the conversion time depends on the capacitor discharging time t_2 (note that t_1 is fixed) and this depends on v_c and, hence, on the input signal value v_s (see equation (*i*)). It follows that, unlike the successive-approximation ADC, the dual-slope ADC has a conversion time that directly depends on the magnitude of the input data sample. This is a disadvantage in a way because, in many applications, one prefers to have a constant conversion rate.

The above discussion assumed that the input signal is positive. For a negative signal, the polarity of the supply voltage v_{ref} must be changed. Furthermore, the sign must be properly represented in the contents of the output register — as for the case of successive-approximation ADC, for example.

Counter ADC

The counter-type ADC has several aspects in common with the successive-approximation ADC. Both are comparison-type (or closed-loop) ADCs. Both use a DAC unit internally to compare the input signal with the converted signal. The difference is that, in a counter ADC, the comparison starts with the LSB and proceeds down. It follows that, in the counter ADC, the conversion time

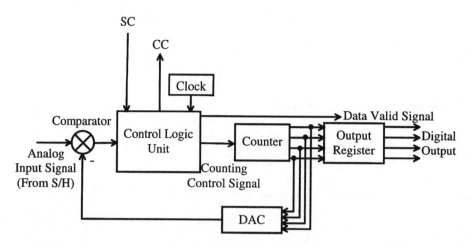

FIGURE 9.20 Counter ADC.

depends on the signal level, because counting (comparison) stops when a match is made, giving shorter conversion times for smaller signal values.

A schematic diagram of a counter ADC is shown in Figure 9.20. Note that this is quite similar to Figure 9.18. Initially, all registers are cleared (all bits and counts are set to 0). As an analog data signal (from the sample and hold circuit) arrives at the comparator, a "start conversion" (SC) pulse is sent to the control logic unit. When the ADC is ready for conversion (i.e., when the data valid signal is on), the control logic unit initiates the counter. Now, the counter sets its count to 1 and the LSB of the DAC register is set to 1 as well. The resulting DAC output is subtracted from the analog input by the comparator. If the comparator output is positive, the count is incremented by one, and this causes the binary number in the DAC register to be incremented by one LSB. The new (increased) output of the DAC is now compared with the input signal. This count incrementing and comparison cycle is carried out until the comparator output becomes zero or negative. At that point, the control logic unit will send out a "conversion complete" (CC) signal and will transfer the contents of the counter to the output register. Then, the data valid signal is also turned on, indicating that the ADC is ready for a new conversion cycle, and the contents of the output register (the digital output) are available to be read by the interfaced digital system.

The count of the counter is available in the binary form, which is compatible with the output register as well as the DAC register. Hence, the count can be transferred directly to these registers. The count when the analog signal is equal to (or slightly less than) the output of the DAC is proportional to the analog signal value. Hence, this count represents the digital output. Again, the sign of the input signal must be properly accounted for in the bipolar operation.

9.4.3 ADC Performance Characteristics

For ADCs that internally use a DAC, the same error sources discussed previously for DACs will apply. Code ambiguity at the output register will not be a problem because the converted digital quantity is transferred instantaneously to the output register. Code ambiguity in the DAC register can still cause error in an ADC that uses a DAC. Conversion time is a major factor, this being much larger for an ADC. In addition to resolution and dynamic range, quantization error will be applicable to an ADC. These considerations that govern the performance of an ADC are discussed below.

Resolution and Quantization Error

The number of bits n in an ADC register determines the resolution and dynamic range of the ADC. For an n-bit ADC, the output register size is n bits. Hence, the smallest possible increment of the digital output is one LSB. The change in the analog input that results in a change of one LSB at the output is the *resolution* of the ADC. The range of digital outputs is from 0 to $2^n - 1$ for the unipolar (unsigned) case. This represents the dynamic range. Hence, as for a DAC, the dynamic range of an *n*-bit ADC is given by the ratio:

$$DR = 2^n - 1 \tag{9.70}$$

or, in decibels,

$$DR = 20\log_{10}(2^n - 1) \text{ dB} \tag{9.71}$$

The *full-scale value* of an ADC is the value of the analog input that corresponds to the maximum digital output.

Suppose that an analog signal within the dynamic range of the ADC is converted. Since the analog input (sample value) has infinitesimal resolution and the digital representation has a finite resolution (one LSB), an error is introduced into the analog-to-digital conversion process. This is known as the *quantization error*. A digital number will increment in constant steps of 1 LSB. If an analog value falls at an intermediate point within a single-LSB step, then there is a quantization error. Rounding off of the digital output can be accomplished as follows. The magnitude of the error, when quantized up, is compared with that when quantized down, say, using two hold elements and a differential amplifier. Then, one retains the digital value corresponding to the lower error magnitude. If the analog value is below the ½-LSB mark, then the corresponding digital value is represented by the value in the beginning of the step. If the analog value is above the ½-LSB mark, then the corresponding digital value is the value at the end of the step. It follows that with this type of rounding off, the quantization error does not exceed ½ LSB.

Monotonicity, Nonlinearity, and Offset Error

Considerations of monotonicity and nonlinearity are important for an ADC as well as for a DAC. The input is an analog signal and the output is digital in the case of ADC. Disregarding quantization error, the digital output of an ADC will increase in constant steps in the shape of an ideal staircase when the analog input is increased from 0 in steps of the device resolution (δy). This is the ideally *monotonic* case. The best straight-line fit to this curve has a slope equal to $1/\delta y$ (LSB per volt). This is the *ideal gain* or *ideal scale factor*. But still there will be an *offset error* of ½ LSB because the best linear fit will not pass through the origin. Adjustments can be made for this offset error.

Incorrect bit transitions can take place in an ADC, due to various errors that might be present and possibly due to circuit malfunctions. The best linear fit under such faulty conditions will have a slope different from the ideal gain. The difference is the gain error. Nonlinearity is the maximum deviation of the output from the best linear fit. It is clear that with perfect bit transitions, in the ideal case, a nonlinearity of 1/2 LSB would be present. Nonlinearities larger than this would result due to incorrect bit transitions. As in the case of DAC, another source of nonlinearity in ADC is circuit nonlinearities that would deform the analog input signal before being converted into the digital form.

ADC Conversion Rate

It is clear that analog to digital conversion is much more time consuming than digital to analog conversion. The conversion time is a very important factor because the rate at which conversion can take place governs many aspects of data acquisition, particularly in real-time applications. For example, the data sampling rate must synchronize with the ADC conversion rate. This, in turn, will determine the Nyquist frequency (half the sampling rate), which is the maximum value of useful frequency present in the sampled signal (see Chapter 4). Furthermore, the sampling rate will dictate storage and memory requirements. Another important consideration related to the ADC conversion rate is the fact that a signal sample must be maintained at that value during the entire process of conversion into the digital form. This would require a hold circuit, and this circuit should be able to perform accurately at the largest possible conversion time for the particular ADC unit.

The time needed for a sampled analog input to be converted into the digital form will depend on the type of ADC. Usually in a comparison-type ADC (which uses an internal DAC), each bit-transition will take place in one clock period Δt. Also, in an integrating (dual-slope) ADC, each clock count will need a time of Δt. On this basis, the following figures can be given for conversion times of the three types of ADC that have been discussed:

1. Successive-approximation ADC: For an n-bit ADC, n comparisons are needed in this case. Hence, the conversion time is given by

$$t_c = n\Delta t \tag{9.72}$$

   in which Δt is the clock period. Note that t_c does not depend on the signal level (analog input) in this case.

2. Dual-slope (integrating) ADC: In this case, conversion time is the time needed for the two counts n_1 and n_2 (see Figure 9.19(c)). Hence,

$$t_c = (n_1 + n_2)\Delta t \tag{9.73}$$

   Note that n_1 is a fixed count; but n_2 is a variable count that represents the digital output, and is proportional to the analog input (signal level). Hence, in this type of ADC, conversion time depends on the analog input level. The largest output for an n-bit converter is $2^n - 1$. Hence, the largest conversion time can be given by

$$t_{c\max} = (n_1 + 2^n - 1)\Delta t \tag{9.74}$$

3. Counter ADC: For a counter ADC, the conversion time is proportional to the number of bit transitions (1 LSB per step) from zero to the digital output n_o. Hence, the conversion time is given by

$$t_c = n_o \Delta t \tag{9.75}$$

   in which n_o is the digital output value (in decimal).

   Note that, in this case also, t_c depends on the magnitude of the input data sample. Since the maximum value of n_o is $2^n - 1$ for an n-bit ADC, the maximum conversion time becomes

$$t_{c\max} = (2^n - 1)\Delta t \tag{9.76}$$

By comparing equations (9.72), (9.74), and (9.76), it can be concluded that the successive-approximation ADC is the fastest of the three types discussed.

The total time taken to convert an analog signal will depend on other factors besides the time taken for conversion from sampled data to digital data. For example, in multiple-channel data acquisition, the time taken to select the channel must be counted. Furthermore, the time needed to sample the data and the time needed to transfer the converted digital data into the output register must also be included. The *conversion rate* for an ADC is actually the inverse of this overall time needed for a conversion cycle. Typically, however, the conversion rate depends primarily on the bit conversion time in the case of a comparison-type ADC, and on the integration time in the case of an integration-type ADC. A typical time period for a comparison step or counting step in an ADC is $\Delta t = 5\mu s$. Hence, for an 8-bit successive-approximation ADC, the conversion time is 40 μs. The corresponding sampling rate would be of the order of (less than) $1/40 \times 10^{-6} = 25 \times 10^3$ samples per second (or 25 kHz). The maximum conversion rate for an 8-bit counter ADC would be about $5 \times (2^8 - 1) = 1275$ μs. The corresponding sampling rate would be of the order of 780 samples per second. Note that this is considerably slow. The maximum conversion time for a dual-slope ADC might be still larger (slower).

9.4.4 Sample-and-Hold (S/H) Circuitry

In typical applications of data acquisition that use analog-to-digital conversion, the analog input to an ADC can be very transient. Furthermore, analog-to-digital conversion is not instantaneous (the conversion time is much larger than the digital-to-analog conversion time). Specifically, the incoming analog signal might be changing at a rate higher than the ADC conversion rate. Then, the input signal value will vary during the conversion period and there will be an ambiguity as to the input value corresponding to a digital output value. Hence, it is necessary to sample the analog input signal and maintain the input to the ADC at this value until the analog-to-digital conversion is complete. In other words, because one is typically dealing with analog signals that can vary at a high speed, it will be necessary to *sample and hold* (S/H) the input signal for each analog-to-digital conversion cycle. Each data sample must be generated and captured by the S/H circuit on the issue of the "start conversion" (SC) control signal, and the captured voltage level must be maintained constant until the "conversion complete" (CC) control signal is issued by the ADC unit.

The main element in an S/H circuit is the holding capacitor. A schematic diagram of a sample-and-hold circuit is shown in Figure 9.21. The analog input signal is supplied through a voltage follower to a solid-state switch. The switch typically uses a field-effect transistor (FET), such as the metal-oxide semiconductor field effect transistor (MOSFET).

The switch is closed in response to a "sample pulse" and is opened in response to a "hold pulse." Both control pulses would be generated by the control logic unit of the ADC. During the time interval between these two pulses, the holding capacitor is charged to the voltage of the sampled input. This capacitor voltage is then supplied to the ADC through a second voltage follower.

The functions of the two voltage followers are now explained. When the FET switch is closed in response to a sample command (pulse), the capacitor must be charged as quickly as possible. The associated time constant (charging time constant) τ_c is given by

$$\tau_c = R_s C \qquad (9.77)$$

where

R_s = source resistance
C = capacitance of the holding capacitor.

Since τ_c must be very small for fast charging, and because C is fixed by the holding requirements (typically, C is of the order of 100 pF where 1 pF = 1×10^{-12} F), a very small source resistance is needed. This requirement is met by the input voltage follower (which is known to have a very low output impedance), thereby providing a very small R_s. Furthermore, because a voltage follower has a unity gain, the voltage at the output of this input voltage follower would be equal to the voltage of the analog input signal, as required.

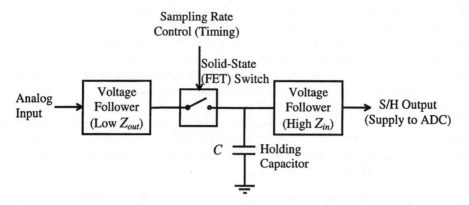

FIGURE 9.21 A sample-and-hold circuit.

Next, once the FET switch is opened in response to a hold command (pulse), the capacitor should not discharge. This requirement is met due to the presence of the output voltage follower. Since the input impedance of a voltage follower is very high, the current through its leads would be almost 0. Because of this, the holding capacitor will have a virtually 0 discharge rate under hold conditions. Furthermore, one would like the output of this second voltage follower to be equal to the voltage of the capacitor. This condition is also satisfied due to the fact that a voltage follower has a unity gain. Hence, the sampling would be almost instantaneous and the output of the S/H circuit would be maintained (almost) constant during the holding period, due to the presence of the two voltage followers. Note that the practical S/H circuits are *zero-order-hold* devices by definition.

9.4.5 MULTIPLEXERS (MUX)

A multiplexer (also known as a *scanner*) is used to select one channel at a time from a bank of signal channels and connect it to a common hardware unit. In this manner, a costly and complex hardware unit can be time-shared among several signal channels. Typically, channel selection is done in a sequential manner at a fixed channel-select rate.

There are two types of multiplexers: *analog multiplexers* and *digital multiplexers*. To scan a group of analog channels, an analog multiplexer is used. Alternatively, a digital multiplexer can be used to sequentially read one data word at a time from a set of digital data words.

The process of distributing a single channel of data among several output channels is known as *demultiplexing*. A demultiplexer (or data distributor) performs the reverse function of a multiplexer (or scanner). A demultiplexer can be used, for example, when the same (processed) signal from a digital computer is needed for several purposes (e.g., digital display, analog reading, and digital plotting).

Multiplexing used in short-distance signal transmission applications (e.g., control and data logging) is usually *time-division multiplexing*. In this method, channel selection is made with respect to time. Hence, only one input channel is connected to the output channel of the multiplexer. This is the method described here. Another method of multiplexing, used particularly in long-distance transmission of several data signals, is known as *frequency-division multiplexing*. In this method, the input signals are *modulated* (e.g., by *amplitude modulation* as discussed previously) using *carrier signals* having different frequencies and are transmitted simultaneously through the same data channel. The signals are separated by *demodulation* at the receiving end.

Analog Multiplexers

Dynamic system monitoring often requires the measurement of several process responses. These signals must be conditioned (e.g., amplification and filtering) and modified in some manner (e.g., analog-to-digital conversion) before being supplied to a common-purpose system such as a digital computer or data logger. Usually, data-modification devices are costly. In particular, it has been noted that ADCs are more expensive than DACs. An expensive option for interfacing several analog signals with a common-purpose system such as a digital computer would be to provide separate data-modification hardware for each signal channel. This method has the advantage of high speed. An alternative, low-cost method is to use an *analog multiplexer* (*analog scanner*) to select one signal channel at a time, sequentially, and connect it to a common signal-modification hardware unit (consisting of amplifiers, filters, S/H, ADC, etc.). In this way, by time-sharing expensive hardware among many data channels, the data acquisition speed is traded off to some extent for significant cost savings. Because very high channel-selection speeds are possible with solid-state switching (e.g., solid-state speeds of the order of 10 MHz), the reduced speed is not a significant drawback in most applications. But, because the cost of hardware components such as an ADC is declining due to advances in solid-state technologies, cost reductions attainable through the use of multiplexing might not be substantial in some applications. Hence, some economic evaluation and engineering judgment is needed in deciding on the use of signal multiplexing for a particular data acquisition and control application.

A schematic diagram of an analog multiplexer is shown in Figure 9.22. The figure represents the general case of N input channels and one output channel. This is called an $N \times 1$ analog multiplexer. Each input channel is connected to the output through a solid-state switch, typically a field-effect transistor (FET) switch. One switch is closed (turned on) at a time. A switch is selected by a digital word that contains the corresponding *channel address*. Note that an n-bit address can assume 2^n digital values in the range of 0 to $2^n - 1$. Hence, a MUX with an n-bit address can handle $N = 2^n$ channels. Channel selection can be done by an external microprocessor that places the address of the channel on the address bus and simultaneously sends a control signal to the MUX to enable the MUX. The address decoder decodes the address and activates the corresponding FET switch. In this manner, channel selection could be done in an arbitrary order and with arbitrary timing, controlled by the microprocessor. In simple versions of multiplexers, the channel selection is made in a fixed order at a fixed speed, however.

Typically, the output of an analog MUX is connected to an S/H circuit and an ADC. Voltage followers can be provided both at the input and the output in order to reduce loading problems. A *differential amplifier* (or *instrumentation amplifier*) can be used at the output to reduce noise problems, particularly to reject common-mode interference, as discussed earlier in this chapter. Note that the channel-select speed must be synchronized with sampling and ADC speeds for each signal channel. The multiplexer speed is not a major limitation because very high speeds (solid-state speeds of 10 MHz or more) are available with solid-state switching.

Digital Multiplexers

Sometimes it is required to select one data word at a time from a set of digital data words, to be fed into a common device. For example, the set of data may be the outputs from a bank of digital transducers (e.g., shaft encoders which measure angular motions) or outputs from a set of ADCs that are connected to a set of analog signal channels. Then the selection of the particular digital output (data word) can be made using *addressing* and *data-bus transfer* techniques that are commonly used in digital systems.

A digital multiplexing (or logic multiplexing) configuration is shown in Figure 9.23. The N registers of the multiplexer hold a set of N data words. The contents of each register might correspond to a response measurement and, hence, will change regularly. The registers can represent separate hardware devices (e.g., output registers of a bank of ADCs) or locations in a computer memory to which data are being transferred (read in) regularly. Each register has a unique binary address. As in the case of an analog MUX, an n-bit address can select (address) 2^n registers. Hence, the number of registers is $N = 2^n$, as before. When the address of the register to be selected is

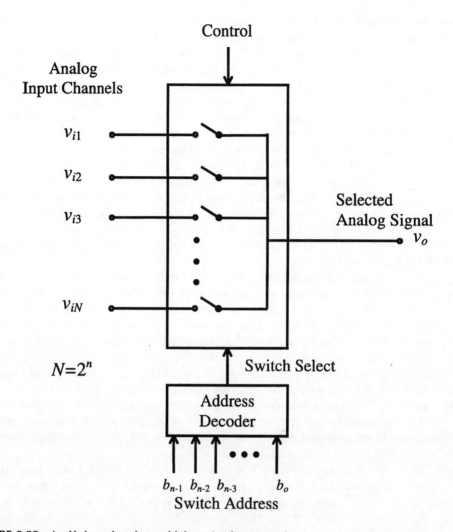

FIGURE 9.22 An N-channel analog multiplexer (analog scanner).

placed on the address bus, it enables the particular register. This causes the contents of that register to be placed on the data bus. Now the data bus is read by the device (e.g., computer) that is time-shared among the N data registers. Another address on the address bus will result in selecting another register and reading the contents of that register as before.

Digital multiplexing is usually faster than analog multiplexing, and has the usual advantages of digital devices: high accuracy, better noise immunity, robustness (no drift and errors due to parameter variations), long-distance data transmission capability without associated errors due to signal weakening, capability to handle very large numbers of data channels, etc. Furthermore, a digital multiplexer can be modified using software, usually without the need for hardware changes. If, however, instead of using an analog multiplexer followed by a single ADC, a separate ADC is used for each analog signal channel and then digital multiplexing is used, it is quite possible for the digital multiplexing approach to be more costly. If, on the other hand, the measurements are already available in the digital form (e.g., as encoder outputs), then digital multiplexing would be very cost effective and most desirable.

The transfer of a digital word from a single data source (e.g., a data bus) into several data registers, to be accessed independently, can be interpreted as *digital demultiplexing*. This is also a straightforward process of digital data transfer and reading.

Signal Conditioning and Modification

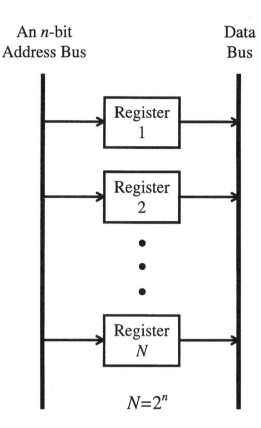

FIGURE 9.23 An $N \times 1$ digital multiplexer.

9.4.6 DIGITAL FILTERS

A filter is a device that eliminates undesirable frequency components in a signal and passes only the desirable frequency components through it. In *analog filtering*, the filter is a *physical dynamic system* — typically an electric circuit. The signal to be filtered is applied (*input*) to this dynamic system. The output of the dynamic system is the filtered signal. It follows that any physical dynamic system can be interpreted as an analog filter.

An analog filter can be represented by a differential equation with respect to time. It takes an analog input signal $u(t)$, that is defined continuously in time t, and generates an analog output $y(t)$. A *digital filter* is a device that accepts a sequence of discrete input values (say, sampled from an analog signal at sampling period Δt):

$$\{u_k\} = \{u_0, u_1, u_2, \ldots\} \tag{9.78}$$

and generates a sequence of discrete output values;

$$\{y_k\} = \{y_0, y_1, y_2, \ldots\} \tag{9.79}$$

Hence, a digital filter is a discrete-time system and it can be represented by a *difference equation*.

An nth order linear difference equation can be written in the form

$$a_0 y_k + a_1 y_{k-1} + \ldots + a_n y_{k-n} = b_0 u_k + b_1 u_{k-1} + \ldots + b_m u_{k-m} \qquad (9.80)$$

This is a *recursive algorithm* in the sense that it generates one value of the output sequence using previous values of the output sequence and all values of the input sequence up to the present time point. Digital filters represented in this manner are termed *recursive digital filters*. There are filters that employ digital processing where a block (a collection of samples) of the input sequence is converted in a one-shot computation into a block of the output sequence. Such filters are not recursive filters. Nonrecursive filters usually employ *digital Fourier analysis*, the *fast Fourier transform* (FFT) algorithm, in particular (see Chapter 4 and Appendix D). We restrict our discussion below to recursive digital filters. This section provides a brief (and nonexhaustive) introduction to the subject of digital filtering.

Software Implementation and Hardware Implementation

In digital filters, signal filtering is accomplished through digital processing of the input signal. The sequence of input data (usually obtained by sampling and digitizing the corresponding analog signal) is processed according to the recursive algorithm of the particular digital filter. This generates the output sequence. This digital output can be converted into an analog signal using a DAC if so desired.

A recursive digital filter is an implementation of a recursive algorithm that governs the particular filtering (e.g., low-pass, high-pass, bandpass, and band-reject). The filter algorithm can be implemented either by *software* or *hardware*. In software implementation, the filter algorithm is programmed into a digital computer. The *processor* (e.g., *microprocessor*) of the computer can process an input data sequence according to the run-time filter program stored in the memory (in *machine code*), to generate the filtered output sequence.

Digital processing of data is accomplished by means of logic circuitry that can perform basic arithmetic operations such as addition. In the software approach, the processor of a digital computer makes use of these basic logic circuits to perform digital processing according to the instructions of a software program stored in the computer memory. Alternatively, a hardware digital processor can be put together to perform a somewhat complex, yet fixed, processing operation. In this approach, the program of computation is said to be in hardware. The hardware processor is then available as an IC chip whose processing operation is fixed and cannot be modified. The logic circuitry in the IC chip is designed to accomplish the required processing function. Digital filters implemented by this hardware approach are termed *hardware digital filters*.

The software implementation of digital filters has the advantage of flexibility; the filter algorithm can be easily modified by changing the software program that is stored in the computer. If, on the other hand, a large number of filters of a particular (fixed) structure is commercially needed, then it is economical to design the filter as an IC chip and replicate the chip in mass production. In this manner, very low-cost digital filters can be produced. A hardware filter can operate at a much faster speed in comparison to a software filter because, in the former case, processing takes place automatically through logic circuitry in the filter chip without having to access the processor, a software program, and various data items stored in the memory. The main disadvantage of a hardware filter is that its algorithm and parameter values cannot be modified, and the filter is dedicated to a fixed function.

9.5 BRIDGE CIRCUITS

A full bridge is a circuit having four *arms* connected in a *lattice* form. Four nodes are formed in this manner. Two opposite nodes are used for excitation (voltage or current supply) of the bridge, and the remaining two opposite nodes provide the bridge output. Further details on bridge circuits and applications are found in Chapter 8.

Signal Conditioning and Modification

A bridge circuit is used to make some form of measurement. Typical measurements include change in resistance, change in inductance, change in capacitance, oscillating frequency, or some variable (stimulus) that causes these. There are two basic methods of making the measurement:

1. Bridge balance method
2. Imbalance output method.

A bridge is said to be balanced when the output voltage is 0. In the bridge-balance method, one starts with a balanced bridge. Then, in making a measurement, since the balance of the bridge will be upset due to the associated variation, it will result in a non-zero output voltage. The bridge can be balanced again by varying one of the arms of the bridge (assuming, of course, that some means is provided for fine adjustments that may be required). The change that is required to restore the balance provides the measurement. In this method, the bridge can be balanced precisely using a servo device.

In the imbalance output method, one usually starts with a balanced bridge, but the bridge is not balanced again after undergoing the change due to the variable that is being measured. Instead, the output voltage of the bridge due to the resulting imbalance is measured and used as an indication of the measurement.

There are many types of bridge circuits. If the supply to the bridge is DC, then one has a *DC bridge*. Similarly, an *AC bridge* has an AC excitation. A *resistance bridge* has only resistance elements in its four arms. An *impedance bridge* has impedance elements consisting of resistors, capacitors, and inductors in one or more of its arms. If the bridge excitation is a constant-voltage supply, it is a *constant-voltage bridge*. If the bridge supply is a constant-current source, one has a *constant-current bridge*.

9.5.1 WHEATSTONE BRIDGE

This is a resistance bridge with a constant DC voltage supply (i.e., a constant-voltage resistance bridge). A Wheatstone bridge is used in strain-gage measurements, and also in force, torque, and tactile sensors that employ strain-gage techniques. Because a Wheatstone bridge is used primarily in the measurement of small changes in resistance, it could be used in other types of sensing applications as well (e.g., in resistance temperature detectors or RTD).

Consider the Wheatstone bridge circuit shown in Figure 9.24(a). The bridge output v_o can be expressed as (see Chapter 8)

$$v_o = \frac{(R_1 R_4 - R_2 R_3)}{(R_1 + R_2)(R_3 + R_4)} v_{ref} \tag{9.81}$$

Note that the bridge-balance requirement is

$$\frac{R_1}{R_2} = \frac{R_3}{R_4} \tag{9.82}$$

Suppose that $R_1 = R_2 = R_3 = R_4 = R$ in the beginning (the bridge is balanced according to equation (9.82)) and then R_1 is increased by δR. For example, R_1 may represent the only active strain gage and the remaining three elements in the bridge are identical dummy elements. Then, in view of equation (9.81), the change in output due to the change δR is given by

$$\delta v_o = \frac{[(R+\delta R)R - R^2]}{(R+\delta R+R)(R+R)} v_{ref} - 0$$

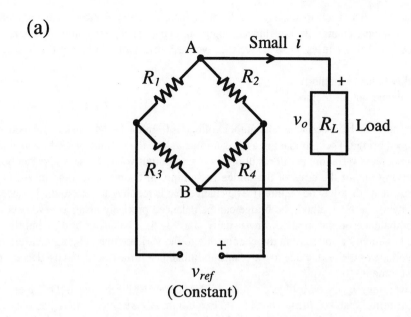

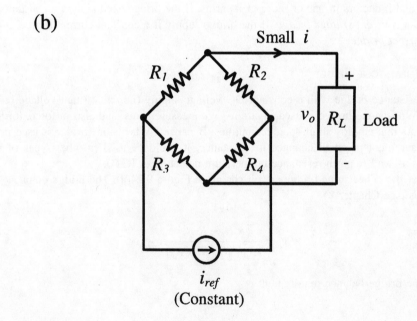

FIGURE 9.24 (a) Wheatstone bridge (the constant-voltage resistance bridge), and (b) the constant-current bridge.

or

$$\frac{\delta v_o}{v_{ref}} = \frac{\delta R/R}{(4 + 2\delta R/R)} \quad (9.83)$$

Note that the output is nonlinear in $\delta R/R$. If, however, $\delta R/R$ is assumed small in comparison to 2, one obtains the linearized relationship:

Signal Conditioning and Modification

$$\frac{\delta v_o}{v_{ref}} = \frac{\delta R}{4R} \tag{9.84}$$

as obtained in Chapter 8.

The error due to linearization, which is a measure of nonlinearity, can be given as the percentage

$$N_P = 100\left(1 - \frac{\text{Linearized output}}{\text{Actual output}}\right)\% \tag{9.85}$$

Hence, from equations (9.83) and (9.84), one obtains

$$N_P = 50\frac{\delta R}{R}\% \tag{9.86}$$

9.5.2 Constant-Current Bridge

When large resistance variations δR are required for a measurement, the Wheatstone bridge may not be satisfactory due to its nonlinearity, as indicated by equation (9.83). The constant-current bridge has less nonlinearity and is preferred in such applications. However, it needs a current-regulated power supply, which is typically more costly than a voltage-regulated power supply.

As shown in Figure 9.24(b), the constant-current bridge uses a constant-current excitation i_{ref} instead of a constant-voltage supply. Note that the output equation for the constant-current bridge can be determined from equation (9.81) simply by knowing the voltage at the current source. Suppose that this is voltage v_{ref} with the polarity as shown in Figure 9.24(a). Now, since the load current is assumed small (high-impedance load), the current through R_2 is equal to the current through R_1 and is given by

$$\frac{v_{ref}}{(R_1 + R_2)}$$

Similarly, current through R_4 and R_3 is given by

$$\frac{v_{ref}}{(R_3 + R_4)}$$

Accordingly,

$$i_{ref} = \frac{v_{ref}}{(R_1 + R_2)} + \frac{v_{ref}}{(R_3 + R_4)}$$

or

$$v_{ref} = \frac{(R_1 + R_2)(R_3 + R_4)}{(R_1 + R_2 + R_3 + R_4)} i_{ref} \tag{9.87}$$

Substituting equation (9.87) in (9.81), one obtains the output equation for the constant-current bridge; thus,

$$v_o = \frac{(R_1 R_4 - R_2 R_3)}{(R_1 + R_2 + R_3 + R_4)} i_{ref} \tag{9.88}$$

Note that the bridge-balance requirement is again given by equation (9.82).

To estimate the nonlinearity of a constant-current bridge, suppose that $R_1 = R_2 = R_3 = R_4 = R$ in the beginning and R_1 is changed by δR while the other resistors remain inactive. Again, R_1 will represent the active element (sensing element) and may correspond to an active strain gage. The change in output δv_o is given by

$$\delta v_o = \frac{[(R+\delta R)R - R^2]}{(R+\delta R + R + R + R)} i_{ref} = 0$$

or

$$\frac{\delta v_o}{R i_{ref}} = \frac{\delta R / R}{(4 + \delta R / R)} \tag{9.89}$$

By comparing the denominator on the RHS of this equation with equation (9.83), one observes that the constant-current bridge is more linear. Specifically, using the definition given by equation (9.85), the *percentage nonlinearity* can be expressed as

$$N_P = 25 \frac{\delta R}{R} \% \tag{9.90}$$

It is noted that the nonlinearity is halved by using a constant-current excitation instead of a constant-voltage excitation.

9.5.3 Bridge Amplifiers

The output from a resistance bridge is usually very small in comparison to the reference, and it must be amplified in order to increase the voltage level to a useful value (e.g., in system monitoring or data logging). A *bridge amplifier* is used for this purpose. This is typically an *instrumentation amplifier* or a *differential amplifier*. The bridge amplifier is modeled as a simple gain K_a that multiplies the bridge output.

Half-Bridge Circuits

A half-bridge can be used in some applications that require a bridge circuit. A half-bridge has only two arms, and the output is tapped from the mid-point of the two arms. The ends of the two arms are excited by a positive voltage and a negative voltage. Initially, the two arms have equal resistances so that, nominally, the bridge output is 0. One of the arms has the active element. Its change in resistance results in a non-zero output voltage. It is noted that the half-bridge circuit is somewhat similar to a potentiometer circuit.

A half-bridge amplifier consisting of a resistance half-bridge and an output amplifier is shown in Figure 9.25. The two bridge arms have resistances R_1 and R_2, and the amplifier uses a feedback

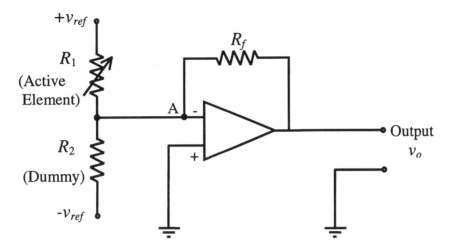

FIGURE 9.25 A half-bridge with an output amplifier.

resistance R_f. To get the output equation, one can use the two basic facts for an unsaturated op-amp: the voltages at the two leads are equal (due to high gain) and the current in both leads is 0 (due to high input impedance). Hence, voltage at node A is 0 and the current balance equation at node A is

$$\frac{v_{ref}}{R_1} + \frac{(-v_{ref})}{R_2} + \frac{v_o}{R_f} = 0$$

This gives

$$v_o = R_f \left(\frac{1}{R_2} - \frac{1}{R_1} \right) v_{ref} \quad (9.91)$$

Now, suppose that initially $R_1 = R_2 = R$, and the active element R_1 changes by δR. The corresponding change in output is

$$\delta v_o = R_f \left(\frac{1}{R} - \frac{1}{R + \delta R} \right) v_{ref} - 0$$

or

$$\frac{\delta v_o}{v_{ref}} = \frac{R_f}{R} \frac{\delta R/R}{(1 + \delta R/R)} \quad (9.92)$$

Note that R_f/R is the amplifier gain. Now, in view of equation (9.85), the percentage nonlinearity of the half-bridge circuit is

$$N_p = 100 \frac{\delta R}{R} \% \quad (9.93)$$

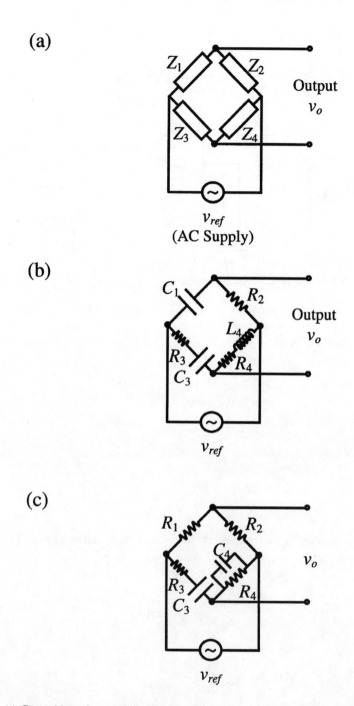

FIGURE 9.26 (a) General impedance bridge; (b) Owen bridge; and (c) Wien-bridge oscillator.

It follows that the nonlinearity of a half-bridge circuit is worse than for the Wheatstone bridge.

9.5.4 Impedance Bridges

An impedance bridge contains general impedance elements Z_1, Z_2, Z_3, and Z_4 in its four arms, as shown in Figure 9.26(a). The bridge is excited by an AC supply v_{ref}. Note that v_{ref} would represent a carrier signal, and the output v_o has to be demodulated if a transient signal representative of the

Signal Conditioning and Modification

variation in one of the bridge elements is needed. Impedance bridges can be used, for example, to measure capacitances in *capacitive sensors* and changes of inductance in *variable-inductance sensors* and *eddy current sensors*. Also, impedance bridges can be used as oscillator circuits. An oscillator circuit could serve as a constant-frequency source of a signal generator (in vibration testing) or it could be used to determine an unknown circuit parameter by measuring the oscillating frequency.

By analyzing using frequency-domain concepts it is seen that the frequency spectrum of the impedance-bridge output is given by

$$v_o(\omega) = \frac{(Z_1 Z_4 - Z_2 Z_3)}{(Z_1 + Z_2)(Z_3 + Z_4)} v_{ref}(\omega) \qquad (9.94)$$

This reduces to equation (9.81) in the DC case of a Wheatstone bridge. The balanced condition is given by

$$\frac{Z_1}{Z_2} = \frac{Z_3}{Z_4} \qquad (9.95)$$

The bridge balance equation may be used to measure an unknown circuit parameter in the bridge. Consider now two examples.

Owen Bridge

The Owen bridge shown in Figure 9.26(b) can be used to measure inductance L_4 or capacitance C_3, by the bridge-balance method. To derive the necessary equation, note that the voltage-current relation for an inductor is

$$v = L \frac{di}{dt} \qquad (9.96)$$

and for a capacitor it is

$$i = C \frac{dv}{dt} \qquad (9.97)$$

It follows that the voltage/current transfer function (in the Laplace domain) for an inductor is

$$\frac{v(s)}{i(s)} = Ls \qquad (9.98)$$

and, that for a capacitor is

$$\frac{v(s)}{i(s)} = \frac{1}{Cs} \qquad (9.99)$$

Accordingly, the impedance of an inductor element at frequency ω is

$$Z_L = j\omega L \tag{9.100}$$

and the impedance of a capacitor element at frequency ω is

$$Z_c = \frac{1}{j\omega C} \tag{9.101}$$

Applying these results for the Owen bridge, one obtains

$$Z_1 = \frac{1}{j\omega C_1}$$

$$Z_2 = R_2$$

$$Z_3 = R_3 + \frac{1}{j\omega C_3}$$

$$Z_4 = R_4 + j\omega L_4$$

in which ω is the excitation frequency. Now, from equation (9.95),

$$\frac{1}{j\omega C_1}(R_4 + j\omega L_4) = R_2\left(R_3 + \frac{1}{j\omega C_3}\right)$$

By equating the real parts and the imaginary parts of this equation, one obtains the two equations

$$\frac{L_4}{C_1} = R_2 R_3$$

and

$$\frac{R_4}{C_1} = \frac{R_2}{C_3}$$

Hence,

$$L_4 = C_1 R_2 R_3 \tag{9.102}$$

and

$$C_3 = C_1 \frac{R_2}{R_4} \tag{9.103}$$

It follows that L_4 and C_3 can be determined with the knowledge of C_1, R_2, R_3, and R_4 under balanced conditions. For example, with fixed C_1 and R_2, an adjustable R_3 could be used to measure the variable L_4, and an adjustable R_4 could be used to measure the variable C_3.

Signal Conditioning and Modification

Wien-Bridge Oscillator

Now consider the Wien-bridge oscillator shown in Figure 9.26(c). For this circuit,

$$Z_1 = R_1$$

$$Z_2 = R_2$$

$$Z_3 = R_3 + \frac{1}{j\omega C_3}$$

$$\frac{1}{Z_4} = \frac{1}{R_4} + J\omega C_4$$

Hence, from equation (9.95), the bridge-balance requirement is

$$\frac{R_1}{R_2} = \left(R_3 + \frac{1}{j\omega C_4}\right)\left(\frac{1}{R_4} + j\omega C_4\right)$$

Equating the real parts yields

$$\frac{R_1}{R_2} = \frac{R_3}{R_4} + \frac{C_4}{C_3} \qquad (9.104)$$

and equating the imaginary parts yields

$$0 = \omega C_4 R_3 - \frac{1}{\omega C_3 R_4}$$

Hence,

$$\omega = \frac{1}{\sqrt{C_3 C_4 R_3 R_4}} \qquad (9.105)$$

Equation (9.105) implies that the circuit is an oscillator whose natural frequency is given by this equation, under balanced conditions. If the frequency of the supply is equal to the natural frequency of the circuit, large-amplitude oscillators will take place. The circuit can be used to measure an unknown resistance (e.g., in strain-gage devices) by first measuring the frequency of the bridge signals at resonance (natural frequency). Alternatively, an oscillator that is excited at its natural frequency can be used as an accurate source of periodic signals (signal generator).

9.6 LINEARIZIING DEVICES

Nonlinearity is present in any physical device, to varying levels. If the level of nonlinearity in a system (component, device, or equipment) can be neglected without exceeding the error tolerance, then the system can be assumed linear.

In general, a linear system is one that can be expressed as one or more *linear differential equations*. Note that the *principle of superposition* holds for linear systems. Specifically, if the system response to an input u_1 is y_1 and the response to another input u_2 is y_2, then the response to $a_1u_1 + a_2u_2$ would be $a_1y_1 + a_2y_2$.

Nonlinearities in a system can appear in two forms:

1. Dynamic manifestation of nonlinearities
2. Static manifestation of nonlinearities.

The useful operating region of many systems can exceed the frequency range where the frequency response function is flat. The operating response of such a system is said to be *dynamic*. Examples include a typical dynamic system or plant (e.g., automobile, aircraft, chemical process plant, robot), actuator (e.g., hydraulic motor), and controller (e.g., PID control circuitry). Nonlinearities of such systems can manifest themselves in a dynamic form such as the *jump phenomenon* (also known as the *fold catastrophe*), *limit cycles*, and *frequency creation*. Design changes, extensive adjustments, or reduction of the operating signal levels and bandwidths would be necessary, in general, to reduce or eliminate these dynamic manifestations of nonlinearity. In many instances, such changes would not be practical, and one must somehow cope with the presence of these nonlinearities under dynamic conditions. Design changes might involve replacing conventional gear drives with devices such as harmonic drives in order to reduce backlash; replacing nonlinear actuators by linear actuators; and using components that have negligible Coulomb friction and that make small motion excursions.

A wide variety of sensors, transducers, and signal modification devices are expected to operate in the flat region of the frequency-response function. The input/output relation of these types of devices, in the operating range, is expressed (modeled) as a *static curve* rather than a differential equation. Nonlinearities in these devices will manifest themselves in the static operating curve in many forms. These manifestations include *saturation*, *hysteresis*, and *offset*.

In the first category of systems (plants, actuators, and compensators), if nonlinearity is exhibited in the dynamic form, proper modeling and control practices should be employed in order to avoid unsatisfactory degradation of the system performance. In the second category of systems (sensors, transducers, and signal modification devices), if nonlinearities are exhibited in the "static" operating curve, again the overall performance of the system will be degraded. Hence, it is important to "linearize" the output of such devices. Note that in dynamic manifestations, it is not realistic to "linearize" the output because the response is in the dynamic form. The solution in that case is either to minimize nonlinearities by design modifications and adjustments, so that a linear approximation would be valid, or to take the nonlinearities into account in system modeling and control. In the present section, one is not concerned with this aspect; instead, one is interested in the "linearization" of devices in the second category whose operating characteristics can be expressed by static input-output curves.

Linearization of a static device can be attempted by making design changes and adjustments as well, as in the case of dynamic devices. But, because the response is "static," and because one normally deals with an available (fixed) device whose internal hardware cannot be modified, one should consider ways of linearizing the input-output characteristic by modifying the output itself.

Static linearization of a device can be made in three ways:

1. Linearization using digital software
2. Linearization using digital (logic) hardware
3. Linearization using analog circuitry.

In the software approach to linearization, the output of the device is read into a processor with software-programmable memory, and the output is modified according to the program instructions. In the hardware approach, the device output is read by a device having fixed logic circuitry that would process (modify) the data. In the analog approach, a linearizing circuit is directly connected

Signal Conditioning and Modification

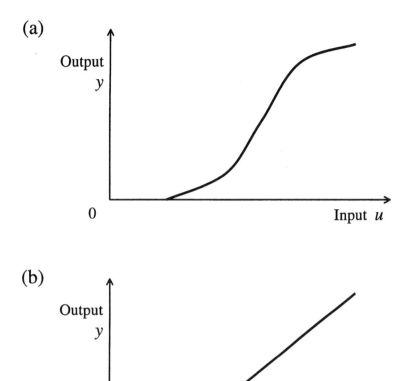

FIGURE 9.27 (a) A general static nonlinear characteristic, and (b) an offset nonlinearity.

at the output of the device so that the output of the linearizing circuit is proportional to the input of the device. These three approaches are discussed in the remainder of this section, heavily emphasizing the analog-circuit approach.

Hysteresis-type static nonlinearity characteristics have the property that the input-output curve is not one-to-one. In other words, one input value may correspond to more than one (static) output value, and one output value may correspond to more than one input value. In the present discussion, disregard these types of nonlinearities. Then, the main concern is with the linearization of a device having a single-valued static response curve that is not a straight line. An example of a typical nonlinear input-output characteristic is shown in Figure 9.27(a). Strictly speaking, a straight-line characteristic with a simple *offset*, as shown in Figure 9.27(b), is also a nonlinearity. In particular, note that superposition does not hold for an input-output characteristic of this type, given by

$$y = ku + c \qquad (9.106)$$

It is very easy, however, to linearize such a device because a simple addition of a DC component will convert the characteristic into the linear form given by

$$y = ku \qquad (9.107)$$

This method of linearization is known as *offsetting*. Linearization is more difficult in the general case where the characteristic curve could be much more complex.

9.6.1 LINEARIZATION BY SOFTWARE

If the nonlinear relationship between the input and output of a nonlinear device is known, the input can be "computed" for a known value of the output. In the software approach of linearization, a processor and memory that can be programmed using software (i.e., a digital computer) is used to compute the input using output data. Two approaches can be used; they are:

1. Equation inversion
2. Table lookup.

In the first method, the nonlinear characteristic of the device is known in the analytic (equation) form:

$$y = f(u) \qquad (9.108)$$

where

u = device input
y = device output.

Assuming that this is a one-to-one relationship, a unique inverse given by the equation

$$u = f^{-1}(y) \qquad (9.109)$$

can be determined. This equation is programmed into the *read-and-write memory* (RAM) of the computer as a *computation algorithm*. When the output values y are supplied to the computer, the processor will compute the corresponding input values u using the instructions (excitable program) stored in the RAM.

In the table lookup method, a sufficiently large number of pairs of values (y,u) are stored in the memory of the computer in the form of a table of ordered pairs. These values should cover the entire operating range of the device. Then, when a value for y is entered into the computer, the processor scans the stored data to check whether that value is present. If so, the corresponding value of u is read and this is the linearized output. If the value of y is not present in the data table, then the processor will *interpolate* the data in the vicinity of the value and will compute the corresponding output. In the *linear interpolation* method, the neighborhood of the data table where the y value falls is fitted with a straight line, and the corresponding u value is computed using this straight line. Higher-order interpolations use nonlinear interpolation curves such as quadratic and cubic polynomial equations (splines).

Note that the equation inversion method is usually more accurate than the table lookup method and it does not need excessive memory for data storage; but it is relatively slow because data are transferred and processed within the computer using program instructions that are stored in the memory and that typically have to be accessed in a sequential manner. The table lookup method is fast. Because the accuracy depends on the amount of stored data values, this is a memory-intensive method. For better accuracy, more data should be stored. But, because the entire data table has to be scanned to check for a given data value, this increase in accuracy is derived at the expanse of speed as well as memory requirements.

Signal Conditioning and Modification

9.6.2 Linearization by Hardware Logic

The software approach of linearization is flexible in the sense that the linearization algorithm can be modified (e.g., improved, changed) simply by modifying the program stored in the RAM. Furthermore, highly complex nonlinearities can be handled by the software method. As mentioned before, the method is relatively slow, however.

In the hardware logic method of linearization, the linearization algorithm is permanently implemented in the *integrated-circuit* (IC) form using appropriate *digital logic circuitry* for data processing, and memory elements (e.g., *flip-flops*). Note that the algorithm and numerical values of parameters (except input values) cannot be modified without redesigning the IC chip, because a hardware device typically does not have *programmable memory*. Furthermore, it will be difficult to implement very complex linearization algorithms by this method; and unless the chips are mass produced for an extensive commercial market, the initial chip development cost will make the production of linearizing chips economically infeasible. In bulk production, however, the per-unit cost will be very small. Furthermore, since the access of stored program instructions and extensive data manipulation are not involved, the hardware method of linearization can be substantially faster than the software method.

A digital linearizing unit having a processor and a read-only memory (ROM) whose program cannot be modified, also lacks the flexibility of a programmable software device. Hence, such a ROM-based device also falls into the category of hardware logic devices.

9.6.3 Analog Linearizing Circuitry

Three types of analog linearizing circuitry can be identified:

1. Offsetting circuitry
2. Circuitry that provides a proportional output
3. Curve shapers.

Each of these categories is described below.

An offset is a nonlinearity that can be easily removed using an analog device. This is accomplished simply by adding a DC offset of equal value to the response, in the opposite direction. Deliberate addition of an offset in this manner is known as *offsetting*. The associated removal of original offset is known as *offset compensation*. There are many applications of offsetting. Unwanted offsets such as those present in the results of ADCs and DACs can be removed by analog offsetting. Constant (DC) error components such as steady-state errors in dynamic systems due to load changes, gain changes, and other disturbances can be eliminated by offsetting. Common-mode error signals in amplifiers and other analog devices can also be removed by offsetting. In measurement circuitry, such as potentiometer (ballast) circuits, where the actual measurement signal is a "change" δv_o in a steady output signal v_o, the measurement can be completely wiped out due to noise. To reduce this problem, first the output should be offset by $-v_o$ so that the net output is δv_o and not $v_o + \delta v_o$. This output is subsequently conditioned by filtering and amplification. Another application of offsetting is the additive change of scale of a measurement — for example, from a relative scale (e.g., velocity) to an absolute scale. In summary, some of the applications of offsetting include:

1. Removal of unwanted offsets and DC components in signals (e.g., in ADC, DAC, signal integration)
2. Removal of steady-state error components in dynamic system responses (e.g., due to load changes and gain changes in Type 0 systems) (*Note*: Type 0 systems are open-loop systems having no free integrators)
3. Rejection of common-mode levels (e.g., in amplifiers and filters)

4. Error reduction when a measurement is an increment of a large steady output level (e.g., in ballast circuits for strain-gage and RTD sensors)
5. Scale changes in an additive manner (e.g., conversion from relative to absolute units or from absolute to relative units).

One can remove unwanted offsets in the simple manner as discussed above. Now consider more complex nonlinear responses that are nonlinear in the sense that the input-output curve is not a straight line. Analog circuitry can be used to linearize these type of responses as well. The linearizing circuit used will generally depend on the particular device and the nature of its nonlinearity. Hence, linearizing circuits of this type often must be discussed with respect to a particular application. For example, such linearization circuits are useful in *transverse-displacement capacitative sensors*. Several useful circuits are described below.

Consider the type of linearization known as *curve shaping*. A *curve shaper* is a linear device whose gain (output/input) can be adjusted so that response curves with different slopes can be obtained. Suppose that a nonlinear device having an irregular (nonlinear) input characteristic is to be linearized. First, one applies the operating input simultaneously to the device and the curve shaper, and the gain of the curve shaper is adjusted such that it closely matches that of the device in a small range of operation. Now, the output of the curve shaper can be utilized for any task that requires the device output. The advantage here is that linear assumptions are valid with the curve shaper, which is not the case for the actual device. When the operating range changes, the curve shaper must be adjusted to the new range. Comparison (calibration) of the curve shaper and the nonlinear device can be done off line and, once a set of gain values corresponding to a set of operating ranges is determined in this manner for the curve shaper, it is possible to completely replace the nonlinear device by the curve shaper. Then the gain of the curve shaper can be adjusted, depending on the actual operating range during system operation. This is known as *gain scheduling*. Note that one can replace a nonlinear device with a linear device (curve shaper) within a multi-component system in this manner without greatly sacrificing the accuracy of the overall system.

9.6.4 OFFSETTING CIRCUITRY

Common-mode outputs and *offsets* in amplifiers and other analog devices can be minimized by including a *compensating resistor* that can provide fine adjustments at one of the input leads. Furthermore, the larger the feedback signal level in a feedback system, the smaller the steady-state error. Hence, steady-state offsets can be reduced by reducing the *feedback resistance* (thereby increasing the feedback signal). Furthermore, since a ballast (potentiometer) circuit provides an output of $v_o + \delta v_o$ and a bridge circuit provides an output of δv_o, the use of a *bridge circuit* can be interpreted as an offset compensation method.

The most straightforward way of offsetting is by using a *differential amplifier* (or a *summing amplifier*) to subtract (or add) a DC voltage to the output of the nonlinear device. The DC level must be variable so that various levels of offset can be provided with the same circuit. This is accomplished using an adjustable resistance at the DC input lead of the amplifier.

An operational-amplifier circuit for offsetting is shown in Figure 9.28. Since the input v_i is connected to the – lead of the op-amp, one obtains an *inverting amplifier*, and the input signal will appear in the output v_o with its sign reversed. This is also a *summing amplifier* because two signals can be added together by this circuit. If the input v_i is connected to the + lead of the op-amp, one has a *noninverting amplifier*.

The DC voltage v_{ref} provides the offsetting voltage. The resistor R_c (compensating resistor) is variable so that different values of offset can be compensated using the same circuit. To obtain the circuit equation, one can write the current balance equation for node A, using the usual assumption

Signal Conditioning and Modification

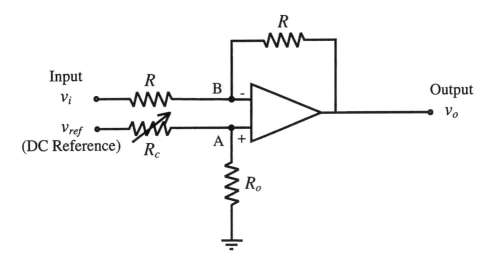

FIGURE 9.28 An inverting amplifier circuit for offset compensation.

that the current through an input lead is 0 for an op-amp (because of very high input impedance); thus,

$$\frac{v_{ref} - v_A}{R_c} = \frac{v_A}{R_o}$$

or

$$v_A = \frac{R_o}{(R_o + R_c)} v_{ref} \qquad (i)$$

Similarly, the current balance at node B gives

$$\frac{v_i - v_B}{R} + \frac{v_o - v_B}{R} = 0$$

or

$$v_o = -v_i + 2v_B \qquad (ii)$$

Since $v_A = v_B$ for the op-amp (because of very high open-loop gain), one can substitute equation (i) in (ii). Then,

$$v_o = -v_i + \frac{2R_o}{(R_o + R_c)} v_{ref} \qquad (9.110)$$

Note the sign reversal of v_i at the output (because this is an inverting amplifier). This is not a problem because polarity can be reversed at input or output in connecting this circuit to other circuitry, thereby recovering the original sign. The important result here is the presence of a constant offset term on the RHS of equation (9.110). This term can be adjusted by picking the proper value for R_c so as to compensate for a given offset in v_i.

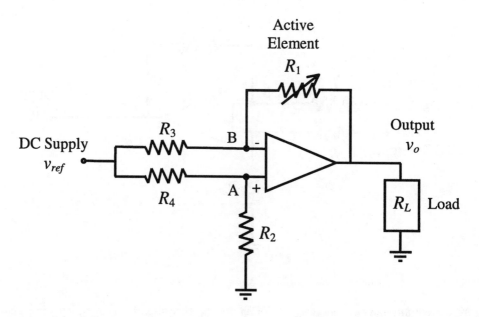

FIGURE 9.29 A proportional-output circuit for an active resistance element (strain gage).

9.6.5 Proportional-Output Circuitry

An operational-amplifier circuit can be employed to linearize the output of a capacitive transverse-displacement sensor. In constant-voltage and constant-current resistance bridges and in a constant-voltage half-bridge, the relation between the bridge output δv_o and the measurand (change in resistance in the active element) is nonlinear. The nonlinearity is least for the constant-current bridge and it is highest for the half-bridge. Since δR is small compared to R, however, the nonlinear relations can be linearized without introducing large errors. But the linear relations are inexact, and are not suitable if δR cannot be neglected in comparison to R. Under these circumstances, the use of a linearizing circuit would be appropriate.

One way to obtain a proportional output from a Wheatstone bridge is to feedback a suitable factor of the bridge output into the bridge supply v_{ref}. Another way is to use the op-amp circuit shown in Figure 9.29. This should be compared with the Wheatstone bridge shown in Figure 9.24(a). Note that R represents the only active element (e.g., an active strain gage).

First, one can show that the output equation for the circuit in Figure 9.29 is quite similar to equation (9.81). Using the fact that the current through an input lead of an unsaturated op-amp can be neglected, one has the following current balance equations for nodes A and B:

$$\frac{v_{ref} - v_A}{R_4} = \frac{v_A}{R_2}$$

$$\frac{v_{ref} - v_B}{R_3} + \frac{v_o - v_B}{R_1} = 0$$

Hence,

$$v_A = \frac{R_2}{(R_2 + R_4)} v_{ref}$$

and

$$v_B = \frac{R_1 v_{ref} + R_3 v_o}{(R_1 + R_3)}$$

Now using the fact $v_A = v_B$ for an op-amp, one obtains

$$\frac{R_1 v_{ref} + R_3 v_o}{(R_1 + R_3)} = \frac{R_2}{(R_2 + R_4)} v_{ref}$$

Accordingly, the circuit output equation is

$$v_o = \frac{(R_2 R_3 - R_1 R_4)}{R_3 (R_2 + R_4)} v_{ref} \qquad (9.111)$$

Note that this relation is quite similar to the Wheatstone bridge equation (9.81). The balance condition (i.e., $v_o = 0$) is again given by equation (9.82).

Suppose that $R_1 = R_2 = R_3 = R_4 = R$ in the beginning (the circuit is balanced), so the $v_o = 0$. Then suppose that the active resistance R_1 is changed by δR (say, due to a change in strain in the strain gage R_1). Then, using equation (9.111), one can write an expression for the change in circuit output as

$$\delta v_o = \frac{\left[R^2 - R(R + \delta R)\right]}{R(R + R)} v_{ref} - 0$$

or

$$\frac{\delta v_o}{v_{ref}} = -\frac{1}{2} \frac{\delta R}{R} \qquad (9.112)$$

By comparing this result with equation (9.83) one observes that the circuit output δv_o is proportional to the measurand δR. Furthermore, note that the sensitivity ($\frac{1}{2}$) of the circuit in Figure 9.29 is double that of a Wheatstone bridge ($\frac{1}{4}$) that has one active element, which is a further advantage of the proportional-output circuit. The sign reversal is not a drawback because it can be accounted for by reversing the load polarity.

Curve-Shaping Circuitry

A curve shaper can be interpreted as an amplifier whose gain is adjustable. A typical arrangement for a curve-shaping circuit is shown in Figure 9.30. The feedback resistance R_f is adjustable by some means. For example, a switching circuit with a bank of resistors (say, connected in parallel through solid-state switches as in the case of weighted-resistor DAC) can be used to switch the feedback resistance to the required value. Automatic switching can be realized using Zener diodes that will start conducting at certain voltage levels. In both cases (external switching by switching pulses or automatic switching using Zener diodes), amplifier gain is variable in discrete steps. Alternatively, a potentiometer can be used as R_f so that the gain can be continuously adjusted (manually or automatically).

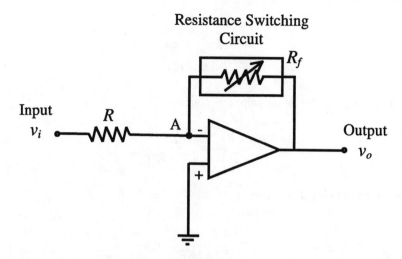

FIGURE 9.30 A curve-shaping circuit.

The output equation for the curve-shaping circuit shown in Figure 9.30 is obtained by writing the current balance at node A, noting that $v_A = 0$; thus,

$$\frac{v_i}{R} + \frac{v_o}{R_f} = 0$$

or

$$v_o = -\frac{R_f}{R} v_i \qquad (9.113)$$

It follows that the gain (R_f/R) of the amplifier can be adjusted by changing R_f.

9.7 MISCELLANEOUS SIGNAL-MODIFICATION CIRCUITRY

In addition to the signal modification devices discussed thus far in this chapter, there are many other types of circuitry that are used for signal modification and related tasks. Examples are phase shifters, voltage-to-frequency converters, frequency-to-voltage converters, voltage-to-current converters, and peak-hold circuits. The objective of the present section is to briefly discuss several such miscellaneous circuits and components that are useful in the instrumentation of dynamic systems.

9.7.1 PHASE SHIFTERS

A sinusoidal signal given by

$$v = v_a \sin(\omega t + \phi) \qquad (9.114)$$

has the following three representative parameters:

v_a = amplitude
ω = frequency
ϕ = phase angle.

Signal Conditioning and Modification

Note that the phase angle represents the *time reference* (starting point) of the signal. The phase angle is an important consideration only when two or more signal components are compared. The Fourier spectrum of a signal is presented as its amplitude (magnitude) and the phase angle with respect to the frequency (also see Chapter 4).

Phase-shifting circuits have many applications. When a signal passes through a system, its phase angle changes due to dynamic characteristics of the system. Consequently, the phase change provides very useful information about the dynamic characteristics of the system. Specifically, for a linear constant-coefficient system, this phase shift is equal to the phase angle of the *frequency-response function (frequency-transfer function)* of the system at that particular frequency. This phase-shifting behavior is, of course, not limited to electrical systems and is equally exhibited by other types of systems, including mechanical vibrating systems. The phase shift between two signals can be determined by converting the signals into the electrical form (using suitable transducers), and shifting the phase angle of one signal through known amounts using a phase-shifting circuit until the two signals are in phase.

Another application of phase shifters is in signal demodulation. For example, one method of amplitude demodulation involves processing the modulated signal together with the carrier signal. This, however, requires the modulated signal and the carrier signal to be in phase. But, usually, since the modulated signal has already transmitted through electrical circuitry having impedance characteristics, its phase angle will have changed. Then, it is necessary to shift the phase angle of the carrier until the two signals are in phase, so that demodulation can be performed accurately. Hence, phase shifters are used in demodulating (e.g., LVDT displacement-sensor outputs).

A phase-shifter circuit, ideally, should not change the signal amplitude while changing the phase angle by a required amount. Practical phase shifters could introduce some degree of amplitude distortion (with respect to frequency) as well. A simple phase-shifter circuit can be constructed using resistance (R) and capacitance (C) elements. A resistance or a capacitor of such an RC circuit is made fine-adjustable so as to obtain a variable phase shifter.

An op-amp-based phase-shifter circuit is shown in Figure 9.31. One can show that this circuit provides a phase shift without distorting the signal amplitude. The circuit equation is obtained by writing the current balance equations at nodes A and B, as usual, noting that the current through the op-amp leads can be neglected due to high input impedance; thus,

$$\frac{v_i - v_A}{R_C} = C \frac{dv_A}{dt}$$

$$\frac{v_i - v_B}{R} + \frac{v_o - v_B}{R} = 0$$

On simplifying and introducing the Laplace variable, one obtains

$$v_i = (\tau s + 1) v_A \qquad (i)$$

and

$$v_B = \frac{1}{2}(v_i + v_o) \qquad (ii)$$

in which the circuit *time constant* τ is given by

$$\tau = R_c C$$

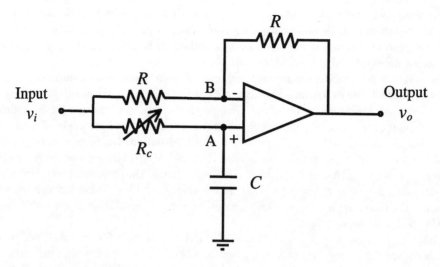

FIGURE 9.31 A phase-shifter circuit.

Since $v_A = v_B$, as a result of very high gain in the op-amp, by substituting equation (*ii*) in (*i*), one obtains

$$v_i = \frac{1}{2}(\tau s + 1)(v_i + v_o)$$

It follows that the transfer function $G(s)$ of the circuit is given by

$$\frac{v_o}{v_i} = G(s) = \frac{(1-\tau s)}{(1+\tau s)} \tag{9.115}$$

It is seen that the magnitude of the *frequency-response function* $G(j\omega)$ is

$$|G(j\omega)| = \frac{\sqrt{1+\tau^2\omega^2}}{\sqrt{1+\tau^2\omega^2}}$$

or

$$|G(j\omega)| = 1 \tag{9.116}$$

and the phase angle of $G(j\omega)$ is

$$\angle G(j\omega) = -\tan^{-1}\tau\omega - \tan^{-1}\tau\omega$$

or

$$\angle G(j\omega) = -2\tan^{-1}\tau\omega = -2\tan^{-1}R_c C\omega \tag{9.117}$$

Signal Conditioning and Modification

As needed, the transfer function magnitude is unity, indicating that the circuit does not distort the signal amplitude over the entire bandwidth. Equation (9.117) gives the *phase lead* of the output v_o with respect to the input v_i. Note that this angle is negative, indicating that actually a *phase lag* is introduced. The phase shift can be adjusted by varying the resistance R_c.

9.7.2 Voltage-to-Frequency Converter (VFC)

A voltage-to-frequency converter generates a periodic output signal whose frequency is proportional to the level of an input voltage. Because such an *oscillator* generates a periodic output according to the voltage excitation, it is also called a *voltage-controlled oscillator* (VCO).

A common type of VFC uses a capacitor. The time needed for the capacitor to be charged to a fixed voltage level will depend on the charging voltage (inversely proportional). Suppose that this voltage is governed by the input voltage. Then, if the capacitor is made to periodically charge and discharge, one obtains an output whose frequency (inverse of the charge-discharge period) is proportional to the charging voltage. The output amplitude will be given by the fixed voltage level to which the capacitor is charged in each cycle. Consequently, one gets a signal with a fixed amplitude, and a frequency that depends on the charging voltage (input).

A voltage-to-frequency converter (or voltage-controlled oscillator) circuit is shown in Figure 9.32(a). The voltage-sensitive switch closes when the voltage across it exceeds a reference level v_s, and it will open again when the voltage across it falls below a lower limit $v_o(0)$. The *programmable unijunction transistor* (PUT) is such a switching device.

Note that the polarity of the input voltage v_i is reversed. Suppose that the switch is open. Then, current balance at node A of the op-amp circuit gives

$$\frac{v_i}{R} = C \frac{dv_o}{dt}$$

As usual v_A = voltage at + lead = 0 because the op-amp has a very high gain; and current through the op-amp leads = 0 because the op-amp has a very high input impedance. The capacitor charging equation can be integrated for a given value of v_i. This gives

$$v_o(t) = \frac{1}{RC} v_i t + v_o(0)$$

The switch will be closed when the voltage across the capacitor $v_o(t)$ equals the reference level v_s. Then the capacitor will be immediately discharged through the closed switch. Hence, the capacitor charging time T is given by

$$v_s = \frac{1}{RC} v_i T + v_o(0)$$

Accordingly,

$$T = \frac{RC}{v_i}\left(v_s - v_o(0)\right) \tag{9.118}$$

The switch will be open again when the voltage across the capacitor drops to $v_o(0)$, and the capacitor will again begin to charge from $v_o(0)$ up to v_s. This charging and instantaneous discharge cycle will

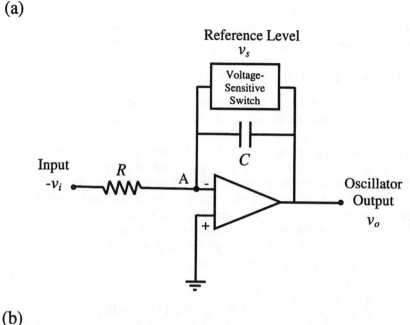

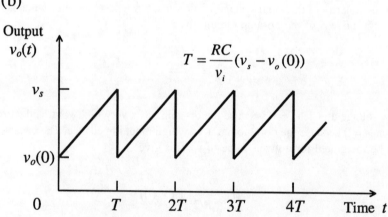

FIGURE 9.32 A voltage-to-frequency converter (voltage-controlled oscillator): (a) circuit, and (b) output signal.

repeat periodically. The corresponding output signal will be as shown in Figure 9.32(b). This is a periodic (*saw-tooth*) wave with period T. The frequency of oscillation of the output ($1/T$) is given by

$$f = \frac{v_i}{RC(v_s - v_o(0))} \tag{9.119}$$

It is seen that the oscillator frequency is proportional to the input voltage v_i. The oscillator amplitude is v_s, which is fixed.

Voltage-controlled oscillators have many applications. One application is in *analog-to-digital conversion*. In the VCO-type analog-to-digital converters, the analog signal is converted into an oscillating signal using a VCO. Then, the oscillator frequency is measured using a *digital counter*. This count, which is available in the digital form, is representative of the input analog signal level. Another application is in *digital voltmeters*. Here, the same method as for ADC is used. Specifically,

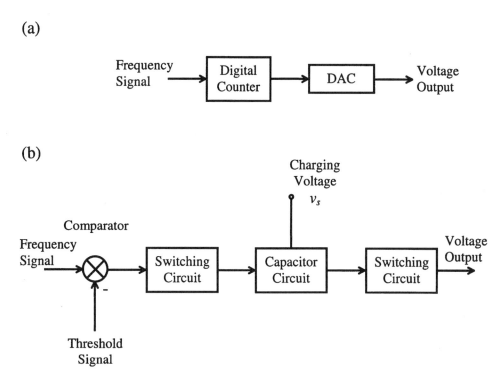

FIGURE 9.33 Frequency-to-voltage converters: (a) digital counter method, and (b) capacitor charging method.

the voltage is converted into an oscillator signal and its frequency is measured using a digital counter. The count can be scaled and displayed to provide the voltage measurement. A direct application of VCO is apparent from the fact that VCO is actually a *frequency modulator* (FM), providing a signal whose frequency is proportional to the input (modulating) signal. Hence, VCO is useful in applications that require frequency modulation. Also, VCO can be used as a signal (wave) generator for variable-frequency applications; for example, excitation inputs for shakers in vibration testing (see Chapters 8 and 10), excitations for frequency-controlled DC motors, and pulse signals for translator circuits of stepping motors.

9.7.3 FREQUENCY-TO-VOLTAGE CONVERTER (FVC)

A frequency-to-voltage converter generates an output voltage whose level is proportional to the frequency of its input signal. One way to obtain an FVC is to use a digital counter to count the signal frequency and then use a digital-to-analog converter (DAC) to obtain a voltage proportional to the frequency. A schematic representation of this type of FVC is shown in Figure 9.33(a).

An alternative FVC circuit is schematically shown in Figure 9.33(b). In this method, the frequency signal is supplied to a comparator along with a threshold voltage level. The sign of the comparator output will depend on whether the input signal level is larger or smaller than the threshold level. The first sign change (−ve to +ve) in the comparator output is used to trigger a switching circuit that will respond by connecting a capacitor to a fixed charging voltage. This will charge the capacitor. The next sign change (+ve to −ve) of the comparator output will cause the switching circuit to short the capacitor, thereby instantaneously discharging it. This charging-discharging process will be repeated in response to the oscillator input. Note that the voltage level to which the capacitor is charged each time will depend on the switching period (charging voltage is fixed), which is in turn governed by the frequency of the input signal. Hence, the output voltage of the capacitor circuit will be representative of the frequency of the input signal. Because the

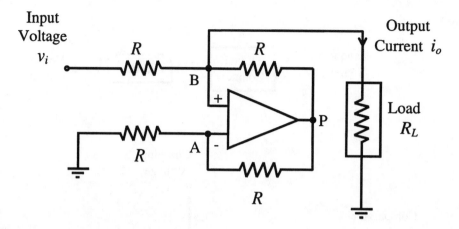

FIGURE 9.34 A voltage-to-current converter.

output is not steady due to the ramp-like charging curve and instantaneous discharge, a smoothing circuit is provided at the output to remove the noise ripples.

Applications of FVC include demodulation of frequency-modulated signals, frequency measurement in mechanical vibration applications, and conversion of pulse outputs in some types of sensors and transducers into analog voltage signals.

9.7.4 Voltage-to-Current Converter (VCC)

Measurement and feedback signals are usually transmitted as current levels in the range of 4 mA to 20 mA rather than as voltage levels. This is particularly useful when the measurement site is not close to the monitoring room. Because the measurement itself is usually available as a voltage, it must be converted into current using a voltage-to-current converter. For example, pressure transmitters and temperature transmitters in operability testing systems provide current outputs that are proportional to the measured values of pressure and temperature.

There are many advantages to transmitting current rather than voltage. In particular, the voltage level will drop due to resistance in the transmitting path, but the current through a conductor will remain unchanged unless the conductor is branched. Hence, current signals are less likely to acquire errors due to signal weakening. Another advantage of using current instead of voltage as the measurement signal is that the same signal can be used to operate several devices in series (e.g., a display, plotter, and signal processor simultaneously), again without causing errors due to signal weakening by the power lost at each device, because the same current is applied to all devices. A voltage-to-current converter (VCC) should provide a current proportional to an input voltage, without being affected by the load resistance to which the current is supplied.

An operational-amplifier-based voltage-to-current converter circuit is shown in Figure 9.34. Using the fact that the currents through the input leads of an unsaturated op-amp can be neglected (due to very high input impedance), one can write the current summation equations for the two nodes A and B; thus,

$$\frac{v_A}{R} = \frac{v_P - v_A}{R}$$

and

$$\frac{v_i - v_B}{R} + \frac{v_P - v_B}{R} = i_o$$

Signal Conditioning and Modification

Accordingly,

$$2v_A = v_P \quad (i)$$

and

$$v_i - 2v_B + v_P = Ri_o \quad (ii)$$

Now, using the fact that $v_A = v_B$ for the op-amp (due to very high gain), one can substitute equation (*i*) in (*ii*); this gives

$$i_o = \frac{v_i}{R} \quad (9.120)$$

where

i_o = output current
v_i = input voltage.

It follows that the output current is proportional to the input voltage, irrespective of the value of the load resistance R_L, as required for a VCC.

9.7.5 Peak-Hold Circuit

Unlike a sample-and-hold circuit that holds every sampled value of the signal, a peak-hold circuit holds only the largest value reached by the signal during the monitored period. Peak holding is useful in a variety of applications. In signal processing for shock and vibration studies, what is known as *response spectra* (e.g., *shock response spectrum*) are determined using a *response spectrum analyzer* that exploits a peak-holding scheme (see Chapter 10). Suppose that a signal is applied to a simple oscillator (a single-degree-of-freedom second-order system with no zeros) and the peak value of the response (output) is determined. A plot of the peak output as a function of the natural frequency of the oscillator, for a specified damping ratio, is known as the response spectrum of the signal for that damping ratio. Peak detection is also useful in machine monitoring and alarm systems. In short, when just one representative value of a signal is needed in a particular application, the peak value would be a leading contender.

Peak detection of a signal can be conveniently done using digital processing. For example, the signal is sampled and the previous sample value is replaced by the present sample value if and only if the latter is larger than the former. By sampling, and then holding one value in this manner, the peak value of the signal is retained. Note that, usually, the time instant at which the peak occurs is not retained.

Peak detection can be done using analog circuitry as well. This is, in fact, the basis of analog spectrum analyzers. A peak-holding circuit is shown in Figure 9.35. The circuit consists of two voltage followers. The first voltage follower has a diode at its output that is forward-biased by the positive output of the voltage follower and reverse-biased by a low-leakage capacitor, as shown. The second voltage follower presents the peak voltage that is held by the capacitor to the circuit output at a low output impedance, without loading the previous circuit stage (capacitor and first voltage follower). To explain the operation of the circuit, suppose that the input voltage v_i is larger than the voltage to which capacitor is charged (v). Since the voltage at the + lead of the op-amp is v_i and the voltage at the − lead is v, the first op-amp will be saturated. Since the differential input

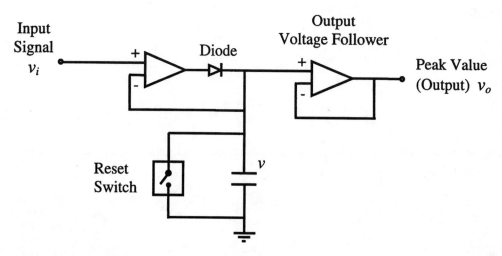

FIGURE 9.35 A peak-holding circuit.

to the op-amp is positive under these conditions, the op-amp output will be positive. The output will charge the capacitor until the capacitor voltage v equals the input voltage v_i. This voltage (call it v_o) is in turn supplied to the second voltage follower, which presents the same value to its output (gain = 1 for a voltage follower), but at a very low impedance level. Note that the op-amp output remains at the saturated value for only a very short time (the time taken by the capacitor to charge). Now suppose that v_i is smaller than v. Then, the differential input of the op-amp will be negative, and the op-amp output will be saturated at the negative saturation level. This will reverse-bias the diode. Hence, the output of the first op-amp will be in open-circuit and, as a result, the voltage supplied to the output voltage follower will still be the capacitor voltage and not the output of the first op-amp. It follows that the voltage level of the capacitor (and hence the output of the second voltage follower) will always be the peak value of the input signal. The circuit can be reset by discharging the capacitor through a solid-state switch that is activated by an external pulse.

9.8 SIGNAL ANALYZERS AND DISPLAY DEVICES

Vibration signal analysis can employ both analog and digital procedures. (Chapter 4 is devoted entirely to this topic.) Since signal analysis results in extracting various useful information from the signal, it is appropriate to consider the topic within the present context of signal modification as well. This section introduces digital signal analyzers that essentially make use of the same techniques that were described previously (in Chapter 4 and Appendix D).

Signal display devices also make use of at least some signal processing. This may involve filtering and change of signal level and format. More sophisticated signal display devices, particularly digital oscilloscopes, can carry out more complex signal analysis functions such as those normally available with digital signal analyzers. Oscilloscopes as well are introduced in the present section, although they can also be treated under vibration instrumentation (Chapter 8).

Signal-recording equipment commonly employed in vibration practice includes digital storage devices such as hard drives, floppy disks, and CD-ROMs, analog devices like tape recorders, strip-chart recorders, and X-Y plotters, and digital printers. Tape recorders are used to record vibration data (transducer outputs) that are subsequently reproduced for processing or examination. Often, tape-recorded waveforms are also used to generate (by replay) signals that drive vibration test exciters (shakers). Tape recorders use tapes made of a plastic material that has a thin coating of a specially treated ferromagnetic substance. During the recording process, magnetic flux proportional to the recorded signal is produced by the recording head (essentially an electromagnet), which magnetizes the tape surface in proportion to the signal variation. Reproduction is the reverse process, whereby an electrical signal is generated at the reproduction head by electromagnetic induction in accordance with the magnetic flux of the magnetized (recorded) tape. Several signal-conditioning circuitries are involved in the recording and reproducing stages. Recording by FM is very common in vibration testing.

Strip-chart recorders are usually employed to plot time histories (i.e., quantities that vary with time), although they can also be used to plot such data as frequency-response functions and response spectra. In these recorders, a paper roll unwinds at a constant linear speed, and the writing head moves across the paper (perpendicular to the paper motion) proportionally to the signal level. There are many kinds of strip-chart recorders, which are grouped according to the type of writing head employed. Graphic-level recorders, which use ordinary paper, employ such heads as ink pens or brushes, fiber pens, and sapphire styli. Visicoders are simple oscilloscopes capable of producing permanent records; they employ light-sensitive paper for this. Several channels of input data can be incorporated with a visicoder. Obviously, graphic-level recorders are generally limited by the number of writing heads available (typically, one or two), but visicoders can have many more input channels (typically 24). Performance specifications of these devices include paper speed, frequency range of operation, dynamic range, and power requirements.

In vibration experimentation, X-Y plotters are generally employed to plot frequency data (e.g., psd, frequency-response functions, response spectra, and transmissibility curves, as defined in Chapters 3 and 4), although they can also be used to plot time-history data. Many types of X-Y plotters are available, most of them using ink pens on ordinary paper. There are also hard-copy units that use laser printing or heat-sensitive paper in conjunction with a heating element as the writing head. The writing head in an X-Y plotter is moved in the X and Y directions on the paper by two input signals that form the coordinates for the plot. In this manner, a trace is made on stationary plotting paper. Performance specifications of X-Y plotters are governed by such factors as paper size; writing speed (inches per second, centimeters per second); deadband (expressed as a percentage of the full scale), which measures the resolution of the plotter head; linearity (expressed as a percentage of full scale), which measures the accuracy of the plot; minimum trace separation (inches, centimeters) for multiple plots on the same axes; dynamic range; input impedance; and maximum input (millivolts per inch, millivolts per centimeter).

Today, the most widespread signal recording device is in fact the digital computer (memory, storage) and printer combination. This, and also the other (analog) devices used in signal recording and display, make use of some signal modification to accomplish their functions. These devices will not be discussed in the present section, however.

9.8.1 SIGNAL ANALYZERS

Modern signal analyzers employ digital techniques of signal analysis, as described in Chapter 4, to extract useful information that is carried by the signal. Digital Fourier analysis using fast Fourier transform (FFT) is perhaps the single common procedure used in the vast majority of signal analyzers (see Appendix D). As noted before, Fourier analysis will produce the frequency spectrum of a time signal. It should be clear, therefore, why the terms digital signal analyzer, FFT analyzer, frequency analyzer, spectrum analyzer, and digital Fourier analyzer are synonymous to some extent, as used in commercial instrumentation literature.

A signal analyzer typically has two (dual) or more (multiple) input signal channels. To generate results such as frequency response (transfer) functions, cross-spectra, coherence functions, and cross-correlation functions, one needs at least two data signals and, hence, a dual-channel analyzer.

In hardware analyzers, digital circuitry rather than software is used to carry out the mathematical operations. Clearly, they are very fast but less flexible (in terms of programmability and functional capability) for this reason. Digital signal analyzers, regardless of whether they use the hardware or the software approach, employ some basic operations. These operations, carried out in sequence, are:

1. Anti-alias filtering (analog)
2. Analog-to-digital conversion (i.e., signal sampling)
3. Truncation of a block of data and multiplication by a window function
4. FFT analysis of the block of data.

These operations were explained in Chapter 4. Also noted are the following facts: If the sampling period of the analog-to-digital convertor (ADC) is ΔT (i.e., the sampling frequency is $1/\Delta T$), then the Nyquist frequency $f_c = \dfrac{1}{2\Delta T}$. This Nyquist frequency is the upper limit of the useful frequency content of the sampled signal. The cutoff frequency of the anti-aliasing filter should be set at f_c or less. If there are N data samples in the block of data used in the FFT analysis, the corresponding record length is $T = N\Delta T$. Then, the spectral lines in the FFT results are separated at a frequency spacing of $\Delta F = 1/T$. In view of the Nyquist frequency limit, there will be only $N/2$ useful spectral lines in the FFT result.

Strictly speaking, a real-time signal analyzer should analyze a signal instantaneously and continuously as the signal is received by the analyzer. This is usually the case with an analog signal analyzer. But, in digital signal analyzers, which are usually based on digital Fourier analysis, a block of data (i.e., N samples of record length T) is analyzed together to produce $N/2$ useful spectral lines (at frequency spacing $1/T$). This is, then, not a truly real-time analysis. But for practical purposes, if the speed of analysis is sufficiently fast, the analyzer can be considered real-time, which is usually the case with hardware analyzers and also modern, high-speed, software analyzers.

The bandwidth B of a digital signal analyzer is a measure of its speed of signal processing. Specifically, for an analyzer that uses N data samples in each block of signal analysis, the associated processing time can be given by

$$T_c = \frac{N}{B} \tag{9.121}$$

Note that the larger the B, the smaller the T_c. Then, the analyzer is considered a real-time one if the analysis time (T_c) of the data record is less than the generation time ($T = N\Delta T$) of the data record. Hence, one requires that

$$T_c < T$$

or

$$\frac{N}{B} < T$$

or

$$\frac{N}{B} < N\Delta T$$

or

$$\frac{1}{\Delta T} < B \tag{9.122}$$

In other words, a real-time analyzer has a bandwidth greater than its sampling rate.

Signal Conditioning and Modification

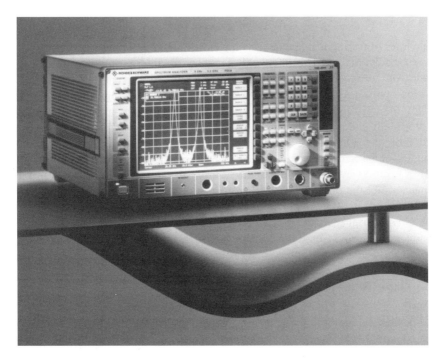

FIGURE 9.36 A digital spectrum analyzer. (Copyright 1999, Tektronix, Inc. All Rights Reserved. With permission)

A multi-channel digital signal analyzer can analyze one or more signals simultaneously and generate (and display) results such as Fourier spectra, power spectral densities, cross-spectral densities, frequency-response functions, coherence functions, autocorrelations, and cross-correlations. They are able to perform high-resolution analysis on a small segment of the frequency spectrum of a signal. This is termed *zoom analysis*. Essentially, in this case, the spectral line spacing ΔF is decreased while keeping the number of lines (N) and hence the number of time data samples the same. That means the record length ($T = 1/\Delta F$) must be increased in proportion for zoom analysis. A photo of a modern signal analyzer is give in Figure 9.36.

9.8.2 OSCILLOSCOPES

An oscilloscope is used to observe one or two signals separately or simultaneously. Amplitude, frequency, and phase information of the signals can be obtained using an oscilloscope. In this sense, it is a signal modification as well as a measurement (monitoring) and display device. Both analog and digital oscilloscopes are available. A typical application of ocilloscopes is to observe (monitor) experimental data such as vibration signals of machinery as obtained from transducers. They are also useful in observing and examining vibration test results, such as frequency-response plots, psd curves, and response spectra. Typically, only temporary records are available on an analog oscilloscope screen. The main component of an analog oscilloscope is the cathode-ray tube (CRT), which consists of an electron gun (cathode) that deflects an electron ray according to the input-signal level. The oscilloscope screen has a coating of electron-sensitive material, so that the electron ray that impinges on the screen leaves a temporary trace on it. The electron ray sweeps across the screen horizontally, so that waveform traces can be recorded and observed. Usually, two input channels are available. Each input can be observed separately, or the variation in an input can be observed against those of the other. In this manner, signal phasing can be examined. Several sensitivity settings for the input-signal amplitude scale (in the vertical direction) and sweep-speed selections are available on the panel.

Triggering

The voltage level of the input signal deflects the electron gun in proportion to the vertical (y-axis) direction on the CRT screen. This alone will not show the time evolution of the signal. The true time variation of the signal is achieved by means of a saw-tooth signal that is generated internally in the oscilloscope and used to move the electron gun in the horizontal (x-axis) direction. As the name implies, the saw-tooth signal increases linearly in amplitude up to a threshold value, and then suddenly drops to 0, and repeats this cycle again. In this manner, the observed signal is repetitively swept across the screen and a trace of it can be observed as a result of the temporary retention of the illumination of the electron gun on the fluorescent screen. The saw-tooth signal can be controlled (triggered) in several ways. For example, the *external trigger* mode uses an external signal from another channel (not the observed channel) to generate and synchronize the saw-tooth signal. In the *line trigger* mode, the saw-tooth signal is synchronized with the AC line supply (60 Hz or 50 Hz). In the *internal trigger* mode, the observed signal (which is used to deflect the electron beam in the y direction) itself is used to generate (synchronize) the saw-tooth signal. Because the frequency and the phase of the observed signal and the trigger signal are perfectly synchronized in this case, the trace on the oscilloscope screen will appear stationary. Careful observation of a signal can be made in this manner.

Lissajous Patterns

Suppose that two signals x and y are provided to the two channels of an oscilloscope. If they are used to deflect the electron beam in the horizontal and vertical directions, respectively, a pattern known as Lissajous pattern will be observed on the oscilloscope screen. Useful information about the amplitude and phasing of the two signals can be observed by means of these patterns. Consider sine waves x and y. Several special cases of Lissajous patterns are given below:

1. *Same frequency, same phase*:
   Here,

$$x = x_o \sin(\omega t + \phi)$$

$$y = y_o \sin(\omega t + \phi)$$

   Then,

$$\frac{x}{x_o} = \frac{y}{y_o}$$

   which gives a straight-line trace with a positive slope, as shown in Figure 9.37(a).

2. *Same frequency, 90° out-of-phase*:
   Here,

$$x = x_o \sin(\omega t + \phi)$$

$$y = y_o \sin(\omega t + \phi + \pi/2)$$

$$= y_o \cos(\omega t + \phi)$$

   Then,

Signal Conditioning and Modification

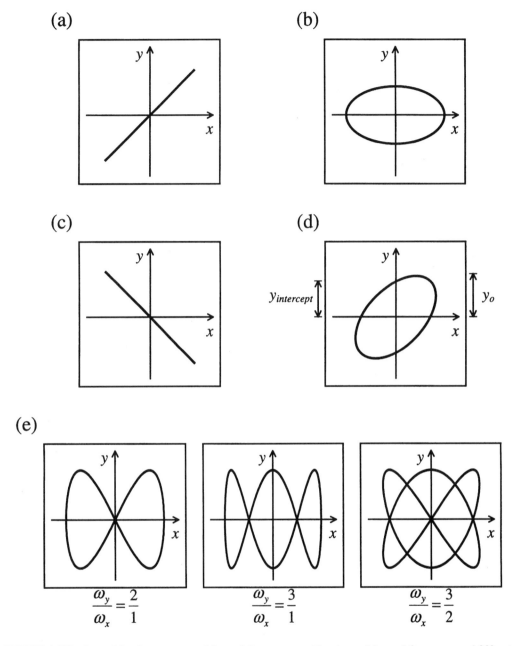

FIGURE 9.37 Some Lissajous patterns: (a) equal frequency and in-phase; (b) equal frequency and 90° out-of-phase; (c) equal frequency and 180° out-of-phase; and (d) equal frequency and θ out-of-phase.

$$\left(\frac{x}{x_o}\right)^2 + \left(\frac{y}{y_o}\right)^2 = 1$$

which gives an ellipse, as shown in Figure 9.37(b).

3. *Same frequency, 180° out-of-phase*:
   Here,

$$x = x_o \sin(\omega t + \phi)$$
$$x = y_o \sin(\omega t + \phi + \pi)$$
$$= -y_o \sin(\omega t + \phi)$$

Hence,

$$\frac{x}{x_o} + \frac{y}{y_o} = 0$$

which corresponds to a straight line with a negative slope, as shown in Figure 9.37(c).

4. *Same frequency, θ out-of-phase*:

$$x = x_o \sin(\omega t + \phi)$$
$$y = y_o \sin(\omega t + \phi + \theta)$$

When $\omega t + \phi = 0$, $y = y_{intercept} = y_o \sin\theta$
Hence,

$$\sin\theta = \frac{y_{intercept}}{y_o}$$

In this case, one obtains a tilted ellipse as shown in Figure 9.37(d). The phase difference θ is obtained from the Lissajous pattern.

5. *Integral Frequency Ratio*:

$$\frac{\omega_y}{\omega_x} = \frac{\text{Number of } y - \text{peaks}}{\text{Number of } x - \text{peaks}}$$

Three examples are shown in Figure 9.37(e).

$$\frac{\omega_y}{\omega_x} = \frac{2}{1}, \quad \frac{\omega_y}{\omega_x} = \frac{3}{1}, \quad \text{and} \quad \frac{\omega_y}{\omega_x} = \frac{3}{2}$$

Note: The above observations are true for narrow-band signals as well. Broad-band random signals produce scattered (irregular) Lissajous patterns.

Digital Oscilloscopes

The basic uses of a digital oscilloscope are quite similar to those of a traditional analog oscilloscope. The main differences stem from the manner in which information is represented and processed "internally" within the oscilloscope. Specifically, a digital oscilloscope first samples a signal that arrives at one of its input channels and stores the resulting digital data within a memory segment. This is essentially a typical analog-to-digital conversion (ADC) operation. This digital data can be processed to extract and display the necessary information. The sampled data and the processed

Signal Conditioning and Modification

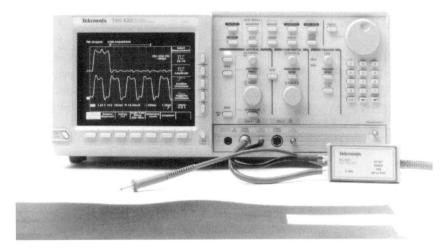

FIGURE 9.38 A digital oscilloscope. (Copyright 1999, Tektronix, Inc. All Rights Reserved. With permission.)

information can be stored on a floppy disk, if needed, for further processing using a digital computer. Also, some digital oscilloscopes have the communication capability so that the information can be displayed on a video monitor or printed to provide a hard copy.

A typical digital oscilloscope has four channels so that four different signals can be acquired (sampled) into the oscilloscope and displayed. Also, it has various triggering options so that the acquisition of a signal can be initiated and synchronized by means of either an internal or an external trigger. Apart from the typical capabilities that are possible with an analog oscilloscope, a digital oscilloscope can automatically provide other useful features, including the following:

1. Automatic scaling of the acquired signal
2. Computation of signal features such as frequency, period, amplitude, mean, root-mean-square (rms) value, and rise time
3. Zooming into regions of interest of a signal record
4. Averaging of multiple signal records
5. Enveloping of multiple signal records
6. Fast Fourier transform (FFT) capability, with various window options and anti-aliasing.

These various functions are menu selectable. Typically, a channel of the incoming data (signal) is selected first, and then an appropriate operation on the data is chosen from the menu (through menu buttons). A modern digital oscilloscope is shown in Figure 9.38.

PROBLEMS

9.1 A linear variable differential transformer (LVDT) is a displacement sensor commonly used in machine monitoring. Consider a digital vibration monitoring loop that uses an LVDT measurement. Typically, the LVDT is energized by a DC power supply. An oscillator provides an excitation signal in the kilohertz range to the primary windings of the LVDT. The secondary winding segments are connected in series opposition. An AC amplifier, demodulator, low-pass filter, amplifier, and ADC are used in the monitoring path. Using a schematic diagram, show the various hardware components in the monitoring loop and indicate their functions.

At the null position of the LVDT stroke, there was a residual voltage. A compensating resistor was used to eliminate this voltage. Indicate the connections for this compensating resistor.

9.2 Today, *machine vision* is used in many industrial tasks, including process monitoring. In an industrial system based on machine vision, an imaging device such as a charge-coupled-device (CCD) camera is used as the sensing element. The camera provides an image (picture) of a scene related to the industrial process (the measurement), to an image processor. The computed results from the image processor are used to determine the necessary information about the process (plant).

A CCD camera has an image plate consisting of a matrix of metal-oxide-semiconductor field-effect-transistor (MOSFET) elements. The electrical charge held by each MOSFET element is proportional to the intensity of light falling on the element. The output circuit of the camera has a charge-amplifier-like device (capacitor-coupled) that is supplied by each MOSFET element. The MOSFET element to be connected to the output circuit at a given instant is determined by the control logic that systematically scans the matrix of MOSFET elements. The capacitor circuit provides a voltage that is proportional to the charge in each MOSFET element.

a. Draw a schematic diagram for a process monitoring system based on machine vision, that uses a CCD camera. Indicate the necessary signal modification operations at various stages in the monitoring loop, showing whether analog filters, amplifiers, ADC, and DAC are needed and, if so, at which locations.

An image can be divided into *pixels* (or picture elements) for representation and subsequent processing. A pixel has a well-defined coordinate location in the picture frame, relative to some reference coordinate frame. In a CCD camera, the number of pixels per image frame is equal to the number of CCD elements in the image plate. The information carried by a pixel (in addition to its location) is the *photointensity* (or *gray level*) at the image location. This number must be expressed in the digital form (using a certain number of bits) for digital image processing. The need for very large data-handling rates is a serious limitation on a real-time controller that uses machine vision.

b. Consider a CCD image frame of the size 488×380 pixels. The refresh rate of the picture frame is 30 frames per second. If 8 bits are needed to represent the gray level of each pixel, what is the associated data (baud) rate?

c. Discuss whether you prefer hardware processing or programmable-software-based processing in a process monitoring system based on machine vision.

9.3 Usually, an operational amplifier circuit is analyzed making use of the following two assumptions:
i. The potential at the + input lead is equal to the potential at the − input lead.
ii. The current through each of the two input leads is zero.
Explain why these assumptions are valid under unsaturated conditions of an op-amp.

An amateur electronics enthusiast connects an op-amp without a feedback element to a circuit. Even when there is no signal applied to the op-amp, the output was found to oscillate between +12 V and −12 V once the power supply is turned on. Give a reason for this behavior.

An operational amplifier has an open-loop gain of 5×10^5 and a saturated output of ± 14 V. If the noninverting input is -1 µV, and the inverting input is $+0.5$ µV, what is the output? If the inverting input is 5 µV and the noninverting input is grounded, what is the output?

9.4 Define the following terms in connection with an operational amplifier:
a. Offset current
b. Offset voltage (at input and output)

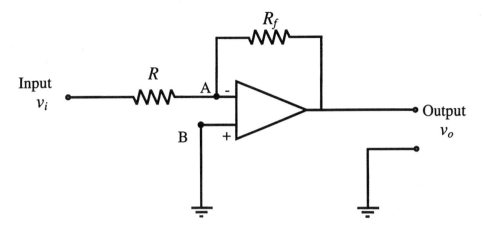

FIGURE P9.5 An amplifier circuit.

    c. Unequal gains
    d. Slew rate.
    Give typical values for these parameters. The open-loop gain and the input impedance of an op-amp are known to vary with frequency and are known to drift (with time). Still, the op-amp circuits are known to behave very accurately. What is the main reason for this?

9.5 What is a voltage follower? Discuss the validity of the following statements:
    a. A voltage follower is a current amplifier.
    b. A voltage follower is a power amplifier.
    c. A voltage follower is an impedance transformer.
    Consider the amplifier circuit shown in Figure P9.5. Determine an expression for the voltage gain K_v of the amplifier in terms of the resistances R and R_f. Is this an inverting amplifier or a noninverting amplifier?

9.6 The speed of response of an amplifier can be represented using three parameters: bandwidth, rise time, and slew rate. For an idealized linear model (transfer function), it can be verified that the rise time and the bandwidth are independent of the size of the input and the DC gain of the system. Since the size of the output (under steady conditions) can be expressed as the product of the input size and the DC gain, it is seen that rise time and bandwidth are independent of the amplitude of the output for a linear model.

    Discuss how slew rate is related to bandwidth and rise time of a practical amplifier. Usually, amplifiers have a limiting slew rate value. Show that bandwidth decreases with the output amplitude in this case.

    A voltage follower has a slew rate of 0.5 V·µs^{-1}. If a sinusoidal voltage of amplitude 2.5 V is applied to this amplifier, estimate the operating bandwidth. If, instead, a step input of magnitude 5 V is applied, estimate the time required for the output to reach 5 V.

9.7 Define the following terms:
    a. Common-mode voltage
    b. Common-mode gain
    c. Common-mode rejection ratio (CMRR).
What is a typical value for the CMRR of an op-amp? Figure P9.7 shows a differential amplifier circuit with a flying capacitor. The switch pairs A and B are turned on and off alternately during operation. For example, first the switches denoted by A are turned on (closed) with the switches B off (open). Next, the switches A are open and the switches B

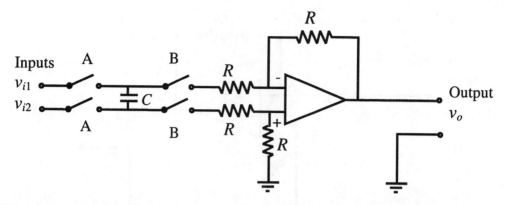

FIGURE P9.7 A differential amplifier with a flying capacitor for common-mode rejection.

are closed. Explain why this arrangement provides good common-mode rejection characteristics.

9.8 Compare the conventional (textbook) meaning of system stability and the practical interpretation of instrument stability.

An amplifier is known to have a temperature drift of 1 mV·°C^{-1} and a long-term drift of 25 µV per month. Define the terms *temperature drift* and *long-term drift*. Suggest ways to reduce drift in an instrument.

9.9 Electrical isolation of a device (or circuit) from another device (or circuit) is very useful in instrumentation practice. An isolation amplifier can be used to achieve this. This provides a transmission link that is almost "one way," and avoids loading problems. In this manner, damage in one component due to an increase in signal levels in the other components — perhaps due to shorts, malfunctions, noise, high common-mode signals, etc. — can be reduced. An isolation amplifier can be constructed from a transformer and a demodulator with other auxiliary components such as filters and amplifiers. Draw a schematic diagram for an isolation amplifier and explain the operation.

9.10 What are passive filters? List several advantages and disadvantages of passive (analog) filters in comparison to active filters.

A simple way to construct an active filter is to start with a passive filter of the same type and add a voltage follower to the output. What is the purpose of such a voltage follower?

9.11 Give one application each for the following types of analog filters:
   a. Low-pass filter
   b. High-pass filter
   c. Bandpass filter
   d. Notch filter.

Suppose that several single-pole active filter stages are cascaded. Is it possible for the overall (cascaded) filter to possess a resonant peak? Explain.

9.12 The Butterworth filter is said to have a "maximally flat magnitude." Explain what is meant by this. Give another characteristic that is desired from a practical filter.

9.13 An active filter circuit is given in Figure P9.13.
   a. Obtain the input-output differential equation for the circuit.
   b. What is the filter-transfer function?
   c. What is the order of the filter?
   d. Sketch the magnitude of the frequency-transfer function, and state what type of filter it represents.
   e. Estimate the cutoff frequency and the roll-off slope.

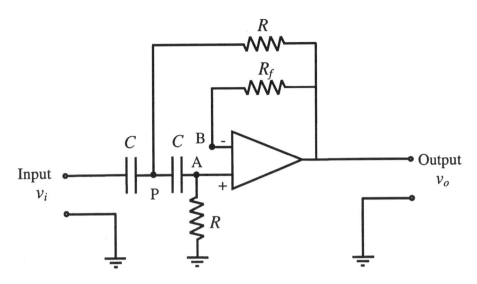

FIGURE P9.13 An active filter circuit.

9.14 What is meant by modulation, modulating signal, carrier signal, modulated signal, and demodulation? Explain the following types of signal modulation, giving an application for each case:
   a. Amplitude modulation
   b. Frequency modulation
   c. Phase modulation
   d. Pulse-width modulation
   e. Pulse-frequency modulation
   f. Pulse-code modulation.

   How could the sign of the modulating signal be accounted for during demodulation in each of these types of modulation?

9.15 Give two situations where amplitude modulation is intentionally introduced; and in each situation, explain how amplitude modulation would be beneficial. Also, describe two devices where amplitude modulation might be naturally present. Could the fact that amplitude modulation is present be exploited to one's advantage in these two natural situations? Explain.

9.16 A vibration monitoring system for a ball bearing of a rotating machine is schematically shown in Figure P9.16(a). It consists of an accelerometer to measure the bearing vibration, and an FFT analyzer to compute the Fourier spectrum of the vibration signal. This spectrum is examined over a period of 1 month after installation of the rotating machine in order to detect any degradation in the bearing performance. An interested segment of the Fourier spectrum can be examined with high resolution using the "zoom analysis" capability of the FFT analyzer. The magnitude of the original spectrum and that of the spectrum determined 1 month later, in the same zoom region, are shown in Figure P9.16(b).
   a. Estimate the operating speed of the rotating machine and the number of balls in the bearing.
   b. Do you suspect any bearing problems?

9.17 Explain the following terms:
   a. Phase-sensitive demodulation
   b. Half-wave demodulation
   c. Full-wave demodulation.

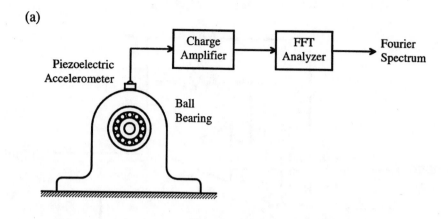

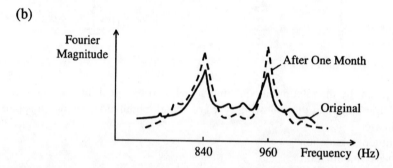

FIGURE P9.16. (a) A vibration monitoring system for a ball bearing, and (b) a zoomed Fourier spectrum.

When vibrations in rotating machinery such as gear boxes, bearings, turbines, and compressors are monitored, it is observed that a peak of the spectral magnitude curve does not usually occur at the frequency corresponding to the forcing function (e.g., tooth meshing, ball or roller hammer, blade passing). But, instead, two peaks occur on the two sides of this frequency. Explain the reason for this observation.

9.18 Define the following terms in relation to an analog-to-digital converter:
 a. Resolution
 b. Dynamic range
 c. Full-scale value
 d. Quantization error.

9.19 Single-chip amplifiers with built-in compensation and filtering circuits are becoming popular for signal-conditioning tasks associated with data acquisition and machine monitoring. Signal processing such as integration that would be needed to convert, say, an accelerometer into a velocity sensor, can also be accomplished in the analog form using an IC chip. What are advantages of such signal-modification chips in comparison to the conventional analog signal-conditioning hardware that employ discrete circuit elements and separate components to accomplish various signal-conditioning tasks?

9.20 Compare the three types of bridge circuits: constant-voltage bridge; constant-current bridge; and half-bridge, in terms of nonlinearity, effect of change in temperature, and cost.
 Obtain an expression for the percentage error in a half-bridge circuit output due to an error δv_{ref} in the voltage supply v_{ref}. Compute the percentage error in the output if the voltage supply has a 1% error.

9.21 The Maxwell bridge circuit is shown in Figure P9.21. Obtain the conditions for a balanced Maxwell bridge in terms of the circuit parameters R_1, R_2, R_3, R_4, C_1, and L_4. Explain how this circuit can be used to measure a variation in C_1 or L_4.

Signal Conditioning and Modification

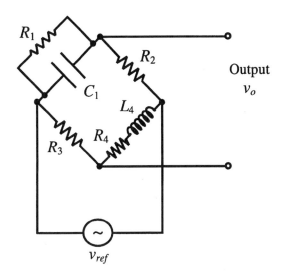

FIGURE P9.21 The Maxwell bridge.

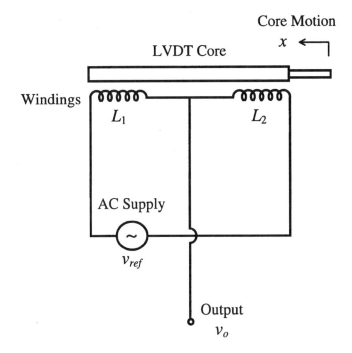

FIGURE P9.22 A half-bridge circuit for an LVDT.

9.22 The standard LVDT (linear variable differential transducer or transformer) arrangement has a primary coil and two secondary coil segments connected in series opposition. Alternatively, some LVDTs use a bridge circuit to produce their output. An example of a half-bridge circuit for an LVDT is shown in Figure P9.22. Explain the operation of this arrangement. Extend this idea to a full-impedance bridge for LVDT measurement.

9.23 The output of a Wheatstone bridge is nonlinear with respect to the variations in the bridge resistance. This nonlinearity is negligible for small changes in resistance. For large variations in resistance, however, some method of calibration or linearization should be employed. One way to linearize the bridge output is to positive feedback the output voltage signal into the bridge supply using a feedback operational amplifier. Consider the Wheatstone bridge circuit shown in Figure 9.24(a). Initially, the bridge is balanced with $R_1 = R_2 = R_3 = R_4 = R$. Then, the resistor R_1 is varied to $R + \delta R$. Suppose that the bridge output δv_o is fed back (positive) with a gain of 2 into the bridge supply v_{ref}. Show that this will linearize the bridge equation.

9.24 The noise in an electrical circuit can depend on the nature of the coupling mechanism. In particular, the following types of coupling can be identified:
   a. Conductive coupling
   b. Inductive coupling
   c. Capacitive coupling
   d. Optical coupling.

Compare these four types of coupling in terms of the nature and level of noise that is fed through or eliminated in each case. Discuss ways to reduce noise that is fed through in each type of coupling.

The noise due to variation in ambient light can be a major problem in optically coupled systems. Briefly discuss a method that could be used in an optically coupled device in order to make the device immune to variation in the ambient light level.

10 Vibration Testing

Vibration testing is usually performed by applying a vibratory excitation to a test object and monitoring the structural integrity and performance of the intended function of the object. The technique can be useful in several stages of (1) *design development*, (2) *production*, and (3) *utilization* of a product. In the initial design stage, the design weaknesses and possible improvements can be determined through vibration testing of a preliminary design prototype or a partial product. In the production stage, the quality of workmanship of the final product can be evaluated using both destructive and non-destructive vibration testing. A third application, termed *product qualification*, is intended to determine the adequacy of a product of good quality for a specific application (e.g., the seismic qualification of a nuclear power plant) or a range of applications.

The technology of vibration testing has rapidly evolved since World War II, and the technique has been successfully applied to a wide spectrum of products — ranging from small printed circuit boards and microprocessor chips to large missiles and structural systems. Until recently, however, much of the signal processing required in vibration testing was performed through analog methods. In these methods, the measured signal is usually converted into an electric signal, which in turn is passed through a series of electrical or electronic circuits to achieve the required processing. Alternatively, motion or pressure signals could be used in conjunction with mechanical or hydraulic (e.g., fluidic) circuits to perform analog processing. Today's complex test programs require the capability of fast and accurate processing of large numbers of measurements. The performance of analog signal analyzers is limited by hardware costs, size, data handling capacity, and computational accuracy. Digital processing, for the synthesis and analysis of vibration test signals and for the interpretation and evaluation of test results, began to replace the classical analog methods in the late 1960s. Today, special-purpose digital analyzers with real-time digital Fourier analysis capability (see Chapters 4 and 9, and Appendix D) are commonly used in vibration testing applications. The advantages of incorporating digital processing in vibration testing include the flexibility and convenience with respect to the type of the signal that can be analyzed and the complexity of the nature of processing that can be handled, increased speed of processing, accuracy and reliability, reduction in operational costs, practically unlimited repeatability of processing, and reduction in overall size and weight of the analyzer.

Vibration testing is usually accomplished using a shaker apparatus as shown by the schematic diagram in Figure 10.1. Theest object is secured to the shaker table in a manner representative of its installation during actual use (service). In-service operating conditions are simulated while the shaker table is actuated by applying a suitable input signal. Shakers of different types, with electromagnetic, electromechanical, or hydraulic actuators are available, as discussed in Chapter 8. The shaker device may depend on the test requirement, availability, and cost. More than one signal may be required to simulate three-dimensional characteristics of the vibration environment. The test input signal is either stored on an analog magnetic tape or generated in real-time by a signal generator. The capability of the test object or a similar unit to withstand a "predefined" vibration environment is evaluated by monitoring the dynamic response (accelerations, velocities, displacements, strains, etc.) and functional operability variables (e.g., temperatures, pressures, flow rates, voltages, currents). Analysis of the response signals will aid in detecting existing defects or impending failures in various components of the test equipment. The control sensor output is useful in several ways — particularly in feedback control of the shaker, frequency band equalization in real-time of the excitation signal, and synthesizing of future test signals.

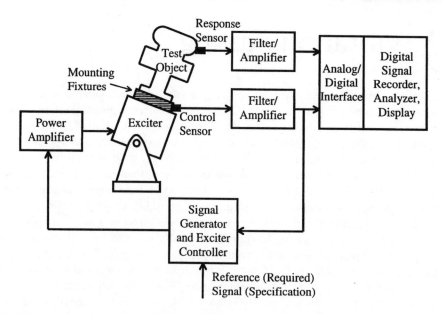

FIGURE 10.1 A typical vibration testing arrangement.

The excitation signal is applied to the shaker through a shaker controller, which usually has a built-in power amplifier. The shaker controller compares the "control sensor" signal, from the shaker-test-object interface, with the reference excitation signal from the signal generator. The associated error is used to control the shaker motion so as to push this error to 0. This is termed "equalization." Hence, a shaker controller serves as an equalizer as well.

The signals monitored from the test object include *test response signals* and *operability signals*. The former category of signals provides the dynamic response of the test object, and can include velocities, accelerations, and strains. The latter category of signals is used to check whether the test object performs in-service functions (i.e., it operates properly) during the test excitation, and can include flow rates, temperatures, pressures, currents, voltages, and displacements. The signals can be recorded in a computer or a digital oscilloscope for subsequent analysis. Also, by using an oscilloscope or a spectrum analyzer, some analysis can be done online, and the results are displayed immediately.

The most uncertain part of a vibration test program is the simulation of the test input. For example, the operating environment of a product such as an automobile is not deterministic and will depend on many random factors. Consequently, it is not possible to generate a single test signal that can completely represent various operating conditions. As another example, in seismic qualification of an equipment, the primary difficulty stems from the fact that the probability of accurately predicting the recurrence of an earthquake at a given site during the design life of the equipment is very small, and that of predicting the nature of the ground motions if an earthquake were to occur is even smaller. In this case, the best that one could do would be to make a conservative estimate of the nature of the ground motions due to the strongest earthquake that is reasonably expected. The test input should have (1) *amplitude*, (2) *phasing*, (3) *frequency content*, and (4) *damping* characteristics comparable to the expected vibration environment, if satisfactory representation is to be achieved. A frequency domain representation (see Chapters 3 and 4) of the test inputs and responses, in general, can provide better insight regarding their characteristics in comparison to a time domain representation (namely, a time history). Fortunately, frequency domain information can be derived from time domain data using Fourier transform techniques.

In vibration testing, Fourier analysis is used in three principal ways: (1) to determine the frequency response of the test object by means of prescreening tests; (2) to represent the vibration environment by its Fourier spectrum or its power spectral density so that a test input signal can be generated to represent it; and (3) to monitor the Fourier spectrum of the response at key locations in the test object and at control locations of the test table and use the information diagnostically or in controlling the exciter.

The two primary steps of a vibration testing scheme are:

Step 1: Specify the test requirements.
Step 2: Generate a vibration test signal that conservatively satisfies the specifications of step 1.

10.1 REPRESENTATION OF A VIBRATION ENVIRONMENT

A complete knowledge of the vibration environment in which a device is operating is not available to the test engineer or the test program planner. The primary reason for this is that the operating environment is a random process. When performing a vibration test, however, either a deterministic or a random excitation can be employed to meet the test requirements. This is known as the *test environment*.

Based on the vibration-testing specifications or product qualification requirements, the test environment should be developed to have the required characteristics of (1) intensity (amplitude); (2) frequency content (effect on the test-object resonances and the like); (3) decay rate (damping); and (4) phasing (dynamic interactions). Usually, these parameters are chosen to conservatively represent the worst possible vibration environment that is reasonably expected during the design life of the test object. So long as this requirement is satisfied, it is not necessary for the test environment to be identical to the operating vibration environment.

In vibration testing, the excitation input (test environment) can be represented in several ways. The common representations are by (1) time signal, (2) response spectrum, (3) Fourier spectrum, and (4) power spectral density function. Once the required environment is specified by one of these forms, the test should be conducted either by directly employing them to drive the exciter or by using a more conservative excitation when the required environment cannot be exactly reproduced.

10.1.1 TEST SIGNALS

Vibration testing can employ both random and deterministic signals as test excitations. Regardless of its nature, the test input should conservatively meet the specified requirements for that test.

Stochastic versus Deterministic Signals

Consider a seismic time-history record. Such a ground-motion record is not stochastic. It is true that earthquakes are random phenomena and the mechanism by which the time history was produced is a random process. Once a time history is recorded, however, it is known completely as a curve of response value versus time (a deterministic function of time). Therefore, it is a deterministic set of information. However, it is also a "sample function" of the original stochastic process (the earthquake) by which it was generated. Hence, very valuable information about the original stochastic process itself can be determined by analyzing this sample function on the basis of the *ergodic hypothesis* (see Section 10.1.3 on stochastic representation). Some might think that an irregular time-history record corresponds to a random signal. It should be remembered that some random processes produce very smooth signals. As an example, consider the sine wave given by $a\sin(\omega t + \phi)$. Assume that the amplitude a and the frequency ω are deterministic quantities, and the phase angle ϕ is a random variable. This is a random process. Every time this particular random process is activated, a sine wave is generated that has the same amplitude and frequency, but generally a different phase angle. Nevertheless, the sine wave will always appear as smooth as a deterministic sine wave.

In a vibration testing program, if one uses a recorded time history to derive the exciter, it would be a deterministic signal, even if it was originally produced by a random phenomenon such as an earthquake. Also, if one uses a mathematical expression for the signal in terms of completely known (deterministic) parameters, it is again a deterministic signal. If the signal is generated by some random mechanism (computer simulation or physical) in real-time, however, and if that signal is used as the excitation in the vibration test simultaneously as it is being generated, then one has a truly random excitation. Also, if one uses a mathematical expression (with respect to time) for the excitation signal and some of its parameters are not known numerically, and the values are assigned to them during the test in a random manner, one has a truly random test signal.

10.1.2 Deterministic Signal Representation

In vibration testing, time signals that are completely predefined can be used as test excitations. They should be capable, however, of subjecting the test object to the specified levels of intensity, frequency, decay rate (and phasing in the case of simultaneous multiple test excitations).

Deterministic excitation signals (time histories) used in vibration testing are divided into two broad categories: single-frequency signals and multifrequency signals.

Single-Frequency Signals

Single-frequency signals have only one predominant frequency component at a given time. For the entire duration, however, the frequency range covered is representative of the frequency content of the vibration environment. For seismic-qualification purposes, for example, this range should be at least 1 Hz to 33 Hz. Some typical single-frequency signals used as excitation inputs in vibration testing of equipment are shown in Figure 10.2. The signals shown in the figure can be expressed by simple mathematical expressions. This is not a requirement, however. It is quite acceptable to store a very complex signal in a storage device and subsequently use it in the procedure. In picking a particular time history, one should give proper consideration to its ease of reproduction and the accuracy with which it satisfies the test specifications. Next, the acceleration signals shown in Figure 10.2 are described mathematically.

Sine Sweep

One obtains a sine sweep by continuously varying the frequency of a sine wave. Mathematically,

$$u(t) = a\sin[\omega(t)t + \phi] \tag{10.1}$$

The amplitude a and the phase angle ϕ are usually constants, and the frequency $\omega(t)$ is a function of time. Both linear and exponential variations of frequency over the duration of the test are in common usage, but exponential variations are more common. For the linear variation (see Figure 10.3):

$$\omega(t) = \omega_{min} + (\omega_{max} - \omega_{min})\frac{t}{T_d} \tag{10.2}$$

where

ω_{min} = lowest frequency in the sweep
ω_{max} = highest frequency in the sweep
T_d = duration of the sweep.

Vibration Testing

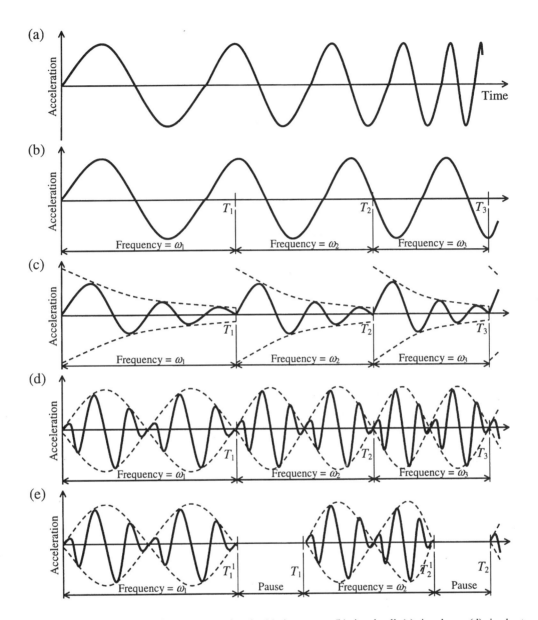

FIGURE 10.2 Typical single-frequency test signals: (a) sine sweep, (b) sine dwell, (c) sine decay, (d) sine beat, and (e) sine beat with pause.

For the exponential variation (see Figure 10.3):

$$\log\left[\frac{\omega(t)}{\omega_{min}}\right] = \frac{t}{T_d}\log\left[\frac{\omega_{max}}{\omega_{min}}\right] \quad (10.3)$$

or

$$\omega(t) = \omega_{min}\left[\frac{\omega_{max}}{\omega_{min}}\right]^{t/T_d} \quad (10.4)$$

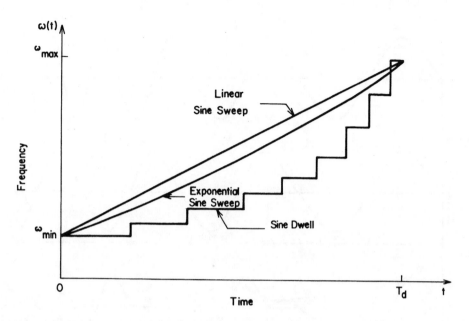

FIGURE 10.3 Frequency variation in some single-frequency vibration test signals.

This variation is sometimes incorrectly called logarithmic variation. This confusion arises because of its definition using equation (10.3) instead of equation (10.4). It is actually an inverse logarithmic (i.e., exponential) variation. Note that the logarithm in equation (10.3) can be taken to any arbitrary base. If base 10 is used, the frequency increments are measured in *decades* (multiples of 10); if base 2 is used, the frequency increments are measured in *octaves* (multiples of 2). Thus, the number of decades in the frequency range from ω_1 to ω_2 is given by $\log_{10}(\omega_2/\omega_1)$; for example, with $\omega_1 = 1$ rad·s^{-1} and $\omega_2 = 100$ rad·s^{-1}, $\log_{10}(\omega_2/\omega_1) = 2$, which corresponds to two decades. Similarly, the number of octaves in the range ω_1 to ω_2 is given by $\log_2(\omega_2/\omega_1)$. Then, with $\omega_1 = 2$ rad·s^{-1} and $\omega_2 = 32$ rad·s^{-1} we have $\log_2(\omega_2/\omega_1) = 4$, a range of four octaves. Note that these quantities are ratios and have no physical units. The foregoing definitions can be extended to smaller units; for example one-third octave represents increments of $2^{1/3}$. Thus, if one starts with 1 rad·s^{-1}, and increments the frequency successively by one-third octave, one obtains 1, $2^{1/3}$, $2^{2/3}$, 2, $2^{4/3}$, $2^{5/3}$, 2^2, etc. It is clear, for example, that there are four one-third octaves in the frequency range from $2^{2/3}$ to 2^2. Note that ω is known as the angular frequency (or radian frequency) and is usually measured in the units of radians per second (rad·s^{-1}). The more commonly used frequency is the cyclic frequency, which is denoted by f. This is measured in hertz (Hz), which is identical to cycles per second (cps). It is clear that

$$f = \frac{\omega}{2\pi} \qquad (10.5)$$

because there are 2π radians in one cycle.

So that all important vibration frequencies of the test object (or its model) are properly excited, the sine sweep rate should be as slow as is feasible. Typically, one octave per minute or slower rates are employed.

Sine Dwell

Sine-dwell signal is the discrete version of a sine sweep. The frequency is not varied continuously, but is incremented by discrete amounts at discrete time points. This is shown graphically in Figure 10.3. Mathematically, for the *r*th time interval, the dwell signal is

Vibration Testing

$$u(t) = a\sin(\omega_r t + \phi_r) \quad \text{for} \quad T_{r-1} \leq t \leq T_r \tag{10.6}$$

$$r = 1, 2, \ldots, n$$

in which ω_r, a, and ϕ_r are kept constant during the time interval (T_{r-1}, T_r). The frequency can be increased by a constant increment, or the frequency increments can be made bigger with time (exponential-type increment). The latter procedure is more common. Also, the dwelling-time interval is usually made smaller as the frequency is increased. This is logical because, as the frequency increases, the number of cycles that occur during a given time also increases. Consequently, steady-state conditions can be achieved in a shorter time.

Sine-dwell signals can be specified using either a graphical form (see Figure 10.3) or tabular form, giving the dwell frequencies and corresponding dwelling-time intervals. The amplitude is usually kept constant for the entire duration $(0, T_d)$, but the phase angle ϕ_r might have to be changed with each frequency increment in order to maintain the continuity of the signal.

Decaying Sine

Actual transient vibration environments (e.g., seismic ground motions) decay with time as the vibration energy is dissipated by some means. This decay characteristic is not present, however, in sine-sweep and sine-dwell signals. Sine-decay representation is a sine dwell with decay (see Figure 10.2). For an exponential decay, the counterpart of equation (10.6) can be written as

$$u(t) = a\exp(-\lambda_r t)\sin(\omega_r t + \phi_r) \quad T_{r-1} \leq t \leq T_r \tag{10.7}$$

The damping parameter (inverse of the time constant) λ is typically increased with each frequency increment in order to represent the increased decay rates of a dynamic environment (or increased modal damping) at higher frequencies.

Sine Beat

When two sine waves having the same amplitude but different frequencies (that are closer together) are mixed (added or subtracted) together, a sine beat is obtained. This signal is considered a sine wave having the average frequency of the two original waves, which is amplitude-modulated by a sine wave of frequency equal to half the difference of the frequencies of the two original waves. The amplitude modulation produces a transient effect that is similar to that caused by the damping term in the sine-decay equation (10.7). The sharpness of the peaks becomes more prominent when the frequency difference of the two frequencies is made smaller.

Consider two cosine waves having frequencies $(\omega_r + \Delta\omega_r)$ and $(\omega_r - \Delta\omega_r)$ and the same amplitude $a/2$. If the first signal is subtracted from the second (i.e., added with a 180° phase shift from the first wave), one obtains

$$u(t) = \frac{a}{2}\left[\cos(\omega_r - \Delta\omega_r)t - \cos(\omega_r + \Delta\omega_r)t\right] \tag{10.8}$$

By straightforward use of trigonometric identities, one obtains

$$u(t) = a(\sin\omega_r t)(\sin\Delta\omega_r t) \quad T_{r-1} \leq t \leq T_r \tag{10.9}$$

This is a sine wave of amplitude a and frequency ω_r modulated by a sine wave of frequency $\Delta\omega_r$. Sine-beat signals are commonly used as test excitation inputs in vibration testing. Usually, the ratio $\omega_r/\Delta\omega_r$ is kept constant. A typical value used is 20, in which case one gets 10 cycles per beat. Here, the cycles refer to the cycles at the higher frequency ω_r, and a beat occurs at each half cycle of the smaller frequency $\Delta\omega_r$. Thus, a beat is identified by a peak of amplitude a in the modulated wave, and the beat frequency is $2\Delta\omega_r$.

As in the case of a sine dwell, the frequency ω_r of a sine-beat excitation signal is incremented at discrete time points T_r so as to cover the entire frequency interval of interest $(\omega_{min}, \omega_{max})$. It is common practice to increase the size of the frequency increment and decrease the time duration at a particular frequency, for each frequency increment, just as is done for the sine dwell. The reasoning for this is identical to that given for sine dwell. The number of beats for each duration is usually kept constant (typically at a value over 7). A sine-beat signal is shown in Figure 10.2(d).

Sine Beat with Pauses

If one includes pauses between sine-beat durations, one obtains a sine-beat signal with pauses. Mathematically,

$$
\begin{aligned}
u(t) &= a(\sin\omega_r t)(\sin\Delta\omega_r t) && \text{for } T_{r-1} \leq t \leq T'_r \\
&= 0 && \text{for } T'_r \leq t \leq T_r
\end{aligned}
\quad (10.10)
$$

This situation is shown in Figure 10.2(e). When a sine-beat signal with pauses is specified as a test excitation, one must give the frequencies, corresponding time intervals, and corresponding pause times. Typically, the pause time is also reduced with each frequency increment.

The single-frequency signal relations described in this section are summarized in Table 10.1.

Multifrequency Signals

In contrast to single-frequency signals, multifrequency signals usually appear irregular and will have more than one predominant frequency component at a given time. Some common examples of multifrequency signals include aerodynamic disturbances, actual earthquake records, and simulated road disturbance signals used in automotive dynamic tests.

Actual Excitation Records

Typically, actual excitation records such as overhead guideway vibrations are sample functions of random processes. By analyzing these deterministic records, however, characteristics of the original stochastic processes can be established, provided that the records are sufficiently long. This is possible because of the *ergodic hypothesis*. Results thus obtained are not quite accurate because the actual excitation signals are usually nonstationary random processes and hence are not quite ergodic. Nevertheless, the information obtained by Fourier analysis is useful in estimating the amplitude, phase, and frequency-content characteristics of the original excitation. In this manner, one can pick a past excitation record that can conservatively represent the design-basis excitation for the object that needs to be tested.

Excitation time histories can be modified to make them acceptably close to a design-basis excitation by using spectral-raising and spectral-suppressing methods. In spectral-raising procedures, a sine wave of required frequency is added to the original time history to improve its capability of excitation at that frequency. The sine wave should be properly phased such that the time of maximum vibratory motion in the original time history is unchanged by the modification. Spectral suppressing is achieved essentially by using a narrow band-reject filter for the frequency band that needs to be removed. Physically, this is realized by passing the time-history signal through a linearly

TABLE 10.1
Typical Single-Frequency Signals Used in Vibration Testing

Single-Frequency Acceleration Signal	Mathematical Expression
Sine sweep	$u(t) = a \sin[\omega(t)t + \phi]$
	$\omega(t) = \omega_{min} + (\omega_{max} - \omega_{min})t/T_d$ (linear)
	$\omega(t) = \omega_{min} \left(\dfrac{\omega_{max}}{\omega_{min}}\right)^{t/T_d}$ (exponential)
Sine dwell	$u(t) = a \sin(\omega_r t + \phi_r) \quad T_{r-1} \leq t \leq T_r,\ r = 1, 2, \ldots, n$
Decaying sine	$u(t) = a \exp(-\lambda_r t) \sin(\omega_r t + \phi_r) \quad T_{r-1} \leq t \leq T_r$
	$r = 1, 2, \ldots, n$
Sine beat	$u(t) = a(\sin \omega_r t)(\sin \Delta\omega_r t) \quad T_{r-1} \leq t \leq T_r$
	$r = 1, 2, \ldots, n \quad \omega_r / \Delta\omega_r = $ constant
Sine beat with pauses	$u(t) = a(\sin \omega_r t)(\sin \Delta\omega_r t) \quad$ for $T_{r-1} \leq t \leq T'_r$
	$= 0 \quad$ for $T'_r \leq t \leq T_r$

damped oscillator that is tuned to the frequency to be rejected and connected in series with a second damper. Damping of this damper is chosen to obtain the required attenuation at the rejected frequency.

Simulated Excitation Signals

Random-signal-generating algorithms can be easily incorporated into digital computers. Also, physical experiments can be developed that have a random mechanism as an integral part. A time history from any such random simulation, once generated, is a sample function. If the random phenomenon is accurately programmed or physically developed so as to conservatively represent a design-basis excitation, a signal from such a simulation can be employed in vibration testing. Such test signals are usually available either as analog records on magnetic tapes or as digital records on a computer disk. Spectral-raising and spectral-suppressing techniques, mentioned earlier, can also be considered as methods of simulating vibration test excitations.

Before concluding this section, it is worthwhile to point out that all test excitation signals considered in this section are oscillatory. Although the single-frequency signals considered may possess little resemblance to actual excitations on a device during operation, they can be chosen to possess the required decay, magnitude, phase, and frequency-content characteristics. During vibration testing, these signals, if used as excitations, will impose reversible stresses and strains to the test object, whose magnitudes, decay rates, and frequencies are representative of those that would be experienced during actual operation during the design life of the test object.

10.1.3 Stochastic Signal Representation

To generate a truly stochastic signal, a random phenomenon must be incorporated into the signal-generating process. The signal must be generated in real-time, and its numerical value at a given time is unknown until that time instant is reached. A stochastic signal cannot be completely specified in advance, but its statistical properties can be prespecified. There are many ways of obtaining random processes, including physical experimentation (e.g., by tossing a coin at equal time steps and assigning a value to the magnitude over a given time step depending on the outcome of the toss), observation of processes in nature (such as outdoor temperature), and digital-computer simulation. The last procedure is the one commonly used in signal generation associated with vibration testing.

Ergodic Random Signals

A *random process* is a signal that is generated by some random (stochastic) mechanism. Each time the mechanism is operated, a different signal (sample function) is usually generated. The likelihood of any two sample functions becoming identical is governed by some probabilistic law. The random process is denoted by $X(t)$, and any sample function by $x(t)$. It should be remembered that no numerical computations can be made on $X(t)$ because it is not known for certain. Its Fourier transform, for example, can be written as an analytical expression but cannot be computed. Once a sample function $x(t)$ is generated, however, any numerical computation can be performed on it, because it is a completely known function of time. This important difference might be somewhat confusing.

At any given time t_1, $X(t_1)$ is a random variable that has a certain probability distribution. Consider a well-behaved function $f\{X(t_1)\}$ of this random variable (which is also a random variable). Its expected value (statistical mean) is denoted $E[f\{X(t_1)\}]$. This is also known as the *ensemble average*, because it is equivalent to the average value at t_1 of a collection (ensemble) of a large number of sample functions of $X(t)$.

Now consider the function $f\{x(t)\}$ of one sample function $x(t)$ of the random process. Its temporal (time) mean is expressed by

$$\lim_{T\to\infty} \frac{1}{2T} \int_{-T}^{T} f\{x(t)\} dt$$

Now, if

$$E[f\{X(t_1)\}] = \lim_{T\to\infty} \frac{1}{2T} \int_{-T}^{T} f\{x(t)\} dt \tag{10.11}$$

then the random signal is said to be *ergodic*. Note that the right-hand side of equation (10.11) does not depend on time. Hence, the left-hand side should also be independent of the time point t_1.

As a result of this relation (known as the *ergodic hypothesis*), one can obtain the properties of a random process merely by performing computations using one of its sample functions. Ergodic hypothesis is the bridge linking the stochastic domain of expectations and uncertainties and the deterministic domain of real records and actual numerical computations. Digital Fourier computations, such as correlation functions and spectral densities, would not be possible for random signals if not for this hypothesis.

Stationary Random Signals

If the statistical properties of a random signal $X(t)$ are independent of the time point considered, it is stationary. In particular, $X(t_1)$ will have a probability density that is independent of t_1, and the joint probability of $X(t_1)$ and $X(t_2)$ will depend only on the time difference $t_2 - t_1$. Consequently, the mean value $E[X(t)]$ of a stationary random signal is independent of t, and the autocorrelation function defined by

$$E[X(t)X(t+\tau)] = \phi_{xx}(\tau) \tag{10.12}$$

depends on τ and not on t. Note that ergodic signals are always stationary, but the converse is not always true.

Consider Parseval's theorem:

$$\int_{-\infty}^{\infty} x^2(t)dt = \int_{-\infty}^{\infty} |X(f)|^2 df \tag{10.13}$$

where $X(f)$ is the Fourier integral transform of $x(t)$.

This can be interpreted as an energy integral and its value is usually infinite for random signals. An appropriate measure for a random signal is its power. This is given by its root-mean-square (rms) value $E[X(t)^2]$. Power spectral density (psd) $\Phi(f)$ is the Fourier transform of the autocorrelation function $\phi(\tau)$; and similarly, the latter is the inverse Fourier transform of the former. Hence,

$$\phi_{xx}(\tau) = \int_{-\infty}^{\infty} \Phi_{xx}(f) \exp(j2\pi f \tau) df \tag{10.14}$$

Now, from equations (10.12) and (10.14), one obtains

$$\text{rms value} = E[X(t)^2] = \phi_{xx}(0) = \int_{-\infty}^{\infty} \Phi_{xx}(f) df \tag{10.15}$$

It follows that the rms value of a stationary random signal is equal to the area under its psd curve.

Independent and Uncorrelated Signals

Two random signals $X(t)$ and $Y(t)$ are independent if their joint probability distribution is given by the product of the individual distributions. A special case is the uncorrelated signals, which satisfy

$$E[X(t_1)Y(t_2)] = E[X(t_1)]E[Y(t_2)] \tag{10.16}$$

Consider the stationary case with mean values

$$\mu_x = E[X(t)] \tag{10.17}$$

$$\mu_y = E[Y(t)] \tag{10.18}$$

The *autocovariance functions* are given by

$$\varphi_{xx}(\tau) = E[\{X(t) - \mu_x\}\{X(t+\tau) - \mu_x\}] = \phi_{xx}(\tau) - \mu_x^2 \tag{10.19}$$

$$\varphi_{yy}(\tau) = E[\{Y(t) - \mu_y\}\{Y(t-\tau) - \mu_y\}] = \phi_{yy}(\tau) - \mu_y^2 \tag{10.20}$$

and the *cross-covariance function* is given by

$$\varphi_{xy}(\tau) = E\left[\{X(t)-\mu_x\}\{Y(t-\tau)-\mu_y\}\right] = \phi_{xy}(\tau) - \mu_x\mu_y \qquad (10.21)$$

For uncorrelated signals [equation (10.16)],

$$\phi_{xy}(\tau) = \mu_x\mu_y \qquad (10.22)$$

and, from equation (10.21), it follows that,

$$\varphi_{xy}(\tau) = 0 \qquad (10.23)$$

The *correlation-function coefficient* is defined by

$$\rho_{xy}(\tau) = \frac{\varphi_{xy}(\tau)}{\sqrt{\varphi_{xx}(0)\varphi_{yy}(0)}} \qquad (10.24)$$

which satisfies

$$-1 \le \rho_{xy}(\tau) \le 1 \qquad (10.25)$$

For uncorrelated signals, $\rho_{xy}(\tau) = 0$. This function measures the degree of correlation of the two signals.

The correlation of two random signals $X(t)$ and $Y(t)$ is measured in the frequency domain by its *ordinary coherence function*:

$$\gamma_{xy}^2(f) = \frac{|\Phi_{xy}(f)|^2}{\Phi_{xx}(f)\Phi_{yy}(f)} \qquad (10.26)$$

which satisfies the condition

$$0 \le \gamma_{xy}^2(f) \le 1 \qquad (10.27)$$

Transmission of Random Excitations

When the excitation input to a system is a random signal, the corresponding system response will also be random. Consider the system shown by the block diagram in Figure 10.4(a). The response of the system is given by the convolution integral:

$$Y(t) = \int_{-\infty}^{\infty} h(t_1)U(t-t_1)dt_1 \qquad (10.28)$$

in which the response psd is given by the Fourier transform:

$$\Phi_{yy}(f) = \mathcal{F}\left\{E[Y(t)Y(t+\tau)]\right\} \qquad (10.29)$$

Vibration Testing

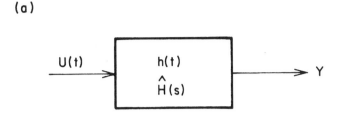

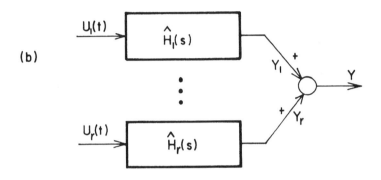

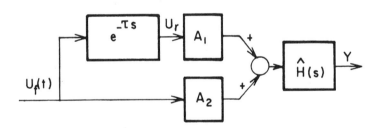

FIGURE 10.4 Combined response of a system to various random excitations: (a) system excited by a single input, (b) response to several random excitations, and (c) response to a delayed excitation.

Now, using equation (10.28) in equation (10.29), in conjunction with the definition of Fourier transform (see Chapter 4), one can write:

$$\Phi_{yy}(f) = \int_{-\infty}^{\infty} d\tau \exp(-j2\pi f\tau) E\left[\int_{-\infty}^{\infty} dt_1 h(t_1) U(t-t_1) \times \int_{-\infty}^{\infty} dt_2 h(t_2) U(t+\tau-t_2)\right]$$

which can be expressed as

$$\Phi_{yy}(f) = \int_{-\infty}^{\infty} dt_1 h(t_1) \int_{-\infty}^{\infty} dt_2 h(t_2) \int_{-\infty}^{\infty} d\tau \exp(-j2\pi f\tau) \times \phi_{uu}(\tau+t_1-t_2)$$

Now, by letting $\tau' = \tau + t_1 - t_2$, one can write

$$\Phi_{yy}(f) = \left[\int_{-\infty}^{\infty} h(t_1)\exp(j2\pi f t_1)dt_1\right]\left[\int_{-\infty}^{\infty} h(t_2)\exp(-j2\pi f t_2)dt_2\right] \times \left[\int_{-\infty}^{\infty} \phi_{uu}(\tau')\exp(-j2\pi f\tau')d\tau'\right]$$

Note that $U(t)$ is assumed to be *stationary*.

Next, since the frequency-response function is given by the Fourier transform of the impulse response function (see Chapters 2 and 3), one obtains

$$\Phi_{yy}(f) = H^*(f)H(f)\Phi_{uu}(f) \tag{10.30}$$

in which $H^*(f)$ is the complex conjugate of $H(f)$. Alternatively, if $|H(f)|$ denotes the magnitude of the complex quantity, one can write

$$\Phi_{yy}(f) = |H(f)|^2 \Phi_{uu}(f) \tag{10.31}$$

Using equation (10.30) or equation (10.31), one can determine the psd of the system response from the psd of the excitation if the system-frequency response function is known.

In a similar manner, it can be shown that the *cross-spectral density function* can be expressed as

$$\Phi_{uy}(f) = H(f)\Phi_{uu}(f) \tag{10.32}$$

Now consider r stationary, independent, random excitations $U_1, U_2, \ldots, U_r$ (which are assumed to have zero-mean values, without loss of generality) applied to r subsystems, having transfer functions $\hat{H}_1(s), \hat{H}_2(s), \ldots, \hat{H}_r(s)$ as shown in Figure 10.4(b). The total response Y consists of the sum of individual responses $Y_1, Y_2, \ldots, Y_r$. It can be shown that $Y_1, Y_2, \ldots, Y_r$ are also stationary, independent, zero-mean, random processes. By definition, then

$$\phi_{yy}(\tau) = E\left[\{Y_1(t)+\ldots+Y_r(t)\}\{Y_1(t+\tau)+\ldots+Y_r(t+\tau)\}\right] \tag{10.33}$$

Now, for independent, zero-mean Y_i, equation (10.33) becomes

$$\phi_{yy}(\tau) = E[Y_1(t)Y_1(t+\tau)] + \ldots + E[Y_r(t)Y_r(t+r)] \tag{10.34}$$

Since Y_i are stationary, one has

$$\phi_{yy}(\tau) = \phi_{y_1 y_1}(\tau) + \cdots + \phi_{y_r y_r}(\tau) \tag{10.35}$$

On Fourier transformation, one obtains

$$\Phi_{yy}(f) = \Phi_{y_1 y_1}(f) + \cdots + \Phi_{y_r y_r}(f) \tag{10.36}$$

In view of equation (10.31), it can be written

Vibration Testing

$$\Phi_{yy}(f) = \sum_{i=1}^{r} |H_i(f)|^2 \Phi_{u_i u_i}(f) \qquad (10.37)$$

from which the response psd can be determined if the input psd values are known.

If all inputs $U_i(t)$ have identical probability distributions (e.g., when they are generated by the same mechanism), the corresponding psd's will be identical. (Note that this does not imply that the inputs are equal. They could be dependent, independent, correlated, or uncorrelated.) In this case, equation (10.37) becomes

$$\Phi_{yy}(f) = \left[\sum_{i=1}^{r} |H_i(f)|^2\right] \Phi_{uu}(f) \qquad (10.38)$$

in which $\Phi_{uu}(f)$ is the common input psd.

Finally, consider the linear combination of two excitations $U_f(t)$ and $U_r(t)$, with the latter excitation delayed in time by τ but otherwise identical to the former. This situation is shown in Figure 10.4(c). From Laplace transform tables, it is seen that the Laplace transforms of the two signals are related by

$$U_r(s) = \exp(-\tau s) U_f(s) \qquad (10.39)$$

From equation (10.39), it follows that [see Figure 10.4(c)]

$$Y(s) = \left(A_1 \exp(-\tau s) + A_2\right) H(s) U_f(s) \qquad (10.40)$$

Consequently,

$$\Phi_{yy}(f) = \left|\left(A_1 \exp(-j2\pi f \tau) + A_2\right) H(f)\right|^2 \Phi_{uu}(f) \qquad (10.41)$$

From this result, the net response can be determined when the phasing between the two excitations is known. This has applications, for example, in determining the response of a vehicle to road disturbances at the front and rear wheels.

10.1.4 Frequency-Domain Representations

In this section, the Fourier spectrum method and the power spectral density method of representing a test excitation are discussed. These are frequency-domain representations. It is advisable to review Chapters 3 and 4 first in order to learn the necessary fundamentals.

Fourier Spectrum Method

Since the time domain and the frequency domain are related through Fourier transformation, a time signal can be represented by its Fourier spectrum. In vibration testing, a required Fourier spectrum can be given as the test specification. Then, the actual input signal used to excite the test object should have a Fourier spectrum that envelops the required Fourier spectrum. Generation of a signal to satisfy this requirement might be difficult. Usually, digital Fourier analysis (see Appendix D) of the control sensor signal is necessary to compare the actual (test) Fourier spectrum with the required Fourier spectrum. If the two spectra do not match in a certain frequency band, the error (i.e., the

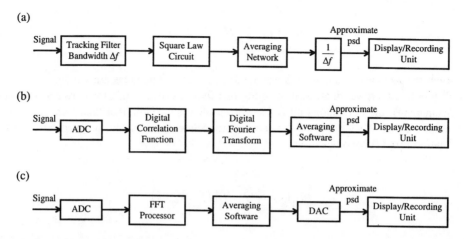

FIGURE 10.5 Some methods of psd determination: (a) filtering, squaring, and averaging method; (b) using autocorrelation function; and (c) using direct FFT.

difference in the two spectra) is fed back to correct the situation. This process is known as *frequency-band equalization*. Also, the sample step of the time signal in the digital Fourier analysis should be adequately small to cover the required frequency range of interest in that particular vibration testing application. Advantages of using digital Fourier analysis in vibration testing include flexibility and convenience with respect to the type of the signal that can be analyzed, availability of complex processing capabilities, increased speed of processing, accuracy and reliability, reduction in the test cost, practically unlimited repeatability of processing, and reduction in overall size and weight of the analyzer.

Power Spectral Density Method

The operational vibration environment of equipment is usually random. Consequently, a stochastic representation of the test excitation appears to be suitable for a majority of vibration testing situations. One way of representing a stationary random signal is by its power spectral density (psd). As noted before, the numerical computation of psd is not possible, however, unless the ergodicity is assumed for the signal. Using the ergodic hypothesis, one can compute the psd of a random signal simply by using one sample function (one record) of the signal.

Three methods of determining the psd of a random signal are shown in Figure 10.5. From Parseval's theorem [equation (10.13)], one sees that the mean square value of a random signal can be obtained from the area under the psd curve. This suggests the method shown in Figure 10.5(a) for estimating the psd of a signal. The mean square value of a sample of the signal in the frequency band Δf having a certain center frequency is obtained by first extracting the signal components in the band and then squaring them. This is done for several samples and averaged to get a high accuracy. It is then divided by Δf. By repeating this for a range of center frequencies, an estimate for the psd is obtained.

In the second scheme, shown in Figure 10.5(b), correlation function is first computed digitally. Its Fourier transform (by fast Fourier transform, or FFT as outlined in Appendix D) gives an estimate of the psd.

In the third scheme, shown in Figure 10.5(c), the psd is computed directly using FFT. Here, the Fourier spectrum of the sample record is computed and the psd is estimated directly, without first computing the autocorrelation function.

In these numerical techniques of computing psd, a single sample function would not give the required accuracy, and averaging of results for a number of sample records is usually needed. In real-time digital analysis, the running average and the current estimate are normally computed. In the running average, it is desirable to give a higher weighting to the more recent estimates. The

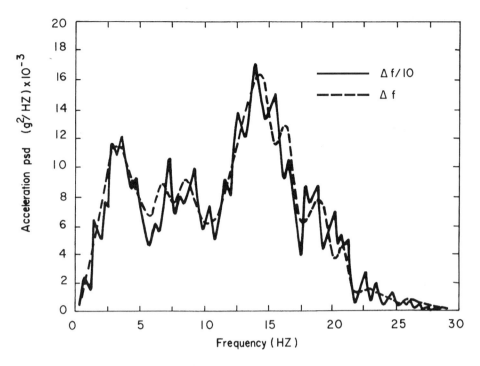

FIGURE 10.6 Effect of filter bandwidth on psd results.

fluctuations in the psd estimate about the local average could be reduced by selecting a larger filter bandwidth Δf (see Figure 10.6) and a large record length T. A measure of this fluctuation is given by

$$\varepsilon = \frac{1}{\sqrt{\Delta f T}} \qquad (10.42)$$

It should be noted that increasing Δf results in reduction of the precision of the estimates while improving the appearance. To offset this, T should be increased further, or averaging should be done for several sample records.

Generating a test input signal with a psd that satisfactorily compares with the required psd could be a tedious task if manually attempted by mixing various signal components. A convenient method is to use an automatic multiband equalizer. By this means, the mean amplitude of the signal in each small frequency band of interest can be made to approach the spectrum of the specified vibration environment (see Figure 10.7). Unfortunately, this type of random-signal vibration testing may be more costly than testing with deterministic signals.

10.1.5 Response Spectrum

Response spectra are commonly used to represent signals associated with vibration testing. A given signal has a certain fixed response spectrum, but many different signals can have the same response spectrum. For this reason, as will be clear shortly, the original signal cannot be reconstructed from its response spectrum (unlike in the case of a Fourier spectrum). This is a disadvantage; but the physical significance of a response spectrum makes it a good representation for a test signal.

If a given signal is applied to a single-degree-of-freedom oscillator (of a specific natural frequency), and the response of the oscillator (mass) is recorded, one can determine the maximum (peak) value of that response. Suppose that the process is repeated for a number of different

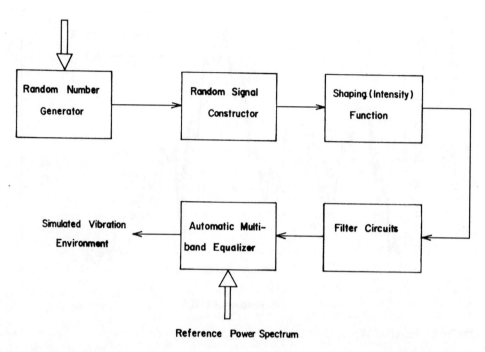

FIGURE 10.7 Generation of a specified random vibration environment.

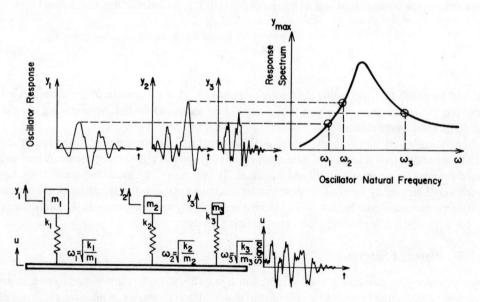

FIGURE 10.8 Definition of the response spectrum of a signal.

oscillators (having different natural frequencies) and then the peak response values thus obtained are plotted against the corresponding oscillator natural frequencies. This procedure is shown schematically in Figure 10.8. For an infinite number of oscillators (or for the same oscillator with continuously variable natural frequency), one obtains a continuous curve that is called the response spectrum of the given signal. It is obvious, however, that the original signal cannot be completely

Vibration Testing

determined from the knowledge of its response spectrum alone. In Figure 10.8, for example, another signal, when passed through a given oscillator, might produce the same peak response.

Note that it is assumed the oscillators to be undamped; that is, the response spectrum obtained using undamped oscillators corresponds to the damping value $\zeta = 0$. If all the oscillators are damped, however, and have the same damping ratio ζ, the resulting response spectrum will correspond to that particular value of ζ. It is therefore clear that ζ is also a parameter in the response-spectrum representation. One should specify the damping value as well when representing a signal by its response spectrum.

Displacement, Velocity, and Acceleration Spectra

It is clear that a motion signal can be represented by the corresponding displacement, velocity, or acceleration. First consider a displacement signal $u(t)$. The corresponding velocity signal is $\dot{u}(t)$, and the acceleration is $\ddot{u}(t)$.

Now consider an undamped simple oscillator, subjected to a support displacement $u(t)$, as shown in Figure 10.9. As usual, assuming that the displacements are measured with respect to a static equilibrium configuration, the gravity effect (which is balanced by the static deflection of the spring) can be ignored. Then, the equation of motion is given by

$$m\ddot{y}_d = k(u - y_d) \tag{10.43}$$

or

$$\ddot{y}_d + \omega_n^2 y_d = \omega_n^2 u(t) \tag{10.44}$$

where the (undamped) natural frequency is given by

$$\omega_n = \sqrt{\frac{k}{m}} \tag{10.45}$$

Suppose that the support (displacement) excitation $u(t)$ is a unit impulse $\delta(t)$. Then, the corresponding (displacement) response y is called the impulse-response function, as discussed in Chapter 2, and is denoted by $h(t)$. It is known that $h(t)$ is the inverse Laplace transform (with zero initial conditions) of the transfer function of the system (10.44), as given by (see Chapter 3)

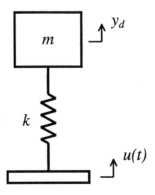

FIGURE 10.9 Undamped simple oscillator subjected to a support motion excitation.

$$H(s) = \frac{\omega_n^2}{\left(s^2 + \omega_n^2\right)} \tag{10.46}$$

The impulse-response function (to an impulsive support excitation) for an undamped, single-degree-of-freedom oscillator having natural frequency ω_n is given by

$$h(t) = \omega_n \sin \omega_n t \tag{10.47}$$

The displacement response $y_d(t)$ of this oscillator, when excited by the displacement signal $u(t)$, is given by the convolution integral

$$y_d(t) = \omega_n \int_0^\infty u(\tau) \sin \omega_n (t - \tau) d\tau \tag{10.48}$$

The "velocity" response of the same oscillator, when excited by the velocity signal $\dot{u}(t)$, is given by

$$y_v(t) = \omega_n \int_0^\infty \dot{u}(\tau) \sin \omega_n (t - \tau) d\tau \tag{10.49}$$

and the "acceleration" response, when excited by the acceleration signal $\ddot{u}(t)$, is

$$y_a(t) = \omega_n \int_0^\infty \ddot{u}(\tau) \sin \omega_n (t - \tau) d\tau \tag{10.50}$$

These results immediately follow from equation (10.44). Specifically, differentiate equation (10.44) once to get

$$\ddot{y}_v = \omega_n^2 y_v = \omega_n^2 \dot{u}(t) \tag{10.51}$$

and differentiate again to get

$$\ddot{y}_a = \omega_n^2 y_a = \omega_n^2 \ddot{u}(t) \tag{10.52}$$

in which

$$y_v = \frac{dy_d}{dt} \tag{10.53}$$

$$y_a = \frac{d^2 y_d}{dt^2} = \frac{dy_v}{dt} \tag{10.54}$$

Vibration Testing

If the peak value of $y_d(t)$ is plotted against ω_n, one obtains the *displacement-spectrum* curve of the displacement signal $u(t)$. If the peak value of $y_v(t)$ is plotted against ω_n, one gets the *velocity-spectrum* curve of the displacement signal $u(t)$. If the peak value of $y_a(t)$ is plotted against ω_n, one obtains the *acceleration-spectrum* curve of the displacement signal $u(t)$. Now consider equation (10.49). Integration by parts gives

$$y_v(t) = [\omega_n u(\tau) \sin \omega_n(t-\tau)]_0^\infty + \omega_n^2 \int_0^\infty u(\tau) \cos \omega_n(t-\tau) d\tau \tag{10.55}$$

The initial and final conditions for $u(t)$ are assumed to be 0. It follows that the first term in equation (10.55) vanishes. The second term is $\omega_n [y_d(t + \pi/(2\omega_n)) - \tau]$, which is clear by noting that $\sin \omega_n(t + \pi/(2\omega_n) - \tau)$ is equal to $\cos \omega_n(t-\tau)$; thus,

$$y_v(t) = -\omega_n y_d\left(t + \frac{\pi}{2\omega_n}\right) \tag{10.56}$$

If one integrates equation (10.50) by parts twice, and applies the end conditions as before, one obtains

$$y_a(t) = -\omega_n^2 y_d(t) \tag{10.57}$$

By taking the peak values of response time histories, it is seen from equations (10.56) and (10.57) that

$$v(\omega_n) = \omega_n d(\omega_n) \tag{10.58}$$

$$a(\omega_n) = \omega_n^2 d(\omega_n) \tag{10.59}$$

in which $d(\omega_n)$, $v(\omega_n)$, and $a(\omega_n)$ represent the displacement spectrum, the velocity spectrum, and the acceleration spectrum, respectively, of the displacement time history $u(t)$. It follows from equations (10.58) and (10.59) that

$$a(\omega_n) = \omega_n v(\omega_n) \tag{10.60}$$

Response-Spectra Plotting Paper

Response spectra are usually plotted on a frequency–velocity coordinate plane or on a frequency–acceleration coordinate plane. Values are normally plotted in logarithmic scale, as shown in Figure 10.10. First, consider the axes shown in Figure 10.10(a). Obviously, constant velocity lines are horizontal for this coordinate system. From equation (10.58), constant-displacement lines correspond to

$$v(\omega_n) = c\omega_n$$

By taking logarithms of both sides, one obtains

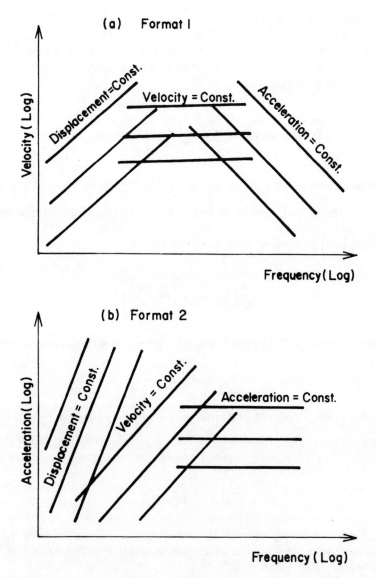

FIGURE 10.10 Response-spectra plotting formats: (a) frequency–velocity plane, and (b) frequency–acceleration plane.

$$\log v(\omega_n) = \log \omega_n + \log c$$

It follows that the constant-displacement lines have a +1 slope on the logarithmic frequency–velocity plane. Similarly, from equation (10.60), constant-acceleration lines correspond to

$$\omega_n v(\omega_n) = c$$

Hence,

$$\log v(\omega_n) = -\log \omega_n + \log c$$

Vibration Testing

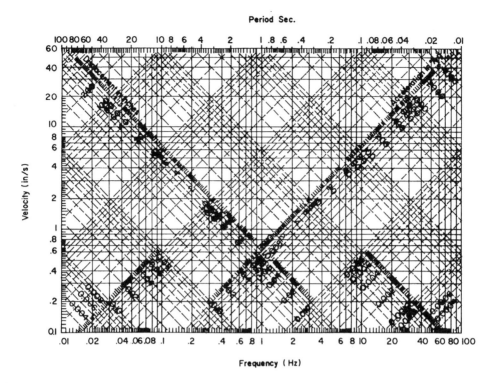

FIGURE 10.11 Response-spectra plotting sheet or nomograph (frequency–velocity plane).

It follows that the constant-acceleration lines have a -1 slope on the logarithmic frequency–velocity plane. Similarly, it can be shown from equations (10.59) and (10.60) that, on the logarithmic frequency–acceleration plane [Figure 10.10(b)], constant-displacement lines have a $+2$ slope, and constant-velocity lines have a $+1$ slope.

On the frequency–velocity plane, a point corresponds to a specific frequency and a specific velocity. The corresponding displacement at the point is obtained [equation (10.58)] by dividing the velocity value by the frequency value at that point. The corresponding acceleration at that point is obtained [equation (10.60)] by multiplying the particular velocity value by the frequency value. Any types of units can be used for displacement, velocity, and acceleration quantities. A typical logarithmic frequency–velocity plotting sheet is shown in Figure 10.11. Note that the sheet is already graduated on constant deplacement, velocity, and acceleration lines. Also, a period axis (period = 1/cyclic frequency) is given for convenience in plotting. A plot of this type is called a *nomograph*.

Zero–Period Acceleration

Frequently, response spectra are specified in terms of accelerations rather than velocities. This is particularly true in vibration testing associated with product qualification, because typical operational disturbance records are usually available as acceleration time histories. Of course, no information is lost because the logarithmic frequency–acceleration plotting paper can be graduated for velocities and displacements as well. It is therefore clear that an acceleration quantity (peak) on a response spectrum has a corresponding velocity quantity (peak) and a displacement quantity (peak). In vibration testing, however, the motion variable that is in common usage is acceleration. *Zero-period acceleration* (ZPA) is an important parameter that characterizes a response spectrum. It should be remembered, however, that zero-period velocity or zero-period displacement can be similarly defined.

Zero-period acceleration is defined as the acceleration value (peak) at zero period (or infinite frequency) on a response spectrum. Specifically,

$$\text{ZPA} = \lim_{\omega_n \to \infty} a(\omega_n) \qquad (10.61)$$

Consider the damped simple oscillator equation (for support motion excitation):

$$\ddot{y} + 2\zeta\omega_n \dot{y} + \omega_n^2 y = \omega_n^2 u(t) \qquad (10.62)$$

By differentiating equation (10.62) throughout, once or twice, it is seen, as in equations (10.51) and (10.52), that if u and y initially refer to displacements, then the same equation is valid when both of them refer to either velocities or accelerations. Consider the case in which u and y refer to input acceleration and response acceleration, respectively. For a sinusoidal signal $u(t)$ given by

$$u(t) = A \sin \omega t \qquad (10.63)$$

the response $y(t)$, neglecting the transient components (i.e., the steady-state value), is given by

$$y(t) = A \frac{\omega_n^2}{\sqrt{\left(\omega_n^2 - \omega^2\right)^2 + 4\zeta^2 \omega_n^2 \omega^2}} \sin(\omega t + \phi) \qquad (10.64)$$

Hence, the acceleration response spectrum, given by $a(\omega_n) = [y(t)]_{\max}$, for a sinusoidal signal of frequency ω and amplitude A is

$$a(\omega_n) = A \frac{\omega_n^2}{\sqrt{\left(\omega_n^2 - \omega^2\right)^2 + 4\zeta^2 \omega_n^2 \omega^2}} \qquad (10.65)$$

A plot of this response is shown in Figure 10.12. Note that $a(0) = 0$. Also,

$$\text{ZPA} = \lim_{\omega_n \to \infty} a(\omega_n) = A \qquad (10.66)$$

It is worth observing that at the point $\omega_n = \omega$ (i.e., when the excitation frequency ω is equal to the natural frequencies ω_n of the simple oscillator), one has $a(\omega_n) = A/(2\zeta)$, which corresponds to an amplification by a factor of $1/(2\zeta)$ over the ZPA value.

Uses of Response Spectra

In vibration testing, response-spectra curves are employed to specify the dynamic environment to which the test object is required to be subjected. This specified response spectrum is known as the *required response spectrum* (RRS). In order to conservatively satisfy the test specification, the response spectrum of the actual test input excitation, known as the *test response spectrum* (TRS), should envelop the RRS. Note that when response spectra are used to represent excitation input signals in vibration testing, the damping value of the hypothetical oscillators used in computing the response spectrum has no bearing on the actual damping that is present in the test object. In

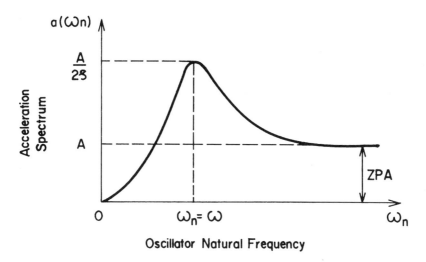

FIGURE 10.12 Response spectrum and ZPA of a sine signal.

this application, the response spectrum is merely a representation of the shaker input signal, and therefore does not depend on system damping.

Another use of response spectra is in estimating the peak value of the response of a multi-degree-of-freedom or distributed-parameter system when it is excited by a signal whose response spectrum is known. To understand this concept, one should recall the fact that, for a multi-degree-of-freedom or truncated (approximated) distributed-parameter system having distinct natural frequencies, the total responses can be expressed as a linear combination of the individual modal responses (see Chapters 5 and 6). Specifically, the response $y(t)$ can be written

$$y(t) = \sum_{i=1}^{r} \alpha_i a(\omega_i) \exp\left[\frac{-\zeta_i \omega_i t}{\sqrt{1-\zeta_i^2}}\right] \sin(\omega_i t + \phi_i) \tag{10.67}$$

in which $a(\omega_i)$ are the amplitude contributions from each mode (simple oscillator equation), with "damped" natural frequency ω_i. Hence, $a(\omega_i)$ corresponds to the value of the response spectrum at frequency ω_i. The linear combination parameters α_i depend on the modal-participation factors and can be determined from system parameters. Since the peak values of all terms in the summation on the right-hand side of equation (10.67) do not occur at the same time, one observes that

$$[y(t)]_{peak} \leq \sum_{i=1}^{r} \alpha_i a(\omega_i) \tag{10.68}$$

It follows that the right-hand side of the inequality (10.68) is a conservative upperbound estimate (i.e., the absolute sum) for the peak response of the multi-degree-of-freedom system. Some prefer to make the estimate less conservative by taking the *square root of sum of the squares* (SRSS):

$$[y(t)]_{SRSS} = \left[\sum_{i=1}^{r} \alpha_i^2 a^2(\omega_i)\right]^{1/2} \tag{10.69}$$

The latter method, however, has the risk of giving an estimate that is less than the true value. Note that, in this application, the damping value associated with the response spectrum is directly related to modal damping of the system. Hence, the response spectrum $a(\omega_i)$ should correspond to the same damping ratio as that of the mode considered within the summation of the inequality (10.68). If all modal damping ratios ζ_i are identical or nearly so, the same response spectrum can be used to compute all terms in the inequality (10.68). Otherwise, different response-spectra curves should be used to determine each quantity $a(\omega_i)$, depending on the applicable modal damping ratio ζ_i.

10.1.6 Comparison of Various Representations

This section states some major advantages and disadvantages of the four representations of the vibration environment that have been discussed.

Time-signal representation has several advantages. It can be employed to represent either deterministic or random vibration environments. It is an exact representation of a single excitation event. Also, when performing multi-excitation (multiple shaker) vibration testing, phasing between the various inputs can be conveniently incorporated by simply delaying each excitation with respect to the others. There are also disadvantages to time-signal representation. Because each time history represents just one sample function (single event) of a random environment, it may not be truly representative of the actual vibration excitation. This can be overcome by using longer signals, which, however, will increase the duration of the test, which is limited by test specifications. If the random vibration is truly ergodic (or at least stationary), this problem will not be as serious. Furthermore, the problem does not arise when testing with deterministic signals. An extensive knowledge of the true vibration environment for which the test object is subjected is necessary, however, in order to conclude that it is stationary or that it can be represented by a deterministic signal. In this sense, time-signal representation is difficult to implement.

The *response-spectrum method* of representing a vibration environment has several advantages. It is relatively easier to implement. Because the peak response of a simple oscillator is used in its definition, it is representative of the peak response or structural stress of simple dynamic systems; hence, there is a direct relation to the behavior of the physical object. An upper bound for the peak response of a multi-degree-of-freedom system can be conveniently obtained by the method outlined in the earlier section on uses of response spectra. Also, by considering the envelope of a set of response spectra at the same damping value, it is possible to use a single response spectrum to conservatively represent more than one excitation event. The method also has disadvantages. It employs deterministic signals in its definition. Sample functions (single events) of random vibrations can be used, however. It is not possible to determine the original vibration signal from the knowledge of its response spectrum because it uses the peak value of response of a simple oscillator (more than one signal can have the same response spectrum). Thus, a response spectrum cannot be considered a complete representation of a vibration environment. Also, characteristics such as the transient nature and the duration of the excitation event cannot be deduced from the response spectrum. For the same reason, it is not possible to incorporate information on excitation-signal phasing into the response-spectrum representation. This is a disadvantage in multiple excitation testing.

Fourier spectrum representation has advantages. Because the signal corresponding to the actual dynamic environment can be obtained by inverse transformation, it has the same advantages as for the time-signal representation. In particular, since a Fourier spectrum is generally complex, phasing information of the test excitation can be incorporated into the Fourier spectra in multiple excitation testing. Furthermore, by considering an envelope Fourier spectrum (like an envelope response spectrum), it can be employed to conservatively represent more than one vibration environment. Also, it gives frequency-domain information (such as resonances), which is very useful in vibration testing situations. The disadvantages of the Fourier spectrum representation include the following. It is a deterministic representation, but, as in the response-spectrum method, a sample function (single event) of a random vibration can be represented by its Fourier spectrum. Transient effects

Vibration Testing

BOX 10.1 Random Testing Versus Sine Testing

Advantages of random testing:
1. more realistic representation of the true environment
2. many frequencies are applied simultaneously
3. all resonances, natural frequencies, and mode shapes are excited simultaneously.

Disadvantages of random testing:
1. needs more power for testing
2. control is more difficult
3. more costly.

Advantages of sine testing:
Appropriate for —
1. fatigue testing of products that operate primarily at a known speed (frequency) under in-service conditions
2. detecting sensitivity of a device to a particular excitation frequency
3. detecting resonances, natural frequencies, modal damping, and mode shapes
4. calibration of vibration sensors and control systems.

Disadvantages of sine testing:
1. usually not a good representation of the true dynamic environment
2. because vibration energy is concentrated at one frequency, it can cause failures that would not occur in service (particularly single-resonance failures)
3. since only one mode is excited at a time, it can hide multiple-resonance failures that might occur in service.

---

and event duration are hidden in this representation. Also, it is somewhat difficult to implement because complex procedures of multiband equalization may be necessary in the signal synthesis associated with this representation.

Power spectral density representation has the following advantages. It takes into account the random nature of a vibration environment. As in response-spectrum and Fourier spectrum representations, by taking an envelope psd, it can be used to conservatively represent more than one environment. It can display important frequency-domain characteristics, such as resonances. Its disadvantages include the following. It is an exact representation only for truly stationary or ergodic random environments. In nonstationary situations, as in seismic ground motions, significant error can result. Also, it is not possible to obtain the original sample function (dynamic event) from its psd. Hence, transient characteristics and duration of the event are not known from its psd. Because mean square values, not peak values, are considered, psd representation is not structural-stress related. Furthermore, because psd functions are real (not complex), one cannot incorporate phasing information into them. This is a disadvantage in multiple excitation testing situations, but this problem can be overcome by considering either the cross-spectrum (which is complex) or the cross-correlation in each pair of test excitations. Random vibration testing is compared with sine testing (single-frequency, deterministic excitations) in Box 10.1. A comparison of various representations of test excitations is given in Box 10.2.

In practice, generation of an excitation signal for vibration testing may not follow any one of the analytical procedures exclusively and may incorporate a combination of them. For example, combination of sine-beat signals of different frequencies, with random phasing is one practical approach to the generation of multi-frequency, pseudo-random excitation signal. This approach is summarized in Box 10.3.

BOX 10.2 Comparison of Test Excitation Representations

Property	Time Signal	Response Spectrum	Fourier Spectrum	Power Spectral Density
True representation of a deterministic environment?	Yes	Yes	Yes	No
True representation of a random environment?	One sample function	One sample function	One sample function	Yes
Frequency-time reversible?	Yes	No	Yes	No
Signal phasing possible for multi-axis testing?	Yes	No	Yes	No
Good representation of peak amplitude/stress events?	Yes	Yes	No	No
Explicit accounting for modal responses?	No	Yes	Yes	Yes

BOX 10.3 Test Signal Generation

Steps:
1. Generate a set of sine beats at discrete frequencies of interest for the vibration test, and having specified amplitudes.
2. Phase shift (time shift) the signal components from step 1 according to a random number generator.
3. Sum the signal components from step 2.

10.2 PRETEST PROCEDURES

The selection of a test procedure for vibration testing of an object should be based on technical information regarding the test object and its intended use. Vendors usually prefer to use more established, conventional testing methods and generally are reluctant to incorporate modifications and improvements. This is primarily due to economic reasons, convenience, testing-time limitation, availability of the equipment and facilities (test-lab limitations), and similar factors. Regulatory agencies, however, usually modify their guidelines from time to time, and some of these requirements are mandatory.

Before conducting a vibration test on a test object, it is necessary to follow several pretest procedures. Such procedures are necessary in order to conduct a meaningful test. Some important pretest procedures include:

1. Understanding the purpose of the test
2. Studying the service functions of the test object
3. Information acquisition on the test object
4. Test-program planning
5. Pretest inspection of the test object
6. Resonance search to gather dynamic information about the test object
7. Mechanical aging of the test object.

In the following sections, each of the first five items of these procedures are discussed to emphasize how they can contribute to a meaningful test. The last two items will be considered separately in Section 10.3 on testing procedures.

10.2.1 Purpose of Testing

As noted previously, vibration testing is useful in various stages of (1) design and development, (2) production and quality assurance, and (3) qualification and utilization of a product. Depending on the outcome of a vibration test, design modifications or corrective actions can be recommended for a preliminary design or a partial product. To determine the most desirable location (in terms of minimal noise and vibration) for the compressor in a refrigerator unit, for example, a resonance-search test can be employed. As another example, vibration testing can be employed to determine vibration-isolation material requirements in structures for providing adequate damping. Such tests fall into the first category of system development tests. They are beneficial for the designer and the manufacturer in improving the quality of performance of the product. Government regulatory agencies do not usually stipulate the requirements for this category of tests, but they might stipulate minimal requirements for safety and performance levels of the final product, which could indirectly affect the development-test requirements. Custom-made items are exceptions, for which the customers could stipulate the design-test requirements.

For special-purpose products, it might also be necessary to conduct a vibration test on the final product before its installation for service operation. For mass-produced items, it is customary to select representative samples from each batch of the product for these tests. The purpose of such testing is to detect any inferiorities in the workmanship or in the materials used. These tests fall into the second category — quality assurance tests. These usually consist of a standard series of routine tests that are well established for a given product.

Distribution qualification and seismic qualification of devices and components are good examples of the use of the third category — qualification tests. A high-quality product such as a valve actuator, for example, which is thoroughly tested in the design-development stage and at the final production stage, will need further dynamic tests or analysis if it is to be installed in a nuclear power plant. The purpose in this instance is to determine whether the product (valve actuator) would be crucial for system safety-related functions. Government regulatory agencies usually stipulate basic requirements for qualification tests. These tests are necessarily application oriented. The vendor or the customer might employ more elaborate test programs than those stipulated by the regulatory agency, but at least the minimum requirements set by the agency should be met before commissioning the plant.

The purpose of any vibration test should be clearly understood before incorporating it into a test program. A particular test might be meaningless under some circumstances. If it is known, for example, that no resonances below 35 Hz exist in a particular equipment piece that requires seismic qualification, then it is not necessary to conduct a resonance search because the predominant frequency content in seismic excitations occurs below 35 Hz. If, however, the test serves a dual purpose, such as mechanical aging in addition to resonance detection, then it may still be conducted even if there are no resonances in the predominant frequency range of excitation.

If testing is performed on one test item selected from a batch of products to ensure the quality of the entire batch or to qualify the entire batch, it is necessary to establish that all items in the batch are of identical design. Otherwise, testing of all items in the batch might be necessary unless some form of design similarity can be identified. "Qualification by similarity" is done in this manner.

The nature of the vibration testing that is employed will usually be governed by the test purpose. Single-frequency tests, using deterministic test excitations, for example, are well suited for design-development and quality-assurance applications. The main reason for this choice is that the test input excitations can be completely defined; consequently, a complete analysis can be performed with relative ease, based on existing theories and dynamic models. Random or multifrequency tests are more realistic in qualification tests, however, because under typical service conditions, the dynamic environments to which an object is subjected are random and multifrequency in nature (e.g., seismic disturbances, ground-transit road disturbances, aerodynamic disturbances). Because random-excitation tests are relatively more expensive and complex in terms of signal generation and data processing, single-frequency tests might also be employed in qualification tests. Under some circumstances, single-frequency testing could add excessive conservatism to the test excitation. It is known, for example, that single-frequency tests are justified in the qualification of line-mounted equipment (i.e., equipment mounted on pipelines, cables, and similar "line" structures), which can encounter in-service disturbances that are amplified because of resonances in the mounting structure.

10.2.2 Service Functions

For product qualification by testing, it is required that the test object remain functional and maintain its structural integrity when subjected to a certain prespecified dynamic environment. In seismic qualification of equipment, for example, the dynamic environment is an excitation that adequately represents the amplitude, phasing, frequency content, and transient characteristics (decay rate and signal duration) of the motions at the equipment-support locations, caused by the most severe seismic disturbance that is expected, with a reasonable probability, during the design life of the equipment. Monitoring the proper performance of in-service functions (functional operability monitoring) of a test object during vibration testing could be crucial in the qualification decision.

The intended service functions of the test object should be clearly defined prior to testing. For active equipment, functional operability is necessary during vibration testing. For passive equipment, however, only the structural integrity needs to be maintained during testing.

Active Equipment

Equipment that should perform a mechanical motion (e.g., valve closure, relay contact) or that produces a measurable signal (e.g., electrical signal, pressure, temperature, flow) during the course of performing its intended functions is termed active equipment. Some examples of active equipment are valve actuators, relays, motors, pumps, transducers, control switches, and data recorders.

Passive Equipment

Passive equipment typically performs containment functions and consequently should maintain a certain minimum structural strength or pressure boundary. Such equipment usually does not perform mechanical motions or produce measurable response signals, but it may have to maintain displacement tolerances. Some examples of passive equipment are piping, tanks, cables, supporting structures, and heat exchangers.

Functional Testing

When defining intended functions of an object for test purposes, the following information should be gathered for each active component of the object that will be tested:

1. The maximum number of times a given function should be performed during the design life of the equipment
2. The best achievable precision (or monitoring tolerance) for each functional-operability parameter and the time duration for which a given precision is required
3. Mechanisms and states of malfunction or failure
4. Limits of the functional-operability parameters (electrical signals, pressures, temperatures, flow rates, mechanical displacements and tolerances, relay chatter, etc.) that correspond to a state of malfunction or failure

It should be noted that, under a state of malfunction, the object would not perform the intended function properly. Under a failure state, however, the object would not perform its intended function at all.

For objects consisting of an assembly of several crucial components, it should be determined how a malfunction or failure of one component could result in malfunction or failure of the entire unit. In such cases, any hardware redundancy (i.e., when component failure does not necessarily cause unit failure) and possible interactive and chain effects (such as failure in one component overloading another, which could result in subsequent failure of the second component, and so on) should be identified. In considering functional precision, it should be noted that high precision usually means increased complexity of the test procedure. This is further complicated if a particular level of precision is required at a prescribed instant.

It is common practice for the test object supplier (customer) to define the functional test, including acceptance criteria and tolerances for each function, for the benefit of the test engineer. This information eventually is used in determining acceptance criteria for the tests of active equipment. Complexity of the required tests also depends on the precision requirements for the intended functions of the test object.

Examples of functional failure are sensor and transducer (measuring instrumentation) failure, actuator (motors, valves, etc.) failure, chatter in relays, gyroscopic and electronic-circuit drift, and discontinuity of electrical signals because of short-circuiting. It should be noted that functional failures caused by mechanical excitation are often linked with the structural integrity of the test object. Such functional failures are primarily caused in two ways: (1) when displacement amplitude exceeds a certain critical value once or several times, or (2) when vibrations of moderate amplitudes occur for an extended period of time. Functional failures in the first category include, for example, short-circuiting, contact errors, instabilities, and nonlinearities (in relays, amplifier outputs, etc.). Such failures are usually reversible, so that when the excitation intensity drops, the system would function normally. In the second category, slow degradation of components would occur because of aging, wear, and fatigue, which could cause drift, offset, etc. and subsequent malfunction or failure. This kind of failure is usually irreversible. It must be emphasized that the first category of functional failure can be better simulated using high-intensity single-frequency testing and shock testing, and the second category by multifrequency or broadband random testing and low-intensity single-frequency testing.

For passive devices, a damage criterion should be specified. This can be expressed in terms of parameters such as cumulative fatigue, deflection tolerances, wearout limits, pressure drops, and leakage rates. Often, damage or failure in passive devices can be determined by visual inspection and other nondestructive means.

10.2.3 Information Acquisition

In addition to information concerning service functions, as discussed in the previous section, and dynamic characteristics determined from a resonance search, as will be discussed later in this chapter, there are other characteristics of the test object that need to be studied in the development of a vibration testing program. In particular, there are characteristics that cannot be described in exact quantitative terms. In determining the value of equipment, for example, the monetary value (or cost) might be relatively easy to estimate, whereas it may be very difficult to assign a dollar value to its significance under service conditions. One reason for this could be that the particular piece of equipment alone might not determine the proper operation of a complex system. Interaction of a particular unit with other subsystems in a complex operation would determine the importance attached to it and, hence, its value. In this sense, the true value of a test object is a relatively complex consideration. The service function of the test object is also an important consideration in determining its value. The value of a test object is important in planning a test program because the cost of a test program and the effort expended therein are governed mainly by this factor.

Many features of a test object that are significant in planning a test program can be deduced from the manufacturer's data for the particular object. The following information is representative:

1. Drawings (schematic or to scale when appropriate) of principal components and the whole assembly, with the manufacturer's name, identification numbers, and dimensions clearly indicated
2. Materials used, design strengths, fatigue life, etc. of various components, and factors determining the structural integrity of the unit
3. Component weight and total weight of the unit
4. Design ratings, capacities, and tolerances for in-service operation of each crucial component
5. Description of intended functions of each component and of the entire unit, clearly indicating the parameters that determine functional operability of the unit
6. Interface details (intercomponent as well as for the entire assembly), including in-service mounting configurations and mounting details
7. Details of the probable operating site or operating environment (particularly with respect to the excitation events if product qualification is intended)
8. Details of any previous testing or analysis performed on that unit or a similar one.

Scale drawings and component-weight information describe the size and geometry of the test object. This information is useful in determining the following:

1. The locations of sensors (accelerometers, strain gages, strobocopes, and the like) for monitoring dynamic response of the test object during tests
2. The necessary ratings for vibration test (shaker) apparatus (power, force, stroke, bandwidth, etc.)
3. The degree of dynamic interaction between the test object and the test apparatus
4. The level of coupling between various degrees of freedom and modal interactions in the test object
5. The assembly level of the test object (e.g., whether it can be treated as a single component, as a subsystem consisting of several components, or as an independent, stand-alone system).

In general, as the size and the assembly level increase, the tests becomes increasingly complex and difficult to perform. To test heavy, complex test objects, we would need a large test apparatus with high power ratings and the capability of multiple excitation locations. In this case, the number of operability parameters that are monitored and the number of observation (sensor) locations will also increase.

Interface Details

The dynamics of a piece of equipment depend on the way the equipment is attached to its support structure. In addition to the mounting details, equipment dynamic response is also affected by other interfacing linkages, such as wires, cables, conduits, pipes, and auxiliary instrumentation. In vibration testing of equipment, such interface characteristics should be simulated appropriately. Dynamics of the test fixture and the details of the test object–fixture interface are very important considerations that affect the overall dynamics of the test object. If interface characteristics are not properly represented during testing, a non-uniform test could result, in which case some parts of the test object would be overtested and other parts undertested. This situation can bring about failures that are not representative of the failures that could take place in actual service. In effect, the testing could become meaningless if interface details are not simulated properly.

Vibration Testing

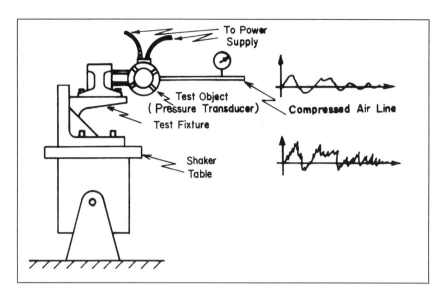

FIGURE 10.13 Influence of test fixture on the test excitation signal.

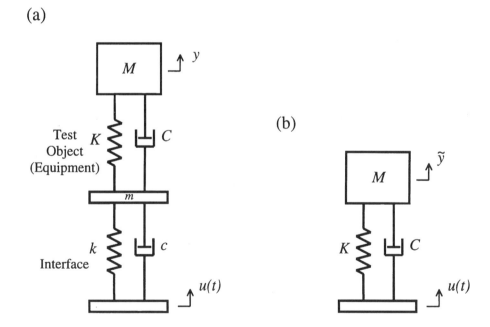

FIGURE 10.14 A simplified model to study the effect of interface dynamics: (a) with interface dynamics, and (b) without interface dynamics.

The test fixture is a structure attached to the shaker table and used to mount the test object (see Figure 10.13). Test fixture dynamics can significantly modify the shaker-table motion before reaching the test object. Such modifications include filtering of the shaker motion and introduction of auxiliary (cross-axis) motions. In the test setup shown in Figure 10.13, for example, the direct motion will be modified to some extent by fixture dynamics. In addition, some transverse and rotational motion components will be transmitted to the test object by the test fixture because of its overhang.

To minimize interface dynamic effects in vibration testing situations, an attempt should be made to (1) make the test fixture as light and as rigid as is feasible; (2) simulate in-service mounting conditions at the test object–fixture interface; and (3) simulate other interface linkages, such as cables, conduits, and instrumentation, to represent in-service conditions. Very often, the design of a proper test fixture can be a costly and time-consuming process. A tradeoff is possible by locating the control sensors (accelerometers) at the mounting locations of the test object and, then, using the error between the actual and the desired excitations through feedback to control the mounting-location excitations during testing.

Effect of Neglecting Interface Dynamics

Consider a simplified model in order to study some important effects of neglecting interface dynamics. In the model shown in Figure 10.14, the equipment and the mounting interface are modeled separately as single-degree-of-freedom systems. Capital letters are used to denote the equipment parameters (mass M, stiffness K, and damping coefficient C). When mounting interface dynamics are included, the model appears as is in Figure 10.14(a). When the mounting interface dynamics are neglected, one obtains the single-degree-of-freedom model shown in Figure 10.14(b). Note that, in the latter case, the shaker motion $u(t)$ is directly applied to the equipment mounts; whereas, in the former case, it is applied through the interface. If the equipment response in the two cases is denoted by y and $\tilde{y}$, respectively, it can be shown by considering the system-frequency transfer functions $Y(\omega)/U(\omega)$ and $\tilde{Y}(\omega)/U(\omega)$ that

$$\frac{\tilde{Y}(\omega)}{Y(\omega)} = \frac{Ms^2}{\left(Ms^2 + Cs + K\right)} \frac{(Cs+K)}{(cs+k)} + \frac{\left(ms^2 + cs + k\right)}{(cs+k)} \tag{10.70}$$

with $s = j\omega$. The following nondimensional parameters are defined:

$$\text{Mass ratio } \alpha = \frac{m}{M} \tag{10.71}$$

$$\text{Natural frequency ratio } \beta = \frac{\omega_n}{\Omega_n} \tag{10.72}$$

$$\text{Normalized excitation frequency } \bar{\omega} = \frac{\omega}{\Omega_n} \tag{10.73}$$

in which the natural frequency (undamped) of the equipment is

$$\Omega_n = \sqrt{\frac{K}{M}} \tag{10.74}$$

and the mounting interface natural frequency is

Vibration Testing

$$\omega_n = \sqrt{\frac{k}{m}} \tag{10.75}$$

Then, equation (10.70) can be written

$$\frac{\tilde{Y}(\omega)}{Y(\omega)} = \frac{\left(\beta^2 + 2j\zeta\beta\overline{\omega} - \overline{\omega}^2\right)}{\left(\beta^2 + 2j\zeta\beta\overline{\omega}\right)} - \frac{\overline{\omega}(1 + 2jZ\overline{\omega})}{\alpha\left(\beta^2 + 2j\zeta\beta\overline{\omega}\right)\left(1 + 2jZ\overline{\omega} - \overline{\omega}^2\right)} \tag{10.76}$$

in which ζ and Z denote the damping ratios of the interface and the equipment, respectively.

The ratio $\tilde{Y}(\omega)/Y(\omega)$ is representative of the equipment-response amplification when interface dynamic effects are neglected (removed) for a harmonic excitation.

Figure 10.15 shows eight curves, corresponding to equation (10.76), for the parameter combinations given in Table 10.2. Interpretation of the results becomes easier when peak values of the response ratios are compared for various parameter combinations. Sample results are given Table 10.2.

Effects of Damping

By comparing cases 1, 2, 3, and 4 in Table 10.2 with cases 5, 6, 7 and 8, respectively, one sees that increasing the interface damping has reduced the peak response (a favorable effect), irrespective of the values of the interface mass and natural frequency (α and β values).

Effects of Inertia

By comparing cases 2, 3, 6, and 7 with cases 4, 1, 8, and 5, respectively, one sees that the interface inertia has the effect of decreasing dynamic interaction by making the response of the sytem with an interface closer to that of the system without an interface, irrespective of the interface damping and natural frequency.

Effect of Natural Frequency

By comparing cases 2, 4, 6, and 8 with cases 3, 1, 7, and 5, respectively, one sees that increasing the interface natural frequency has a favorable effect in decreasing dynamic interactions, irrespective of the interface damping and inertia.

TABLE 10.2
Response Amplification Caused by Neglecting Interface Dynamics

Case (Curve No.)	Parameter Combination				Peak Value of Response Ratio
	ζ	α	β	Z	
1	0.1	2.0	2.0	0.1	1.11
2	0.1	0.5	0.5	0.1	38.80
3	0.1	0.5	2.0	0.1	2.77
4	0.1	2.0	0.5	0.1	10.80
5	0.2	2.0	2.0	0.2	0.89
6	0.2	0.5	0.5	0.2	18.40
7	0.2	0.5	2.0	0.2	1.71
8	0.2	2.0	0.5	0.2	5.98

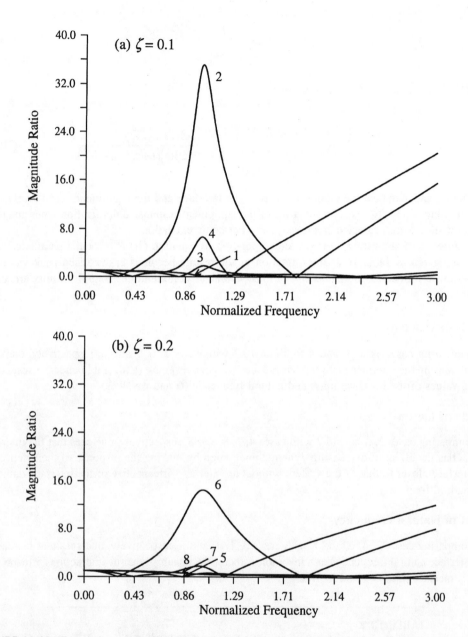

FIGURE 10.15 Response amplification when interface dynamic interactions are neglected.

Effect of Excitation Frequency

All the response plots (see Figure 10.15) diverge to ∞ as $\bar{\omega}$ increases. This indicates that, at very high excitation frequencies, dynamic testing results could become meaningless because of the interactions with interface dynamics.

It can be concluded that, to reduce dynamic interactions caused by a mechanical interface, one should (1) increase interface damping as much as is feasible; (2) increase interface mass as much as is feasible; (3) increase interface natural frequency as much as is feasible; and (4) avoid testing at relatively high frequencies of excitation. It should be noted that, in the foregoing analysis and discussion, the mechanical interface was considered to include test fixtures and the shaker table as well.

Other Effects of Interface

The type of vibration test used sometimes depends on the mechanical interface characteristics. An example is the testing of line-mounted equipment. Single-frequency testing is preferred for such equipment so as to add a certain degree of conservatism, because, as a result of interface resonances, line-mounted equipment could be subjected to higher levels of narrow-band excitation through the support structure.

In vibration testing of multicomponent equipment cabinets, it is customary to test the empty cabinet first, with the components replaced by dummy weights, and then to test the individual components separately, using different test excitations depending on the component locations and their mounting characteristics. Mechanical interface details of individual components are important in such situations. As a result, interface information is an important constituent of the pretest information that is collected for a test object.

Most of the interface data, particularly information related to size and geometry (e.g., mass, dimensions, configurations, and locations), can be gathered simply by observing the test object and using scale drawings supplied by the manufacturer. Size and number of anchor bolts used or weld thickness, for example, can be obtained in this manner. When analysis is also used to augment testing, however, it is often necessary to know the loads transmitted (forces, moments, etc.), relative displacements, and stiffness values at the mechanical interface under in-service conditions. These must be determined by tests, by analysis (static or dynamic) of a suitable model, or from manufacturer's data.

10.2.4 TEST-PROGRAM PLANNING

The test program to which a test object is subjected depends on several factors, including:

1. The objectives and specific requirements of the test
2. In-service conditions, including equipment-mounting features, vibration environment, and specifications of the test environment
3. The nature of the test object, including complexity, assembly level, and functional-operability parameters to be monitored
4. Test-laboratory capabilities, available testing apparatus, past experience, conventions, and established practices of testing.

Some of these factors are based on solid technical reasons, whereas others depend on economics, convenience, and personal likes and dislikes.

Initially, it is not necessary to develop a detailed test procedure; this is required only at the stage of actual testing. In the initial stage, it is only necessary to select the appropriate test method, based on factors such as those listed at the beginning of this section. Before conducting the tests, however, a test procedure should be prepared in sufficient detail. In essence, this is a pretest requirement.

Objectives and specific requirements of a test depend on such considerations as whether the test is conducted at the design stage, the quality-control stage, or the utilization stage. The objective of a particular test could be to verify the outcome of a previously conducted test. In that case, it is necessary to assess the adequacy of one or a series of tests conducted at an earlier time (e.g., when the specifications and government regulations were less stringent). Often, this can be done by analysis alone. Some testing might be necessary at times, but it usually is not necessary to repeat the entire test program. If the previous tests were conducted for the frequency range 1 Hz to 25 Hz, for example, and the present specifications require a wider range of 1 Hz to 35 Hz, it might be adequate merely to demonstrate (by analysis or testing) that there are no significant resonances in the test object in the 25- to 35-Hz range.

If it is necessary to qualify the test object for several different dynamic environments, a generic test that represents (conservatively, but without the risk of overtesting) all these environments can be used. For this purpose, special test-excitation inputs must be generated, taking into account the variability of the excitation characteristics under the given set of environments. Alternatively, several tests might be conducted if the dynamic environments for which the test object is to be qualified are significantly different. Operating-basis earthquake (OBE) tests and safe-shutdown earthquake (SSE) tests in seismic qualification of nuclear power plant equipment, for example, represent two significantly different test conditions. Consequently, they cannot be represented by a single test. When qualifying an equipment piece for several geographic regions or locations, however, one might be able to combine all OBE tests into a single test and all SSE tests into another single test.

Another important consideration in planning a test program is the required accuracy for the test, including the accuracy for the excitation inputs, response and operability measurements, and analysis. This is related to the "value" of the test object and the objectives of the test.

When it is required to evaluate or qualify a group of equipment by testing a sample, it is first necessary to establish that the selected sample unit is truly representative of the entire group. When the items in the batch are not identical in all respects, some conservatism can be added to the tests to minimize the possibility of incorrect qualification decision. It might be necessary to test more than one sample unit in such situations.

When planning a test procedure, one should clearly identify the standards, government regulations, and specifications that are applicable to a particular test. The pertinent sections of the applicable documents should be noted, and proper justification should be given if the tests deviate from regulatory agency requirements.

Excitation input that is employed in a vibration test depends on the in-service vibration environment of the test object. The number of tests needed will also depend on this to some extent. Test orientation depends mainly on the mounting features and the mechanical interface details of the test object under in-service conditions. Mounting features might govern the nature of the test excitations used for a particular test.

Two distinct mounting types can be identified for most equipment: (1) *Line-mounted equipment* and (2) *floor-mounted equipment*. Line-mounted equipment is equipment that is mounted upright or hanging from pipelines, cables, or similar line structures that are not rigid. Generally, devices such as valves, nozzles, valve actuators, and transducers are considered line-mounted equipment. Any equipment that is not line-mounted is considered floor-mounted. The supporting structure is considered relatively rigid in this case. Examples of such mounting structures are floors, walls, and rigid frames. Typical examples of floor-mounted equipment include motors, compressors, and cabinets of relays and switchgear.

Wide-band floor disturbances are filtered by line structures. Consequently, the environmental disturbances to which line-mounted equipment is subjected will generally be narrow-band disturbances. Accordingly, vibration testing of line-mounted equipment is best performed using narrow-band random test excitations or single-frequency deterministic test excitations. Relatively higher test intensities might be necessary for line-mounted equipment because any low-frequency resonances that might be present in the mounting structure (which is relatively flexible in this case) could amplify the excitations before reaching the equipment.

Floor-mounted equipment often requires relatively wide-band random test excitations. As an example, consider a pressure transducer mounted on (1) a rigid wall, (2) a rigid I-section frame, (3) a pressurized gasline, or (4) a cabinet. In cases (1) and (2), wide-band random excitations with response spectra approximately equal to the floor-response spectra could be employed for vibration testing of the pressure transducer. For cases (3) and (4), however, flexibility of the support structure should be taken into consideration in developing the required response spectra (RRS) specifications for vibration testing. In case (3), a single-frequency deterministic test, such as a sine-beat test or a sine-dwell test, can be employed, giving sufficient attention to testing at the equipment-resonant frequencies. In case (4), single-frequency tests can also be employed if the cabinet is considerably

Vibration Testing

flexible and not rigidly attached to a rigid structure (a floor or a wall). Alternatively, a wide-band test on the cabinet itself, with the pressure transducer mounted on it, can be used.

Size, complexity, assembly level, and related features of a test object can significantly complicate and extend the test procedure. In such cases, testing the entire assembly might not be practical and testing of individual components or subassemblies might not be adequate because, in the in-service dynamic environment, the motion of a particular component could be significantly affected by the dynamics of other components in the assembly, the mounting structure, and other interface subsystems.

Functional operability parameters to be monitored during testing should be predetermined. They depend on the purpose of the test, the nature of the test object, and the availability and characteristics of the sensors that are required to monitor these parameters. Malfunction or failure criteria should be related in some way to the monitored operability parameters; that is, each operability parameter should be associated with one or several components in the test object that are crucial to its operation.

The decision of whether to perform an active test (e.g., whether a valve should be cycled during the test) and determination of the actuation time requirements (e.g., the number of times the valve is cycled and at what instants during the test) should be made at this stage. The loading conditions for the test (i.e., in-service loading simulation) should also be defined.

An important nontechnical factor that determines the nature of a vibration test is the availability of hardware (test apparatus) in the test laboratory. This is especially true when nonconventional vibration tests are required. Some specifications require three-degree-of-freedom test inputs, for example, but most test laboratories have only one-degree-of-freedom or two-degree-of-freedom test machines. When two-degree-of-freedom or one-degree-of-freedom tests are used in place of three-degree-of-freedom tests, it is first required to determine what additional orientations of the test object should be tested in order to add the required conservatism. Also, it should be verified by analysis or testing that the modified series of tests does not cause significant undertesting or overtesting of certain parts of the test object. Otherwise, some other form of justification should be provided for replacing the test.

Test plans prepared in the pretest stage should include an adequate description of the following important items:

1. Test purpose
2. Test-object details
3. Test environment, specifications, and standards
4. Functional operability parameters and failure or malfunction criteria
5. Pretest inspection
6. Aging requirements
7. Test outline
8. Instrumentation requirements
9. Data-processing requirements
10. Methods of evaluation of the test results.

Testing of Cabinet-Mounted Equipment

In vibration testing of cabinet- or panel-mounted equipment, the following is standard procedure.

Step 1: Test the cabinet or panel with equipment replaced by a dummy weight.
Step 2: Obtain the cabinet response at equipment-mounting locations and, based on these observations, develop the required vibration environment for testing (the RRS) the equipment.
Step 3: Test the equipment separately, using the excitations developed in step 2.

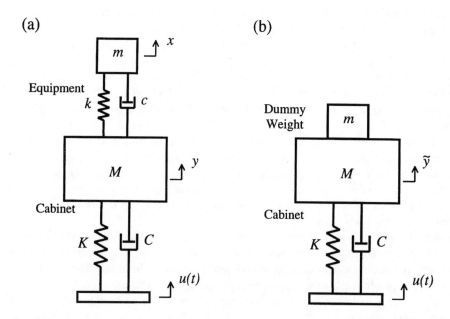

FIGURE 10.16 A simplified model for (a) an equipment cabinet test system, and (b) a dummy-weight cabinet test system.

This procedure may not be satisfactory if there is a considerable degree of dynamic interaction between the equipment and the mounting cabinet. This could be illustrated using a simplified model to represent cabinet-mounted equipment. The cabinet and the equipment are represented separately by single-degree-of-freedom systems, as shown in Figure 10.16. Cabinet parameters are represented by capital letters and equipment parameters by lower-case letters. The cabinet response when the equipment is replaced by a dummy weight of equal mass, is denoted by $\tilde{y}(t)$. The test excitation applied to the cabinet base is denoted by $u(t)$. It can be shown that the frequency-response ratio in the two cases is given by

$$\frac{\tilde{Y}(\omega)}{Y(\omega)} = \frac{\left[ms^2(cs+k)\right]/\left(ms^2+cs+k\right) + Ms^2 + Cs + K}{(M+m)s^2 + Cs + K} \qquad (10.77)$$

with $s = j\omega$. Using the nondimensional parameters defined by equations (10.71) through (10.75), one obtains

$$\frac{\tilde{Y}(\omega)}{Y(\omega)} = \frac{(j\overline{\omega})^2 \alpha(2\zeta\beta j\overline{\omega} + \beta^2) + (j\overline{\omega})^2 + 2Zj\overline{\omega} + 1}{(1+\alpha)(j\overline{\omega})^2 + 2Zj\overline{\omega} + 1} \qquad (10.78)$$

in which ζ denotes the equipment damping ratio and Z denotes the cabinet damping ratio.

The ratio $\tilde{Y}(\omega)/Y(\omega)$ represents the amplification in the cabinet response when the equipment is replaced by a dummy weight for a harmonic excitation. Figure 10.17 shows eight curves obtained for the ζ, α, β, and Z combinations, as given in Table 10.2. Notice that the most suitable response condition is obtained in curve 6, where the response of the actual system is the closest to the system with dummy weight. It can be concluded that a dummy test procedure for cabinet-mounted equipment is satisfactory when the equipment inertia and natural frequency are small in comparison to the values for the cabinet. Also, increasing the damping level has a favorable effect on test results.

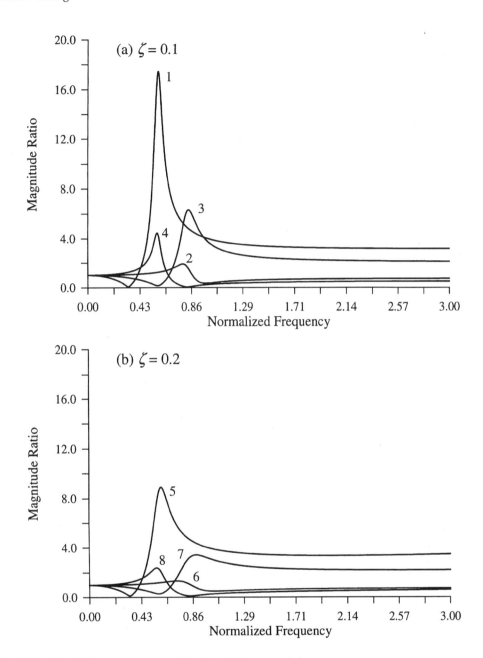

FIGURE 10.17 Cabinet response amplification in dummy-weight tests.

10.2.5 Pretest Inspection

Pretest inspection of a test object is important for at least two major reasons. First, if the equipment supplied for testing is different from the piece of equipment or the group of equipment that is required to be qualified, then these differences must be carefully observed and recorded in sufficient detail. In particular, deviations in the model number, mounting features, and other mechanical interface details, geometry, size, and significant dynamic features should be recorded. Second, before testing, the test object should be inspected for any damage, deficiencies, or malfunctions. Structural integrity usually can be determined by visual inspection alone. To determine malfunctions by operability monitoring, however, the test object must be actuated and the operating environment should be simulated.

If the equipment supplied for testing is not identical to that required to be tested, adequate justification must be provided for the differences to guarantee that the objectives of the test can be achieved by testing the equipment that is supplied. Otherwise, the test should be abandoned pending the arrival of the correct test object.

If any structural failure or operation malfunction is noted during pretest inspection, no corrective action should be taken by the test laboratory personnel unless those actions are notified to and fully authorized by the supplier of the test object. Otherwise, the test should be abandoned and the customer should be promptly notified of the anomalies.

It is important that the functional operability pretest inspection be performed in the same functional environment as that experienced under normal in-service conditions. When monitoring functional operability parameters, it is necessary to guarantee that the monitoring instrumentation meets the required accuracy. Instrumentation data should be provided to the customer for review. This ensures that the observed malfunction is real and not a false alarm caused by a malfunction in the monitoring instrumentation and channels. The monitoring-equipment accuracy should be higher than that required for the operability parameter itself.

Justification is needed if some components in the test object were not actuated and monitored during pretest inspection. Also, the warm-up period and the total time of actuation should be justified. In particular, if the proper operation of the equipment is governed by the continuity of a parameter (such as an electrical signal), the time duration of monitoring should be noted. If, however, the proper function is governed by a change of state (such as opening or closing a valve, switch, or relay), the number of cycles of actuation is important.

10.3 TESTING PROCEDURES

Vibration testing may involve pretesting prior to the main tests. The objectives of pretesting can be (1) exploratory, in order to obtain dynamic information such as natural frequencies, mode shapes, and damping about the test object; or (2) preconditioning, in order to age or pass the infant mortality stage (see Appendix E) so that the main test would be realistic and correspond to normal operating conditions. This section describes both pretesting and main testing in an integrated manner.

10.3.1 Resonance Search

Vibration test programs usually require a resonance-search pretest. This is typically carried out at a lower excitation intensity than that used for the main test in order to minimize the damage potential (overtesting). The primary objective of a resonance-search test is to determine resonant frequencies of the test object. More elaborate tests are employed, however, to determine mode shapes and modal damping ratios (see Chapter 11) in addition to resonant frequencies. Such frequency-response data on the test object are useful in planning and conducting the main test.

Frequency-response data are usually available as a set of complex frequency-response functions. There are tests that determine the frequency-response functions of a test object, and simpler tests are available to determine resonant frequencies alone. Some of the uses of frequency-response data include:

1. A knowledge of the resonant frequencies of the test object is important in conducting the main test. More attention should be given, for example, when performing a main test in the vicinity of resonant frequencies. In the resonance neighborhoods, lower sweep rates should be used if a sine sweep is used in the main test, and larger dwell periods

Vibration Testing

should be used if a sine dwell is part of the main test. Frequency-response data give the most desirable frequency range for conducting main tests.
2. From frequency-response data, it is possible to determine the most desirable test excitation directions and the corresponding input intensities.
3. The degree of nonlinearity and the time variance in system parameters of the test object can be estimated by conducting more than one frequency-response test at different excitation levels. If the deviation in the frequency-response functions thus obtained is sufficiently small, then a linear, time-invariant dynamic model is considered satisfactory in the analysis of the test object.
4. If no resonances are observed in the test object over the frequency range of interest, as determined by the operating environment for a given application, then a static analysis will be adequate to qualify the test object.
5. A set of frequency-response functions can be considered a dynamic model for the test specimen (see Chapter 11). This model can be employed in further studies of the test specimen by analytical means.

10.3.2 Methods of Determining Frequency-Response Functions

Three methods of determining frequency-response functions are outlined here.

Fourier Transform Method

If $y(t)$ is the response at location B of the test object, when a transient input $u(t)$ is applied at location A, then the frequency-response function $H(f)$ between locations A and B is given by the ratio of the Fourier integral transforms of the output $y(t)$ and the input $u(t)$:

$$H(f) = \frac{Y(f)}{U(f)} \qquad (10.79)$$

In particular, if $u(t)$ is a unit impulse, then $U(f) = 1$, and hence, $H(f) = Y(f)$.

Spectral Density Method

If the input excitation is a random signal, the frequency-response function between the input point and the output point can be determined as the ratio of the cross-spectral density $\Phi_{uy}(f)$ of the input $u(t)$ and the output $y(t)$, and the power spectral density $\Phi_{uu}(f)$ of the input:

$$H(f) = \frac{\Phi_{uy}(f)}{\Phi_{uu}(f)} \qquad (10.80)$$

Harmonic Excitation Method

If the input signal is sinusoidal (harmonic) with frequency f, the output will also be sinusoidal with frequency f at steady state, but with a change in the phase angle. Then, the frequency-response function is obtained as a magnitude function and a phase-angle function. The magnitude $|H(f)|$ = steady-state amplification of the output signal, and the phase angle $\angle H(f)$ = steady-state phase lead of the output signal. This pair of curves — the magnitude plot and the phase angle plot — is called a *Bode plot* or *Bode diagram*.

10.3.3 RESONANCE-SEARCH TEST METHODS

There are three basic types of resonance-search test methods. They are categorized according to the nature of the excitation used in the test: specifically, (1) impulsive excitation, (2) initial displacement, or (3) forced vibration. The first two categories are free-vibration tests; that is, response measurements are made on free decay of the test object following a momentary (initial) excitation. Typical tests belonging to each of these categories are described in the following subsections.

Hammer (Bump) Test and Drop Test

In resonance search by the impulsive-excitation method, an impulsive force (a large magnitude of force acting over a very short duration) is applied at a suitable location of the test object, and the resulting transient response of the object is observed, preferably at several locations. This is equivalent to applying an initial velocity to the test object and letting it vibrate freely. By Fourier analysis of the response data, it is possible to obtain the resonant frequencies, corresponding mode shapes, and modal damping.

Hammer tests and drop tests belong to the impulsive-excitation category. A schematic diagram of the hammer test arrangement is shown in Figure 10.18. A schematic diagram of the drop test arrangement is shown in Figure 10.19. The angle of swing of the hammer or the drop height of the object determines the intensity of the applied impulse. Alternatively, the impulse can be generated by explosive cartridges (for relatively large structures) located suitably in the test object, or by firing small projectiles at the test object. The response is monitored at several locations of the test object. The response at the point of application of the impulse is always monitored. Response analysis can be done in real-time, or the response can be recorded for subsequent analysis. A major concern in these tests is making sure that all significant resonances in the required frequency range are excited under the given excitation. Several tests for different configurations of the test object might be necessary to achieve this.

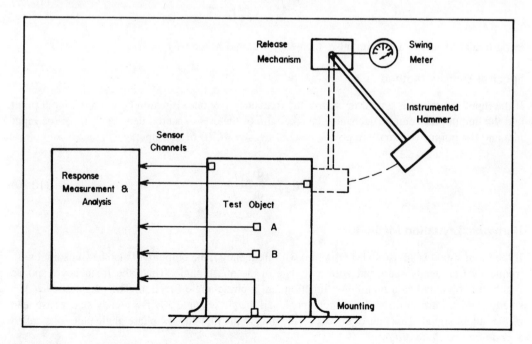

FIGURE 10.18 Schematic diagram of a hammer test arrangement.

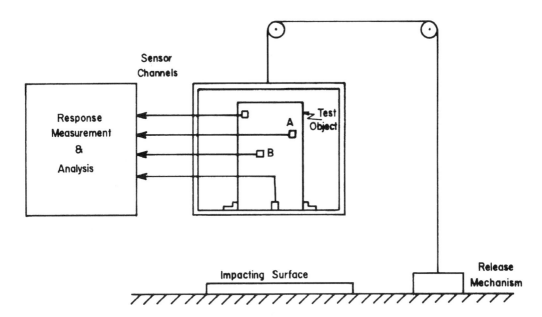

FIGURE 10.19 Schematic diagram of a drop test arrangement.

Proper selection of the response-monitoring locations is also important in obtaining meaningful test results. By changing the impulsive-force intensity and repeating the test, any significant nonlinear (or time-variant-parameter) behavior of the test object can be determined. A common practice is to monitor the impulsive force signal during impact. In this way, poor impacts (e.g., low-intensity impacts, multiple impacts caused by bouncing back of the hammer) can be detected, and the corresponding test results can be rejected. The impacting surface as well, may be instrumented to measure the test signal.

Pluck Test

Resonance search on a test object can be performed by applying a displacement initial condition (rather than a velocity initial condition, as in impulsive tests) to a suitably mounted test object and measuring its subsequent response at various locations as it executes free vibrations. By properly selecting the locations and the magnitudes of the initial displacements, it is sometimes possible to excite various modes of vibration, provided that these modes are reasonably uncoupled.

The pluck test is the most common test that uses the initial-displacement method. A schematic diagram of the test setup is shown in Figure 10.20. The test object is initially deflected by pulling it with a cable. When the cable is suddenly released, the test object will undergo free vibrations about its static-equilibrium position. The response is observed for several locations of the test object and analyzed to obtain the required parameters.

In Figures 10.18 through 10.20, the frequency-response function between two locations (A and B, for example) is obtained by analyzing the corresponding two signals, using either the Fourier transform method [equation (10.79)] or the spectral-density method, [equation (10.80)]. These frequency-domain techniques will automatically provide the natural-frequency and modal-damping information. Alternatively, modal damping can be determined using time-domain methods — for example, by evaluating the logarithmic decrement of the response after passing it through a filter having a center frequency adjusted to the predetermined natural frequency of the test object for that mode. Accuracy of the estimated modal-damping value can be improved significantly by such filtering methods.

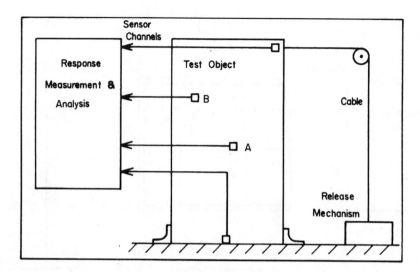

FIGURE 10.20 Diagram of a pluck test arrangement.

Often, the most difficult task in a natural-frequency search is the excitation of a single mode. If two natural frequencies are close together, modal interactions of the two will be present inevitably in the response measurements. Because of the closeness of the frequencies, the response curve will display a beat phenomenon, as shown in Figure 10.21, which makes it difficult to determine damping by the logarithmic-decrement method. It is difficult to distinguish between decay caused by damping and rapid drop-off caused by beating. In this case, one of the frequency components must be filtered out, using a very narrow bandpass filter, before computing damping.

The required testing time for the impulsive-excitation and initial-displacement test methods is relatively small in comparison to forced-vibration test durations. For this reason, these former (free-vibration) tests are often preferred in preliminary (exploratory) testing before conducting the main tests. Directions and locations of impact or initial displacements should be properly chosen, however, so that as many significant modes as possible are excited in the desired frequency range. If the impact is applied at a node point (see Chapters 5 and 6) of a particular mode, for example, it will be virtually impossible to detect that mode from the response data. Sometimes, a large number of monitoring locations will be necessary to accurately determine mode shapes of the test object. This depends primarily on the size and dynamic complexity of the test object and the particular mode number. This, in turn, necessitates the use of more sensors (accelerometers and the like) and recorder channels. If a sufficient number of monitoring channels is not available, the test will have to be repeated each time, using a different set of monitoring locations. Under such circumstances, it is advisable to keep one channel (monitoring location) unchanged and to use it as the reference channel. In this manner, any deviations in the test-excitation input can be detected for different tests and properly adjusted or taken into account in subsequent analysis (e.g., by normalizing the response data).

Shaker Tests

A convenient method of resonance search is through the use of continuous excitation. A forced excitation, which typically is a sinusoidal signal or a random signal, is applied to the test object by means of a shaker, and the response is continuously monitored. The test setup is shown schematically in Figure 10.22. For sinusoidal excitations, signal amplification and phase shift over a range of excitations will determine the frequency-response function. For random excitations, equation (10.80) can be used to determine the frequency-response function.

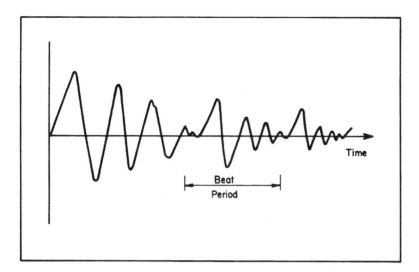

FIGURE 10.21 Beat phenomenon resulting from interaction of closely spaced modes.

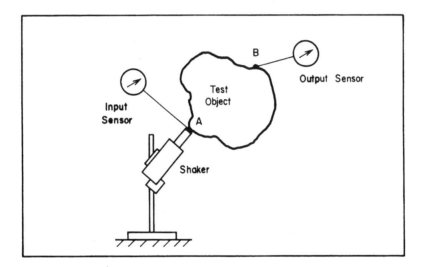

FIGURE 10.22 Schematic diagram of a shaker test for resonance search.

One or several portable exciters (shakers), or a large shaker table similar to that used in the main vibration test, can be employed to excite the test object. The number and the orientations of the shakers, and the mounting configurations and monitoring locations of the test object, should be chosen depending on the size and complexity of the test object, the required accuracy of the resonance-search results, and the modes of vibration that need to be excited. The shaker test method has the advantage of being able to control the nature of the test-excitation input (e.g., frequency content, intensity, sweep rate), although it might be more complex and costly. The results from shaker tests are relatively more accurate and more complete.

Test objects usually display a change in resonant frequencies when the shaker amplitude is increased. This is caused by inherent nonlinearities in complex structural systems. Usually, the change appears as a spring-softening effect, which results in lower resonant frequencies at higher shaker amplitudes. If this nonlinear effect is significant, resonant frequencies for the main test level cannot be accurately determined using a resonance search at low intensity. Some form of extrapolation of the test results, or analysis using an appropriate dynamic model, may be necessary in this case to determine the resonant-frequency information that is required to perform the main test.

10.3.4 MECHANICAL AGING

Before performing a qualification test, it is usually necessary to age the test object to put it into a condition that represents the state following its operation for a predetermined period under in-service conditions. In this manner, it is possible to reduce the probability of burn-in failure (infant mortality) during testing. Some tests, such as design development tests and quality assurance tests, do not require prior aging.

The nature and degree of aging that is required depends on such factors as the intended function of the test object, the operating environment, and the purpose of the dynamic test. In qualification tests, it might be necessary to demonstrate that the test object still has adequate capability to withstand an extreme dynamic environment toward the end of its design life (i.e., the period in which it can be safely operated without requiring corrective action). In such situations, it is required to age the test object to an extreme deterioration state, representing the end of the design life of the test object.

Test objects are aged by subjecting them to various environmental conditions (e.g., high temperatures, radiation, humidity, vibrations). Usually, it is not practical to age the equipment at the same rate as it would age under a normal service environment. Consequently, accelerated aging procedures are used to reduce the test duration and cost. Furthermore, the operating environment may not be fully known at the testing stage. This makes the simulation of the true operating environment virtually impossible. Usually, accelerated aging is done in sequence by subjecting the test equipment to the various environmental conditions one at a time. Under in-service conditions, however, these effects occur simultaneously, with the possibility of interactions between different effects. Therefore, when sequential aging is employed, some conservatism should be added. The type of aging used should be consistent with the environmental conditions and operating procedures of the specific application of the test object. Often, these conditions are not known in advance. In that case, standardized aging procedures should be used.

The main concern in this subsection is mechanical aging, although other environmental conditions can significantly affect the dynamic characteristics of a test object. The two primary mechanisms of mechanical aging are *material fatigue* and *mechanical wearout*. The former mechanism plays a primary role if in-service operation consists of cyclic loading over relatively long periods of time. Wearout, however, is a long-term effect caused by any type of relative motion between components of the test object. It is very difficult to analyze component wearout, even if only the mechanical aspects are considered (i.e., neglecting the effects of corrosion, radiation, and the like). Some mechanical wearout processes resemble fatigue aging, however, depending simultaneously on the number of cycles of load applications and the intensity of the applied load. Consequently, only the cumulative damage phenomenon, which is related to material fatigue, is usually treated in the literature.

Although mechanical aging is often considered a pretest procedure (e.g., a resonance-search test), it actually is part of the main test. If, in a dynamic qualification program, the test object malfunctions during mechanical aging, this amounts to failure in the qualification test. Furthermore, exploratory tests, such as resonance-search tests, are sometimes conducted at higher intensities than what is required to introduce mechanical aging into the test object. Some analytical concepts that are useful are given in Appendix E.

Equivalence for Mechanical Aging

It is usually not practical to age a test object under its normal operating environment, primarily because of time limitations and the difficulty in simulating the actual operating environment.

Vibration Testing

Therefore, it may be necessary to subject the test object to an accelerated aging process in a dynamic environment of higher intensity than that present under normal operating conditions.

Two aging processes are said to be equivalent if the final aged condition attained by the two processes is identical. This is virtually impossible to realize in practice, particularly when the object and the environment are complex and the interactions of many dynamic causes have to be considered. In this case, a single most severe aging effect is used as the standard for comparison to establish the equivalence. The equivalence should be analyzed in terms of both the intensity and the nature of the dynamic excitations used for aging.

Excitation-Intensity Equivalence

A simplified relationship between the dynamic-excitation intensity U and the duration of aging T that is required to attain a certain level of aging, keeping the other environmental factors constant, can be given as

$$T = \frac{c}{U^r} \tag{10.81}$$

in which c is a proportionality constant and r is an exponent. These parameters depend on such factors as the nature and sequence of loading and characteristics of the test object. It follows from equation (10.81) that, by increasing the excitation intensity by a factor n, the aging duration can be reduced by a factor of n^r. In practice, however, the intensity–time relationship is much more complex, and caution should be exercised when using equation (10.81). This is particularly so if the aging is caused by multiple dynamic factors of varying characteristics that are acting simultaneously. Furthermore, there is usually an acceptable upper limit to n. It is unacceptable, for example, to use a value that would produce local yielding or any such irreversible damage to the equipment that is not present under normal operating conditions.

It is not necessary to monitor functional operability during mechanical aging. Furthermore, it might happen that, during accelerated aging, the equipment malfunctions but, when the excitation is removed, it operates properly. This type of reversible malfunction is acceptable in accelerated aging.

The time to attain a given level of aging is usually related to the stress level at a critical location of the test object. Because this critical stress can be related, in turn, to the excitation intensity, the relationship given by equation (10.81) is justified.

Dynamic-Excitation Equivalence

Equivalence of two dynamic excitations that have different time histories can be expressed by using methods employed to represent dynamic excitations (e.g., response spectrum, Fourier spectrum, and power spectral density). If the maximum (peak) excitation is the factor that primarily determines aging in a given system under a particular dynamic environment, then response-spectrum representation is well suited for establishing the equivalence of two excitations. If, however, the frequency characteristics of the excitation are the major determining factor for mechanical aging, then Fourier spectrum representation is favored for establishing the equivalence of two deterministic excitations, and power spectral density representation is suited for random excitations. When two excitation environments are represented by their respective psd functions $\Phi_1(\omega)$ and $\Phi_2(\omega)$, and if the significant frequency range for the two excitations is (ω_1, ω_2), then the degree of aging under the two excitations can be compared using the ratio

$$\frac{A_1}{A_2} = \frac{\int_{\omega_1}^{\omega_2} \Phi_1(\omega) d\omega}{\int_{\omega_1}^{\omega_2} \Phi_2(\omega) d\omega} \tag{10.82}$$

where A denotes a measure of aging. If the two excitations have different frequency ranges of interest, a range consisting of both ranges can be selected for the integrations in equation (10.82).

Cumulative Damage Theory

Miner's linear cumulative damage theory can be used to estimate the combined level of aging resulting from a set of excitation conditions. Consider m excitations acting separately on a system. Suppose that each of these excitations produces a unit level of aging in N_1, N_2, ..., N_m loading cycles, respectively, when acting separately. If, in a given dynamic environment, n_1, n_2, ..., n_m loading cycles, respectively, from the m excitations have actually been applied to the system (possibly all excitations were acting simultaneously), the level of aging attained can be given by

$$A = \sum_{i=1}^{m} \frac{n_i}{N_i} \tag{10.83}$$

The unit level of aging is achieved, theoretically, when $A = 1$. Equation (10.83) corresponds to *Miner's linear cumulative damage theory*.

Because of various interactive effects produced by different loading conditions, when some or all of the m excitations act simultaneously, it is usually not necessary to have $A = 1$ under the combined excitation to attain the unit level of aging. Furthermore, it is extremely difficult to estimate N_i, $i = 1, ..., m$. For such reasons, the practical value of A in equation (10.83) to attain a unit level of aging can vary widely (typically, from 0.3 to 3.0).

10.3.5 TRS Generation

A vibration test can be specified by a required response spectrum (RRS). Then, the response spectrum of the actual excitation signal — that is, the *test response spectrum* (TRS) — should envelop the RRS during testing. It is customary for the purchaser (owner of the test object) to provide the test laboratory with a multichannel FM tape or some form of stored signal containing the components of the excitation signal that should be used in the test. Alternatively, the purchaser can request that the test laboratory generate the required signal components under the purchaser's supervision. If sine beats are combined to generate the test excitations, each FM tape should be supplemented by tabulated data, giving the channel number, the beat frequencies (hertz) in that channel, and the amplitude (g) of each sine-beat component. The RRS curve that is enveloped by the particular input should also be specified.

The excitation signal applied to the shaker-table actuator is generated by combining the contents of each channel in an appropriate ratio so that the response spectrum of the excitation that is actually felt at the mounting locations of the test object (the TRS) satisfactorily matches the RRS supplied to the test laboratory. Matching is performed by passing the contents of each channel through a variable-gain amplifier and mixing the resulting components according to variable proportions. This is known as equalization. These operations are performed by a waveform mixer. The adjustment of the amplifier gains is done by trial and error. The phase of the individual signal components should be maintained during the mixing process.

Each channel can contain a single-frequency component (such as sine beat) or a multifrequency signal of fixed duration (e.g., 20 s). If the RRS is complex, each channel may have to carry a multifrequency signal to achieve close matching of the TRS with the RRS. Also, a large number of channels might be necessary. The test excitation signal is generated continuously by repetitively playing the FM tape loop of fixed duration.

In product qualification, response spectra are usually specified in units of acceleration due to gravity (g). Consequently, the contents in each channel of the test-input FM tape represent acceleration motions. For this reason, the signal from the waveform mixer must be integrated twice before using it to drive the shaker table. The actuator of the exciter is driven by this displacement signal, and its control may be done by feedback from a displacement sensor. But, if the control sensor is an accelerometer, as typical, double intergration of that signal will be needed as well. In typical test facilities, a double integration unit is built into the shaker system. It is then possible to use any type of signal (displacement, velocity, acceleration) as the excitation input and to decide simultaneously on the number of integrations necessary. If the input signal is a velocity time history, for example, one integration should be chosen, etc.

The tape speed should be specified (e.g., 7.5 in·s^{-1}, 15 in·s^{-1}) when signals recorded on tapes are provided to generate input signals for vibration testing. This is important, so that the frequency content of the signal is not distorted. Speeding up of the tape has the effect of scaling up each frequency component in the signal. It also has the effect, however, of filtering out very high frequency components in the signal. If the excitation signals are available as digital records, then a digital-to-analog converter (DAC) is needed to convert them into analog signals.

10.3.6 Instrument Calibration

The test procedure normally stipulates accuracy requirements and tolerances for various critical instruments that are used in testing. It is desirable that these instruments have current calibration records that are agreeable to an accepted standard. Instrument manufacturers usually provide these calibration records. Accelerometers, for example, may have calibration records for several temperatures (e.g., –65°F, 75°F, 350°F) and for a range of frequencies (e.g., 1 Hz to 1000 Hz). Calibration records for accelerometers are given in both voltage sensitivity (mV·g^{-1}) and charge sensitivity (pC·g^{-1}), along with percentage deviation values. These tolerances and peak deviations for various test instruments should be provided for the purchaser's review before they are used in the test apparatus.

From the tolerance data for each sensor or transducer, it is possible to estimate peak error percentages in various monitoring channels in the test setup, particularly in the channels used for functional operability monitoring. The accuracy associated with each channel should be adequate to measure expected deviations in the monitored operability parameter.

It is good practice to calibrate sensor or transducer units, such as accelerometers and associated auxiliary devices, daily or after each test. These calibration data should be recorded using various scales when a particular instrument has multiple scales, and for various instrument settings.

10.3.7 Test-Object Mounting

When a test object is being mounted on a shaker table, care should be taken to simulate all critical interface features under normal installed conditions for the intended operation. This should be done as accurately as is feasible. Critical interface requirements are those that could significantly affect the dynamics of the test object. If the mounting conditions in the test setup significantly deviate from those under installed conditions for normal operation, adequate justification should be provided to show that the test is conservative (i.e., the motions produced under the test mounting conditions are more severe). In particular, local mounting that would not be present under normal installation conditions should be avoided in the test setup.

In simulating in-service interface features, the following details should be considered as a minimum:

1. Test orientation of the test object should be its in-service orientation, particularly with respect to the direction of gravity (vertical), available degrees of freedom, and mounting locations.
2. Mounting details at the interface of the test object and the mounting fixture should represent in-service conditions with respect to the number, size, and strength of welds, bolts and nuts, and other hold-down hardware.
3. Additional interface linkages, such as wires, cables, conduits, pipes, instrumentation (dials, meters, gauges, sensors, transducers, etc.), and their supporting brackets should be simulated at least in terms of mass and stiffness, and preferably in terms of size as well.
4. Any dynamic effects of adjacent equipment cabinets, and supporting structures under in-service conditions, should be simulated or taken into account in analysis.
5. Operating loads, such as those resulting from fluid flow, pressure forces, and thermal effects, should be simulated if they appear to significantly affect test object dynamics. In particular, the nozzle loads (fluid) should be simulated in magnitude, direction, and location.

The required mechanical interface details of the test object are obtained by the test laboratory at the information-acquisition stage. Any critical interface details that are simulated during testing should be included in the test report.

At least three control accelerometers should be attached to the shaker table in the neighborhood of the mounting location of the test object. One control accelerometer measures the excitation-acceleration component applied to the test object in the vertical direction. The other two measure the excitation-acceleration components in two horizontal directions at right angles. The two horizontal (control) directions are chosen to be along the two major freedom-of-motion directions (or dynamic principal axes) of the test object. Engineering judgment should be used in deciding these principal directions of high response in the test object. Often, geometric principal axes are used. The control accelerometer signals are passed through a response-spectrum analyzer (or a suitably programmed digital computer) to compute the TRS in the vertical and two horizontal directions that are perpendicular.

Vibration tests generally require monitoring of the dynamic response at several critical locations of the test object. In addition, the tests may call for determining the mode shapes and natural frequencies of the test object. For this purpose, a sufficient number of accelerometers should be attached to various key locations in the test object. The test procedure (document) should carry a sketch of the test object, indicating the accelerometer locations. Also, the type of accelerometers employed, their magnitudes and directions of sensitivity, and the tolerances should be included in the final test report.

10.3.8 Test-Input Considerations

In vibration testing, a significant effort goes into the development of test excitation inputs. Not only the nature, but also the number and directions of the excitations can have a significant effect on the outcomes of a test. This is so because the excitation characteristics determine the nature of a test.

Test Nomenclature

A common practice in vibration testing is to apply synthesized vibration excitations to a test object that is appropriately mounted on a shaker table. Customarily, only translatory excitations, as generated by linear actuators, are employed. Nevertheless, the resulting motion of the test object usually consists of rotational components as well. A typical vibration environment might consist of three-dimensional motions, however. The specification of a three-dimensional test environment is a complex task, even after omitting the rotational motions at the mounting locations of the test object. Furthermore, practical vibration environments are random and their representation with sufficient accuracy can be done only in a probabilistic sense.

Very often, the type of testing used is governed mainly by the capabilities of the test laboratory to which the contract is granted. Test laboratories conduct tests using their previous experience and engineering judgment. Making extensive improvements to existing tests can be very costly and time-consuming, and this is not warranted from the point of view of the customer or the vendor. Regulatory agencies usually allow simpler tests if sufficient justification can be provided that indicates that a particular test is conservative with respect to regulatory requirements.

Complexity of a shaker table apparatus is governed primarily by the number of actuators employed and the number of independent directions of simultaneous excitation it is capable of producing. Terminology for various tests is based on the number of independent directions of excitation used in the test. It is advantageous to standardize this terminology to be able to compare different test procedures. Unfortunately, the terminology used to denote different types of tests usually depends on the particular test laboratory and the specific application. Attempts to standardize various test methods have become tedious, partly because of the lack of a universal nomenclature for dynamic testing. A justifiable grouping of test configurations is presented in this section. Figure 10.23 illustrates the various test types.

In test nomenclature, the degree of freedom refers to the number of directions of independent motions that can be generated simultaneously by means of independent actuators in the shaker table. According to this concept, three basic types of tests can be identified:

1. Single-degree-of-freedom (or rectilinear) testing, in which the shaker table employs only one exciter (actuator), producing test-table motions along the axis of that actuator. The actuator may not necessarily be in the vertical direction.
2. Two-degree-of-freedom testing, in which two independent actuators oriented at right angles to each other are employed. The most common configuration consists of a vertical actuator and a horizontal actuator. Theoretically, the motion of each actuator can be specified independently.
3. Three-degree-of-freedom testing, in which three actuators, oriented at mutually right angles, are employed. A desirable configuration consists of a vertical actuator and two horizontal actuators. At least theoretically, the motion of each actuator can be specified independently.

It is common practice to specify the directions of excitation with respect to the geometric principal axes of the test object. This practice is somewhat questionable, primarily because it does not take into account the flexibility and inertia distributions of the object. Flexibility and inertia elements in the test object have a significant influence on the level of dynamic coupling present in a given pair of directions. In this respect, it is more appropriate to consider *dynamic principal axes* rather than geometric principal axes of the test object. One useful definition is in terms of eigenvectors of an appropriate three-dimensional frequency-response function matrix that takes into account the response at every critical location in the test object. The only difficulty in this method is that prior frequency-response testing or analysis is needed to determine the test input direction. For practical purposes, the vertical axis (direction of gravity) is taken as one principal axis.

The single-degree-of-freedom (rectilinear) test configuration has three subdivisions, based on the orientation of the vibration exciter (actuator) with respect to the principal axes of the test object. It is assumed that one principal axis of the test object is the vertical axis and that the three principal axes are mutually perpendicular. The three subdivisions are as follows:

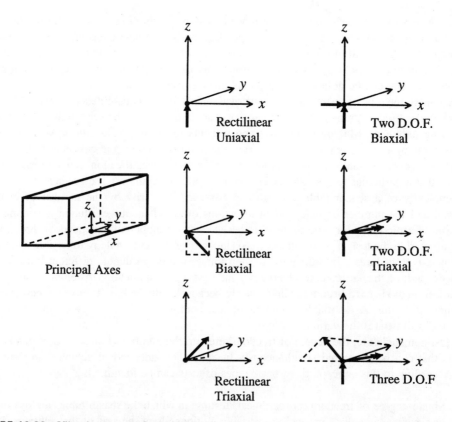

FIGURE 10.23 Vibration test configurations.

1. Rectilinear uniaxial testing, in which the single actuator is oriented along one of the principal axes of the test object.
2. Rectilinear biaxial testing, in which the single actuator is oriented on the principal plane containing the vertical and one of the two horizontal principal axes. The actuator is inclined to both principal axes in the principal plane.
3. Rectilinear triaxial testing, in which the single actuator is inclined to all three orthogonal principal axes of the test object.

The two-degree-of-freedom test configuration has two subdivisions, based on the orientation of the two actuators with respect to the principal axes of the test object, as follows:

1. Two-degree-of-freedom biaxial testing, in which one actuator is directed along the vertical principal axis and the other along one of the two horizontal principal axes of the test object.
2. Two-degree-of-freedom triaxial testing, in which one actuator is positioned along the vertical principal axis and the other actuator is horizontal but inclined to both horizontal principal axes of the test object.

Testing with Uncorrelated Excitations

Simultaneous excitations in three orthogonal directions often produce responses (accelerations, stresses, etc.) that are very different from what is obtained by vectorially summing the responses to separate excitations acting one at a time. This is primarily because of the nonlinear, time-variant nature of test specimens and test apparatus, their dynamic coupling, and the randomness of excitation signals. If these effects are significant, it is theoretically not possible to replace a three-degree-of-freedom test, for example, by a sequence of three single-degree-of-freedom tests. In practice, however, some conservatism can be incorporated into two-degree-of-freedom and single-degree-of-freedom tests to account for these effects. These tests with added conservatism can be employed when three-degree-of-freedom testing is not feasible. It should be clear by now that rectilinear triaxial testing is generally not equivalent to three-degree-freedom testing, because the former merely applies an identical excitation in all three orthogonal directions, with scaling factors (direction cosines). One obvious drawback of rectilinear triaxial testing is that the input excitation in a direction at right angles to the actuator is theoretically 0, and the excitation is maximum along the actuator. In three-degree-of-freedom testing using uncorrelated random excitations, however, no single direction has a zero excitation at all times, and also the probability is 0 that the maximum excitation occurs along a fixed direction at all times.

Three-degree-of-freedom testing is mentioned infrequently in the literature on vibration testing. A major reason for this lack of three-degree-of-freedom testing might be the practical difficulty in building test tables that can generate truly uncorrelated input motions in three orthogonal directions. The actuator interactions caused by dynamic coupling through the test table and mechanical constraints at the table supports are primarily responsible for this. Another difficulty arises because it is virtually impossible to synthesize perfectly uncorrelated random signal to drive the actuators. Two-degree-of-freedom testing is more common. In this case, the test must be repeated for a different orientation of the test object (e.g., with a 90° rotation about the vertical axis) unless some form of dynamic-axial symmetry is present in the test object.

Test programs frequently specify uncorrelated excitations in two-degree-of-freedom testing for the two actuators. This requirement lacks solid justification because two uncorrelated excitations applied at right angles do not necessarily produce uncorrelated components in a different pair of orthogonal directions, unless the mean square values of the two excitations are equal. To demonstrate this, consider the two uncorrelated excitations u and v shown Figure 10.24. The components u' and v' in a different pair of orthogonal directions, obtained by rotating the original coordinates through and angle θ in the counterclockwise direction, is given by:

$$u' = u\cos\theta + v\sin\theta \tag{10.84}$$

$$v' = -u\sin\theta + v\cos\theta \tag{10.85}$$

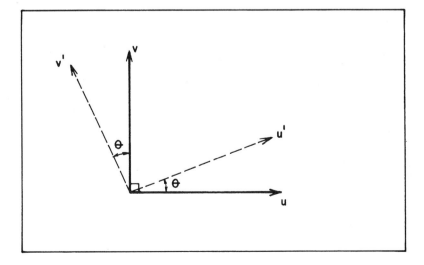

FIGURE 10.24 Effect of coordinate transformation on correlation.

Without loss of generality, one can assume that u and v have zero means. Then, u' and v' will also have zero means. Furthermore, since u and v are uncorrelated, one has

$$E(uv) = E(u)E(v) = 0 \tag{10.86}$$

From equations (10.84) and (10.85), one obtains

$$E(u'v') = E[(u\cos\theta + v\sin\theta)(-u\sin\theta + v\cos\theta)]$$

which, when expanded and substituted with equation (10.86), becomes

$$E(u'v') = \sin\theta\cos\theta\left[E(v^2) - E(u^2)\right] \tag{10.87}$$

Since θ is any general angle, the excitation components u' and v' become uncorrelated if and only if

$$E(v^2) = E(u^2) \tag{10.88}$$

This is the required result. Nevertheless, a considerable effort of digital Fourier analysis is expended by vibration testing laboratories to determine the degree of correlation in test signals employed in two-degree-of-freedom testing.

Symmetrical Rectilinear Testing

Single-degree-of-freedom (rectilinear) testing, which is performed with the test excitation applied along the line of symmetry with respect to an orthogonal system of three principal axes of the test object mainframe, is termed symmetrical rectilinear testing. In product qualification literature, this test is often referred to as the 45° test. The direction cosines of the input orientation are $\left(1/\sqrt{3},\ 1/\sqrt{3},\ 1\sqrt{3}\right)$ for this test configuration. The single-actuator input intensity is amplified by a factor of $\sqrt{3}$ in order to obtain the required excitation intensity in the three principal directions. Note that symmetrical rectilinear testing falls into the category of rectilinear triaxial testing, as defined earlier. This is one of the widely used testing configurations in seismic qualification, for example.

Geometry versus Dynamics

In vibration testing, the emphasis is on the dynamic behavior rather than the geometry of the equipment. For a simple three-dimensional body that has homogeneous and isotropic characteristics, it is not difficult to correlate the geometry to its dynamics. A symmetrical rectilinear test makes sense for such systems. The equipment one comes across is often much more complex, however. Furthermore, one's interest is not merely in determining the dynamics of the mainframe of the equipment; one is more interested in the dynamic reliability of various critical components located within the mainframe. Unless one has some previous knowledge of the dynamic characteristics in various directions of the system components, it is not possible to draw a direct correlation between the geometry and the dynamics of the tested equipment.

Vibration Testing

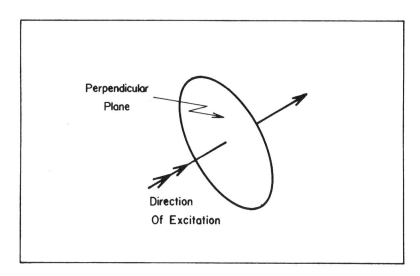

FIGURE 10.25 Illustration of the limitation of a single rectilinear test.

Some Limitations

In a typical symmetrical rectilinear test, one is dealing with black-box equipment whose dynamics are completely unknown. The excitation is applied along the line of symmetry of the principal axes of the mainframe. A single test of this type does not guarantee excitation of all critical components located inside the equipment. Figure 10.25 illustrates this further. Consider the plane perpendicular to the direction of excitation. The dynamic effect caused by the excitation is minimal along any line on this plane. (Any dynamic effect on this plane is caused by dynamic coupling among different body axes.) Accordingly, if there is a component (or several components) inside the equipment whose direction of sensitivity lies on this perpendicular plane, the single excitation might not adequately excite that component. Since one is dealing with a black box, one does not know the equipment dynamics beforehand. Hence, there is no way of identifying the existence of such unexcited components. When the equipment is put into service, a vibration of sufficient intensity can easily overstress this component along its direction of sensitivity and may bring about component failure. It is apparent that at least three tests, performed in three orthogonal directions, are necessary to guarantee excitation of all components, regardless of their direction of sensitivity.

A second example is given in Figure 10.26. Consider a dual-arm component with one arm sensitive in the O-O direction and the second arm sensitive in the P-P direction. If component failure occurs when the two arms are in contact, a single excitation in either the O-O direction or the P-P direction will not bring about component failure. If the component is located inside a black box, such that either the O-O direction or the P-P direction is very close to the line of symmetry of the principal axes of the mainframe, a single symmetrical rectilinear test will not result in system malfunction. This may very well be true, because one does not have a knowledge of component dynamics in such cases. Again, under service conditions, a vibration of sufficient intensity can produce an excitation along the A-A direction, subsequently causing system malfunction.

A further consideration in using rectilinear testing is dynamic coupling between the directions of excitation. In the presence of dynamic coupling, the sum of individual responses of the test object resulting from four symmetrical rectilinear tests is not equal to the response obtained when the excitations are applied simultaneously in the four directions. Some conservatism should be introduced when employing rectilinear testing for objects having a high level of dynamic coupling between the test directions. If the test-object dynamics are restrained to only one direction under normal operating

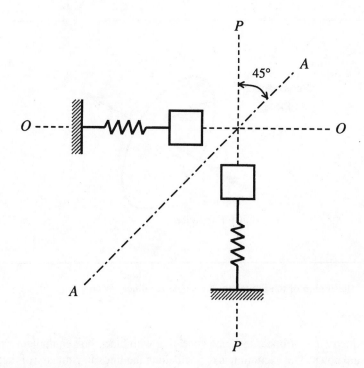

FIGURE 10.26 Illustrative example of the limitation of several rectilinear tests.

conditions, however, then rectilinear testing becomes more realistic and can be used without applying any conservatism.

Testing of Black Boxes

When the equipment dynamics are unknown, a single rectilinear test does not guarantee proper testing of the equipment. To ensure excitation of every component within the test object that has directional sensitivities, three tests should be carried out along three independent directions. The first test is carried out with a single horizontal excitation, for example. The second test is performed with the equipment rotated through 90° about its vertical axis, and using the same horizontal excitation. The last test is performed with a vertical excitation.

Alternatively, if symmetrical rectilinear tests are preferred, four such tests should be performed for four equipment orientations (e.g., an original test, a 90° rotation, a 180° rotation, and a 270° rotation about the vertical axis). These tests also ensure excitation of all components that have directional sensitivities. This procedure might not be very efficient, however. The shortcoming of this series of four tests is that some of the components would be overtested. It is clear from Figure 10.27, for example, that the vertical direction is excited by all four tests. The method has the advantage, however, of simplicity of performance.

Phasing of Excitations

The main purpose of rotating the test orientation in rectilinear testing is to ensure that all components within the equipment are excited. Phasing of different excitations also plays an important role, however, when several excitations are used simultaneously. To explore this concept further, it should be noted that a random input applied in the A-B direction or in the B-A direction has the same frequency and amplitude (spectral) characteristics. This is clear because the psd of u = psd of $(-u)$ and the autocorrelation of u = autocorrelation of $(-u)$. Hence, it is seen that, if the test is performed

Vibration Testing

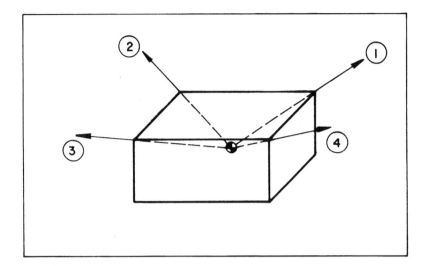

FIGURE 10.27 Directions of excitation in a sequence of four rectilinear tests.

along the A-B direction, it is of no use to repeat the test in the B-A direction. It should be understood, however, that the situation is different when several excitations are applied simultaneously.

The simultaneous action of u and v is not the same as the simultaneous action of $-u$ and v (see Figure 10.28). The simultaneous action of u and v is the same, however, as the simultaneous action of $-u$ and $-v$. Obviously, this type of situation does not arise when there are no simultaneous excitations, as in rectilinear testing.

Testing a Gray or White Box

When some information regarding the true dynamics of the test object is available, it is possible to reduce the number of necessary tests. In particular, if the equipment dynamics are completely known, then a single test will be adequate. The best direction for excitation of the system in Figure 10.26, for example, is A-A. (Note that A-A can be lined up in any arbitrary direction inside the equipment housing. In such a situation, a knowledge of the equipment dynamics is crucial.) This also indicates that it is very important to accumulate and use any past experience and data on the dynamic behavior of similar equipment. Any test that does not use some previously known information regarding the equipment is a blind test, and it cannot be optimal. As more and more information becomes available, better and better tests can be conducted.

Overtesting in Multitest Sequences

It is well known that increasing the test duration increases aging of the test object, because of prolonged stressing and load cycling of various components. This is the case when a test is repeated one or more times at the same intensity as that prescribed for a single test. The symmetrical rectilinear test requires four separate tests at the same excitation intensity as that prescribed for a single test. As a result, the equipment becomes subjected to overtesting, at least in certain directions. The degree of overtesting is small if the tests are performed in only three orthogonal directions. In any event, a certain amount of dynamic coupling is present in the test-object structure; and, to minimize overtesting in these sequential tests, a smaller intensity than that prescribed for a single test should be employed. The value of the intensity-reduction factor clearly depends on the characteristics of the test object, the degree of reliability expected, and the intensity value itself. More research is necessary to develop expressions for intensity-reduction factors for various test objects.

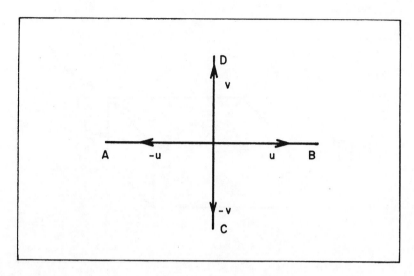

FIGURE 10.28 Significance of excitation phasing in two-degree-of-freedom testing.

10.4 PRODUCT QUALIFICATION TESTING

Vibration testing is used in product qualification. Here, the objective is to test the adequacy of a product of good quality for a specific use, in a typical operating environment. Clearly, the nature of the testing and the test requirements, including test specifications, will depend on the type of application and the class of product. This section considers just two types of product qualification: distribution qualification and seismic qualification. Procedures of vibration testing for other types of qualification will be similar.

10.4.1 Distribution Qualification

The term "distribution qualification" is used to denote the process by which the ability of a product to withstand a clearly defined distribution environment is established. Dynamic effects on the product due to handling loads, characteristics of packaging, and excitations under various modes of transportation (truck, rail, air, and ocean) must be properly represented in the test specifications used for distribution qualification. If a product fails a qualification test, corrective measures and subsequent requalification are necessary prior to commercial distribution. Product redesign, packaging redesign, and modification of existing shipping procedures might be required to meet qualification requirements.

Often, the necessary improvements can be determined by analyzing data from prior tests. Proper distribution qualification will result in improved product quality (and associated reliability and performance), reduced wastage and inventory problems, cost-effective packaging, reduced shipping and handling costs, and reduced warranty and service costs.

Random testing can more accurately represent vibrations in distribution environments; some inherent characteristics make it superior to sine testing. A sine test is a single-frequency test; thus, only one frequency is applied to a test object at a given instant. As a result, failure modes caused by the simultaneous excitation of two or more modes of vibration cannot be realized by sine testing, at least under steady excitations. In random testing, on the other hand, many frequencies are simultaneously applied to the test object. Conditions are thus more conducive to multiple-mode excitations and associated complex failures. A comparison of testing with four types of excitation signals is given in Table 10.3.

TABLE 10.3
Comparison of Test Types

	Sine Testing	Random Testing	Narrow-Band Random Sweep	Sweep on Wide-Band Random
Simultaneous multimodal (multiresonant) excitation possible?	No	Yes	No	Yes
Test duration	Long	Short	Long	Moderate
Power requirements	Low	High	Low	High
Represents a random environment?	No	Yes	Yes	Yes
Test system cost	Low	High	Moderate to high	High
Overtesting possibility	High	Low	High	Low

Drive-Signal Generation

The first step in drive-signal synthesis is to assign independent and identically distributed, random phase angles to the digitized spectral magnitude (spectral lines) of the drive spectrum. The number of lines chosen is consistent with the fast Fourier transform algorithm that is employed (see Appendix D) and the desired numerical accuracy. The inverse Fourier transform is obtained from the resulting discrete, complex Fourier spectrum. In general, the signal so obtained would be neither ergodic nor Gaussian.

Stationarity can be attained by randomly shifting the signal with respect to time and summing the results. The resulting signal would be weakly ergodic as well. Ergodicity is improved by increasing the duration of the signal. To obtain Gaussianity, sufficiently large numbers of time-shifted signals must be summed as dictated by the central limit theorem. Furthermore, because the magnitude of a Gaussian signal almost always remains within three times its standard deviation (99.7% of the time), Gaussianity can be imposed simply by windowing the time-shifted signal. The amplitude of the window function is governed by the required standard deviation of the drive signal. Unwanted frequency components introduced as a result of sharp end transitions in each time-shifted signal component can be suppressed by properly shaping the window. This process introduces a certain degree of non-stationarity into the synthesized signal, particularly if the windowed signal segments are joined end to end to generate the drive signal. A satisfactory way to overcome this problem is to introduce a high overlap from one segment to the next. Because the processing time increases in proportion to the degree of overlap, however, a compromise must be reached.

In summary, for a given drive spectral magnitude, the drive signal can be synthesized as follows:

1. Assign independent, identically distributed, random phase values to the drive-spectral lines.
2. Perform an inverse Fourier transform of the resulting spectrum using FFT.
3. Generate a set of independent and identically distributed time-shift values.
4. Perform a time-shift of the signal obtained in step 2 using the values from step 3.
5. Window the time-shifted signals.
6. Join the windowed signals with a fixed overlap.

The resulting digital drive signal is converted into an analog signal using a digital-to-analog converter (DAC) and passed through a low-pass filter to remove any unwanted frequency components (see Chapter 9) before it is used to drive the shaker. This procedure is illustrated in Figure 10.29.

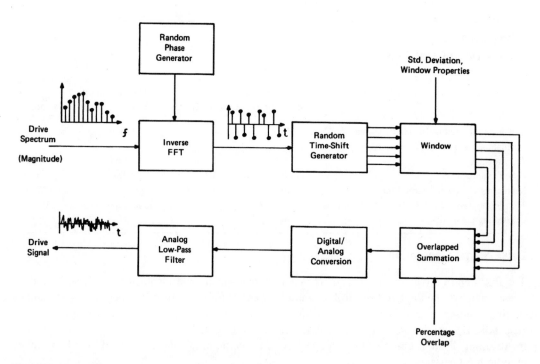

FIGURE 10.29 The synthesis of a random drive signal.

Distribution Spectra

The distribution environment to which a product is subjected depends on several factors. In particular, one must consider (1) the nature and severity of handling prior to and during shipment, (2) the mode of transportation (truck, rail, air cargo, ship), (3) geographic factors, (4) environmental conditions, (5) characteristics of the protective packaging used, and (6) dynamic characteristics of the product itself. These factors are complex and essentially random in nature. Laboratory simulation of such an environment is difficult even if a combination of several types of tests — for example, vibration, shock, drop, and thermal cycling — is employed. A primary difficulty arises from the requirement that test specifications should be simple yet accurately represent the true environment. The test must also be repeatable to allow standardization of the test procedure and to facilitate evaluation and comparison of test data. Finally, testing must be cost-effective.

During transportation, a package is subjected to multi-degree-of-freedom excitations that can include rectilinear and rotational excitations at more than one location simultaneously. However, test machines are predominantly single-axis devices that generate excitations along a single direction. Thus, any attempt to duplicate a realistic distribution environment in a laboratory setting can prove futile.

An alternative might be to use trial shipments. But, because of the random nature of the distribution environment, many such trials would be necessary before the data would be meaningful. Trial shipments are thus not appealing from a cost-benefit point of view and also because test control and data acquisition would be difficult. Data from trial shipments are extremely useful, however, in developing qualification-test specifications and in improving existing laboratory test procedures.

A more realistic goal of testing would be to duplicate possible failure and malfunction modes without actually reproducing the distribution environment. This, in fact, is the underlying principle of testing for distribution qualification. For example, sine tests can reproduce some types of failure caused during shipment although the test signal does not resemble the actual dynamic environment; however, a random test is generally superior.

Vibration Testing

Test specifications are expressed in terms of distribution spectra in distribution qualification where random testing is used. Specification development begins with a sufficient collection of realistic data. Sources of data include field measurements during trial shipments, computer simulations (e.g., Monte Carlo simulations), and previous specifications for similar products and environments. For best results, all possible modes of transportation, excitation levels, and handling severities should be included. The data, expressed as psd, must be reduced to a common scale — particularly with respect to the duration of excitation — for comparison purposes. Scaling can be accomplished by applying a similarity law based on a realistic damage criterion. For example, a similarity law might relate the excitation duration and the psd level such that the value of a suitable damage function would remain constant. Time-dependent damage criteria are developed primarily on the basis of fatigue-strength characteristics of a test product.

Due to nonlinearities of the environment, spectral characteristics (frequency content) will change with the excitation level. If such changes are significant, one should properly account for them. The influence of environmental conditions (temperature and humidity, for example) must be considered as well. The psd curves conditioned in this manner are plotted on a log–log plane to establish an envelope curve. This curve represents the worst composite environment that is typically expected. The envelope is then fitted with a small number of straight-line segments.

At this point, the psd curve should be scaled so that the root-mean-square (rms) value is equal to that before the straight-line segments were fitted. The resulting psd curve can be used as the test specification. Test duration can be established from the time-scaling criterion. If the corresponding test duration is excessively long, thereby making the test impractical, the test duration should be shortened by increasing the test level according to a realistic similarity criterion.

Product overtesting can be significant only if one reference spectrum is used to represent all possible distribution environments. Shipping procedures should thus be classified into several groups, depending on the dynamic characteristics of the shipping environment; and a representative reference spectrum should be determined for each group. In addition, reference spectra should be modified and classified according to product type if a range of products with significantly diverse dynamic characteristics is being qualified. At the testing stage, a reference spectrum must be chosen from a spectral data base, depending on the product type and the applicable shipping procedures. Alternatively, a general composite spectrum can be developed by assigning weights to a chosen set of reference spectra and computing the weighted sum.

Vibration levels in land vehicles and aircraft can range up to several kilohertz (kHz). Ships are known to have lower levels of excitation. In general, the energy content in vibrations experienced during the distribution of computer products is known to remain within 20 Hz. Consequently, the test specification spectra (reference spectra) used in distribution qualification are usually limited to this bandwidth. The typical specification curve shown in Figure 10.30 can be specified simply from the coordinates of the break points of the psd curve. Intermediate values can be determined easily because the break points are joined by straight-line segments on a log–log plane.

The area beneath the psd curve gives the required mean-square value of the test excitation. The square root of this value is the rms value; it is specified along with the psd curve, although it can be determined directly from the psd curve. An acceptable tolerance band for the control spectrum — usually ±3 dB — is also specified. Test duration should be supplied with the test specification.

Test Procedures

Dynamic test systems, with digital control, are easy to operate. In menu-driven systems, a routine or mode is activated by picking the appropriate item from a menu that is displayed on the CRT screen. The system then asks for necessary data, and then necessary parameter values are entered into the system. Lower and upper rms limits for test abort levels, breakpoint coordinates of the reference spectrum, and test duration are typically supplied by the user. The tolerance bands for

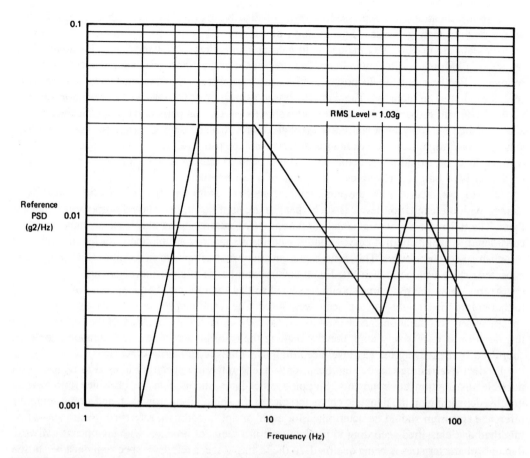

FIGURE 10.30 A reference spectrum for the distribution qualification of personal computers.

test spectrum equalization and the accelerometer sensitivities are also entered. More than one test setup can be stored; then a number is assigned to denote each test.

Preprogrammed tests can be modified, in the edit mode, using a similar procedure. Any preprogrammed test can be carried out simply by entering the corresponding test number. Computed results such as psd curves and transmissibility functions are stored for future evaluation. If desired, these results can be displayed, printed, or plotted with proper annotations and scales while the test is in progress.

The main steps of a typical test procedure are as follows:

1. Carefully examine the test object and record obvious structural defects, abnormalities, and hazardous or unsafe conditions.
2. Perform a functional test (i.e., operate the product) according to specifications, and record any malfunctions and safety hazards.
   Note: The test can be abandoned at this stage if the test object is defective.
3. Mount the test object rigidly on the shaker table, so that the loading points and the excitation axis are consistent with standard shipping conditions and the specified test sequence.
4. Perform an exploratory test at half the specified rms level (one fourth the specified psd level); monitor the response of the test package at critical locations including the control sensor location.
5. Perform the full-level test for the specified duration. Record the response data.
6. Change the orientation in accordance with the specified test sequence and repeat the test.

Vibration Testing

7. After the test sequence is completed, carefully inspect the test object and record any structural defects, abnormalities, and safety hazards.
8. Conduct a functional test and record any malfunctions, failures, and safety problems.

An exploratory test at a fraction of the specified test level is required for new product models that are being tested for the first time, or for older models that have been subjected to major design modifications. Three mutually perpendicular axes are usually tested, including the primary orientation (vertical axis) that is used for shipping. If product handling during distribution is automated, it is adequate to test only the primary axis.

When multiple tests are required, the test sequence is normally stipulated. If the test sequence is not specified, it can be chosen such that the least-severe orientation (orientation least likely to fail) is tested first. The test is repeated successively for the remaining orientations, ending with the most severe one. The rationale is that with this choice of test sequence, the aging of the most severe direction will be maximized, thereby making the test more reliable.

The test report should contain the following:

1. Description of the test object: Serial number, size (dimensions and weight), product function (e.g., system unit, hard drive, power supply, printer, keyboard, mouse, monitor, and floppy disk drive of a personal computer), and packaging particulars. Descriptive photos are useful.
2. Test plan: Usually standard and attached to the report as an appendix.
3. Test setup: Test orientations, sensor (accelerometer) locations, details of mounting fixtures, and a brief description of the test apparatus. Photos can be included.
4. Test procedure: A standard attachment that is usually given according to corporate specifications.
5. Test results: Ambient conditions in the laboratory (e.g., temperature, humidity), pretest observations (e.g., defects, abnormalities, malfunctions), test data (e.g., reference spectrum, equalized control spectrum, drive spectrum, response time histories and corresponding spectra, transmissibility plots, coherence plots), and post-test observations.
6. Comments and recommendations: General comments regarding the test procedure and test item and recommendations for improving the test, product, or packaging.

Names and titles of the personnel who conducted the test would be given in the test report, with appropriate signatures, dates, and location of the test facility.

Tests for distribution qualification can be conducted on both packaged products and those without any protective packaging, although it is the packaged product that is shipped. The reference spectra used in the two cases are usually not the same, however. The spectrum used for testing a product without protective packaging is generally less severe. Response spectra used for testing an unpackaged product should reflect the excitations experienced by the product during packaging.

10.4.2 Seismic Qualification

It is often necessary to determine whether a given piece of equipment is capable of withstanding a preestablished seismic environment in a specific application. This process is known as *seismic qualification*. Electric utility companies, for example, should qualify their equipment for seismic capability before installing it in earthquake-prone geographic localities. Also, safety-related equipment in nuclear power plants requires seismic qualification. Regulatory agencies usually specify the general procedures to follow in seismic qualification.

Seismic qualification by testing is appropriate for complex equipment, but, in such cases, the equipment size is a limiting factor. For large systems that are relatively simple to model, qualification by analysis is suitable. Often, however, both testing and analysis are needed in the qualification of a given piece of equipment. Seismic qualification of equipment by testing is accomplished by applying a dynamic excitation, by means of a shaker, to the equipment (which is suitably mounted on a test table), and then monitoring structural integrity and functional operability of the equipment. Special attention should be given to the development of the dynamic-test environment, mounting features, the operability variables that should be monitored, the method of monitoring functional operability and structural integrity, and the acceptance criteria used to decide qualification.

In monitoring functional operability, the test facility would normally require auxiliary systems to load (i.e., to apply loading conditions on) the test object or to simulate in-service operating conditions. Such systems include actuators, dynamometers, electrical-load and control-signal circuitry, fluid flow and pressure loads, and thermal loads. In seismic qualification by analysis, a suitable model is first developed for the equipment, and static or dynamic analysis (or computer analysis) is performed under an analytically defined dynamic environment. The analytical dynamic environment is developed on the basis of the specified dynamic environment for seismic qualification. By analytically and/or computationally determining system response at various locations, and by checking for such crucial parameters as relative deflections, stresses, and strains, qualification can be established.

Stages of Seismic Qualification

Consider the construction of a nuclear power plant. In this context, the plant owner is the customer. Actual construction of the plant is done by the plant builder, who is directly responsible to the customer concerning all equipment purchased from the equipment supplier or vendor. The vendor is often the equipment manufacturer as well. The equipment can be purchased by the customer and handed over to the plant builder or directly purchased by the plant builder. Accordingly, the purchaser can be the plant builder or the plant owner. A regulatory agency might stipulate seismic-excitation capability requirements for the equipment used in the plant, or the regulatory agency might specify the qualification requirements for various categories of equipment. The customer is directly responsible to the regulatory agency for adherence to these stipulations. The vendor, however, is responsible to the plant builder and the customer for the seismic capability of the equipment. The vendor can perform seismic qualification on the equipment according to required specifications. More often, however, the vendor hires the services of a test laboratory, which is the contractor, for seismic qualification of equipment in the plant. The qualification procedure and the report, which are usually developed by the test laboratory by adhering to the qualification requirements, can be reviewed by a reviewer, who is hired by the plant builder or the customer. A flowchart for test-object movement and for associated information interactions between various groups in a qualification program, is illustrated in Figure 10.31.

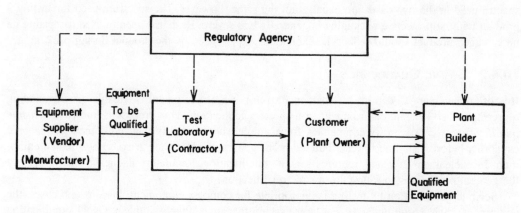

FIGURE 10.31 Test-object movement and information interactions in seismic qualification.

Vibration Testing

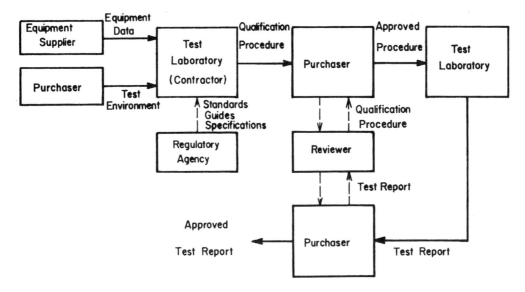

FIGURE 10.32 Information flow in a seismic qualification program.

A basic step in any qualification program is the preparation of a *qualification procedure*. This is a document that describes in sufficient detail such particulars as the tests that will be conducted on the test object, pretest procedures, the nature of test-input excitations and the method of generating these signals, inspection and response-monitoring procedures during testing, definitions of equipment malfunction, and qualification criteria. If analysis is also used in the qualification, the analytical methods and computer programs that will be used should be described adequately in the qualification procedure. The qualification procedure is prepared by the test laboratory (contractor); equipment particulars are obtained from the vendor or the purchaser; and the test environment for which the equipment will be qualified is usually supplied by the purchaser.

Before the qualification tests are conducted, the test procedure is submitted to the purchaser for approval. The purchaser normally hires a reviewer to determine whether the qualification procedure satisfies the requirements of both the regulatory agency and the purchaser. There can be several stages of revision of the test procedure until it is finally accepted by the purchaser on the recommendation of the reviewer.

The approved qualification procedure is sent to the test laboratory, and qualification is performed according to it. The test laboratory prepares a qualification report, which also includes the details of static or dynamic analysis when incorporated. The qualification report is sent to the purchaser for evaluation. The purchaser might obtain the services of an authority to review the qualification report. The report might have to be revised, and even analysis and tests might have to be repeated before the final decision is made on the qualification of the equipment. Information flow in a typical qualification program is shown in Figure 10.32.

10.4.3 Test Preliminaries

Seismic qualification tests are usually conducted by one of two methods, depending on whether single-frequency or multifrequency excitation inputs are employed in the main tests. The two test categories are (1) single-frequency tests and (2) multifrequency tests. The second test method is more common in seismic qualification by testing, although the first method is used under some conditions, depending on the nature of the test object and its mounting features (e.g., line-mounted versus floor-mounted equipment). Typically, multifrequency excitations are preferred in qualification tests, and single-frequency excitations are favored in design-development and quality-assurance tests.

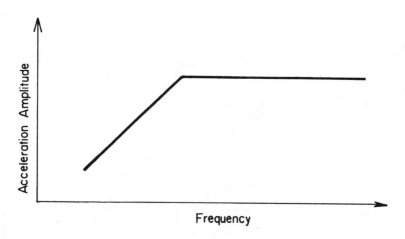

FIGURE 10.33 A typical required input motion (RIM) curve.

In single-frequency testing, the amplitude of the excitation input is specified by a required input motion (RIM) curve, similar to that shown in Figure 10.33. If single-frequency dwells (e.g., sine dwell, sine beat) are employed, the excitation input is applied to the test object at a series of selected frequency values in the frequency range of interest for that particular test environment. In such cases, dwell times (and number of beats per cycle, when sine beats are employed) at each frequency point should be specified.

If a single-frequency sweep (such as a sine sweep) is employed as the excitation signal, the sweep rate should be specified. When the single-frequency test-excitation is specified in this manner, the tests are conducted very much like multifrequency tests.

Multifrequency tests are normally conducted employing the response spectra method to represent the test-input environment. Basically, the test object is excited using a signal whose response spectrum, known as the test response spectrum (TRS), envelops a specified response spectrum, known as the required response spectrum (RRS). Ideally, the TRS should equal the RRS, but it is practically impossible to achieve this condition. Hence, multifrequency tests are conducted using a TRS that envelops the RRS so that, in significant frequency ranges, the two response spectra are nearly equal (see Figure 10.34). Excessive conservatism, however, which would result in overtesting, should be avoided. It is usually acceptable to have TRS values below the RRS at a few frequency points.

The RRS is part of the data supplied to the test laboratory prior to the qualification tests being conducted. Two types of RRS are provided, representing (1) the operating-basis earthquake (OBE) and (2) the safe-shutdown earthquake (SSE). The response spectrum of the OBE represents the most severe motions produced by an earthquake under which the equipment being tested would remain functional without undue risk of malfunction or safety hazard. If the equipment is allowed to operate at a disturbance level higher than the OBE level for a prolonged period, however, there will be a significant risk of malfunction.

The response spectrum of the SSE represents the most severe motions produced by an earthquake that the equipment being tested could safely withstand while the entire nuclear power plant is being shut down. Prolonged operation (i.e., more than the duration of one earthquake), however, could result in equipment malfunction; in other words, equipment is designed to withstand only one SSE in addition to several OBEs.

A typical seismic qualification test would first subject the equipment to several OBE-level excitations, primarily for aging the equipment mechanically to its end-of-design-life condition, and then would subject it to one SSE-level excitation. When providing RRS test specifications, it is customary to supply only the SSE requirement. The OBE requirement is then taken as a fraction (typically, 0.5 or 0.7) of the SSE requirement.

Vibration Testing

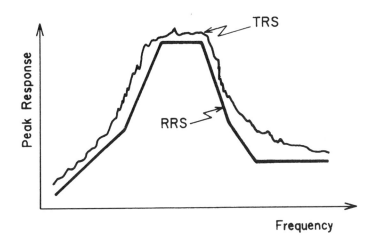

FIGURE 10.34 The TRS enveloping the RRS in a multifrequency test.

Test response spectra corresponding to the excitation signals are generated by the test laboratory during testing. The purchaser usually supplies the test laboratory with an FM tape containing frequency components that should be combined in some ratio to generate the test-input signal.

Qualification tests are conducted according to the test procedure approved and accepted by the purchaser. The main steps of seismic qualification testing are outlined in the following subsections.

Single-Frequency Testing

Seismic ground motions usually pass through various support structures before they are eventually transmitted to equipment. In seismic qualification of that equipment by testing, one should in theory apply to it the actual excitations felt by it — and not the seismic ground motions. In an ideal case, the shaker-table motion should be equivalent to the seismic response of the supporting structure at the point of attachment of the equipment.

The supporting structure would have a particular frequency-response function between the ground location and the equipment-support location (see Figure 10.35). Consequently, it can be considered a filter that modifies seismic ground motions before they reach the equipment mounts. In particular, the components of the ground motion that have frequencies close to a resonant frequency of the supporting structure will be felt by the equipment at a relatively higher intensity. Furthermore, the ground motion components at very high frequencies will be almost entirely filtered out by the structure. If the frequency response of the supporting structure is approximated by a lightly damped simple oscillator, then the response felt by the equipment will be almost sinusoidal, with a frequency equal to the resonant frequency of the structure.

When the supporting structure has a very sharp resonance in the significant frequency range of the dynamic environment (e.g., 1 Hz to 35 Hz for seismic ground motions), it follows from the previous discussion that it is desirable to use a short-duration single-frequency test in seismic qualification of the equipment. Equipment that is supported on pipelines (valves, valve actuators, gauges, etc.) falls into this category. Such equipment is termed *line-mounted equipment*.

The resonant frequency of the supporting structure is usually not known at the time of the seismic qualification test. Consequently, single-frequency testing must be performed over the entire frequency range of interest for that particular dynamic environment.

Another situation in which single-frequency testing is appropriate arises when the test object (equipment) itself does not have more than one sharp resonance in the frequency range of interest. In this case, the most prominent response of the test object occurs at its resonant frequency, even when the dynamic environment is an arbitrary excitation. Consequently, a single-frequency exci-

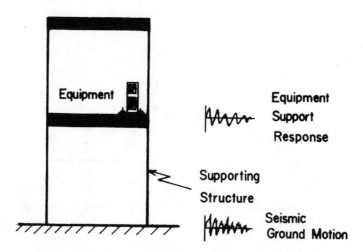

FIGURE 10.35 Schematic representation of the filtering of seismic ground motions by a supporting structure.

tation would yield conservative test results. Equipment that has more than one predominant resonance can employ single-frequency testing, provided that each resonance corresponds to a dynamic degree of freedom (e.g., one resonance along each dynamic principal axis) and that cross-coupling between these degrees of freedom is negligible.

In summary, single-frequency testing can be used if one or more of the following conditions are satisfied:

1. The supporting structure has one sharp resonance in the frequency range of interest (line-mounted equipment is included).
2. The test object does not have more than one sharp resonance in the frequency range of interest.
3. The test object has a resonance in each degree of freedom, but the degrees of freedom are uncoupled (for which adequate verification should be provided in the test procedure).
4. The test object can be modeled as a simple dynamic system (such as a simple oscillator), for which adequate justification or verification should be provided.

Usually, the required SSE excitation level for a single-frequency test over a frequency range is specified by a curve such as the one shown in Figure 10.33. This curve is known as the required input motion (RIM) magnitude curve. The OBE excitation level is usually taken as a fraction (typically, 0.5 or 0.7) of the RIM values given for the SSE. For a sine-sweep test, the sweep rate and the number of sweeps in the test should also be specified. Typically, the sweep rate for seismic qualification tests is less than one octave per minute. One sweep, from the state of rest to the maximum frequency in the range and back to the state of rest, is normally carried out in an SSE test (e.g., 1 Hz to 35 Hz to 1 Hz). Several sweeps (typically five) are performed in an OBE test.

In an SSE sine-dwell test, the dwell time for each dwell frequency should be specified. The dwell-frequency intervals should not be high (typically, a half octave or less). For an OBE test, the dwell times are longer (typically five times longer) than those specified for an SSE test.

For an SSE test using sine beats, the minimum number of beats and the minimum duration of excitation (with or without pauses) at each test frequency should be specified. In addition, the pause time for each test frequency should be specified when sine beats with pauses are employed. For an OBE test, the duration of excitation should be increased (as in a sine-dwell test).

The dwell time at each test frequency should be adequate to perform at least one functional-operability test. Furthermore, a dwell should be carried out at each resonant frequency of the test

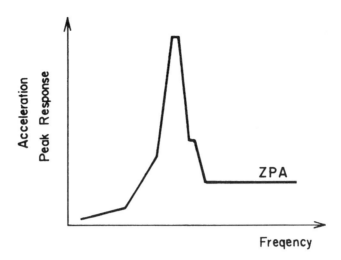

FIGURE 10.36 A typical RRS for a narrow-band excitation test.

object as well as at those frequencies that are specified. Total duration of an SSE test should be representative of the duration of the strong-motion part of a standard safe-shutdown earthquake.

Sometimes, narrow-band random excitations may be used in situations where single-frequency testing is recommended. Narrow-band random signals are those that have their power concentrated over a narrow frequency band. Such a signal can be generated for test-excitation purposes by passing a random signal through a narrow bandpass filter. By tuning the filter to different center frequencies in narrow bands, the test-excitation frequency can be varied during testing. This center frequency of the filter should be swept up and down over the desired frequency range at a reasonably slow rate (e.g., 1.0 octave per minute) during the test. Thus, a multifrequency test with a sharp frequency-response spectrum (RRS), as typified in Figure 10.36, is adequate in cases where single-frequency testing is recommended. A requirement that must be satisfied by the test-excitation signal in this case is that its amplitude should be equal to or greater than the zero-period acceleration of the RRS for the test.

Multifrequency Testing

When equipment is mounted very close to the ground under its normal operating conditions, or if its supporting structure and mounting can be considered rigid, then seismic ground motions will not be filtered significantly before they reach the equipment mounts. In this case, the seismic excitations felt by the equipment will retain broadband characteristics. Multifrequency testing is recommended for seismic qualification of such equipment.

Whereas single-frequency tests are specified by means of an RIM curve along with the test duration at each frequency (or sweep rates), multifrequency tests are specified by means of an RRS curve. The test requirement in multifrequency testing is that the response spectrum of the test excitation (the TRS), which is felt by the equipment mounts, should envelop the RRS. Note that all frequency components of the test excitation are applied simultaneously to the test object, in contrast to single-frequency testing, in which, at a given instant, only one significant frequency component is applied.

When random excitations are employed in multifrequency testing, enveloping of the RRS by the TRS can be achieved by passing the random signal produced by a signal generator through a spectrum shaper. As the analyzing frequency bandwidth (e.g., one-third octave bands, one-sixth octave bands) decreases, the flexibility of shaping the TRS improves. A real-time spectrum analyzer (or a personal computer) can be used to compute and display the TRS curve corresponding to the

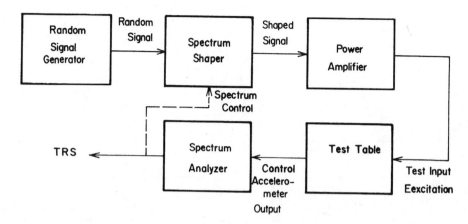

FIGURE 10.37 Matching of the TRS with the RRS in multifrequency testing.

control accelerometer signal (see Figure 10.37). By monitoring the displayed TRS, it is possible to adjust the gains of the spectrum-shaper filter so as to obtain the desired TRS that will envelop the RRS.

Most test laboratories generate their multifrequency excitation signals by combining a series of sine beats that have different peak amplitudes and frequencies. Using the same method, many other signal types, such as decaying sinusoids, can be superimposed to generate the required multifrequency excitation signal. A combination of signals of different types can also be employed to produce a desired test input. A commonly used combination is a broadband random signal and a series of sine beats. In this combination, the random signal is adjusted to have a response spectrum that will envelop the broadband portion of the RRS without much conservatism. The narrow-band peaks of the RRS that generally will not be enveloped by such a broad-band response spectrum will be covered by a suitable combination of sine beats.

By employing such mixed composite signals, it is possible to envelop the entire RRS without having to increase the amplitude of the test excitation to a value that is substantially higher than the ZPA of the RRS. One important requirement in multifrequency testing is that the amplitudes of the test excitation be equal to or greater than the ZPA of the RRS.

10.4.4 GENERATION OF RRS SPECIFICATIONS

Seismic qualification of an object is usually specified in terms of a required response spectrum (RRS). The excitation input that is used in seismic-qualification analysis and testing should conservatively satisfy the RRS; that is, the response spectrum of the actual excitation input should envelop the RRS (without excessive conservatism, of course).

For equipment that is intended to be installed in a building or on some other supporting structure, the RRS generally cannot be obtained as the response spectrum of a modified seismic ground-motion time history. The supporting structure usually introduces an amplification effect and a filtering effect on seismic ground motions. This amplification factor alone can be as high as 3. Some of the major factors that determine the RRS for a particular seismic qualification test are:

1. the nature of the building that will be qualified
2. the dynamic characteristics of the building or structure and the location (elevation and the like) where the object is expected to be installed
3. the in-service mounting orientation and support characteristics of the object

4. the nature of the seismic ground motions in the geographic region where the object is to be installed
5. the test severity and conservatism required by the purchaser or the regulatory agency.

The basic steps in developing the RRS for a specific seismic qualification application include:

1. the development of representative safe-shutdown earthquake (SSE) ground-motion time histories for the building (or support structure) location
2. the development of a suitable building (or support structure) model
3. the response analysis of the building model, using the time histories obtained in step 1
4. the development of response spectra for various critical locations in the building (or support structure), using the response time histories obtained in step 3
5. the normalization of the response spectra obtained in step 4 to unity ZPA (i.e., dividing by their individual ZPA values)
6. the identification of the similarities in the set of normalized response spectra obtained in step 5 and grouping them into a small number of groups
7. the representation of each similar group by a response spectrum consisting of straight-line segments that envelop all members in the group, giving a normalized RRS for each group
8. the determination of scale factors for various locations in the building for use in conjunction with the corresponding normalized RRS curves.

Representative strong-motion earthquake time histories (SSEs) are developed by suitably modifying actual seismic ground-motion time histories that have been observed in that geographic location (or a similar one), or by using a random-signal-generation (simulation) technique or any other appropriate method. These time histories are available as either digital or analog records, depending on the way in which they are generated. If computer simulation is used in their development, a statistical representation of the expected seismic disturbances in the particular geographic region (using geological features in the region, seismic activity data, and the like) should be incorporated in the algorithm. The intensity of the time histories can be adjusted, depending on the required test severity and conservatism.

The normalized response spectra are grouped so that those spectra that have roughly the same shape are put in the same group. In this manner, relatively few groups of normal response spectra (normalized) are obtained. Then, the response spectra that belong to each group are plotted on the same graph paper. Next, straight-line segments are drawn to envelop each group of response spectra. This procedure results in a normalized RRS for each group of analytical response spectra.

The RRS used for a particular seismic qualification scheme is obtained as follows. First, the normalized RRS — corresponding to the location in the building where the object would be installed — is selected. The normalized RRS curve is then multiplied by the appropriate scaling factor. The scaling factor normally consists of the product of the actual ZPA value under SSE conditions at that location (as obtained from the analytical response spectrum at that location, for example) and a factor of safety that depends on the required test severity and conservatism.

Actually, three RRS curves corresponding to the vertical, east-west, and north-south directions might be needed, even for single-degree-of-freedom seismic qualification tests, because, by mounting three control accelerometers in these three directions, triaxial monitoring can be accomplished. If only one control accelerometer is used in the test, then only one RRS curve is used. In this case, the resultant of the three orthogonal RRS curves should be used. One way to obtain the resultant RRS curve is to apply the square-root of the sum of squares (SRSS) method to the three orthogonal components. Alternatively, the envelope of the three orthogonal RRS curves is obtained and multiplied by a safety factor (greater than unity).

Note that more than one building or even many different geographic locations can be included in the described procedure for developing RRS curves. The resulting RRS curves are then valid for the collection of buildings or geographic locations considered. When the generality of an RRS curve is extended in this manner, the test conservatism increases. This also will result in an RRS curve with a much broader band.

In a particular seismic qualification project, in practice, only a few normalized RRS curves are employed. In conjunction with these RRS curves, a table of data is provided that identifies the proper RRS curves and the scaling factors that should be used for different physical locations (e.g., elevations) in various buildings that are situated at several geographic locations.

PROBLEMS

10.1 For electric capacitors, suppose that the test voltage intensity k is related to the duration of the test T through the relationship

$$T \propto \frac{1}{k^p}$$

where p is a parameter that depends on such factors as the particular capacitor used and the environmental conditions. If the intensity for a single test procedure has been prescribed as k_s, determine the intensity for a test sequence involving four tests.

10.2 Consider a test object that has symmetry (dynamic as well as geometric) about vertical planes through the two horizontal principal axes (the x-axis and y-axis in Figure P10.2). In this case, symmetrical rectilinear testing in what directions will produce identical results?

10.3 Consider a test object that has dynamic and geometric symmetry about the vertical plane through one horizontal principal axis (the y-axis in Figure P10.3). What are the directions of symmetry? What sets of rectilinear vibration testing would you suggest for this object?

10.4 Rectilinear testing is the most widespread method employed in seismic qualification. Two-degree-of-freedom testing is employed in some situations, however; but this depends on the availability of appropriate shaker tables. In this type of testing, if the two excitations are random and statistically independent (or at least uncorrelated), suggest a sequence of tests.

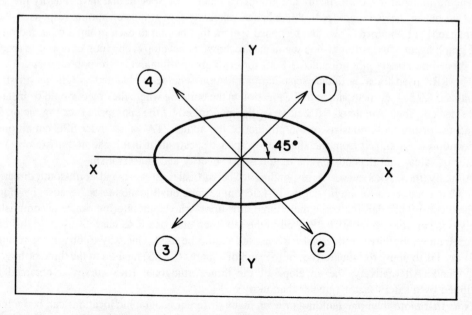

FIGURE P10.2 An object that has two orthogonal planes of symmetry.

Vibration Testing

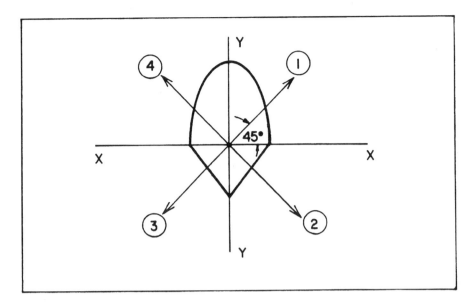

FIGURE P10.3 An object that has one plane of symmetry.

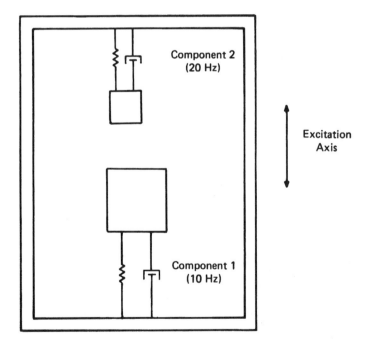

FIGURE P10.6 A device that can malfunction under random excitations but not under sine excitations.

10.5 Compare sine testing and random testing, giving advantages and disadvantages. Sketch a typical excitation signal for each case, giving the probability density function of the random signal.

10.6 Consider the device shown in Figure P10.6. Two components with fundamental natural frequencies at 10 Hz and 20 Hz are mounted such that their axes of sensitivity coincide. Suppose that a functional failure occurs when the two components come in contact. Discuss failure possibility under (a) sine testing, and (b) random testing.

TABLE P10.9
Random Vibration Tests for a Product Development Application

Vibration Test	RMS Value of Excitation (g)	Peak Value of the Excitation psd (g²·Hz⁻¹)	Minimum Times the Random Vibration is Applied	Minimum Duration of Vibration (min)	Vibration Axes
A	2.7	0.01	1	60	Major horizontal axis
B	6.0	0.05	2	30	Major horizontal axis
C	3.2	0.01	1	15	All three
D	5.8	0.02	1	15	All three
E	4.9	0.01	1	15	All three
F	6.3	0.04	2	5	All three

10.7 Who provides the specifications for a vibration test of a piece of equipment?

10.8 Give typical frequency ranges for vibration testing of dynamic equipment in the following applications:
   a. Military avionics (fighter airplanes, etc.)
   b. Human comfort equipment (mainly household applications)
   c. Vehicle ride quality
   d. Distribution qualification of computers and hardware
   e. Precision machine tools
   f. Seismic qualification of nuclear power plant equipment.

10.9 Table P10.9 lists several random vibration tests in the frequency range of 0 Hz to 500 Hz, in an application related to product development. Compare important characteristics of these tests.

10.10 a. Define the following terms:
   1. Reference spectrum
   2. Drive spectrum
   3. Control spectrum
   in relation to vibration testing.
   b. What do you mean by spectrum equalization?
   c. Give a simple algorithm that can be used in shaker control for spectrum equalization.

10.11 Define the following terms:
   a. Octave
   b. Decade.
      i. What is a one-third octave?
      ii. How many decades are there in 100 Hz?
      iii. How many decades correspond to a frequency change from 0.1 rad·s⁻¹ to 10.0 rad·s⁻¹?
      iv. How many octaves correspond to a frequency change from 1 Hz to $\sqrt{2}$ Hz?
      v. What are the dimensions of decades?
   Hint: Some of the questions are posed incorrectly.

10.12 What is a decibel? What are the advantages of using this unit in vibration data presentation?

10.13 Give several advantages of using logarithmic axes in plotting (presenting) vibration test data.

10.14 Discuss how vibration monitoring can be used in the prediction of failure in gear mechanisms. Would you use an acceleration signal or a velocity signal for this task?

10.15 An electromagnetic shaker has the following testing capabilities:

   Maximum displacement amplitude (stroke) = 5.0 cm
   Maximum velocity amplitude = 150.0 cm·s⁻¹
   Maximum acceleration amplitude = 100.0 g

a. Would it be possible to obtain all three of these peak performance capabilities simultaneously for this shaker? Explain.

b. On log–log paper that is used for plotting response spectra, having frequency-velocity axes, mark the capabilities of the shaker and the feasible test region. (This is called a nomograph.)

c. If the shaker operates at a displacement amplitude of 5.0 cm and a velocity amplitude of 150.0 cm·s^{-1}, simultaneously, what is the corresponding maximum acceleration amplitude that would be possible? Also, what is the corresponding test frequency?

d. If the shaker operates at a velocity amplitude of 150.0 cm·s^{-1} and an acceleration amplitude of 100.0 g, simultaneously, what is the corresponding stroke that would be possible? What is the corresponding test frequency?

10.16 Draw a typical schematic diagram for a vibration test arrangement, showing the main components and instrumentation. Describe the function of the following components:

a. Piezoelectric accelerometer
b. Charge amplifier
c. Vibration meter
d. Phase meter
e. Power amplifier
f. Shaker
g. Sine-random generator
h. Low-pass filter
i. High-pass filter
j. Bandpass filter
k. Tunable filter
l. Tracking filter
m. Spectrum analyzer
n. Oscilloscope.

10.17 a. Vibration testing of a device primarily involves application of a test excitation to the device and measuring the resulting response at one or more key locations of the device. Identify four general areas (not specific applications) where vibration testing is used in practice.

b. A piezoelectric accelerometer is a motion sensor that is widely used in vibration measurement. Describe its principle of operation. Why is it that the signal generated by a piezoelectric accelerometer cannot be directly used without proper signal conditioning, for the purposes of recording, analysis, and control? What type of signal conditioning device is commonly used with a piezoelectric accelerometer?

c. The operating capability (ratings) of an electrodynamic (electromagnetic) shaker (exciter) is shown in Figure P10.17, on the frequency–velocity plane, to a log–log scale. Given this information, one engineering student comments that it is practically useless because it is the acceleration versus frequency capability that matters for an exciter, and not the velocity versus frequency capability. A brilliant student who has recently taken an undergraduate course in mechanical vibrations objects to this statement, saying that the given information can be easily converted to a rating curve on the frequency–acceleration plane, to a log–log scale using the units Hz and m·s^{-2}. You are that student.

  i. Compute the coordinates at the two break points between the straight-line segments of this acceleration rating curve. Sketch this (acceleration versus frequency) curve.

  ii. What is the displacement limit (in units of cm), and the acceleration limit when testing a 4-kg object (in units of g, the acceleration due to gravity), for this shaker?

  iii. Suppose that a 4-kg object is tested at 15 Hz. What are the limits of shaker head displacement (cm), velocity (m·s^{-1}), and acceleration (g) for this test? If a 5-kg object is tested at 15 Hz, how would these limiting values change (an approximate estimate will be adequate)?

10.18 a. Electrodynamic shakers are commonly used in the dynamic testing of products. One possible configuration of a shaker/test-object system is shown in Figure P10.18(a). A simple, linear, lumped-parameter model of the mechanical system is shown in Figure P10.18 (b). Note that the driving motor is represented by a torque source T_m. Also, the following parameters are indicated:

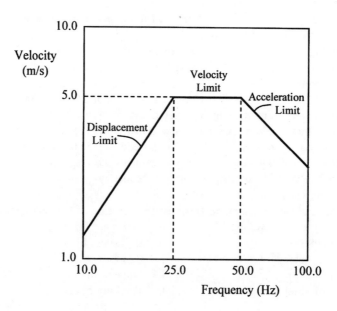

FIGURE P10.17 The rating curve of an electrodynamic exciter.

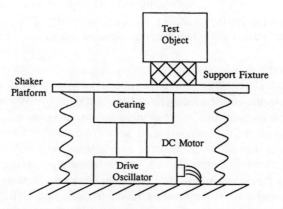

FIGURE P10.18(a) A dynamic-testing system.

J_m = equivalent moment of inertia of motor rotor, shaft, coupling, gears, and shaker platform
r_1 = pitch circle radius of the gear wheel attached to the motor shaft
r_2 = pitch circle radius of the gear wheel rocking the shaker platform
ℓ = lever arm from the rocking gear center to the support location of the test object
m_L = equivalent mass of the test object and support fixture
k_L = stiffness of the support fixture
b_L = equivalent viscous damping constant of the support fixture
k_s = stiffness of the suspension system of the shaker table
b_s = equivalent viscous damping constant of the suspension system.

Note that, since the inertia effects are lumped into equivalent elements, it can be assumed that the shafts, gearing, platform, and support fixtures are light. The following variables are of interest:

Vibration Testing

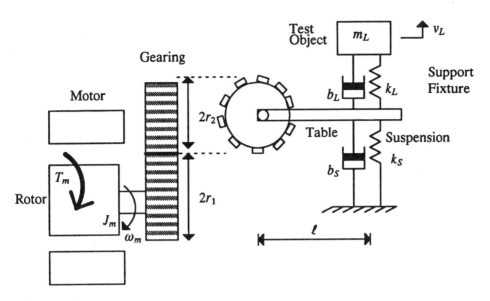

FIGURE P10.18(b) A model of the dynamic testing system.

TABLE P10.20(a)
Capabilities of Five Commercial Control Systems for Vibration Test Shakers

System	A	B	C	D	E
Random test	Yes	Yes	Yes	Yes	Yes
Sine test	Yes	Yes	Optional	No	Yes
Transient and shock tests	Yes	Yes	Optional	No	Yes
Hydraulic shaker	O.K.	O.K.	O.K.	O.K.	?O.K.
Preprogrammed test setups	Max. 63	Max. 25	Max. 99	10 per disk	?
Amplitude scheduling	32 levels and duration	Min. start: −25 dB; Min. step: 0.25 dB; Can pick step durations	10 Levels over 60 dB	0.5 dB steps; Can pick no. of steps and rate	No
On-line reference modification	Yes	No	Yes	No	No
Use of measured spectra as reference	Yes	Yes (measurement — pass feature)	Yes	No	No
Transmissibility	Yes	Measurement option	Yes	No	Yes
Coherence	Yes	Measurement option	No	No	Yes
Correlation	Yes	No	No	No	Yes
Shock response spectrum	Yes	Yes	Optional	No	Yes
Sine on random	Yes	Sine bursts	Optional	No	No
Random on random	Yes	No	Optional	No	No

ω_m = angular speed of the drive motor
v_L = vertical speed of motion of the test object
f_L = equivalent dynamic force (in spring k_L) of the support fixture
f_s = equivalent dynamic force (in spring k_s) of the suspension system.

a. Obtain an expression for the motion ratio

$$r = \frac{\text{Vertical movement of the shaker table at the test object support location}}{\text{Angular movement of the drive motor shaft}}$$

b. Using $x = [\omega_m, f_s, f_L, v_L]^T$ as the state vector, $u = [T_m]$ as the input, and $y = [v_L, f_L]^T$ as the output vector, obtain a complete state-space model for the system.

10.19 What is meant by the term "distribution qualification" of a product? What type of test excitations would be appropriate in distribution qualification procedures? Give possible difficulties in generating such test excitations.

10.20 a. Table P10.20(a) lists the capabilities of five commercial control systems that can be used for shaker control in random vibration testing of products. Compare the five systems. In your discussion, describe the meanings of the terms (capabilities) when necessary.

b. Table P10.20(b) summarizes important hardware characteristics of the five control systems. By defining the terms when necessary, comparatively discuss the various characteristics of the five systems.

c. Table 10.20(c) gives some important specifications of the five control systems. Defining the terms where necessary, discuss the significance of these specifications.

TABLE P10.20(b)
Important Hardware Characteristics of Five Systems for Shaker Control

System	A	B	C	D	E
Reference spectrum break points	40	32	50	? 10	45
Spectrum resolution (number of spectral lines)	Can pick 100, 200, 400, 600, 800 lines	Can pick 64, 128, 256, 512 lines (optional 1024 lines)	Can pick 100, 200, 400, 800 lines	200 lines; 10 Hz spacing	Pick any number: 10–1000 lines (optional 2048 lines
Nature or random drive signal	?	Gaussian, periodic pseudo-random	Gaussian	Gaussian	Pseudo-random
Measured signal averaging	RMS, peak-hold	Arithmetic peak-hold	True power	Peak pick	No
Operator interface	Keyboard, menu-driven RS 232	Keyboard, push button, dialog, set-up	Keyboard, push button, dialog, menu-driven	Keyboard 10 soft keys, dialog	Keyboard
Output devices	CRT screen, hard copy, video print, digital plot	Standard or graphics terminal, X-Y record printer, digital plotter	Graphics terminal, video hard copier, digital plotter, X-Y recorder	Like IBM PC, monochrome-9" Epson printer	Graphics terminal, printer, hard copy, X-Y plotter
Memory	128K	64K	128K	64K	32K Std; 64K Option
Mass storage	Floppy drive 0.5 MB; hard drive 10 MB	One floppy drive, 256K	Hard + floppy 8M, 20M, 30M	Two floppy drives 360K each	Two floppy drives 256K each
Number of measurements (control input) channels	2 Standard; 16 optional	2 Standard; 4 optional; multiplexer optional	1 Standard; 16, 31 optional	1 Standard; 4 optional	?
Number of controller output channels	One	One	One	One	One

Vibration Testing

TABLE P10.20(c)
Specifications of Five Shaker Control Systems

System	A	B	C	D	E
Accelerometer signal (controller input)	±125 mV to ±8 V full scale	10 mV rms to ±8 V max.; typically > 500 mV rms	Max. 10 V peak, 3.5 V, rms	1 to 1000 mV/g user picked	Not given
Controller output signal	2.4 V rms (random); 20 V peak to peak (sine and random)	20 V P-P max.; 50 mA max.	20 V P-P max.	10 V peak, 3V rms	Not given
Input frequency ranges	Random: DC to 200 Hz, 500 Hz, 1 kHz, 2 kHz, 3 kHz, 4 kHz, 8 kHz; Sine: 1 Hz-8 Hz; Shock: 10 Hz–125 Hz, 312 Hz–5 kHz	Seven ranges Max. freq.: 500 Hz–5 kHz, min. freq. = 1 line	100 Hz, 500 Hz, 1 kHz, 2 kHz, 4 kHz, 5 kHz, 10 kHz	10–2000 Hz	10–5000 Hz
Control loop time	2.1 s (2 kHz, 200 lines)	0.3 s, 64 lines; 0.9 s, 256 lines; 3 s, 1024 lines (2 kHz)	4 s, 100 lines; 8 s, 200 lines (2 kHz)	2 s	2.5 s for 256 lines at 2 kHz
Equalization time for 10-dB range	Within ±3 dB in two loops	2 or 3 loops	Within ±1 dB in one loop	Within ±1 dB in 6 s	Not given.
Resolution	12 Bits	12 Bits			12 Bits
Dynamic range	65 dB	65 dB	72 dB	60 dB	
Control accuracy	±1 dB at Q=30, ±2dB at Q=50 (100 Hz Resonance at 1 octave/min.)	±1 dB (at 90% Confidence)	±1 dB Over 72 dB	±1 dB	±1 dB (at 95% confidence)
Sine sweep rate	OK	0.1–100 oct/min (log) 1 Hz–100 kHz/min (linear)	0.1–100 oct/min max.; 0.1 Hz–6 kHz/min^{-1}	N/A	0.001–10 oct/min; 1–6000 Hz/ min

Which of these five shaker control systems would be suitable for use in distribution qualification of products?

10.21 A conventional electrodynamic exciter (shaker) that is used in vibration testing has a shaker head that is suspended on a rigid housing through a flexible diaphragm. The shaker head is excited by the electromagnetic force (as in the case of a DC motor) generated in the drive coil, which is wound around the core of the head.

Consider a shaker of this type, with a test object mounted on it. An accelerometer is mounted on the shaker head. The coil is excited by a known transient current from the drive amplifier. This current is proportional to the electromagnetic force that is generated. The acceleration of the shaker head is measured. The frequency-response function between the drive force and the head acceleration is computed using a spectrum analyzer. Its magnitude is found to have two prominent resonances that are separated by a flat region. Explain this characteristic shape of the frequency response of the shaker, indicating the sources of the two resonances.

10.22 Although shaker tables capable of very high payloads have been reported (e.g., a test object capacity of 1000 tons, table size 15 m × 15 m, three translational and three rotational degrees of freedom, testing frequency range 0 to 30 Hz, stroke 0.2 m, velocity 0.75 m·s^{-1} for the shaker table in Tadotsu Island, Japan), it is quite difficult to carry out shaker-table tests on large civil engineering structures (buildings, bridges, etc.). Several other testing procedures are also employed in testing large test objects. One approach is to make use of natural excitations (e.g., aerodynamic forces) and monitor the response of the structure at several locations. Another is to excite the structure using several portable exciters (shakers) at strategic locations and directions (degrees of freedom), assuming that each exciter has its own controller in generating the excitation. If this second approach could be carried out quite accurately, there would not be a need for large-scale table testing. Clearly, there are difficulties that limit the use of multiple shakers in large-scale testing. Discuss some of these potential problems.

10.23 The size and geometry of a test object, and the mounting characteristics of the test object and its fixtures, have direct implications in vibration testing using shakers. Indicate several effects of these on a test routine.

11 Experimental Modal Analysis

Experimental modal analysis, basically, is a procedure of "experimental modeling." The primary purpose is to develop a dynamic model for a mechanical system using experimental data. In this sense, experimental modal analysis is similar to "model identification" in control system practice, and can utilize somewhat related techniques of "parameter estimation." It is the nature of the developed model that can distinguish experimental modal analysis (EMA) from other conventional procedures of model identification. Specifically, EMA produces a *modal model* that consists of

1. Natural frequencies
2. Modal damping ratios
3. Mode shape vectors

as the primary result. Once a modal model is known, standard results of modal analysis can be used to extract an *inertia (mass) matrix*, a *damping matrix*, and a *stiffness matrix*, which constitute a complete dynamic model for the experimental system, in the time domain. The modal analysis of lumped-parameter systems is covered in Chapter 5, and that of distributed-parameter systems in Chapter 6. Vibration testing and signal analysis are studied in Chapters, 4, 8, 9, and 10. These chapters should be reviewed for the necessary background prior to reading the present chapter.

Since experimental modal analysis produces a modal model (and in some cases, a complete time-domain dynamic model) for a mechanical system form test data of the system, its uses can be extensive. In particular, EMA is useful in

1. Design
2. Diagnosis
3. Control

of mechanical systems, primarily with regard to vibration. In the area of design, the following three approaches that utilize EMA should be mentioned:

1. Component modification
2. Modal response specification
3. Substructuring.

In component modification, one can modify (i.e., add, remove, or vary) inertia (mass), stiffness, and damping parameters in a mechanical system and determine the resulting effect on the modal response (natural frequencies, damping ratios, and mode shapes) of the system. In modal response specification, one can establish the best changes, from the design point of view, in system parameters (inertia, stiffness, and damping values and their degrees of freedom), in order to give a "specified" (prescribed) change in the modal response. In substructuring, two or more subsystem models are combined using dynamic interfacing components, and the overall model is determined. Some of the subsystem models used in this manner can be of analytical origin (e.g., finite element models).

Diagnosis of problems (faults, performance degradation, component deterioration, impending failure, etc.) of a mechanical system requires *condition monitoring* of the system, and analysis and evaluation of the monitored information. Often, analysis involves extraction of modal parameters using monitored data. Diagnosis may involve the establishment of changes (both gradual and sudden), patterns, and trends in these system parameters.

Control of a mechanical system may be based on modal analysis. Standard and well-developed techniques of modal control are widely used in mechanical system practice. In particular, vibration control, both active and passive, may use modal control (see Chapter 12). In this approach, the system is first expressed as a modal model. Then, control excitations, parameter adaptations, etc. are established that would result in a specified (derived) behavior in various modes of the system. Of course, techniques of experimental modal analysis are commonly used here, both in obtaining a modal model from test data, and in establishing modal excitations and parameter changes that are needed to realize a prescribed behavior in the system.

The standard steps of experimental modal analysis are:

1. Obtain a suitable (admissible) set of test data, consisting of forcing excitations and motion responses, for various pairs of degrees of freedom of the test object.
2. Compute the frequency transfer functions (frequency response functions) of the pairs of test data using Fourier analysis. Digital Fourier analysis using fast Fourier transform (FFT) is the standard way of accomplishing this. Either software-based (computer) equipment or hardware-based instrumentation can be used.
3. Curve fit analytical transfer functions to the computed transfer functions. Determine natural frequencies, damping ratios, and residues for various modes in each transfer function.
4. Compute mode shape vectors.
5. Compute inertia (mass) matrix M, stiffness matrix K, and damping matrix C.

Some variations of these steps may be possible in practice, and step 5 is omitted in many situations. The present chapter focuses on some of the standard techniques and procedures associated with the process of experimental modal analysis. The first step in generating test data is not discussed here, as it is extensively covered elsewhere (see Chapters 8 and 10).

11.1 FREQUENCY-DOMAIN FORMULATION

Frequency-domain analysis of vibrating systems is very useful in a wide variety of applications. The analytical convenience of frequency domain methods results from the fact that differential equations in the time domain become algebraic equations in the frequency domain. Once the necessary analysis is performed in the frequency domain, it is often possible to interpret the results without having to transform them back to the time domain through inverse Fourier transformation. In the context of the present chapter, frequency-domain representation is particularly important because it is the frequency-transfer functions that are used for extracting the necessary modal parameters.

For the convenience of notation, the frequency-domain results are developed using the Laplace variable s. As usual, the straightforward substitution of $s = j\omega$, or $s = j2\pi f$, gives the corresponding frequency–domain results.

11.1.1 Transfer Function Matrix

Consider a linear mechanical system that is represented by

$$M\ddot{y} + C\dot{y} + Ky = f(t) \tag{11.1}$$

where

$f(t)$ = forcing excitation vector (nth order column)
y = displacement response vector (nth order column)
m = mass (inertia) matrix ($n \times n$)

C = damping (linear viscous) matrix ($n \times n$)
K = stiffness matrix ($n \times n$).

If the assumption of *proportional damping* is made, it is seen in Chapter 5 that the coordinate transformation

$$y = \Psi q \qquad (11.2)$$

decouples equations (11.1) into the canonical form of modal equations

$$\overline{M}\ddot{q} + \overline{C}\dot{q} + \overline{K}q = \Psi^T f(t) \qquad (11.3)$$

where

Ψ = modal matrix ($n \times n$) of n independent modal vector vectors $[\psi_1, \psi_2, ..., \psi_n]$
$\overline{M}$ = diagonal matrix of modal masses M_i
$\overline{C}$ = diagonal matrix of modal damping constants C_i
$\overline{K}$ = diagonal matrix of modal stiffnesses K_i

Specifically,

$$\overline{M} = \Psi^T M \Psi \qquad (11.4)$$

$$\overline{C} = \Psi^T C \Psi \qquad (11.5)$$

$$\overline{K} = \Psi^T K \Psi \qquad (11.6)$$

If the modal vectors are assumed to be *M*-normal, then

$$M_i = 1$$

$$K_i = \omega_i^2$$

and, furthermore, one can express C_i in the convenient form

$$C_i = 2\zeta_i \omega_i$$

where

ω_i = undamped natural frequency
ζ_i = modal damping ratio.

By Laplace transformation of the response canonical equations of modal motion (11.3), assuming zero initial conditions, one obtains

$$\begin{bmatrix} s^2 + 2\zeta\omega_1 s + \omega_1^2 & & & 0 \\ & s^2 + 2\zeta\omega_2 s + \omega_2^2 & & \\ & & \ddots & \\ 0 & & & s^2 + 2\zeta\omega_n s + \omega_n^2 \end{bmatrix} Q(s) = \Psi^T F(s) \qquad (11.7)$$

Laplace transforms of the modal response (or generalized coordinate) vector $q(t)$ and the forcing excitation vector $f(t)$ are denoted by the column vectors $Q(s)$ and $F(s)$, respectively. The square matrix on the left-hand side of equation (11.7) is a diagonal matrix. Its inverse is obtained by inverting the diagonal elements. Consequently, the following modal transfer relation results:

$$Q(s) = \begin{bmatrix} G_1 & & & 0 \\ & G_2 & & \\ & & \ddots & \\ 0 & & & G_n \end{bmatrix} \Psi^T F(s) \tag{11.8}$$

in which, the diagonal elements are the damped simple-oscillator transfer functions:

$$G_i(s) = \frac{1}{\left[s^2 + 2\zeta_i \omega_i s + \omega_i^2\right]} \quad \text{for } i = 1, 2, \ldots, n \tag{11.9}$$

Note that ω_i, the ith undamped natural frequency (in the time domain), is only approximately equal to the frequency of the ith *resonance* of the transfer function (in the frequency domain) as given by

$$\omega_{ri} = \sqrt{1 - 2\zeta_i^2}\, \omega_i \tag{11.10}$$

As discussed before, and clear from equation (11.10), the approximation improves for decreasing modal damping. Consequently, in most applications of experimental modal analysis, the resonant frequency is taken equal (approximately) to the natural frequency for a given mode.

From the time-domain coordinate transformation (11.2), the Laplace-domain coordinate transformation relation is obtained as:

$$Y(s) = \Psi Q(s) \tag{11.11}$$

Substitute equation (11.8) into (11.11); thus,

$$Y(s) = \Psi \begin{bmatrix} G_1 & & & 0 \\ & G_2 & & \\ & & \ddots & \\ 0 & & & G_n \end{bmatrix} \Psi^T F(s) \tag{11.12}$$

Equation (11.12) is the excitation-response (input-output) transfer relation. It is clear that the $n \times n$ *transfer function matrix G* for the n-degree-of-freedom system is given by

$$G(s) = \Psi \begin{bmatrix} G_1 & & & 0 \\ & G_2 & & \\ & & \ddots & \\ 0 & & & G_n \end{bmatrix} \Psi^T \tag{11.13}$$

Notice in particular that $G(s)$ is a symmetric matrix; specifically,

$$G^T(s) = G(s) \tag{11.14}$$

which should be clear from the matrix transposition property $(ABC)^T = C^T B^T A^T$.

Experimental Modal Analysis

An alternative version of equation (11.13) that is extensively used in experimental modal analysis can be obtained using the partitioned form (or assembled form) of the modal matrix in equation (11.13). Specifically,

$$G(s) = [\psi_1, \psi_2, \ldots, \psi_n] \begin{bmatrix} G_1 & & & 0 \\ & G_2 & & \\ & & \ddots & \\ 0 & & & G_n \end{bmatrix} \begin{bmatrix} \psi_1^T \\ \psi_2^T \\ \vdots \\ \psi_n^T \end{bmatrix} \quad (11.15)$$

On multiplying out the last two matrices on the RHS of equation (11.15), term by term, the following intermediate result is obtained:

$$G(s) = [\psi_1, \psi_2, \ldots, \psi_n] \begin{bmatrix} G_1 \psi_1^T \\ G_2 \psi_2^T \\ \vdots \\ G_n \psi_n^T \end{bmatrix}$$

Note that G_i are scalars, while ψ_i are column vectors. The two matrices in this product can now be multiplied out to obtain the matrix sum:

$$G(s) = G_1 \psi_1 \psi_1^T + G_2 \psi_2 \psi_2^T + \cdots + G_n \psi_n \psi_n^T$$
$$= \sum_{r=1}^{n} G_r \psi_r \psi_r^T \quad (11.16)$$

in which ψ_r is the rth modal vector that is normalized with respect to the mass matrix. Notice that each term $\psi_r \psi_r^T$ in the summation (11.16) is an $n \times n$ matrix with the element corresponding to its ith row and kth column being $(\psi_i \psi_k)_r$. The ikth element of the transfer matrix $G(s)$ is the transfer function $G_{ik}(s)$, which determines the transfer characteristics between the response location i and the excitation location k. From equation (11.16), this is given by

$$G_{ik}(s) = \sum_{r=1}^{n} G_r (\psi_i \psi_k)_r$$
$$= \sum_{r=1}^{n} \frac{(\psi_i \psi_k)_r}{[s^2 + 2\zeta_r \omega_r s + \omega_r^2]} \quad (11.17)$$

where $s = j\omega = j2\pi f$ in the frequency domain. Note that $(\psi_i)_r$ is the ith element of the rth modal vector, and is a scaler quantity. Similarly, $(\psi_i \psi_k)_r$ is the product of the ith element and the kth element of the rth modal vector, and is also a scalar quantity. This is the numerator of each modal transfer function within the RHS summation of equation (11.17), and is the "residue" of the pole (eignevalue) of that mode.

Equation (11.17) is useful in experimental modal analysis. Essentially, one starts by determining the *residues* $(\psi_i \psi_k)_r$ of the poles in an admissible set of measured transfer functions. One can determine the modal vectors in this manner. In addition, by analyzing the measured transfer functions, *modal damping ratios* ζ_i and the *natural frequencies* ω_i can be estimated. From these results, an estimate for the time-domain model (i.e., the matrices M, K, and C) can be determined.

11.1.2 PRINCIPLE OF RECIPROCITY

By the symmetry of transfer matrix, as given by equation (11.14) it follows that

$$G_{ik}(s) = G_{ki}(s) \tag{11.18}$$

This fact is further supported by equation (11.17). This symmetry can be interpreted as *Maxwell's principle of reciprocity*. To understand this further, consider the complete set of transfer relations given by equations (11.12) and (11.13):

$$\begin{aligned} Y_1(s) &= G_{11}(s)F_1(s) + G_{12}(s)F_2(s) + \cdots + G_{1n}(s)F_n(s) \\ Y_2(s) &= G_{21}(s)F_1(s) + G_{22}(s)F_2(s) + \cdots + G_{2n}(s)F_n(s) \\ &\vdots \\ Y_n(s) &= G_{n1}(s)F_1(s) + G_{n2}(s)F_2(s) + \cdots + G_{nn}(s)F_n(s) \end{aligned} \tag{11.19}$$

Note that the diagonal elements G_{11}, G_{22}, ..., G_{nn} are *driving-point transfer functions* (or auto-transfer functions), and the rest are *cross-transfer functions*. Suppose that a single excitation $F_k(s)$ is applied at the kth degree of freedom with all the other excitations set to 0. The resulting response at the ith degree of freedom is given by

$$Y_i(s) = G_{ik}(s)F_k(s) \tag{11.20}$$

Similarly, when a single excitation $F_i(s)$ is applied at the ith degree of freedom, the resulting response at the kth degree of freedom is given by

$$Y_k(s) = G_{ki}(s)F_i(s) \tag{11.21}$$

In view of the symmetry that is indicated by equation (11.18), it follows from (11.20) and (11.21) that if the two separate excitations $F_k(s)$ and $F_i(s)$ are identical, then the corresponding responses $Y_i(s)$ and $Y_k(s)$ also become identical. In other words, the response at the ith degree of freedom due to a single force at the kth degree of freedom is equal to the response at the kth degree of freedom when the same single force is applied at the ith degree of freedom. This is the frequency-domain version of the principle of reciprocity.

EXAMPLE 11.1

Consider the two-degree-of-freedom system shown in Figure 11.1. Assume that the excitation forces $f_1(t)$ and $f_2(t)$ act at the y_1 and y_2 degrees of freedom, respectively. The equations of motion are given by

$$\begin{bmatrix} m & 0 \\ 0 & m \end{bmatrix}\ddot{y} + \begin{bmatrix} c & 0 \\ 0 & c \end{bmatrix}\dot{y} + \begin{bmatrix} 2k & -k \\ -k & 2k \end{bmatrix}y = f(t) \tag{i}$$

Experimental Modal Analysis

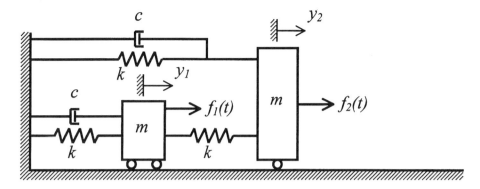

FIGURE 11.1 A vibrating system with proportional damping.

It has been noted in Chapter 5 that this system has proportional damping (specifically, it is clear that C is proportional to M) and, hence, possesses the same real modal vectors as for the undamped system. First obtain the transfer matrix in the direct manner. By taking the Laplace transform (with zero initial conditions) of the equations of motion (*i*), one obtains

$$\begin{bmatrix} ms^2 + cs + 2k & -k \\ -k & ms^2 + cs + 2k \end{bmatrix} Y(s) = F(s) \tag{ii}$$

Hence, in the relation $Y(s) = G(s)F(s)$, the transfer matrix G is given by

$$G(s) = \begin{bmatrix} ms^2 + cs + 2k & -k \\ -k & ms^2 + cs + 2k \end{bmatrix}^{-1}$$

$$= \frac{1}{\left[(ms^2 + cs + 2k)^2 - k^2\right]} \begin{bmatrix} ms^2 + cs + 2k & k \\ k & ms^2 + cs + 2k \end{bmatrix} \tag{iii}$$

The characteristic polynomial $\Delta(s)$ of the system is

$$\Delta(s) = (ms^2 + cs + 2k)^2 - k^2 = (ms^2 + cs + k)(ms^2 + cs + 3k) \tag{iv}$$

and is common to the denominator of all four transfer functions in the matrix. Specifically,

$$G_{11}(s) = G_{22}(s) = \frac{ms^2 + cs + 2k}{\Delta(s)} \tag{v}$$

$$G_{12}(s) = G_{21}(s) = \frac{k}{\Delta(s)} \tag{vi}$$

What this result implies is that the characteristic equation characterizes the entire system (particularly the natural frequencies and damping ratios), and no matter what transfer function is measured (or analyzed), the same natural frequencies and modal damping are obtained.

One can put these transfer functions into the partial fraction form. For example,

$$\frac{ms^2 + cs + 2k}{(ms^2 + cs + k)(ms^2 + cs + 3k)} = \frac{A_1 s + A_2}{(ms^2 + cs + k)} + \frac{A_3 s + A_4}{(ms^2 + cs + 3k)} \quad (vii)$$

By comparing the numerator coefficients, one finds that $A_1 = A_3 = 0$ (this is the case when the modes are *real*; with complex modes, $A_1 \neq 0$ and $A_3 \neq 0$ in general) and $A_2 = A_4 = 1/2$. These results are summarized below.

$$G_{11}(s) = G_{22}(s) = \frac{1/(2m)}{\left[s^2 + 2\zeta_1 \omega_1 s + \omega_1^2\right]} + \frac{1/(2m)}{\left[s^2 + 2\zeta_2 \omega_2 s + \omega_2^2\right]} \quad (viii)$$

$$G_{12}(s) = G_{21}(s) = \frac{1/(2m)}{\left[s^2 + 2\zeta_1 \omega_1 s + \omega_1^2\right]} - \frac{1/(2m)}{\left[s^2 + 2\zeta_2 \omega_2 s + \omega_2^2\right]} \quad (ix)$$

where

$$\omega_1 = \sqrt{k/m}, \; \omega_2 = \sqrt{3k/m}, \; \zeta_1 = c/\sqrt{4mk}, \text{ and } \zeta_2 = c/\sqrt{12mk}$$

By comparing the *residues* (numerators) of these expressions with relation (11.17), one can determine the *M*-normal modal vectors. Specifically,

by examining G_{11}: $(\psi_1^2)_1 = \frac{1}{2m}$, $(\psi_1^2)_2 = \frac{1}{2m}$

by examining G_{12}: $(\psi_1 \psi_2)_1 = \frac{1}{2m}$, $(\psi_1 \psi_2)_2 = -\frac{1}{2m}$

One needs consider only two admissible transfer functions (e.g., G_{11} and G_{12}, or G_{11} and G_{21}, or G_{12} and G_{22}, or G_{21} and G_{22}) in order to completely determine the modal vectors. Specifically, one obtains

$$\begin{bmatrix} \psi_1 \\ \psi_2 \end{bmatrix}_1 = \begin{bmatrix} 1/\sqrt{2m} \\ 1/\sqrt{2m} \end{bmatrix} \quad \text{and} \quad \begin{bmatrix} \psi_1 \\ \psi_2 \end{bmatrix}_2 = \begin{bmatrix} 1/\sqrt{2m} \\ -1/\sqrt{2m} \end{bmatrix}$$

Note that the modal masses are unity for these modal vectors. Also, there is an arbitrariness in the sign. As usual, this problem is overcome by making the first element of each modal vector positive. These modal vectors agree with the results obtained in a previous example for the same system (see Example 5.8 in Chapter 5).

□

11.2 EXPERIMENTAL MODEL DEVELOPMENT

It has been noted that the process of extracting modal data (natural frequencies, modal damping, and mode shapes) from measured excitation-response data is termed "experimental modal analysis." Modal testing and the analysis of test data are the two main steps of experimental modal analysis. Information obtained through experimental modal analysis is useful in many applications, including

Experimental Modal Analysis

validation of analytical models for dynamic systems, fault diagnosis in machinery and equipment, on-site testing for requalification to revised regulatory specifications, and design development of mechanical systems.

In the present development, it is assumed that the test data are available in the frequency domain as a set of transfer functions. In particular, suppose that an *admissible* set of transfer functions is available. The actual process of constructing or computing these frequency-transfer functions from measured excitation-response (input-output) test data (in the time domain) is known as *model identification* in the frequency domain. This step should precede the actual modal analysis in practice. Numerical analysis (or curve-fitting) is the basic tool used for this purpose, and it will be discussed in a later section.

The basic result used in experimental modal analysis is equation (11.17) with $s = j\omega$ or $s = j2\pi f$ for the frequency-transfer functions. For convenience, however, the following notation is used:

$$G_{ik}(\omega) = \sum_{r=1}^{n} \frac{(\psi_i \psi_k)_r}{\left[\omega_r^2 - \omega^2 + 2j\zeta_r \omega_r \omega\right]} \quad (11.22)$$

or, equivalently,

$$G_{ik}(f) = \sum_{r=1}^{n} \frac{(\psi_i \psi_k)_r}{4\pi^2 \left[f_1^2 - f^2 + 2j\zeta_r f_r f\right]} \quad (11.23)$$

where ω and f are used in place of $j\omega$ and $j2\pi f$ in the function notation $G(\ )$. As already observed in Example 11.1, it is not necessary to measure all n^2 transfer functions in the $n \times n$ transfer function matrix G in order to determine the complete modal information. Due to the symmetry of G, it follows that, at most, only $1/2 n(n + 1)$ transfer functions are needed. In fact, it can be "shown by construction" (i.e., in the process of developing the method itself) that only n transfer functions are needed. These n transfer functions cannot be chosen arbitrarily, however, although there is a wide choice for the admissible set of n transfer functions. A convenient choice would be to measure any one *row* or any one *column* of the transfer-function matrix. It should be clear from the following development that any set of transfer functions that spans all n degrees of freedom of the system would be an admissible set, provided that only one auto-transfer function is included in the minimal set. Hence, for example, all the transfer functions on the main diagonals, or on the main cross-diagonal of G, do not form an admissible set.

Suppose that the kth column (G_{ik}, $i = 1, 2, \ldots, n$) of the transfer-function matrix is measured by applying a single forcing excitation at the kth degree of freedom and measuring the corresponding responses at all n degrees of freedom in the system. The main steps in extracting the modal information from this data are given below.

Step 1: Curve-fit the (measured) n transfer functions to expressions of the form given by equation (11.22). In this manner, determine the natural frequencies ω_r, the damping ratios ζ_r, and the residues $(\psi_i \psi_k)_r$ for the set of modes $r = 1, 2, \ldots$

Step 2: The residues of a diagonal transfer function (i.e., point transfer functions or auto-transfer function) G_{kk} are $(\psi_k^2)_1, (\psi_k^2)_2, \ldots, (\psi_k^2)_n$. From these, determine the kth row of the modal matrix $(\psi_k)_1, (\psi_k)_2, \ldots, (\psi_k)_n$. Note that M-normality is assumed. Still, the modal vectors are arbitrary up to a multiplier of -1. Hence, one can choose this row to have all positive elements.

Step 3: The residues of a nondiagonal transfer function (i.e., cross-transfer function) $G_{k+i,k}$ are $(\psi_{k+i} \psi_k)_1, (\psi_{k+i} \psi_k)_2, \ldots, (\psi_{k+i} \psi_k)_n$. By substituting the values obtained in step 2 into these values, determine the $k+i$th row of the modal matrix $(\psi_{k+i})_1, (\psi_{k+i})_2, \ldots, (\psi_{k+i})_n$.

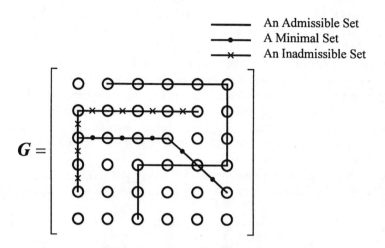

FIGURE 11.2 A non-minimal admissible set, a minimal set, and an inadmissible set of possible transfer function measurements.

The complete modal matrix Ψ is obtained by repeating this step for $i = 1, 2, \ldots, n - k$ and $i = -1, -2, \ldots, -k + 1$. Note that the associated modal vectors are M-normal.

The procedure just outlined for determining the modal matrix verifies, by construction, that only n transfer functions are needed to extract the complete modal information. It further reveals that it is not essential to perform the transfer function measurements in a row fashion or column fashion. A diagonal element (i.e., a *point transfer function*, or an *auto-transfer function*) should always be measured. The remaining $n - 1$ transfer functions have to be off diagonal, but otherwise can be chosen arbitrarily, provided that all n degrees of freedom are spanned either as an excitation point or measurement location (or both). This guarantees that no symmetric transfer function elements are included. This defines a *minimal set* of transfer function measurements. An *admissible set* of more than n transfer functions can be measured in practice so that redundant measurements would be available in addition to the minimal set that is required. Such redundant data are useful for checking the accuracy of the modal estimates. Examples for an admissible (nonminimal) set, a minimal set, and an inadmissible set of transfer function matrix elements are shown schematically in Figure 11.2. Note that the inadmissible set in this example contains 8 transfer function measurements, but the 6th degree of freedom is not covered by this set. On the other hand, a minimal set requires only six transfer functions.

11.2.1 Extraction of the Time-Domain Model

Once the complete modal information is extracted by modal analysis, it is possible — at least in theory — to determine a time-domain model (M, K, and C matrices) for the system. To obtain the necessary equations, first premultiply by $(\Psi^T)^{-1}$ and postmultiply by Ψ^{-1} the equations (11.4), (11.5), and (11.6) to get

$$M = \left(\Psi^T\right)^{-1} \overline{M} \Psi^{-1} \tag{11.24}$$

where $\overline{M} = I$ = identity matrix

$$K = \left(\Psi^T\right)^{-1} \overline{K} \Psi^{-1} \tag{11.25}$$

Experimental Modal Analysis

$$C = \left(\Psi^T\right)^{-1} \overline{C} \Psi^{-1} \tag{11.26}$$

Since the modal matrix Ψ is nonsingular, because M is assumed nonsingular in the dynamic models that are used here (i.e., each degree of freedom has an associated mass, or the system does not possess static modes), the inverse transformations given by equations (11.24) to (11.26) are feasible. It appears, however, that two matrix inversions are needed for each result. Since $\overline{M}, \overline{K}$, and $\overline{C}$ matrices are diagonal, their inverse is given by inverting the diagonal elements. This fact can be used to obtain each result through just one matrix inversion.

Equations (11.24) to (11.26) are written as

$$M = \left(\Psi \overline{M}^{-1} \Psi^T\right)^{-1} \tag{11.27}$$

$$K = \left(\Psi \overline{K}^{-1} \Psi^T\right)^{-1} \tag{11.28}$$

$$C = \left(\Psi \overline{C}^{-1} \Psi^T\right)^{-1} \tag{11.29}$$

Note that for the present M-normal case

$$\overline{M}^{-1} = I \tag{11.30}$$

$$\overline{K}^{-1} = \text{diag}\left[1/\omega_1^2, 1/\omega_2^2, \ldots, 1/\omega_n^2\right] \tag{11.31}$$

$$\overline{C}^{-1} = \text{diag}\left[1/(2\zeta_1\omega_1), 1/(2\zeta_2\omega_2), \ldots, 1/(2\zeta_n\omega_n)\right] \tag{11.32}$$

By substituting equations (11.30) to (11.32) into equations (11.27) to (11.29), one obtains the relations that can be used in computing the time-domain model:

$$M = \left(\Psi \Psi^T\right)^{-1} \tag{11.33}$$

$$K = \left(\Psi \begin{bmatrix} 1/\omega_1^2 & & & 0 \\ & 1/\omega_2^2 & & \\ & & \ddots & \\ 0 & & & 1/\omega_n^2 \end{bmatrix} \Psi^T\right)^{-1} \tag{11.34}$$

$$C = \left(\Psi \begin{bmatrix} 1/(2\zeta_1\omega_1) & & & 0 \\ & 1/(2\zeta_2\omega_2) & & \\ & & \ddots & \\ 0 & & & 1/(2\zeta_n\omega_n) \end{bmatrix} \Psi^T\right)^{-1} \tag{11.35}$$

Alternatively, only one matrix inversion (that of Ψ) is needed for all three matrices if one uses

$$(\Psi^T)^{-1} = (\Psi^{-1})^T$$

Then,

$$M = (\Psi^{-1})^T \overline{M} \Psi^{-1} \qquad (11.36)$$

$$K = (\Psi^{-1})^T \overline{K} \Psi^{-1} \qquad (11.37)$$

$$C = (\Psi^{-1})^T \overline{C} \Psi^{-1} \qquad (11.38)$$

The main steps of experimental modal analysis are summarized in Box 11.1. In practice, frequency-response data are relatively less accurate at higher resonances. Some of the main sources of error include:

1. Aliasing distortion in the frequency domain due to finite sampling rate of data will distort high-frequency results during digital computation (see Chapter 4).
2. Inadequate spectral-line resolution (or frequency resolution) and frequency coverage (bandwidth) can introduce errors at high-frequency resonances. The frequency resolution is fixed both by the signal record length (T) and the type of time window used in digital Fourier analysis, but the resonant peaks are sharper for higher frequencies. Frequency coverage depends on the data sampling rate.
3. Low signal-to-noise ratio (SNR) at high frequencies, in part due to noise and poor dynamic range of equipment, and in part due to low signal levels, will result in data measurement errors. Signal levels are usually low at high frequencies because inertia in a mechanical system acts as a low-pass filter $\left(\dfrac{1}{m\omega^2}\right)$.
4. Computations involving high-order matrices (multiplication, inversion, etc.) in complex systems with many degrees of freedom, will lead to numerical errors.

It is customary, therefore, to extract modal information only for the first several modes. In that case, it is not possible to accurately recover the mass, stiffness, and damping matrices. Even if these matrices were computed, their accuracy would be questionable due to their sensitivity to the factors listed above.

11.3 CURVE-FITTING OF TRANSFER FUNCTIONS

Parameter estimation in vibrating systems can be interpreted as a technique of *experimental modeling*. This process requires experimental data in a suitable form — preferably excitation-response data and often represented as a set of transfer functions in the frequency domain. Parameter estimation using measured response data is termed *model identification*, or simply identification, in the literature on systems and control. A parameter estimation procedure that involves frequency-transfer functions, and which is particularly useful in experimental modal analysis, is presented.

Experimental Modal Analysis

BOX 11.1 Main Steps of Experimental Modal Analysis

1. Measure an admissible set of excitation (u) and response (y) signals. (Cover all dof; one response measurement should be for the excitation location.)
2. Group the signals, assign windows, and filter the signals.
3. Compute transfer functions using FFT and the spectral-density method:

$$G_{uy} = \Phi_{uy}/\Phi_{uu}$$

4. Compute ordinary coherence functions:

$$\gamma_{uy}^2 = \frac{|\Phi_{uy}|^2}{\Phi_{uu}\Phi_{yy}}$$

   and choose the accurate transfer functions on this basis (γ_{uy} close to 1 $\Rightarrow$ accept; γ_{uy} close to 0 $\Rightarrow$ reject).
5. Curve-fit n admissible transfer functions to expressions:

$$G_{ik} = \sum_{r=1}^{n} \frac{(\psi_i \psi_k)_r}{[\omega_r^2 - \omega^2 + 2j\zeta_r \omega_r \omega]}$$

   Hence, extract:
   Residues $(\psi_i \psi_k)_r \Rightarrow$ mode shapes vectors ψ_r, which are M-normal
   Natural frequencies (undamped) ω_r
   Modal damping ratios (viscous) ζ_r
6. Form the modal matrix $\Psi = [\psi_1, \psi_2, ..., \psi_n]$.
   Compute Ψ^{-1}.
7. Modal mass matrix $\overline{M} = I$

   Modal stiffness matrix $\overline{K} = \text{diag}[\omega_1^2, \omega_2^2, ..., \omega_n^2]$

   Modal damping matrix $\overline{C} = \text{diag}[2\zeta_1\omega_1, 2\zeta_2\omega_2, ..., 2\zeta_n\omega_n]$
8. Compute the system model:
   Mass matrix $M = (\Psi^{-1})\Psi^{-1}$
   Stiffness matrix $K = (\Psi^{-1})^T \overline{K} \Psi^{-1}$
   Damping matrix $C = (\Psi^{-1})\overline{C}\Psi^{-1}$

11.3.1 Problem Identification

Transfer functions that are computed from measured time histories using digital Fourier analysis (e.g., fast Fourier transform, or FFT) cannot be directly used in modal analysis computations. The data must be available as analytical transfer functions. Therefore, it is important to represent the computed transfer functions by suitable analytical expressions. This is done, in practice, either by curve-fitting a suitable transfer function model into the computed data or by simplified methods such as "peak picking." Accordingly, this convention of data is an experimental modeling technique.

Identification of transfer function models from measured data is an essential step in experimental modal analysis. Apart from that, it has other important advantages. In particular, analytical transfer function plots clearly identify system resonances and generate numerical values for the corresponding parameters (resonant frequencies, damping, phase angles, and magnitudes) in a convenient manner. This form represents a significant improvement over the crude transfer function plots that are normally far less presentable and rather difficult to interpret.

11.3.2 SINGLE-DEGREE-OF-FREEDOM AND MULTI-DEGREE-OF-FREEDOM TECHNIQUES

Several single degree-of-freedom (dof) techniques exist for extracting analytical parameters from experimental transfer functions. In particular, the methods of curve fitting (circle fitting) and peak picking are considered here. In a single-dof method, only one resonance is considered at a time.

Single-degree-of-freedom curve fitting or, more correctly, *single-resonance curve fitting* is the term used to denote any curve-fitting procedure that fits a quadratic (second-order) transfer function into each resonance in the measured transfer function, one at a time. In the case of closely spaced modes (or closely spaced resonances), the associated error can be very large. The accuracy is improved if expressions of higher order than quadratic are used for this purpose, but unacceptable errors can still exist. In peak picking, each resonance of experimental transfer function data is examined individually, and resonant frequency, and the damping constant corresponding to that resonance, are determined by comparing with an analytical single-dof transfer function.

In multi-degree-of-freedom curve fitting or, more appropriately, *multiresonance curve-fitting*, all resonances (or modes) of importance are considered simultaneously and fitted into an analytical transfer function of suitable order. This method is generally more accurate but computationally more demanding than the single-resonance method. In choosing between the single-resonance and multiresonance methods, the required accuracy should be weighed against the cost and speed of computation.

11.3.3 SINGLE-DEGREE-OF-FREEDOM PARAMETER EXTRACTION IN THE FREQUENCY DOMAIN

The theory of curve-fitting by a circle (i.e., circle fitting) for each resonance of an experimentally determined transfer function is presented first. Next, the peak picking method will be described.

Circle-Fit Method

It can be shown that the *mobility* transfer function (velocity/force) of a single-degree-of-freedom (dof) system with linear *viscous damping*, when plotted on the *Nyquist plane* of real axis and imaginary axis for the frequency transfer function, is a circle. Similarly, it can be shown that the *receptance* or *dynamic flexibility* or compliance transfer function (displacement/force) of a single-dof system with *hysteretic damping*, when plotted on the Nyquist plane, is also a circle. Note that for hysteretic damping, the damping constant (in the time domain) is not actually a constant but is inversely proportional to the frequency of motion (see Chapter 7). But, in the frequency domain, the damping term will be independent of frequency, in this case. The fact that such circle representations are possible for transfer functions of a single-dof system, can be used in fitting a circle to a transfer function that is computed from experimental data. This will lead to determining the analytical parameters for the transfer function. This approach is illustrated now, through analytical development.

Case of Viscous Damping

Consider a single-dof system with linear, viscous damping, as given by

$$m\ddot{y} + c\dot{y} + ky = f(t) \tag{11.39}$$

Experimental Modal Analysis

where m, k, and c are the mass, stiffness, and the damping constant of the system, respectively; $f(t)$ is the excitation force; and y is the displacement response. Equation (11.39) can be expressed in the standard form:

$$\ddot{y} + 2\zeta\omega_n\dot{y} + \omega_n^2 y = \frac{1}{m}f(t) \qquad (11.40)$$

$$\text{Receptance } \frac{Y(s)}{F(s)} = \frac{1}{m[s^2 + 2\zeta\omega_n s + \omega_n^2]} \quad \text{with } s = j\omega \qquad (11.41)$$

$$\text{Mobility } \frac{V(s)}{F(s)} = \frac{sY(s)}{F(s)} = \frac{s}{m[s^2 + 2\zeta\omega_n s + \omega_n^2]} \quad \text{with } s = j\omega \qquad (11.42)$$

Consider the mobility (velocity/force) transfer function given by

$$G(s) = \frac{s}{s^2 + 2\zeta\omega_n s + \omega_n^2} \qquad (11.43)$$

where the constant parameter m in equation (11.42) has been omitted, without loss of generality. In the frequency domain ($s = j\omega$), then

$$G(j\omega) = \frac{j\omega}{\omega_n^2 - \omega^2 + 2j\zeta\omega_n\omega} \qquad (11.44)$$

Multiply the numerator and the denominator of $G(j\omega)$ in equation (11.44) by the complex conjugate of the denominator (i.e., $\omega_n^2 - \omega^2 - 2j\zeta\omega_n\omega$). Then, the denominator is converted to the square of its original magnitude, as given by

$$\Delta = \left(\omega_n^2 - \omega^2\right)^2 + \left(2\zeta\omega_n\omega\right)^2 \qquad (11.45)$$

and the frequency-transfer function (11.44) is converted into the form

$$G(j\omega) = \frac{j\omega}{\Delta}\left[\omega_n^2 - \omega^2 - 2j\zeta\omega_n\omega\right] \qquad (11.46)$$

$$G(j\omega) = Re + jIm, \quad \text{where} \quad Re = \frac{2\zeta\omega_n\omega^2}{\Delta} \quad \text{and} \quad Im = \frac{\omega}{\Delta}\left(\omega_n^2 - \omega^2\right) \qquad (11.47)$$

Now, one can write

$$Re - \frac{1}{4\zeta\omega_n} = \frac{8\zeta^2\omega_n^2\omega^2 - 4\zeta^2\omega_n^2\omega^2 - (\omega_n^2 - \omega^2)^2}{4\zeta\omega_n\Delta}$$

$$= \frac{4\zeta^2\omega_n^2\omega^2 - (\omega_n^2 - \omega^2)^2}{4\zeta\omega_n\Delta}$$

Hence, in view of equation (11.47), one obtains

$$\left[Re - \frac{1}{4\zeta\omega_n}\right]^2 + Im^2 = \left[\frac{4\zeta^2\omega_n^2\omega^2 - (\omega_n^2 - \omega^2)^2}{4\zeta\omega_n\Delta}\right]^2 + \frac{\omega^2(\omega_n^2 - \omega^2)^2}{\Delta^2}$$

$$= \frac{16\zeta^4\omega_n^4\omega^4 - 8\zeta^2\omega_n^2\omega^2(\omega_n^2 - \omega^2)^2 + (\omega_n^2 - \omega^2)^4 + 16\zeta^2\omega_n^2\omega^2(\omega_n^2 - \omega^2)^2}{16\zeta^2\omega_n^2\Delta^2}$$

$$= \frac{\left[4\zeta^2\omega_n^2\omega^2 + (\omega_n^2 - \omega^2)^2\right]^2}{16\zeta^2\omega_n^2\Delta^2} = \frac{\Delta^2}{16\zeta^2\omega_n^2\Delta^2} = \frac{1}{16\zeta^2\omega_n^2}$$

$$= R^2$$

It follows that the transfer function $G(j\omega)$ represents a circle in the real-imaginary plane, with the following properties:

$$\text{Circle radius } R = \frac{1}{4\zeta\omega_n} \qquad (11.48)$$

$$\text{Circle center} = \left(\frac{1}{4\zeta\omega_n}, 0\right) \qquad (11.49)$$

Now, one can reintroduce the constant parameter m back into the transfer function, as in equation (11.42). Then,

$$\text{Circle radius } R = \frac{1}{4\zeta\omega_n m}, \qquad \text{Circle center} = \left(\frac{1}{4\zeta\omega_n m}, 0\right) \qquad (11.50)$$

A sketch of this circle is shown in Figure 11.3(a). As mentioned before, the plane formed by the real and imaginary parts of $G(j\omega)$ as the Cartesian x and y axes, respectively, is the *Nyquist plane*. The plot of $G(j\omega)$ on this plane is the *Nyquist diagram*. It follows that the Nyquist diagram of the mobility function (11.42) [or (11.44)] is a circle.

Case of Hysteretic Damping
Consider a single-dof system with *hysteretic damping*. The equation motion is given by

Experimental Modal Analysis

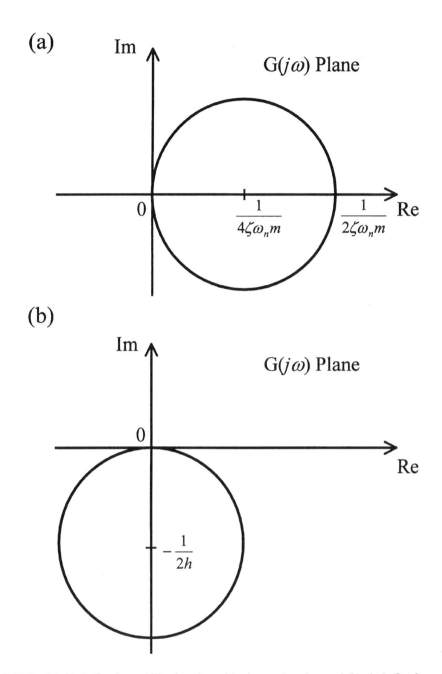

FIGURE 11.3 (a) Circle fit of a mobility function with viscous damping, and (b) circle fit of a receptance function with hysteretic damping.

$$m\ddot{y} + \frac{h}{\omega}\dot{y} + ky = f(t) \quad \text{with } f(t) = f_0 \sin \omega t \tag{11.51}$$

Note the frequency-dependent damping constant, with the hysteretic damping parameter h, in the time domain. The receptance function $G(j\omega)$ is given by

$$G(j\omega) = \frac{1}{(k - m\omega^2) + jh}$$

Note that the damping term jh is independent of frequency in the frequency domain, for this case. As for the case of viscous damping, one can easily show that the Nyquist plot of this transfer function is a circle with

$$\text{Radius} = \frac{1}{2h} \quad \text{and} \quad \text{Center} = \left(0, \frac{1}{2h}\right) \tag{11.52}$$

A sketch of the resulting circle is shown in Figure 11.3(b).

In general, for a multi-dof viscous-damped system, one has the "mobility" function

$$G_{ik}(j\omega) = j\omega \sum_{r=1}^{n} \frac{(\psi_i \psi_k)_r}{[\omega_r^2 - \omega^2 + 2j\zeta_r \omega_r \omega]} \tag{11.53}$$

If the resonances are not closely spaced, one can assume that near each resonance (r)

$$G_{ik} = \text{Constant offset (complex)} + \text{Single - dof mobility function}$$

$$= \text{Constant offset} + \left[\frac{j\omega(\psi_i \psi_k)_r}{\omega_r^2 - \omega^2 + 2j\zeta_r \omega_r \omega}\right] \tag{11.54}$$

One can curve-fit each resonance r to a circle this way and thereby extract the $(\psi_i \psi_k)_r$ value (the residue) from the radius of the circle fit.

Note: This method will lead to larger errors if the resonances are closely spaced and, consequently, if significant modal interactions are present.

Peak Picking Method

This is also a single-dof method in view of the fact that each resonance of an experimentally determined transfer function is considered separately. The approach is to compare the resonance region with an analytical transfer function of a damped single-dof system. One of three types of transfer functions — receptance, mobility, or acceleration — as listed in Table 11.1, can be used for this purpose. Note that when the level of damping is small, it can be assumed (approximately) that the resonance is at the undamped natural frequency $\omega_n = \sqrt{k/m}$. Substituting this value for ω in each of the frequency transfer functions, one can determine the transfer function value at resonance, denoted by $G_{\text{peak}}(j\omega)$. It is noted from Table 11.1 that, in general, this function value depends on the damping constant and the natural frequency. Because ω_n is known directly from the peak location of the transfer function, it is possible to compute the damping constant c (or damping ratio ζ) by first determining the corresponding peak magnitude.

Specifically, from Table 11.1, it is clear that one should pick the imaginary part of the frequency-transfer function for receptance or acceleration data, and the real part of the transfer function for mobility data. Then, one picks the peak value of the chosen part of the transfer function, and the frequency at the peak.

Peak picking is good for cases where modes are well separated and lightly damped. It does not work when the system is highly damped (or overdamped), or when the damping is 0 (infinite peak). It is a quick approach that is appropriate for preliminary evaluations and troubleshooting.

TABLE 11.1
Some Frequency Transfer Functions Used in Peak Picking

Single-dof System		$m\ddot{y} + c\dot{y} + ky = f(t)$	
Receptance (dynamic flexibility, compliance)	$G^r(j\omega) = \dfrac{\text{Displacement } y}{\text{Force } f}$	$\dfrac{1}{ms^2 + cs + k}$	with $s = j\omega$
Mobility	$G^m(j\omega) = \dfrac{\text{Velocity } v}{\text{Force } f}$	$\dfrac{s}{ms^2 + cs + k}$	with $s = j\omega$
Accelerance	$G^a(j\omega) = \dfrac{\text{Acceleration } a}{\text{Force } f}$	$\dfrac{s^2}{ms^2 + cs + k}$	with $s = j\omega$
Resonant peak $G_{\text{peak}}(j\omega)$ (occurs approximately at $\omega = \omega_n$ for light damping)		$G^r_{\text{peak}} = -\dfrac{j}{c\omega_n}$ $G^m_{\text{peak}} = \dfrac{1}{c}$ $G^a_{\text{peak}} = \dfrac{j\omega_n}{c}$	

11.3.4 Multi-Degree-of-Freedom Curve Fitting

A general multi-resonance curve-fitting method is now presented; the corresponding single-resonance method should also be clear from this general procedure. Note that many different versions of problem formulation and algorithm development are possible for least-squares curve fitting, but the results should be essentially the same. The method presented here is a frequency-domain method, as one is dealing in this chapter with experimentally determined frequency-transfer functions. In a comparable time-domain method, a suitable analytical expression of the complex exponential form is fitted into the experimental impulse response function obtained by the inverse Fourier transformation of a measured transfer function. That method inherits additional error due to truncation (*leakage*) and finite sampling rate (*aliasing*) during the inverse FFT (see Chapter 4 and Appendix D).

Formulation of the Method

The objective of the present multi-resonance (multi-dof) curve-fitting procedure is to fit the computed (measured) transfer function data into an analytical expression of the form:

$$G(s) = \frac{[b_0 + b_1 s + \cdots + b_m s^m]}{[a_0 + a_1 s + \cdots + a_{p-1} s^{p-1} + s^p]} \quad \text{for } m \leq p \quad (11.55)$$

The data for curve fitting would be the N complex transfer function values $[G_1, G_2, \ldots, G_N]$ computed at discrete frequencies $[\omega_1, \omega_2, \ldots, \omega_N]$. Typically, if 1024 samples of time history were used in the FFT computations to determine the transfer function, one would have 512 valid spectral lines of transfer function data. But, near the high-frequency end, these data values become excessively distorted due to the aliasing error; only a part of the 512 spectral lines is usable, typically the first 400 lines. In that case, one has $N = 400$. This value can be doubled by doubling the FFT block size (to 2048 words in the buffer), thereby doubling the record length or the sampling rate. It is acceptable to leave out part of the computed transfer function, not for poor accuracy but because that part falls outside the frequency band of interest in that particular modal analysis problem. A less wasteful practice would be to pick the sampling rate of the measured time history data to reflect the highest frequency of interest in the modal analysis.

The (complex) error in the estimated value at each frequency point (spectral line) ω_i is given by

$$\tilde{e}_i = G(\omega_i) - G_i$$

$$= \frac{[b_0 + b_1 s_i + \cdots + b_m s_i^m]}{[a_0 + a_1 s_i + \cdots + a_{p-1} s_i^{p-1} + s^p]} - G_i \quad \text{with } s = j\omega \quad (11.56)$$

The characteristic equation of the dynamic model is given by

$$\Delta(s) = a_0 + a_1 s + \cdots + a_{p-1} s^{p-1} + s^p = 0 \quad (11.57)$$

Its roots are the eigenvalues of the system. For damped oscillatory systems, they occur in complex conjugates with negative real parts (*Note*: $p = 2 \times$ number of degrees of freedom, in typical cases). For systems with *rigid body modes* (see Chapter 5), 0 eigenvalues will also be present. But because there is some damping in the system and because the lowest frequency that is tested and analyzed is normally greater than 0 even for systems with rigid body modes, one obtains

$$\Delta(j\omega) \neq 0 \quad (11.58)$$

in the frequency range of interest. Hence, the estimation error given by equation (11.56) can be expressed as

$$e_i = [b_0 + b_1 s_i + \cdots + b_m s_i^m] - G_i [a_0 + a_1 s_i + \cdots + a_{p-1} s_i^{p-1} + s_i^p] \quad \text{with } s = j\omega \quad (11.59)$$

The quadratic error function is given by the sum of the squares of magnitude error for all discrete frequency points used in modal analysis; thus,

$$J = \sum_{i=1}^{N} |e_i|^2 = \sum_{i=1}^{N} e_i^* e_i \quad (11.60)$$

Note that [ ]* denotes the complex conjugate. Complex conjugation is achieved by simply replacing $(j\omega)$ by $(-j\omega)$ in equation (11.59). It follows that equation (11.60) can be written as

$$J = \sum_{i=1}^{N} \Big\langle \big[b_0 + b_1(-j\omega_i) + \cdots + b_m(-j\omega_i)^m - G_i^* \{a_0 + a_1(-j\omega_i) + \cdots + a_{p-1}(-j\omega_i)^{p-1} + (-j\omega_i)^p\} \big] \\ \big[b_0 + b_1(j\omega_i) + \cdots + b_m(j\omega_i)^m - G_i \{a_0 + a_1(j\omega_i) + \cdots + a_{p-1}(j\omega_i)^{p-1} + (j\omega_i)^p\} \big] \Big\rangle \quad (11.61)$$

Experimental Modal Analysis

The basis of the least-squares curve-fitting method of parameter estimation is to pick the transfer function parameters b_i ($i = 0, 1, \ldots, m$) and a_i ($i = 0, \ldots, p - 1$) such that the quadratic error function J is a minimum. Analytically, this requires that

$$\frac{\partial J}{\delta b_k} = 0 \quad k = 0, 1, \ldots, m \tag{11.62}$$

$$\frac{\partial J}{\delta a_k} = 0 \quad k = 0, 1, \ldots, p-1 \tag{11.63}$$

Note that equations (11.62) and (11.63) correspond to $m + p + 1$ linear equations in the $m + p + 1$ unknowns b_i ($i = 0, 1, \ldots, m$) and a_i ($i = 0, \ldots, p - 1$). A well-defined solution exists to this set of *nonhomogeneous equations*, provided that the equations are *linearly independent*, which is guaranteed if the determinant of the coefficients of the unknown parameters does not vanish. It is a good practice to check for linear independence of the set of $m + p + 1$ equations using this determinant condition prior to performing further computations to solve the equations. The solution approach itself is primarily computational in nature and is not presented here. Figure 11.4 shows a result of multi-dof curve fitting on an experimental frequency-transfer function, as collected from a civil engineering structure. Note the close match of the magnitude but not the phase angle. This analysis resulted in the resonant frequency and damping ratio values that are given in Table 11.2. Also note how the damping ratio decreases with the mode number.

11.3.5 A Comment on Static Modes and Rigid Body Modes

Some test systems may possess static modes, and rigid body modes (see Chapter 5) under rare circumstances. Static modes arise in analytical models if one fails to assign an inertia (mass) element for every degree of freedom. Rigid body modes arise in analytical models if proper restraints are not provided for the inertia elements. In practice, however, static modes arise if a coordinate is assigned to a dof that actually does not exist, or if some parts of the physical system are relatively very light with stiff restraints (i.e., very high natural frequencies); and rigid body modes arise in the presence of relatively very heavy components restrained by very flexible elements (i.e., very low natural frequencies). Note that the assumed transfer function (11.55) allows for both these extremes. Specifically, if static modes are present, it is necessary that the transfer function can be expressed as a sum of a constant term (static mode) and an ordinary transfer function (without a static mode). Hence, it will approach a non-zero constant value as the frequency ω increases. This requires that $m = p$. If rigid body modes are present, the characteristic polynomial $\Delta(s)$ of the model should have a factor s^2. This corresponds to $a_0 = a_1 = 0$.

11.3.6 Residue Extraction

The estimated transfer function as given by equation (11.55) is in the form of the ratio of two polynomials; the *rational fraction form*. This has to be converted into the partial fraction form given by equation (11.17) in order to extract the residues $(\psi_i \psi_k)_r$ that are needed for determining the mode shapes. For this, the natural frequencies ω_r and the modal damping ratios ζ_r should be computed first. These are given by the roots of the characteristic equation (11.57) as

$$\lambda_r, \lambda_r^* = \zeta_r \omega_r \pm j\sqrt{1 - \zeta_r^2}\,\omega_r; \quad r = 1, 2, \ldots, n \tag{11.64}$$

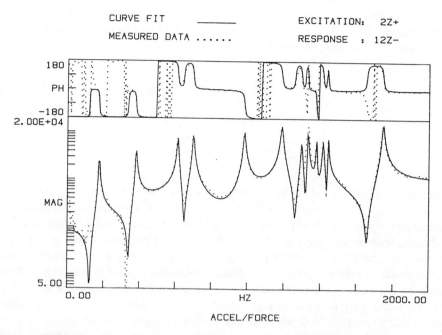

FIGURE 11.4 An example of multi-dof curve fitting on experimental data.

TABLE 11.2
Extracted Parameters in an Example of Experimental Modal Analysis

Mode No.	Resonant Frequency (Hz)	Damping Ratio (viscous)
1	1.773×10^2	1.170×10^{-2}
2	3.829×10^2	8.149×10^{-3}
3	6.145×10^2	6.033×10^{-3}
4	7.018×10^2	5.931×10^{-3}
5	9.839×10^2	4.580×10^{-3}
6	1.190×10^3	3.676×10^{-3}

Once these eigenvalues are known, by solving equation (11.57) using the estimated values for a_0, a_1, ..., a_{p-1}, it is a straightforward task to compute the quadratic factors

$$\Delta_r(s) = s^2 + 2\zeta_r \omega_r s + \omega_r^2 \qquad r = 1, 2, ..., n \tag{11.65}$$

Note from equation (11.17) that

$$(\psi_i \psi_r) = [G_{ik}(s) \cdot \Delta_r(s)]_{s=\lambda_r} \tag{11.66}$$

assuming distinct eigenvalues. This is true because when the partial fraction form is multiplied by $\Delta_r(s)$, it will cancel out with the denominator of the partial fraction corresponding to the rth mode, leaving its residue. Then, when s is set equal to λ_r, all the remaining partial fraction terms will vanish due to the fact that $\Delta_r(\lambda_r) = 0$, provided that the eigenvalues are distinct. Since $G_{ik}(s)$ are known from the estimated transfer functions, the residues can be computed using equation (11.66).

Experimental Modal Analysis

BOX 11.2 Curve Fitting of Transfer Function Data

Single-Resonance Curve Fitting:
A. Viscous Damping:
   1. Compute the mobility (velocity/force) function near resonance
   2. Scale the data
   3. Curve-fit to a circle in the Nyquist plane (Argand diagram).
B. Hysteretic Damping:
   1. Compute receptance (displacement/force) function near resonance
   2. Scale the data
   3. Curve-fit to a circle in the Nyquist plane (Argand diagram).

Multi-Resonance Curve Fitting:
   1. Compute a transfer function over the entire frequency range
   2. Scale the data
   3. Curve-fit to a general polynomial ratio with static and rigid body modes.

Finally, the mode shapes are determined using the procedure outlined earlier. Some curve-fitting approaches are summarized in Box 11.2.

11.4 LABORATORY EXPERIMENTS

Testing and analysis are important in the practice of mechanical vibration and are integral in experimental modal analysis. This section describes two experiments in the category of modal testing. One experiment deals with a lumped-parameter system and the other with a distributed-parameter (or continuous) system. Both experiments have direct practical implications and have been used in an established undergraduate course in mechanical vibrations.

11.4.1 LUMPED-PARAMETER SYSTEM

A schematic representation of a prototype unit used in a laboratory for modal testing is shown in Figure 11.5. A view of the experimental system is shown in Figure 11.6. The system is a crude representation of an engine unit that is supported on flexible mounts and subjected to unbalance forces and moments.

The test object is assumed to consist of lumped elements of inertia, stiffness, and damping. The rectangular metal box, which represents the engine housing, is mounted on four springs and damping elements at the four corners. Inside the box are two pairs of identical and meshed gears that are driven by a single DC motor. Each gear has two slots at diametrically opposite locations in order to place the eccentric masses. Various types of unbalance excitations can be generated by placing the four eccentric masses at different combinations of locations on the gear wheels.

The drive motor is operated by a DC power supply with a speed control knob. The motor speed (and hence the gear speed) is measured using an optical encoder that is mounted on the drive shaft. It generates pulses as the encoder disk rotates with the shaft, in proportion to the angle of rotation. The pulse frequency of the encoder determines the shaft speed. A pair of accelerometers with magnetic bases are mounted on the top of the engine box. The locations that are used for this purpose are indicated in Figure 11.5. Figure 11.6 shows, from left to right, the following components of the experimental system:

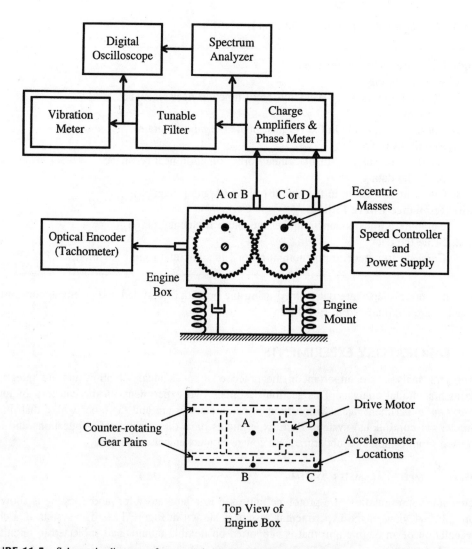

FIGURE 11.5 Schematic diagram of an experimental setup for modal testing in a laboratory.

1. Digital spectrum analyzer (The equipment in the background of this unit is not related to the equipment.)
2. A combined instrument panel consisting of a vibration meter, a tunable bandpass filter, and a unit consisting of a conditioning amplifier and a phase meter
3. Power supply for the instrument panel, placed on top of the panel
4. Engine unit with two accelerometers mounted on top surface of the housing
5. Digital oscilloscope placed on a shelftop immediately above the engine unit
6. DC power supply and speed controller combination for the drive motor.

The phase meter measures the phase difference between two input signals. The tunable filter is a bandpass filter, and it can be tuned by a fine-adjustment dial so that a signal in a very narrow band (i.e., harmonic signal) can be filtered and measured. The vibration meter measures the magnitude (peak or rms value) of a signal. The choice of a displacement value (i.e., double integration), a velocity value (a single integration), or an acceleration value (no integration) is available, and can be selected using a knob.

Experimental Modal Analysis

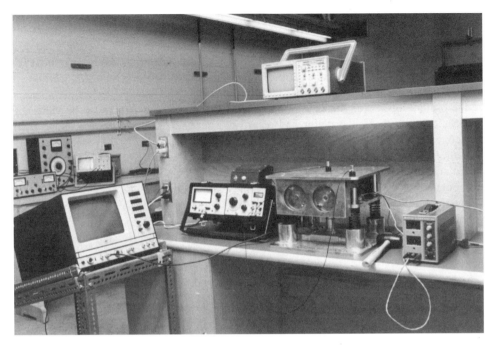

FIGURE 11.6 A view of an experimental setup for modal testing. (Courtesy of the University of British Columbia. With permission.)

By placing the eccentric masses at various locations on the gear wheels, different modes can be excited. For example, if all four eccentric masses are placed at the vertical radius location above the rotating axis, that will generate a net harmonic force in the vertical direction as the motor is driven. This will excite the heave (up and down) mode of the engine box. If the two masses on a meshed pair are placed at the vertical radius location below the rotating axis while the masses on the other meshed pair are placed vertically above the rotating axis, then it will result in a net moment (pitch) about a central horizontal axis of the engine box. This will excite the pitch mode, and so on. For a given arrangement of eccentric masses, two tests can be carried out: one in the frequency domain and the other in the time domain.

Frequency-Domain Test

Choose the "displacement" setting of the vibration meter. Start the motor and maintain the speed at a low value (e.g., 4 Hz). Tune the filter, using its dial, until the vibration meter reading becomes the largest. The tuned frequency will be, in the ideal case, equal to the motor speed. Record the motor speed (i.e., the excitation frequency) and the magnitude of the displacement response. Increase the motor speed in 1-Hz steps and repeat the measurements, up to a reasonably high frequency, covering at least one resonance (i.e., 25 Hz). Reduce the speed in steps of 1 Hz and repeat the measurements. Take some more measurements in the neighborhood of each resonance using smaller frequency steps. Plot the data, as a frequency spectrum, after compensating for the fact that the amplitude of the excitation force increases with the square of the drive speed (hence, divide the vibration magnitudes by square of the frequency). This experiment can be used, for example, to measure mode shapes, resonant frequencies, and damping ratios (by the half-power bandwidth method). The analytical details are found in Chapters 3 and 7.

Time-Domain Tests

A test can be conducted, by using the logarithmic decrement method, to determine the damping ratio corresponding to a particular mode. Here, first pick the eccentric mass arrangement so as to excite the desired mode. Then increase the motor speed and then fine-tune the operation at the desired resonance. Maintain a steady speed at this condition, and observe the accelerometer signal using the oscilloscope, while making sure that at least ten complete cycles can be viewed on the screen. Suddenly, turn off the motor and record the decay of the acceleration signal using the oscilloscope. Analytical details are found in Chapters 2 and 7.

Another test that can be carried out is an impact (hammer) test. Here, use the spectrum analyzer to record and analyze the vibration response of the engine box, through an accelerometer. Gently tap the engine box in different critical directions (e.g., at points A, B, C, and D in the vertical direction, in Figure 11.5; or in the horizontal direction on the side of the engine box in the neighborhood of these points) and acquire the vibration signal using the spectrum analyzer. Process the signal using the spectrum analyzer, obtain the resonant frequencies, and compare them with those obtained from sine testing.

11.4.2 Distributed-Parameter System

All real-life vibrating systems have continuous components. However, one often makes distributed-parameter assumptions, depending on the properties and the operating frequency range of the vibrating system. When a lumped-parameter approximation is not adequate, a distributed-parameter analysis will be needed. Modal testing and comparison with analytic results can validate an analytical model.

The response of a distributed-parameter system will depend on the boundary conditions (supporting conditions) as well as the initial conditions. For forced excitations, the response will depend on the nature of the excitation as well. Natural frequencies and mode shapes are system characteristics and will depend on the boundary conditions, but not on the initial conditions and forcing excitations. This subject is discussed in Chapter 6.

Consider the experimental setup schematically represented in Figure 11.7. A view of the setup is shown in Figure 11.8. The device that is tested is a ski. For analytical purposes, it can be approximated as a thin beam (see Chapter 6 for the Bernoulli-Euler beam model). The objective of the test is to determine the natural frequencies and mode shapes of the ski. Because the significant frequency range of the excitation forces on a ski, during use, is below 15 Hz, it is advisable to determine the modal information in the frequency range of about double the operating range (i.e., 0 Hz to 30 Hz). In particular, in the design of a ski, natural frequencies below 15 Hz should be avoided, while keeping the unit as light and strong as possible. These are conflicting design requirements. It follows that modal testing can play an important role in the design development of a ski.

Consider the experimental setup sketched in Figure 11.7. The ski is firmly supported at its middle, on the electrodynamic shaker. Two accelerometers are mounted on either side of the support and are movable along the ski. The accelerometer signals are acquired and conditioned using charge amplifiers. The two signals are observed in the x-y mode of the digital oscilloscope so that both the amplitudes and the phase difference can be measured. The sine-random signal generator is set to the sine mode so that a harmonic excitation is generated at the shaker head. The shape of the motion can be observed in slow motion by illuminating the ski with the hand-held stroboscope, with the strobe frequency set to within about ± 1 Hz of the excitation frequency.

In the experimental system shown in Figure 11.8, one observes, from left to right, the following components:

Experimental Modal Analysis

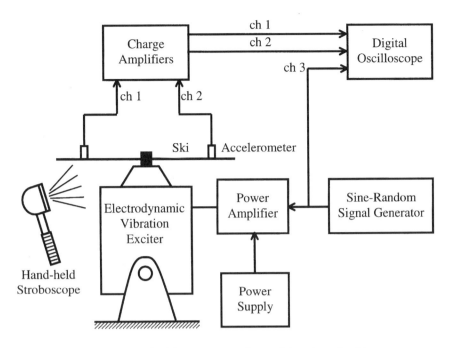

FIGURE 11.7 Schematic diagram of a laboratory setup for modal testing of a ski.

FIGURE 11.8 A view of the experimental system for modal testing of a ski. (Courtesy of the University of British Columbia. With permission.)

1. Electrodynamic shaker with the ski mounted on its exciter head; two accelerometers are mounted on the ski
2. Hand-held stroboscope, placed beside the shaker
3. Power amplifier for driving the shaker, placed on top of the side table

4. Two charge amplifiers placed on top of the power amplifier and connected to the accelerometers
5. Sine-random signal generator, placed on the tabletop, next to the amplifier
6. Digital oscilloscope
7. Static load-deflection measurement device for determining the modulus of rigidity (EI) of the ski (placed on the floor below the intstrument table).

Prior to modal testing, the modulus of rigidity of the ski is determined by supporting it on the two smooth end pegs of the loading structure, and loading at the mid-span using incremental steps of 500-g weights up to 4.0 kg, placed on a scale pan that is suspended at the mid-span of the ski. The mid-span deflection of the ski is measured using a spring-loaded dial gage that is mounted on the loading structure. If the mid-span stiffness (force/deflection) as measured in this manner is k, it is known that the modulus of rigidity is

$$EI = \frac{kl^3}{48} \qquad (11.67)$$

where l is the length between the support points of the ski. Note that this formula is for a simply supported ski, which is the case in view of the smooth supporting pegs. Also, weigh the ski and then compute m = mass per unit length. With this information, the natural frequencies and mode shapes can be computed for various end conditions, as discussed in Chapter 6. In particular, compute this modal information for the following supporting conditions:

1. Free-free
2. Clamped at the center

Next, perform modal testing using the experimental setup and compare the results with those computed using the analytical formulation.

The natural frequencies (actually, resonant frequencies, which are almost equal to the natural frequencies in the present case of light damping) can be determined by increasing the frequency of excitation in small steps using the sine generator and noting the frequency values at which the amplitudes of the accelerometer signals reach local maxima, as observed on the oscilloscope screen. A mode shape is measured as follows: first detect the corresponding natural frequency as above; while maintaining the shaker excitation at this frequency, place the accelerometer near the shaker head, and then move the other accelerometer from one end of the ski to the other in small steps of displacement and observe the amplitude ratio and the phase difference of the two accelerometer signals, using the oscilloscope. Note that in-phase signals mean that the motions of the two points are in the same direction, and the out-of-phase signals mean that the motions are in opposite directions. The mode shapes can be verified by observing the modal vibrations in slow motion, using the stroboscope, as indicated before. Node points are the vibration-free points. They can be detected from the mode shapes. In particular, a tiny piece of paper will remain stationary at a node while making large jumps on either side of the node. Also, the phase angle of the vibration signal, as measured by an accelerometer, will jump by 180° if the accelerometer is carefully moved across a node point.

11.5 COMMERCIAL EMA SYSTEMS

Commercially available *experimental modal analysis* (EMA) systems typically consist of an FFT analyzer, a modal analysis processor, a graphics terminal, and a storage device. Digital plotters, channel selectors, hard copy units, and other accessories can be interfaced, and the operation of

Experimental Modal Analysis

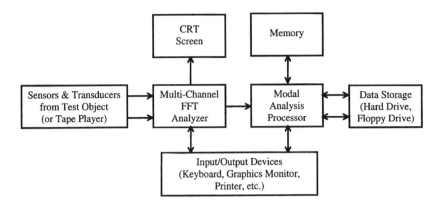

FIGURE 11.9 The configuration of a commercial experimental modal analysis system.

the overall system can be coordinated through a host computer to enhance its capability. The selection of hardware for a particular application should address specific objectives as well as hardware capabilities. Software selection is equally important. Proper selection of an EMA system is difficult unless the underlying theory is understood — in particular, determination of transfer functions via FFT analysis; extraction of natural frequencies, modal damping ratios, and mode shapes from transfer function data; and the construction of mass, stiffness, and damping matrices from modal data should be considered. The underlying theory has been presented in this chapter. The present section describes the features of a typical experimental modal analysis system.

11.5.1 System Configuration

The extraction of modal parameters from dynamic test data is essentially a two-step procedure consisting of (1) FFT analysis and (2) modal analysis.

In the first step, appropriate frequency transfer functions are computed and stored. These raw transfer functions form the input data for the subsequent modal analysis, yielding modal parameters (natural frequencies, damping ratios, and mode shapes) and a linear differential equation model for the dynamic system (test object).

FFT Analysis Options

The basic hardware configuration of a commercial modal analysis system is shown in Figure 11.9. Notice that the FFT analyzer forms the front end of the system. The excitation signal and the response measurements can be transmitted on-line to the FFT analyzer (through charge amplifiers for piezoelectric sensors); many signals can be transmitted simultaneously in the multiple-channel case. Alternatively, all measurements can first be recorded on a multiple-track FM tape and subsequently fed into the analyzer through a multiplexer. In the first case, it would be necessary to take the FFT analyzer to the test site; an FM tape recorder is needed at the test site in the second case.

Through advances in microelectronics and LSI technology, the FFT analyzer has rapidly evolved into a powerful yet compact instrument that is often smaller in size than the conventional tape recorder used in vibration data acquisition; either device can be used in the field with equal convenience. On-site FFT analysis, however, allows one to identify and reject unacceptable measurements (e.g., low signal levels and high noise components) during data acquisition, so that alternative data that might be needed for a complete modal identification can be collected without having to repeat the test at another time. The main advantage of the FM tape method is that data are available in analog form, free of quantization error (digital word-size dependent), aliasing distortion (data sampling-rate dependent), and signal truncation error (data block-size dependent).

Sophisticated analog filtering is often necessary, however, to remove extraneous noise entering from the recording process (e.g., line noise and tape noise), as well as from the measurement process (e.g., sensor and amplifier noise).

The analog-to-digital converter (ADC) is normally an integral part of the analyzer (see Chapter 9). The raw transfer functions, once computed, are stored on a floppy disk (or hard disk) as the "transfer function file." This constitutes the input data file for modal extraction. Some analyzers, instead, compute power spectral densities with respect to the excitation signal, and store these in the data file. From these data, it is possible to instantly compute coherence functions, transfer functions, and other spectral information using keyboard commands. Another procedure has been to compute Fourier spectra of all signals and store them as raw data, from which other spectral functions can be conveniently computed. Most analyzers have small CRT screens to display spectral results. Low-coherent transfer functions are detected by analytical or visual monitoring, and are automatically discarded.

In principle, the same processor can be used for both FFT analysis and modal analysis. Some commercial modal analysis systems use a plug-in programmable FFT card in a common processor cage. Historically, however, the digital FFT analyzer was developed as a stand-alone hardware unit to be used as a powerful measuring instrument, rather than just as a data processor, in a wide variety of applications. Uses include measurement of resonant frequencies and damping in vibration isolation applications, measurement of phase lag between two signals, estimation of signal noise levels, identification of the sources of noise in measured signals, and measurement of correlation in a pair of signals. Because of this versatility, most modal analysis systems do come with a standard FFT analyzer unit as the front end, and a separate computer for modal analysis.

Modal Analysis Components

In addition to the transfer function file, the modal analysis processor needs geometric information about the test object; typically, coordinates of the mass points and directions of the degrees of freedom. This information is stored in a "geometry file." The results of modal analysis are usually stored in two separate files: the "parameter file," containing natural frequencies, modal damping ratios, mass matrix, stiffness matrix, and damping matrix; and the "mode shape file," containing mode shape vectors that are used for graphics display and printout. Individual modes can be displayed on the CRT screen of the graphics monitor, either as a static trace or in animated (dynamic) form. The graphics monitor and printer are standard components of the system. The entire system can be interfaced with other peripheral I/O devices using an IEEE-488 interface bus or the somewhat slower serial RS-232 interface. For example, the overall operation can be coordinated, and further processing done, using a host computer. A desktop (personal) computer can be substituted for the modal analysis processor, graphics monitor, and storage devices in the standard system, resulting in a reasonable reduction of the overall cost as well. An alternative configuration that is particularly useful in data transfer and communication from remote test sites uses a voice-grade telephone line and a modem coupler to link the FFT analyzer to the main processor.

PROBLEMS

11.1 Describe the equipment needed to obtain test data for use in experimental modal analysis. Describe hardware and software components in a commercial modal analysis system.

11.2 List the main steps of experimental modal analysis, starting with the measurement of force-response time histories and ending with the computation of the mass, stiffness, and damping matrices.

Consider a six-degree-of-freedom system. The transfer-function matrix $G(j\omega)$ is schematically represented as in Figure P11.2. How many transfer function measurements are needed in order to extract the M, K, and C matrices? Two possibilities are marked as

Experimental Modal Analysis

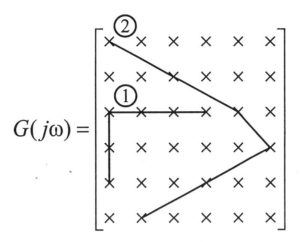

FIGURE P11.2 Example sets of transfer function measurements.

<u>1</u> and <u>2</u> in solid lines. Is there an unacceptable set here? Is there an acceptable set? Explain your answers?

11.3 Explain under what conditions the circle fit method can be used to extract modal parameters in multi-degree-of-freedom (multi-resonance) transfer functions.

Consider a single-degree-of-freedom vibrating system with mass m, stiffness k, and hysteretic damping constant h. The system is supported on a rigid floor and a force f is applied to the mass element. The response (output) is the displacement of the mass. Write the time-domain (differential equation) model and the frequency-domain (transfer function) model for the system.

Show that the Nyquist plot of this transfer function (receptance) is a circle.

11.4 In the frequency domain, receptance = displacement/force; mobility = velocity/force; accelerance = acceleration/force. Table P11.4 gives normalized expressions for these three frequency transfer functions in the standard case of a single-degree-of-freedom mechanical system with

a. Viscous damping
b. Hysteretic damping.

Note that $r = \omega/\omega_n$, where ω is the excitation frequency and ω_n is the undamped natural frequency; ζ is the damping ratio in the case of viscous damping; and $d = h/k$, where h = hysteretic damping parameter and k = system stiffness.

Generate the Nyquist plots (i.e., real part vs. imaginary part) of the frequency-transfer functions as r varies from 0 to 50, for the two cases $\zeta = 0.1$ and $d = 0.2$. Discuss the nature of these plots.

11.5 Consider a standard single-degree-of-freedom vibrating system that is given by

$$m\ddot{y} + b\dot{y} + ky = f(t)$$

in which m, b, and k are the inertia, viscous damping constant, and the stiffness; y is the displacement response; and $f(t)$ is the forcing excitation.

a. What is the mobility function, in the frequency domain, as a function of frequency ω?
b. For small frequencies of excitation, what is the slope of the mobility magnitude plot in a log–log scale (i.e., in dB per decade)?
c. For large frequencies of excitation, what is the slope of the mobility magnitude plot in dB per decade?

TABLE P11.4
Normalized Frequency Response Functions for Single-dof Curve Fitting

Frequency Response Function	With Viscous Damping	With Hysteretic Damping
Receptance	$\dfrac{1}{1-r^2+2j\zeta r}$	$\dfrac{1}{1-r^2+jd}$
Mobility	$\dfrac{jr}{1-r^2+2j\zeta r}$	$\dfrac{jr}{1-r^2+jd}$
Accelerance	$\dfrac{-r^2}{1-r^2+2j\zeta r}$	$\dfrac{-r^2}{1-r^2+jd}$

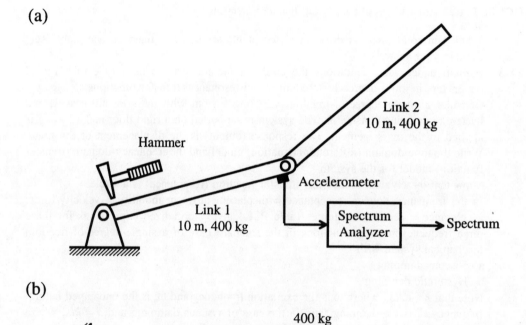

FIGURE P11.6 (a) An instrumented two-link space robot, and (b) an equivalent approximate model.

d. Suppose that a mobility plot of a vibrating system was made near a resonance, using experimental data. If the points very close to the resonance do not have an adequate resolution, how would you estimate the resonant frequency from this data? Assume that the other possible modes of the system are located far away from the present resonance.

11.6 A two-link space robot with two identical links, each with length 10 m and mass 400 kg, was tested for its natural frequencies using the arrangement illustrated in Figure P11.6(a). First, the two joints were locked. A hammer impact was made in the lateral direction

Experimental Modal Analysis

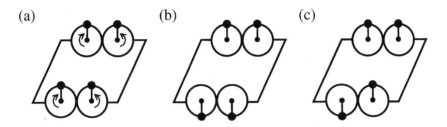

FIGURE P11.7 Three possible configurations of counter-rotating eccentric masses.

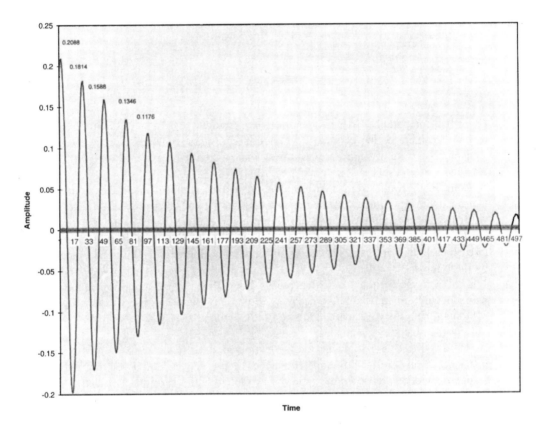

FIGURE P11.8 A free decay signal (approximately 11 Hz) of a mechanical system.

near the first joint, and the resulting response near the second joint, in the lateral direction, was measured using an accelerometer. The response signal was processed using a spectrum analyzer. The resulting primary resonance was found to be at 0.35 Hz. This result is verified analytically in the following manner. First, the robot system is approximated as a cantilever of length 10 m and mass 400 kg, with a lumped end mass 400 kg, as shown in Figure P11.6(b). Next, the fundamental natural frequency is computed using the standard formula for a Bernoulli-Euler beam (see Table 6.3)

$$\omega_1 = \lambda_1^2 \sqrt{\frac{EI}{m}}$$

where

l = length of the cantilever = 10.0 m
m = mass per unit length = 40.0 kg·m^{-1}
EI = modulus of rigidity of the cantilever
     = 8.25×10^5 N·m² (given)
λ_1 = mode shape parameter for mode 1.

The following two approaches are used:
a. Assume a heavy uniform beam with a lumped end mass with end mass/beam mass ratio = 1.0. In this case, it is known that the exact solution is $\lambda_1 l = 1.2479$.
b. Compute the fundamental natural frequency of a uniform cantilever of length 10 m and mass 400 kg, without an end mass. In this case, $\lambda_1 l = 1.8751$. Also, the lateral stiffness at the free end of a cantilever is known to be

$$k = \frac{f}{y_0} = \frac{3EI}{l^3}$$

From this, compute the equivalent end mass m_{eq} that will give the same natural frequency as that obtained for the cantilever without an end mass. Finally, compute the natural frequency for a light cantilever of stiffness k and a combined end mass $m_l + m_{eq}$, where m_l = mass of 2nd link = 400 kg.

Compare the results from the two methods (a) and (b), with that obtained experimentally.

11.7 Consider the experimental system shown in Figure 11.5, consisting of an engine unit with unbalance excitations. Three possible configurations for locating the eccentric masses are shown in Figure P11.7. Give the directions of the resulting excitation forces in each case, and discuss what modes of vibration of the engine unit would be excited by these forces.

11.8 A test object was excited in one of its modes of vibration using a shaker. Then, the shaker was suddenly turned off and the decaying response of the test object was recorded. The resulting time history is shown in Figure P11.8. Estimate the equivalent viscous damping ratio of the object in this mode.

A sine dwell test was conducted on the test object in a frequency band containing the resonance, using sufficiently small frequency steps. The magnitude of the frequency-response function (response/excitation) was plotted and the resonant peak was established, which corresponds to the same mode of vibration as for the previous test. The half-power frequencies were found from the response spectrum. They are $f_1 = 10.2$ Hz and $f_2 = 11.0$ Hz. Estimate the viscous damping ratio using this information, and comment on the accuracy of the result.

11.9 Three mode shapes of a beam, as determined analytically using the Bernoulli-Euler model, are shown in Figure P11.9. Guess the boundary conditions of the beam. What are the practical difficulties that one would encounter in experimental determination of these mode shapes?

11.10 A modal test was performed on an aluminum I-beam. A mode shape that was determined is shown in Figure P11.10 (a). Next, a known mass (1 kg) was attached at one corner of an end plate and the test was repeated. The resulting mode shape is shown in Figure P11.10(b). The frequency-response function between the exciter location and the response location, determined using a shaker test, is shown in Figure P11.10(c) for the I-beam with an extra mass. Identify which curve corresponds to the beam with the extra mass. Indicate an application where experimental procedures of this type are useful.

Experimental Modal Analysis

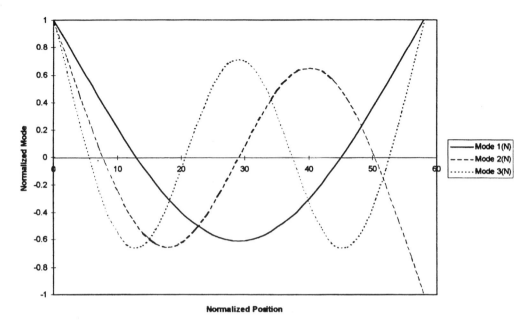

FIGURE P11.9 Shapes of the first three vibrating modes of a beam.

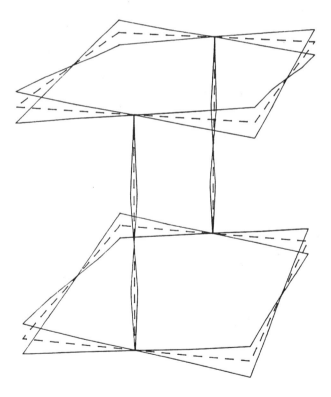

FIGURE P11.10(a) A mode shape of an aluminum I-beam.

11.11 A vibration test was carried out by exciting a mechanical system at the degree of freedom 1 and then measuring the response at the degrees of freedom 1, 2, and 3. This procedure is schematically shown in Figure P11.11. The frequency-response functions

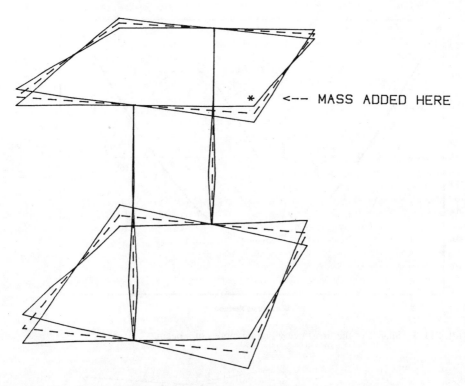

FIGURE P11.10(b) The mode shape when an extra mass is attached.

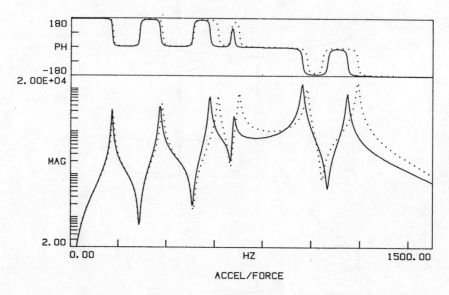

FIGURE P11.10(c) The frequency response function between an excitation–response location pair.

G_{11}, G_{21}, and G_{31} are computed from the test results and the undamped natural frequencies ω_i, the modal damping ratios ζ_i, and the residues R_{ij} are determined, as given in Table P11.11.

Determine:
a. the mode shape vectors of the first three modes
b. the modal matrix

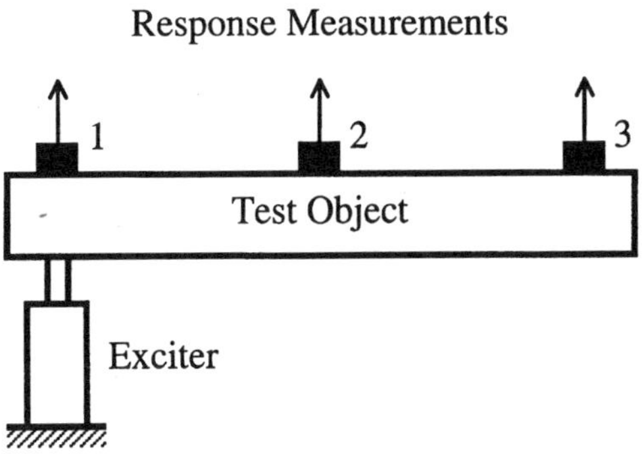

Experimental Modal Analysis

Response Measurements

FIGURE P11.11 Schematic representation of a modal testing procedure.

TABLE P11.11
Results from an Experimental Modal Analysis

Mode Number	Undamped Natural Frequency (rad·s^{-1})	Damping Ratio (viscous)	Residues of G_{11}	G_{21}	G_{31}
1	1.726×10^2	1.137×10^{-2}	0.53	−0.44	0.73
2	3.730×10^2	7.910×10^{-3}	0.38	0.62	−0.65
3	5.796×10^2	5.658×10^{-3}	0.16	0.24	0.41

TABLE P11.12
Comparative Data for Four Modal Analysis Systems

Description	System A	System B	System C	System D
Number of weighting window options available	3	10	5	4
Analyzer data channels	2	2	2	2
Max. degrees of freedom per analysis	750 @ 20 modes	450	725 @ 5 modes	750
Max. number of modes analyzed	50 @ 250 dof	20	10 (typical)	64
Multi-degree-of-freedom curve fitting	Yes	Yes	No	Yes
FFT resolution (usable spectral lines/512)	400	400	400	400
Zoom analysis capability in FFT	Yes	Yes	Yes	Optional
Statistical error-band analysis	No	No	No	No
Static mode-shape extremes	Yes	Yes	Yes	Yes
Animated graphics capability	Yes	Yes	Yes	Yes
Color graphics capability	No	Yes	No	No
Hidden-line display	No	No	No	No
Color printing	No	No	Yes	No
Structural mass and stiffness matrices	No	No	No	Yes
Approximate cost	$30,000	$20,000	$25,000	$50,000

    c. the modal mass, modal stiffness, and modal damping matrices.

Indicate how the mass, stiffness, and damping matrices of the system can be determined from these results. What is a shortcoming of the resulting dynamic model?

11.12 The first step in selecting a modal analysis system for a particular application is to understand the specific needs of that application. For industrial applications of modal testing, the following requirements are typically adequate:

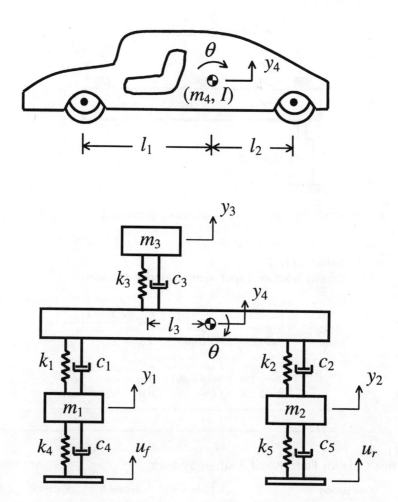

FIGURE P11.15 A planar model of a vehicle.

    i. acceptance of a wide range of measured signals having a variety of transient and frequency band characteristics
    ii. capability of handling up to 300 degrees of freedom of measured data in a single analysis
    iii. FFT with frequency resolution of at least 400 spectral lines per 512
    iv. zoom analysis capability
    v. capability of performing statistical error-band analysis
    vi. static display and plot of mode-shape extremes
    vii. animated (dynamic) display of mode shapes
    viii. color graphics
    ix. hidden-line display
    x. color printing with high line resolution
    xi. capability of generating an accurate time-domain model (mass, stiffness, and damping matrices).

The capabilities of four representative modal analysis systems are summarized in Table P11.12. Give a comparative evaluation of these systems.

11.13 Experimental modal analysis is capable of determining a complete dynamic model (i.e., the mass, stiffness, and damping matrices) of a system. What is the basic requirement with respect to the number of responses that are measured from the test object, in realizing this complete model?

Experimental Modal Analysis

In addition to the extensive computational effort needed for a complete model identification, list several reasons why such an exercise could become futile in practice.

11.14 Experimental modal analysis (EMA) is carried out on "real" systems that have some energy dissipation (damping). However, a fundamental assumption that is made in EMA is the existence of normal (real) modes. Consider the damped system represented by

$$M\ddot{y} + C\dot{y} + Ky = f(t)$$

in the usual notation. Denote the modal matrix by Ψ, where the mode shapes are M-normal.

a. Show that $M^{-1} = \Psi\Psi^T$

b. Show that the damped system and the undamped system have the same (real, normal) modes if and only if the two matrices $M^{-1}C$ and $M^{-1}K$ commute, for a general K and a non-singular M.

11.15 A five-degree-of-freedom planar model of an automobile with a passenger seat is shown in Figure P11.15. Lumped masses m_1 through m_4, body moment of inertia I, lumped stiffnesses k_1 through k_5, and the corresponding damping constants c_1 through c_5 are used in the model as shown. Vertical displacements at various degrees of freedom are denoted by y_1 through y_4. Vehicle pitch angle is denoted by θ. Displacement inputs at the front and rear wheels are u_f and u_r, respectively.

The following parameter values are given:

$i =$	1	2	3	4	5
m_i (lb)	100	100	300	5000	—
k_i (lb·in^{-1})	500	500	100	1500	1500
c_i (lb·s·in^{-1})	25	25	15	5	5
l_i (in)	50	75	10	—	—

a. Obtain the mass matrix, stiffness matrix, and damping matrix of the system.

b. Compute undamped natural frequencies and damped natural frequencies of the system. Compare the values and discuss whether the model is realistic.

c. Determine all five mode shapes of the system, assuming that real modes exist. Is this assumption true for the given damping matrix? Explain your answer.

11.16 A normalized transfer function obtained in an experimental procedure is given by

$$G(j\omega) = \frac{\omega^4 - \omega^2 + 1 - j(\omega^3 - \omega)}{\omega^2(1 - \omega^2 + 0.4j\omega)}$$

Discuss the nature of the physical system on the basis of this information. In particular, comment on the nature of the modes. Determine the natural frequencies and damping ratios of the oscillatory modes.

12 Vibration Design and Control

It has been pointed out that there are desirable and undesirable types and situations of mechanical vibration. This chapter discusses ways of either eliminating or reducing the undesirable effects of vibration. Undesirable vibrations are those that cause human discomfort and hazards, structural degradation and failure, performance deterioration and malfunction of machinery and processes, and various other problems. General approaches to vibration mitigation can be identified from the dynamic systems point of view.

Consider the schematic diagram of a vibratory system shown in Figure 12.1. Forcing excitations $f(t)$ to the mechanical system S cause the vibration responses y. The objective here is to suppress y to a level that is acceptable. Clearly, there are three general ways of doing this.

1. *Isolation*: Suppress the excitations of vibration. This method primarily deals with f.
2. *Design modification*: Modify or redesign the mechanical system so that, for the same levels of excitation, the resulting vibrations are acceptable. This method deals with S.
3. *Control*: Absorb or dissipate the vibrations, using external devices, through implicit or explicit sensing and control. This method primarily deals with y.

Within each of these three categories, several approaches can be used to achieve the objective of vibration mitigation. Essentially, all these approaches involve designing (either complete redesign or incremental design modification) of the system on the one hand, and controlling the vibration through external means (passive or active devices) on the other. The analytical basis for many such approaches was presented in previous chapters. Further analytical procedures will be given in the present chapter. Note that removal of faults (e.g., misalignments and malfunctions by repair or parts replacement) can also remove vibrations. This may fall into any of the three categories listed above, but primarily into the second category of modifying S.

The category of *vibration isolation* involves "isolating" a mechanical system (S) from vibration excitations (f) so that the excitation signals are "filtered" out or dissipated prior to reaching the system. The use of properly designed suspension systems, mounts, and damping layers falls within this category. The category of *design modification* involves making changes to the components and the structure of a mechanical system according to a set of specifications and design guidelines. Balancing of rotating machinery, and structural modification through modal analysis and design techniques, fall into this category. The category of *control* involves either passive devices (which do not use external power) such as dynamic absorbers and dampers, or active control devices (which need external power for operation). In the passive case, the control device implicitly senses the vibration response and dissipates it (as in the case of a damper), or absorbs and stores its energy where it is slowly dissipated (as in the case of a dynamic absorber). In the active case, the vibrations y are explicitly sensed through *sensors* and *transducers*; what forces should be acted on the system to counteract and suppress vibrations are determined by a controller; and the corresponding forces or torques are applied to the system through one or more *actuators*.

Note that there may be some overlap in the three general categories of vibration mitigation mentioned above. For example, the addition of a mount (category 1) can also be interpreted as a design modification (category 2) or as incorporating a passive damper (category 3). It should be noted as well that the general approach commonly known as that *source alteration* may fall into either category 1 or category 2. The purpose in this case is to alter or remove the source of vibration. The source could be either external (e.g., road irregularities that result in vehicle vibrations), a category 1 problem; or internal (imbalance or misalignment in rotating devices that results in periodic

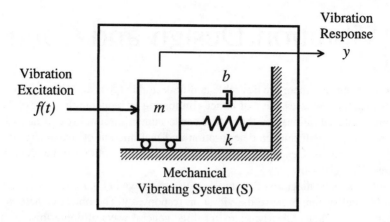

FIGURE 12.1 A vibrating mechanical system.

forces, moments, and vibrations), a category 2 problem. It can be more difficult for a system user to alter external vibration sources (e.g., resurfacing the roadways) than to modify the internal sources (e.g., balancing of rotating machinery). Furthermore, the external source of vibration can be quite random and also may not be accessible at all for alteration (e.g., aerodynamic forces on an aircraft). The present chapter will address some useful topics on the design for vibration suppression and the control of vibration. Typically, a set of vibration specifications is given as simple threshold values (bounds) or frequency spectra, and the goal is to either design or control the system so as to meet these specifications.

SHOCK AND VIBRATION

Sometimes, response to shock loads are considered separately from response to vibration excitations for the purpose of design and control of mechanical systems. For example, shock isolation and vibration isolation are treated under different headings in some literature. This is actually not necessary. Although vibration analysis predominantly involves periodic excitations and responses, transient and random oscillations (vibrations) are also commonly found in practice. The frequency band of the latter two types of signals is much broader than that of a simple periodic signal. A shock signal is transient by definition, and has a very short duration (in comparison to the predominant time constants of the mechanical system to which the shock load is applied). Hence, it will possess a wide band of frequencies. Consequently, frequency-domain techniques are still applicable. Time-domain techniques are particularly suited to dealing with transient signals in general, and shock signals in particular. In that context, a shock excitation can be treated as an impulse whose effect is to instantaneously change the velocity of an inertia element. Then, in the time domain, a shock load can also be treated as an initial-velocity excitation of an otherwise free (unforced) system.

12.1 SPECIFICATION OF VIBRATION LIMITS

Design and control procedures of vibration have the primary objective of ensuring that, under normal operating conditions, the system of interest does not encounter vibration levels that exceed the specified values. In this context, then, the ways of specifying vibration limits become important. This section will present some common ways of vibration specification.

Vibration Design and Control

12.1.1 Peak Level Specification

Vibration limits for a mechanical system can be specified either in the time domain or in the frequency domain. In the time domain, the simplest specification is the peak level of vibration (typically acceleration in units of g, the acceleration due to gravity). Then, the techniques of isolation, design, or control should ensure that the peak vibration response of the system does not exceed the specified level. In this case, the entire time interval of operation of the system is monitored and the peak values are checked against the specifications. Note that in this case, it is the instantaneous peak value at a particular time instant that is of interest, and what is used in representing vibration is an instantaneous amplitude measure rather than an average amplitude or an energy measure.

12.1.2 RMS Value Specification

The root-mean-square (rms) value of a vibration signal $y(t)$ is given by the square root of the average (mean value) of the squared signal:

$$y_{rms} = \left[\frac{1}{T} \int_0^T y^2 dt \right]^{\frac{1}{2}} \qquad (12.1)$$

Note that by squaring the signal, its sign is eliminated and essentially the energy level of the signal is used. The period T over which the squared signal is averaged will depend on the problem and the nature of the signal. For a periodic signal, one period is adequate for averaging. For transient signals, several time constants (typically four times the largest time constant) of the vibrating system will be sufficient. For random signals, a value that is as large as feasible should be used.

In the method of rms value specification, the rms value of the acceleration response (typically, acceleration in gs) is computed using equation (12.1) and is then compared with the specified value. In this method, instantaneous bursts of vibration do not have a significant effect because they are filtered out as a result of the integration. It is the average energy or power of the response signal that is considered. The duration of exposure enters into the picture indirectly and in an undesirable manner. For example, a highly transient vibration signal can have a damaging effect in the beginning; but the larger the T that is used in equation (12.1), the smaller the computed rms value. Hence, the use of a large value for T in this case would lead to diluting or masking the damage potential. In practice, the longer the exposure to a vibration signal, the greater the harm caused by it. Hence, when using specifications such as peak and rms values, they have to be adjusted according to the period of exposure. Specifically, larger levels of specification should be used for longer periods of exposure.

12.1.3 Frequency-Domain Specification

It is not quite realistic to specify the limitation to vibration exposure of a complex dynamic system by just a single threshold value. Usually, the effect of vibration on a system depends on at least the following three parameters of vibration:

1. Level of vibration (peak, rms, power, etc.)
2. Frequency content (range) of excitation
3. Duration of exposure to vibration.

This is particularly true because the excitations that generate the vibration environment may not necessarily be a single-frequency (sinusoidal) signal and may be broad-band and random; and

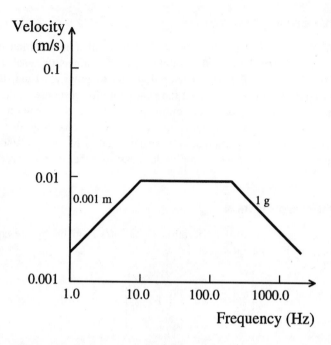

FIGURE 12.2 Operating vibration specification (nomograph) for a machine.

furthermore, the response of the system to the vibration excitations will depend on its frequency-transfer function, which determines its resonances and damping characteristics. Under these circumstances, it is desirable to provide specifications in a *nomograph*, where the horizontal axis gives frequency (Hz) and the vertical axis could represent a motion variable such as displacement (m), velocity (m·s⁻¹), or acceleration (m·s⁻² or g). It is not very important which of these motion variables represents the vertical axis of the nomograph. This is true because, in the frequency domain,

$$\text{Velocity} = j\omega \times \text{Displacement}$$

$$\text{Acceleration} = j\omega \times \text{Velocity}$$

and one form of motion can be easily converted into one of the remaining two motion representations. In each of the forms, assuming that the two axes of the nomograph are graduated in a logarithmic scale, the constant displacement, constant velocity, and constant acceleration lines are straight lines.

Consider a simple specification of machinery vibration limits as given by the following values:

$$\text{Displacement limit (peak)} = 0.001 \text{ m}$$

$$\text{Velocity limit} = 0.01 \text{ m} \cdot \text{s}^{-1}$$

$$\text{Acceleration limit} = 1.0 \text{ g}$$

This specification can be represented in a velocity vs. frequency nomograph (log–log) as in Figure 12.2.

Usually, such simple specifications in the frequency domain are not adequate. As noted before, the system behavior will vary, depending on the excitation frequency range. For example, motion sickness in humans might be predominant in low frequencies in the range of 0.1 Hz to 0.6 Hz, and passenger discomfort in ground transit vehicles might be most serious in the frequency range of 4

Vibration Design and Control

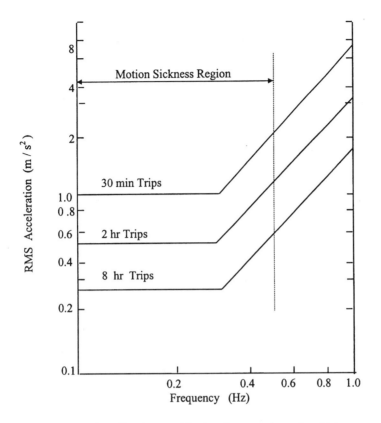

FIGURE 12.3 A severe-discomfort vibration specification for ground transit vehicles.

Hz to 8 Hz for vertical motion and 1 Hz to 2 Hz for lateral motion. Also, for any dynamic system, particularly at low damping levels, the neighborhoods of resonant frequencies should be avoided and, hence, should be specified by low vibration limits in the resonant regions. Furthermore, the duration of vibration exposure should be explicitly accounted for in specifications. For example, Figure 12.3 presents a ride comfort specification for a ground transit vehicle, where lower vibration levels are specified for longer trips.

Before leaving this section, it should be noted that the specifications of concern in the present context of design and control are *upper bounds* of vibration. The system should perform below (within) these specifications under normal operating conditions. Test specifications, as discussed in Chapter 10, are *lower bounds*. The test should be conducted at or above these vibration levels so that the system will meet the test specifications. Some considerations of vibration engineering are summarized in Box 12.1.

12.2 VIBRATION ISOLATION

The purpose of vibration isolation is to "isolate" the system of interest from vibration excitations by introducing an *isolator* in between them. Examples of isolators are machine mounts and vehicle suspension systems. Two general types of isolation can be identified:

1. Force isolation (related to force transmissibility)
2. Motion isolation (related to motion transmissibility).

BOX 12.1 Vibration Engineering

Vibration Mitigation Approaches:
- Isolation (buffers system from excitation)
- Design modification (modifies the system)
- Control (senses vibration and applies a counteracting force: passive/active).

Vibration Specification:
- Peak and rms values
- Frequency-domain specs on a nomograph
  —Vibration levels
  —Frequency content
  —Exposure duration.

  Note:

  $$|\text{Velocity}| = \omega \times |\text{Displacement}|$$

  $$|\text{Acceleration}| = \omega \times |\text{Velocity}|$$

Limiting Specifications:
- Operation (design) specs: Specify upper bounds
- Testing specs: Specify lower bounds.

In *force isolation*, vibration forces that would be ordinarily transmitted directly from a source to a supporting structure (isolated system) are filtered out by an isolator through its flexibility (spring) and dissipation (damping) so that part of the force is routed through an inertial path. Clearly, the concepts of *force transmissibility* are applicable here. In *motion isolation*, vibration motions that are applied at a moving platform of a mechanical system (isolated system) are absorbed by an isolator through its flexibility and dissipation so that the motion transmitted to the system of interest is weakened. The concepts of *motion transmissibility* are applicable in this case. The design problem in both cases is to select applicable parameters for the isolator so that the vibrations entering the system are below specified values within a frequency band of interest (the operating frequency range).

Force transmissibility and motion transmissibility were studied in Chapter 3, but the main concepts are revisited here. Figure 12.4(a) gives a schematic model of force transmissibility through an isolator. Vibration force at the source is $f(t)$. In view of the isolator, the source system (with impedance Z_m) is made to move at the same speed as the isolator (with impedance Z_s). This is a parallel connection of impedances, as noticed in Chapter 3. Hence, the force $f(t)$ is split so that part of it is taken up by the inertial path (broken line) of Z_m, and only the remainder (f_s) is transmitted through Z_s to the supporting structure, which is the isolated system. As derived in Chapter 3, force transmissibility is

$$T_f = \frac{f_s}{f} = \frac{Z_s}{Z_m + Z_s} \tag{12.2}$$

Figure 12.4(b) gives a schematic model of motion transmissibility through an isolator. Vibration motion $v(t)$ of the source is applied through an isolator (with impedance Z_s and mobility M_s) to the isolated system (with impedance Z_m and mobility M_m). The resulting force is assumed to transmit

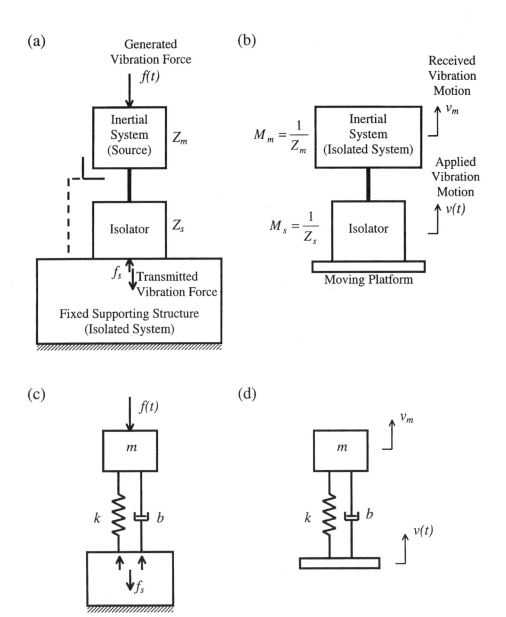

FIGURE 12.4 (a) Force isolation; (b) motion isolation; (c) force isolation example; and (d) motion isolation example.

directly from the isolator to the isolated system and, hence, these two units are connected in series (see Chapter 3). Consequently, one obtains the motion transmissibility:

$$T_m = \frac{v_m}{v} = \frac{M_m}{M_m + M_s} = \frac{Z_s}{Z_s + Z_m} \quad (12.3)$$

It is noticed, according to these two models,

$$T_f = T_m \quad (12.4)$$

As a result, the concepts of force transmissibility and motion transmissibility can usually be studied using just one common transmissibility function T.

Simple examples of force isolation and motion isolation are shown in Figure 12.4(c) and (d). As derived in Chapter 3, for both cases, the transmissibility function is given by

$$T = \frac{k + bj\omega}{\left(k - m\omega^2 + bj\omega\right)} \tag{12.5}$$

where ω is the frequency of vibration excitation. Note that the model (12.5) is not restricted to sinusoidal vibrations. Any general vibration excitation can be represented by a Fourier spectrum, which is a function of frequency ω. Then, the response vibration spectrum is obtained by multiplying the excitation spectrum by the transmissibility function T. The associated design problem is to select the isolator parameters k and b to meet the specifications of isolation.

Equation (12.5) can be expressed as

$$T = \frac{\omega_n^2 + 2j\zeta\omega_n\omega}{\left(\omega_n^2 - \omega^2 + 2j\zeta\omega_n\omega\right)} \tag{12.6}$$

where

$\omega_n = \sqrt{k/m}$ = undamped natural frequency of the system

$\zeta = \dfrac{b}{2\sqrt{km}}$ = damping ratio of the system.

Equation (12.6) can be written in the nondimensional form

$$T = \frac{1 + 2j\zeta r}{1 - r^2 + 2j\zeta r} \tag{12.7}$$

where the nondimensional excitation frequency is defined as

$$r = \omega/\omega_n$$

The transmissibility function has a phase angle as well as magnitude. In practical applications, it is the level of attenuation of the vibration excitation that is of primary importance, rather than the phase difference between the vibration excitation and the response. Accordingly, the transmissibility magnitude

$$|T| = \sqrt{\frac{1 + 4\zeta^2 r^2}{\left(1 - r^2\right)^2 + 4\zeta^2 r^2}} \tag{12.8}$$

is of interest. It can be shown that $|T| < 1$ for $r > \sqrt{2}$, which corresponds to the isolation region. Hence, the isolator should be designed such that the operating frequencies ω are greater than $\sqrt{2}\,\omega_n$. Furthermore, a threshold value for $|T|$ would be specified, and the parameters k and b of the isolator should be chosen so that $|T|$ is less than the specified threshold in the operating frequency range (which should be given). This procedure can be illustrated using an example.

Vibration Design and Control

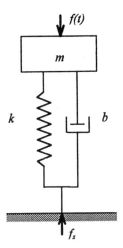

FIGURE 12.5 A simplified model of a machine tool and its supporting structure.

EXAMPLE 12.1

A machine tool and its supporting structure are modeled as the simple mass–spring–damper system shown in Figure 12.5.
  a. Draw a mechanical-impedance circuit for this system in terms of the impedances of the three elements: mass (m), spring (k), and viscous damper (b).
  b. Determine the exact value of the frequency ratio r in terms of the damping ratio ζ, at which the force transmissibility magnitude will peak. Show that for small ζ, this value is $r = 1$.
  c. Plot $|T_f|$ versus r for the interval $r = [0, 5]$, with one curve for each of the five ζ values 0.0, 0.3, 0.7, 1.0, and 2.0 on the same plane. Discuss the behavior of these transmissibility curves.
  d. From part (c), determine for each of the five ζ values, the excitation frequency range with respect to ω_n, for which the transmissibility magnitude is
     i. Less than 1.05
     ii. Less than 0.5.
  e. Suppose that the device in Figure 12.5 has a primary, undamped natural frequency of 6 Hz and a damping ratio of 0.2. It is required that the system has a force transmissibility magnitude of less than 0.5 for operating frequency values greater than 12 Hz. Does the existing system meet this requirement? If not, explain how one should modify the system to meet this requirement.

SOLUTION

a. Here, the elements m, b, and k are in parallel, with a common velocity v across them, as shown in Figure 12.6. In the circuit, $Z_m = mj\omega$, $Z_b = b$, and $Z_k = \dfrac{k}{j\omega}$

The force transmissibility is

$$T_f = \frac{F_s}{F} = \frac{F_s/V}{F/V} = \frac{Z_s}{Z_s + Z_0} = \frac{Z_b + Z_k}{Z_m + Z_b + Z_k} \qquad (i)$$

Substituting the element impedances, one obtains

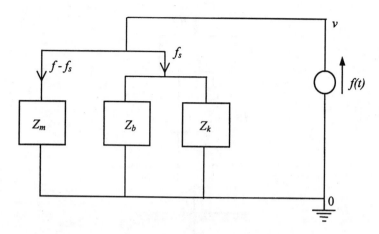

FIGURE 12.6 The mechanical impedance circuit of the force isolation problem.

$$T_f = \frac{b + \dfrac{k}{j\omega}}{mj\omega + b + \dfrac{k}{j\omega}} = \frac{bj\omega + k}{-\omega^2 m + bj\omega + k} = \frac{j\omega b/m + k/m}{-\omega^2 + j\omega b/m + k/m} \quad (ii)$$

The last expression is obtained by dividing the numerator and the denominator by m. Now use the fact that

$$\frac{k}{m} = \omega_n^2 \quad \text{and} \quad \frac{b}{m} = 2\zeta\omega_n$$

and divide (ii) throughout by ω_n^2 to obtain

$$T_f = \frac{\omega_n^2 + 2j\zeta\omega_n\omega}{\omega_n^2 - \omega^2 + 2j\zeta\omega_n\omega} = \frac{1 + 2j\zeta r}{1 - r^2 + 2j\zeta r} \quad (iii)$$

The transmissibility magnitude is

$$|T_f| = \sqrt{\frac{1 + 4\zeta^2 r^2}{(1-r^2)^2 + 4\zeta^2 r^2}} \quad (iv)$$

where $r = \omega/\omega_n$ is the normalized frequency.

b. To determine the peak point of $|T_f|$, differentiate the expression within the square-root sign in (iv) and equate to 0:

$$\frac{\left[(1-r^2)^2 + 4\zeta^2 r^2\right]8\zeta^2 r - \left[1 + 4\zeta^2 r^2\right]\left[2(1-r^2)(-2r) + 8\zeta^2 r\right]}{\left[(1-r^2)^2 + 4\zeta^2 r^2\right]^2} = 0$$

Vibration Design and Control

Hence,

$$4r\left\{\left[(1-r^2)^2 + 4\zeta^2 r^2\right]2\zeta^2 + \left[1+4\zeta^2 r^2\right]\left[1-r^2-2\zeta^2\right]\right\} = 0$$

which simplifies to

$$r(2\zeta^2 r^4 + r^2 - 1) = 0$$

the roots are

$$r = 0 \quad \text{and} \quad r^2 = \frac{-1 \pm \sqrt{1+8\zeta^2}}{4\zeta^2}$$

The root $r = 0$ corresponds to the initial stationary point at zero frequency. That does not represent a peak. Taking only the positive root for r^2 and then its positive square-root, the peak point of the transmissibility magnitude is given

$$r = \frac{\left[\sqrt{1+8\zeta^2} - 1\right]^{1/2}}{2\zeta} \quad (v)$$

For small ζ, Taylor series expansion gives

$$\sqrt{1+8\zeta^2} \approx 1 + \frac{1}{2} \times 8\zeta^2 = 1 + 4\zeta^2$$

With this approximation, equation (v) evaluates to 1. Hence, for small damping, the transmissibility magnitude will have a peak at $r = 1$ and, from equation (iv), its value is

$$|T_f| \approx \frac{\sqrt{1+4\zeta^2}}{2\zeta} \approx \frac{1 + \frac{1}{2} \times 4\zeta^2}{2\zeta}$$

or

$$|T_f| \approx \frac{1}{2\zeta} + \zeta \approx \frac{1}{2\zeta} \quad (12.9)$$

c. The five curves of $|T_f|$ verses r for $\zeta = 0, 0.3, 0.7, 1.0,$ and 2.0 are shown in Figure 12.7. Note that these curves use the exact expression (iv).
From the curves, one observes the following:

1. There is always a non-zero frequency value at which the transmissibility magnitude will peak. This is the resonance.
2. For small ζ, the peak transmissibility magnitude is obtained at approximately $r = 1$. As ζ increases, this peak point shifts to the left (i.e., a lower value for peak frequency).

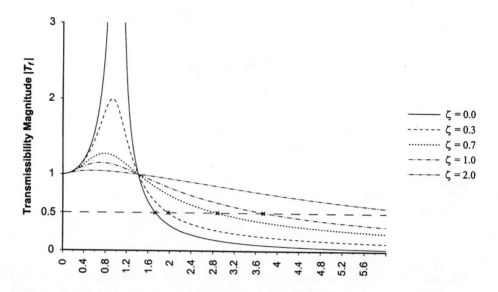

FIGURE 12.7 Transmissibility curves for a simple oscillator model.

  3. The peak magnitude decreases with increasing ζ.
  4. All the transmissibility curves pass through the magnitude value 1.0 at the same frequency $r = \sqrt{2}$.
  5. The isolation (i.e., $|T_f| < 1$) is given by $r > \sqrt{2}$. In this region, $|T_f|$ increases with ζ.
  6. The transmissibility magnitude decreases as r increases, in the isolation region.

d. From the curves in Figure 12.7 or numerically, one obtains
   - For $|T_f| < 1.05$; $r > \sqrt{2}$ for all ζ
   - For $|T_f| < 0.5$; $r > 1.73$, 1.964, 2.871, 3.77, and 7.075 for $\zeta = 0.0, 0.3, 0.7, 1.0,$ and 2.0, respectively.by
e. One needs

$$\sqrt{\frac{1+4\zeta^2 r^2}{(1-r^2)^2 + 4\zeta^2 r^2}} < \frac{1}{2}$$

or

$$\frac{1+4\zeta^2 r^2}{(1-r^2)^2 + 4\zeta^2 r^2} < \frac{1}{4}$$

or

$$4 + 16\zeta^2 r^2 < (1-r^2)^2 + 4\zeta^2 r^2$$

or

$$r^4 - 2r^2 - 12\zeta^2 r^2 - 3 > 0$$

For $\zeta = 0.2$ and $r = 12/6 = 2$, the LHS expression computes to

$$2^4 - 2 \times 2^2 - 12 \times (0.2)^2 \times 2^2 - 3 = 3.08 > 0$$

Hence, the requirement is met. In fact, since, for $r = 2$,

$$\text{LHS} = 2^4 - 2 \times 2^2 - 12 \times 2^2 \zeta^2 - 3 = 5 - 48\zeta^2$$

it follows that the requirement would be met for

$$5 - 48\zeta^2 > 0$$

or

$$\zeta < \sqrt{\frac{5}{48}} = 0.32$$

If the requirement was not met (e.g., if $\zeta = 0.4$), the option would be to reduce damping.

□

12.2.1 Design Considerations

The *level of isolation* is defined as $1 - T$. It was noted that in the isolation region ($r > \sqrt{2}$), the transmissibility decreases (hence, the level of isolation increases) as the damping ratio ζ decreases. Thus, the best conditions of isolation are given by $\zeta = 0$. This is not feasible in practice, but one should maintain ζ as small as possible. For small ζ in the isolation region, equation (12.8) can be approximated by

$$T = \frac{1}{(r^2 - 1)} \qquad (12.10)$$

Note that T is real in this case of $\zeta \cong 0$; and is also positive because $r > \sqrt{2}$. But in the general case, T can denote the magnitude of the transmissibility function. Substitute

$$r^2 = \omega^2/\omega_n^2 = \omega^2 m/k$$

One obtains

$$k = \frac{\omega^2 m T}{(1 + T)} \qquad (12.11)$$

This equation can be used to determine the design stiffness of the isolator for a specified level of isolation $(1 - T)$ in the operating frequency range $\omega > \omega_0$, for a system of known mass (including the isolator mass). Often, the static deflection δ_s of the spring is used in design procedures and is given by

BOX 12.2 Vibration Isolation

Transmissibility (force/force or motion/motion):

$$|T| = \sqrt{\frac{1+4\zeta^2 r^2}{(1-r^2)^2 + 4\zeta^2 r^2}} \cong \frac{1}{(r^2-1)} \quad \text{for } r > 1 \text{ and small } \zeta$$

Properties:

1. $T_{peak} \cong \dfrac{\sqrt{1+4\zeta^2}}{2\zeta} \cong \dfrac{1}{2\zeta}$ for small ζ

2. T_{peak} occurs at $r_{peak} = \dfrac{\left[\sqrt{1+8\zeta^2}-1\right]^{1/2}}{2\zeta} \cong 1$ for small ζ

3. All $|T|$ curves coincide at $r = \sqrt{2}$ for all ζ

4. Isolation region: $r > \sqrt{2}$

5. In isolation region:
   $|T|$ decreases with r (i.e., better isolation at higher frequencies)
   $|T|$ increases with ζ (i.e., better isolation at lower damping).

Design Formulas:

Level of isolation $= 1 - T$

Isolator stiffness $k = \dfrac{\omega^2 m T}{(1+T)}$

where m = system mass
ω = operating frequency

Static deflection $\delta_s = \dfrac{mg}{k} = (1+T)\dfrac{g}{\omega^2 T}$

$$\delta_s = \frac{mg}{k} \tag{12.12}$$

Substituting equation (12.12) in (12.11), one obtains

$$\delta_s = (1+T)\frac{g}{\omega^2 T} \tag{12.13}$$

Since the isolation region is $\omega > \sqrt{2}\,\omega_n$, it is desirable to make ω_n as small as possible so as to obtain the widest frequency range of operation. This is achieved by making the isolator as soft as possible (k as low as possible). However, there are limits to this from the points of view of structural strength, stability, static deflection, and availability of springs. Then, m can be increased by adding an inertia block as the base of the system, which is then mounted on the isolator spring (with a damping layer) or an air-filled pneumatic mount. The inertia block will also lower the centroid of the system, thereby providing added desirable effects of stability and a reduction of rocking motions and noise transmission. For improved load distribution, instead of just one spring

Vibration Design and Control

of design stiffness k, a set of n springs each with stiffness k/n and uniformly distributed under the inertia block should be used.

Another requirement for good vibration isolation is low damping. Usually, metal springs have very low damping (typically ζ less than 0.01). On the other hand, higher damping is needed to reduce resonant vibrations that will be encountered during start-up and shut-down conditions when the excitation frequency will vary and pass through the resonances. Also, vibration energy must be effectively dissipated even under steady operating conditions. Isolation pads made of damping material such as cork, natural rubber, and neoprene can be used for this purpose. They can provide damping ratios of the order of 0.01.

The basic design steps for a vibration isolator, in force isolation, are as follows:

1. The required level of isolation $(1 - T)$ and the lowest frequency of operation (ω_0) are specified. The mass of the vibration source (m) is known.
2. Use equation (12.11) with $\omega = \omega_0$ to compute the required stiffness k of the isolator.
3. If the resulting component k is not satisfactory, increase m by introducing an inertia block and recompute k.
4. Distribute k over several springs.
5. Introduce a mounting pad of known stiffness and damping. Modify k and b accordingly, and compute T using equation (12.8). If the specified T is exceeded, modify the isolator parameters as appropriate and repeat the design cycle.

Some relations that are useful in design for vibration isolation are given in Box 12.2.

EXAMPLE 12.2

Consider a motor and fan unit of a building ventilation system, weighing 50 kg and operating in the speed range of 600 to 3,600 rpm. Since offices are located directly underneath the motor room, a 90% vibration isolation is desired. A set of mounting springs, each having a stiffness of 100 N·cm^{-1}, is available. Design an isolation system to mount the motor fan unit on the room floor.

SOLUTION

For an isolation level of 90%, the required force transmissibility is $T = 0.1$. The lowest frequency of operation is $\omega = \dfrac{600}{60} \times 2\pi$ rad·s^{-1}. First, try four mounting points. The overall spring stiffness is $k = 4 \times 100 \times 10^2$ N·m^{-1}. Substituting in equation (12.11).

$$4 \times 100 \times 100 = \frac{(10 \times 2\pi)^2 \times m \times 0.1}{1.1}$$

gives $m = 111.5$ kg. Since the mass of the unit is 50 kg, one should use an inertia block of mass 61.5 kg or more.

□

12.2.2 VIBRATION ISOLATION OF FLEXIBLE SYSTEMS

The simple model shown in Figures 12.4(c) and (d) might not be adequate in the design of vibration isolators for sufficiently flexible systems. A model that is more appropriate in this situation is shown in Figure 12.8. Note that the vibration isolator has an inertia block of mass m in addition to damped

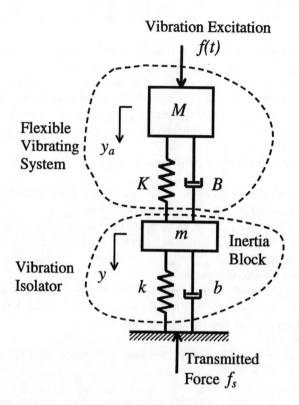

FIGURE 12.8 A model for vibration isolation of a flexible system.

flexible mounts of stiffness k and damping constant b. The vibrating system itself has a stiffness K and damping constant B in addition to its mass M.

In the absence of K, B, and the inertia block (m) as in Figure 12.4(c), the vibrating system becomes a simple inertia (M). Then, y_a and y are the same, and the equation of motion is

$$M\ddot{y} + b\dot{y} + ky = f(t) \tag{12.14}$$

with the force transmitted to the support structure, f_s, given by

$$f_s = b\dot{y} + ky \tag{12.15}$$

The force transmissibility in this case is

$$T_{\text{inertial}} = \frac{f_s}{f} = \frac{bs + k}{Ms^2 + bs + k} \quad \text{with } s = j\omega \tag{12.16}$$

For the flexible system and isolator shown in Figure 12.8, the equations of motion are

$$M\ddot{y}_a + B(\dot{y}_a - \dot{y}) + K(y_a - y) = f(t) \tag{12.17}$$

$$m\ddot{y} + B(\dot{y} - \dot{y}_a) + K(y - y_a) + b\dot{y} + ky = 0 \tag{12.18}$$

Hence, in the frequency domain, one has

$$(Ms^2 + Bs + K)y_a - (Bs + K)y = f \qquad (12.19)$$

$$[ms^2 + (B+b)s + K + k]y = (Bs + K)y_a \quad \text{with } s = j\omega \qquad (12.20)$$

Substitute equation (12.20) into (12.19) for eliminating y_a to obtain

$$\left\{(Ms^2 + Bs + K)\frac{[ms^2 + (B+b)s + K + k]}{(Bs + K)} - (Bs + K)\right\}y = f$$

which simplifies to

$$\left\{\frac{Ms^2[ms^2 + (B+b)s + K + k] + (Bs + K)(ms^2 + bs + k)}{(Bs + K)}\right\}y = f \qquad (12.21)$$

The force transmitted to the supporting structure is still given by equation (12.15). Hence, the transmissibility with the flexible system is

$$T_{\text{flexible}} = \frac{(Bs + K)(bs + k)}{\{Ms^2[ms^2 + (B+b)s + K + k] + (Bs + K)(ms^2 + bs + k)\}} \quad \text{with } s = j\omega \quad (12.22)$$

From equations (12.16) and (12.22), the transmissibility magnitude ratio is

$$\frac{T_{\text{flexible}}}{T_{\text{inertial}}} = \left|\frac{(Bs + K)(Ms^2 + bs + k)}{Ms^2[ms^2 + (B+b)s + K + k] + (Bs + K)(ms^2 + bs + k)}\right| \quad \text{with } s = j\omega \quad (12.23)$$

or

$$\frac{T_{\text{flexible}}}{T_{\text{inertial}}} = \left|\frac{(Ms^2 + bs + k)}{Ms^2(ms^2 + bs + k)/(Bs + K) + Ms^2 + ms^2 + bs + k}\right| \quad \text{with } s = j\omega \quad (12.24)$$

In the nondimensional form:

$$\frac{T_{\text{flexible}}}{T_{\text{inertial}}} = \left|\frac{1 - r^2 + 2j\zeta_b r}{-r^2(1 - r_m r^2 + 2j\zeta_b r)/(r_\omega^2 + 2j\zeta_a r_\omega r) + 1 - (1 + r_m)r^2 + 2j\zeta_b r}\right| \qquad (12.25)$$

where

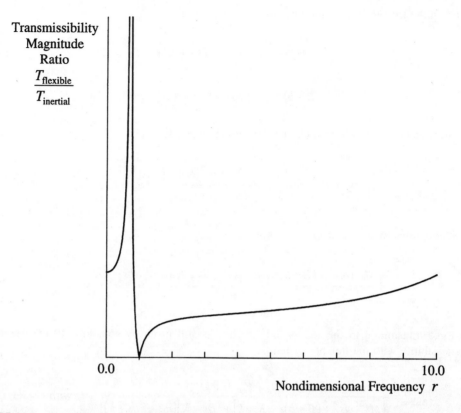

FIGURE 12.9 The effect of system flexibility on the transmissibility magnitude in the undamped case (mass ratio = 1.0; natural frequency ratio = 10.0).

$$r = \frac{\omega}{\sqrt{k/M}}$$

$$r_m = \frac{m}{M}$$

$$r_\omega = \frac{\sqrt{K/M}}{\sqrt{k/M}} = \sqrt{\frac{K}{k}}$$

$$\zeta_a = \frac{B}{2\sqrt{KM}}$$

$$\zeta_b = \frac{b}{2\sqrt{kM}}$$

Again, the design problem of vibration isolation is to select the parameters r_m, r_ω, ζ_a, and ζ_b so that the required level of vibration isolation is realized for an operating frequency range of r.

A plot of equation (12.25) for the undamped case with $r_m = 1.0$ and $r_\omega = 10.0$ is given in Figure 12.9. Generally, the transmissibility ratio will be 0 at $r = 1$ (the resonance of the inertial system), and there will be two values of r (the resonances of the flexible system) for which the ratio will become infinity, in the undamped case. The latter two neighborhoods should be avoided under steady operating conditions.

Vibration Design and Control

12.3 BALANCING OF ROTATING MACHINERY

Many practical devices that move contain rotating components. Examples are wheels of vehicles, shafts and gear transmissions of machinery, belt drives, motors, turbines, compressors, fans, and rollers. An unbalance (imbalance) is created in a rotating part when its center of mass does not coincide with the axis of rotation. The reasons for this *eccentricity* include the following:

1. Inaccurate production procedures (machining, casting, forging, assembly, etc.)
2. Wear and tear
3. Loading conditions (mechanical)
4. Environmental conditions (thermal loads and deformation)
5. Use of inhomogeneous and anisotropic material (that does not have a uniform density distribution)
6. Component failure
7. Addition of new components to a rotating device.

For a component of mass m and eccentricity e, and rotating at angular speed ω, the centrifugal force generated is $me\omega^2$. Note the quadratic variation with ω. This rotating force can be resolved into two orthogonal components that will be sinusoidal with frequency ω. It follows that harmonic forcing excitations are generated due to the unbalance, which can generate undesirable vibrations and associated problems.

Problems caused by unbalance include wear and tear, malfunction and failure of components, poor quality of products, and undesirable noise. The problem becomes increasingly important due to the present trend of developing high-speed machinery. It is estimated that the speed of operation of machinery has doubled during the past 50 years. This means that the level of unbalance forces may have quadrupled during the same period, causing more serious vibration problems.

An unbalanced rotating component can be balanced by adding or removing material to or from the component. One needs to know both the magnitude and location of the balancing masses to be added or removed. The present section will address the problem of component balancing for vibration suppression.

Note that the goal is to remove the source of vibration — namely, the mass eccentricity — typically by adding one or more balancing mass elements. Two methods are available:

1. Static (single-plane) balancing
2. Dynamic (two-plane) balancing.

The first method concerns balancing of planar objects (e.g., pancake motors, disks) whose longitudinal dimension along the axis of rotation is not significant. The second method concerns balancing of objects that have a significant longitudinal dimension. Both methods are discussed.

12.3.1 STATIC BALANCING

Consider a disk rotating at angular velocity ω about a fixed axis. Suppose that the mass center of the disk has an eccentricity e from the axis of rotation, as shown in Figure 12.10(a). Place a fixed coordinate frame x-y at the center of rotation. The position $\bar{e}$ of the mass center in this coordinate frame can be represented as:

1. A position vector rotating at angular speed ω, or
2. A complex number, with x-coordinate denoting the real part and y-coordinate denoting the imaginary part.

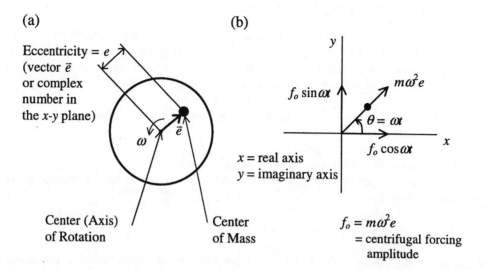

FIGURE 12.10 (a) Unbalance in a rotating disk due to mass eccentricity, and (b) rotating vector (phasor) of centrifugal force due to unbalance.

The centrifugal force due to the mass eccentricity is also a vector in the direction of $\bar{e}$, but with a magnitude $f_o = m\omega^2 e$, as shown in Figure 12.10(b). It is seen that harmonic excitations result in both x and y directions, given by $f_o\cos\omega t$ and $f_o\sin\omega t$, respectively, where $\theta = \omega t$ = orientation of the rotating vector with respect to the x-axis. To balance the disk, one should add a mass m at $-\bar{e}$. But we don't know the value of m and the location of $\bar{e}$.

Balancing Approach

1. Measure the amplitude V_u and the phase angle ϕ_1 (e.g., by the signal from an accelerometer mounted on the bearing of the disk) of the unbalance centrifugal force, with respect to some reference.
2. Mount a known mass (trial mass) M_t at a known location on the disk. Suppose that its own centrifugal force is given by the rotating vector $\bar{V}_w$, and the resultant centrifugal force due to both the original unbalance and the trial mass is $\bar{V}_r$.
3. Measure the amplitude V_r and the phase angle ϕ_2 of the resultant centrifugal force, as in step 1, with respect to the same phase reference.

A vector diagram showing the centrifugal forces $\bar{V}_u$ and $\bar{V}_w$ due to the original unbalance and the trial-mass unbalance, respectively, is shown in Figure 12.11. The resultant unbalance is $\bar{V}_r = \bar{V}_u + \bar{V}_w$. Note that $-\bar{V}_u$ represents the centrifugal force due to the balancing mass that is needed. So, if one determines the angle ϕ_b in Figure 12.11, it will give the orientation of the balancing mass. Furthermore, suppose that the balancing mass is M_b and it is mounted at an eccentricity equal to that of the trial mass M_t. Then,

$$\frac{M_b}{M_t} = \frac{V_u}{V_w}$$

One needs to determine the ratio V_u/V_w and the angle ϕ_b. These values can be derived as follows:

$$\phi = \phi_2 - \phi_1 \tag{12.26}$$

The cosine rule gives

Vibration Design and Control

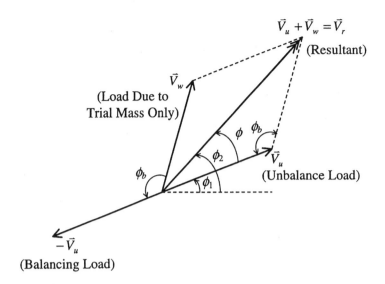

FIGURE 12.11 A vector diagram of the single-plane (static) balancing problem.

$$V_w^2 = V_u^2 + V_r^2 - 2V_u V_r \cos \phi \qquad (12.27)$$

This will provide V_w since V_u, V_r, and ϕ are known. Apply the cosine rule again:

$$V_r^2 = V_u^2 + V_w^2 - 2V_u V_w \cos \phi_b$$

Hence,

$$\phi_b = \cos^{-1}\left[\frac{V_u^2 + V_w^2 - V_r^2}{2V_u V_w}\right] \qquad (12.28)$$

Note: One might think that because ϕ_1 is measured, that one exactly knows where $\vec{V}_u$ is. This is not so, because one does not know the reference line with respect to which ϕ_1 is measured. One only knows that this reference is kept fixed (through strobe synchronization of the body rotation) during measurements. Hence, one needs to know ϕ_b, which gives the location of $-\vec{V}_u$ with respect to the known location of $\vec{V}_w$ on the disk.

12.3.2 COMPLEX NUMBER/VECTOR APPROACH

Again, suppose that the imbalance is equivalent to a mass of M_b that is located at the same eccentricity (radius) r as the trial mass M_t. Define complex numbers (mass location vectors in a body frame):

$$\vec{M}_b = M_b \angle \theta_b \qquad (12.29)$$

$$\vec{M}_t = M_t \angle \theta_t \qquad (12.30)$$

as shown in Figure 12.12.
  Associated force vectors are:

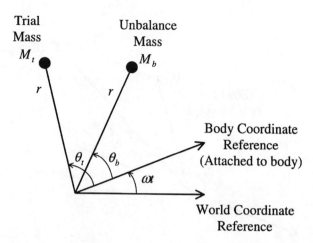

FIGURE 12.12 Rotating vectors of mass location.

$$\vec{V}_u = \omega^2 r e^{j\omega t} M_b \angle \theta_b \tag{12.31}$$

$$\vec{V}_w = \omega^2 r e^{j\omega t} M_t \angle \theta_t \tag{12.32}$$

or

$$\vec{V}_u = \vec{A}\vec{M}_b \tag{12.33}$$

$$\vec{V}_w = \vec{A}\vec{M}_t \tag{12.34}$$

where $\vec{A} = \omega^2 r e^{j\omega t}$ is the conversion factor or "influence coefficient" (complex) from the mass to the resulting dynamic force (rotating). This factor is the same for both cases because r is the same. What is needed is to determine $\vec{M}_b$.

From equation (12.33),

$$\vec{M}_b = \frac{\vec{V}_u}{\vec{A}} \tag{12.35}$$

Substitute equation (12.34):

$$\vec{M}_b = \frac{\vec{V}_u}{\vec{V}_w} \cdot \vec{M}_t \tag{12.36}$$

But, since

$$\vec{V}_r = \vec{V}_u + \vec{V}_w \tag{12.37}$$

one has

Vibration Design and Control

$$\vec{M}_b = \frac{\vec{V}_u}{(\vec{V}_r - \vec{V}_u)} \cdot \vec{M}_t \qquad (12.38)$$

Since one knows $\vec{M}_t$ and one measures $\vec{V}_u$ and $\vec{V}_r$ to the same scaling factor, one can compute $\vec{M}_b$ using equation (12.38). Locate the balancing mass at $-\vec{M}_b$ (with respect to the body frame).

EXAMPLE 12.3

Consider the following experimental steps:

Measured: Accelerometer amplitude (oscilloscope reading) of 6.0 with a phase lead (with respect to a strobe signal reference that is synchronized with the rotating body frame) of 50°.
Added: Trial mass $M_t = 20$ gm at angle 180° (with respect to a body reference radius).
Measured: Accelerometer amplitude of 8.0 with a phase lead of 60° (with respect to the synchronized strobe signal).

Determine the magnitude and location of the balancing mass.

SOLUTION

Method 1:
The data given:

$$\phi = 60° - 50° = 10°$$

$$V_u = 6.0$$

$$V_r = 8.0$$

Hence, from equation (12.27),

$$V_w = \sqrt{6^2 + 8^2 - 2 \times 6 \times 8 \cos 10°} = 2.37$$

Balancing mass:

$$M_b = \frac{6.0}{2.37} \times 20 = 50.63 \text{ gm}$$

Equation (12.28) gives

$$\phi_b = \cos^{-1}\left[\frac{6^2 + 2.37^2 - 8^2}{2 \times 6 \times 2.37}\right] = \cos^{-1}(-0.787)$$

$$= 142° \text{ or } 218°$$

Pick the result $0° \leq \phi_b \leq 180°$, as is clear from the vector diagram shown in Figure 12.11. Hence,

$$\phi_b = 142°$$

But,

$$\bar{M}_t = 20\angle 180° \text{ gm}$$

It follows that

$$-\bar{M}_b = 50.63\angle(180° + 142°) \text{ gm}$$
$$= 50.63\angle 322° \text{ gm}$$

Method 2:
Given:

$$\bar{M}_t = 20\angle 180° \text{ gm}$$
$$\bar{V}_u = 6.0\angle 50°$$
$$\bar{V}_r = 8.0\angle 60°$$

Then from equation (12.38), one obtains

$$\bar{M}_b = \frac{6.0\angle 50°}{(8.0\angle 60° - 6.0\angle 50°)} 20\angle 180° \text{ gm}$$

First, compute:

$$8.0\angle 60° - 6.0\angle 50° = (8.0\cos 60° + j8.0\sin 6.0°) - (6.0\cos 50° + j6.0\sin 50°)$$
$$= (8.0\cos 60° - 6.0\cos 50°) + j(8.0\sin 60° - 6.0\sin 50°)$$
$$= 0.1433 + j2.332 = 2.336\angle 86.48°$$

Hence,

$$\bar{M}_b = \frac{6.0\angle 50°}{2.336\angle 86.48°} 20\angle 180° = \frac{6.0 \times 20}{2.336} \angle(50° + 180° - 86.48°)$$
$$= 51\angle 143.5° \text{ gm}$$

The balancing mass should be located at

$$-\bar{M}_b = 51\angle 323.5° \text{ gm}$$

Note: This angle is measured from the same body reference as for the trial mass.

□

Vibration Design and Control

12.3.3 Dynamic (Two-Plane) Balancing

Consider, instead of an unbalanced disk, an elongated rotating object, supported at two bearings, as shown in Figure 12.13. In this case, in general, there may not be an equivalent single unbalanced force at a single plane normal to the shaft axis. To show this, recall that a system of forces can be represented by a single force at a specified location and a couple (two parallel forces that are equal and opposite). If this single force (resultant force) is 0, one is left with only a couple. The couple cannot be balanced by a single force.

All the unbalance forces at all the planes along the shaft axis can be represented by an equivalent single unbalance force at a specified plane and a couple. If this equivalent force is 0, then to balance the couple, one needs two equal and opposite forces at two different planes.

On the other hand, if the couple is 0, then a single force in the opposite direction at the same plane of the resultant unbalance force will result in complete balancing. But this unbalance plane may not be reachable, even if it is known, for the purpose of adding the balancing mass.

In the present (two-plane) balancing problem, the balancing masses are added at the two bearing planes so that both the resultant unbalance force and couple are balanced.

It is clear from Figure 12.13 that even a sole unbalance mass $\vec{M}_b$ at a single unbalance plane can be represented by two unbalance masses $\vec{M}_{b1}$ and $\vec{M}_{b2}$ at the bearing planes 1 and 2. In the presence of an unbalance couple as well, one can simply add two equal and opposite forces at the planes 1 and 2 so that its couple is equal to the unbalance couple. Hence, a general unbalance can be represented by the two unbalance masses $\vec{M}_{b1}$ and $\vec{M}_{b2}$ at planes 1 and 2, as shown in Figure 12.13. As for the single-plane balancing problem, the resultant unbalance forces at the two bearings (that would be measured by the accelerometers at 1 and 2) are:

$$\vec{V}_{u1} = \vec{A}_{11}\vec{M}_{b1} + \vec{A}_{12}\vec{M}_{b2} \tag{12.39}$$

$$\vec{V}_{u2} = \vec{A}_{21}\vec{M}_{b1} + \vec{A}_{22}\vec{M}_{b2} \tag{12.40}$$

Suppose that a trial mass of $\vec{M}_{t1}$ (at a known location with respect to a known body reference line) was added at plane 1. The resulting unbalance forces at the two bearings would then be:

$$\vec{V}_{r11} = \vec{A}_{11}\left(\vec{M}_{b1} + \vec{M}_{t1}\right) + \vec{A}_{12}\vec{M}_{b2} \tag{12.41}$$

$$\vec{V}_{r21} = \vec{A}_{21}\left(\vec{M}_{b1} + \vec{M}_{t1}\right) + \vec{A}_{22}\vec{M}_{b2} \tag{12.42}$$

Next, suppose that a trial mass of $\vec{M}_{t2}$ (at a known location with respect to a known body reference line) was added at plane 2, after removing $\vec{M}_{t1}$. The resulting unbalance forces at the two bearings would be:

$$\vec{V}_{r12} = \vec{A}_{11}\vec{M}_{b1} + \vec{A}_{12}\left(\vec{M}_{b2} + \vec{M}_{t2}\right) \tag{12.43}$$

$$\vec{V}_{r22} = \vec{A}_{21}\vec{M}_{b1} + \vec{A}_{22}\left(\vec{M}_{b2} + \vec{M}_{t2}\right) \tag{12.44}$$

The following subtractions of equations are made now:

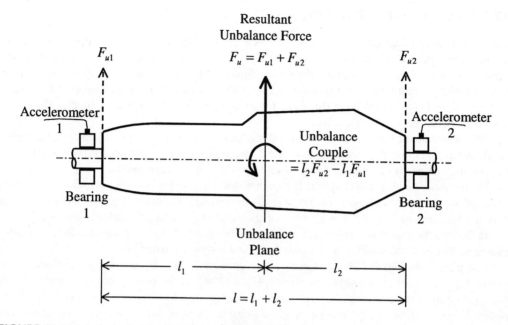

FIGURE 12.13 A dynamic (two-plane) balancing problem.

(12.41) – (12.39):

$$\vec{V}_{r11} - \vec{V}_{u1} = \vec{A}_{11}\vec{M}_{t1} \quad \text{or} \quad \vec{A}_{11} = \frac{\vec{V}_{r11} - \vec{V}_{u1}}{\vec{M}_{t1}} \tag{12.45}$$

(12.42) – (12.40):

$$\vec{V}_{r21} - \vec{V}_{u2} = \vec{A}_{21}\vec{M}_{t1} \quad \text{or} \quad \vec{A}_{21} = \frac{\vec{V}_{r21} - \vec{V}_{u2}}{\vec{M}_{t1}} \tag{12.46}$$

(12.43) – (12.39):

$$\vec{V}_{r12} - \vec{V}_{u1} = \vec{A}_{12}\vec{M}_{t2} \quad \text{or} \quad \vec{A}_{12} = \frac{\vec{V}_{r12} - \vec{V}_{u1}}{\vec{M}_{t2}} \tag{12.47}$$

(12.44) – (12.40):

$$\vec{V}_{r22} - \vec{V}_{u2} = \vec{A}_{22}\vec{M}_{t2} \quad \text{or} \quad \vec{A}_{22} = \frac{\vec{V}_{r22} - \vec{V}_{u2}}{\vec{M}_{t2}} \tag{12.48}$$

Hence, generally,

$$\vec{A}_{ij} = \frac{\vec{V}_{rij} - \vec{V}_{ui}}{\vec{M}_{tj}} \tag{12.49}$$

Vibration Design and Control

These parameters A_{ij} are called *influence coefficients*, and are complex numbers.

Next, in equations (12.39) and (12.40), eliminate each equality $\vec{M}_{b2}$ and $\vec{M}_{b1}$ separately, to determine the other; thus,

$$\vec{A}_{22}\vec{V}_{u1} - \vec{A}_{12}\vec{V}_{u2} = \left(\vec{A}_{22}\vec{A}_{11} - \vec{A}_{12}\vec{A}_{21}\right)\vec{M}_{b1}$$

$$\vec{A}_{21}\vec{V}_{u1} - \vec{A}_{11}\vec{V}_{u2} = \left(\vec{A}_{21}\vec{A}_{12} - \vec{A}_{11}\vec{A}_{22}\right)\vec{M}_{b2}$$

or

$$\vec{M}_{b1} = \frac{\vec{A}_{22}\vec{V}_{u1} - \vec{A}_{12}\vec{V}_{u2}}{\left(\vec{A}_{22}\vec{A}_{11} - \vec{A}_{12}\vec{A}_{21}\right)} \quad (12.50)$$

$$\vec{M}_{b2} = \frac{\vec{A}_{21}\vec{V}_{u1} - \vec{A}_{11}\vec{V}_{u2}}{\left(\vec{A}_{21}\vec{A}_{12} - \vec{A}_{11}\vec{A}_{22}\right)} \quad (12.51)$$

Substitute equations (12.45) through (12.48) in (12.50) and (12.51) to determine $\vec{M}_{b1}$ and $\vec{M}_{b2}$. Balancing masses that should be added are $-\vec{M}_{b1}$ and $-\vec{M}_{b2}$ in planes 1 and 2, respectively, at the same eccentricity as the corresponding trial masses.

The single-plane and two-plane balancing approaches are summarized in Box 12.3.

EXAMPLE 12.4

Suppose that the following measurements are obtained.

Without Trial Mass:
Accelerometer at 1: Amplitude = 10.0; Phase lead = 55°.
Accelerometer at 2: Amplitude = 7.0; Phase lead = 120°.

With Trial Mass 20 gm at Location 270° of Plane 1:
Accl. 1:   Ampl. = 7.0; Phase lead = 120°.
Accl. 2:   Ampl. = 5.0; Phase lead = 225°.

With Trial Mass 25 gm at Location 180° of Plane 2:
Accl. 1:   Ampl. = 6.0; Phase lead = 120°.
Accl. 2:   Ampl. = 12.0; Phase lead = 170°.

Determine the magnitude and orientation of the necessary balancing masses in planes 1 and 2 in order to completely balance (dynamic) the system.

SOLUTION

In the phasor notation, we can represent the given data as follows:

BOX 12.3 Balancing of Rotating Components

Static or Single-Plane Balancing
(Balances a single equivalent dynamic force)

Experimental Approach:
1. Measure magnitude (V) and phase (ϕ), with respect to a marked body reference line (that is kept fixed by strobe light), of accelerometer signal at bearing,
    a. without trial mass: (V_u, ϕ_1) or $\vec{V}_u = V_u \angle \phi_1$
    b. with trial mass M_t: (V_r, ϕ_2) or $\vec{V}_r = V_r \angle \phi_2$
2. Compute balancing mass M_b and its location with respect to M_t
3. Remove M_t and add M_b at determined location.

Computation Approach 1:

$$V_w = \left[V_u^2 + V_r^2 - 2V_u V_r \cos(\phi_2 - \phi_1)\right]^{1/2}$$

$$\phi_b = \cos^{-1}\left[\frac{V_u^2 + V_w^2 - V_r^2}{2V_u V_w}\right] \quad \text{and} \quad M_b = \frac{V_u}{V_w} M_t$$

Locate M_b at ϕ_b from M_t at the same eccentricity as M_t.

Computation Approach 2:

Unbalance mass phasor $\vec{M}_b = \dfrac{\vec{V}_u}{(\vec{V}_r - \vec{V}_u)} \vec{M}_t$ where $\vec{M}_t = M_t \angle \theta_t$ (Trial mass phasor)

Locate balancing mass at $-\vec{M}_b$ and at the same eccentricity as $\vec{M}_t$.

Dynamic or Two-Plane Balancing
(Balances an equivalent dynamic force and a couple)

Experimental Approach:
1. Measure $\vec{V}_{ui}$ at bearings $i = 1, 2$, without a trial mass.
2. Measure $\vec{V}_{rij}$ at bearings $i = 1, 2$, with only one trial mass $\vec{M}_{tj}$ at $j = 1, 2$
3. Compute unbalance mass phasor $\vec{M}_{bi}$ in planes $i = 1, 2$
4. Remove trial mass and place balancing masses $-\vec{M}_{bi}$ in planes $i = 1, 2$.
    Note: Measurements made while a body reference is kept fixed at the same location, using strobe light.

Computations:

Influence coefficients: $\vec{A}_{ij} = (\vec{V}_{rij} - \vec{V}_{ui})/\vec{M}_{tj}$

Unbalance mass phasors:

$$\vec{M}_{b1} = \frac{\vec{A}_{22}\vec{V}_{u1} - \vec{A}_{12}\vec{V}_{u2}}{(\vec{A}_{22}\vec{A}_{11} - \vec{A}_{12}\vec{A}_{21})} \quad \text{and} \quad \vec{M}_{b2} = \frac{\vec{A}_{21}\vec{V}_{u1} - \vec{A}_{11}\vec{V}_{u2}}{(\vec{A}_{21}\vec{A}_{12} - \vec{A}_{11}\vec{A}_{22})}$$

Vibration Design and Control

$$\vec{V}_{u1} = 10.0\angle 55°; \qquad \vec{V}_{u2} = 7.0\angle 120°$$

$$\vec{V}_{r11} = 7.0\angle 120°; \qquad \vec{V}_{r21} = 5.0\angle 225°$$

$$\vec{V}_{r12} = 6.0\angle 120°; \qquad \vec{V}_{r22} = 12.0\angle 170°$$

$$\vec{M}_{t1} = 20\angle 270° \text{ gm}; \qquad \vec{M}_{t2} = 25\angle 180° \text{ gm}$$

From equations (12.45) through (12.48), one obtains

$$\vec{A}_{11} = \frac{7.0\angle 120° - 10.0\angle 55°}{20\angle 270°}; \qquad \vec{A}_{21} = \frac{5.0\angle 225° - 7.0\angle 120°}{20\angle 270°}$$

$$\vec{A}_{12} = \frac{6.0\angle 120° - 10.0\angle 55°}{25\angle 180°}; \qquad \vec{A}_{22} = \frac{12.0\angle 170° - 7.0\angle 120°}{25\angle 180°}$$

These phasors are computed as below.

$$\vec{A}_{11} = \frac{(7.0\cos 120° - 10\cos 55°) + j(7\sin 120° - 10\sin 55°)}{20\angle 270°} = \frac{-9.235 - j2.129}{20\angle 270°}$$

$$= \frac{9.477\angle 193°}{20\angle 270°} = 0.474\angle -77°$$

$$\vec{A}_{21} = \frac{(5\cos 225° - 7\cos 120°) + j(5\sin 225° - 7\sin 120°)}{20\angle 270°} = \frac{-7.036 - j9.6}{20\angle 270°}$$

$$= \frac{11.9\angle 234°}{20\angle 270°} = 0.595\angle -36°$$

$$\vec{A}_{12} = \frac{(6\cos 120° - 10\cos 55°) + j(6\sin 120° - 10\sin 55°)}{25\angle 180°} = \frac{-8.736 - j3.0}{25\angle 180°}$$

$$= \frac{9.237\angle 199°}{25\angle 180°} = 0.369\angle 19°$$

$$\vec{A}_{22} = \frac{(12\cos 170° - 7\cos 120°) + j(12\sin 170° - 7\sin 120°)}{25\angle 180°} = \frac{-8.318 - j4.0}{25\angle 180°}$$

$$= \frac{9.23\angle 205.7°}{25\angle 180°} = 0.369\angle 25.7°$$

Next, the denominators of the balancing mass phasors (in equations (12.50) and (12.51)) are computed as:

$$\vec{A}_{22}\vec{A}_{11} - \vec{A}_{12}\vec{A}_{21} = (0.369\angle 25.7° \times 0.474\angle -77°) - (0.369\angle 19° \times 0.595\angle -36°)$$

$$= 0.1749\angle -51.3° - 0.2196\angle -17°$$

$$= (0.1749\cos 51.3° - 0.2196\cos 17°) - j(0.1749\sin 51.3° - 0.2196\sin 17°)$$

$$= -0.1 - j0.0723$$

$$= 0.1234\angle 216°$$

and hence,

$$-(\vec{A}_{22}\vec{A}_{11} - \vec{A}_{12}\vec{A}_{21}) = 0.1234\angle 36°$$

Finally, the balancing mass phasors are computed using equations (12.50) and (12.51) as:

$$\vec{M}_{b1} = \frac{0.369\angle 25.7° \times 10\angle 55° - 0.369\angle 19° \times 7.0\angle 120°}{0.1234\angle 216°}$$

$$= \frac{3.69\angle 80.7° - 2.583\angle 139°}{0.1234\angle 216°}$$

$$= \frac{(3.69\cos 80.7° - 2.583\cos 139°) + j(3.69\sin 80.7° - 2.5838\sin 139°)}{0.1234\angle 216°}$$

$$= \frac{2.546 + j1.947}{0.1234\angle 216°} = \frac{3.205\angle 37.4°}{0.1234\angle 216°} = 26\angle -178.6°$$

$$\vec{M}_{b2} = \frac{0.595\angle -36° \times 10\angle 55° - 0.474\angle -77° \times 7.0\angle 120°}{0.1234\angle 36°}$$

$$= \frac{5.95\angle 19° - 3.318\angle 43°}{0.1234\angle 36°}$$

$$= \frac{(5.95\cos 19° - 3.318\cos 43°) + j(5.95\sin 19° - 3.318\sin 43°)}{0.1234\angle 36°}$$

$$= \frac{3.2 + j0.326}{0.1234\angle 36°} = \frac{1.043\angle 5.8°}{0.1234\angle 36°} = 8.45\angle -30.0°$$

Finally, we have

$$-\vec{M}_{b1} = 26\angle 1.4° \text{ gm}; \quad -\vec{M}_{b2} = 8.45\angle 150° \text{ gm}$$

□

12.3.4 Experimental Procedure of Balancing

The experimental procedure for determining the balancing masses and locations for a rotating system should be clear from the analytical developments and examples given above. The basic steps are: (1) determine the magnitude and the phase angle of accelerometer signals at the bearings with and without trial masses at the bearing planes; (2) using this data, compute the necessary

Vibration Design and Control

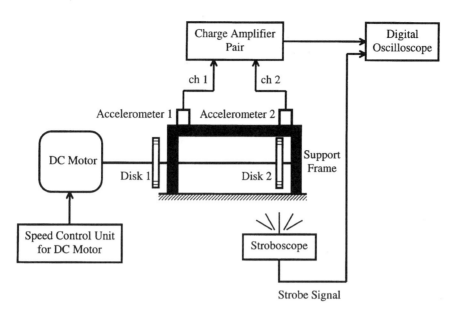

FIGURE 12.14 Schematic arrangement of a rotor balancing experiment.

balancing masses (magnitude and location) at the bearing planes; (3) place the balancing masses; and (4) check whether the system is balanced. If not, repeat the balancing cycle.

A laboratory experimental setup for two-plane balancing is schematically shown in Figure 12.14. A view of the system is shown in Figure 12.15. The two disks rigidly mounted on the shaft, are driven by a DC motor. The drive speed of the motor is adjusted by the manual speed controller. The shaft bearings (two) are located very close to the disks, as shown in Figure 12.14. Two accelerometers are mounted on the top of the bearing housing so that the resulting vertical accelerations can be measured. The accelerometer signals are conditioned using the two-channel charge amplifier, and read and displayed through two channels of the digital oscilloscope. The output of the stroboscope (tachometer) is used as the reference signal with respect to which the phase angles of the accelerometer signals are measured.

In Figure 12.15, the items of equipment are seen, from left to right, as follows. The first item is the two-channel digital oscilloscope. Next is the manual speed controller, with control knob, for the DC motor. The pair of charge amplifiers for the accelerometers is situated next. The strobelight unit (strobe-tacho) is placed on top of the common housing of the charge-amplifier pair. The two-disk rotor system with the drive motor is shown as the last item to the right. Also, note the two accelerometers (seen as small vertical projections) mounted on the bearing frame of the shaft, directly above the two bearings.

In determining an unbalance load, the accelerator readings must be taken with respect to a body reference on the rotating object. Since this reference must always be fixed, prior to reading the oscilloscope data, the strobe-tacho should be synchronized with the disk rotation with respect to both frequency and phase. This is achieved as follows. Note that all the readings are taken with the same rotating speed, which is adjusted by the manual speed controller. Make a physical mark (e.g., black spot in a white background) on one of the disks. Aim the strobe flash at this disk. As the motor speed is adjusted to the required fixed value, the strobe flash is synchronized such that the mark on the disk "appears" stationary at the same location (e.g., at the uppermost location of the circle of rotation). This ensures not only that the strobe frequency is equal to the rotating speed of the disk, but also that the same phase angle reference is used for all readings of accelerometer signals.

The two disks have slots at locations for which the radius is known and for which the angular positions with respect to a body reference line (a radius representing the 0° reference line) are

FIGURE 12.15 A view of the experimental setup for two-plane balancing. (Courtesy of the University of British Colombia. With permission.)

clearly marked. Known masses (typically bolts and nuts of known mass) can be securely mounted in these slots. Readings obtained through the oscilloscope are:

1. Amplitude of each accelerometer signal
2. Phase lead of the accelerometer signal with respect to the synchronized and reference-fixed strobe signal (*Note*: a phase lag should be represented by a negative sign in the data.)

The measurements taken and the computations made in the experimental procedure should be clear from Example 12.4.

12.4 BALANCING OF RECIPROCATING MACHINES

A reciprocating mechanism has a slider that moves rectilinearly back and forth along some guideway. A piston-cylinder device is a good example. Often, reciprocating machines contain rotatory components in addition to the reciprocating mechanisms. The purpose would be to either convert a reciprocating motion to a rotary motion (as in the case of an automobile engine), or to convert a rotary motion to a reciprocating motion (as in the opto-slider mechanism of a photocopier). No matter what type of reciprocating machine is employed, it is important to remove the vibratory excitations that arise, in order to realize the standard design goals of smooth operation, accuracy, low noise, reliability, mechanical integrity, and extended service life. Naturally, reciprocating mechanisms with rotary components are more prone to unbalance than purely rotary components, in view of their rotational asymmetry. Removing the "source of vibration" by proper balancing of the machine would be especially appropriate in this situation.

Vibration Design and Control

12.4.1 Single-Cylinder Engine

A common practical example of a reciprocating machine with integral rotary motion is the internal combustion (IC) engine of an automobile. A single-cylinder engine is sketched in Figure 12.16. Note the nomenclature of the components. The reciprocating motion of the piston is transmitted through the connecting rod and crank, into a rotatory motion of the crank shaft. The crank, as sketched in Figure 12.16, has a counterbalance mass, the purpose of which is to balance the rotary force (centrifugal). In this analysis, this counterbalance mass will be ignored because the goal is to determine the unbalance forces and ways to balance them.

Clearly, both the connecting rod and the crank have distributed mass and moment of inertia. To simplify the analysis, the following approximations are made:

1. Represent the crank mass by an equivalent lumped mass at the crank pin (equivalence can be based on either centrifugal force or kinetic energy).
2. Represent the mass of the connecting rod by two lumped masses: one at the crank pin and the other at the crosshead (piston pin).

The piston itself has a significant mass, which is also lumped at the crosshead. Hence, the equivalent system has a crank and a connecting rod, both of which are considered massless, with a lumped mass m_c at the crank pin and another lumped mass m_p at the piston pin (crosshead).

Furthermore, under normal operation, the crankshaft rotates at a constant angular speed (ω). Note that this steady speed is realized not by natural dynamics of the system, but by proper speed control, which is a topic outside the scope of the present treatment.

It is a simple matter to balance the lumped mass m_c at the crank pin. Simply place a countermass m_c at the same radius in the radially opposite location (or a mass in inverse proportion to the radial distance from the crankshaft, still in the radially opposite direction). This explains the presence of the countermass in the crank shown in Figure 12.16. Once complete balancing of the rotating inertia (m_c) is achieved in this manner, what remains to be done to realize complete elimination of the effect of the vibration source on the crankshaft is the compensation for the forces and moments on the crankshaft that result from (1) the reciprocating motion of the lumped mass m_p, and (2) time-varying combustion (gas) pressure in the cylinder. Both types of forces act on the piston in the direction of its reciprocating (rectilinear) motion. Hence, their influence on the crankshaft can be analyzed in the same way, except that the combustion pressure is much more difficult to determine.

The foregoing discussion justifies the use of the simplified model shown in Figure 12.17 for analyzing the balancing of a reciprocating machine. The characteristics of this model are as follows:

1. A light crank OC of radius r rotates at constant angular speed ω about O, which is the origin of the x-y coordinate frame.
2. A light connecting rod CP of length l is connected to the connecting rod at C and to the piston at P with frictionless pins. Since the rod is light and the joints are frictionless, the force f_c supported by it will act along its length (assume that the force f_c in the connecting rod is compressive, for the purpose of sign convention). The connecting rod makes an angle ϕ with OP (the negative x-axis).
3. A lumped mass m_p is present at the piston. A force f acts at P in the negative x-direction. This can be interpreted as either the force due to the gas pressure in the cylinder, or the inertia force $m_p a$, where a is the acceleration of m_p in the positive x-direction (by D'Alembert's principle). These two cases of forcing will be considered separately.
4. A lateral reaction force f_l acts on the piston by the cylinder wall, in the positive y-direction.

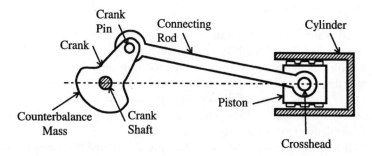

FIGURE 12.16 A single-cylinder reciprocating engine.

Note again that the lumped mass m_c at C is not included in the model of Figure 12.17 because it is assumed to be completely balanced by a countermass in the crank. Furthermore, the lumped mass m_p includes not just the mass of the piston, but also part of the inertia of the connecting rod.

There are no external forces at C. Furthermore, the only external forces at P are f and f_l, where f is interpreted as either the inertia force in m_p or the gas force on the piston. Hence, there should be equal and opposite forces at the crank shaft O, as shown in Figure 12.17, to support the forces acting at P. Now to determine f_l, proceed as follows:

Equilibrium at P gives

$$f = f_c \cos \phi$$

$$f_l = f_c \sin \phi$$

Hence,

$$f_l = f \tan \phi \tag{12.52}$$

This lateral force f_l acting at both O and P, albeit in the opposite directions, forms a couple $\tau = xf_l$, or in view of equation (12.52),

$$\tau = xf \tan \phi \tag{12.53}$$

This couple acts as a torque on the crankshaft. It follows that, once the rotating inertia m_c at the crank is completely balanced by a countermass, the load at the crankshaft is due only to the piston load f, and it consists of (1) a force f in the direction of the piston motion (x), and (2) a torque $\tau = xf\tan\phi$ in the direction of rotation of the crank shaft (i.e., about z).

The means of removing f at the crankshaft, which is discussed below, will also remove τ in many cases. Hence, only the approach of balancing f will be discussed here.

12.4.2 BALANCING THE INERTIA LOAD OF THE PISTON

Now, consider the inertia force f due to m_p. Here,

$$f = m_p a \tag{12.54}$$

where a is the acceleration $\ddot{x}$, with the coordinate x locating the position P of the piston (in other words, $OP = x$). Notice from Figure 12.17 that

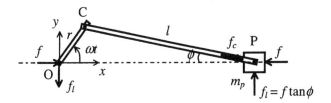

FIGURE 12.17 The model used to analyze balancing of a reciprocating engine.

$$x = r\cos\omega t + l\cos\phi \qquad (12.55)$$

But,

$$r\sin\omega t = l\sin\phi \qquad (12.56)$$

Hence, from trigonometry,

$$\cos\phi = \left[1 - \left(\frac{r}{l}\right)^2 \sin^2\omega t\right]^{\frac{1}{2}} \qquad (12.57)$$

which can be expanded up to the first term of Taylor series as

$$\cos\phi \cong 1 - \frac{1}{2}\left(\frac{r}{l}\right)^2 \sin^2\omega t \qquad (12.58)$$

This approximation is valid because l is usually several time larger than r and, hence, $(r/l)^2$ is much smaller than unity. Next, in view of

$$\sin^2\omega t = \frac{1}{2}[1 - \cos 2\omega t] \qquad (12.59)$$

one has

$$\cos\phi \cong 1 - \frac{1}{4}\left(\frac{r}{l}\right)^2 [1 - \cos 2\omega t] \qquad (12.60)$$

Substitute equation (12.60) in (12.55) to get, approximately,

$$x = r\cos\omega t + \frac{l}{4}\left(\frac{r}{l}\right)^2 \cos 2\omega t + l - \frac{l}{4}\left(\frac{r}{l}\right)^2 \qquad (12.61)$$

Differentiate equation (12.61) twice with respect to t to obtain the acceleration

$$a = \ddot{x} = -r\omega^2 \cos\omega t - l\left(\frac{r}{l}\right)^2 \omega^2 \cos 2\omega t \qquad (12.62)$$

Hence, from equation (12.54), the inertia force at the piston (and its reaction at the crankshaft) is

$$f = -m_p r\omega^2 \cos\omega t - m_p l\left(\frac{r}{l}\right)^2 \omega^2 \cos 2\omega t \qquad (12.63)$$

It follows that the inertia load of the reciprocating piston exerts a vibratory force on the crankshaft, which has a *primary component* of frequency ω and a smaller *secondary component* of frequency 2ω, where ω is the angular speed of the crank. The primary component has the same form as that created by a rotating lumped mass at the crank pin. But, unlike the case of a rotating mass, this vibrating force acts only in the x direction (there is no $\sin\omega t$ component in the y direction) and, hence, cannot be balanced by a rotating countermass. Similarly, the secondary component cannot be balanced by a countermass rotating at double the speed. The means employed to eliminate f are to use multiple cylinders whose connecting rods and cranks being connected to the crankshaft with their rotations properly phased (delayed) so as to cancel out the effects of f.

12.4.3 Multicylinder Engines

A single-cylinder engine generates a primary component and a secondary component of vibration load at the crankshaft, and they act in the direction of piston motion (x). Since there is no complementary orthogonal component (y), it is inherently unbalanced and cannot be balanced using a rotating mass. It can be balanced, however, using several piston-cylinder units, with their cranks properly phased along the crankshaft. This method of balancing multicylinder reciprocating engines is addressed now.

Consider a single cylinder whose piston inertia generates a force f at the crankshaft, in the x direction, given according to equation (12.63) by

$$f = f_p \cos\omega t + f_s \cos 2\omega t \qquad (12.64)$$

Note that the primary and secondary forcing amplitudes f_p and f_s, respectively, are as given by equation (12.63). Suppose that there is a series of cylinders in parallel, arranged along the crankshaft, and the crank of cylinder i makes an angle α_i with the crank of cylinder 1, in the direction of rotation, as schematically shown in Figure 12.18(a). Hence, force f_i on the crankshaft (in the x direction, shown as vertical in Figure 12.18) due to cylinder i is

$$f_i = f_p \cos(\omega t + \alpha_i) + f_s \cos(2\omega t + 2\alpha_i) \qquad \text{for } i = 1, 2, \ldots, \text{ with } \alpha_1 = 0 \qquad (12.65)$$

Not only the cranks need to be properly phased, but the cylinders should be properly spaced along the crankshaft as well, in order to obtain the necessary balance. Now consider two examples.

Two-Cylinder Engine

Consider the two-cylinder case, as schematically shown in Figure 12.18(b) where the two cranks are in radially opposite orientations (i.e., 180° out of phase). In this case, $\alpha_2 = \pi$. Hence,

$$f_1 = f_p \cos\omega t + f_s \cos 2\omega t \qquad (12.66)$$

$$\begin{aligned} f_2 &= f_p \cos(\omega t + \pi) + f_s \cos(2\omega t + 2\pi) \\ &= -f_p \cos\omega t + f_s \cos 2\omega t \end{aligned} \qquad (12.67)$$

Vibration Design and Control

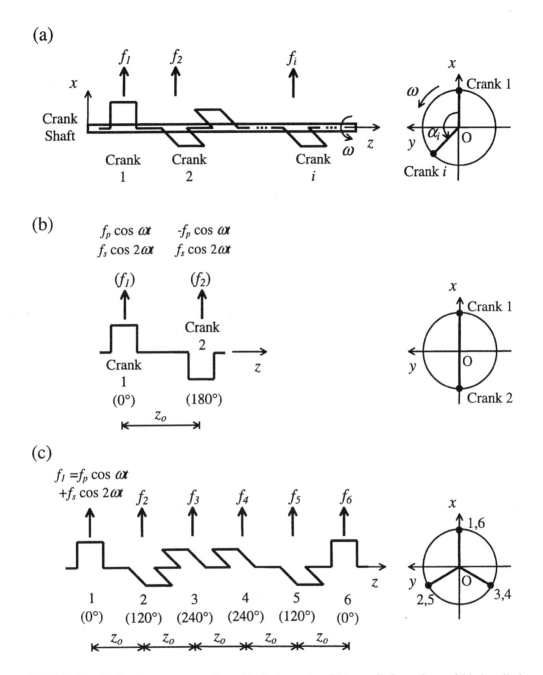

FIGURE 12.18 (a) Crank arrangement of a multicylinder engine; (b) two-cylinder engine; and (c) six-cylinder engine (balanced).

It follows that the primary force components cancel out; however, they form a couple $z_0 f_p \cos\omega t$, where z_0 is the spacing of the cylinders. This causes a bending moment on the crankshaft, and it will not vanish unless the two cylinders are located at the same point along the crankshaft. Furthermore, the secondary components are equal and additive to $2f_s \cos 2\omega t$. This resultant component acts at the mid-point of the crankshaft segment between the two cylinders. There is no couple due to the secondary components, however.

Six-Cylinder Engine

Consider the six-cylinder arrangement schematically shown in Figure 12.18(c). Here, the cranks are arranged such that $\alpha_2 = \alpha_5 = 2\pi/3$, $\alpha_3 = \alpha_4 = 4\pi/3$, and $\alpha_1 = \alpha_6 = 0$. Furthermore, the cylinders are equally spaced, with spacing z_0. In this case,

$$f_1 = f_6 = f_p \cos\omega t + f_s \cos 2\omega t \qquad (i)$$

$$f_2 = f_5 = f_p \cos(\omega t + 2\pi/3) + f_s \cos(2\omega t + 4\pi/3) \qquad (ii)$$

$$f_3 = f_4 = f_p \cos(\omega t + 4\pi/3) + f_s \cos(2\omega t + 8\pi/3) \qquad (iii)$$

Now, use the fact that

$$\cos\theta + \cos\left(\theta + \frac{2\pi}{3}\right) + \cos\left(\theta + \frac{4\pi}{3}\right) = 0 \qquad (iv)$$

which can be proved either by straightforward trigonometric expansion, or using geometric interpretation (i.e., three sides of an equilateral triangle, the sum of whose components in any direction vanishes). The relation (*iv*) holds for any θ, including $\theta = \omega t$ and $\theta = 2\omega t$. Furthermore, $\cos(2\omega t + 8\pi/3) = \cos\left(2\omega t + \frac{2\pi}{3}\right)$. Then, from equations (*i*) through (*iii*), one concludes that

$$f_1 + f_2 + f_3 + f_4 + f_5 + f_6 = 0 \qquad (12.68)$$

This means that the lateral forces on the crankshaft that are exerted by the six cylinders will completely balance. Furthermore, by taking moments about the location of crank 1 of the crankshaft, one obtains

$$(z_0 + 4z_0)\left[f_p \cos\left(\omega t + \frac{2\pi}{3}\right) + f_s \cos\left(2\omega t + \frac{4\pi}{3}\right)\right]$$

$$+ (2z_0 + 3z_0)\left[f_p \cos\left(\omega t + \frac{4\pi}{3}\right) + f_s \cos\left(2\omega t + \frac{8\pi}{3}\right)\right] \qquad (v)$$

$$+ 5z_0\left[f_p \cos\omega t + f_s \cos 2\omega t\right]$$

which also vanishes in view of relation (*iv*). Hence, the set of six forces is in complete equilibrium; and as a result, there will be neither a reaction force nor a bending moment on the bearings of the crankshaft from these forces.

Also, it can be shown that the torques $x_i f_i \tan\phi_i$ on the crankshaft due to this set of inertial forces f_i will also add to 0, where x_i is the distance from the crankshaft to the piston of the *i*th cylinder and ϕ_i is the angle ϕ of the connecting rod of the *i*th cylinder. Hence, this six-cylinder configuration is in complete balance with respect to the inertial loads.

EXAMPLE 12.5

An eight-cylinder in-line engine (with identical cylinders that are placed in parallel along a line) has its cranks arranged according to the phasing angles 0°, 180°, 90°, 270°, 270°, 90°, 180°, and 0° on the crankshaft. The cranks (cylinders) are equally spaced, with spacing z_0. Show that this engine is balanced with respect to primary and secondary components of reaction forces and bending moments of inertial loading on the bearings of the crankshaft.

SOLUTION

The sum of the reaction forces on the crankshaft are:

$$2 \times \left[f_p \cos \omega t + f_s \cos 2\omega t + f_p \cos(\omega t + \pi) + f_s \cos(2\omega t + 2\pi) + f_p \cos\left(\omega t + \frac{\pi}{2}\right) + f_s(2\omega t + \pi) \right.$$
$$\left. + f_p \cos\left(\omega t + \frac{3\pi}{2}\right) + f_s \cos(2\omega t + 3\pi) \right]$$

$$= 2\left[f_p \cos \omega t - f_p \cos \omega t - f_p \sin \omega t + f_p \sin \omega t + f_s \cos 2\omega t + f_s \cos 2\omega t - f_s \cos 2\omega t - f_s \cos 2\omega t \right]$$

$$= 0$$

Hence, both primary forces and secondary forces are balanced. The moment of the reaction forces about the crank 1 location of the crankshaft:

$$(z_0 + 6z_0)\left[f_p \cos(\omega t + \pi) + f_s \cos(2\omega t + 2\pi) \right] + (2z_0 + 5z_0)\left[f_p \cos\left(\omega t + \frac{\pi}{2}\right) + f_s \cos(2\omega t + \pi) \right]$$

$$+ (3z_0 + 4z_0)\left[f_p \cos\left(\omega t + \frac{3\pi}{2}\right) + f_s \cos(2\omega t + 3\pi) \right] + 7z_0\left[f_p \cos \omega t + f_s \cos 2\omega t \right]$$

$$= 7z_0\left[-f_p \cos \omega t + f_s \cos 2\omega t - f_p \sin \omega t - f_s \cos 2\omega t + f_p \sin \omega t - f_s \cos 2\omega t + f_p \cos \omega t + f_s \cos 2\omega t \right]$$

$$= 0$$

Hence, both primary bending moments and secondary bending moments are balanced as well. Thus, the engine is completely balanced.

$\square$

The formulas applicable in balancing reciprocating machines are summarized in Box 12.4.

Before leaving the topic of balancing the inertial loading at the piston, it should be noted that in the configuration considered above, the cylinders are placed in parallel along the crankshaft. These are termed *in-line engines*. Their resulting forces f_i act in parallel along the shaft. In other configurations, such as V6 and V8, the cylinders are placed symmetrically around the shaft; in such cases, the cylinders (and their inertial forces that act on the crankshaft) are not parallel. Then, a complete force balance can be achieved without even having to phase the cranks and, furthermore, the bending moments of the forces can be reduced by placing the cylinders nearly at the same location along the crankshaft. Complete balancing of the combustion/pressure forces is possible as well with such an arrangement.

BOX 12.4 Balancing of Reciprocating Machines

Single cylinder engine:
Inertia force at piston (and its reaction on crankshaft)

$$f = -m_p r\omega^2 \cos\omega t - m_p l \left(\frac{r}{l}\right)^2 \omega^2 \cos 2\omega t$$

$$= f_p \cos\omega t + f_s \cos 2\omega t$$

where

ω = rotating speed of crank
m_p = equivalent lumped mass at piston
r = crank radius
l = length of connecting rod
f_p = amplitude of the primary unbalance force (frequency ω)
f_s = amplitude of the secondary unbalance force (frequency 2ω).

Multicylinder engine:

Net unbalance reaction force on the crankshaft = $\sum_{i=1}^{n} f_i$

Net unbalance moment on the crankshaft = $\sum_{i=1}^{n} z_i f_i$

where

$f_i = f_p \cos(\omega t + \alpha_i) + f_s \cos(2\omega t + 2\alpha_i)$
α_i = angular position of the crank of ith cylinder, with respect to a body (rotating) reference (i.e., crank phasing angle)
z_i = position of the ith crank along the crankshaft, measured from a reference point on the shaft
n = Number of cylinders (assumed identical).

Note: For a completely balanced engine, both the net unbalance force and the net unbalance moment should vanish.

12.4.4 COMBUSTION/PRESSURE LOAD

In the balancing approach presented thus far, the force f on the piston represents the inertia force due to the equivalent reciprocating mass. Its effect on the crankshaft is an equal reaction force f in the lateral direction (x) and a torque $\tau = xf\tan\phi$ about the shaft axis (z). The balancing approach that is used employs a series of cylinders so that their reaction forces f_i on the crankshaft from an equilibrium set. Then no net reaction or bending moment is transmitted to the bearings of the shaft. The torques τ_i can also be balanced by the same approach — which is the case, for example, in the six-cylinder engine.

There is another important force that acts along the direction of piston reciprocation. It is the drive force due to gas pressure in the cylinder (created, for example, by combustion of the fuel–air

Vibration Design and Control

mixture of an internal combustion engine). This force can be analyzed as before, by denoting it as f. However, several important observations should be made first.

1. The combustion force f is not sinusoidal of frequency ω. It is reasonably periodic, but the shape of a force period is complex and depends on the firing/fuel-injection cycle and the associated combustion process.
2. The reaction forces f_i on the crankshaft, which are generated from cylinders i, should be balanced to avoid the transmission of reaction forces and bending moments to the shaft bearings (and hence, to the supporting frame — the vehicle). But the torques τ_i in this case are in fact the drive torques. Obviously, they are the desired output of the engine and should not be balanced unlike the inertia torques.

In view of these observations, the analysis that was done for balancing the inertia forces cannot be directly used here. But, approaches similar to the use of multiple cylinders can be employed for reducing the gas-force reactions. This is a rather difficult problem, mainly in view of the complexity of the combustion process itself. In practice, much of the left-over effects of the ignition cycle are suppressed by properly designed engine mounts. Experimental investigations have indicated that in a properly balanced engine unit, much of the vibration that is transmitted through the engine mounts is caused by the engine firing cycle (internal combustion) rather than the reciprocating inertia (sinusoidal components of frequency ω and 2ω). Hence, active mounts, where stiffness can be varied according to the frequency of excitation, are being considered to reduce engine vibrations in the entire range of operating speeds (say, 500 to 2500 rpm).

12.5 WHIRLING OF SHAFTS

The previous two sections focused on the vibration excitations caused on rotating shafts and their bearings due to some form of mass eccentricity. Methods of balancing these systems so as to eliminate the undesirable effects were presented as well. One limitation of the given analysis is the assumption that the rotating shaft is rigid and hence does not deflect from its axis of rotation due to the unbalance excitations. In practice, however, rotating shafts are made lighter than the components they carry (rotors, disks, gears, etc.) and will undergo some deflection due to the unbalance loading. As a result, the shaft will bow out, and this will further increase the mass eccentricity and associated unbalance excitations and gyroscopic forces of the rotating elements (disks, rotors, etc.). The nature of damping of rotating machinery, which is rather complex and incorporates effects of rotation at bearings, structural deflections, and lateral speeds, will further affect the dynamic behavior of the shaft under these conditions. In this context, the topic of whirling of rotating shafts becomes quite relevant.

Consider a shaft that is driven at a constant angular speed ω (e.g., using a motor or some other actuator). The central axis of the shaft (passing through its bearings) will bow out. This deflected axis itself will rotate, and this rotation is termed *whirling* or *whipping*. The whirling speed is not necessarily equal to the drive speed ω (at which the shaft rotates about its axis with respect to a fixed frame). But, when the whirling speed is equal to ω, the condition is called *synchronous whirl*, and the associated deflection of the shaft can be quite excessive and damaging.

To develop an analytical basis for whirling, consider a light shaft supported on two bearings and carrying a disk of mass m in between the bearings, as shown in Figure 12.19(a). Note that C is the point on the disk at which it is mounted on the shaft. Originally, in the neutral configuration when the shaft is not driven ($\omega = 0$), the point C coincides with point O on the axis joining the two bearings. If the shaft were rigid, the points C and O would continue to coincide during motion. The mass center (centroid, or center of gravity for constant g) of the disk is denoted as G in Figure 12.19. During motion, C will move away from O due to the shaft deflection. The whirling speed (speed of rotation of the shaft axis) is the speed of rotation of the radial line OC with respect

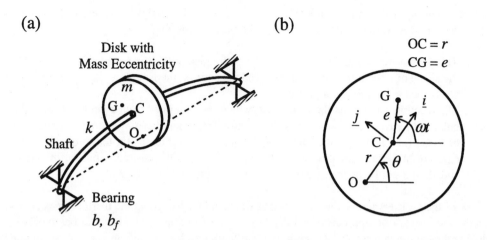

FIGURE 12.19 (a) A whirling shaft carrying a disk with mass eccentricity, and (b) end view of the disk and whirling shaft.

to a fixed reference. Denoting the angle of OC with respect to a fixed reference as θ, the whirling speed is $\dot\theta$. This is explained in Figure 12.19(b) where an end view of the disk is given under deflected conditions. The constant drive speed ω of the shaft is the speed of the shaft spin with respect to a fixed reference and is the speed of rotation of the radial line CG with respect to the fixed horizontal line shown in Figure 12.19(b). Hence, the angle of shaft spin is ωt, as measured with respect to this line. The angle of whirl, θ, is also measured from the direction of this fixed line, as shown.

12.5.1 Equations of Motion

Under practical conditions, the disk moves entirely in a single plane. Hence, its complete set of equations of motion consists of two equations for translatory (planar) motion of the centroid (with lumped mass m) and one equation for rotational motion about the fixed bearing axis. The latter equation depends on the motor torque that derives the shaft at constant speed ω, and is not of interest in the present context. So, one can limit the development to the two translatory equations of motion. The equations can be written either in a Cartesian coordinate system (x,y) or a polar coordinate system (r,θ). The polar coordinate system is used here.

Consider a coordinate frame (i, j) that is fixed to the disk with its i axis lying along OC as shown in Figure 12.19(b). Note that the angular speed of this frame is $\dot\theta$ (about k axis that is orthogonal to i and j). Hence, as is well known, one has

$$\frac{di}{dt} = \dot\theta j \quad \text{and} \quad \frac{dj}{dt} = -\dot\theta i \qquad (12.69)$$

The position vector of the mass point G from O is

$$\overrightarrow{OG} = r_G = \overrightarrow{OC} + \overrightarrow{CG} = ri + e\cos(\omega t - \theta)i + e\sin(\omega t - \theta)j \qquad (12.70)$$

The velocity vector v_G of the mass point G can be obtained simply by differentiating equation (12.70), with the use of (12.69). But, this can be simplified because ω is constant. Here, line CG has a velocity $e\omega$ perpendicular to it, about C. This can be resolved along the axes i and j. Hence, the velocity of G relative to C is

Vibration Design and Control

$$v_{G/C} = -e\omega\sin(\omega t - \theta)i + e\omega\cos(\omega t - \theta)j$$

The velocity of point C is

$$v_C = \frac{d}{dt}ri = \dot{r}i + r\frac{di}{dt} = \dot{r}i + r\dot{\theta}j$$

Hence, the velocity of G, which is given by $v_G = v_C + v_{G/C}$, can be expressed as

$$v_G = \dot{r}i + r\dot{\theta}j - e\omega\sin(\omega t - \theta)i + e\omega\cos(\omega t - \theta)j \tag{12.71}$$

Similarly, the acceleration of C is

$$a_C = \frac{d}{dt}v_C = \frac{d}{dt}\left[\dot{r}i + r\dot{\theta}j\right] = \ddot{r}i + \dot{r}\dot{\theta}j + \dot{r}\dot{\theta}j + r\ddot{\theta}j - r\dot{\theta}^2 i = \left(\ddot{r} - r\dot{\theta}^2\right)i + \left(r\ddot{\theta} + 2\dot{r}\dot{\theta}\right)j$$

Also, since line CG rotates at constant angular speed ω about C, the point G has only a radial (centrifugal) acceleration $e\omega^2$ along GC, about C. This can be resolved along i and j as before. Hence, the acceleration of G relative to C is

$$a_{G/C} = -e\omega^2\cos(\omega t - \theta)i - e\omega^2\sin(\omega t - \theta)j$$

It follows that the acceleration of point G, given by $a_G = a_C + a_{G/C}$, can be expressed as

$$a_G = \left(\ddot{r} - r\dot{\theta}^2\right)i + \left(r\ddot{\theta} + 2\dot{r}\dot{\theta}\right)j - e\omega^2\cos(\omega t - \theta)i - e\omega^2\sin(\omega t - \theta)j \tag{12.72}$$

The forces acting on the disk are as follows:

Restraining elastic force due to lateral deflection of the shaft $= -kri$

Viscous damping force (proportional to the velocity of C) $= -b\dot{r}i - br\dot{\theta}j$

In addition, there is a frictional resistance at the bearing, which is proportional to the reaction and, hence, the shaft deflection r, and also depends on the spin speed ω. The following approximate model is used:

Bearing friction force $= -b_f r\omega j$

Here,

k = lateral deflection stiffness of the shaft at the location of the disk
b = viscous damping constant for lateral motion of the shaft
b_f = bearing frictional coefficient.

The overall force acting on the disk is

$$f = -(kr + b\dot{r})i - (br\dot{\theta} + b_f r\omega)j \qquad (12.73)$$

The equation of rectilinear motion

$$f = ma_G \qquad (12.74)$$

on using equations (12.72) and (12.73), reduces to the following pair in the i and j directions:

$$-kr - b\dot{r} = m\left[\ddot{r} - r\dot{\theta}^2 - e\omega^2 \cos(\omega t - \theta)\right] \qquad (12.75)$$

$$-br\dot{\theta} - b_f r\omega = m\left[r\ddot{\theta} + 2\dot{r}\dot{\theta} - e\omega^2 \sin(\omega t - \theta)\right] \qquad (12.76)$$

These equations can be expressed as

$$\ddot{r} + 2\zeta_v \omega_n \dot{r} + \left(\omega_n^2 - \dot{\theta}^2\right)r = e\omega^2 \cos(\omega t - \theta) \qquad (12.77)$$

$$r\ddot{\theta} + 2(\zeta_v \omega_n r + \dot{r})\dot{\theta} + 2\zeta_f \omega_n \omega r = e\omega^2 \sin(\omega t - \theta) \qquad (12.78)$$

where the undamped natural frequency of lateral vibration is

$$\omega_n = \sqrt{\frac{k}{m}} \qquad (12.79)$$

and

ζ_v = viscous damping ratio of lateral motion
ζ_f = frictional damping ratio of the bearings.

Equations (12.77) and (12.78), which govern the whirling motion of the shaft-disk system, are a pair of coupled nonlinear equations, with excitations (depending on ω) that are coupled with a motion variable (θ). Hence, a general solution would be rather complex. A relatively simple solution is possible, however, under steady-state whirling.

12.5.2 STEADY-STATE WHIRLING

Under steady-state conditions, the whirling speed $\dot{\theta}$ is constant at $\dot{\theta} = \omega_w$; hence, $\ddot{\theta} = 0$. Also, the lateral deflection of the shaft is constant; hence, $\dot{r} = \ddot{r} = 0$. Then, equations (12.77) and (12.78) become

$$\left(\omega_n^2 - \omega_w^2\right)r = e\omega^2 \cos(\omega t - \theta) \qquad (12.80)$$

$$2\zeta_v \omega_n \omega_w r + 2\zeta_f \omega_n \omega r = e\omega^2 \sin(\omega t - \theta) \qquad (12.81)$$

In equations (12.80) and (12.81), the LHS is independent of t. Hence, the RHS also should be independent of t. For this, one must have

Vibration Design and Control

$$\theta = \omega t - \phi \qquad (12.82)$$

where ϕ is interpreted as the phase lag of whirl with respect to the shaft spin (ω), and should be clear from Figure 12.19(b). It follows from equation (12.82) that, for steady-state whirl, the whirling speed $\dot{\theta} = \omega_w$ is

$$\omega_w = \omega \qquad (12.83)$$

This condition is called *synchronous whirl* because the whirl speed (ω_w) is equal to the shaft spin speed (ω). It follows that under steady state, one should have the state of synchronous whirl. The equations governing steady-state whirl are

$$\left(\omega_n^2 - \omega^2\right)r = e\omega^2 \cos\phi \qquad (12.84)$$

$$2\zeta\omega_n\omega r = e\omega^2 \sin\phi \qquad (12.85)$$

along with equation (12.82), and hence equation (12.83). Here, $\zeta = \zeta_v + \zeta_f$ is the overall damping ratio of the system. Note that the phase angle ϕ and the shaft deflection r are determined from equations (12.84) and (12.85). In particular, squaring these two equations and adding, to eliminate ϕ, one obtains

$$r = \frac{e\omega^2}{\sqrt{\left(\omega_n^2 - \omega^2\right)^2 + \left(2\zeta\omega_n\omega\right)^2}} \qquad (12.86)$$

which is of the form of magnitude of the frequency-transfer function of a simple oscillator, with an acceleration excitation. Divide equation (12.85) by equation (12.84) to get the phase angle:

$$\phi = \tan^{-1}\frac{2\zeta\omega_n\omega}{\left(\omega_n^2 - \omega^2\right)} \qquad (12.87)$$

By simple calculus (differentiate the square of equation (12.86) and equate to 0), one can show that the maximum deflection occurs at the critical spin speed ω_c given by

$$\omega_c = \frac{\omega_n}{\sqrt{1 - 2\zeta^2}} \qquad (12.88)$$

This *critical speed* corresponds to a resonance. For light damping, one has approximately, $\omega_c = \omega_n$. Hence, the critical speed, for low damping, is equal to the undamped natural frequency of bending vibration of the shaft-rotor unit. The corresponding shaft deflection is [see equation (12.86)]

$$r_c = \frac{e}{2\zeta} \qquad (12.89)$$

which is also a good approximation of r at critical speed, with light damping. From equations (12.84), (12.85), and (12.89) one sees that at critical speed (with low damping), $\sin\phi = 1$ and $\cos\phi = 0$, which gives $\phi = \frac{\pi}{2}$. Also, note from equation (12.86) that the steady-state shaft deflection is almost 0 at low speeds, and it approaches e at very high speeds. But, from equation (12.87) for small ω, note that $\tan\phi$ is positive and small, but from equation (12.85), $\sin\phi$ is positive. This means that ϕ itself is small for small ω. For large ω, from equation (12.86) it is seen that r approaches e. Then from equation (12.87), $\tan\phi$ is small and negative; while from equation (12.85), $\sin\phi$ is positive. Hence, ϕ approaches π for large ω.

It is seen from equation (12.89) that at critical speed, the shaft deflection increases with mass eccentricity and decreases with damping. This observation, along with the given analysis, indicate that the approaches for reducing the damaging effects of whirling are:

1. Eliminate or reduce the mass eccentricity through proper construction practices and balancing
2. Increase damping
3. Increase shaft stiffness
4. Avoid operation near critical speed.

There will be limitations to the use of these approaches — particularly to making the shaft stiffer. Note further that this analysis does not include the mass distribution of the shaft. A Bernoulli-Euler type beam analysis (see Chapter 6) must be incorporated for a more accurate analysis of whirling, for shafts whose mass cannot be accurately represented by a single parameter that is lumped at the location of the rotor. Formulas related to whirling of shafts are summarized in Box 12.5.

EXAMPLE 12.6

The fan of a ventilation system has a normal operating speed of 3600 rpm. The blade set of the fan weighs 20 kg and is mounted in mid-span of a relatively light shaft that is supported on lubricated bearings at its two ends. The bending stiffness of the shaft at the location of the fan is 4.0×10^6 N·m^{-1}. The equivalent damping ratio that acts on the possible whirling motion of the shaft is 0.05. Due to fabrication error, the centroid of the fan has an eccentricity of 1.0 cm from the neutral axis of rotation of the shaft.

   a. Determine the critical speed of the fan system and the corresponding shaft deflection at the location of the fan, at steady state.
   b. What is the steady-state shaft deflection at the fan during normal operation?

The fan was balanced subsequently using a mass of 5 kg. The centroid eccentricity was reduced to 2 mm by this means. What is the shaft deflection at the fan during normal operation now? Comment on the improvement that has been realized.

SOLUTION

   a. The system is lightly damped. Hence, the critical speed is given by the undamped natural frequency; thus,

$$\omega_c \cong \omega_n = \sqrt{\frac{k}{m}} = \sqrt{\frac{4 \times 10^6}{20}} = 447.2 \text{ rad} \cdot \text{s}^{-1}$$

The corresponding shaft deflection is

$$r_c = \frac{e}{2\zeta} = \frac{1.0}{2 \times 0.05} = 10.0 \text{ cm}$$

Vibration Design and Control

BOX 12.5 Whirling of Shafts

Whirling: A shaft spinning at speed ω about its axis can bend due to flexure. The bent (bowed out) axis will rotate at speed ω_w. This is called whirling.

Equations of motion:

$$\ddot{r} + 2\zeta_v \omega_n \dot{r} + \left(\omega_n^2 - \dot{\theta}^2\right)r = e\omega^2 \cos(\omega t - \theta)$$

$$r\ddot{\theta} + 2\left(\zeta_v \omega_n r + \dot{r}\right)\dot{\theta} + 2\zeta_f \omega_n \omega r = e\omega^2 \sin(\omega t - \theta)$$

where (r, θ) = polar coordinates of shaft deflection at the mounting point of lumped mass

- e = eccentricity of the lumped mass from the spin axis of shaft
- $\dot{\theta} = \omega_w$ = whirling speed
- ω = spin speed of shaft
- $\omega_n = \sqrt{k/m}$ = natural frequency of bending vibration of shaft
- k = bending stiffness of shaft at lumped mass
- m = lumped mass
- ζ_v = damping ratio of bending motion of shaft
- ζ_f = damping ratio of shaft bearings.

Steady-state whirling (synchronous whirl):

Here, whirling speed ($\dot{\theta}$ or ω_w) is constant and equals the shaft spin speed ω (i.e., $\omega_w = \omega$ for steady-state whirling).

Shaft deflection at lumped mass $r = \dfrac{e\omega^2}{\left[\left(\omega_n^2 - \omega^2\right)^2 + \left(2\zeta\omega_n\omega\right)^2\right]^{1/2}}$

Phase angle between shaft deflection (r) and mass eccentricity (e)

$$\phi = \tan^{-1} \dfrac{2\zeta\omega_n\omega}{\left(\omega_n^2 - \omega^2\right)}$$

where $\zeta = \zeta_v + \zeta_f$

Note: For small spin speeds ω, r and ϕ are small. For large ω, $r \cong e$ and $\phi \cong \pi$.

Critical speed:

Spin speed

$$\omega_c = \dfrac{\omega_n}{\sqrt{1 - 2\zeta^2}} \quad \text{for small } \zeta$$

$\phi = \pi/2$

b. Operating speed $\omega = \dfrac{3600}{60} \times 2\pi = 377 \text{ rad} \cdot \text{s}^{-1}$. Using equation (12.86), the corresponding shaft deflection, at steady state, is

$$r = \dfrac{1.0 \times (377)^2}{\left[(447.2^2 - 377^2)^2 + (2 \times 0.05 \times 447.2 \times 377)^2\right]^{1/2}} = 2.36 \text{ cm}$$

After balancing, the new eccentricity $e = 0.2$ cm.
The new natural frequency (undamped) is

$$\omega_n = \sqrt{\dfrac{4 \times 10^6}{25}} = 400 \text{ rad} \cdot \text{s}^{-1}$$

The corresponding shaft deflection during steady-state operation, is

$$r = \dfrac{0.2 \times (377)^2}{\left[(400^2 - 377^2)^2 + (2 \times 0.05 \times 400 \times 377)^2\right]^{1/2}} = 1.216 \text{ cm}$$

Note that although the eccentricity has been reduced by a factor of 5, by balancing, the operating deflection of the shaft has been reduced only by a factor of less than 2. The main reason for this is the fact that the operating speed is close to the critical speed. Methods of improving the performance include: changing the operating speed; using a smaller mass to balance the fan; using more damping; and making the shaft stiffer. Some of these methods may not be feasible. Operating speed is determined by the task requirements. A location may not be available that is sufficiently distant to place a balancing mass that is appropriately small. Increased damping will increase heat generation, cause bearing problems, and will also reduce the operating speed. Replacement or stiffening of the shaft may require too much modification to the system and added cost. A more preferable alternative would be to balance the fan by removing some mass. This will move the critical frequency (natural frequency) away from the operating speed rather than closer to it, while reducing the mass eccentricity at the same time. For example, suppose that a mass of 3 kg is removed from the fan, which results in an eccentricity of 2.0 mm. The new natural/critical frequency is

$$\sqrt{\dfrac{4 \times 10^6}{17}} = 485.1 \text{ rad} \cdot \text{s}^{-1}$$

The corresponding shaft deflection during steady operation is

$$r = \dfrac{0.2 \times (377)^2}{\left[(485.1^2 - 377^2)^2 + (2 \times 0.05 \times 485.1 \times 377)^2\right]^{1/2}} = 0.3 \text{ cm}$$

In this case, the deflection has been reduced by a factor of 8.

□

Vibration Design and Control

12.5.3 SELF-EXCITED VIBRATIONS

It has been pointed out that equations (12.77) and (12.78), which represent the general whirling motion of a shaft, are nonlinear and coupled. A further characteristic of these equations is the fact that the motion variables (r and θ) occur as (nonlinear) products of the excitation (ω). Such systems are termed self-excited. Note that, in general (before reaching the steady state), the response variables r and θ will exhibit vibratory characteristics in view of the presence of the excitation functions $\cos(\omega t - \theta)$ and $\sin(\omega t - \theta)$. Hence, a whirling shaft can exhibit self-excited vibrations. Because the excitation forces directly depend on the motion itself, it is possible that a continuous energy flow into the system could occur. That will result in a steady growth of the motion amplitudes and represents an *unstable* behavior.

A simple example of self-excited vibration is provided by a simple pendulum whose length is time variable. Although the system is stable when the length is fixed, it can become unstable under conditions of variable length. Practical examples of self-excited vibrations with possible exhibition of instability are: flutter of aircraft wings due to coupled aerodynamic forces, wind-induced vibrations of bridges and tall structures, galloping of ice-covered transmission lines due to air flow-induced vibrations, and chattering of machine tools due to friction-related excitation forces. Proper design and control methods, as discussed in this chapter, are important in suppressing self-excited vibrations.

12.6 DESIGN THROUGH MODAL TESTING

Modal analysis, modal testing, and experimental modal analysis (EMA) are topics that have been covered in Chapters 5, 6, 10, and 11. In particular, EMA involves extracting modal parameters (natural frequencies, modal damping ratios, mode shapes) of a mechanical system through testing (notably through excitation-response data) and then developing a dynamic model of the system (mass, stiffness, and damping matrices) on that basis. The techniques of EMA are directly useful in modeling, as mentioned, and model validation (i.e., verification of the accuracy of an existing model that was obtained, for example, through analytical modeling). Apart from these uses, EMA is also a versatile tool for design development. In the present context of "design for vibration," EMA can be employed in the design and design modification of mechanical systems with the goal of achieving desired performance under vibrating conditions. The present section will introduce this approach.

In applying experimental modal analysis for design development of a mechanical system, three general approaches are employed:

1. Component modification
2. Modal response specification
3. Substructuring.

The method of *component modification* allows one to modify (i.e., add, remove, or vary) physical parameters (inertia, stiffness, damping) in a mechanical system and to determine the resulting effect on the modal response (natural frequencies, damping ratios, and mode shapes) of the system. The method of *modal response specification* provides the capability to establish the best changes, from the design viewpoint, in system parameters (inertia, stiffness, damping values, and associated directions) in order to realize a specified change in the modal response. In the technique of *substructuring*, two or more subsystem models are combined using proper components of interfacing (interconnection), and the overall model of the integrated system is determined. Some of the subsystems used in this approach could be of analytical or computational origin (e.g., finite element models). It should be clear how these methods can be used in the design development of a mechanical system for proper vibration performance. The first method is essentially a trial-and-error technique of incremental design. Here, some appropriate parameters are changed and the resulting modal behavior is determined. If the resulting performance is not satisfactory, further changes are made in discrete steps until an acceptable performance (with regard to natural frequencies, response magnification factors, etc.) is achieved. The second method is clearly a direct design approach, where first the design specifications are developed in terms of modal characteristics, and then the design procedure will generate the size and type of the physical parameters that would enable the system to meet the specifications. In the third method, first a suitable set of subsystems is designed, so as to meet performance characteristics of each subsystem. Then, these subsystems are linked through suitable mechanical interfacing components, and the performance of the overall system is determined to verify acceptance. In this manner, a complex system can be designed through the systematic design of its subsystems.

12.6.1 Component Modification

The method of component modification involves changing a mass, stiffness, or damping element in the system and determining the corresponding dynamic response — particularly the natural frequencies, modal damping ratios, and mode shapes. This is relatively straightforward because a single modal analysis or modal test (EMA) will give the required information. Since what is achieved in a single step of component modification might not be acceptable as an appropriate design (e.g., a natural frequency might be too close to a significant frequency component of a vibration excitation), a number of modifications may be necessary before reaching a suitable design. For such incremental procedures, modal analysis would be more convenient and cost effective than EMA because in the latter case, physical modification and retesting would be needed, while the former involves the same computational steps as before, but with a new set of parameter values.

As an illustration, consider an aluminum I-beam which has a number of important modes of vibration including bending and torsional modes. Figure 12.20(a) shows the 4th mode shape of vibration at natural frequency 678.4 Hz. The dotted line in Figure 12.20(b) shows the transfer function magnitude when the beam is excited at some location in the vertical direction and the response is measured in the vertical direction, at some other location, with neither of the locations being node points. The curve shows the first six natural frequencies.

Next, a lumped mass is added to the top flange at the shown location. The corresponding transfer function magnitude is shown by the solid curve in Figure 12.20(b). Note that all the natural frequencies have decreased due to the added mass, but the effect is larger for higher modes. Similarly, mode shapes will also change. If the new modes are not satisfactory (e.g., a particular natural frequency has not shifted enough), further modification and evaluation will be required.

Consider a mechanical vibrating system whose free response y is described by

$$M\ddot{y} + Ky = 0 \tag{12.90}$$

Damping has been ignored for simplicity, but the following discussion can be extended to a damped system as well (quite directly, for the case of proportional damping). If the mass matrix M and the stiffness matrix K are modified by δM and δK, respectively, the corresponding response, and also the natural frequencies and mode shapes, will be different from those of the original system in general. To illustrate this, let the modal matrix (the matrix whose columns are the independent mode shape vectors, as discussed in Chapters 5 and 11) of the original system be Ψ. Then, using the modal transformation

$$y = \Psi q \tag{12.91}$$

equation (12.90) can be expressed in the canonical form, with modal generalized coordinates q, as

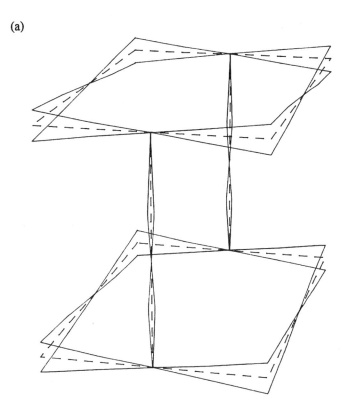
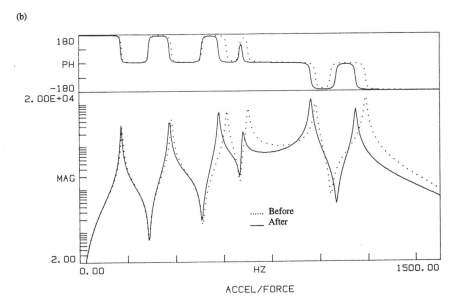

FIGURE 12.20 An example of component modification: (a) the shape of mode 4 prior to modification, and (b) transfer function magnitude and phase before and after modification.

$$\overline{M}\ddot{q} + \overline{K}q = 0 \qquad (12.92)$$

where

$$\Psi^T M \Psi = \overline{M} = \text{diag}[M_1, M_2, \ldots, M_n] \qquad (12.93)$$

$$\Psi^T K \Psi = \overline{K} = \text{diag}[K_1, K_2, \ldots, K_n] \qquad (12.94)$$

If the same transformation (12.91) is used for the modified system

$$(M + \delta M)\ddot{y} + (K + \delta K)y = 0 \qquad (12.95)$$

one obtains

$$(\overline{M} + \Psi^T \delta M \Psi)\ddot{q} + (\overline{K} + \Psi^T \delta K \Psi)q = 0 \qquad (12.96)$$

Since both $\Psi^T \delta M \Psi$ and $\Psi^T \delta K \Psi$ are not diagonal matrices in general, Ψ would not remain the modal matrix for the modified system. Furthermore, the original natural frequencies $\omega_i = \sqrt{K_i/M_i}$ will change due to the component modification. For the special case of proportional modifications (δM proportional to M and δK proportional to K), the mode shapes will not change. But, the natural frequencies will change in general.

The reverse problem is the modal response specification. Here, a required set of modal parameters (ω_{ir} and ψ_{ir}) is specified and the necessary changes δM and δK to meet the specifications must be determined. Note that the solution is not unique in general, and is more difficult than the direct problem. In this case, initially a sensitivity analysis can be performed to determine the directions and magnitudes of the modal shift for a particular physical parameter shift. Then, the necessary magnitudes of physical shift, to achieve the specified modal shift, are estimated on that basis. The corresponding modifications are made and the modified system is analyzed/tested to check whether it is within specification. If not, further cycles of modification should be performed.

EXAMPLE 12.7

As an illustrative example of component modification, consider the familiar problem of a two-degree-of-freedom system as shown in Figure 12.21. It was shown in Chapter 5 that the squared nondimensional natural frequencies $r_i^2 = (\omega_i/\omega_0)^2$ of the systems are given by

$$r_1^2, r_2^2 = \frac{1}{2\alpha}\{\alpha + \beta + \alpha\beta\}\left\{1 \pm \sqrt{1 - \frac{4\alpha\beta}{(\alpha + \beta + \alpha\beta)^2}}\right\}$$

where $\omega_0 = \sqrt{k/m}$. Also, it was shown that the mode shapes, as given by the ratio of the displacement of mass 2 to that of mass 1 at a natural frequency, are:

$$\left(\frac{\psi_2}{\psi_1}\right)_i = \frac{1 + \beta - r_i^2}{\beta} \qquad \text{for mode } i$$

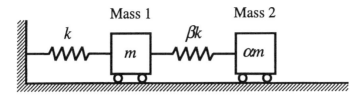

FIGURE 12.21 A two-degree-of-freedom example.

Consider a system with $\alpha = 0.5$ and $\beta = 0.5$. By direct computation, one can show that $r_1 = 0.71$ and $r_2 = 1.41$ for this case. Estimate the modification of β (the relative stiffness of the second spring) that would be necessary to shift the system natural frequencies to approximately $r_1 = 0.8$ and $r_2 = 2.0$. Check the corresponding shift in mode shapes.

SOLUTION

For $\alpha = 0.5$ and $\beta = 0.5$, direct substitution yields $r_1 = 0.71$ and $r_2 = 1.41$ with $(\psi_2/\psi_1)_1 = 2.0$ and $(\psi_2/\psi_1)_2 = -1.0$. Now consider an incremental change in β of 0.1. Then, $\beta = 0.6$. The corresponding natural frequencies are computed as

$$r_i^2 = \frac{1}{2 \times 0.5}\{0.5 + 0.6 + 0.5 \times 0.6\}\left\{1 \mp \left[1 - \frac{4 \times 0.5 \times 0.6}{(0.5 + 0.6 + 0.5 \times 0.6)^2}\right]^{1/2}\right\}$$

$$= 0.528,\ 2.272$$

Hence,

$$r_1,\ r_2 = 0.727,\ 1.507$$

This step can be interpreted as a way of establishing the *sensitivity* of the system to the particular component modification. Clearly, the problem of modification is not linear in general. But, as a first approximation, assume a linear variation of r_i^2 with β, and make modifications according to

$$\frac{\delta\beta}{\delta\beta_0} = \frac{\delta r_i^2}{\delta r_{i0}^2} \tag{12.97}$$

where, the subscript 0 refers to the initial trial variation ($\delta\beta_0 = 0.1$). Equation (12.97) is intuitively satisfying in view of the nature of the physical problem and the fact that for a single-dof problem, squared frequency varies with k_0. Then, one has

For Mode 1:

$$\frac{\delta\beta}{0.1} = \frac{0.8^2 - 0.71^2}{0.727^2 - 0.71^2} = 5.634$$

or

$$\delta\beta = 0.56$$

For Mode 2:

$$\frac{\delta\beta}{0.1} = \frac{2^2 - 1.41^2}{1.507^2 - 1.41^2} = 7.09$$

or

$$\delta\beta = 0.709$$

So, one should use $\delta\beta = 0.71$, which is the larger of the two. This corresponds to

$$\beta = 0.5 + 0.71 = 1.21$$

The natural frequencies are computed as usual:

$$r_1^2, r_2^2 = \frac{1}{2 \times 0.5}\{0.5 + 1.21 + 0.5 \times 1.21\}\left\{1 \mp \left[1 - \frac{4 \times 0.5 \times 1.21}{(0.5 + 1.21 + 0.5 \times 1.21)^2}\right]^{1/2}\right\}$$

$$= 0.60, \ 4.03$$

or

$$r_1, r_2 = 0.78, \ 2.01$$

In view of the nonlinearity of the problem, this shift in frequencies is satisfactory. The corresponding mode shapes are

$$\left(\frac{\psi_2}{\psi_1}\right)_1 = \frac{(1+1.21) - 0.6}{1.21} = 1.33$$

$$\left(\frac{\psi_2}{\psi_1}\right)_2 = \frac{(1+1.21) - 4.03}{1.21} = -1.50$$

It follows that as the stiffness of the second spring is increased, the motions of the two masses become closer in Mode 1. Furthermore, in Mode 2, the node point becomes closer to Mass 1. Note a limitation of this particular component modification. As $\beta \to \infty$, the two masses become rigidly linked, giving a frequency ratio of $r_1 = \sqrt{k/(m+\alpha m)}/\sqrt{k/m} = 1/\sqrt{1+\alpha} = 1/\sqrt{1.5} = 0.816$, with $r_2 \to \infty$. Hence, it is unreasonable to expect a frequency ratio that is closer to this value of r_1 by a change in β alone.

□

12.6.2 Substructuring

For large and complex mechanical systems with many components, the approach of substructuring can make the process of "design for vibration" more convenient and systematic. In this approach, the system is first divided into a convenient set of subsystems that are more amenable to testing and/or analysis. The subsystems are then separately modeled and designed through the approaches

Vibration Design and Control

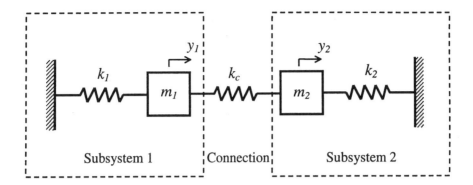

FIGURE 12.22 An example of substructuring.

of modal analysis and testing, along with any other convenient approaches (e.g., finite element technique). Note that the performance of the overall system depends on the interface conditions that link the subsystems, as well as the characteristics of the individual subsystems. Hence, it is not possible to translate the design specifications for the overall system into those for the subsystems, without taking the interface conditions into account. The overall system is *assembled* from the designed subsystems, using *compatibility* requirements at the assembly locations together with dynamic equations of the interconnecting components such as spring-mass-damper units or rigid linkages. If the assembled system does not meet the design specifications, then modifications should be made to one or more of the subsystems and interfacing (assembly) linkages, and the procedure should be repeated. Hence, the main steps of using the approach of substructuring for vibration design of a complex system are as follows:

1. Divide the mechanical system into convenient subsystems (substructuring) and represent the interconnection points of subsystems by forces/moments.
2. Develop models for the subsystems through analysis, modal testing, and other standard procedures.
3. Design the subsystems so that their performance is well within the performance specifications provided for the overall system.
4. Establish the interconnecting (assembling) linkages for the subsystems, and obtain dynamic equations for them in terms of the linking forces/moments and motions (displacements/rotations).
5. Establish continuity (force balance) and compatibility (motion consistency) conditions at the assembly locations (see Appendix A on modeling).
6. Through matrix methods, eliminate the unknown variables and assemble the overall system.
7. Analyze (or test) the overall system to determine its vibration performance. If satisfactory, stop. If not, make modifications to the systems and/or assembly conditions and repeat steps 4 through 7.

As a simple example, consider two single-degree-of-freedom systems that are interconnected by a spring linkage, as shown in Figure 12.22. The two subsystems can be represented by

$$\begin{bmatrix} m_1 & 0 \\ 0 & m_2 \end{bmatrix} \ddot{y} + \begin{bmatrix} k_1 & 0 \\ 0 & k_2 \end{bmatrix} y = 0$$

and the corresponding natural frequencies are

$$\omega_{s1} = \sqrt{k_1/m_1} \quad \text{and} \quad \omega_{s2} = \sqrt{k_2/m_2}$$

The overall interconnected system is given by

$$\begin{bmatrix} m_1 & 0 \\ 0 & m_2 \end{bmatrix} \ddot{y} + \begin{bmatrix} k_1 + k_c & -k_c \\ -k_c & k_2 + k_c \end{bmatrix} y = 0$$

Its natural frequencies are obtained by solving the characteristic equation

$$\det \begin{bmatrix} k_1 + k_c - \omega^2 m_1 & -k_c \\ -k_c & k_2 + k_c - \omega^2 m_2 \end{bmatrix} = 0$$

or

$$(k_1 + k_c - \omega^2 m_1)(k_2 + k_c - \omega^2 m_2) - k_c^2 = 0$$

which simplifies to

$$\omega^4 - \left[\frac{k_1 + k_c}{m_1} + \frac{k_2 + k_c}{m_2} \right] \omega^2 + \frac{k_1 k_2 + k_c(k_1 + k_2)}{m_1 m_2} = 0$$

The sum of the roots is

$$\omega_1^2 + \omega_2^2 = \frac{k_1 + k_c}{m_1} + \frac{k_2 + k_c}{m_2} > \omega_{s1}^2 + \omega_{s2}^2$$

The product of the roots is

$$\omega_1^2 \omega_2^2 = \frac{k_1 k_2 + k_c(k_1 + k_2)}{m_1 m_2} > \omega_{s1}^2 \omega_{s2}^2$$

This does not mean that both frequencies will increase due to the interconnection. Note that the limit on the lower frequency, as $k_c \to \infty$, is given by that of a single-dof system with mass $m_1 + m_2$ and stiffness $k_1 + k_2$, which is $\sqrt{(k_1 + k_2)/(m_1 + m_2)}$. This value can be larger or smaller than the natural frequency of a subsystem, depending on the relative values of the parameters. Hence, even for this system, exact satisfaction of a set of design natural frequencies would be somewhat challenging, as these frequencies depend on the interconnection as well as the subsystems.

It is noted that substructuring is a design development technique where complex designs can be accomplished through a parallel and separate development of several subsystems and interconnections. Furthermore, through this procedure, dynamic interactions among subsystems can be estimated and potential problems can be detected that will allow redesigning of the subsystems and/or interface linkages prior to building the prototype. Design approaches using experimental modal analysis that can be used in vibration problems are summarized in Box 12.6.

Vibration Design and Control

BOX 12.6 Test-Based Design Approaches for Vibration

1. **Component modification:**
   Modify a component (mass, spring, damper) and determine modal parameters (natural frequencies, damping ratios, mode shapes).
   - Can determine sensitivity to component changes
   - Can check whether a particular change is desirable
2. **Modal Response Specification:**
   Specify a desired modal response (natural frequencies, damping ratios, mode shapes) and determine the "best" component changes (mass, spring, damper) that will realize the modal specs.
   - Can be accomplished by first performing a sensitivity study (as in Item 1)
3. **Substructuring:**
   a. Design subsystems to meet specs (analytically, experimentally, or by a mixed approach).
   b. Establish interconnections between subsystems, and obtain continuity (force balance) and compatibility (motion consistency) at assembly locations.
   c. Assemble the overall system by eliminating unknown variables at interconnections.
   d. Analyze or test the overall system. If satisfactory, stop. Otherwise, make changes to the subsystems and/or interconnections, and repeat the above steps.

12.7 PASSIVE CONTROL OF VIBRATION

Techniques presented thus far in this chapter, for the reduction of effects of mechanical vibration, primarily fall into the categories of vibration isolation and design for vibration. The third category — vibration control — is addressed now. What is characteristic of vibration control is the use of a sensing device to detect the level of vibration in a system, and an actuation (forcing) device to apply a forcing function to the system so as to counteract the effects of vibration. In some such devices, the sensing and forcing functions are implicit and integrated together.

Vibration control can be subdivided into the following two broad categories:

1. Passive control
2. Active control.

Passive control of vibration employs passive controllers. Passive devices, by definition, do not require external power for their operation. The two passive controllers of vibration studied in the present section are vibration absorbers (or dynamic absorbers or Frahm absorbers, named after H. Frahm who first employed the technique for controlling ship oscillations) and dampers. In both types of devices, sensing is implicit and control is done through a force that is generated by the device as a result of its response to the vibration excitation. A dynamic absorber is a mass-spring type mechanism with very little or no damping, which can "absorb" vibration excitations through energy transfer into it, thereby reducing the vibrations of the primary system. The energy received by the absorber will be slowly dissipated due to its own damping. A damper is a purely dissipative device which, unlike a dynamic absorber, directly dissipates the energy received from the system, rather than storing it. Hence, it is a more wasteful device that can also exhibit problems related to wear and thermal effects. However, it has advantages as well over an absorber; for example, a wider frequency of operation.

12.7.1 UNDAMPED VIBRATION ABSORBER

A dynamic vibration absorber (or dynamic absorber, vibration absorber, or Frahm absorber) is a simple mass-spring oscillator with very low damping. An absorber that is tuned to a frequency of vibration of a mechanical system is able to receive a significant portion of the vibration energy from the primary system at that frequency. The resulting vibration of the absorber in effect applies an oscillatory force opposing the vibration excitation of the primary system, and thereby virtually canceling the effect. In theory, then, the vibration of the system can be completely removed, while the absorber itself undergoes vibratory motion. Since damping is quite low in practical vibration absorbers, consider first the case of an undamped absorber.

A vibration absorber can be used for vibration control in two common types of situations, as shown in Figure 12.23. Here, the primary system for which the vibration needs to be controlled is modeled as an undamped, single-degree-of-freedom mass-spring system (denoted by the subscript p). An undamped vibration absorber is also a single-degree-of-freedom mass-spring system (denoted by the subscript a). In the application shown in Figure 12.23(a), the objective of the absorber is to reduce the vibratory response y_p of the primary system as a result of a vibration excitation $f(t)$. But, because the force f_s transmitted to the support structure, due to the vibratory response of the system, is given by

$$f_s = k_p y_p \tag{12.98}$$

the objective of reducing y_p can also be interpreted as one of reducing this transmitted vibratory force (a goal of vibration isolation). In the second type of application, as represented in Figure 12.23(b), the primary system is excited by a vibratory support motion and the objective of the absorber is again to reduce the resulting vibratory motions y_p of the primary system. Note that in both classes of application, the purpose is to reduce the vibratory responses. Hence, static loads (e.g., gravity) are not considered in the analysis.

Development of the equations of motion for the two systems shown in Figure 12.23 is summarized in Table 12.1. Since one is primarily interested in the control of oscillatory responses to oscillatory excitations, the frequency-domain model is particularly useful. Note from Table 12.1 that the transfer function f_s/f of system (a) is simply k times the transfer function y_p/f, and is in fact identical to the transfer function y_p/u of system (b). Hence, the two problems are essentially identical and it suffices to address only one of them.

Before investigating the common transfer function for the two types of problems, one should look closely at the frequency-domain equations for the system shown in Figure 12.23(a):

$$\left(k_p + k_a - \omega^2 m_p\right) y_p - k_a y_a = f \tag{12.99}$$

$$\left(k_a - \omega^2 m_a\right) y_a = k_a y_p \tag{12.100}$$

along with equation (12.98). Here, m_p and k_p are the mass and the stiffness of the primary system, m_a and k_a are the mass and the stiffness of the absorber, f is the excitation amplitude, ω is the excitation frequency, y_p is the primary-mass response, and y_a is the absorber response. Now note from equation (12.100) that if $\omega = \sqrt{k_a/m_a}$, then $y_p = 0$. This means that if the absorber is tuned so that its natural frequency is equal to the excitation frequency (drive frequency), the primary system (ideally) will not undergo any vibratory motion, and hence is perfectly controlled. The reason for this should be clear from equation (12.99) which, when $y_p = 0$ is substituted, gives $k_a y_a = -f$. In other words, a tuned absorber applies to the primary system a spring force that is

Vibration Design and Control

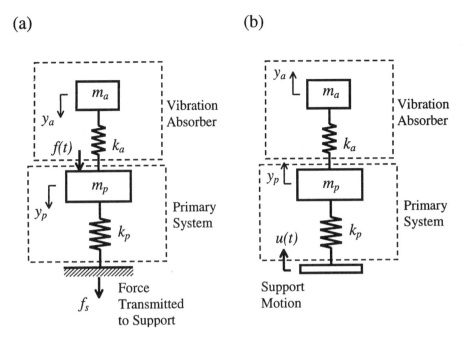

FIGURE 12.23 Two types of applications of a vibration absorber: (a) reduction of the response to forcing excitation (or to reduce the force transmitted to support structure), and (b) reduction of the response to support motion.

exactly equal and opposite to the excitation force, thereby neutralizing the effect. Note that the absorber mass itself moves, albeit at 180° out of phase with the excitation. The frequency of these motions will be ω (same as that of the excitation), and the amplitude is proportional to that of the excitation (f) and inversely proportional to the stiffness of the absorber spring. It follows that a vibration absorber "absorbs" vibration energy from the primary system. Furthermore, note from equation (12.98) that, with a tuned absorber, the vibration force transmitted to the support structure is (ideally) 0 as well. All this information is observed without any mathematical manipulation of the equations of motion.

Keep in mind that one is dealing with vibratory excitations and responses; hence, static loading (such as gravity and spring pre-loads) is not considered (or one investigates responses with respect to the static equilibrium configuration of the system). In summary, one is now able to state the following characteristics of a vibration absorber (undamped):

1. It is effective only for a single excitation frequency (i.e., a sinusoidal excitation).
2. For the best effect, it should be "tuned" such that its natural frequency $\sqrt{k_a/m_a}$ is equal to the excitation frequency.
3. In the case of forcing vibration excitation, a tuned absorber can (ideally) make both the vibratory response of the primary system and the vibratory force transmitted to the support structure 0.
4. In the case of a vibratory support motion, a tuned absorber can make the resulting response of the primary system 0.
5. A dynamic absorber functions by acquiring vibration energy from the primary system and storing it (as kinetic energy of the mass and/or potential energy of the spring) rather than by directly dissipating the energy.

TABLE 12.1
Equations for the Two Types of Absorber Applications

	Absorber Application for the Reduction of Response to a:	
	Forcing Excitation	**Support Motion**
Time-domain equations	$m_p \ddot{y}_p = -k_p y_p - k_a(y_p - y_a) + f(t)$ $m_a \ddot{y}_a = k_a(y_p - y_a)$	$m_p \ddot{y}_p = k_p(u(t) - y_p) - k_a(y_p - y_a)$ $m_a \ddot{y}_a = k_a(y_p - y_a)$
Frequency-domain equations	$(-\omega^2 m_p + k_p + k_a)y_p = k_a y_a + f$ $(-\omega^2 m_a + k_a)y_a = k_a y_p$	$(-\omega^2 m_p + k_p + k_a)y_p = k_a y_a + k_p u$ $(-\omega^2 m_a + k_a)y_a = k_a y_p$
Matrix form	$\begin{bmatrix} k_p + k_a - \omega^2 m_p & -k_a \\ -k_a & k_a - \omega^2 m_a \end{bmatrix} \begin{bmatrix} y_p \\ y_a \end{bmatrix} = \begin{bmatrix} f \\ 0 \end{bmatrix}$	$\begin{bmatrix} k_p + k_a - \omega^2 m_p & -k_a \\ -k_a & k_a - \omega^2 m_a \end{bmatrix} \begin{bmatrix} y_p \\ y_a \end{bmatrix} = k_p \begin{bmatrix} u \\ 0 \end{bmatrix}$
Transfer-function matrix form	$\begin{bmatrix} y_p \\ y_a \end{bmatrix} = \frac{1}{\Delta} \begin{bmatrix} k_a - \omega^2 m_a & k_a \\ k_a & k_p + k_a - \omega^2 m_p \end{bmatrix} \begin{bmatrix} f \\ 0 \end{bmatrix}$	$\begin{bmatrix} y_p \\ y_a \end{bmatrix} = \frac{k_p}{\Delta} \begin{bmatrix} k_a - \omega^2 m_a & k_a \\ k_a & k_p + k_a - \omega^2 m_p \end{bmatrix} \begin{bmatrix} u \\ 0 \end{bmatrix}$
Vibration control transfer function	$\dfrac{f_s}{f} = \dfrac{k_p y_p}{f} = \dfrac{k_p}{\Delta}(k_a - \omega^2 m_a)$	$\dfrac{y_p}{u} = \dfrac{k_p}{\Delta}(k_a - \omega^2 m_a)$
Characteristic polynomial	$\Delta = (k_p + k_a - \omega^2 m_p)(k_a - \omega^2 m_a) - k_a^2$ $= m_p m_a \omega^4 - [k_a(m_p + m_a) + k_p m_a]\omega^2 + k_p k_a$	

6. A dynamic absorber functions by applying a vibration force to the primary system, which is equal and opposite to the excitation force, thereby neutralizing the excitation.
7. The amplitude of motion of the vibration absorber is proportional to the excitation amplitude and is inversely proportional to the absorber stiffness. The frequency of the absorber motion is the same as the excitation frequency.

Now for some formal analysis, consider the transfer function (f_s/f or y_p/u) of an undamped vibration absorber, as given in Table 12.1. Then,

$$G(\omega) = \frac{k_p(k_a - \omega^2 m_a)}{m_p m_a \omega^4 - [k_a(m_p + m_a) + k_p m_a]\omega^2 + k_p k_a} \tag{12.101}$$

As usual, it is convenient to use a nondimensional form in analyzing this frequency-transfer function. To that end, one can define the following nondimensional parameters and frequency variable:

$$\text{Fractional mass of the absorber } \mu = \frac{m_a}{m_p}$$

$$\text{Nondimensional natural frequency of the absorber } \alpha = \frac{\omega_a}{\omega_p}$$

$$\text{Nondimensional excitation (drive) frequency } r = \frac{\omega}{\omega_p}$$

Vibration Design and Control

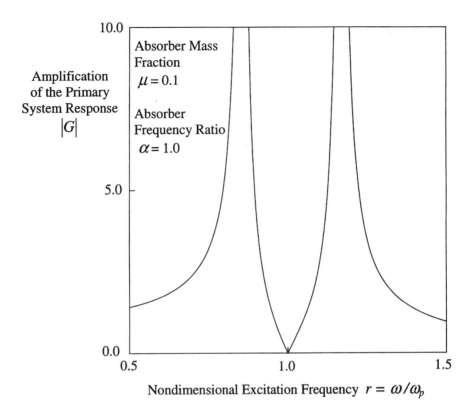

FIGURE 12.24 The effect of an undamped vibration absorber on the vibration response of a primary system.

where

$$\omega_a = \sqrt{\frac{k_a}{m_a}} = \text{natural frequency of the absorber}$$

$$\omega_p = \sqrt{\frac{k_p}{m_p}} = \text{natural frequency of the primary system.}$$

Then, it is straightforward to divide the numerator and the denominator by $k_p k_a$ and carry out simple algebraic manipulations to express the transfer function of equation (12.101) in the nondimensional form as

$$G(r) = \frac{\alpha^2 - r^2}{r^4 - [\alpha^2(1+\mu)+1]r^2 + \alpha^2} \qquad (12.102)$$

For this undamped system, there is no difference between the resonant frequencies (where the magnitude of the transfer function peaks) and the natural frequencies (roots of the characteristic equation), which correspond to the "natural" or free time-response oscillations. These are obtained by solving the characteristic equation

$$r^4 - [\alpha^2(1+\mu)+1]r^2 + \alpha^2 = 0 \qquad (12.103)$$

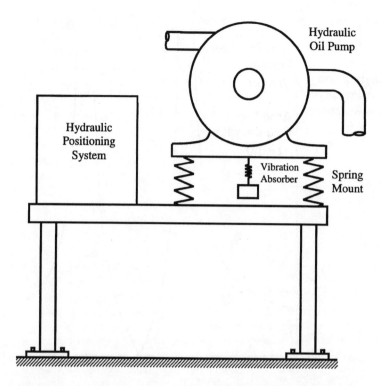

FIGURE 12.25 A hydraulic positioning system with a gear pump.

which gives

$$r_1^2, r_2^2 = \frac{1}{2}\left[\alpha^2(1+\mu)+1\right] \mp \frac{1}{2}\sqrt{\left[\alpha^2(1+\mu)+1\right]^2 - 4\alpha^2} \qquad (12.104)$$

These are squared frequencies, both of which are positive, as clear from equation (12.104). The actual nondimensional natural frequencies are their square roots. The magnitude of the transfer function becomes infinite at either of these two natural/resonant frequencies. Furthermore, it is clear from equation (12.102) that the transfer function magnitude becomes 0 at $r = \alpha$, where the excitation frequency (ω) is equal to the natural frequency of the absorber (ω_a). This fact has been pointed out already. Note further that, in the present undamped case, the transfer function $G(r)$ is real; but it can be either positive or negative. The magnitude referred to is, then, the absolute value of $G(r)$, which is positive. The magnitude plot given in Figure 12.24 shows the resonant and control characteristics of a system with an undamped vibration absorber, as discussed here. Originally, the primary system had a resonance at $r = 1$ (i.e., $\omega = \omega_p$). When the absorber (which also has a resonance at $r = 1$) is added, the original resonance becomes an *antiresonance* with a 0 response. But two new resonances are created: one at $r = 0.854$ and the other at $r = 1.171$, which are on either side of the tuned frequency ($r = 1$) of the absorber.

In view of these two resonances, the effective region of the absorber is limited to a narrow frequency band centering its tuned frequency. Specifically, the absorber is not effective unless $|G| < 1$. The effective frequency band of a vibration absorber can be determined using this condition.

EXAMPLE 12.8

A high-precision, yet high-power, positioning system uses a hydraulic actuator and a valve. The pressurized oil to this hydraulic servo system is provided by a gear-type rotary pump. The pump

Vibration Design and Control

and the positioning system are mounted on the same workbench. The mass of the pump is 25 kg. The normal operating speed of the pump is 3600 rpm. During operation, it was observed that the pump exhibits a vertical resonance at this speed and it affects the accuracy of the position servo system. To control the vibrations of the pump at its operating speed, a vibration absorber of mass 1.25 kg, and tuned to the normal operating speed of the pump, is attached as schematically shown in Figure 12.25. Since the speed of the pump normally fluctuates during operation, determine the speed range within which the vibration absorber is effective. What are the new resonant frequencies of the system? Neglect damping.

SOLUTION

For this problem, the fractional mass $\mu = 1.25/25.0 = 0.05$, and since the absorber is tuned to the resonant frequency of the pump, $\alpha = 1.0$. Hence, from equation (12.103), the characteristic equation of the modified system becomes

$$r^4 - 2.05r^2 + 1 = 0$$

which has roots $r_1 = 0.854$ and $r_2 = 1.171$. It follows that the new resonances are at 0.854×3600 rpm and 1.171×3600 rpm. These are 3074.4 rpm and 4215.6 rpm, which should be avoided. From equation (12.102), the system transfer function is

$$G(r) = \frac{1 - r^2}{\left(r^4 - 2.05r^2 + 1\right)}$$

The effective frequency band of the absorber corresponds to $|G(r)| < 1.0$. Since a sign reversal of $G(r)$ occurs at $r = 1$, one needs to solve both

$$\frac{1 - r^2}{\left(r^4 - 2.05r^2 + 1\right)} = 1 \quad \text{and} \quad -1$$

The first equation give the roots $r = 0$ and 1.025. The second equation gives the roots $r = 0.977$ and 1.45. Hence, the effective frequency band corresponds to $\Delta r = [0.977, 1.025]$. In terms of the operating speed of the pump, then, one has an effective band of 3517.2 rpm to 3690 rpm. So, a speed fluctuation of about ±80 rpm is acceptable.

□

Before leaving the topic of an undamped vibration absorber, recall that the presence of the absorber generates two new resonances on either side of the resonance of the original system (to which the absorber is normally tuned). It is also clear from equation (12.104) that these two resonances become farther and farther apart as the fractional mass μ of the vibration absorber is increased.

12.7.2 DAMPED VIBRATION ABSORBER

Damping is not the primary means by which vibration control is achieved in a vibration absorber. As noted before, the absorber acquires vibration energy from the primary system (and, in return, exerts a force on the system that is equal and opposite to the vibration excitation), thereby suppressing the vibratory motion. The energy received by the absorber must be dissipated gradually and, hence, some damping should be present in the absorber. Furthermore, as one will notice in the following development, the two resonances created by adding the absorber have an infinite

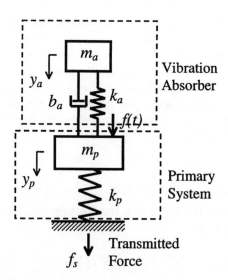

FIGURE 12.26 Primary system with a damped vibration absorber.

magnitude in the absence of damping. Hence, damping has the added benefit of lowering these resonant peaks as well.

The analysis of a vibratory system with a damped absorber is as straightforward as, but somewhat more complex than, that involving an undamped absorber. Furthermore, an extra design parameter — the damping ratio of the absorber — enters into the scene. Consider the model shown in Figure 12.26. Another version of application of a damped absorber, which corresponds to Figure 12.23(b), can be presented as well. But because the two types of application have the same transfer function, it is sufficient to consider Figure 12.26 alone. Again, the transfer function of vibration control can be taken as either y_a/f or f_s/f, the latter being simply k_p times the former. Hence, one can consider the dimensionless case of f_s/f, but the results are equally valid for y_p/f, except that the responses must be converted from force to displacement by dividing by k_p.

There is no need to derive the transfer function anew for the damped system. Simply replace k_a in equation (12.101) by the complex stiffness $k_a + j\omega b_a$, which incorporates the viscous damping constant b_a and the excitation frequency ω. Hence, the transfer function of the damped system is

$$G(\omega) = \frac{k_p\left(k_a + j\omega b_a - \omega^2 m_a\right)}{m_p m_a \omega^4 - \left[\left(k_a + j\omega b_a\right)\left(m_p + m_a\right) + k_p m_a\right]\omega^2 + k_p\left(k_a + j\omega b_a\right)} \quad (12.105)$$

With the parameters defined as before, the nondimensional form of this transfer function is obtained by dividing throughout by $k_p k_a$ and then substituting the appropriate parameters. In particular, use the fact that

$$\frac{b_a}{k_a} = \frac{2b_a}{2\sqrt{k_a m_a}} \cdot \sqrt{\frac{m_a}{k_a}} = \frac{2\zeta_a}{\omega_a} = \frac{2\zeta_a}{\omega_p} \cdot \frac{\omega_p}{\omega_a} = \frac{2\zeta_a}{\alpha \omega_p} \quad (12.106)$$

where the damping ratio ζ_a of the absorber is given by

$$\zeta_a = \frac{b_a}{2\sqrt{k_a m_a}} \quad (12.107)$$

Vibration Design and Control

as usual. Then, follow the same procedure used to derive equation (12.102) from (12.101) to get

$$G(r) = \frac{\alpha^2 - r^2 + 2j\zeta_a \alpha r}{r^4 - \left[\left(\alpha^2 + 2j\zeta_a \alpha r\right)(1+\mu) + 1\right]r^2 + \left(\alpha^2 + 2j\zeta_a \alpha r\right)} \tag{12.108}$$

Note that this final result is equivalent to simply replacing α^2 by $\alpha^2 + 2j\zeta_a \alpha r$ in equation (12.102).

At this juncture, the usual cautionary statement should be made about natural frequencies and resonant frequencies. The undamped natural frequencies are obtained by solving the characteristics equation, with $\zeta_a = 0$. These are the same as before, and given by the square roots of equation (12.104). The damped natural frequencies are obtained by first setting $jr = \lambda$ (hence, $r^2 = -\lambda^2$ and $r^4 = \lambda^4$), and solving the resulting characteristics equation [see the denominator of equation (12.108)]

$$\lambda^4 + 2\zeta_a \alpha(1+\mu)\lambda^3 + \left(\alpha^2 + \alpha^2\mu + 1\right)\lambda^2 + 2\zeta_a \alpha\lambda + \alpha^2 = 0 \tag{12.109}$$

and then taking the imaginary parts of the roots of λ. These depend on ζ_a and are different from those obtained from equation (12.104). The resonant frequencies correspond to the r values where the magnitude of $G(r)$ will peak. These are, generally, not the same as the undamped or damped natural frequencies; but for low damping (small ζ_a compared to 1), these three types of system characteristics frequencies are almost identical.

The magnitude of the transfer function (12.108) is plotted in Figure 12.27 for the case $\mu = 1.0$ and $\alpha = 1.0$, as in Figure 12.24, but for damping ratios $\zeta_a = 0.01, 0.1,$ and 0.5. Note that the curve for $\zeta_a = 0.01$ is very close to that in Figure 12.24 for the undamped case. When ζ_a is large, as shown in the case of $\zeta_a = 0.5$, the two masses m_p and m_a tend to become locked together and appear to behave like a single mass. Then the system tends to act like a single-degree-of-freedom one, and the primary system is modified only in its mass (which increases). Consequently, there is only one resonant frequency that is smaller than that of the original primary system. Furthermore, the effect of a vibration absorber is no longer present, as expected, in this high-damping case.

All three curves in Figure 12.27 pass through the two common points A and B, as shown. This is, in fact, true for all curves corresponding to all values of ζ_a, and particularly for the extreme cases of $\zeta_a = 0$ and $\zeta_a \to \infty$. Hence, these points can be determined as the points of intersection of the transfer function magnitude curves for the limiting cases $\zeta_a = 0$ and $\zeta_a \to \infty$.

Equation (12.102) gives $G(r)$ for $\zeta_a = 0$. Next, from equation (12.108), note that as $\zeta_a \to \infty$, all the terms not containing ζ_a can be neglected. Hence,

$$G(r) = \frac{2j\zeta_a \alpha r}{-2j\zeta_a \alpha r(1+\mu)r^2 + 2j\zeta_a \alpha r}$$

Cancel the common term, and obtain (for $r \neq 0$)

$$G(r) = \frac{1}{1-(1+\mu)r^2} \quad \text{for } \zeta_a \to \infty \tag{12.110}$$

Note that this is the normalized transfer function of a single-degree-of-freedom system of nondimensional natural frequency $1/\sqrt{1+\mu}$. This result confirms the fact that as $\zeta_a \to \infty$, the two masses m_p and m_a become locked together and act as a single mass $(m_p + m_a)$ supported on a spring of

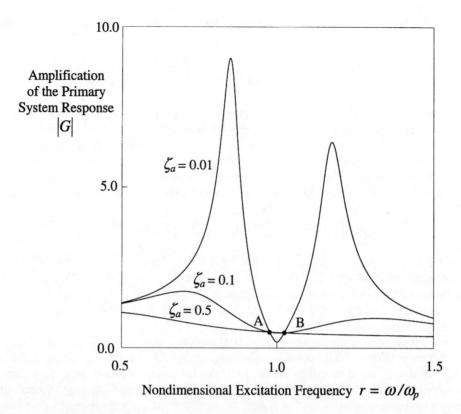

FIGURE 12.27 Vibration amplification (transfer function magnitude) curves for damped vibration absorbers (absorber mass $\mu = 0.1$, absorber resonant frequency $\alpha = 1.0$.

stiffness k_p. Its natural frequency is $\sqrt{k_p/(m_p + m_a)}$, which when normalized with respect to $\sqrt{k_p/m_p}$ becomes $\sqrt{\dfrac{k_p}{(m_p + m_a)} \cdot \dfrac{m_p}{k_p}} = \sqrt{\dfrac{m_p}{(m_p + m_a)}} = \dfrac{1}{\sqrt{1+\mu}}$.

In determining the points of intersection between the functions (12.102) and (12.110), note first that at the first point of intersection (A), the function (12.102) is negative and (12.110) is positive, while the reverse is true for the second point of intersection (B). That means that for either point, the sign of one of the functions should be reversed before equating them; thus,

$$\frac{\alpha^2 - r^2}{r^4 - [\alpha^2(1+\mu)+1]r^2 + \alpha^2} = -\frac{1}{1-(1+\mu)r^2}$$

which gives,

$$(2+\mu)r^4 - 2[\alpha^2(1+\mu)+1]r^2 + 2\alpha^2 = 0 \qquad (12.111)$$

This is the equation whose roots (e.g., r_1 and r_2) give the points A and B. Then one has the sum of the squared roots equal to the negative coefficient of r^2 in the quadratic (in r^2) equation (12.111); thus,

Vibration Design and Control

$$r_1^2 + r_2^2 = \frac{2[\alpha^2(1+\mu)+1]}{(2+\mu)} \tag{12.112}$$

Also, the product of the squared roots is equal to the constant term in the quadratic (in r^2) equation (12.111). Hence,

$$r_1^2 r_2^2 = \frac{2\alpha^2}{(2+\mu)} \tag{12.113}$$

Optimal Absorber Design

It has been pointed out, primarily by J.P. Den Hartog, that an optimal absorber design should not only have equal response magnitudes at the common points of intersection (i.e., have the ordinates of points A and B equal in Figure 12.27), but also the resonances should occur at these points so that some balance and uniformity is achieved in the response amplification in the region surrounding the tuned frequency of the absorber. It is expected that these (intuitive) design conditions would give relations between the parameters α, μ, and ζ_a corresponding to an optimal absorber.

Consider the first requirement of equal transfer function magnitudes at A and B. Since, as noted earlier, these two points do not depend on ζ_a, one can use equation (12.110) to satisfy the requirement. Thus, again keeping in mind the sign reversal of the transfer function between A and B (i.e., as the transfer function passes through the resonance), one obtains

$$\frac{1}{1-(1+\mu)r_1^2} = -\frac{1}{1-(1+\mu)r_2^2}$$

which gives

$$r_1^2 + r_2^2 = \frac{2}{1+\mu} \tag{12.114}$$

Substituting this result (for equal ordinates) in the intersection-point condition (12.112), one obtains

$$\frac{2}{1+\mu} = \frac{2[\alpha^2(1+\mu)+1]}{(2+\mu)}$$

On simplification, one obtains the simple result

$$\alpha = \frac{1}{1+\mu} \tag{12.115}$$

Next, turn to achieving peak magnitudes of the transfer function at the points of intersection (A and B). Unfortunately, when one point peaks, the other does not, in general. As reported by Den Hartog, with straightforward but lengthy analysis, one obtains the following two results:

$$\zeta_a^2 = \frac{\mu\left[3 - \sqrt{\mu/(\mu+2)}\right]}{8(1+\mu)^3} \tag{12.116}$$

for peak at the first intersection point, and

$$\zeta_a^2 = \frac{\mu\left[3 + \sqrt{\mu/(\mu+2)}\right]}{8(1+\mu)^3} \qquad (12.117)$$

for peak at the second intersection point. So, for design purposes, a balance is obtained by taking the average value of the results (12.116) and (12.117), as

$$\zeta_a^2 = \frac{3\mu}{8(1+\mu)^3} \qquad (12.118)$$

Then, equations (12.115) and (12.118) correspond to an optimal vibration absorber. In addition, note that practical requirements and limitations need to be addressed in any design procedure. In particular, since μ is considerably less than unity (i.e., absorber mass is a small fraction of the primary mass), in order to receive the energy of the primary system, the absorber mass should undergo relatively large amplitudes at the operating frequency. The absorber spring needs to be designed accordingly, while meeting the tuning frequency conditions that determine the ratio m_a/k_a.

EXAMPLE 12.9

The air compressor of a wind tunnel weighs 48 kg and normally operates at 2400 rpm. The first major resonance of the compressor unit occurs at 2640 rpm, with severe vibration amplitudes that are quite dangerous. Design a vibration absorber (damped) for installation on the mounting base of the compressor. What are the vibration amplifications of the compressor unit at the new resonances of the modified system? Compare these with the vibration amplitude of the original system in normal operation.

SOLUTION

As usual, one tunes the absorber to the normal operating speed (2400 rpm). Then, the nondimensional resonant frequency of the absorber is given by

$$\alpha = \frac{\omega_a}{\omega_p} = \frac{2400}{2640} = \frac{12}{13}$$

Now, for an optimal absorber, from equation (12.115),

$$\mu = \frac{1}{\alpha} - 1 = \frac{13}{12} - 1 = \frac{1}{12}$$

Hence, the absorber mass $m_a = 48 \times \frac{1}{12}$ kg = 4.0 kg. Then, from equation (12.118), the damping ratio of the absorber is

$$\zeta_a = \left[\frac{3/12}{8(1+1/12)^3}\right]^{1/2} = 0.157$$

Now,

Vibration Design and Control

$$\omega_a = \sqrt{\frac{k_a}{m_a}} = \sqrt{\frac{k_a}{4.0}} = \frac{2400}{60} \times 2\pi = 80\pi \text{ rad} \cdot \text{s}^{-1}$$

Hence,

$$k_a = (80\pi)^2 \times 4.0 = 2.527 \times 10^5 \text{ N} \cdot \text{m}^{-1}$$

Also,

$$\zeta_a = \frac{1}{2} \frac{b_a}{\sqrt{m_a k_a}}$$

Then,

$$b_a = 2 \times 0.157 \sqrt{4.0 \times 2.527 \times 10^5} = 315.7 \text{ N} \cdot \text{s} \cdot \text{m}^{-1}$$

Hence, the damped absorber is designed. Now check its performance. One knows that, in theory, the vibration amplitude at the operating speed should be almost 0 now; but two resonances are created around the operating point. Since damping is small, one can use the undamped characteristic equation (12.103) to compute these resonances:

$$r^4 - \left[\frac{12^2}{13^2}\left(1 + \frac{1}{12}\right) + 1\right]r^2 + \frac{12^2}{13^2} = 0$$

which gives

$$r^4 - \frac{25}{13}r^2 + \frac{12^2}{13^2} = 0$$

The roots of r^2 are 0.692 and 1.231. The (positive) roots of r are: 0.832 and 1.109. These correspond to compressor speeds of (multiply r by 2640 rpm) 2196 rpm and 2929 rpm. Although they are approximately at −10% and +20% of the operating speed, the first resonance will be encountered during start-up and shut-down conditions. To determine the corresponding vibration amplifications (force/force), use equation (12.108) which, when the undamped characteristic equation is substituted into the denominator, becomes

$$G(r) = \frac{\alpha^2 - r^2 + 2j\zeta_a \alpha r}{\left[-2j\zeta_a \alpha r(1+\mu)r^2 + 2j\zeta_a \alpha r\right]} = \frac{1 - j(\alpha^2 - r^2)/(2\zeta_a \alpha r)}{1 - (1+\mu)r^2} \quad (12.119)$$

Substitute the resonant frequencies $r_1 = 0.832$ and $r_2 = 1.109$ to obtain $|G(r_1)| = 4.223$ and $|G(r_2)| = 4.634$. Without the absorber, one can approximate the system by a simple undamped oscillator with transfer function

$$G_p(r) = \frac{1}{1 - r^2}$$

The corresponding vibration amplification at the operating speed is:

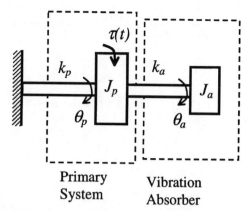

FIGURE 12.28 The application of a rotary vibration absorber.

$$|G_p(r_0)| = \frac{1}{|1-12^2/13^2|} = 6.76$$

It is observed that the resonant vibrations, after adding the absorber, are smaller than even the operating vibrations of the original system. Hence, the design is satisfactory. Note that the force/force transfer functions were used. To get the displacement/force transfer functions, divide by k_p. However,

$$\sqrt{\frac{k_p}{m_p}} = \frac{2640}{60} \times 2\pi = 88\pi \text{ rad} \cdot \text{s}^{-1}$$

Hence,

$$k_p = (88\pi)^2 \times 48 \text{ N} \cdot \text{m}^{-1} = 3.67 \times 10^6 \text{ N} \cdot \text{m}^{-1} = 3.67 \times 10^3 \text{ N} \cdot \text{mm}^{-1}$$

So, the amplitude of operating vibrations of the original system is equal to

$$\frac{6.76}{3.67 \times 10^3} \text{ mm} \cdot \text{N}^{-1} = 1.84 \times 10^{-3} \text{ mm} \cdot \text{N}^{-1}$$

The amplitudes of the resonant vibrations of the modified system are

$$\frac{4.223}{3.67 \times 10^3} \text{ mm} \cdot \text{N}^{-1} \text{ and } \frac{4.634}{3.67 \times 10^3} \text{ mm} \cdot \text{N}^{-1}, \text{ or } 1.15 \times 10^{-3} \text{ mm} \cdot \text{N}^{-1} \text{ and } 1.26 \times 10^{-3} \text{ mm} \cdot \text{N}^{-1}$$

□

Vibration absorbers are simple and passive devices commonly used in the control of narrow-band vibrations (limited to a very small interval of frequencies). Applications are found in vibration suppression of transmission wires (e.g., a stockbridge damper, which simply consists of a piece of cable carrying two masses at its ends), consumer appliances, automobile engines, and industrial machinery. Before leaving the topic, it should be stated that the concepts presented for a rectilinear vibration absorber can be directly extended to a rotary vibration absorber. A schematic representation of a rotary vibration absorber is shown in Figure 12.28. This model corresponds to vibration

Vibration Design and Control

BOX 12.7 Vibration Control

Passive control (no external power):
1. Dampers
   a. A dissipative approach (thermal problems, mechanical degradation)
   b. Useful over a wide frequency band.
2. Vibration absorbers (dynamic absorbers, Frahm absorbers)
   a. Absorbs energy from vibrating system and applies a counteracting force
   b. Useful over a very narrow frequency band (near the tuned frequency)
   c. Absorber executes large motions.

Undamped absorber design:

$$\text{Transfer function of system with absorber} = \frac{\alpha^2 - r^2}{r^4 - \left[\alpha^2(1+\mu)+1\right]r^2 + \alpha^2}$$

where

→ μ = absorber mass/primary system mass
→ α = absorber natural frequency/primary system natural frequency
→ r = excitation frequency/primary system natural frequency.

The most effective operating frequency $r_{op} = \alpha$.
Avoid the two resonances.

Optimal damped absorber design:
To get the transfer fucntion, replace α^2 by $\alpha^2 + 2j\zeta_a \alpha r$ in the undamped case.

→ Mass ratio $\mu = \dfrac{1}{\alpha} - 1$

Damping ratio $\zeta_a = \dfrac{3\mu}{8(1+\mu)^3}$

Active control (needs external power):
1. Measure vibration response using sensors/transducers.
2. Apply control forces to vibrating system through actuators, according to a suitable control algorithm.

---

force excitations [compare with Figure 12.23(a)]. The case of rotational support-motion excitations [see Figure 12.23(b)] can be addressed as well, which has essentially the same transfer function. Approaches of vibration control are summarized in Box 12.7.

12.7.3 Vibration Dampers

As discussed above, vibration absorbers are simple and effective passive devices that are used in vibration control. They have the added advantage of being primarily non-dissipative. The main disadvantage of a vibration absorber is the fact that it is effective only over a very narrow band of

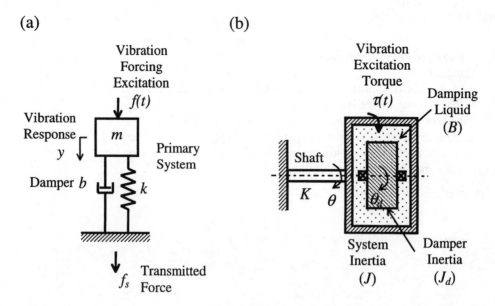

FIGURE 12.29 (a) A system with a linear viscous damper, and (b) a rotary system with a Houdaille damper.

frequencies enclosing its resonant frequency (tuned frequency). When passive vibration control over a wide band of frequencies is required, a damper would be a good choice.

Vibration dampers are dissipative devices. They accomplish the function of vibration control through direct dissipation of the vibration energy of the primary (vibrating) system. As a result, however, there will be substantial heat generation, as well as associated thermal problems and component wear. Consequently, methods of cooling (e.g., use of a fan, coolant circulation, and thermal conduction blocks) may be required in some special situations.

Consider a vibrating system that is modeled as an undamped single-degree-of-freedom mass-spring system (simple oscillator). The magnitude of the excitation-response transfer function will have a resonance with a theoretically infinite magnitude in this case. Operation in the immediate neighborhood of such a resonance would be destructive. Adding a simple viscous damper, as shown in Figure 12.29(a), will correct the situation. The equation of motion (about the static equilibrium position) is

$$m\ddot{y} + b\dot{y} + ky = f(t) \tag{12.120}$$

with the dynamic force that is transmitted through the support base (f_s) given by

$$f_s = ky + b\dot{y} \tag{12.121}$$

Hence, the transfer function between the forcing excitation f and the vibration response y is

$$\frac{y}{f} = \frac{1}{k - \omega^2 m + j\omega b} \tag{12.122}$$

and that between the forcing excitation and the force transmitted to the support structure is

$$\frac{f_s}{f} = \frac{k + j\omega b}{k - \omega^2 m + j\omega b} \tag{12.123}$$

Using the nondimensional frequency variable $r = \omega/\omega_n$, where $\omega_n = \sqrt{k/m}$ is the undamped natural frequency of the system, and the damping ratio $\zeta = b/(2\sqrt{km})$, one can express equations (12.122) and (12.123) in the form:

$$\frac{y}{f} = \frac{1}{k(1 - r^2 + 2j\zeta r)} \tag{12.124}$$

$$\frac{f_s}{f} = \frac{1 + 2j\zeta r}{(1 - r^2 + 2j\zeta r)} \tag{12.125}$$

When vibration control of the primary system is desired, one can use the transfer function (12.124); and when force transmissibility is the primary consideration, one can use (12.125). Furthermore, it is convenient to use the transfer function (12.124) in the nondimensional form:

$$\frac{ky}{f} = G(r) = \frac{1}{(1 - r^2 + 2j\zeta r)} \tag{12.126}$$

The magnitude of this transfer function is plotted in Figure 12.30 for several values of damping ratio. Note how the addition of significant levels of damping can considerably lower the resonant peak, and furthermore flatten the overall response. This example illustrates the broad-band nature of the effect of a damper; but, unlike a vibration absorber, it is not possible with a simple damper to bring the vibration levels to a theoretical 0. However, a damper is able to bring the response uniformly close to the static value (unity in Figure 12.30).

Another common application of a damper is found where it is connected through a free inertia element. Such an arrangement, for a rotational system, is known as the Houdaille damper, and is modeled as in Figure 12.30(b). The equations of motion are

$$J\ddot{\theta} + B(\dot{\theta} - \dot{\theta}_d) + K\theta = \tau(t) \tag{12.127}$$

$$J_d\ddot{\theta}_d + B(\dot{\theta}_d - \dot{\theta}) = 0 \tag{12.128}$$

In this case, the transfer function between the vibratory excitation torque τ and the response angle θ is given by

$$\frac{\theta}{\tau} = \frac{B + J_d j\omega}{KB - B(J + J_d)\omega^2 - J_d J j\omega^3 + KJ_d j\omega} \tag{12.129}$$

Again, use the normalized form of $K\theta/\tau$ to obtain

$$\frac{K\theta}{\tau} = G(r) = \frac{2\zeta + jr\mu}{2\zeta[1 - (1+\mu)r^2] + jr\mu(1 - r^3)} \tag{12.130}$$

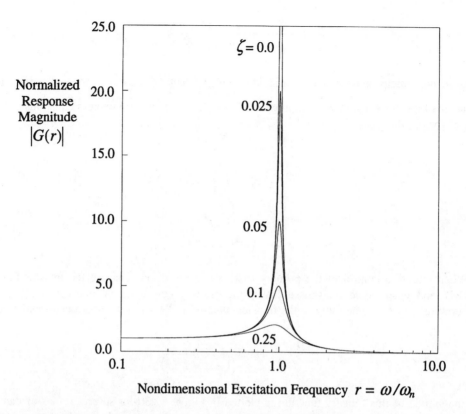

FIGURE 12.30 Frequency response of a system containing a linear damper.

where

$r = \omega/\omega_n$

$\zeta = B/(2\sqrt{KJ})$

$\mu = J_d/J$

$\omega_n = \sqrt{K/J}$

Note the two extreme cases: (1) when $\zeta = 0$, the system becomes the original undamped system, as expected; and (2) when $\zeta \to \infty$, again, the system becomes an undamped simple oscillator, but with a lower natural frequency of $r = 1/\sqrt{1+\mu}$, instead of $r = 1$ that was present in the original system. This is to be expected because as $\zeta \to \infty$, the two inertia elements become locked together and act as a single combined inertia $J + J_d$. Clearly, in these two extreme systems, the effect of damping is not present. Optimal damping occurs somewhere in between, as is clear from the curves of response magnitude shown in Figure 12.31 for the case of $\mu = 0.2$.

Proper selection of the nature and values of damping is crucial to the use of a damper in vibration control. Damping in physical systems is known to be nonlinear and frequency dependent, as well as time-variant and dependent on the environment (e.g., temperature). Models are available for different types of damping, but these are only models or approximate representations. In practice, depending on such considerations as the type of damper used, nature of the system, specific application, and the speed of operation, a particular model (linear viscous, hysteric, Coulomb, Stribeck, quadratic aerodynamic, etc.) may be valid. In addition to the simple linear theory

Vibration Design and Control

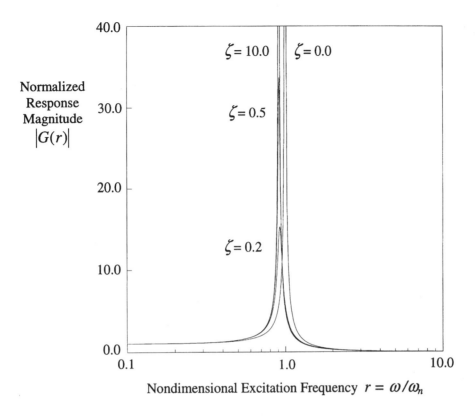

FIGURE 12.31 Response curves for a rotary system with Houdaille damper of inertia ratio $\mu = 0.2$.

of viscous damping, as used in the present section, specific properties of physical damping should be taken into consideration in practical designs. The topic of damping has been addressed in more detail in Chapter 7.

12.8 ACTIVE CONTROL OF VIBRATION

Passive control of vibration is relatively simple and straightforward. It is also known to be robust, reliable, and economical, but it has limitations. Note that the control force generated in a passive device depends entirely on the natural dynamics. Once the device is designed (i.e., after the parameter values for mass, damping constant, stiffness, location, etc. are chosen), it is not possible to adjust the control forces that are naturally generated by it in real-time. Furthermore, in a passive device, there is no supply of power from an external source. Hence, even the magnitudes of the control forces cannot be changed from their natural values. Because a passive device senses the response of the system implicitly as an integral process of the overall dynamics of the system, it is not always possible to directly target the control action at particular responses (e.g., particular modes). This can result in incomplete control, particularly in complex and high-order (e.g., distributed-parameter) systems. These shortcomings of passive control can be overcome using active control, where the system responses are directly sensed using sensor-transducer devices; and on that basis, control actions of specific desired values are applied to desired locations/modes of the system.

12.8.1 ACTIVE CONTROL SYSTEM

A schematic diagram of an active control system is shown in Figure 12.32. The mechanical dynamic system for which vibrations need to be controlled is the *plant* or *process*. The device that generates

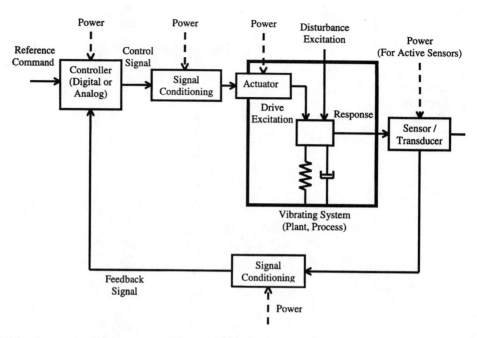

FIGURE 12.32 A system for active control of vibration.

the signal (or command) according to some scheme (or control law) and controls vibrations of the plant is called the *controller*. The plant and the controller are the two essential components of a *control system*. Usually, the plant must be monitored and its response must be measured using sensors, for feedback into the controller. Then, the controller compares the sensed signal with a desired response as specified externally, and uses the error to generate a proper control signal. In this manner, one has a *feedback control system*. In the absence of a sensor and feedback, what one has is an open-loop control system. In *feed-forward control*, the excitation (i.e., input signal) — not the response (i.e., output signal) — is measured and used (i.e., fed forward into the controller) for generating the control signal. Both feedback and feed-forward schemes can be used together in the same control system.

The actuator that receives a control signal and drives the plant can be an integral part of the plant (e.g., the motor that drives the blade of a saw), or it can be specifically added as an external component for the control actuation (e.g., a piezoelectric or electromagnetic actuator for controlling blade vibrations of a saw). In the former case, in particular, proper signal conditioning is needed to convert the control signal to a form that is compatible with the existing actuator. In the latter case, both the controller and the actuator must be developed in parallel for integration into the plant. In digital control, the controller is a digital processor. The control signal is in the digital form in this case and typically must be converted into the analog form prior to use in the actuator. Hence, digital-to-analog conversion (DAC) is a form of signal conditioning that is useful here. Furthermore, the analog signal that is generated might have to be filtered and amplified to an appropriate level for use in the actuator. It follows that filters and amplifiers are signal-conditioning devices that are useful in vibration control (see Chapters 8 and 9). In *software control*, the control signal is generated by a computer that functions as the digital controller. In *hardware control*, the control signal is generated very rapidly by digital hardware, without using software programs. Alternatively, *analog control* can be used, where the control signal is generated directly using analog circuitry. In this case as well, the controller is quite fast and, furthermore, it does not require DAC. Note that the actuator may need high levels of power. Also, the controller and associated signal conditioning will require some power.

Vibration Design and Control

This need for an external power source for control particularly distinguishes active control from passive control. Some important aspects of control instrumentation and signal conditioning were presented in Chapters 8 and 9.

In a feedback control system, sensors are used to measure the plant response that enables the controller to determine whether the plant operates properly. A sensor unit that "senses" the response can automatically convert (transduce) this "measurement" into a suitable form. In examples discussed in Chapter 8, a piezoelectric accelerometer senses acceleration and converts it into an electric charge; an electromagnetic tachometer senses velocity and converts it into a voltage; and a shaft encoder senses a rotation and converts it into a sequence of voltage pulses. Hence, the terms *sensor* and *transducer* are used interchangeably to denote a sensor-transducer unit. The signal generated in this manner might need conditioning before feeding into the controller. For example, the charge signal from a piezoelectric accelerometer must be converted to a voltage signal of appropriate level using a charge amplifier, and then it must be digitized using an analog-to-digital converter (ADC) for use in a digital controller. Furthermore, filtering may be needed to remove measurement noise. Hence, signal conditioning is usually required between the sensor and the controller, as well as the controller and the actuator. External power will be required to operate active sensors (e.g., potentiometer), whereas passive sensors (e.g., electromagnetic tachometer) employ self-generation and do not need an external power source. As before, external power might be needed for conditioning the sensor signals. See Chapters 8 and 9 for details. Finally, as indicated in Figure 12.32, a vibrating system may have unknown disturbance excitations that can make the control problem particularly difficult. Removing such excitations at the source level is desirable, through proper design or vibration isolation, as previously discussed in this chapter. But in the context of control, if these disturbances can be measured or, some information about them is available, they can be compensated for within the controller itself. This is, in fact, the approach of feed-forward control.

12.8.2 Control Techniques

The purpose of a vibration controller is to excite (activate) a vibrating system so as to control its vibration response in a desired manner. In the present context of active feedback control, the controller uses measured response signals and compares them with their desired values in its task of determining an appropriate action. The relationship that generates the control action from a measured response (and also using a desired value for the response) is called a *control law*. Sometimes, a *compensator* (analog or digital; hardware or software) is employed to improve the system performance or to enhance the controller so that the task of control becomes easier. But, for purposes herein, one can consider a compensator as an integral part of the controller and, hence, a distinction between the two is not made.

Various control laws, both linear and nonlinear, have been developed for practical applications. Many of them are suitable in vibration control. A comprehensive presentation of all such control laws is outside the scope of this book. Several linear control laws that are common and representative of what is available, are presented here. These techniques are based on a linear representation (linear model) of the vibrating system (plant). Even when the overall operating range of a plant (e.g., robotic manipulator) is nonlinear, it is often possible to linearize the vibration response (e.g., link vibrations and joint vibrations of a robot) about a reference configuration (e.g., robot trajectory). Then, these linear control techniques will still be suitable, although the overall dynamics of the system is nonlinear.

State-Space Models

In applying many types of control techniques, it is convenient to represent the vibrating system (plant) by a state-space model. This is simply a set of first-order ordinary differential equations that can be coupled, nonlinear, and have time-varying parameters (time-variant models). The concept of state has been discussed in Chapters 2 and 5 and Appendix A. The discussion here is limited to linear and time-invariant state-space models. Such a model is expressed as

$$\dot{x} = Ax + Bu \qquad (12.131)$$

$$y = Cx + Du \qquad (12.132)$$

where

- $x = [x_1, x_2, \ldots, x_n]^T$ = state vector (nth order column)
- $u = [u_1, u_2, \ldots, u_r]^t$ = input vector (rth order column)
- $y = [y_1, y_2, \ldots, y_m]^t$ = output vector (mth order column)
- A = system matrix ($n \times n$ square)
- B = input gain matrix ($n \times r$)
- C = measurement gain matrix ($m \times n$)
- D = feed-forward gain matrix ($m \times r$).

Usually, for vibrating systems, it is possible to make $D = 0$, and hence one can drop this matrix in the sequel. Furthermore, a state variable x_i need not have a direct physical meaning, but an output variable y_j should have some physical meaning and, in typical situations, should be measurable as well. The input variables are the "control variables" and are the ones used for controlling the system (plant). The output variables are the "controlled variables;" they correspond to the system response and are measured for feedback control.

It can be verified that the eigenvalues of the system matrix A occur in complex conjugates of the form $-\zeta_i \omega_i \pm j\sqrt{1-\zeta_i^2}\,\omega_i$ in the damped oscillatory case, or as $\pm j\omega_i$ in the undamped case, where ω_i is the ith undamped natural frequency of the system and ζ_i is the corresponding damping ratio (of the ith mode). The mathematical verification requires some linear algebra. An intuitive verification can be made since equation (12.131) is an equivalent model for a system having the traditional mass-spring-damper model

$$M\ddot{y} + C\dot{y} + Ky = f(t) \qquad (12.133)$$

where

- M = mass matrix
- C = damping matrix
- K = stiffness matrix
- $f(t)$ = forcing input vector
- y = displacement response vector.

Both models (12.131) and (12.133), being equivalent, should have the same characteristic equation, which by its roots determines the natural frequencies and modal damping ratios. This must be the case because one is looking at just two different mathematical representations of the same system. Hence, the parameters of its dynamics, such as ω_i and ζ_i should remain unchanged. In fact, the state-space model (12.131) is not unique, and different versions of state vectors and corresponding models are possible. Of course, all of them should have the same characteristic polynomial (and hence, the same ω_i and ζ_i). One such state-space model can be derived from equation (12.133) as follows.

Vibration Design and Control

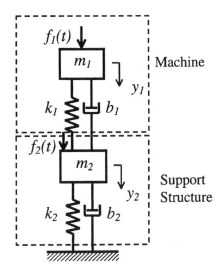

FIGURE 12.33 A model of a machine mounted on a support structure.

Define the state vector as

$$x = \begin{bmatrix} y \\ \dot{y} \end{bmatrix} \quad \text{and} \quad u = f(t) \tag{12.134}$$

Then, since (for nonsingular M, as required) equation (12.133) can be written as

$$\ddot{y} = -M^{-1}Ky - M^{-1}C\dot{y} + M^{-1}f(t) \tag{12.135}$$

One obtains

$$\dot{x} = \begin{bmatrix} 0 & I \\ -M^{-1}K & -M^{-1}C \end{bmatrix} x + \begin{bmatrix} 0 \\ M^{-1} \end{bmatrix} u \tag{12.136}$$

This is a state-space model that is equivalent to the conventional model (12.133), and can be shown to have the same characteristic equation. The development of a state-space model for a vibrating system can be illustrated using an example.

EXAMPLE 12.10

Consider a machine mounted on a support structure, modeled as in Figure 12.33. Using the excitation forces $f_1(t)$ and $f_2(t)$ as the inputs and the displacements y_1 and y_2 of the masses m_1 and m_2 as the outputs, develop a state-space model for this system.

SOLUTION

Assume that the displacements are measured from the static equilibrium positions of the masses. Hence, the gravity forces do not enter into the formulation. Newton's second law is applied to the two masses; thus,

$$m_1 \ddot{y}_1 = f_1 - k_1(y_1 - y_2) - b_1(\dot{y}_1 - \dot{y}_2)$$

$$m_2\ddot{y}_2 = f_2 - k_1(y_2 - y_1) - b_1(\dot{y}_2 - \dot{y}_1) - k_2 y_2 - b_2 \dot{y}_2$$

The following state variables are defined:

$x_1 = y_1$; $x_2 = \dot{y}_1$; $x_3 = y_2$; $x_4 = \dot{y}_2$. Also, the input vector is $u = [f_1\ f_2]^T$ and the output vector is $y = [y_1\ y_2]^T$. Then,

$$\dot{x}_1 = x_2$$

$$m_1 \dot{x}_2 = u_1 - k_1(x_1 - x_3) - b_1(x_2 - x_4)$$

$$\dot{x}_3 = x_4$$

$$m_2 \dot{x}_4 = u_2 - k_1(x_3 - x_1) - b_1(x_4 - x_2) - k_2 x_3 - b_2 x_4$$

Accordingly, the state-space model is given by equations (12.131) and (12.132), with

$$A = \begin{bmatrix} 0 & 1 & 0 & 0 \\ -k_1/m_1 & -b_1/m_1 & k_1/m_1 & b_1/m_1 \\ 0 & 0 & 0 & 1 \\ k_1/m_2 & b_1/m_2 & -(k_1 + k_2)/m_2 & -(b_1 + b_2)/m_2 \end{bmatrix}$$

$$B = \begin{bmatrix} 0 & 0 \\ 1/m_1 & 0 \\ 0 & 0 \\ 0 & 1/m_2 \end{bmatrix}$$

$$C = \begin{bmatrix} 1 & 0 & 0 & 0 \\ 0 & 0 & 1 & 0 \end{bmatrix}$$

and

$$D = 0$$

Also, note that the system can be expressed as

$$\begin{bmatrix} m_1 & 0 \\ 0 & m_2 \end{bmatrix} \ddot{y} + \begin{bmatrix} b_1 & -b_1 \\ -b_1 & (b_1 + b_2) \end{bmatrix} \dot{y} + \begin{bmatrix} k_1 & -k_1 \\ -k_1 & (k_1 + k_2) \end{bmatrix} y = f(t)$$

Its characteristic equation can be expressed as the determinant equation

$$\det \begin{bmatrix} m_1 s^2 + b_1 s + k_1 & -b_1 s - k_1 \\ -b_1 s - k_1 & m_2 s^2 + (b_1 + b_2)s + (k_1 + k_2) \end{bmatrix} = 0$$

It can be verified through direct expansion of the determinants that this equation is equivalent to the characteristic equation of the matrix A, as given by $\det(\lambda I - A) = 0$, or

$$\det\begin{bmatrix} \lambda & -1 & 0 & 0 \\ k_1/m_1 & \lambda+b_1/m_1 & -k_1/m_1 & -b_1/m_1 \\ 0 & 0 & \lambda & -1 \\ -k_1/m_2 & -b_1/m_2 & (k_1+k_2)/m_2 & \lambda+(b_1+b_2)/m_2 \end{bmatrix} = 0$$

□

Note that in the present context, x and y represent the vibration response of the plant, and the control objective is to reduce these to 0. Some common control techniques that can achieve this goal are given below.

Position and Velocity Feedback

In this technique, the position and velocity of each degree of freedom are measured and fed into the system with sign reversal (negative feedback) and amplification by a constant gain. Since velocity is the derivative of position and the gains are constant (i.e., proportional), this method falls into the general category of proportional-plus-derivative (PD or PPD) control. In this approach, it is tacitly assumed that the degrees of freedom are uncoupled. Then, control gains are chosen so that the degrees of freedom in the controlled system are nearly uncoupled, thereby justifying the original assumption. To explain this control method, suppose that a degree-of-freedom of a vibrating system is represented by

$$m\ddot{y} + b\dot{y} + ky = u(t) \tag{12.137}$$

where y is the displacement (position) of the degree-of-freedom and u is the excitation input applied. Now suppose that u is generated according to the (active) control law

$$u = -k_c y - b_c \dot{y} + u_r \tag{12.138}$$

where k_c is the position feedback gain and b_c is the velocity feedback gain. The implication here is that the position y and the velocity $\dot{y}$ are measured and fed into the controller, which in turn generates u according to equation (12.138). Also, u_r is some reference input that is provided externally to the controller. Then, substituting equation (12.138) into (12.137), one obtains

$$m\ddot{y} + (b+b_c)\dot{y} + (k+k_c)y = u_r \tag{12.139}$$

The closed-loop system (the controlled system) now behaves according to equation (12.139). The control gains b_c and k_c can be chosen somewhat arbitrarily (subject to the limitations of the physical controller, signal conditioning circuitry, the actuator, etc.) and might even be negative. In particular, by increasing b_c, the damping of the system can be increased; and similarly, by increasing k_c, the stiffness (and the natural frequency) of the system can be increased. Although a passive spring and damper with stiffness k_c and damping constant b_c can accomplish the same task, once the devices are chosen, it will not be possible to conveniently change their parameters. Furthermore, it will not be possible to make k_c or b_c negative in this case of passive physical devices. The method of PPD control is simple and straightforward, but the assumption of linear uncoupled degrees of freedom places a limitation on its general use.

Linear Quadratic Regulator (LQR) Control

This is an optimal control technique. Consider a vibrating system that is represented by the linear state-space model

$$\dot{x} = Ax + Bu \qquad (12.131)$$

Assume that all the states x are measurable and all the system modes are controllable. Then, use the constant-gain feedback control law:

$$u = Kx \qquad (12.140)$$

The choice of parameter values for the feedback gain matrix K is infinite. Thus, one can use this freedom to minimize the cost function:

$$J = \frac{1}{2}\int_t^\infty [x^T Q x + u^T R u]d\tau \qquad (12.141)$$

This is the time integral of a quadratic function in both state and input variables, and the optimization goal can be interpreted as bringing x down to 0 (regulating x to $\mathbf{0}$), but without spending a rather high control effort; hence, the name linear quadratic regulator (LQR). Also, Q and R are weighting matrices, with the former being at least positive semi-definite and the latter positive definite. Typically, Q and R are chosen as diagonal matrices with positive diagonal elements whose magnitudes are decided based on the degree of relative emphasis that should be made on various elements of x and u. It is well-known that K, which minimizes the cost function (12.141), is given by

$$K = -R^{-1} B^T K_r \qquad (12.142)$$

where K_r is the positive-definite solution of the matrix Riccati algebraic equation

$$K_r A + A^T K_r - K_r B R^{-1} B^T K_r + Q = 0 \qquad (12.143)$$

It is also known that the resulting closed-loop control system is stable. Furthermore, the minimum (optimal) value of the cost function (12.141) is given by

$$J_m = \frac{1}{2} x^T K_r x \qquad (12.144)$$

where x is the present value of the state vector. A major computational burden of the LQR method lies in the solution equation (12.143). Other limitations of the technique arise due to the need for measuring all the state variables (which may be relaxed to some extent). Furthermore, although the stability of the controlled system is guaranteed, the level of stability that is achieved (i.e., stability margin or the level of modal damping) cannot be directly specified. Also, robustness of the control system, in the presence of model errors, unknown disturbances, etc., may be questionable. Besides, the cost function incorporates an integral over an infinite time duration, which does not typically reflect the practical requirement of rapid vibration control.

Modal Control

The LQR control technique has the serious limitation of not being able to directly achieve specified levels of modal damping, which can be an important goal in vibration control. The method of modal control accomplishes this objective through *pole placement*, where poles (eignevalues) of

Vibration Design and Control

the controlled system are placed at specified values. Specifically, consider the plant (12.131) and the feedback control law (12.140). Then, the closed-loop system is given by

$$\dot{x} = (A + BK)x \qquad (12.145)$$

It is well known that if the plant (A, B) is controllable, then a control gain matrix K can be chosen that will arbitrarily place the eigenvalues of the closed-loop system matrix $A + BK$. That means, under the given assumptions, the modal control technique can not only assign the modal damping, but also the damped natural frequencies at specified values. The assumptions given above are quite stringent, but they can be relaxed to some degree. A shortcoming, however, of this method is the fact that it does not place a restriction on the control effort, as, for example, the LQR technique does, in achieving a specified level of modal control.

12.8.3 Active Control of Saw Blade Vibration

Saw blades are thin-plate-like distributed (continuous) mechanical systems. They exhibit vibrations, with theoretically an infinite number of modes. The mode shapes of the first four modes of vibration of a circular blade, as obtained from a finite element analysis, are shown in Figures 12.34(a) through (d). Also, as the speed of rotation increases, the tension effect due to the associate centrifugal forces also increases. Hence, just like in the case of a beam in tension and undergoing flexural vibrations (as analyzed in Chapter 6), the natural frequencies will increase with speed. Table 12.2 gives the natural frequencies of the first five modes of vibration of a circular saw, at different rotation speeds in the range of 0 to 4200 rpm.

Circular saws are widely used in the wood machining industry. Operating speeds of these saws can range from 600 to 4000 rpm. The blade diameter can be in the range of 35.0 to 150.0 cm, and the blade thickness can be from 2.5 to 5.5 mm. Vibrations in saw blades can have several detrimental effects. First, the quality of the cut (sawing accuracy and finish) will degrade with vibration. Second, the wood recovery will decrease due to inaccurate cuts; a wastage level of 7% is typical due to blade vibration. Third, because approximately 12% of the processed wood ends up as sawdust, the related problems will be exacerbated due to inaccurate cuts caused by blade vibrations. Deterioration of the saw-blade life and damaging effects on the saw, due to vibration, can be significant as well.

Finer cuts of high quality require thinner blades. Also, increased product quality and productivity will call for higher blade speeds, and will make the speed become closer to the fundamental natural frequency of the blade. Both these requirements of modern wood processing will potentially increase the likelihood of vibration problems. It follows that vibration control of saw blades can be quite beneficial in wood processing. A project has been undertaken by George Wang and associates at the National Research Council of Canada, in active control of saw blade vibration. In the present example, their experimental setup is outlined. A schematic diagram of the experimental setup is given in Figure 12.35(a), and a view of the laboratory system is shown in Figure 12.35(b).

Since the blade rotates during normal operation, both the sensors and actuators for the blade itself should preferably be of the non-contact type. Of course, it is possible to monitor speeds, forces, and movements at the shaft and bearings of the blade using sensors and transducers such as optical encoders, tachometers, accelerometers, and strain-gages (see Chapter 8); but for the direct measurement of blade vibration, proximity sensors are preferred, as used in the system of Figure 12.35. Magnetic-induction type or eddy-current type proximity sensors and optical sensors (e.g., lasers) can be employed for this purpose, as indicated in Chapter 8, with their particular advantages and disadvantages. Notably, eddy-current proximity sensors can be used at high frequencies and can also be easily compensated for environmental effects, using a bridge circuit. They require a relatively small sensing area and, hence, the sensor signal will not be affected by the motion of other parts of the blade system. This is not the case for a magnetic-induction proximity

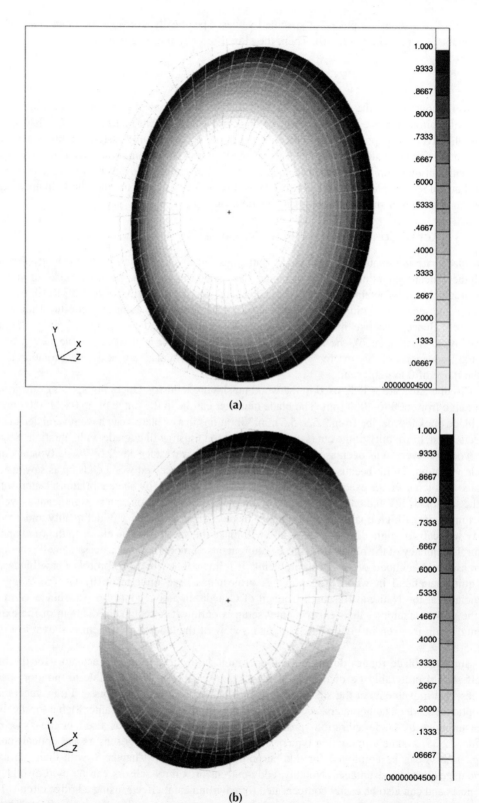

FIGURE 12.34 Mode shapes of vibration of a circular saw blade: (a) Mode 1, and (b) Mode 2. (Courtesy of Dr. George Wang; taken from Wang, G. et al., National Research Council, Integrated Manufacturing Technologies Institute, Vancouver, Canada, pp. 5, 8, 25–28, May 1998. With permission.)

Vibration Design and Control

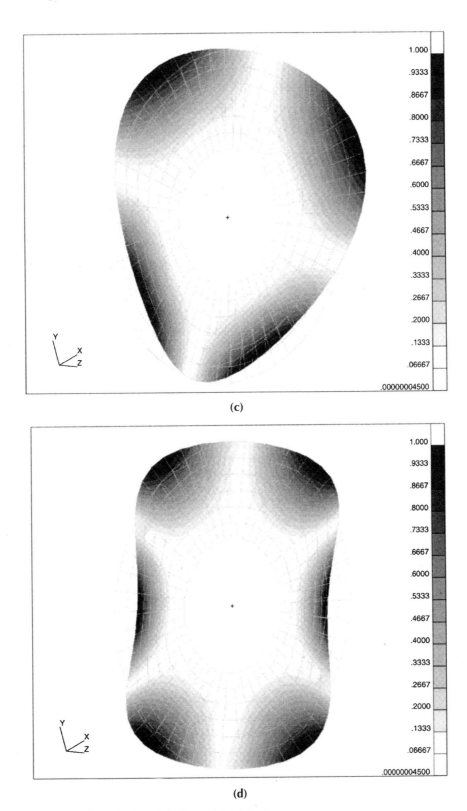

FIGURE 12.34 (*continued*) (c) Mode 3, and (d) Mode 4.

TABLE 12.2
Natural Frequencies of Vibration of a Circular Saw Blade (Diameter = 45.7 cm, Thickness = 2.5 mm)

Blade Speed (rpm)	Natural Frequency (Hz) for Mode No.					
	1	2	3	4	5	6
0	96.06	96.91	110.03	154.91	235.97	347.52
600	96.92	97.86	111.13	155.99	236.95	348.41
1500	101.32	102.71	116.73	161.59	242.06	353.04
2400	108.99	111.14	126.48	171.51	251.28	361.49
3300	119.29	122.39	139.48	185.04	264.17	373.48
4200	131.55	135.74	154.93	201.45	280.22	388.69

(Courtesy of Dr. George Wang; taken from Wang, G. et al., National Research Council, Integrated Manufacturing Technologies Institute, Vancouver, Canada, pp. 5, 8, 25–28, May 1998. With permission.)

sensor. Optical sensors are convenient, but their measurements will be affected by environmental lighting, surface material and texture of the object, dust, etc. Electrodynamic type actuators, which operate on the same principle as an electric motor or a shaker used in vibration testing, are used in the system of Figure 12.35. These do not have to be in contact with the blade and can be accurately controlled through a digital controller and a drive amplifier.

In Figure 12.35, a signal generator is used to apply to the blade a vibration excitation of known magnitude and frequency characteristics, through an electrodynamic actuator. But, in the practical situation, it is the blade rotation that excites various modes of vibration of the blade, and the actuators are used only for the purpose of active control. A spectrum analyzer is used for off-line frequency analysis, of the sensor signals. Again, in a practical active control system, typically, the sensor signals are not analyzed off-line, but rather conditioned and directly used in the controller.

The task of active control involves sensing a vibration component and then, using an actuator, applying an excitation force to counteract the measured vibrations. This seemingly simple task can become quite difficult in complex, continuous-parameter systems such as rotating saw blades, which have an infinite number of modes. For example, both the value of the control excitation and the location of the actuator can be varied to realize the most desirable vibration performance. The mode shapes and natural frequencies of the system should be taken into account. For example, if a sensor or an actuator is located at a node of a particular mode, it will not be possible to accurately control that mode. Control techniques such as those mentioned previously can be employed in generating the control signal from the sensor signals. The issues involved are quite similar to those that arise in the vibration control of a beam. This topic will be addressed in the next section.

12.9 CONTROL OF BEAM VIBRATIONS

A beam is a distributed-parameter system that has, in theory, an infinite number of modes of vibration, with associated mode shapes and natural frequencies. In this sense, it is an "infinite order" system, with infinite degrees of freedom. Hence, the computation of modal quantities and associated control inputs can be quite complex. Fortunately, however, only a few modes need be retained in a dynamic model without sacrificing a great deal of accuracy, and thereby, facilitating simpler control. Some concepts of controlling vibrations in a beam are considered in this section. The present treatment is intended as an illustration of the relevant techniques and is not meant to be exhaustive. These techniques can be extended to other types of continuous systems, such as beams with different boundary conditions and plates. Since the control techniques that were outlined

Vibration Design and Control

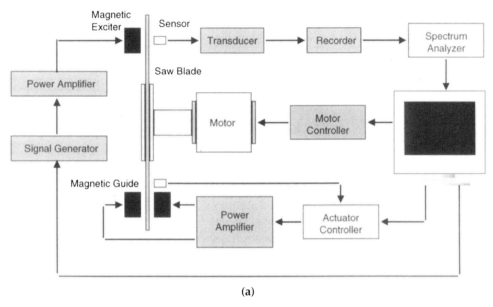

(a)

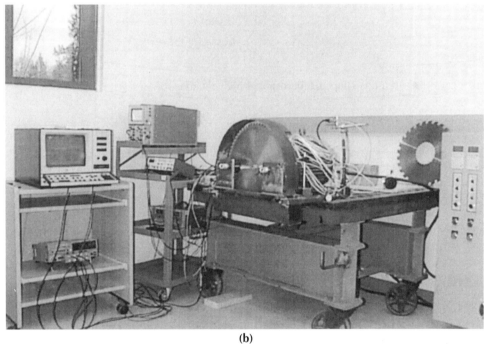

(b)

FIGURE 12.35 An active control system for saw blade vibration: (a) schematic diagram, and (b) experimental setup. (Courtesy of Dr. George Wang; taken from Wang, G. et al., National Research Council, Integrated Manufacturing Technologies Institute, Vancouver, Canada, pp. 5, 8, 25–28, May 1998. With permission.)

previously depend on a model, the procedure of obtaining a state-space model for a beam is illustrated first.

12.9.1 STATE-SPACE MODEL OF BEAM DYNAMICS

Consider a Bernoulli-Euler type beam with Kelvin-Voigt type internal (material) damping, as discussed in Chapters 6 and 7. The beam equation can be expressed as

$$ELv(x,t) + E^*L\frac{\partial v(x,t)}{\partial t} + \rho A(x)\frac{\partial^2 v(x,t)}{\partial t^2} = f(x,t) \tag{12.146}$$

in which L is the partial differential operator given by

$$L = \frac{\partial^2 I(x)}{\partial x^2}\frac{\partial^2}{\partial x^2} \tag{12.147}$$

and

$f(x,t)$ = distributed force excitation per unit length of beam
$v(x,t)$ = displacement response at location x along the beam at time t
$I(x)$ = second moment of area of the beam cross section about the neutral axis
E = Young's modulus of the beam material
E^* = Kelvin-Voigt material damping parameter.

Note that a general beam with non-uniform characteristics is assumed and, hence, the variations of $I(x)$ and $\rho A(x)$ with x are retained in the formulation.

Using the approach of modal expansion, the response of the beam can be expressed by

$$v(x,t) = \sum_{i=1}^{\infty} Y_i(x) q_i(t) \tag{12.148}$$

where $Y_i(x)$ is the ith mode shape of the beam, which satisfies

$$LY_i(x) = \frac{\rho A(x)}{E}\omega_i^2 Y_i(x) \tag{12.149}$$

and ω_i is the ith undamped natural frequency. The orthogonality condition for this general case of a non-uniform beam is

$$\int_{x=0}^{l} \rho A Y_i Y_j dx = 0 \quad \text{for } i \neq j \tag{12.150}$$
$$= \alpha_j \quad \text{for } i = j$$

Suppose that the forcing excitation on the beam is a set of r point forces $u_k(t)$ located at $x = l_k$, $k = 1, 2, \ldots, r$. Then,

$$f(x,t) = \sum_{k=1}^{r} u_k \delta(x - l_k) \tag{12.151}$$

where $\delta(x - l_i)$ is the *Dirac delta function*. Now substitute equations (12.148) and (12.151) into (12.146); use (12.149); multiply throughout by $Y_j(x)$; and integrate over $x[0,l]$ using (12.150). This gives

Vibration Design and Control

$$\ddot{q}_j + \gamma_j \dot{q}_j(t) + \omega_j^2 q_j = \frac{1}{\alpha_j} \sum_{k=1}^{r} u_k Y_j(l_k) \quad \text{for } j = 1, 2, \ldots \quad (12.152)$$

where

$$\gamma_j = \frac{E^*}{E} \omega_j^2 \quad (12.153)$$

Now, define the state variables x_j according to

$$x_{2j-1} = \omega_j q_j$$
$$x_{2j} = \dot{q}_j \quad \text{for } j = 1, 2, \ldots \quad (12.154)$$

Then, assuming that only the first m modes are retained in the expansion, the state equations are:

$$\dot{x}_{2j-1} = \omega_j x_{2j}$$
$$\dot{x}_{2j} = -\omega_j x_{2j-1} - \gamma_j x_{2j} + \frac{1}{\alpha_j} \sum_{k=1}^{r} u_k Y_j(l_k) \quad \text{for } j = 1, 2, \ldots, m \quad (12.155)$$

This can be put in the matrix-vector form of a state-space model:

$$\dot{x} = Ax + Bu \quad (12.131)$$

where

$$A = \begin{bmatrix} 0 & \omega_1 & & & 0 \\ -\omega_1 & -\gamma_1 & \ddots & 0 & \omega_m \\ 0 & & & -\omega_m & -\gamma_m \end{bmatrix}_{n \times n} \quad (12.156)$$

and

$$B = \begin{bmatrix} 0 & & 0 \\ Y_1(l_1)/\alpha_1 & \cdots & Y_1(l_r)/\alpha_1 \\ \vdots & & \vdots \\ 0 & & 0 \\ Y_m(l_1)/\alpha_m & \cdots & Y_m(l_r)/\alpha_m \end{bmatrix}_{n \times r} \quad (12.157)$$

with $n = 2m$, where m is the number of modes retained in the modal expansion. Note that as the number of modes used in this model increases, the accuracy increases and, simultaneously, the computational effort needed for the control problem increases as well, because of the proportional increase of the system order. At some point, the potential improvement in accuracy, by further increasing the model size, will be insignificant in comparison with added computational burden. Hence, a balance must be struck in this tradeoff of modal truncation.

12.9.2 CONTROL PROBLEM

The state-space model (12.131) for the beam dynamics, with matrices (12.156) and (12.157), is known to be *controllable*. Hence, it is possible to determine a constant-gain feedback controller $u = Kx$ that minimizes a quadratic-integral cost function of the form (12.141). Also, a similar controller can be determined that places the eigenvalues of the system at specified locations, thereby achieving not only specified levels of modal damping but also a specified set of natural frequencies. However, there is a practical obstacle to achieving such an active controller. Note that in the model (12.156) and (12.157), the state variables are proportional to the modal variables q_i and their time derivatives $\dot{q}_i$. They are not directly measurable. What can be measured, normally, are the displacements and velocities at a set of discrete locations along the beam. Let these locations (s) be denoted by $p_1, p_2, \ldots, p_s$. Then, in view of the modal expansion (12.148), the measurements can be expressed as

$$v(p_j,t) = \sum_{i=1}^{m} Y_i(p_j) q_i(t)$$

$$\dot{v}(p_j,t) = \sum_{i=1}^{m} Y_i(p_j) \dot{q}_i(t) \quad \text{for } j = 1, 2, \ldots, s$$

(12.158)

Now, define the output (measurement) vector y according to

$$y = \left[v(p_1,t),\ \dot{v}(p_1,t),\ \ldots,\ v(p_s,t),\ \dot{v}(p_s,t) \right]^T \tag{12.159}$$

Then, in view of equations (12.158) and the definitions (12.154) of the state variable, one can write

$$y = Cx \tag{12.160}$$

with

$$C = \begin{bmatrix} Y_1(p_1)/\omega_1 & 0 & \cdots & Y_m(p_1)/\omega_m & 0 \\ 0 & Y_1(p_1) & \cdots & 0 & Y_m(p_1) \\ \vdots & \vdots & \cdots & \vdots & \vdots \\ Y_1(p_s)/\omega_1 & 0 & \cdots & Y_m(p_s)/\omega_m & 0 \\ 0 & Y_1(p_s) & \cdots & 0 & Y_m(p_s) \end{bmatrix}_{2s \times n} \tag{12.161}$$

Hence, what is possible is an active controller of the form

$$u = Hy \tag{12.162}$$

which is an output feedback controller. Then, in view of equation (12.160), one has

$$u = HCx \tag{12.163}$$

Vibration Design and Control

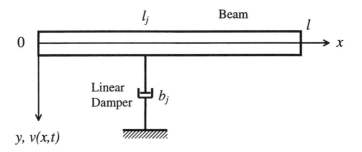

FIGURE 12.36 Use of linear dampers in beam vibration control.

This is not the same as complete-state feedback $u = Kx$, where K can take any real value, and hence, the LQR solution (12.142) or the complete pole placement solution cannot be applied directly. In equation (12.163), only H can be arbitrarily chosen, and C is completely determined according to equation (12.161). The resulting product HC will not usually correspond to either the LQR solution or the complete pole assignment solution. Still, the output feedback controller (12.162) can provide satisfactory performance. However, a sufficient number of displacement and velocity sensors (s) must be used in conjunction with a sufficient number of actuators (r) for active control. This will increase the system complexity and cost. Furthermore, due to added components and their active nature, the reliability of fault-free operation may degrade somewhat. A satisfactory alternative would be to use passive control devices such as dampers and dynamic absorbers. The use of dampers is discussed in Section 12.9.3. The approach using dynamic absorbers follows from this (except for the need of additional state variables to represent the dynamics of the absorbers) and is left as an exercise. Before leaving the present discussion, however, note that in the matrices B and C given by equations (12.157) and (12.161), both the actuator locations l_i and the sensor locations p_j are variable. Hence, there exists an additional design freedom (or optimization parameters) in selecting the sensor and actuator locations for achieving satisfactory control.

12.9.3 Use of Linear Dampers

Now consider the use of a discrete set of linear dampers for controlling beam vibration. Suppose that r linear dampers with damping constants b_j are placed at locations $l_j, j = 1, 2, \ldots, r$ along the beam, as schematically shown in Figure 12.36. The damping forces are given by

$$u_j = -b_j \dot{v}(l_j, t) \qquad \text{for } j = 1, 2, \ldots, r \tag{12.164}$$

By substituting the truncated modal expansion (m modes)

$$\dot{v}(l_j, t) = \sum_{i=1}^{m} Y_i(l_j) \dot{q}_i(t) \tag{12.165}$$

one obtains, in view of equation (12.154), the passive feedback control action

$$u = -Kx \tag{12.166}$$

with

$$K = \begin{bmatrix} 0 & bY_1(l_1) & \cdots & 0 & b_1Y_m(l_1) \\ \vdots & \vdots & \cdots & \vdots & \vdots \\ 0 & b_rY_1(l_r) & \cdots & 0 & b_rY_m(l_r) \end{bmatrix}_{r \times n} \qquad (12.167)$$

By substituting equation (12.166) into (12.131), one obtains the closed-loop system equation

$$\dot{x} = (A - F)x = A_c x \qquad (12.168)$$

where $F = BK$ and is given by

$$F = \begin{bmatrix} 0 & 0 & \cdots & 0 & 0 \\ 0 & \sum b_i Y_{11}(l_i)/\alpha_1 & \cdots & 0 & \sum b_i Y_{1m}(l_i)/\alpha_1 \\ \vdots & \vdots & \cdots & \vdots & \vdots \\ 0 & 0 & \cdots & 0 & 0 \\ 0 & \sum b_i Y_{m1}(l_i)/\alpha_m & \cdots & 0 & \sum b_i Y_{mm}(l_i)/\alpha_m \end{bmatrix}_{n \times n} \qquad (12.169)$$

with

$$Y_{ij}(x) = Y_i(x) Y_j(x) \qquad (12.170)$$

The controller design in this case involves the selection of the damping constants b_i and the damper locations l_j so as to achieve the required performance. This can be achieved, for example, by seeking to make the eigenvalues of the closed-loop system matrix A_c reach a set of desired values, hence giving the desired modal damping and natural frequency characteristics. But in view of the fact that the structure of the F matrix is fixed as given in equation (12.169), this is not equivalent to complete state feedback (and not even complete output feedback). Hence, it will not be possible, in general, to place the poles of the system exactly at the desired locations; i.e., exact pole assignment may fail.

Design Example

In realizing a desirable modal response of a beam using a set of linear dampers, one can seek to minimize a cost function of the form

$$J = \text{Re}(\lambda - \lambda_d)^T Q \, \text{Re}(\lambda - \lambda_d) + \text{Im}(\lambda - \lambda_d)^T R (\lambda - \lambda_d) \qquad (12.171)$$

where λ are the actual eigenvalues of the closed-loop system matrix (A_c) and λ_d are the desired eigenvalues that will give the required modal performance (damping ratios and natural frequencies). Re denotes the real part and Im denotes the imaginary part. Weighting matrices Q and R which are real and diagonal with positive diagonal elements, should be chosen to relatively weight various eigenvalues. This allows one to emphasize some eigenvalues over others, with real parts and imaginary parts weighting separately.by

Various computational algorithms are available for minimizing the cost function (12.171). The details are beyond the scope of this book; only an example result is presented here. Consider a uniform, simply supported 12×5 American Standard beam with the following pertinent specifications: $E = 2 \times 10^8$ kPa (29 × 10^6 psi), $\rho A = 47$ kg·m^{-1} (2.6 lb·in^{-1}), length $l = 15.2$ m (600 in.), $I = 9 \times 10^{-5}$ m^4 (215.8 in.^{4}). The internal damping parameter for the jth mode of vibration is given by

TABLE 12.3
Eigenvalues of the Open-Loop (Uncontrolled) Beam

Mode	Eigenvalue (rad·s⁻¹) (Multiply by 26.27)
1	$-0.000126 \pm j\, 1.0$
2	$-0.000776 \pm j\, 4.0$
3	$-0.002765 \pm j\, 9.0$
4	$-0.007453 \pm j\, 16.0$
5	$-0.016741 \pm j\, 25.0$
6	$-0.033.75 \pm j\, 36.0$

$$E^*(\omega_j) = (g_1/\omega_j) + g_2 \qquad (12.172)$$

in which ω_j is the jth undamped natural frequency given

$$\omega_j = (j\pi/l)^2 \sqrt{EI/\rho A} \qquad (12.173)$$

The numerical values used for the damping parameters are $g_1 = 88 \times 10^4$ kPa (12.5×10^4 psi) and $g_2 = 3.4 \times 10^4$ kPa·s (5×10^3 psi·s). For the present problem, $Y_j(x) = \sqrt{2}\, \sin(j\pi x/l)$ and $\alpha_j = \rho A l$ for all j.

First, ω_j and γ_j are computed using equations (12.173) and (12.153), respectively, along with (12.172). Next, the open-loop system matrix A is formed according to equation (12.156) and its eigenvalues are computed. These are listed in Table 12.3, scaled to the first undamped natural frequency (ω_1). Note that in view of the very low levels of internal material damping of the beam, the actual natural frequencies, as given by the imaginary parts of the eigenvalues, are almost identical to the undamped natural frequencies.

Next, attempt to place the real parts of all the (scaled) eigenvalues at -0.20, while exercising no constraint on the imaginary parts (i.e., damped natural frequencies) by using (1) single damper, and (2) two dampers. In the cost function (12.171), the first three modes are more heavily weighted than the remaining three. Initial values of the damper parameters are $b_1 = b_2 = 0.1$ lbf·s·in⁻¹ (17.6 N·s·m⁻¹) and the initial locations $l_1/l = 0.0$ and $l_2/l = 0.5$. At the end of the numerical optimization, using a modified gradient algorithm, the following optimized values are obtained:

(1) *Single-damper control*:

$b_1 = 36.4$ lbf·s·in.⁻¹ $(6.4 \times 10^3$ N·s·m⁻¹$)$

$l_1/l = 0.3$

The corresponding normalized eigenvalues (of the closed-loop system) are given in Table 12.4.

(b) *Two-damper control*:

$b_1 = 22.8$ lbf·s·in.⁻¹ $(4.0 \times 10^3$ N·s·m⁻¹$)$

$b_2 = 12.1$ lbf·s·in.⁻¹ $(2.1 \times 10^3$ N·s·m⁻¹$)$

$l_1/l = 0.25, \; l_2/l = 0.43$

The corresponding normalized eigenvalues are given in Table 12.5.

TABLE 12.4
Eigenvalues of the Beam with an Optimal Single Damper

Mode	Eigenvalue (rad·s^{-1}) (Multiply by 26.27)
1	$-0.225 \pm j\, 0.985$
2	$-0.307 \pm j\, 3.955$
3	$-0.037 \pm j\, 8.996$
4	$-0.119 \pm j\, 15.995$
5	$-0.355 \pm j\, 24.980$
6	$-0.158 \pm j\, 35.990$

TABLE 12.5
Eigenvalues of the Beam with Optimized Two Dampers

Mode	Eigenvalue (rad·s^{-1}) (Multiply by 26.27)
1	$-0.216 \pm j\, 0.982$
2	$-0.233 \pm j\, 3.974$
3	$-0.174 \pm j\, 8.997$
4	$-0.079 \pm j\, 15.998$
5	$-0.145 \pm j\, 24.999$
6	$-0.354 \pm j\, 35.989$

It would be highly optimistic to expect perfect assignment of all real parts at -0.2. However, note that good levels of damping have been achieved for all modes except for Mode 3 in the single-damper control and for Mode 4 in the two-damper control. In any event, since the contribution of the higher modes toward the overall response is relatively smaller, it is found that the total response (say, at point $x = l/12$) is well damped in both cases of control.

PROBLEMS

12.1 In general terms, outline the procedures of:
   a. System design for vibration
   b. Vibration control in a system.
   What are the differences and similarities of the two procedures?

12.2 On a velocity versus frequency nomograph, to log–log scale, mark suitable operating (design) vibration regions for the following applications:
   a. Ground transportation (30-minute trips)
   b. Ground transportation (8-hour trips)
   c. Tool-workpiece region of milling machines
   d. Automobile transmissions
   e. Lateral vibration of a building tower
   f. Pile drivers in civil engineering constructions (bridges)
   g. Forging machines
   h. Concrete drilling machines
   i. Delicate robotic experimentation in a space station
   j. Compact-disk players.

Vibration Design and Control

12.3 Indicate similarities and differences in the specifications of vibration environment for:
   a. Product operation
   b. Product testing.

12.4 The following are procedures for achieving a desired vibration performance from a mechanical system. Categorize them into vibration isolation, vibration design modification, and vibration control.
   a. Shock absorbers in an automobile
   b. Stiffening crossbars added to a structure
   c. Rotary damper placed on the shaft of a rotating machine
   d. Dynamic absorber mounted on the casing of a delicate instrument
   e. Elastomeric mounts of an exhaust fan
   f. Spring mounts of a heavy engine
   g. Inertia block at a machine base
   h. Stabilizer suspensions of power transmission lines
   i. Helical spoilers of tall incinerators and chimneys
   j. Vibration sensor-actuator combinations for distributed systems
   k. Active suspension systems of transit vehicles
   l. Balancing of rotating machines through the addition and removal of mass.

12.5 Consider the transmissibility function magnitude of a simple-oscillator mechanical system, as given by:

$$|T| = \left[\frac{1 + 4\zeta^2 r^2}{(1 - r^2)^2 + 4\zeta^2 r^2} \right]^{1/2}$$

Using straightforward differentiation of this function, it can be shown that the peak transmissibility occurs at

$$r_{peak} = \frac{\left[\sqrt{1 + 8\zeta^2} - 1 \right]^{1/2}}{2\zeta}$$

In particular, for $\zeta = 0$, $r_{peak} = 1$. By computing r_{peak} for a range of ζ values from 0.001 to 1.9, show that r_{peak} decreases with ζ.

12.6 A landlord rents a room in his basement to two university students. The room is just beneath the kitchen. Two weeks later, the students complain about the shaking of their ceiling when the dishwasher is operating. The landlord decides to install the dishwasher on four spring mounts in order to achieve a vibration isolation level of 80%. The following data are known:

Mass of the dishwasher = 50 kg
Normal operating speed = 300 rpm

Determine the required stiffness for each of the four spring mounts. What will be the static deflection of the springs?

12.7 Under normal conditions, a washing machine operates at a steady speed in the range of 1200 to 1800 rpm. The weight of the washing machine is 75 kg. It is required to achieve a vibration isolation level of at least 80%, and preferably about 90%. Also, during starting and stopping conditions of the washing machine, the peak (resonant) transmissibility should be about 2.5, but not exceed 3. Design a damped spring mount to achieve these operating requirements. Specifically, determine the stiffness k and damping constant b

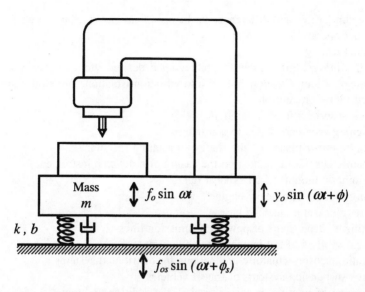

FIGURE P12.8 A milling machine with damped flexible mounts.

of each of the four identical mounts to be incorporated at the base of the washing machine. What is the (undamped) natural frequency of the system?

Hint: Use the approximate design relation (assuming small or 0 damping ratio) and then check adherence to the specifications by using more accurate relations (for a sufficiently large damping ratio).

12.8 A milling machine weighing 500 kg is rigidly mounted on a concrete floor. Loadcells were placed on the base and measurements were made to determine the vertical forces generated by the machine that are transmitted to the floor during operation, in the frequency range of 10 Hz to 60 Hz.

The worst-case amplitude of the transmitted force was found to be 2000 N and the vibrations were nearly sinusoidal. Also, large-amplitude vibratory motion was noticed during start-up and shut-down procedures. To reduce floor vibrations that affect adjoining operations and offices, vibration isolation was found to be required. Furthermore, in order to maintain the machining accuracy during normal operation, the vibratory motion during these steady operating conditions needed to be reduced. The following specifications are given:

Amplitude of vibratory motion at resonace = 1.0 cm or less
Level of vibration isolation under normal operation = 80% (approx.)
Amplitude of vibratory motion under normal operation = 2.0 mm or less.

Design a mounting system to achieve these specifications. A schematic representation of the system is shown in Figure P12.8.

12.9 Consider the flexible vibrating system with a vibration isolator as shown in Figure 12.8. For the case of negligible damping (B and b are neglected) and a unity mass ratio ($m/M = 1$), show that the transmissibility ratio $T_{\text{flexible}}/T_{\text{inertial}}$ of the flexible system to the inertial system (where B, k, and m are absent as in Figure 12.4(c)) is given by

$$\frac{T_{\text{flexible}}}{T_{\text{inertial}}} = \left| \frac{r^2 - 1}{-r^2(r^2 - 1)/r_\omega^2 + 2r^2 - 1} \right|$$

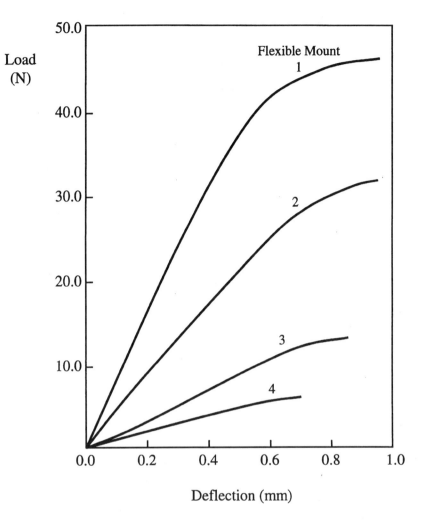

FIGURE P12.10 Static load-deflection characteristics of four spring mounts.

where the nondimensional frequency variable:

$$r = \frac{\omega}{\sqrt{k/M}}$$

and the natural frequency ratio:

$$r_\omega = \frac{\sqrt{K/M}}{\sqrt{k/M}} = \sqrt{\frac{K}{k}}$$

The excitation frequency is ω.

Plot this transmissibility ratio against r in the range $r = 0$ to 10, for a frequency ratio value of $r_\omega = 10.0$.

For the inertial system with an undamped isolator, what is the minimum operating frequency ratio r_{op} for achieving an isolation level of 90%? What is the isolation level of the flexible system at the same operating frequency?

12.10 A high-speed punch press operates at a steady speed in the range of 2400 to 3600 rpm under normal conditions. In order to stop vibration effects of the punch press from affecting other processes and environments, a 75% vibration isolation is sought. Four flexible mounts are available for this purpose and their static characteristics are shown in Figure P12.10. Which of these four mounts would you choose for this application? Perform an analysis to justify your choice. Assume that the damping in the mounts is very small.

12.11 The following data are obtained from an experiment carried out on a disk that is mounted very close to a bearing, in a single-plane balancing problem. Magnitude and location of the trial mass with respect to a body reference line:

$$\vec{M}_t = 13.5 \angle 0° \text{ gm}$$

Magnitude and phase angle (with respect to a body reference) of the accelerometer signal at the bearing, in the absence of the trial mass:

$$\vec{V}_u = 356 \angle 242.2°$$

Magnitude and phase angle (with respect to the same body reference) of the accelerometer signal at the bearing, in the presence of the trial mass:

$$\vec{V}_r = 348 \angle 75.6°$$

Determine the magnitude and orientation of the necessary balancing mass to be mounted on the disk at the same radius as the trial mass in order to completely balance it, after removing the trial mass.

12.12 a. List five causes of unbalance in rotating devices. What are detrimental effects of unbalance? Give some of the ways of eliminating/reducing unbalance.

b. A pancake motor has a disk-like rotor. When rotating at a fixed speed, an accelerometer mounted on the rotor bearing shows an excitation amplitude of 350 mV through a charge amplifier. Also, this signal has a phase lead of 200° with respect to a body reference of the rotor, as determined with respect to a synchronized stroboscope signal. A trial mass of 15 gm was placed on the rotor at a known radius and an angular location of 0° with respect to a body-reference radius that is marked on the rotor surface. Then, the accelerometer signal was found to have an amplitude of 300 mV and a phase lead of 70° with respect to the same synchronized (with respect to both frequency and phase) strobe signal as before.
Determine the magnitude and location of the mass that must be placed at the same radius as the trial mass in order to balance the rotor, after removing the trial mass.

12.13 a. When is two-plane balancing preferred over single-plane balancing? Comment on the terms "static balancing" and "dynamic balancing."

b. A turbine rotor is supported on two bearings at the two ends. Two accelerometers are mounted on the housing of these bearings. The rotor is driven at a fixed speed and the accelerometer signals obtained. Their amplitudes and phase leads, with respect to a strobe signal that is synchronized to a fixed body reference, are found to be:

400 mV and 100° at bearing 1
700 mV and 120° at bearing 2

Vibration Design and Control

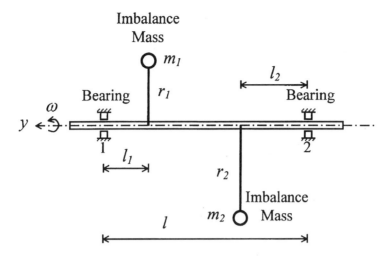

FIGURE P12.14 A dynamic balancing problem.

Next, a trial mass of 20 gm is placed at a known radius and an angular position of 90° with respect to a known body reference, in the balancing plane 1 (close to bearing 1). The new readings of the accelerometer signals are:

350 mV and 140° at bearing 1
600 mV and 130° at bearing 2

Subsequently, this trial mass is removed and a second trial mass of 25 gm is placed at a known radius and an angular position of 30° with respect to a known body reference, in the balancing plane 2 (close to bearing 2). The resulting readings of the accelerometers are:

10 mV and 150° at bearing 1
750 mV and 170° at bearing 2

Determine the magnitudes and locations of the balancing masses that should be placed on planes 1 and 2, at the same radii as the trial masses placed on these planes, after removing the trial masses.

12.14 a. Give four causes of dynamic imbalance in rotating machinery.
   b. What is static balancing and what is dynamic balancing? Assuming that any location/plane of a rotating machine is available for placing a balancing mass, is it possible to completely balance the machine using the static balancing (single-plane) method? Fully justify your answer.
   c. Consider a completely balanced rigid shaft that is supported horizontally on two bearings at distance l apart. A point mass m_1 is attached to the shaft at a distance l_1 from the left bearing using a light, rigid radial arm of length r_1 measured from the rotating axis of the shaft. Similarly, a point mass m_2 is attached to the shaft at a distance l_2 from the right bearing using a light, radial arm of length r_2. The mass m_2, however, is placed in a radially opposite configuration with respect to m_1, as shown in Figure P12.14. The two masses securely rotate with the shaft as a single rigid body, without any deformation. Note that when m_1 is vertically above the shaft axis, m_2 will be vertically below the axis. The angular speed of rotation of the shaft is ω. Suppose that at time $t = 0$, the mass m_1 is vertically above the shaft axis.

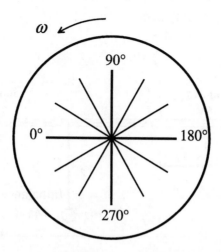

FIGURE P12.15 Angular coordinates for locating the masses in a single-plane balancing problem.

  i. Giving all the necessary steps, derive expressions for the horizontal (x-axis) and vertical (z-axis) components of the reactions on the shaft at the left bearing (1) and the right bearing (2). Specifically, obtain R_{x1}, R_{z1}, R_{x2}, and R_{z2} in terms of the given parameters (m_1, m_2, r_1, r_2, l_1, l_2, and l) of the system, for a general time instant t.
  ii. Using the planes at the two bearings as the balancing planes, determine the magnitude and orientation of the balancing masses m_{b1} and m_{b2} that should be placed on these planes at a specified radius r, in order to dynamically balance the system. Give the orientation of the balancing masses with respect to the orientation of the mass m_1.
  iii. If $m_1 = 1$ kg, $m_2 = 2$ kg, $l = 1.0$ m, $l_1 = 0.2$ m, $l_2 = 0.3$ m, $r_1 = 0.1$ m, $r_2 = 0.2$ m, and $r = 0.1$ m, compute the balancing masses and their orientations.
  d. Would it be possible to balance this problem by the single-plane (static) method? Explain your answer. Also, consider the special case where $m_1 = m_2 = m$ and $r_1 = r_2 = r$.

12.15 Consider the problem of single-plane balancing. It should be clear that the angular position for locating the trial and balancing masses should be measured in the same direction as the angular velocity of the rotating disk, in using the equation

$$\vec{M}_b = \frac{\vec{V}_u}{\left(\vec{V}_r - \vec{V}_u\right)} \vec{M}_t$$

where the phase angles of the accelerometer signals $\vec{V}_u$ and $\vec{V}_r$ are taken as phase leads in the usual notation.

  a. In a laboratory-experimental setup, the disk was found to be graduated as shown in Figure P12.15, while the angular speed ω was in the indicated direction (counterclockwise). What interpretations must be made on the experimental data in this case, in using the above equation for computing the balancing mass?
  b. With an experimental setup of the above type, the following data were obtained: Without a trial mass, the amplitude and the phase lead (with respect to the strobe signal) of the accelerometer signal were 36.1 mV and 209.3°.

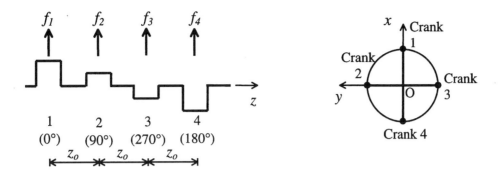

FIGURE P12.17 A possible crank arrangement of a four-cylinder engine.

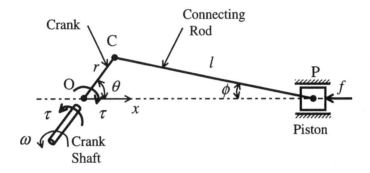

FIGURE P12.18 A single-cylinder model used in load analysis.

With a trial mass of 10.4 gm placed at the location of 130°, the amplitude and the phase lead of the accelerometer signal were 38.7 mV and 247.5°. Determine the magnitude and the location of the balancing mass, with the trial mass removed.

12.16 Although it is possible to accurately compute the magnitude and the location of the necessary balancing mass for a rotating component, in practice it may not be possible to achieve a perfect balance, particularly in a single trial. Give reasons for this situation.

12.17 Examine the statement "the most difficult part of balancing the inertial loading of a reciprocating engine is the removal of effects due to the equivalent reciprocating mass."

Consider a four-cylinder engine where the engines are placed in parallel (in-line) and equally spaced (z_0) with their cranks phased at the angles 0°, 90°, 270°, and 180°, in sequence, with respect to a rotating reference. Show that, for this engine, the inertial loading on the crankshaft, due to the reciprocating masses, has the following characteristics:

i. Primary components (frequency ω) of the forces are balanced.
ii. Primary components of the bending moments are not balanced.
iii. Secondary components (frequency 2ω) of the forces are balanced.
iv. Secondary components of the bending moments are balanced.

Here, ω is the angular speed of the crankshaft. The crank arrangement is shown in Figure P12.17.

12.18 Clearly justify the assumption of massless crank and connecting rod in the balancing analysis of a reciprocating engine. Also, justify the assumption that the resultant end force at each end of a connecting rod acts along the length of the rod.

Suppose that a force f acts on the piston of a single-cylinder engine, as shown in Figure P12.18. Note that f can represent either the inertia force of the equivalent reciprocating mass, or the force due to gas pressure in the cylinder. As a result, a torque τ is

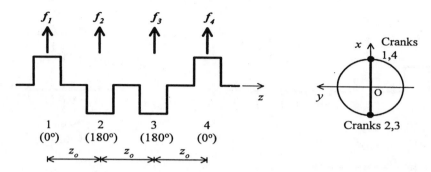

FIGURE P12.19 An alternative crank configuration of a four-cylinder engine.

applied on the crankshaft in the direction of its rotation, and an equal reaction torque τ is applied by the crankshaft on the crank, in the opposite direction. Use the principle of virtual work, with justification of its use, to show that $\tau = (r\cos\theta + l\cos\phi)f\tan\phi$, where r and l are the lengths of the crank and the connecting rod, respectively; θ is the inclination of the crank; and ϕ is the inclination of the connecting rod, to the line connecting the crankshaft and the piston, as shown.

12.19 In the analysis of inertial load balancing in reciprocating engines, the inertia of the connecting rod is frequently represented by two lumped masses at its two ends, joined by a massless rod. The end masses are chosen such that the vector sum of their inertia forces is equal to the inertia force of the mass of the rod assumed to be concentrated at the centroid. What are the limitations of this model?

A four-cylinder in-line engine where the cylinders are equally spaced and the cranks placed at the angles 0°, 180°, 180°, and 0°, sequentially, is schematically shown in Figure P12.19. Show that in this case the inertial loading on the crankshaft, due to the reciprocating masses, is such that
  i. primary components (with frequency ω) of the forces and bending moments are completely balanced.
  ii. secondary components (with frequency 2ω) of the forces and bending moments are not balanced.
Note that ω is the angular speed of the crankshaft.

12.20 In a multicylinder engine, the inertia force of each reciprocating mass causes on the crankshaft a lateral reaction force in the direction of reciprocation and a torque in the direction of rotation of the shaft. For proper operation of the engine, all these reaction forces, bending moments, and torques should be balanced. Similarly, the piston force due to gas pressure in the cylinder causes a reaction force, a bending moment, and a torque in the crankshaft. It is desirable to balance the reaction forces and bending moments, but not the torques, in this case.
Explain why:
  i. Balancing of the gas-pressure load is much more difficult than that of the inertial load
  ii. Torque on the crank shaft due to the gas-pressure load should not be balanced.
Consider a six-cylinder in-line engine with the crank orientations 0°, 120°, 240°, 240°, 120°, and 0°, in sequence. Check whether the torques generated on the crankshaft due to the inertia forces of the reciprocating masses are balanced.

12.21 Consider a light shaft that is supported by bearings at its ends and carries a rotor in its mid-span. The shaft is driven at an angular speed of ω. The magnitude of the shaft deflection at the rotor, in whirling motion, at steady state is given by

Vibration Design and Control

$$r = \frac{e\omega^2}{\left[\left(\omega_n^2 - \omega^2\right)^2 + \left(2\zeta\omega_n\omega^2\right)\right]^{1/2}}$$

where

e = eccentricity of the rotor centroid from the axis of rotation of the shaft
ω_n = undamped natural frequency of bending vibration of the shaft-rotor system
ζ = equivalent damping ratio for whirling motion.

Show that the peak value of r occurs at a shaft speed of

$$\omega = \frac{\omega_n}{\sqrt{1 - 2\zeta^2}}$$

12.22 A light shaft that is supported on two bearings carries a fly wheel at its mid-span. The centroid of the flywheel has an eccentricity e with respect to the axis of rotation of the shaft. The effective damping ratio in whirling motion of the shaft is ζ. The bending stiffness of the shaft at its midspan is k. What is the reaction at each bearing when the shaft system rotates at its critical speed?

12.23 An experimental procedure for determining the equivalent damping ratio that is provided by a pair of bearings on a shaft in whirling motion is as follows. A radial arm with a lumped mass of 1.0 kg is rigidly attached at the mid-span so that the eccentricity of the mass from the axis of rotation of the shaft is 10.0 cm. The shaft is driven at the normal operating speed of 2400 rpm, and the average reaction at the two bearings is measured using load cells. It was found to be 4.56×10^3 N. Using strobe lighting with manually adjustable frequency, directed axially from one end of the shaft, the mid-span deflection of the shaft is measured approximately using a background scale, while the strobe frequency is synchronized with the operating speed of the shaft. This reading was found to be 4.7 cm. Also, using similar means, the angle between the radial arm and the direction of bending (bowing) of the shaft was measured approximately. This was found to be 20°. The bending stiffness of the shaft at mid-span was measured while the system was stationary, by applying a known load and measuring the deflection. It was determined to be $k = 2.0 \times 10^5$ N·m^{-1}. Estimate the damping ratio.

12.24 A turbine rotor has a mass of 50 kg and is supported on a light shaft with end bearings. The bending stiffness of the shaft at the rotor location is 3.0×10^6 N·m^{-1}. The centroid of the rotor has an eccentricity of 2.0 cm from the axis of rotation of the shaft. The normal operating speed of the turbine is 3600 rpm. The equivalent damping ratio of the system in whirling is 0.1.
   a. What is the critical speed of rotation of the system?
   b. What is the shaft deflection at the rotor during normal operation?
   c. A mass of 1 kg is added to the rotor in order to achieve a better balance. By what factor should the eccentricity be reduced by this means in order to reduce the shaft deflection by a factor of 10 during normal operation?

12.25 A student proceeds to determine the growth of the deflection of a whirling shaft that carries a rotor in its mid-span and is supported on bearings at its ends as follows:
   1. Assume synchronous whirl (steady whirling rotation at the same speed as the shaft spin) so that $\dot\theta = \omega$ and $\theta = \omega t - \phi$, where ϕ is the phase lag between the whirl and the spin.
   2. The equations of motion [see equations (12.77) and (12.78)] become

$$\ddot{r} + 2\zeta_v \omega_n \dot{r} + (\omega_n^2 - \omega^2)r = e\omega^2 \cos\phi \qquad (i)$$

$$\dot{r} + \zeta\omega_n r = \frac{1}{2} e\omega \sin\phi \qquad (ii)$$

Solve equation (*ii*) with $r = 0$ at $t = 0$ to get

$$r = \frac{1}{2}\frac{e\omega}{\zeta\omega_n}\sin\phi \left[1 - e^{-\zeta\omega_n t}\right] \qquad (iii)$$

3. Substitute equation (*iii*) in (*i*). Set the overall coefficient of $e^{-\zeta\omega_n t}$ to 0, as $e^{-\zeta\omega_n t} \neq 0$ in general. Then set the rest of the terms to 0. In this manner, obtain ϕ and hence the time variation of r.

Do you agree with this approach? If so, provide justification. If not, give reasons why the approach might fail.

12.26 Explain how experimental modal analysis can be used in the design of a mechanical system for proper performance under vibration. What are limitations of its use? In substructuring, if the linkages of the subsystem have inertial elements that are not negligible, what additional issues should be addressed in a vibration design procedure?

12.27 Consider two single-degree-of freedom subsystems (k_1, m_1) and (k_2, m_2) that are interconnected by a spring element of stiffness k_c, as shown in Figure 12.22. Use $\omega_0 = \sqrt{k_1/m_1}$ in nondimensionalizing the frequencies according to $r_i = \omega_i/\omega_0$. Suppose that $k_2/k_1 = 7.0$ and $m_2/m_1 = 1.0$. Design the interconnection element k_c so that the two natural frequencies satisfy the condition $r_i^2 \geq 2.0$ for $i = 1$ and 2.

12.28 Determine an expression for the separation interval between the two dominant resonant frequencies of a vibrating (primary) system once a vibration absorber is added. Show that this expression, with respect to the original resonance of the primary system, can be expressed as

$$r_2 - r_1 = \left[\alpha^2(1+\mu) + 1 - 2\alpha\right]^{1/2}$$

where $r_i = \omega_i/\omega_p$, $\alpha = \omega_a/\omega_p$, and $\mu = m_a/m_p$, and

ω_i = a newly created resonant frequency
ω_p = original resonant frequency of the primary system
ω_a = resonant frequency of the vibration absorber
m_p = mass of the primary system
m_a = mass of the vibration absorber.

For a system with $\alpha = 1.0$, $\mu = 0.1$, and $\omega_p = 120\pi$ rad·s^{-1}, compute the frequency interval of the two resonances.

12.29 An induction motor weighing 10 kg is mounted on a relatively light structure and is used to drive a conveyor at a steady speed. A schematic diagram is given in Figure P12.29. The normal operating speed of the motor is 2400 rpm, as required for driving the conveyor. When the motor speed was slowly increased, a significant vertical resonance was found at 3000 rpm. With the intention of mitigating this problem and further reducing its spill-over effect at the normal operating speed, a technician installs a vibration absorber

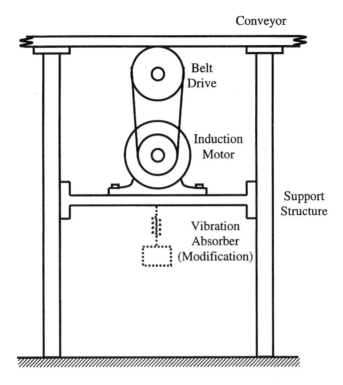

FIGURE P12.29 A conveyor motor.

on the support frame of the motor. The absorber is tuned to 2700 rpm (a value in between the operating speed and the original resonance). The absorber mass is 1.0 kg. Determine whether the addition of the vibration absorber mitigates the problem. In particular, answer the following questions:

a. What are the main resonant frequencies of the modified system when the vibration absorber is added?
b. What is the effective speed range of operation of the modified system?
c. What was the magnitude of vibration amplification of the original system at the operating speed, and what is it after the modification?

Neglect damping in this analysis.

12.30 An undamped vibration absorber generates two resonances for which separation is equal to β times the tuned frequency (resonant frequency) of the absorber. Obtain an expression for the fractional mass μ of the absorber in terms of β and the nondimensional resonant frequency α of the absorber (with respect to the primary frequency).

A machine of mass 100 kg has a significant resonance at 2400 rpm. The normal operating speed is 2200 rpm. Design an undamped vibration absorber, tuned to the operating speed, such that the two generated resonances are at least 20% apart with respect to the operating frequency.

12.31 The tubes of a steam generator in a nuclear power plant facility exhibited significant wear and tear due to vibration. Vibration monitoring and signal analysis showed that under normal operation, the tube vibration was narrow-band, and limited to a very small interval near 30 Hz. Furthermore, vibration testing indicated that the primary significant resonance of the steam generator occurs at 32 Hz. The mass of the steam generator is 50 kg. Design a damped vibration absorber for the system. Check the magnitude of the

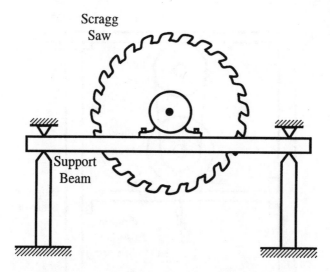

FIGURE P12.34 A wood-cutting saw mounted on a beam.

operating vibration in the modified system and compare it with the performance before modification.

12.32 Compare and contrast a simple linear damper and a dynamic absorber as vibration control devices.

The exhaust fan of a pulp and paper mill operates at 450 rpm. At the time of installation of the fan on its support structure, a static deflection of 1.0 cm was experienced by the structure. During normal operation, the amplitude of vertical vibrations of the fan was found to be 3.0 mm. During start-up and shut-down of the fan, it exhibited a vertical resonance with vibration amplitude 2.0 cm. Estimate the damping ratio of the fan-support-structure system.

12.33 Using the damped simple oscillator model of a system, justify that at low frequencies of excitation, system dynamics are primarily determined by its stiffness characteristics; and at high frequencies of excitation it is primarily determined by its inertia characteristics. Furthermore, justify that near the resonance, it is the damping that primarily determines the dynamic characteristics of the system.

It is also known that a dynamic absorber can be quite effective in vibration control in the neighborhood of a system resonance. But, typically, a dynamic absorber is a low-damping device that can function properly even without any damping — at least in theory. Is this a contradiction in view of the previous observation about the dominance of damping in determining dynamics near a resonance?

12.34 a. In comparison with lumped-parameter systems, what are some of the difficulties that arise in the vibration control of a distributed-parameter (i.e., continuous) system?

Also, give several reasons for considering active control to be more difficult than passive control in the vibration reduction of distributed-parameter systems.

b. A scragg saw commonly used in cutting relatively small diameter logs of wood is firmly supported at the mid-span of a beam of length $l = 2$ m, as schematically shown in Figure P12.34. The saw weighs 20 kg and normally operates at a steady speed of 600 rpm. The beam is simply supported at its ends and has the following parameter values:

I (2nd moment of area of cross section about the neutral axis) $= 1.0 \times 10^{-7}$ m^4
ρA (mass per unit length) $= 5.0$ kg·m^{-1}
E (Young's modulus) $= 2.0 \times 10^{11}$ N·m^{-2}

(*Note*: 1 N·m^{-2} = 1 Pa = 1 × 10^{-3} kPa)

Estimate the fundamental natural frequency of the overall system (the saw and the supporting beam) by first determining the equivalent mass of the beam at the mid-span in the first mode of vibration. Design an optimal (and damped) vibration absorber, to be tuned to the normal operating speed of the saw and mounted at the mid-span of the supporting beam.

Note: The equivalent stiffness (force/displacement) at the mid-span of a simply supported beam is given by $\dfrac{48\,EI}{l^3}$.

12.35 The shock absorbers of an automobile are primarily damping devices, with springs provided by the suspension system. Explain why conventional dynamic absorbers are not suitable in this application.

Consider the use of dynamic absorbers to control vibration of a beam. Formulate a state-space model for this problem, as shown in Figure P12.35. Specifically, determine the model for the case of a single dynamic absorber. Note that $x = l_j$ is the location of the jth absorber along the beam, and s_j is the displacement of the mass of this absorber. Also, m_j, k_j, and b_j are the mass, stiffness, and damping constant, respectively, of the jth absorber.

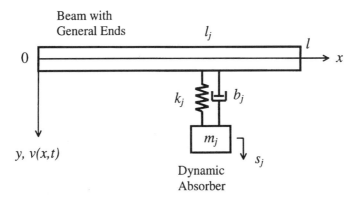

FIGURE P12.35 Use of dynamic absorbers for vibration control of a beam.

Appendix A
Dynamic Models and Analogies

A system may consist of a set of interacted components or elements, each possessing an input-output (or cause-effect, or causal) relationship. A dynamic system is one whose response variables are functions of time, with non-negligible rates of changes. A more formal mathematical definition can be given, but it is adequate here to understand that vibrating systems are dynamic systems. A model is some form of representation of a practical system. An analytical model (or mathematical model) comprises a set of equations, or an equivalent, that approximately represents the system. Sometimes, a set of curves, digital data (table) stored in a computer, and other numerical data — rather than a set of equations — can be termed an analytical model if such data represent the system of interest. A model developed by applying a suitable excitation to a system and measuring the resulting response of the system is called an experimental model. In general, then, models can be grouped into the following categories:

1. Physical models (prototypes)
2. Analytical models
3. Computer (numerical) models
4. Experimental models (using input/output experimental data)

Mathematical definitions for a dynamic system are given with reference to an analytical model of the system; for example, a *state-space model*. In that context, the system and its analytical model are synonymous. In reality, however, an analytical model, or any model for that matter, is an idealization of the actual system. Analytical properties that are established and results that are derived would be associated with the model rather than the actual system, whereas the excitations are applied to and the output responses are measured from the actual system. This distinction should be clearly recognized.

Analytical models are very useful in predicting the dynamic behavior (response) of a system when it is subjected to a certain excitation (input). Vibration is a dynamic phenomenon and its analysis, practical utilization, and effective control require a good understanding of the vibrating system. A recommended way to achieve this is through the use of a suitable model of the system. A model can be employed for designing a mechanical system for proper vibration performance. In the context of vibration testing, for example, analytical models are commonly used to develop test specifications, and also the input signal applied to the shaker, and to study dynamic effects and interactions in the test object, the shaker table, and their interfaces. In product qualification by analysis, a suitable analytical model of the product replaces the test specimen. In vibration control, a dynamic model of the vibrating system can be employed to develop the necessary control schemes (e.g., model-based control).

Analytical models can be developed for mechanical, electrical, fluid, and thermal systems in a rather analogous manner because some clear analogies are present among these four types of systems. In practice, a dynamic system can exist as a combination of two or more of the these various types, and is termed a *mixed system*. In view of the analogy, then, a unified approach can be adopted in analysis, design, and control of these different types of systems and mixed systems. In this context, understanding the analogies in different system types is quite useful in the development and utilization of models. This appendix outlines some basic concepts of modeling. Also, fundamental analogies that exist among different types of dynamic systems are identified.

A.1 MODEL DEVELOPMENT

There are two broad categories of models for dynamic systems: lumped-parameter models and continuous-parameter models. Lumped-parameter models are more commonly employed than continuous-parameter models, but continuous-parameter elements sometimes are included in otherwise lumped-parameter models in order to improve the model accuracy.

In lumped-parameter models, various characteristics in the system are lumped into representative elements located at a discrete set of points in a geometric space. A coil spring, for example, has a mass, an elastic (spring) effect, and an energy-dissipation characteristic, each of which is distributed over the entire coil. In an analytical model, however, these individual distributed characteristics can be approximated by a separate mass element, a spring element, and a damper element, which are interconnected in some parallel-series configuration, thereby producing a lumped-parameter model.

Development of a suitable analytical model for a large and complex system requires a systematic approach. Tools are available to aid this process. Signals (excitations and response) can be represented either in the frequency domain or in the time domain. A time-domain model consists of a set of differential equations. In the frequency domain, a model is represented by a set of transfer functions (or frequency-response functions). Frequency-domain transfer function considerations also lead to the concepts of mechanical impedance, mobility, and transmissibility.

The process of modeling can be made simple by following a systematic sequence of steps. The main steps are summarized below:

1. Identify the system of interest by defining its purpose and system boundaries.
2. Identify or specify the variables of interest. These include inputs (forcing functions or excitations) and outputs (responses).
3. Approximate (or model) various segments (or processes or phenomena) in the system by ideal elements, suitably interconnected.
4. Draw a free-body diagram for the system with suitably isolated components.
5. a. Write constitutive equations (physical laws) for the elements.
   b. Write continuity (or conservation) equations for through variables (equilibrium of forces at joints; current balance at nodes, etc.).
   c. Write compatibility equations for across (potential or path) variables. These are loop equations for velocities (geometric connectivity), voltages (potential balance), etc.
   d. Eliminate auxiliary (unwanted) variables that are redundant and not needed to define the model.
6. Express system boundary conditions and response initial conditions using system variables.

These steps should be self-explanatory or integral with the particular modeling techniques.

A.2 ANALOGIES

Analogies exist among mechanical, electrical, hydraulic, and thermal systems. The basic system elements can be divided into two groups: energy-storage elements and energy-dissipation elements. Table A.1 shows the linear relationships that describe the behavior of translatory mechanical, electrical, thermal, and fluid elements. These relationships are known as constitutive relations. In particular, Newton's second law is considered the constitutive relation for the mass element. The analogy used in Table A.1 between mechanical and electrical elements is known as the force–current analogy. This analogy appears more logical than a force–voltage analogy, as is clear from Table A.2. This follows from the fact that both force and current are *through variables*, which are analogous to fluid flow through a pipe; and both velocity and voltage are *across variables*, which vary across

Appendix A Dynamic Models and Analogies

TABLE A.1
Some Linear Constitutive Relations

	Constitutive Relation for		
	Energy Storage Elements		Energy Dissipating Elements
System Type	A-Type (Across) Element	T-Type (Through) Element	D-Type (Dissipative) Element
Translatory mechanical: v = Velocity f = Force	Mass: $m\dfrac{dv}{dt} = f$ (Newton's second law) m = Mass	Spring: $\dfrac{df}{dt} = kv$ (Hooke's law) k = Stiffness	Viscous damper: $f = bv$ b = Damping constant
Electrical: v = Voltage i = Current	Capacitor: $C\dfrac{dv}{dt} = i$ C = Capacitance	Inductor: $L\dfrac{di}{dt} = v$ L = Inductance	Resistor: $Ri = v$ R = Resistance
Thermal: T = Temperature difference Q = Heat transfer rate	Thermal capacitor: $C_t\dfrac{dT}{dt} = Q$ C_t = Thermal capacitance	None	Thermal Resistor: $R_t Q = T$ R_t = Thermal resistance
Fluid: P = Pressure difference Q = Volume flow rate	Fluid capacitor: $C_f\dfrac{dP}{dt} = Q$ C_f = Fluid capacitance	Fluid inertor: $I_f\dfrac{dQ}{dt} = P$ I_f = Inertance	Fluid resistor: $R_f Q = P$ R_f = Fluid resistance

TABLE A.2
Force–Current Analogy

System Type	Mechanical	Electrical
System-response variables:		
Through variables	Force f	Current i
Across variables	Velocity v	Voltage v
System parameters	m	C
	k	$1/L$
	b	$1/R$

the flow direction, as in the case of pressure. The correspondence between the parameter pairs given in Table A.2 follows from the relations in Table A.1. Note that the rotational mechanical elements possess constitutive relations between torque and angular velocity, which can be treated as a generalized force and a generalized velocity, respectively. In fluid systems as well, basic elements corresponding to capacitance (capacity), inductance (fluid inertia), and resistance (fluid friction) exist. Constitutive relations between pressure difference and mass flow rate can be written for these elements. In thermal systems, generally only two elements — capacitance and resistance — can be identified. Constitutive relations exist between temperature difference and heat transfer rate in this case.

Proper selection of system variables is crucial in developing an analytical model for a dynamic system. A general approach that can be adopted is to use across variables of the *A*-type (or across-type) energy storage elements and the through variables of the *T*-type (or through-type) energy storage elements as system variables (state variables). Note that if any two elements are not independent (e.g., if two spring elements are directly connected in series or parallel), then only a single state variable should be used to represent both elements. Independent variables are not needed for D-type (dissipative) elements because their response can be represented in terms of the state variables of the energy storage elements (A-type and T-type).

A.3 MECHANICAL ELEMENTS

Here, one uses velocity (across variable) of each independent mass (A-type element) and force (through variable) of each independent spring (T-type element) as system variables (state variables). The corresponding constitutive equations form the "shell" of an analytical model. These equations will directly lead to a *state-space model* of the system.

A.3.1 Mass (Inertia) Element

Constitutive equation (Newton's second law):

$$m\frac{dv}{dt} = f \tag{A.1}$$

Since power $= fv$, the energy of the element is given by

$$E = \int fv dt = \int m\frac{dv}{dt} v dt = \int mv dv$$

or

$$\text{Energy } E = \frac{1}{2}mv^2 \tag{A.2}$$

This is the well-known *kinetic energy*. Now, integrating equation (A.1),

$$v(t) = v(0^-) + \frac{1}{m}\int_{0^-}^{t} f dt \tag{A.3}$$

By setting $t = 0^+$, one sees that

$$v(0^+) = v(0^-) \tag{A.4}$$

unless an infinite force f is applied to m. Hence, one can state the following:

1. Velocity can represent the state of an inertia element. This is justified first because, from equation (A.3), the velocity at any time t can be completely determined with the knowledge of the initial velocity and the applied force, and because, from equation (A.2), the energy of an inertia element can be represented in terms of v alone.
2. Velocity across an inertia element cannot change instantaneously unless an infinite force/torque is applied to it.
3. A finite force cannot cause an infinite acceleration. A finite instantaneous change (step) in velocity will need an infinite force. Hence, v is a natural output (or state) variable and f is a natural input variable for an inertia element.

A.3.2 Spring (Stiffness) Element

Constitutive equation (Hook's law):

$$\frac{df}{dt} = kv \tag{A.5}$$

Note that the conventional force-deflection Hooke's law has been differentiated in order to be consistent with the variable (velocity) that is used with the inertia element.

As before, the energy is

$$E = \int fv\,dt = \int f\frac{1}{k}\frac{df}{dt}dt = \int \frac{1}{k}f\,df$$

or

$$\text{Energy } E = \frac{1}{2}\frac{f^2}{k} \tag{A.6}$$

This is the well-known (elastic) *potential energy*.
Also,

$$f(t) = f(0^-) + k\int_{0^-}^{t} v\,dt \tag{A.7}$$

and hence,

$$f(0^+) = f(0^-) \tag{A.8}$$

unless an infinite velocity is applied to the spring element. In summary,

1. Force can represent the state of a stiffness (spring) element. This is justified because the force of a spring at any general t can be completely determined with the knowledge of the initial force and the applied velocity, and also because the energy of a spring element can be represented in terms of f alone.
2. Force through a stiffness element cannot change instantaneously unless an infinite velocity is applied to it.
3. Force f is a natural output (state) variable and v is a natural input variable for a stiffness element.

A.4 ELECTRICAL ELEMENTS

Here, one uses voltage (across variable) of each independent capacitor (*A*-type element) and current (through variable) of each independent inductor (*T*-type element) as system (state) variables.

A.4.1 Capacitor Element

Constitutive equation:

$$C\frac{dv}{dt} = i \tag{A.9}$$

Since power is iv, the energy is

$$E = \int iv\,dt = \int C\frac{dv}{dt}v\,dt = \int Cv\,dv$$

or

$$\text{Energy } E = \frac{1}{2}Cv^2 \qquad (A.10)$$

This is the *electrostatic energy* of a capacitor.
Also,

$$v(t) = v(0^-) + \frac{1}{C}\int_{0^-}^{t} i\,dt \qquad (A.11)$$

Hence, for a capacitor,

$$v(0^+) = v(0^-) \qquad (A.12)$$

unless an infinite current is applied to a capacitor. In summary,

1. Voltage is an appropriate response variable (or state variable) for a capacitor element.
2. Voltage across a capacitor cannot change instantaneously unless an infinite current is applied.
3. Voltage is a natural output variable and current is a natural input variable for a capacitor.

A.4.2 Inductor Element

Constitutive equation:

$$L\frac{di}{dt} = v \qquad (A.13)$$

$$\text{Energy } E = \frac{1}{2}Li^2 \qquad (A.14)$$

This is the *electromagnetic energy* of an inductor.
Also,

$$i(t) = i(0^-) + \frac{1}{L}\int_{0^-}^{t} v\,dt \qquad (A.15)$$

Hence, for an inductor,

$$i(0^+) = i(0^-) \qquad (A.16)$$

unless an infinite voltage is applied. In summary,

1. Current is an appropriate response variable (or state variable) for an inductor.
2. Current through an inductor cannot change instantaneously unless an infinite voltage is applied.
3. Current is a natural output variable and voltage is a natural input variable for an inductor.

A.5 THERMAL ELEMENTS

Here, the across variable is temperature (T) and the through variable is the heat transfer rate (Q). The thermal capacitor is the A-type element. There is no T-type element in a thermal system. The reason is clear. There is only one type of energy (thermal energy) in a thermal system, whereas there are two types of energy in mechanical and electrical systems.

A.5.1 THERMAL CAPACITOR

Consider a material control volume V, of density ρ, and specific heat c. Then, for a net heat transfer rate Q into the control volume, one obtains

$$Q = \rho v c \frac{dT}{dt} \tag{A.17}$$

or

$$C_t \frac{dT}{dt} = Q \tag{A.18}$$

where $C_t = \rho v c$ is the thermal capacitance of the control volume.

A.5.2 THERMAL RESISTANCE

There are three basic processes of heat transfer:

1. Conduction
2. Convection
3. Radiation.

There is a thermal resistance associated with each process, given by their constitutive relations as given below.

Conduction: $$Q = \frac{kA}{\Delta x} T \tag{A.19}$$

where

k = conductivity
A = area of cross section of the heat conduction element
Δx = length of heat conduction with a temperature drop of T.

The conductive resistance is

$$R_k = \frac{\Delta x}{kA} \tag{A.20}$$

Convection:
$$Q = h_c A T \quad (A.21)$$

where

h_c = convection heat transfer coefficient
A = area of heat convection surface with temperature drop of T.

The convective resistance is

$$R_c = \frac{1}{h_c A} \quad (A.22)$$

Radiation:
$$Q = \sigma F_E F_A A \left(T_1^4 - T_2^4\right) \quad (A.23)$$

where

σ = Stefan-Boltzmann constant
F_E = effective emmisivity of the radiation source (of temperature T_1)
F_A = shape factor of the radiation receiver (of temperature T_2)
A = effective surface area of the receiver.

This corresponds to a nonlinear thermal resistor.

A.6 FLUID ELEMENTS

Here, one uses pressure (across variable) of each independent fluid capacitor (A-type element) and volume flow rate (through variable) of each independent fluid inertor (T-type element) as system (state) variables.

A.6.1 FLUID CAPACITOR

The heat transfer rate is

$$C_f \frac{dP}{dt} = Q \quad (A.24)$$

Note that a fluid capacitor stores potential energy (a "fluid spring") unlike the mechanical A-type element (inertia), which stores kinetic energy.

For a liquid control volume V of bulk modulus β, the fluid capacitance is given by

$$C_{\text{bulk}} = \frac{V}{\beta} \quad (A.25)$$

For an isothermal (constant temperature, slow-process) gas of volume V and pressure P, the fluid capacitance is

$$C_{\text{comp}} = \frac{V}{P} \quad (A.26)$$

Appendix A Dynamic Models and Analogies

For an adiabatic (zero heat transfer, fast process) gas, the capacitance is

$$C_{\text{comp}} = \frac{V}{kP} \tag{A.27}$$

where

$$k = \frac{c_p}{c_v} \tag{A.28}$$

which is the ratio of specific heats at constant pressure and constant volume.

For an incompressible fluid in a container of flexible area A and stiffness k, the capacitance is

$$C_{\text{elastic}} = \frac{A^2}{k} \tag{A.29}$$

Note: For a fluid with bulk modulus, the equivalent capacitance would be

$$C_{\text{bulk}} + C_{\text{elastic}}$$

For an incompressible fluid column with an area of cross section A and density ρ, the capacitance is

$$C_{\text{grav}} = \frac{A}{\rho g} \tag{A.30}$$

A.6.2 Fluid Inertor

$$I_f \frac{dQ}{dt} = P \tag{A.31}$$

This represents a *T*-type element. However, it stores kinetic energy, unlike the mechanical *T*-type element (spring), which stores potential energy. For a flow with uniform velocity distribution across an area A and over a length segment Δx, the fluid inertance is given by

$$I_f = \rho \frac{\Delta x}{A} \tag{A.32}$$

For a non-uniform velocity distribution,

$$I_f = \alpha \rho \frac{\Delta x}{A} \tag{A.33}$$

where a correction factor α has been introduced. For a flow of circular cross section with a parabolic velocity distribution, use $\alpha = 2.0$.

A.6.3 Fluid Resistance

For the approximate, linear case:

$$P = R_f Q \tag{A.34}$$

The more general, nonlinear case is given by

$$P = K_R Q^n \tag{A.35}$$

where K_R and n are parameters of the nonlinearity. For viscous flow through a uniform pipe, one obtains

$$R_f = 128\mu \frac{\Delta x}{\pi d^4} \tag{A.36}$$

for a circular cross-section of diameter d, and

$$R_f = 12\mu \frac{\Delta x}{w b^3} \tag{A.37}$$

for a rectangular cross section of height b that is much smaller than its width w. Also, μ is the absolute viscosity (or dynamic viscosity) of the fluid, and is related to the kinematic viscosity υ through

$$\mu = \upsilon \rho \tag{A.38}$$

A.6.4 Natural Oscillations

Mechanical systems can produce natural oscillatory responses (or free vibrations) because they can possess two types of energy (kinetic and potential). When one type of stored energy is converted to the other type repeatedly, back and forth, the resulting response is oscillatory. Of course, some of the energy will dissipate (through the dissipative mechanism of damper) and the free natural oscillations will decay as a result. Similarly, electrical circuits and fluid systems can exhibit natural oscillatory responses due to the presence of two types of energy storage mechanism, where energy can "flow" back and forth repeatedly between the two types of elements. However, thermal systems have only one type of energy storage element (A-type) with only one type of energy (thermal energy). Hence, purely thermal systems cannot naturally produce oscillatory responses unless forced by external means, or integrated with other types of systems (e.g., fluid systems).

A.7 STATE-SPACE MODELS

More than one variable might be needed to represent the response of a dynamic system. There also could be more than one input variable. A time-domain analytical model is a set of differential equations relating the response variables to the input variables. This set of system equations is generally coupled, so that more than one response variable appears in each differential equation. A particularly useful time-domain representation for a dynamic system is a state-space model. In this representation (state-space representation), an nth-order system is represented by n first-order differential equations, which generally are coupled. This is of the general form:

Appendix A Dynamic Models and Analogies

$$\frac{dq_1}{dt} = f_1(q_1, q_2, \ldots, q_n, r_1, r_2, \ldots, r_m, t)$$

$$\frac{dq_2}{dt} = f_2(q_1, q_2, \ldots, q_n, r_1, r_2, \ldots, r_m, t) \quad (A.39)$$

$$\vdots$$

$$\frac{dq_n}{dt} = f_n(q_1, q_2, \ldots, q_n, r_1, r_2, \ldots, r_m, t)$$

The n state variables can be expressed as the state vector

$$q = [q_1, q_2, \ldots, q_n]^T \quad (A.40)$$

which is a column vector. Note that $[\ ]^T$ denotes the transpose of a matrix or vector. The space formed by all possible state vectors of a system is the state space. At this stage, one may wish to review the concepts of linear algebra given in Appendix C.

The state vector of a dynamic system is a least set of variables that is required to completely determine the state of the system at all instants of time. They may or may not have a physical interpretation. The state vector is not unique; many choices are possible for a given system. Output (response) variables of a system can be completely determined from any such choice of state variables. Since the state vector is a least set, a given state variable cannot be expressed as a linear combination of the remaining state variables in that state vector. One suitable choice of state variables is the across variables of the independent A-type energy-storage elements and through variables of the independent T-type energy-storage elements.

The m variables $r_1, r_2, \ldots, r_m$ in equations (A.39) are input variables, and they can be expressed as the input vector:

$$r = [r_1, r_2, \ldots, r_m]^T \quad (A.41)$$

Now, equation (A.39) can be written in the vector notation

$$\dot{q} = f(q, r, t) \quad (A.42)$$

When t is not present explicitly in the function f, the system is said to be *autonomous*.

A.7.1 LINEARIZATION

Equilibrium states of the dynamic system given by equation (A.42), correspond to

$$\dot{q} = 0 \quad (A.43)$$

Consequently, the equilibrium states $\bar{q}$ are obtained by solving the set of n algebraic equations

$$f(q, r, t) = 0 \quad (A.44)$$

for a special steady input $\bar{r}$. Usually, a system operates in the neighborhood of one of its equilibrium states. This state is known as its *operating point*. The steady state of a dynamic system is also an equilibrium state.

Suppose that a slight excitation is given to a dynamic system that is operating at an equilibrium state. If the system response builds up and deviates further from the equilibrium state, the equilibrium state is said to be *unstable*. If the system returns to the original operating point, the equilibrium state is *stable*. If it remains at the new state without either returning to the equilibrium state or building up the response, the equilibrium state is said to be *neutral*.

To study the stability of various equilibrium states of a nonlinear dynamic system, it is first necessary to linearize the system model about these equilibrium states. Linear models are also useful in analyzing nonlinear systems when it is known that the variations of the system response about the system operating point are small in comparison to the maximum allowable variation (dynamic range). Equation (A.42) can be linearized for small variations δq and δr about an equilibrium point $(\bar{q}, \bar{r})$ by employing up to only the first derivative term in the Taylor series expansion of the nonlinear function f. The higher-order terms are negligible for small δq and δr. This method yields

$$\delta \dot{q} = \frac{\partial f}{\partial q}(\bar{q},\bar{r},t)\delta q + \frac{\partial f}{\partial r}(\bar{q},\bar{r},t)\delta r \qquad (A.45)$$

The state vector and input vector for the linearized system are denoted by

$$\delta q = x = [x_1, x_2, \ldots, x_n]^T \qquad (A.46)$$

$$\delta r = u = [u_1, u_2, \ldots, u_m]^T \qquad (A.47)$$

The linear *system matrix* $A(t)$ and the *input gain matrix* $B(t)$ are defined as

$$A(t) = \frac{\partial f}{\partial q}(\bar{q},\bar{r},t) \qquad (A.48)$$

$$B(t) = \frac{\partial f}{\partial r}(\bar{q},\bar{r},t) \qquad (A.49)$$

Then, the linear state model can be expressed as

$$\dot{x} = Ax + Bu \qquad (A.50)$$

If the dynamic system is a constant-parameter system, or if it can be assumed as such for the time period of interest, then A and B become constant matrices.

A.7.2 Time Response

Time variation of the state vector of a linear, constant-parameter dynamic system can be obtained using the Laplace transform method. The Laplace transform of equation (A.50) is given by

$$sX(s) - x(0) = AX(s) + BU(s) \qquad (A.51)$$

Consequently,

$$x(t) = \mathcal{L}^{-1}(sI - A)^{-1}x(0) + \mathcal{L}^{-1}(sI - A)^{-1}BU(s) \qquad (A.52)$$

Appendix A Dynamic Models and Analogies

in which I denotes the identity (unit) matrix. Note that $\mathcal{L}^{-1}$ denotes the inverse Laplace transform operator. The square matrix $(sI - A)^{-1}$ is known as the *resolvent matrix*. Its inverse Laplace transform is the *state-transition matrix*:

$$\Phi(t) = \mathcal{L}^{-1}(sI - A)^{-1} \tag{A.53}$$

It can be shown that $\Phi(t)$ is equal to the matrix exponential

$$\Phi(t) = \exp(At) = I + At + \frac{1}{2!}A^2 t^2 + \ldots \tag{A.54}$$

The state-transition matrix can be analytically determined as a closed-form matrix function by the direct use of inverse transformation on each term of the resolvent matrix, using equation (A.53), or as a series solution using equation (A.54). One can reduce the infinite series given in equation (A.54) into a finite matrix polynomial of order $n - 1$ using the Cayley-Hamilton theorem. This theorem states that a matrix satisfies its own characteristic equation. The characteristic polynomial of A can be expressed as

$$\begin{aligned}\Delta(\lambda) &= \det(A - \lambda I) \\ &= a_n \lambda^n + a_{n-1} \lambda^{n-1} + \cdots + a_0\end{aligned} \tag{A.55}$$

in which det( ) denotes determinant. The notation

$$\Delta(A) = a_n A^n + a_{n-1} A^{n-1} + \cdots + a_0 I \tag{A.56}$$

is used. Then, by the Cayley-Hamilton equation, one obtains

$$\mathbf{0} = a_n A^n + a_{n-1} A^{n-1} + \cdots + a_0 I \tag{A.57}$$

To get a polynomial expansion for $\exp(At)$, one can write

$$\exp(At) = S(A) \cdot \Delta(A) + \alpha_{n-1} A^{n-1} + \alpha_{n-2} A^{n-2} + \cdots + \alpha_0 I \tag{A.58}$$

in which $S(A)$ is an appropriate infinite series. Since $\Delta(A) = \mathbf{0}$ by the Cayley-Hamilton theorem, however, one has

$$\exp(At) = \alpha_{n-1} A^{n-1} + \alpha_{n-2} A^{n-2} + \cdots + \alpha_0 I \tag{A.59}$$

Now, it is just a matter of determining the coefficients $\alpha_0, \alpha_1, \ldots, \alpha_{n-1}$, which are functions of time. This is done as follows. From equation (A.58),

$$\exp(\lambda t) = S(\lambda) \cdot \Delta(\lambda) + \alpha_{n-1} \lambda^{n-1} + \alpha_{n-2} \lambda^{n-2} + \cdots + \alpha_0 \tag{A.60}$$

If $\lambda_1, \lambda_2, \ldots, \lambda_n$ are the eigenvalues of A, however, then, by definition,

$$\Delta(\lambda_i) = \det(A - \lambda_i I) = 0 \quad \text{for } i = 1, 2, \ldots, n \tag{A.61}$$

Thus, from equation (A.60), one obtains

$$\exp(\lambda_i t) = \alpha_{n-1}\lambda_i^{n-1} + \alpha_{n-2}\lambda_i^{n-2} + \cdots + \alpha_0 \qquad \text{for } i = 1, 2, \ldots, n \qquad (A.62)$$

If the eigenvalues are all distinct, equation (A.62) represents a set of n independent algebraic equations from which the n unknowns $\alpha_0, \alpha_1, \ldots, \alpha_{n-1}$ can be determined.

Because the product in the Laplace domain is a convolution integral in the time domain, and vice versa, the second term on the right-hand side of equation (A.52) can be expressed as a matrix convolution integral. This gives

$$x(t) = \Phi(t)x(0) + \int_0^t \Phi(t-\tau)Bu(\tau)d\tau \qquad (A.63)$$

The first part of this solution is the *zero-input response*; the second part is the *zero-state response*.

State variables are not necessarily measurable and generally are not system outputs. A linearized relationship between state variables and system output (response) variables $y(t)$ can be expressed as

$$y(t) = Cx(t) \qquad (A.64)$$

in which the output vector is

$$y = \begin{bmatrix} y_1, y_2, \ldots, y_p \end{bmatrix}^T \qquad (A.65)$$

and C denotes the output (measurement) gain matrix. When $m > 1$ and $p > 1$, the system is said to be a multi-input–multi-output (MIMO) system. Note that, in this case, one has a transfer matrix $H(s)$ given by

$$H(s) = C(sI - A)^{-1}B \qquad (A.66)$$

which satisfies

$$Y(s) = H(s)U(s) \qquad (A.67)$$

Since

$$(sI - A)^{-1} = \frac{\text{adj}(sI - A)}{\det(sI - A)} \qquad (A.68)$$

in which adj( ) denotes the adjoint (see Appendix C). It is seen that the poles, or eigenvalues, of the system (matrix A) are given by the solution of its characteristic equation:

$$\det(sI - A) = 0$$

which should be compared with equation (A.61). If all eigenvalues of A have negative real parts, then the state-transition matrix $\Phi(t)$ in equation (A.63) will be bounded as $t \to \infty$, which means that the linear system is stable.

A.7.3 SOME FORMAL DEFINITIONS

A *state vector* x is a column vector that contains a minimum set of *state variables* $x_1, x_2, \ldots, x_n$ that completely determines the state of a dynamic system. The number of state variables, n, is the *order* of the system. This is typically equal to the number of independent energy storage elements in the system, and is twice the number of degrees of freedom (for a mechanical system).

Property 1: If the state vector $x(t_0)$ at the time t_0 and the input (forcing excitation) $u[t_0, t_1]$ over the time interval $[t_0, t_1]$ are known, where t_1 is any future time, then $x(t_1)$ can be uniquely determined. In other words, a transformation g can be defined such that

$$x(t_1) = g(t_0, t_1, x(t_0), u[t_0, t_1]) \tag{A.69}$$

By this property it should be clear that the order of a dynamic system is equal to the number of independent initial conditions needed to completely determine the system response.

Note that according to the *causality* of dynamic systems, future states cannot be determined unless the inputs up to that future time are known. This means that the transformation g is *nonanticipative*.

Each forcing function $u[t_0, t_1]$ defines a *state trajectory*. The *n-dimensional vector space* formed by all possible state trajectories is known as the *state space*.

Property 2: If the state $x(t_1)$ and the input $u(t_1)$ are known at any time t_1, the system response (output) vector $y(t_1)$ can be uniquely determined. This can be expressed as

$$y(t_1) = h(t_1, x(t_1), u(t_1)) \tag{A.70}$$

Note that the transformation h has no *memory* in the sense that the response at a previous time cannot be determined through the knowledge of the present state and input. Note also that, in general, system outputs are not identical to the states although the former can be uniquely determined by the latter.

A.7.4 ILLUSTRATIVE EXAMPLE

A torsional dynamic model of a pipeline segment is shown in Figure A.1(a). Free-body diagrams in Figure A.1(b) show internal torques acting at sectioned inertia junctions for free motion. A state model is obtained using the generalized velocities (angular velocities Ω_i) of the independent inertia elements and the generalized forces (torques T_i) of the independent elastic (torsional spring) elements as state variables. The minimum set of states that is required for a complete representation determines the system order. There are two inertia elements and three spring elements — a total of five energy-storage elements. The three springs are not independent, however. The motion of two springs completely determines the motion of the third. This indicates that the system is a fourth-order system. One obtains the model as follows.

Newton's second law gives

$$I_1 \dot{\Omega}_1 = -T_1 + T_2$$

$$I_2 \dot{\Omega}_2 = -T_2 - T_3$$

Hooke's law gives

$$\dot{T}_1 = k_1 \Omega_1$$

$$\dot{T}_2 = k_2 (\Omega_2 - \Omega_1)$$

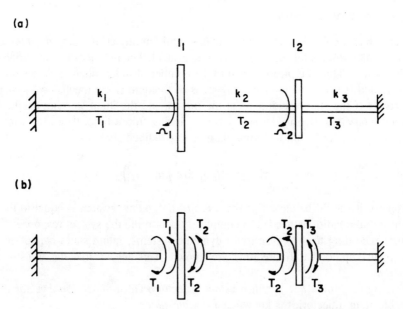

FIGURE A.1 (a) Dynamic model of a pipeline segment, and (b) free-body diagrams.

Torque T_3 is determined in terms of T_1 and T_2, using the displacement relation (compatibility) for the inertia I_2:

$$\frac{T_1}{k_1} + \frac{T_2}{k_2} = \frac{T_3}{k_3}$$

The state vector is chosen as

$$x = [\Omega_1, \Omega_2, T_1, T_2]^T$$

The corresponding system matrix is

$$A = \begin{bmatrix} 0 & 0 & -\dfrac{1}{I_1} & \dfrac{1}{I_1} \\ 0 & 0 & -\dfrac{1}{I_2}\left(\dfrac{k_3}{k_1}\right) & -\dfrac{1}{I_2}\left(1+\dfrac{k_3}{k_2}\right) \\ k_1 & 0 & 0 & 0 \\ -k_2 & k_1 & 0 & 0 \end{bmatrix}$$

The output-displacement vector is

$$y = \left[\frac{T_1}{k_1}, \frac{T_1}{k_1} + \frac{T_2}{k_2}\right]^T$$

which corresponds to the following output-gain matrix:

$$C = \begin{bmatrix} 0 & 0 & \dfrac{1}{k_1} & 0 \\ 0 & 0 & \dfrac{1}{k_1} & \dfrac{1}{k_2} \end{bmatrix}$$

For the special case given by $I_1 = I_2 = I$ and $k_1 = k_3 = k$, the system eigenvalues are

$$\lambda_1, \overline{\lambda}_1 = \pm j\omega_1 = \pm j\sqrt{\dfrac{k}{I}}$$

$$\lambda_2, \overline{\lambda}_2 = \pm j\omega_2 = \pm j\sqrt{\dfrac{k + 2k_2}{I}}$$

and the corresponding eigenvectors are

$$X_1, \overline{X}_1 = R_1 \pm jI_1 = \dfrac{\alpha_1}{2}[\omega_1,\ \omega_1,\ \mp jk_1,\ 0]^T$$

$$X_2, \overline{X}_2 = R_2 \pm jI_2 = \dfrac{\alpha_2}{2}[\omega_2,\ -\omega_2,\ \mp jk_1,\ \pm 2jk_2]^T$$

The modal contributions to the displacement vector are

$$Y_1 = \begin{bmatrix} 1 \\ 1 \end{bmatrix} \alpha_1 \sin \omega_1 t$$

and

$$Y_2 = \begin{bmatrix} 1 \\ -1 \end{bmatrix} \alpha_2 \sin \omega_2 t$$

The mode shapes are given by the vectors $S_1 = [1, 1]^T$ and $S_2 = [1, -1]^T$, which are sketched in Figure A.2. In general, each modal contribution introduces two unknown parameters, α_i and ϕ_i, into the free response (homogeneous solution), where ϕ_i are the phase angles associated with the sinusoidal terms. For an n-degree-of-freedom (order-$2n$) system, this results in $2n$ unknowns, which require the $2n$ initial conditions $x(0)$. Further developments of modal analysis for lumped-parameter systems are found in Chapter 5.

A.7.5 Causality and Physical Realizability

Consider a dynamic system represented by the single input-output differential equation:

$$\dfrac{d^n y}{dt^n} + a_{n-1} \dfrac{d^{n-1} y}{dt^{n-1}} + \ldots + a_0 y = b_m \dfrac{d^m u}{dt^m} + b_{m-1} \dfrac{d^{m-1} u}{dt^{m-1}} + \ldots + b_0 u \qquad (A.71)$$

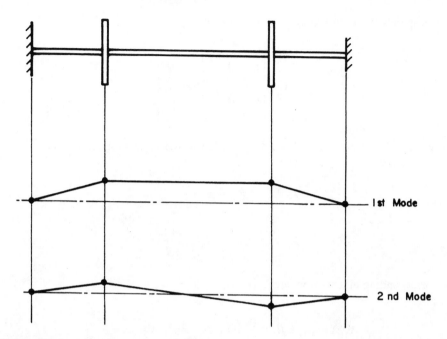

FIGURE A.2 Mode shapes of the pipeline segment.

where y and u are the output and the input, respectively. The causality (cause-effect) of this system should dictate that u is the input and y is the output. Its transfer function is given by

$$G(s) = \frac{N(s)}{\Delta(s)} = \frac{b_m s^m + b_{m-1} s^{m-1} + \ldots + b_0}{s^n + a_{n-1} s^{n-1} + \ldots + a_0} \quad (A.72)$$

Note that n is the *order*, $\Delta(s)$ is the characteristic polynomial, and $N(s)$ is the numerator polynomial of the system.

Suppose that $m > n$. Then, if one integrates equation (A.71) n times, one obtains y and its integrals on the LHS, but the RHS will contain at least one derivative of u. Since the derivative of a step function is an impulse, this implies that a finite change in input will result in an infinite change in the response. This is not physically realizable. It follows that a physically realizable system cannot have a numerator order greater than the denominator order in its transfer function. If, in fact, $m > n$, then what this means physically is that y should be the system input and u should be the system output. In other words, the causality should be reversed in this case. Furthermore, for a physically realizable system, a simulation block diagram can be established using integrals $\left(\dfrac{1}{s}\right)$ alone, without the need for derivatives (s). Note that pure derivatives are physically not realizable. If $m > n$, the simulation block diagram will need at least one derivative for linking u to y. That will not be physically realizable, again, because it would imply the possibility of producing an infinite response by a finite input. In other words, feed-forward paths with pure derivatives will not be needed in a simulation block diagram of a physically realizable system.

Appendix B
Newtonian and Lagrangian Mechanics

A vibrating system can be interpreted as a collection of mass particles. In the case of distributed systems, the number of particles is infinite. The flexibility and damping effects can be introduced as forces acting on these particles. It follows that Newton's second law for a mass particle forms the basis of describing vibratory motions.

System equations can be obtained directly by applying Newton's second law to each particle. It is convenient, however, to use Lagrange's equations for this purpose, particularly when the system is relatively complex. A variational principle known as Hamilton's principle, which can be established from Newton's second law, is the starting point in the derivation of Lagrange's equations.

This appendix outlines some useful results of dynamics, in both Newtonian and Lagrangian approaches. The Newtonian approach uses forces, torques, and motions, which are vectors. Hence, it is important to deal with *vector mechanics* in the Newtonian approach. The Lagrangian approach is based on energy, which is a scalar quantity. Scalar energies can be expressed in terms of vectorial positions and velocities.

The subject of dynamics deals with forces (torques) and motions. The study of motion alone belongs to the subject of kinematics. This appendix starts with vectorial kinematics, and then addresses Newtonian mechanics (dynamics) and finally Lagrangian dynamics.

B.1 VECTOR KINEMATICS

B.1.1 Euler's Theorem (See Figure B.1)

Every displacement of a rigid body can be represented by a single rotation θ about some axis (of unit vector υ).

Important Corollary

Rotations cannot be represented by vectors unless they are infinitesimally small.

Proof

Give a small rotation $\delta\theta = \delta\theta\upsilon$ about the υ-axis, the corresponding displacement (of point P) is $\delta r = \delta\theta \times r$.

The new position (of P) is $r_1 = r + \delta\theta \times r$.

Give another small rotation $\delta\phi = \delta\phi\mu$ about the μ-axis; the new position (of P) is $r_2 = r_1 + \delta\phi \times r_1$.

The combined displacement

$$\Delta r = r_2 - r$$

$$= r_1 + \delta\phi \times r_1 - r = r + \delta\theta \times r + \delta\phi \times [r + \delta\theta \times r] - r$$

$$= (\delta\theta + \delta\phi) \times r + \underbrace{\delta\phi \times (\delta\theta \times r)}_{o(\delta^2)}$$

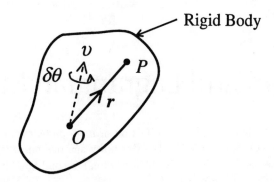

FIGURE B.1 Single rotation of a rigid body about an axis.

The second term can be neglected for small rotations.
Hence,

$$\Delta r = (\delta\theta + \delta\phi) \times r \quad \text{Q.E.D.} \tag{B.1}$$

B.1.2 Angular Velocity and Velocity at a Point of a Rigid Body (See Figure B.2)

From Figure B.1, it is clear that

Angular velocity
$$\omega = \lim_{\delta t \to 0} \frac{\delta\theta}{\delta t} \tag{B.2}$$

Note: Since this definition uses small rotations $\delta\theta$, it follows that ω is always a vector.

The velocity of P relative to O (see Figure B.1) is given by

$$v_{P/O} = \lim_{\delta t \to 0} \frac{\delta r}{\delta t} = \lim_{\delta t \to 0} \frac{\delta\theta \times r}{\delta t} = \omega \times r \tag{B.3}$$

Theorem

For a rigid body, ω is unique (does not vary from point to point on the body).

Proof

Suppose that, on the contrary, angular velocities ω_1 and ω_2 are associated with points C_1 and C_2 (Figure B.2). Then,

$$v_{P/O} = v_{C_1/O} + v_{P/C_1} = v_{C_1/O} + \omega_1 \times r$$

Also,

$$v_{P/O} = v_{C_1/O} + v_{C_2/C_1} + v_{P/C_2} = v_{C_1/O} + \omega_1 \times (r - s) + \omega_2 \times s$$

Hence,

$$\omega_1 \times r = \omega_1 \times (r - s) + \omega_2 \times s$$

Appendix B Newtonian and Lagrangian Mechanics

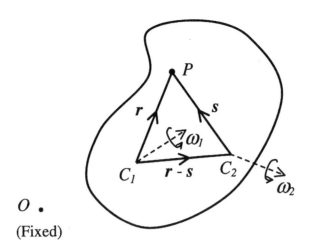

FIGURE B.2 Uniqueness of the angular velocity of a rigid body.

It follows that

$$(\boldsymbol{\omega}_1 - \boldsymbol{\omega}_2) \times \boldsymbol{s} = 0$$

Since $(\boldsymbol{\omega}_1 - \boldsymbol{\omega}_2)$ is not parallel to s in general, because P is an arbitrary point, one has

$$\boldsymbol{\omega}_1 = \boldsymbol{\omega}_2 \quad \text{Q.E.D.}$$

B.1.3 Rates of Unit Vectors Along Axes of Rotating Frames (See Figure B.3)

General Result

Suppose that i_1, i_2, i_3 are unit vectors along the three orthogonal axes of a frame rotating at $\boldsymbol{\omega}$. The rates are the velocities of these vectors about the origin of the frame. Hence, from equation (B.3), one obtains

$$\frac{di_1}{dt} = \boldsymbol{\omega} \times i_1, \qquad \frac{di_2}{dt} = \boldsymbol{\omega} \times i_2, \qquad \frac{di_3}{dt} = \boldsymbol{\omega} \times i_3 \tag{B.4}$$

Some special cases (natural frames of reference) are considered below.

Cartesian Coordinates

Suppose that the frame is free to move independently in the x, y, and z directions only. This corresponds to a translatory motion (no rotation). The rates of the unit vectors in this moving frame are 0; that is, $\boldsymbol{\omega} = 0$

$$\frac{di_x}{dt} = \frac{di_y}{dt} = \frac{di_z}{dt} = 0 \quad \text{(Cartesian)} \tag{B.5}$$

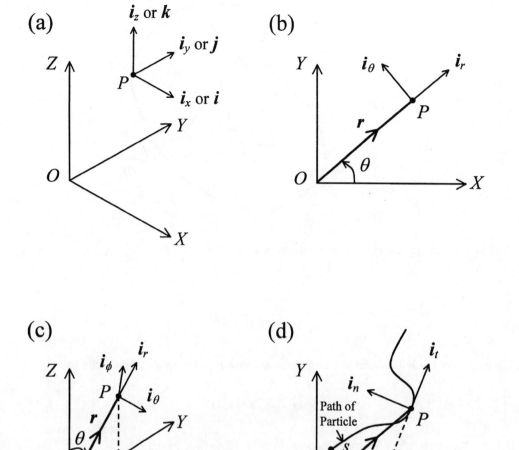

FIGURE B.3 Some natural coordinate frames: (a) Cartesian coordinates; (b) polar coordinates (2-D); (c) spherical polar coordinates; and (d) tangential-normal coordinates (2-D).

Polar Coordinates (2-D)

Suppose that the frame is free to move independently in the r and θ directions, but an increment in the r direction (i.e., δr) causes no rotation. Hence,

$$\boldsymbol{\omega} = \dot{\theta} \boldsymbol{i}_z$$

From equation (B.4),

Appendix B Newtonian and Lagrangian Mechanics

$$\frac{di_r}{dt} = \dot{\theta}i_\theta$$

(2 - D polar) (B.6)

$$\frac{di_\theta}{dt} = -\dot{\theta}i_r$$

Spherical Polar Coordinates

The coordinates are (r, θ, ϕ) as in Figure B.3. To find natural ω for the frame:

Give δr to frame	$\Rightarrow$	No rotation
Give $\delta\theta$ to frame	$\Rightarrow$	Rotation $\delta\theta i_\phi$
Give $\delta\phi$ to frame	$\Rightarrow$	Rotation $\delta\phi i_z$

Hence,

$$\omega = \dot{\theta}i_\phi + \dot{\phi}i_z$$

But $i_z = \cos\theta i_r - \sin\theta i_\theta$
Hence,

$$\omega = \dot{\phi}\cos\theta i_r - \dot{\phi}\sin\theta i_\theta + \dot{\theta}i_\phi \tag{B.7}$$

From equation (B.4),

$$\frac{di_r}{dt} = \dot{\theta}i_\theta + \dot{\phi}\sin\theta i_\phi$$

$$\frac{di_\theta}{dt} = -\dot{\theta}i_r + \dot{\phi}\cos\theta i_\phi \quad \text{(Spherical polar)} \tag{B.8}$$

$$\frac{di_\phi}{dt} = -\dot{\phi}\sin\theta i_r - \dot{\phi}\cos\theta i_\theta$$

Tangential-Normal (Intrinsive) Coordinates (2-D)

The coordinates are (s, ψ) as in Figure B.3. Note that s is the *curvilinear distance* along the *path* of the particle (from a reference point P_o), and ψ is the angle of slope of the path. To find ω of the natural frame,

| Give δs | $\Rightarrow$ | No rotation |
| Give $\delta\psi$ | $\Rightarrow$ | Rotation $\delta\psi i_z$ |

Hence,

$$\dot{\omega} = \dot{\psi}i_z$$

The derivatives of the tangential and normal unit vectors i_t and i_n are given by

$$\frac{di_t}{dt} = \dot{\psi} i_n$$
$$\frac{di_n}{dt} = -\dot{\psi} i_t$$
(Tangential - normal) (B.9)

Note: The vector cross-product of the two vectors

$$a = a_x i + a_y j + a_z k$$
$$b = b_x i + b_y j + b_z k$$

can be obtained by expanding the determinant

$$a \times b = \begin{vmatrix} i & j & k \\ a_x & a_y & a_z \\ b_x & b_y & b_z \end{vmatrix}$$ (B.10)

B.1.4 Acceleration Expressed in Rotating Frames

Spherical Polar Coordinates

Position: $\quad r = r i_r$

Velocity: $\quad v = \dfrac{dr}{dt} = \dot{r} i_r + r \dfrac{di_r}{dt}$

Substitute equation (B.8), and obtain

$$v = \dot{r} i_r + (r\dot{\theta}) i_\theta + (r \sin\theta \dot{\phi}) i_\phi$$ (B.11)

Acceleration $a = \dfrac{dv}{dt}$. Hence, by differentiating equation (B.11) and using (B.8), one obtains

$$a = \begin{bmatrix} a_r \\ a_\theta \\ a_\phi \end{bmatrix} = \begin{bmatrix} \ddot{r} - r\dot{\theta}^2 - r\sin^2\theta \dot{\phi}^2 \\ \dfrac{1}{r}\dfrac{d}{dt}(r^2 \dot{\theta}) - r\sin\theta\cos\theta \dot{\phi}^2 \\ \dfrac{1}{r\sin\theta}\dfrac{d}{dt}(r^2 \sin^2\theta \dot{\phi}) \end{bmatrix}$$ (B.12)

Tangential-Normal Coordinates (2-D) (See Figure B.4)

The velocity is always tangential to the path. Hence,

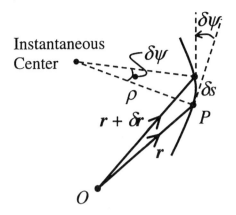

$$\text{Radius of curvature } \rho = \frac{ds}{d\psi}$$

$$\text{Hence, } \frac{d\psi}{dt} = \frac{1}{\rho}\frac{ds}{dt} = \frac{v}{\rho}$$

FIGURE B.4 Velocity representation in tangential-normal coordinates.

$$v = \frac{d\mathbf{r}}{dt} = \frac{ds}{dt}\mathbf{i}_t = v\mathbf{i}_t \tag{B.13}$$

By differentiating equation (B.13) and using (B.9), one obtains

$$\mathbf{a} = \begin{bmatrix} a_t \\ a_n \end{bmatrix} = \begin{bmatrix} \dot{v} \\ v\dot{\psi} \end{bmatrix} = \begin{bmatrix} v\dfrac{dv}{ds} \\ \dfrac{v^2}{\rho} \end{bmatrix} \tag{B.14}$$

See Figure B.4.

$$\text{Note}: \quad \frac{dv}{dt} = \frac{dv}{ds}\frac{ds}{dt} = \frac{v\,dv}{ds}$$

$$\frac{d\psi}{dt} = \frac{d\psi}{ds}\frac{ds}{dt} = \frac{v}{\rho}$$

$$\rho = \frac{ds}{d\psi} = \text{radius of curvature}$$

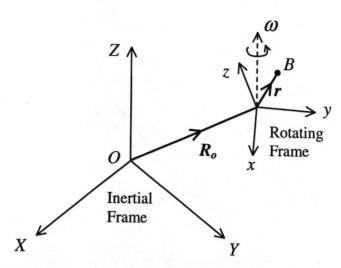

FIGURE B.5 Representation of a vector with respect to a rotating frame.

B.2 NEWTONIAN (VECTOR) MECHANICS

B.2.1 Frames of Reference Rotating at Angular Velocity ω (See Figure B.5)

Newton's second law holds with respect to (w.r.t.) an inertial frame of reference (normally a frame fixed on the earth's surface or moving at constant velocity). The rate of change of vector B w.r.t. an inertial frame is related to the rate of change w.r.t. a frame rotating at ω by

$$\frac{dB}{dt} = \left(\frac{\partial B}{\partial t}\right)_{rel} + \omega \times B \tag{B.15}$$

From equation (B.15), the following results can be obtained for velocity v and acceleration a of point B (see Figure B.5).

$$v = \underbrace{\frac{dR_o}{dt} + \omega}_{v_{frm}} + v_{rel} \tag{B.16}$$

$$a = \underbrace{\frac{d^2 R_o}{dt^2} + \dot{\omega} \times r + \overbrace{\omega \times (\omega \times r)}^{\text{centripetal}}}_{a_{frm}} + \underbrace{2\omega \times v_{rel}}_{\text{Coriolis}} + a_{rel} \tag{B.17}$$

Note: v_{frm} and a_{frm} are velocity and acceleration, respectively, of a point just beneath P and fixed to rotating frame.

Appendix B Newtonian and Lagrangian Mechanics

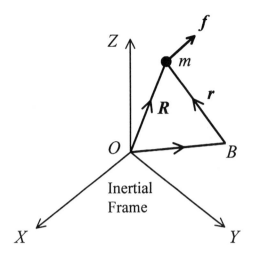

FIGURE B.6 Vector motion of a particle in space.

B.2.2 Newton's Second Law for a Particle of Mass m (see Figure B.6)

$$f = \frac{dp}{dt} \tag{B.18}$$

$$= ma$$

(Linear momentum principle)

Note: Linear momentum $p = m\dfrac{dR}{dt}$

Cross multiply equation (B.18) by r. The torque about B is

$$\tau_B = r \times f = r \times \frac{dp}{dt} = \frac{d}{dt}(r \times p) - \frac{dr}{dt} \times p$$

Now,

$$r \times p = h_B = \text{angular momentum about } B$$

Also,

$$\frac{dr}{dt} \times p = \left(\frac{dR}{dt} - v_B\right) \times m\frac{dR}{dt} = -mv_B \times \frac{dR}{dt} = -v_B \times p$$

This cross-product vanishes if either B is fixed ($v_B = 0$) or B moves parallel to the velocity of m. Hence, the angular momentum principle:

$$\tau_B = \frac{dh_B}{dt} + v_B \times p \qquad \text{in general} \tag{B.19a}$$

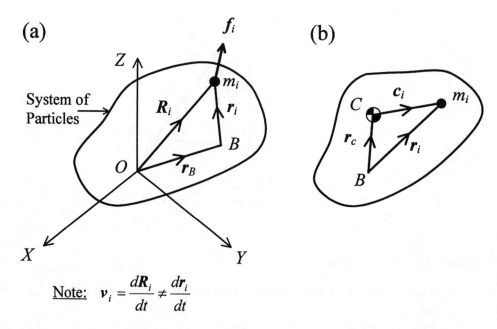

FIGURE B.7 (a) Dynamics of a system of particles, and (b) dynamics with respect to the centroid.

$$\tau_B = \frac{dh_B}{dt} \quad \text{if} \quad \begin{cases} B \text{ fixed, or} \\ B \text{ moves parallel to the velocity of } m \end{cases} \quad \text{(B.19b)}$$

B.2.3 Second Law for a System of Particles — Rigidly or Flexibly Connected (See Figure B.7)

$$F = \frac{dP}{dt}$$

$$= Ma_C \quad \text{if constant mass} \quad \text{(B.20)}$$

(Linear momentum principle)

Note: $F = \sum f_i$ = resultant external force

$P = \sum m_i v_i = M v_C$ = total linear momentum

C = centroid of the particles

Now, using the procedure outlined before for a single particle, and summing the results, one obtains

$$\underbrace{\sum r_i \times f_i}_{\tau_B} = \underbrace{\sum \frac{d}{dt}(r_i \times p_i)}_{\frac{dH_B}{dt}} - \sum \frac{dr_i}{dt} \times p_i$$

But

Appendix B Newtonian and Lagrangian Mechanics

$$\sum \frac{dr_i}{dt} \times p_i = \sum \left(\frac{dR_i}{dt} - v_B \right) \times \left(m_i \frac{dR_i}{dt} \right) = -v_B \times \sum m_i \frac{dR_i}{dt} = -v_B \times P$$

$$= -Mv_B \times v_C = 0 \quad \text{if} \begin{cases} v_B = 0, \text{ or} \\ B \text{ moves parallel to } C \\ (\text{special case}; B \equiv C) \end{cases}$$

Hence, we have the angular momentum principle

$$\tau_B = \frac{d}{dt} H_B + v_B \times P \qquad \text{in general} \tag{B.21a}$$

$$\tau_B = \frac{d}{dt} H_B \qquad \text{if} \begin{cases} B \text{ fixed, or} \\ B \text{ moves parallel to } C \\ (\text{special case}; B \equiv C) \end{cases} \tag{B.21b}$$

Note: (See Figure B.7(b))

$$H_B = \sum r_i \times m_i v_i = \sum (r_c + c_i) \times m_i v_i$$

$$= r_c \times \underbrace{\sum m_i v_i}_{P} + \underbrace{\sum c_i \times m_i v_i}_{H_c}$$

Hence,

$$H_B = H_c + r_c \times P \tag{B.22}$$

B.2.4 Rigid Body Dynamics (See Figure B.8) — Inertia Matrix and Angular Momentum

Note: Equation (B.20) for a system of particles is also the convenient form of the linear momentum principle for rigid bodies. But a more convenient form of equation (B.21) is possible using ω — the angular velocity of the rigid body.

Angular momentum about O is a given by $H_o = \sum r_i \times m_i \frac{dr_i}{dt} = \sum m_i r_i \times (\omega \times r_i)$

Now, using the cross-product relation

$$a \times (b \times c) = b(a \cdot c) - c(a \cdot b)$$

one obtains

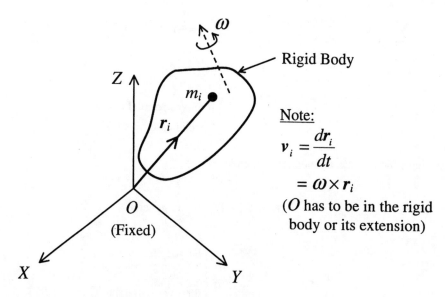

FIGURE B.8 Motion of a rigid body with respect to a fixed frame.

$$H_o = [I]_o \omega$$
(Angular momentum about fixed point O) \hfill (B.23a)

where the *inertia matrix* is

$$[I]_o = \begin{bmatrix} I_{xx} & I_{xy} & I_{xz} \\ & I_{yy} & I_{yz} \\ \text{Symmetric} & & I_{zz} \end{bmatrix} \begin{cases} = \sum m_i \left(r_i^2 [1] - r_i r_i^T \right) & \text{Discrete body} \\ = \int dm \left(r^2 [1] - r r^T \right) & \text{Continuous body} \end{cases} \quad (B.24)$$

Note: In Cartesian coordinates, $r_i^T = [x_i, y_i, z_i]$, and $r_i = |r_i| = \sqrt{x_i^2 + y_i^2 + z_i^2}$

$$I_{xx} = \sum m_i \left(r_i^2 - x_i^2 \right) = \int dm \left(r^2 - x^2 \right)$$

$$I_{xy} = -\sum m_i x_i y_i = -\int dm\, xy$$

etc.

Angular momentum about the centroid:

$$H_c = \sum c_i \times m_i \frac{dr_i}{dt}$$

$$= \sum m_i c_i \times \left(\frac{dr_c}{dt} + \frac{dc_i}{dt} \right) = \underbrace{\left(\sum m_i c_i \right)}_{=0 \text{ by definition of centroid}} \times \frac{dr_c}{dt} + \sum m_i c_i \times \frac{dc_i}{dt}$$

Now

Appendix B Newtonian and Lagrangian Mechanics

$$\sum m_i c_i \times \frac{dc_i}{dt} = \sum m_i c_i \times (\omega \times c_i) \qquad \text{for rigid body}$$

$$= [I]_c \omega \qquad \text{as for equation (B.23)}$$

Hence,

$$H_c = [I]_c \omega \qquad (B.23b)$$

(Angular momentum about centroid)

Equations (B.23a) and (B.23b) can be written

$$H_B = [I]_B \omega \qquad \begin{cases} B \text{ fixed, or} \\ B \text{ is the centroid} \end{cases} \qquad (B.23)$$

B.2.5 Manipulation of Inertia Matrix (See Figure B.9)

Parallel Axis Theorem — Translational Transformation of [*I*]

$$r_c = \begin{bmatrix} a \\ b \\ c \end{bmatrix} = \text{position of centroid w.r.t. a frame at } B$$

If the axes of the two frames are parallel,

$$[I]_B = [I]_C + M \begin{bmatrix} (b^2 + c^2) & (-ab) & (-ac) \\ & (c^2 + a^2) & (-bc) \\ \text{Symmetric} & & (a^2 + b^2) \end{bmatrix} \qquad (B.25)$$

Rotational Transformation of [*I*]

If a nonsingular square matrix [*C*] satisfies

$$[C][C]^T = [1] \qquad (B.26)$$

then it is an *orthogonal matrix*, and the transformation of coordinates from *r* to *r'* through

$$r' = [C]r \qquad (B.27)$$

is called an *orthogonal transformation*. It can be verified using equation (B.24) that

$$[I'] = [C][I][C]^T \qquad (B.28)$$

Principal Directions (Eigenvalue Problem)

Principal directions ≡ Directions in which angular momentum is parallel to the angular velocity. Then,

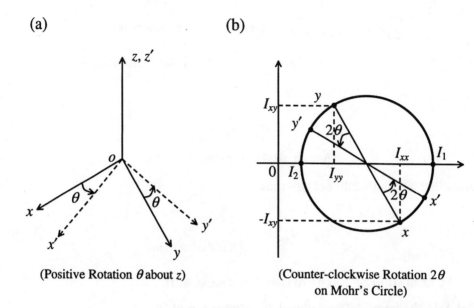

FIGURE B.9 (a) Planar coordinate transformation (b) Mohr's circle for moments of inertia.

$$H = \lambda \omega$$

Substitute equation (B.23): $[I]\omega = \lambda \omega$

If the corresponding direction vector is u (i.e., $\omega = \omega u$), then

$$[I]u = \lambda u \tag{B.29}$$

Equation (B.29) represents an *eigenvalue problem*. The nontrivial solutions for u are *eigenvectors* and represent *principal directions*. Since $[I]$ is symmetric, three independent, *real* solutions (u_1, u_2, u_3) exist for u. The corresponding values of λ are I_1, I_2, and I_3. These are termed *principal moments of inertia*. The matrix of normalized eigenvectors

$$[C] = \begin{bmatrix} u_1^T \\ u_2^T \\ u_3^T \end{bmatrix} \tag{B.30}$$

is an orthogonal matrix. The orthogonal transformation of coordinates

$$r_p = [C]r \tag{B.31}$$

rotates the frame into the principal directions and hence diagonalizes the inertia matrix

$$[C][I][C]^T = \begin{bmatrix} I_1 & & 0 \\ & I_2 & \\ 0 & & I_3 \end{bmatrix} \tag{B.32}$$

Appendix B Newtonian and Lagrangian Mechanics

Mohr's Circle

In the two-dimensional (2-D) case of rigid bodies, the principal directions and principal moments of inertia can be determined conveniently using Mohr's circle. In Figure B.9(b), the direct inertias I_{xx} and I_{yy} are read on the horizontal axis, and the cross inertia I_{xy} on the vertical axis. If the inertia matrix is given in the (x, y, z) frame, two diametrically opposite points of the circle are known. These determine the Mohr's circle. The inertia matrix in any other frame (x', y', z'), rotated by angle θ about the common z-axis in the positive direction, is obtained by moving through 2θ counter closewise on the circle. This procedure also determines the principal moments I_1 and I_2, and the principal direction.

B.2.6 Euler's Equations (for a Rigid Body Rotating at ω)

First consider a general body (rigid or not) for which angular momentum about a point B is expressed in terms of directions of a frame that rotates at ω. Then, from equation (B.15), one obtains

$$\frac{d}{dt} H_B = \left(\frac{\partial H_B}{\partial t}\right)_{rel} + \omega \times H_B \tag{B.33}$$

Now consider a rigid body rotating at ω. If the body frame is oriented in the principal directions, then from equation (B.23), one obtains

$$H_B = \begin{bmatrix} I_1 & & 0 \\ & I_2 & \\ 0 & & I_3 \end{bmatrix} \begin{bmatrix} \omega_1 \\ \omega_2 \\ \omega_3 \end{bmatrix} = \begin{bmatrix} I_1\omega_1 \\ I_2\omega_2 \\ I_3\omega_3 \end{bmatrix} \quad \text{if} \begin{cases} B \text{ fixed, or} \\ B \text{ moving parallel to } C \\ (\text{special case } B = C) \end{cases}$$

From equation (B.19b), the Euler's equations (B.34) are obtained:

$$\tau_B = \begin{bmatrix} I_1\dot{\omega}_1 - (I_2 - I_3)\omega_2\omega_3 \\ I_2\dot{\omega}_2 - (I_3 - I_1)\omega_3\omega_1 \\ I_3\dot{\omega}_3 - (I_1 - I_2)\omega_1\omega_2 \end{bmatrix} \quad \begin{array}{l} \bullet \text{ For principal directions only} \\ \bullet \text{ } B \text{ fixed or moving parallel to centroid} \end{array} \tag{B.34}$$

Note:
$$\left(\frac{\partial H_B}{\partial t}\right)_{rel} = [I_1\dot{\omega}_1, \ I_2\dot{\omega}_2, \ I_3\dot{\omega}_3]^T$$

(because I_1, I_2, I_3 are constant relative to the body frame)

B.2.7 Euler's Angles

Consider two Cartesian frames F and F' in different orientations. One can rotate F to coincide with F' in three steps, as follows:

Step 1: Rotate by angle ψ about the z-axis. The orthogonal transformation matrix for this rotation is

$$[\psi] = \begin{bmatrix} \cos\psi & \sin\psi & 0 \\ -\sin\psi & \cos\psi & 0 \\ 0 & 0 & 1 \end{bmatrix} \tag{B.35}$$

Step 2: Rotate by angle θ about the new y-axis. The orthogonal transformation matrix for this rotation is

$$[\theta] = \begin{bmatrix} \cos\theta & 0 & -\sin\theta \\ 0 & 1 & 0 \\ \sin\theta & 0 & \cos\theta \end{bmatrix} \tag{B.36}$$

Step 3: Rotate by angle ϕ about the new x-axis. The orthogonal transformation matrix for this rotation is:

$$[\phi] = \begin{bmatrix} 1 & 0 & 0 \\ 0 & \cos\phi & \sin\phi \\ 0 & -\sin\phi & \cos\phi \end{bmatrix} \tag{B.37}$$

The same point P can be expressed as vector r in F, or vector r' in F'. It follows that the two are related through

$$r' = [\phi][\theta][\psi]r \tag{B.38}$$

In the Euler angle representation, the angular velocity can be considered to consist of the following components:

- $\dot{\psi}$ about the z-axis of the original frame
- $\dot{\theta}$ about the y-axis of the intermediate frame
- $\dot{\phi}$ about the x-axis of the final frame.

It follows that the angular velocity expressed in F' is

$$\begin{aligned}\boldsymbol{\omega}' &= [\phi][\theta][\psi]\begin{bmatrix} 0 \\ 0 \\ \dot{\psi} \end{bmatrix} + [\phi][\theta]\begin{bmatrix} 0 \\ \dot{\theta} \\ 0 \end{bmatrix} + [\phi]\begin{bmatrix} \dot{\phi} \\ 0 \\ 0 \end{bmatrix} \\ &= [\phi][\theta]\begin{bmatrix} 0 \\ 0 \\ \dot{\psi} \end{bmatrix} + [\phi]\begin{bmatrix} 0 \\ \dot{\theta} \\ 0 \end{bmatrix} + \begin{bmatrix} \dot{\phi} \\ 0 \\ 0 \end{bmatrix}\end{aligned} \tag{B.39a}$$

Also, angular velocity expressed in F is

$$\boldsymbol{\omega} = \begin{bmatrix} 0 \\ 0 \\ \dot{\psi} \end{bmatrix} + [\psi]^T \begin{bmatrix} 0 \\ \dot{\theta} \\ 0 \end{bmatrix} + [\psi]^T [\theta]^T \begin{bmatrix} \dot{\phi} \\ 0 \\ 0 \end{bmatrix} \tag{B.40a}$$

The resulting expressions are:

$$\boldsymbol{\omega}' = \begin{bmatrix} \dot{\phi} - \dot{\psi}\sin\theta \\ \dot{\theta}\cos\phi + \dot{\psi}\sin\phi\cos\theta \\ \dot{\psi}\cos\phi\cos\theta - \dot{\theta}\sin\phi \end{bmatrix} \tag{B.39b}$$

$$\boldsymbol{\omega} = \begin{bmatrix} \dot{\phi}\cos\theta\cos\psi - \dot{\theta}\sin\psi \\ \dot{\phi}\cos\theta\sin\psi + \dot{\theta}\cos\psi \\ \dot{\psi} - \dot{\phi}\sin\theta \end{bmatrix} \tag{B.40b}$$

Once $\boldsymbol{\omega}$ is expressed in this manner, v and a can be expressed in terms of Euler's angles (ψ, θ, ϕ) and their time derivatives. This is the basis of using Euler angles to write equations of motion. *Note*: The set of Euler angles described here is known as the (3, 2, 1) set or Type I Euler angles. Other combinations are possible. For example, if the first rotation is about the *x*-axis, the second rotation about the new *y*-axis, and the final rotation about the new *x*-axis, then one has the (1, 2, 1) set.

B.3 LAGRANGIAN MECHANICS

B.3.1 Kinetic Energy and Kinetic Coenergy

Consider a single particle.

Kinetic energy
$$T = \int \boldsymbol{v} \cdot d\boldsymbol{p} \tag{B.41}$$

Kinetic coenergy
$$T^* = \int \boldsymbol{p} \cdot d\boldsymbol{v} \tag{B.42}$$

Note: $T + T^* = \boldsymbol{v} \cdot \boldsymbol{p}$

In classical mechanics, the *constitutive relation* between velocity v and linear momentum p is *linear*:

$$\boldsymbol{p} = m\boldsymbol{v} \tag{B.43}$$

Hence,

$$T = T^* = \frac{1}{2}m\mathbf{v}\cdot\mathbf{v} = \frac{1}{2}mv^2 \quad \text{(for a particle)} \tag{B.44}$$

Note: For this reason, it is not necessary to distinguish between T and T^*. But in Lagrangian mechanics, traditionally T^* is retained.

For a system of particles,

$$T^* = \frac{1}{2}\sum m_i \mathbf{v}_i \cdot \mathbf{v}_i \tag{B.45}$$

For a rigid body rotating at $\boldsymbol{\omega}$ (see Figure B.8),

$$T^* = \frac{1}{2}\sum m_i (\boldsymbol{\omega}\times\mathbf{r}_i)\cdot(\boldsymbol{\omega}\times\mathbf{r}_i)$$

Now, using $(\mathbf{a}\times\mathbf{b})\cdot\mathbf{c} = (\mathbf{b}\times\mathbf{c})\cdot\mathbf{a} = (\mathbf{c}\times\mathbf{a})\cdot\mathbf{b}$

$$T^* = \frac{1}{2}\underbrace{\sum m_i[\mathbf{r}_i \times(\boldsymbol{\omega}\times\mathbf{r}_i)]}_{\mathbf{H}_o}\cdot\boldsymbol{\omega}$$

See equation (B.23a). One can write,

$$T^* = \frac{1}{2}\boldsymbol{\omega}^T \mathbf{H}_o = \frac{1}{2}\boldsymbol{\omega}^T[I]_o\boldsymbol{\omega} \tag{B.46}$$

(For a rigid body rotating about fixed O)

In Figure B.7(b), suppose that B is fixed. Premultiply equation (B.22) by $1/2\,\boldsymbol{\omega}^T$ (or take the dot product with $1/2\,\boldsymbol{\omega}$) and obtain

$$T^* = \frac{1}{2}\boldsymbol{\omega}^T\mathbf{H}_c + \frac{1}{2}\boldsymbol{\omega}\cdot(\mathbf{r}_c\times\mathbf{P}) = \underbrace{\frac{1}{2}\boldsymbol{\omega}^T[I]_c\boldsymbol{\omega}}_{\text{From eq. (B.23)}} + \frac{1}{2}\underbrace{\overbrace{(\boldsymbol{\omega}\times\mathbf{r}_c)}^{\mathbf{v}_c}\cdot\overbrace{\mathbf{P}}^{m\mathbf{v}_c}}_{\text{Using vector identity}}$$

Hence, with v_c denoting the magnitude $|\mathbf{v}_c|$,

$$T^* = \frac{1}{2}Mv_c^2 + \frac{1}{2}\boldsymbol{\omega}^T[I]_c\boldsymbol{\omega} \tag{B.47}$$

(For a rigid body with centroid C)

B.3.2 Work and Potential Energy

When the points of application $\mathbf{r}_i$ of a set of forces $\mathbf{f}_i$ move by increments $\delta\mathbf{r}_i$, the incremental work done is given by

Appendix B Newtonian and Lagrangian Mechanics

$$\delta W = \sum f_i \cdot \delta r_i \tag{B.48}$$

Forces can be divided into *conservative forces* and *nonconservative forces*. The work done by conservative forces can be given by the *decrease* in potential energy (this is one definition for potential energy). Hence,

$$\delta V = -(\delta W)_{\text{conserv.}} \tag{B.49}$$

Note that conservative forces are nondissipative forces (e.g., spring force, gravity).

Examples

1. For masses m_i located at elevations y_i in gravity (acceleration g),

$$V = \sum m_i g y_i \tag{B.50}$$

2. For springs of stiffness k_i stretched through x_i, the potential energy (elastic) is

$$V = \frac{1}{2}\sum k_i x_i^2 \tag{B.51}$$

B.3.3 Holonomic Systems, Generalized Coordinates, and Degrees of Freedom

Holonomic constraints can be represented entirely by algebraic relations of the motion variables. Dynamic systems having holonomic constraints only are termed *holonomic systems*. For any system (holonomic or non-holonomic), the number of *degrees of freedom* (n) equals the minimum number of *incremental (variational)* generalized coordinates ($\delta q_1, \delta q_2, \ldots, \delta q_n$) required to completely describe any general small motion (without violating the constraints). The number of (nonincremental) generalized coordinates required to describe large motions may be greater than n in general. But, for holonomic systems:

Number of dof = Number of independent generalized coordinates.

B.3.4 Hamilton's Principle

For a holonomic system, the Lagrangian L is given by

$$L = T^* - V \tag{B.52}$$

Consider the variation integral

$$\delta H = \int_{t_0}^{t_f} \left[\delta L + \sum_{j=1}^{n} Q_j \delta q_j \right] dt \tag{B.53}$$

in which Q_j are the *nonconservative* generalized forces corresponding to the generalized coordinates q_j. For a motion trajectory, t_0 and t_f are the initial and the final times, respectively. Hamilton's principle states that this trajectory corresponds to a *natural motion* of the system if and only if $\delta H = 0$ for arbitrary δq_j about the trajectory.

B.3.5 Lagrange's Equations

Note that L is a function of q_j and $\dot{q}_j$ in general, because V is a function of q_j and T^* is a function of q_j and $\dot{q}_j$. Hence,

$$L = L(q_1, q_2, \ldots, q_n, \dot{q}_1, \dot{q}_2, \ldots, \dot{q}_n) \tag{B.54}$$

Then, it follows from Hamilton's principle [equation (B.53) with $\delta H = 0$], that:

$$\frac{d}{dt}\left(\frac{\partial L}{\partial \dot{q}_j}\right) - \frac{\partial L}{\partial q_j} = Q_j \qquad j = 1, 2, \ldots, n \tag{B.55}$$

These are termed Lagrange's equations, and represent a complete set of equations of motion. *Note*: Newton's equations of motion *are equivalent to* the Lagrange's equations.

To determine Q_j, give an incremental motion δq_j to the system with the other coordinates fixed, and determine the work done δW_j. Then,

$$\delta W_j = Q_j \delta q_j \tag{B.56}$$

This gives Q_j.

Example

Figure B.10 shows a simplified model that can be used to study the mechanical vibrations that are excited by the control-loop disturbances in a single-link robot arm. The length of the arm is l, the mass is M, and the moment of inertia about the joint is I. The gripper hand (end effector) is modeled as a mass m connected to the arm through a spring of stiffness k. The joint has an effective viscous damping constant c for rotary motions. The motor torque (applied at the joint) is $\tau(t)$.

Generalized Coordinates

This is a two-degree-of-freedom holonomic system. The angle of rotation θ of the arm and spring deflection x from the unstretched position are chosen as the generalized coordinates.

Generalized Nonconservative Forces

Keeping x fixed, increment θ by $\delta\theta$. The corresponding incremental work due to *nonconservative forces* is $\delta W_\theta = \tau(t)\delta\theta - c\dot{\theta}\delta\theta$.

Note that the damping torque $c\dot{\theta}$ acts opposite to the increment $\delta\theta$. Thus, the generalized force is given by

$$F_\theta = \tau(t) - c\dot{\theta}$$

Keeping θ fixed, increment x by δx. The corresponding incremental work due to nonconservative forces is $\delta W_x = 0$.
Hence, $F_x = 0$.

Appendix B Newtonian and Lagrangian Mechanics

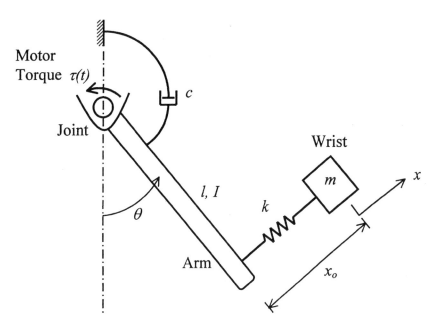

FIGURE B.10

Lagrangian

The total kinetic coenergy (= kinetic energy in these Newtonian systems) is

$$T^* = \frac{1}{2}I\dot{\theta}^2 + \frac{1}{2}m(l\dot{\theta} + \dot{x})^2$$

Note: $(l\dot{\theta} + \dot{x})^2$ is not exact (there is a nonlinear term that is neglected).
The potential energy (due to gravity and spring) is given by:

$$V = Mg\left(-\frac{1}{2}l\cos\theta\right) + mg\left(-l\cos\theta + (x+x_0)\sin\theta\right) + \frac{1}{2}kx^2$$

Note that the centroid of the arm is assumed to be halfway along the link. It follows that the Lagrangian is given by:

$$L = \frac{1}{2}I\dot{\theta}^2 + \frac{1}{2}m(l\dot{\theta} + \dot{x})^2 + \left(\frac{M}{2} + m\right)gl\cos\theta - mg(x+x_0)\sin\theta - \frac{1}{2}kx^2$$

Lagrange's Equations

From equation (B.55) one obtains

For θ: $\dfrac{d}{dt}\left[I\dot{\theta} + m(l\dot{\theta} + \dot{x})l\right] - \left[-\left(\dfrac{M}{2} + m\right)gl\sin\theta - mg(x+x_0)\cos\theta\right] = \tau(1) - c\dot{\theta}$

For x: $\dfrac{d}{dt}\left[m(l\dot{\theta} + \dot{x})\right] - [mg\sin\theta - kx] = 0$

If the steady-state configuration of the link is assumed to be vertical, for small departures from this position (e.g., due to control loop disturbances), the equations of motion are obtained by linearizing for small θ (i.e., $\sin\theta = \theta$, $\cos\theta = 1$). The corresponding equations are

$$\begin{bmatrix} I+ml^2 & ml \\ ml & m \end{bmatrix} \begin{bmatrix} \ddot{\theta} \\ \ddot{x} \end{bmatrix} + \begin{bmatrix} c & 0 \\ 0 & 0 \end{bmatrix} \begin{bmatrix} \dot{\theta} \\ \dot{x} \end{bmatrix} + \begin{bmatrix} \left(\frac{1}{2}M+m\right)gl & mg \\ mg & k \end{bmatrix} \begin{bmatrix} \theta \\ x \end{bmatrix} = \begin{bmatrix} \tau(t) - mgx_0 \\ 0 \end{bmatrix}$$

Note: For equilibrium at $\theta = 0$, one needs $\tau = mgx_0$. In other words, this term represents the static torque needed at the motor joint in order to maintain equilibrium at the ($\theta = 0$, $x = 0$) position. If the system was symmetric, $\tau_{static} = 0$.

Note also that this set of equations of motion is of the form

$$M\ddot{y} + C\dot{y} + Ky = f(t)$$

and the matrices M, C, and K are symmetric. The mass matrix M is not diagonal in the present formulation. It is possible, however, to make it diagonal simply by eliminating the $\ddot{x}$ term in the first equation of motion and the $\ddot{\theta}$ term in the second equation of motion through straightforward algebraic manipulation. Natural frequencies and mode shapes of the undamped system can be determined in the usual manner (see Chapter 5). If the feedback gain of the control loop is such that at least a pair of eigenvalues is complex and has an imaginary part that is approximately equal to a natural frequency of the structural system, then that mode can be excited in an undesirable manner even by a slight disturbance in the control force. Such situations can be avoided by modifying the control system (e.g., by changing the gains) or the structural system (for example, by adding damping, and by changing stiffness and mass) so that the structural natural frequencies would not be near the resonances of the feedback control system. Related issues of design and control for vibration are addressed in Chapter 12.

Appendix C
Review of Linear Algebra

Linear algebra, the algebra of *sets*, *vectors*, and *matrices*, is useful in the study of mechanical vibration. In practical vibrating systems, interactions among various components are inevitable. There are many response variables associated with many excitations. It is thus convenient to consider all excitations (inputs) simultaneously as a single variable, and also all responses (outputs) as a single variable. The use of *linear algebra* makes the analysis of such a system convenient. The subject of linear algebra is complex and is based on a rigorous mathematical foundation. This appendix reviews the basics of *vectors* and *matrices*, which form the foundation of linear algebra.

C.1 VECTORS AND MATRICES

In the analysis of vibrating systems, vectors and matrices will be useful in both time and frequency domains. First, consider the time-domain formulation of a vibration problem. For a single-degree-of-freedom system with a single forcing excitation $f(t)$ and a corresponding single displacement response y, the dynamic equation is

$$m\ddot{y} + c\dot{y} + ky = f(t) \tag{C.1}$$

Note that, in this single-dof case, the quantities f, y, m, c, and k are *scalars*. If the system has n degrees of freedom, with excitation forces $f_1(t)$, $f_2(t)$, ..., $f_n(t)$ and associated displacement responses y_1, y_2, ..., y_n, then the equations of motion can be expressed as

$$\boldsymbol{M}\ddot{\boldsymbol{y}} + \boldsymbol{C}\dot{\boldsymbol{y}} + \boldsymbol{K}\boldsymbol{y} = \boldsymbol{f}(t) \tag{C.2}$$

in which

$$\boldsymbol{y} = \begin{bmatrix} y_1 \\ y_2 \\ \vdots \\ y_n \end{bmatrix} = \text{displacement vector (}n\text{th - order column vector)}$$

$$\boldsymbol{f} = \begin{bmatrix} f_1 \\ f_2 \\ \vdots \\ f_n \end{bmatrix} = \text{forcing excitation vector (}n\text{th - order column vector)}$$

$$\boldsymbol{M} = \begin{bmatrix} m_{11} & m_{12} & \cdots & m_{1n} \\ m_{21} & m_{22} & \cdots & m_{2n} \\ \vdots & & & \\ m_{n1} & m_{n2} & \cdots & m_{nn} \end{bmatrix} = \text{mass matrix (}n \times n\text{ square matrix)}$$

$$C = \begin{bmatrix} c_{11} & c_{12} & \cdots & c_{1n} \\ c_{21} & c_{22} & \cdots & c_{2n} \\ \vdots & & & \\ c_{n1} & c_{n2} & \cdots & c_{nn} \end{bmatrix} = \text{damping matrix } (n \times n \text{ square matrix})$$

$$K = \begin{bmatrix} k_{11} & k_{12} & \cdots & k_{1n} \\ k_{21} & k_{22} & \cdots & k_{2n} \\ \vdots & & & \\ k_{n1} & k_{n2} & \cdots & k_{nn} \end{bmatrix} = \text{stiffness matrix } (n \times n \text{ square matrix})$$

In this manner, vectors and matrices are introduced into the formulation of a multi-degree-of-freedom vibration problem. Further, vector-matrix concepts will enter into the picture in subsequent analysis; for example, in modal analysis, as discussed in Chapters 5 and 11.

Next consider the frequency-domain formulation. In the single-degree-of-freedom case, the system equation can be given as

$$y = Gu \qquad (C.3)$$

where

u = frequency spectrum (fourier spectrum) of the forcing excitation (input)
y = frequency spectrum (Fourier spectrum) of the response (output)
G = frequency-transfer function (frequency-response function) of the system.

The quantities u, y, and G are *scalars* because each one is a single quantity, and not a collection of several quantities.

Next consider a two-degree-of-freedom system having two excitations u_1 and u_2, and two responses y_1 and y_2; y_i now depends on both u_1 and u_2. It follows that one needs four transfer functions to represent all the excitation-response relationships that may exist in this system. One can use the four transfer functions (G_{11}, G_{12}, G_{21}, and G_{22}). For example, the transfer function G_{12} relates the excitation u_2 to the response y_1. The associated two equations that govern the system are:

$$\begin{aligned} y_1 &= G_{11}u_1 + G_{12}u_2 \\ y_2 &= G_{21}u_1 + G_{22}u_2 \end{aligned} \qquad (C.4)$$

Instead of considering the two excitations (two inputs) as two separate quantities, one can consider them as a single "vector" u having the two components u_1 and u_2. As before, one can write this as a "column" vector:

$$u = \begin{bmatrix} u_1 \\ u_2 \end{bmatrix}$$

Alternately, one can write a "row" vector as

$$u = \begin{bmatrix} u_1, u_2 \end{bmatrix}$$

It is common to use the column-vector representation. Similarly, one can express the two outputs y_1 and y_2 as a vector y. Consequently, the column vector is given by

$$y = \begin{bmatrix} y_1 \\ y_2 \end{bmatrix}$$

and the row vector by

$$y = [y_1, y_2]$$

It should be kept in mind that the order in which the components (or elements) are given is important because the vector $[u_1, u_2]$ is not equal to the vector $[u_2, u_1]$. In other words, a vector is an "ordered" collection of quantities.

Summarizing, one can express a collection of quantities, in an orderly manner, as a single vector. Each quantity in the vector is known as a *component* or an *element* of the vector. What each component means will depend on the particular situation. For example, in a dynamic system, it can represent a quantity such as voltage, current, force, velocity, pressure, flow rate, temperature, or heat transfer rate. The number of components (*elements*) in a vector is called the *order*, or *dimension*, of the vector.

Next, the concept of a matrix is introduced, using the frequency-domain example given above. Note that one needs four transfer functions to relate the two excitations to the two responses. Instead of considering these four quantities separately, one can express them as a single matrix G having four *elements*. Specifically, the *transfer function matrix* for the present example is

$$G = \begin{bmatrix} G_{11} & G_{12} \\ G_{21} & G_{22} \end{bmatrix}$$

Note that the matrix has two rows and two columns. Hence, the *size* or *order* of the matrix is 2×2. Since the number of rows is equal to the number of columns in this example, one has a *square matrix*. If the number of rows is not equal to the number of columns, one has a rectangular matrix. Actually, a matrix can be interpreted as a collection of vectors. Hence, in the previous example, the matrix G is an assembly of the two column vectors

$$\begin{bmatrix} G_{11} \\ G_{21} \end{bmatrix} \quad \text{and} \quad \begin{bmatrix} G_{12} \\ G_{22} \end{bmatrix}$$

or, alternatively, an assembly of the two row vectors

$$[G_{11}, G_{12}]$$

and

$$[G_{21}, G_{22}]$$

C.2 VECTOR-MATRIX ALGEBRA

The advantage of representing the excitations and the responses of a vibrating system as the vectors u and y, and the transfer functions as the matrix G, is clear from the fact that the excitation-response (input-output) equations can be expressed as the single equation

$$y = Gu \qquad (C.5)$$

instead of the collection of scalar equations (C.4).

Hence, the response vector y is obtained by "premultiplying" the excitation vector u by the transfer function matrix G. Of course, certain rules of vector-matrix multiplication have to be agreed on in order that this single equation is consistent with the two scalar equations given by equations (C.4). Also, one must agree on rules for the addition of vectors or matrices.

A vector is a special case of a matrix. Specifically, a third-order column vector is a matrix having three rows and one column. Hence, it is a 3×1 matrix. Similarly, a third-order row vector is a matrix having one row and three columns. Accordingly, it is a 1×3 matrix. It follows that one only needs to know matrix algebra, and the vector algebra will follow from the results for matrices.

C.2.1 Matrix Addition and Subtraction

Only matrices of the same size can be added. The result (sum) will also be a matrix of the same size. In matrix addition, one adds the corresponding elements (i.e., the elements at the same position) in the two matrices, and write the results at the corresponding places in the resulting matrix.

As an example, consider the 2×3 matrix

$$A = \begin{bmatrix} -1 & 0 & 3 \\ 2 & 6 & -2 \end{bmatrix}$$

and a second matrix

$$B = \begin{bmatrix} 2 & 1 & -5 \\ 0 & -3 & 2 \end{bmatrix}$$

The sum of these two matrices is given by

$$A + B = \begin{bmatrix} 1 & 1 & -2 \\ 2 & 3 & 0 \end{bmatrix}$$

The order in which the addition is done is immaterial. Hence,

$$A + B = B + A \qquad (C.6)$$

In other words, matrix addition is *commutative*.

Matrix subtraction is defined just like matrix addition, except the corresponding elements are subtracted (or sign changed and added). An example is given below:

$$\begin{bmatrix} -1 & 2 \\ 3 & 0 \\ -4 & 1 \end{bmatrix} - \begin{bmatrix} 4 & 2 \\ 2 & -1 \\ -3 & 0 \end{bmatrix} = \begin{bmatrix} -5 & 0 \\ 1 & 1 \\ -1 & 1 \end{bmatrix}$$

C.2.2 Null Matrix

The *null matrix* is a matrix for which the elements are all zeros. Hence, when one adds a null matrix to an arbitrary matrix, the result is equal to the original matrix. One can define a *null vector* in a similar manner. One can write

$$A + 0 = A \qquad (C.7)$$

As an example, the 2×2 null matrix is:

$$\begin{bmatrix} 0 & 0 \\ 0 & 0 \end{bmatrix}$$

C.2.3 Matrix Multiplication

Consider the product AB of two matrices A and B. One can write this as:

$$C = AB \qquad (C.8)$$

As such, B is *premultiplied* by A or, equivalently, A is *post-multiplied* by B. For this multiplication to be possible, the number of columns in A must be equal to the number of rows in B. Then, the number of rows of the product matrix C is equal to the number of rows in A, and the number of columns in C is equal to the number of columns in B.

The actual multiplication is done by multiplying the elements in a given row (say, the ith row) of A by the corresponding elements in a given column (say, jth column) of B and summing these products. The result is the element c_{ij} of the product matrix C. Note that c_{ij} denotes the element that is common to the ith row and the jth column of matrix C. Thus,

$$c_{ij} = \sum_k a_{ik} b_{kj} \qquad (C.9)$$

As an example, suppose:

$$A = \begin{bmatrix} 1 & 2 & -1 \\ 3 & -3 & 4 \end{bmatrix}$$

$$B = \begin{bmatrix} 1 & -1 & 2 & 4 \\ 2 & 3 & -4 & 2 \\ 5 & -3 & 1 & 0 \end{bmatrix}$$

Note that the number of columns in A is equal to 3, and the number of rows in B is also equal to 3. Hence, one can perform the premultiplication of B by A. For example,

$$c_{11} = 1 \times 1 + 2 \times 2 + (-1) \times 5 = 0$$

$$c_{12} = 1 \times (-1) + 2 \times 3 + (-1) \times (-3) = 8$$

$$c_{13} = 1 \times 2 + 2 \times (-4) + (-1) \times 1 = -7$$

$$c_{14} = 1 \times 4 + 2 \times 2 + (-1) \times 0 = 8$$

$$c_{21} = 3 \times 1 + (-3) \times 2 + 4 \times 5 = 17$$

$$c_{22} = 3 \times (-1) + (-3) \times 3 + 4 \times (-3) = -24$$

etc.

The product matrix is

$$C = \begin{bmatrix} 0 & 8 & -7 & 8 \\ 17 & -24 & 22 & 6 \end{bmatrix}$$

It should be noted that both products AB and BA are not always defined; and even when they are defined, the two results are not equal in general. Unless both A and B are square matrices of the same order, the two product matrices will not be of the same order.

Summarizing, matrix multiplication is *not commutative*:

$$AB \neq BA \tag{C.10}$$

C.2.4 Identity Matrix

An identity matrix (or unity matrix) is a square matrix whose diagonal elements are all equal to 1 and all the remaining (off-diagonal) elements are zeros. This matrix is denoted by I.

For example, the third-order identity matrix is

$$I = \begin{bmatrix} 1 & 0 & 0 \\ 0 & 1 & 0 \\ 0 & 0 & 1 \end{bmatrix}$$

It is easy to see that when any matrix is multiplied by an identity matrix (provided, of course, that the multiplication is possible), the product is equal to the original matrix; thus,

$$AI = IA = A \tag{C.11}$$

C.3 MATRIX INVERSE

An operation similar to scalar division can be defined with regard to the inverse of a matrix. A proper inverse is defined only for a square matrix and, even for a square matrix, an inverse might not exist. The inverse of a matrix is defined as follows:

Suppose that a square matrix A has the inverse B. Then, these must satisfy the equation:

$$AB = I \tag{C.12}$$

or, equivalently,

$$BA = I \tag{C.13}$$

where I is the identity matrix, as previously defined.

The inverse of A is denoted by A^{-1}. The inverse exists for a matrix if and only if the *determinant* of the matrix is non-zero. Such matrices are termed *nonsingular*. The determinant is discussed in section C.3.3; but, before explaining a method for determining the inverse of a matrix, one can verify that

$$\begin{bmatrix} 2 & 1 \\ 1 & 1 \end{bmatrix}$$

is the inverse of

$$\begin{bmatrix} 1 & -1 \\ -1 & 2 \end{bmatrix}$$

To show this, simply multiply the two matrices and show that the product is the second-order unity matrix. Specifically,

$$\begin{bmatrix} 1 & -1 \\ -1 & 2 \end{bmatrix} \begin{bmatrix} 2 & 1 \\ 1 & 1 \end{bmatrix} = \begin{bmatrix} 1 & 0 \\ 0 & 1 \end{bmatrix}$$

or

$$\begin{bmatrix} 2 & 1 \\ 1 & 1 \end{bmatrix} \begin{bmatrix} 1 & -1 \\ -1 & 2 \end{bmatrix} = \begin{bmatrix} 1 & 0 \\ 0 & 1 \end{bmatrix}$$

C.3.1 Matrix Transpose

The transpose of a matrix is obtained by simply interchanging the rows and the columns of the matrix. The transpose of A is denoted by A^T.

For example, the transpose of the 2×3 matrix

$$A = \begin{bmatrix} 1 & -2 & 3 \\ -2 & 2 & 0 \end{bmatrix}$$

is the 3×2 matrix

$$A^T = \begin{bmatrix} 1 & -2 \\ -2 & 2 \\ 3 & 0 \end{bmatrix}.$$

Note that the first row of the original matrix has become the first column of the transposed matrix, and the second row of the original matrix has become the second column of the transposed matrix.

If $A^T = A$, then the matrix A is *symmetric*. Another useful result on the matrix transpose is expressed by

$$(AB)^T = B^T A^T \tag{C.14}$$

It follows that the transpose of a matrix product is equal to the product of the transposed matrices, taken in the reverse order.

C.3.2 Trace of a Matrix

The *trace* of a square matrix is given by the sum of the diagonal elements. The trace of matrix A is denoted by tr(A).

$$\text{tr}(A) = \sum_i a_{ii} \tag{C.15}$$

For example, the trace of the matrix

$$A = \begin{bmatrix} -2 & 3 & 0 \\ 4 & -4 & 1 \\ -1 & 0 & 3 \end{bmatrix}$$

is given by

$$\text{tr}(A) = (-2) + (-4) + 3 = -3$$

C.3.3 Determinant of a Matrix

The *determinant* is defined only for a square matrix. It is a scalar value computed from the elements of the matrix. The determinant of a matrix A is denoted by $\det(A)$ or $|A|$.

Instead of giving a complex mathematical formula for the determinant of a general matrix in terms of the elements of the matrix, one can compute the determinant as follows.
First consider the 2×2 matrix

$$A = \begin{bmatrix} a_{11} & a_{12} \\ a_{21} & a_{22} \end{bmatrix}$$

Its determinant is given by

$$\det(A) = a_{11}a_{22} - a_{12}a_{21}$$

Next consider the 3×3 matrix

$$A = \begin{bmatrix} a_{11} & a_{12} & a_{13} \\ a_{21} & a_{22} & a_{23} \\ a_{31} & a_{32} & a_{33} \end{bmatrix}$$

Its determinant can be expressed as

$$\det(A) = a_{11}M_{11} - a_{12}M_{12} + a_{13}M_{13}$$

where

$$M_{11} = \det \begin{bmatrix} a_{22} & a_{23} \\ a_{32} & a_{33} \end{bmatrix}$$

$$M_{12} = \det \begin{bmatrix} a_{21} & a_{23} \\ a_{31} & a_{33} \end{bmatrix}$$

$$M_{13} = \det \begin{bmatrix} a_{21} & a_{22} \\ a_{31} & a_{32} \end{bmatrix}$$

Note that M_{ij} is the determinant of the matrix obtained by deleting the ith row and the jth column of the original matrix. The quantity M_{ij} is known as the *minor* of the element a_{ij} of the matrix A. If the proper sign is attached to the minor, then depending on the position of the corresponding matrix element, one has a quantity known as the *cofactor*. Specifically, the cofactor C_{ij} corresponding to the minor M_{ij} is given by

$$C_{ij} = (-1)^{i+j} M_{ij} \qquad (C.16)$$

Hence, the determinant of the 3×3 matrix can be given by

$$\det(A) = a_{11} C_{11} + a_{12} C_{12} + a_{13} C_{13}$$

Note that in the two formulas given above for computing the determinant of a 3×3 matrix, one has expanded along the first row of the matrix. The same answer is obtained, however, if one expands along any row or any column. Specifically, when expanded along the ith row, one obtains

$$\det(A) = a_{i1} C_{i1} + a_{i2} C_{i2} + a_{i3} C_{i3}$$

Similarly, if one expands along the jth column, then

$$\det(A) = a_{1j} C_{1j} + a_{2j} C_{2j} + a_{3j} C_{3j}$$

These ideas of computing a determinant can be easily extended to 4×4 and higher-order matrices in a straightforward manner. Hence, one can write

$$\det(A) = \sum_j a_{ij} C_{ij} = \sum_i a_{ij} C_{ij} \qquad (C.17)$$

C.3.4 Adjoint of a Matrix

The *adjoint* of a matrix is the transpose of the matrix whose elements are the cofactors of the corresponding elements of the original matrix. The adjoint of matrix A is denoted by $\mathrm{adj}(A)$.

As an example, in the 3×3 case, one has

$$\mathrm{adj}(A) = \begin{bmatrix} C_{11} & C_{12} & C_{13} \\ C_{21} & C_{22} & C_{23} \\ C_{31} & C_{32} & C_{33} \end{bmatrix}^T$$

$$= \begin{bmatrix} C_{11} & C_{21} & C_{31} \\ C_{12} & C_{22} & C_{32} \\ C_{13} & C_{23} & C_{33} \end{bmatrix}$$

In particular, it is easily seen that the adjoint of the matrix

$$A = \begin{bmatrix} 1 & 2 & -1 \\ 0 & 3 & 2 \\ 1 & 1 & 1 \end{bmatrix}$$

is given by

$$\operatorname{adj}(A) = \begin{bmatrix} 1 & 2 & -3 \\ -3 & 2 & 1 \\ 7 & -2 & 3 \end{bmatrix}^T$$

Accordingly,

$$\operatorname{adj}(A) = \begin{bmatrix} 1 & -3 & 7 \\ 2 & 2 & -2 \\ -3 & 1 & 3 \end{bmatrix}$$

Hence, in general,

$$\operatorname{adj}(A) = \left[C_{ij} \right]^T \tag{C.18}$$

C.3.5 Inverse of a Matrix

At this point, one can define the *inverse* of a square matrix. Specifically,

$$A^{-1} = \frac{\operatorname{adj}(A)}{\det(A)} \tag{C.19}$$

Hence, in the 3×3 matrix example given before, since the adjoint has already been determined, it remains only to compute the determinant in order to obtain the inverse. Now, expanding along the first row of the matrix, the determinant is given by

$$\det(A) = 1 \times 1 + 2 \times 2 + (-1) \times (-3) = 8$$

Accordingly, the inverse is given by

$$A^{-1} = \frac{1}{8} \begin{bmatrix} 1 & -3 & 7 \\ 2 & 2 & -2 \\ -3 & 1 & 3 \end{bmatrix}$$

For two square matrices A and B,

$$(AB)^{-1} = B^{-1} A^{-1} \tag{C.20}$$

BOX C.1 Summary of Matrix Properties

Addition : $A_{m \times n} + B_{m \times n} = C_{m \times n}$

Multiplication : $A_{m \times n} B_{n \times r} = C_{m \times r}$

Identity : $AI = IA = A \Rightarrow I$ is the identity matrix

Note : $AB = 0 \nRightarrow A = 0$ or $B = 0$ in general

Transposition : $C^T = (AB)^T = B^T A^T$

Inverse : $AP = I = PA \Rightarrow A = P^{-1}$ and $P = A^{-1}$

$(AB)^{-1} = B^{-1} A^{-1}$

Community : $AB \neq BA$ in general

Associativity : $(AB)C = A(BC)$

Distributivity : $C(A + B) = CA + CB$

Distributivity : $(A + B)D = AD + BD$

As a final note, if the determinant of a matrix is 0, the matrix does not have an inverse. Then, that matrix is *singular*. Some important matrix properties are summarized in Box C.1.

C.4 VECTOR SPACES

C.4.1 FIELD ($\mathscr{F}$)

Consider a set of scalars. If for any α and β from the set, $\alpha + \beta$ and $\alpha\beta$ are also elements in the set; and if:

1. $\alpha + \beta = \beta + \alpha$ and $\alpha\beta = \beta\alpha$ (Commutativity)
2. $(\alpha + \beta) + \gamma = \alpha + (\beta + \gamma)$ and $(\alpha\beta)\gamma = \alpha(\beta\gamma)$ (Associativity)
3. $\alpha(\beta + \gamma) = \alpha\beta + \alpha\gamma$ (Distributivity)

are satisfied,
and if:

1. Identity elements 0 and 1 exist in the set such that $\alpha + 0 = \alpha$ and $1\alpha = \alpha$
2. Inverse elements exist in the set such that $\alpha + (-\alpha) = 0$ and $\alpha \cdot \alpha^{-1} = 1$

then, the set is a *field*. For example, the set $\mathbb{R}$ of real numbers is a field.

C.4.2 VECTOR SPACE ($\mathscr{L}$)

Properties

1. Vector addition $(x + y)$ and scalar multiplication (αx) are defined.

2. Commutativity: $x + y = y + x$
   Associativity: $(x + y) + z = x + (y + z)$
   are satisfied.
3. Unique null vector **0** and negation $(-x)$ exist such that: $x + 0 = x$
   $x + (-x) = 0.$
4. Scalar multiplication satisfies:

$$\alpha(\beta x) = (\alpha\beta)x \qquad \text{(Associativity)}$$

$$\left.\begin{array}{l}\alpha(x + y) = \alpha x + \beta y \\ (\alpha + \beta)x = \alpha x + \beta x\end{array}\right\} \quad \text{(Distributivity)}$$

$$1x = x, \quad 0x = 0$$

Special Case

Vector space $\mathscr{L}^n$ has vectors with n elements from the field $\mathscr{F}$.
Consider

$$x = \begin{bmatrix} x_1 \\ x_2 \\ \vdots \\ x_n \end{bmatrix}, \quad y = \begin{bmatrix} y_1 \\ y_2 \\ \vdots \\ y_n \end{bmatrix}$$

Then,

$$x + y = \begin{bmatrix} x_1 + y_1 \\ \vdots \\ x_n + y_n \end{bmatrix} = y + x \quad \text{and} \quad \alpha x = \begin{bmatrix} \alpha x_1 \\ \vdots \\ \alpha x_n \end{bmatrix}$$

C.4.3 Subspace $\mathscr{S}$ of $\mathscr{L}$

1. If x and y are in $\mathscr{S}$ then $x + y$ is also in $\mathscr{S}$.
2. If x is in $\mathscr{S}$ and α is in $\mathscr{F}$, then αx is also in $\mathscr{S}$.

C.4.4 Linear Dependence

Consider the set of vectors: $x_1, x_2, \ldots, x_n$. They are linearly independent if any one of these vectors cannot be expressed as a linear combination of one more remaining vectors.
Necessary and sufficient condition for linear independence:

$$\alpha_1 x_1 + \alpha_2 x_2 + \ldots + \alpha_n x_n = 0 \tag{C.21}$$

gives $\alpha = 0$ (trivial solution) as the only solution. For example,

$$x_1 = \begin{bmatrix} 1 \\ 2 \\ 3 \end{bmatrix}$$

$$x_2 = \begin{bmatrix} 2 \\ -1 \\ 1 \end{bmatrix}$$

$$x_3 = \begin{bmatrix} 5 \\ 0 \\ 5 \end{bmatrix}$$

These vectors are not linearly independent because $x_1 + 2x_2 = x_3$.

C.4.5 Basis and Dimension of a Vector Space

1. If a set of vectors can be combined to form any vector in $\mathcal{L}$, then that set of vectors is said to *span* the vector space $\mathcal{L}$ (i.e., a generating system of vectors).
2. If the spanning vectors are all linearly independent, then this set of vectors is a *basis* for that vector space.
3. The number of vectors in the basis = Dimension of the vector space.

Note: The dimension of a vector space is not necessarily the order of the vectors.

For example, consider two intersecting third-order vectors. They will form a basis for the plane (two dimensional) that contains the two vectors. Hence, the dimension of the vector space = 2, but the order of each vector in the basis = 3.

Note: $\mathcal{L}^n$ is spanned by n linearly independent vectors $\Rightarrow \dim(\mathcal{L}^n) = n$

For example,

$$\begin{bmatrix} 1 \\ 0 \\ 0 \\ \vdots \\ 0 \end{bmatrix}, \begin{bmatrix} 0 \\ 1 \\ 0 \\ \vdots \\ 0 \end{bmatrix}, \ldots, \begin{bmatrix} 0 \\ 0 \\ \vdots \\ 0 \\ 1 \end{bmatrix}$$

C.4.6 Inner Product

$$(x, y) = y^H x \tag{C.22}$$

where H denotes the hermitian transpose (i.e., complex conjugate and transpose). Hence $y^H = (y^*)^T$ where $(\ )^*$ denotes complex conjugation.

Note:

1. $(x,x) \geq 0$ and $(x,x) = 0$ if and only if (iff) $x = 0$
2. $(x,y) = (y,x)^*$
3. $(\lambda x, y) = \lambda(x, y)$
   $(x, \lambda y) = \lambda^*(x, y)$
4. $(x, y + z) = (x, y) + (x, z)$

C.4.7 Norm

Properties

$$\|x\| \geq 0 \quad \text{and} \quad \|x\| = 0 \quad \text{iff } x = 0$$

$$\|\lambda x\| = |\lambda|\|x\| \quad \text{for any scalar } \lambda$$

$$\|x + y\| \leq \|x\| + \|y\|$$

For example, the Euclidean norm:
$$\|x\| = \left(\sum_{i=1}^{n} x_i^2\right)^{\frac{1}{2}} \tag{C.23}$$

Unit vector: $\|x\| = 1$

Normalization: $\dfrac{x}{\|x\|} = \hat{x}$

Angle between vectors: We have $\cos\theta = \dfrac{(x,y)}{\|x\|\|y\|} = (\hat{x}, \hat{y})$ (C.24)

where θ is the angle between x and y.

Orthogonal: iff $(x, y) = 0$ (C.25)

Note: n orthogonal vectors in $\mathscr{L}^n$ are linearly independent and span $\mathscr{L}^n$, and form a basis for $\mathscr{L}^n$

C.4.8 Gram-Schmidt Orthogonalization

Given a set of vectors $x_1, x_2, \ldots, x_n$ that are linearly independent in $\mathscr{L}^n$, one can construct a set of orthonormal (orthozonal and normalized) vectors $\hat{y}_1, \hat{y}_2, \ldots, \hat{y}_n$ that are linear combinations of $\hat{x}_i$.

Start: $\hat{y}_1 = \hat{x}_1 = \dfrac{x_1}{\|x_1\|}$

Then: $y_i = x_i - \sum_{j=1}^{i-1}(x_i, \hat{y}_j)\hat{y}_j \quad$ for $i = 1, 2, \ldots, n$

C.4.9 Modified Gram-Schmidt Procedure

In each step, compute new vectors that are orthogonal to the just-computed vector.

Step 1: $\hat{y}_1 = \dfrac{x_1}{\|x_1\|} \quad$ as before

Then: $x_i^{(1)} = x_i - (\hat{y}_1, x_i)\hat{y}_1 \quad$ for $i = 2, 3, \ldots, n$

$\hat{y}_i = \dfrac{x_i^{(1)}}{\|x_i^{(1)}\|} \quad$ for $i = 2, 3, \ldots, n$

and $x_i^{(2)} = x_i^{(1)} - (\hat{y}_2, x_i^{(1)})\hat{y}_2, \quad i = 3, 4, \ldots, n,$ and so on.

C.5 DETERMINANTS

Now one can address several analytical issues of the determinant of a square matrix. Consider the matrix

$$A = \begin{bmatrix} a_{11} & \cdots & a_{1n} \\ \vdots & & \vdots \\ a_{n1} & \cdots & a_{nn} \end{bmatrix}$$

The minor of $a_{ij} = M_{ij}$ = the determinant of the matrix formed by deleting the ith row and the jth column of the original matrix.
Cofactor of $a_{ij} = C_{ij} = (-1)^{i+j} M_{ij}$

cof(A) = Cofactor matrix of A
adj(A) = Adjoint A = (cof A)^T

C.5.1 Properties of Determinant of a Matrix

1. Interchange two rows (columns) $\Rightarrow$ Determinant sign changes
2. Multiply one row (column) by α $\Rightarrow$ αdet( )
3. Add a [$\alpha \times$ row (column)] to a second row (column) $\Rightarrow$ Determinant unchanged
4. Identical rows (columns) $\Rightarrow$ Zero determinant
5. For two square matrices A and B, det(AB) = det(A) det(B)

C.5.2 Rank of a Matrix

Rank A = Number of linearly independent columns
= Number of linearly independent rows
= dim(column space)
= dim(row space)

Here, "dim" denotes the "dimension of."

C.6 SYSTEM OF LINEAR EQUATIONS

Consider the set of linear algebraic equations

$$a_{11}x_1 + a_{12}x_2 + \ldots + a_{1n}x_n = c_1$$

$$a_{21}x_1 + a_{22}x_2 + \ldots + a_{2n}x_n = c_2$$

$$\vdots$$

$$a_{m1}x_1 + a_{m2}x_2 + \ldots + a_{mn}x_n = c_m$$

One needs to solve for $x_1, x_2, \ldots, x_n$.
This problem can be expressed in the vector-matrix form:

$$A_{m \times n} x_n = c_m \qquad B = (A, c)$$

Solution exists iff rank (A,c) = rank (A)

Two cases can be considered:

Case 1: If $m \geq n$ and rank $(A) = n \Rightarrow$ unique solution for x.
Case 2: If $m \leq n$ and rank $(A) = m \Rightarrow$ infinite number of solutions for x;
$x = A^H(AA^H)^{-1}C \Leftarrow$ minimum norm form
Specifically, out of the infinite possibilities, this is the solution that minimizes the norm $x^H x$.

Note that the superscript "H" denotes the "hermitian transpose," which is the transpose of the complex conjugate of the matrix. For example,

$$A = \begin{bmatrix} 1+j & 2+3j & 6 \\ 3-j & 5 & -1-2j \end{bmatrix}$$

Then,

$$A^H = \begin{bmatrix} 1-j & 3+j \\ 2-3j & 5 \\ 6 & -1+2j \end{bmatrix}$$

If the matrix is real, its hermitian transpose is simply the ordinary transpose.

In general, if rank $(A) \leq n \Rightarrow$ infinite number of solutions.
The space formed by solutions $Ax = 0 \Rightarrow$ is called the *null space*.
dim (null space) $= n - k$, where rank $(A) = k$.

REFERENCES

1. Meirovitch, L., *Computational Methods in Structural Dynamics*, Sijthoff & Noordhoff, The Netherlands, 1980.
2. Noble, B., *Applied Linear Algebra*, Prentice-Hall, Inc., Englewood Cliffs, NJ, 1969.

Appendix D
Digital Fourier Analysis and FFT

In the frequency domain, vibration analysis can be carried out using Fourier transform techniques. Three versions of the Fourier transform are available. The frequency content of a periodic signal is conveniently represented by its Fourier Series Expansion (FSE). For nonperiodic (or transient) signals, Fourier-Integral-Transform (FIT) is used. For a discrete sequence of points in a signal (i.e., a set of sampled data), a sequence of discrete data in the frequency domain is obtained using Discrete Fourier Transform (DFT), or "digital" Fourier transform. This can be interpreted as a discrete-data approximation to FIT. Similarly, FSE can be expressed as a special case of FIT. In this sense, the three versions of Fourier transform — FSE, FIT, and DFT — are interrelated.

It should be clear that the DFT is the appropriate version for digital analysis of data, using a computer. The direct use of DFT relations, however, is not computationally efficient because it needs a very large number of operations and liberal use of computer memory. For this reason, Fourier analysis using a digital computer was not considered feasible until 1965. That year, the Fast Fourier Transform (FFT) algorithm was published by Cooley and Tukey. This revolutionized the field of data Fourier analysis by reducing the number of arithmetic operations required for the discrete Fourier transformation of an N-point data sequence by a factor of nearly $2N/\ln_2 N$. Prior to this, the only economical way to perform Fourier analysis of complex time histories was by analog means, where narrow-band analog filters (circuits) were used to extract the frequency components in various frequency bands of interest. Early applications of FFT were limited to off-line computations in a batch mode using software in a large mainframe computer. It was only after the development of large-scale integration (LSI) and the associated microprocessor technology that software-based and dedicated hardware FFT analyzers became cost effective for general applications. The hardware FFT analyzers are particularly suitable in real-time applications. Several stand-alone FFT analyzers were marketed in the late 1970s. Practically unlimited options for frequency (spectral) analysis are available today, through these dedicated analyzers as well as desktop computers.

D.1 UNIFICATION OF THE THREE FOURIER TRANSFORM TYPES

By discrete Fourier transformation of a set of sampled data from a signal, one cannot expect to generate an exact set of points in the analytical Fourier spectrum of the signal. Because of sampling of the signal, some information will be lost. Clearly, one should be able to reduce the error in the computed Fourier spectrum by decreasing the size of the data sample step (ΔT). Similarly, one does not expect to get the exact Fourier series coefficients by discrete Fourier transformation of sampled data from a periodic function. It is very important to study the nature of these errors, which are commonly known as *aliasing distortions*.

D.1.1 Relationship Between DFT and FIT

A fundamental result relating DFT and FIT is established in this section. In view of the FIT relation, the frequency-spectrum values $X_m = X(m \cdot \Delta F)$, $m = 0, \pm 1, \pm 2, \ldots$, sampled at the discrete frequency points of sample step ΔF are given by

$$X_m = \int_{-\infty}^{\infty} x(t)\exp(-j2\pi m\Delta Ft)\,dt$$

$$= \sum_{k=-\infty}^{\infty} \int_{kT}^{(k+1)T} x(t)\exp(-j2\pi mt/T)\,dt \tag{D.1}$$

where $T = 1/\Delta F$. On implementing the change of variable $t \to t + kT$ (i.e., let $t' = t - kT$ and then drop the prime) and interchanging the summation and the integration operations, one obtains

$$X_m = \int_0^T \tilde{x}(t)\exp(-j2\pi mt/T)\,dt \tag{D.2}$$

where

$$\tilde{x}(t) = \sum_{k=-\infty}^{\infty} x(t + kT) \tag{D.3}$$

The fact that $\exp(-j2\pi mk) = 1$ for integers m, k was used in obtaining equation (D.2). Since $\tilde{x}(t)$ is periodic, having the period T, it has an FSE given by

$$\tilde{x}(t) = \frac{1}{T}\sum_{n=-\infty}^{\infty} X_n \exp(j2\pi nt/T) \tag{D.4}$$

which follows from the FSE equation. The sampled values $\tilde{x}_m = \tilde{x}(m \cdot \Delta T)$, $m = 0, \pm 1, \pm 2, \ldots$, at sample steps of ΔT are given by

$$\tilde{x}_m = \frac{1}{T}\sum_{n=-\infty}^{\infty} X_n \exp(j2\pi nm/N)$$

$$= \frac{1}{T}\sum_{k=-\infty}^{\infty} \sum_{n=kN}^{(k+1)N-1} X_n \exp(j2\pi nm/N) \tag{D.5}$$

where $\Delta T = T/N = 1/(N \cdot \Delta F)$. In a manner analogous to the procedure for obtaining equation (D.2), the change of variable $n \to n + kN$ (i.e., let $n' = n - kN$ and then drop the prime) is implemented, and the summation operations are interchanged. This results in

$$\tilde{x}_m = \frac{1}{T}\sum_{n=0}^{N-1} \tilde{X}_n \exp(j2\pi nm/N) \tag{D.6}$$

where

$$\tilde{X}(f) = \sum_{k=-\infty}^{\infty} X(f + kF) \tag{D.7}$$

Note that $\tilde{X}_n = \tilde{X}(n \cdot \Delta F)$, $n = 0, \pm1, \pm2, \ldots$, are the sampled values of the periodic function $\tilde{X}(f)$, having the period F. The frequency parameter $F = N \cdot \Delta F = 1/\Delta T = N/T$ represents the number of samples in a record of unity time duration. It is not possible to extract any information about the frequency spectrum for frequencies $f > F/2 = f_c$ from time-response data sampled at steps of ΔT. The parameter $f_c = 1/(2\Delta T)$ is known as *Nyquist frequency*.

It is evident that by comparing equation (D.6) with the inverse DFT relation that the sequence $\{\tilde{X}_n\} = [\tilde{X}_0, \tilde{X}_1, \ldots, \tilde{X}_{N-1}]$ represents the DFT of the sequence $\{\tilde{x}_m\} = [\tilde{x}_0, x_1, \ldots, x_{N-1}]$. The forward transform is given by

$$\tilde{X}_n = \frac{1}{F}\sum_{m=0}^{N-1} \tilde{x}_m \exp(-j2\pi mn/N) \tag{D.8}$$

In summary, if $X(f)$ is the FIT of $x(t)$, then the N-element sequence $\{\tilde{X}_n\}$ is the DFT of the N-element sequence $\{\tilde{x}_m\}$. The periodic functions $\tilde{x}(t)$ and $\tilde{X}(f)$ are related to $x(t)$ and $X(f)$, respectively, through equations (D.3) and (D.7); $\{\tilde{x}_m\}$ and $\{\tilde{X}_n\}$ being their individual sampled data.

D.1.2 Relationship Between DFT and FSE

A fundamental result relating DFT and FSE will be established in this section. From the FSE equation, it follows that, for a periodic signal $x(t)$ of period T, the sampled values $x_m = x(m \cdot \Delta T)$, $m = 0, \pm1, \pm2, \ldots$, are given by

$$\begin{aligned} x_m &= \frac{1}{T}\sum_{n=-\infty}^{\infty} A_n \exp(j2\pi nm\Delta T/T) \\ &= \frac{1}{T}\sum_{k=-\infty}^{\infty}\sum_{n=kN}^{(k+1)N-1} A_n \exp(j2\pi nm/N) \end{aligned} \tag{D.9}$$

By definition, the sequence $\{x_m\}$ is periodic with N-element periodicity where $N = T/\Delta T$. The procedure for obtaining equation (D.6) is now adopted to obtain

$$x_m = \frac{1}{T}\sum_{n=0}^{N-1} \tilde{A}_n \exp(j2\pi nm/N) \tag{D.10}$$

where

$$\tilde{A}_n = \sum_{k=-\infty}^{\infty} A_{n+kN} \tag{D.11}$$

The sequence $\{\tilde{A}_n\}$ is periodic, with N-element periodicity. By comparing equation (D.10) with the inverse DFT equation, it becomes clear that the N-element sequence $\{\tilde{A}_n\} = [\tilde{A}_0, \tilde{A}_1, \ldots, \tilde{A}_{N-1}]$ is the DFT of the N-element sequence $\{x_m\} = [x_0, x_1, \ldots, x_{N-1}]$. The forward transform is given by

$$\tilde{A}_n = \Delta T \sum_{m=0}^{N-1} x_m \exp(-j2\pi rm/N) \tag{D.12}$$

In summary, if $\{A_n\}$ are the coefficients of the FSE of a periodic signal $x(t)$, then the N-element sequence $\{\tilde{A}_n\}$ is the DFT of the N-element sequence $\{x_m\}$, where $x_m = x(m \cdot \Delta T)$ and $\{\tilde{A}_m\}$ is given by equation (D.11).

D.2 FAST FOURIER TRANSFORM (FFT)

The direct computation of the discrete Fourier transform (DFT) is not recommended, particularly in real-time applications, because of the inefficiency of this procedure. For a sequence of N sampled data points, N^2 complex multiplications and $N/(N-1)$ complex additions are necessary in the direct evaluation of the DFT, assuming that the complex exponential factors $\exp(-j2\pi mn/N)$ are already computed. Many of these arithmetic operations are redundant, however. The Cooley and Tukey algorithm, commonly known as the radix-two fast Fourier transform (FFT) algorithm, is an efficient procedure for computing DFT. Efficiency of the algorithm is achieved by dividing the numerical procedure into several stages so that redundant computations are avoided.

D.2.1 Development of the Radix-Two FFT Algorithm

The multiplicative constant ΔT in the DFT equation is a parameter that was introduced to maintain the consistency with the conventional FIT equation. This constant can be treated as a scaling factor for the final results; or, equivalently, it could be combined with the input data sequence $\{x_n\}$. In any event, the primary computational effort in the DFT equation is directed toward computing the sequence $[A(0), A(1), \ldots, A(N-1)]$ from the data sequence $[a(0), a(1), \ldots, a(N-1)]$, using the relationship

$$A(m) = \sum_{n=0}^{N-1} a(n) W^{mn} \quad \text{for } m = 0, 1, \ldots, N-1 \tag{D.13}$$

where

$$W = \exp(-j2\pi/N) \tag{D.14}$$

The FFT algorithm requires that N be highly composite (i.e., factorizable into many non-unity integers). In particular, for the radix-two algorithm, it is required that $N = 2^r$, where r is a positive integer. If the given data sequence does not satisfy this condition, it must be augmented by a sufficient number of trailing zeros.

A systematic development of the radix-two FFT algorithm is presented now. The integers m and n are expressed in the binary-number-system expansion. Recalling that $0 \leq m \leq N-1$ and $0 \leq n \leq N-1$, one can write

$$\begin{aligned} m &= m_{r-1} 2^{r-1} + m_{r-2} 2^{r-2} + \ldots + m_0 \\ n &= n_{r-1} 2^{r-1} + n_{r-2} 2^{r-2} + \ldots + n_0 \end{aligned} \tag{D.15}$$

in which m_i and n_j take values 0 or 1. Equivalently,

$$\begin{aligned} m &= \text{Binary}(m_{r-1}, m_{r-2}, \ldots, m_0) \\ n &= \text{Binary}(n_{r-1}, n_{r-2}, \ldots, n_0) \end{aligned} \tag{D.16}$$

Next, the indices of the elements $A(\cdot)$ and $a(\cdot)$ in equation (D.13) are expressed by their binary counterparts:

$$A(m_{r-1}, \ldots, m_0) = \sum_{n_0=0}^{1} \sum_{n_1=0}^{1} \cdots \sum_{n_{r-1}=0}^{1} a(n_{r-1}, \ldots, n_0) W^{mn} \tag{D.17}$$

where

$$W^{mn} = W^{(m_{r-1} 2^{r-1} + \ldots + m_0)(n_{r-1} 2^{r-1} + \ldots + n_0)}$$

$$= W^{n_{r-1} 2^{r-1} m_0} W^{n_{r-2} 2^{r-2}(2m_1 + m_0)} W^{n_{r-3} 2^{r-3}(2^2 m_2 + 2m_1 + m_0)} \cdots W^{n_0(m_{r-1} 2^{r-1} + \ldots + m_0)} \tag{D.18}$$

The fact that $W^{2^r} = W^N = 1$ has been used in the foregoing expansion. By defining the intermediate set of sequences $\{A_1(\cdot)\}, \{A_2(\cdot)\}, \{A_r(\cdot)\}$, equation (D.17) can be expressed as the set of equations:

$$A_1(m_0, n_{r-2}, \ldots, n_0) = \sum_{n_{r-1}=0}^{1} a(n_{r-1}, \ldots, n_0) W^{n_{r-1} 2^{r-1} m_0}$$

$$A_2(m_0, m_1, n_{r-3}, \ldots, n_0) = \sum_{n_{r-2}=0}^{1} A_1(m_0, n_{r-2}, \ldots, n_0) W^{n_{r-2} 2^{r-2}(2m_1 + m_0)} \tag{D.19}$$

$$\vdots$$

$$A_r(m_0, m_1, \ldots, m_{r-1}) = \sum_{n_0=0}^{1} A_{r-1}(m_0, \ldots, m_{r-2}, n_0) W^{n_0(m_{r-1} 2^{r-1} + \ldots + m_0)}$$

with

$$A(m_{r-1}, \ldots, m_0) = A_r(m_0, \ldots, m_{r-1}) \tag{D.20}$$

Consequently, the single set of computations given by equation (D.13) for the N-element sequence $\{A(\cdot)\}$ has been replaced by r stages of computations. In each stage, an N-element sequence $\{A_i(\cdot)\}$ must be computed from the immediately preceding N-element sequence $\{A_{i-1}(\cdot)\}$. It soon will be apparent that, as a result of this r-stage factorization, the number of arithmetic operations required has been considerably reduced.

It should be noted that each relationship given in equation (D.19) corresponds to a set of N separate relationships, because the index within the parenthesis of $\{A_i(\cdot)\}$ runs from 0 to $N-1$. In order to observe some important characteristics of the FFT algorithm, the ith relationship of equation (D.19) is examined. Specifically,

$$A_i(m_0, m_1, \ldots, m_{i-1}, n_{r-i-1}, \ldots, n_0)$$

$$= \sum_{n_{r-1}=0}^{1} A_{i-1}(m_0, m_1, \ldots, m_{i-2}, n_{r-i}, \ldots, n_0) W^{n_{r-i} 2^{r-i}(2^{i-1} m_{i-1} + \ldots + m_0)} \qquad i = 1, \ldots, r \tag{D.21}$$

where $A_0(\cdot) = a(\cdot)$. The summation on the right-hand side of equation (D.21) is expanded as

$$A_i(m_0, m_1, \ldots, m_{i-1}, n_{r-i-1}, \ldots, n_0)$$
$$= A_{i-1}(m_0, \ldots, m_{i-2}, 0, n_{r-i-1}, \ldots, n_0) \qquad \text{(D.22)}$$
$$+ A_{i-1}(m_0, \ldots, m_{i-2}, 1, n_{r-i-1}, \ldots, n_0) W^{2^{r-1}(2^{i-1}m_{i-1}+\cdots+m_0)}$$

which involves only one complex multiplication and one complex addition, assuming that the complex exponential terms W^p are precomputed. It is noted that the variable m_{i-1}, which is the binary coefficient of 2^{r-i} in the index of the left-hand-side term $A_i(\cdot)$, does not appear in the binary index of the right-hand-side terms $A_{i-1}(\cdot)$. Consequently, the values of the $A_{i-1}(\cdot)$ terms on the right-hand side remain unchanged as m_{i-1} switches from 0 to 1 in the left-hand-side index. This switch corresponds to a jump in the index of $A_1(\cdot)$ through a value of 2^{r-i}. Accordingly, the computation of, for example, the kth term $A_i(k)$ and the $(k + 2^{r-i})$th term $A_i(k + 2^{r-i})$ of the sequence $\{A_1(\cdot)\}$ in the ith stage involves the same two terms $A_{i-1}(k)$ and $A_{i-1}(k + 2^{r-i})$ of the previous $(i-1)$th sequence $A_{i-1}(\cdot)$. It follows that equation (D.22) takes the more familiar decimal format:

$$A_i(k) = A_{i-1}(k) + A_{i-1}(k + 2^{r-i})W^p$$
$$A_i(k + 2^{r-1}) = A_{i-1}(k) + A_{i-1}(k + 2^{r-i})W^{\tilde{p}} \qquad \text{(D.23)}$$

$$p = 2^{r-i}(2^{i-2}m_{i-2} + \cdots + m_0) \qquad \text{(D.24)}$$

$$\tilde{p} = 2^{r-i}(2^{i-1} + 2^{i-2}m_{i-2} + \cdots + m_0) \qquad \text{(D.25)}$$

Equation (D.24) results when $m_{i-1} = 0$, and equation (D.25) results when $m_{i-1} = 1$. On closer examination, it is evident that $\tilde{p} = N/2 + p$, which follows from $2^{r-i}2^{i-1} = 2^{r-1} = 2^r/2$. Hence,

$$W^{\tilde{p}} = W^{N/2}W^p \qquad \text{(D.26)}$$

From the definition of W [equation (D.14)], however, $W^{N/2} = -1$. Consequently,

$$W^{\tilde{p}} = -W^p \qquad \text{(D.27)}$$

By substituting equation (D.27) in equation (D.23), one obtains

$$A_i(k) = A_{i-1}(k) + A_{i-1}(k + 2^{r-i})W^p$$
$$A_i(k + 2^{r-1}) = A_{i-1}(k) - A_{i-1}(k + 2^{r-i})W^p \qquad \text{(D.28)}$$

for $i = 1, \ldots, r$ and $k = 0, 1, \ldots, 2^{r-i} - 1, 2^{r-i+1}, \ldots$, where p is given by equation (D.24).

The in-place simultaneous computation of the so-called dual terms $A_i(k)$ and $A_i(k + 2^{r-i})$ in the N-term sequence $\{A\cdot(\cdot)\}$ involves just one complex multiplication and two complex additions. As a result, the number of multiplications required has been further reduced by a factor of two. At each stage, the computation of the N-term sequence requires $N/2$ complex multiplications and

Appendix D Digital Fourier Analysis and FFT

N complex additions. Because there are r stages, the radix-two FFT requires a total of $rN/2$ complex multiplications and rN complex additions, in which $r = \ln_2 N$. In other words, the number of multiplications required has been reduced by a factor of $(2N/\ln_2 N)$, and the number of additions has been reduced by a factor of $(N/\ln_2 N)$. For large N, these ratios correspond to a sizable reduction in the computer time required for a DFT. This is a significant breakthrough in real-time digital Fourier analysis.

An insignificant shortcoming of the Cooley-Tukey FFT procedure is evident from equation (D.20). The final sequence $\{A_r(\cdot)\}$ is a scrambled version of the desired transform $\{A(\cdot)\}$. To unscramble the result, it is merely required to interchange the term in the binary location $(m_0, \ldots, m_{r-1})$ with that in the binary location $(m_{r-1}, \ldots, m_0)$. It should be remembered not to duplicate any interchanges while proceeding down the array during the unscrambling procedure. Because an in-place interchange of the elements is performed, there is no necessity for defining a new array. There is an associated saving in computer memory requirements.

D.2.2 THE RADIX-TWO FFT PROCEDURE

The basic steps of the radix-two FFT algorithm are as follows: $N = 2^r$ elements of the data sequence $\{A(\cdot)\}$ are available.

Step 1: Initialize variables. Stage number $i = 1$. Sequence element number $k = 0$.
Step 2: Determine p as follows: From equation (D.24), $p = $ binary $(m_{i-1}, \ldots, m_0, 0, \ldots, 0)_{N\text{bits}}$. From equation (D.21), $k = $ binary $(m_0, m_1, \ldots, m_{i-1}, n_{r-i-1}, \ldots, n_0)$. Shift k register through $(r - i)$ bits to the right, and augment the vacancies by leading zeros. This gives binary $(0, \ldots, 0, m_0, \ldots, m_{i-1})_{N\text{bits}}$. Reverse the bits to obtain p.
Step 3: Compute in place, the dual terms $A_i(k)$ and $A_i(k + 2^{r-i})$, using equation (D.28). *Note*: Since $A_{i-1}(k)$ and $A_{i-1}(k + 2^{r-i})$ are not needed in the subsequent computations, they are destroyed by storing $A_i(k)$ and $A_i(k + 2^{r-i})$ in those locations. As a result, only one array of N elements is needed in the computer memory.
Step 4: Increment $k = k + 1$. If an already-computed dual element is encountered, skip k through 2^{r-i} (i.e., $k = k + 2^{r-i}$). If $k \geq N$, increment $i = i + 1$. If $i > r$, go to step 5. Otherwise, go to step 2.
Step 5: Unscramble the sequence, using equation (D.20), and stop.

D.2.3 ILLUSTRATIVE EXAMPLE

Consider the data sequence $[a(0), a(1), a(2), a(3)]$ of block size $N = 4$. Its DFT sequence $[A(0), A(1), A(2), A(3)]$ is obtained as follows. Note that $r = 2$, and the required complex exponents W^p are available as tabulated data.

Stage 1 ($i = 1$): From equation (D.28), the matrix form of the equations with the indices expressed as binary numbers is

$$\begin{bmatrix} A_1(0,0) \\ A_1(0,1) \\ A_1(1,0) \\ A_1(1,1) \end{bmatrix} = \begin{bmatrix} 1 & 0 & W^0 & 0 \\ 0 & 1 & 0 & W^0 \\ 1 & 0 & -W^0 & 0 \\ 0 & 1 & 0 & -W^0 \end{bmatrix} \begin{bmatrix} a(0,0) \\ a(0,1) \\ a(1,0) \\ a(1,1) \end{bmatrix}$$

Note that the dual jump $2^{r-i} = 2^{2-1} = 2$ for this stage. Furthermore, for $k = 0 = $ binary $(0,0)$, $p = $ binary $(0,0)$; and for $k = 1 = $ binary $(0,1)$, $p = $ binary $(0,0) = 0$.

Stage 2 ($i = 2$):

$$\begin{bmatrix} A_2(0,0) \\ A_2(0,1) \\ A_2(1,0) \\ A_2(1,1) \end{bmatrix} = \begin{bmatrix} 1 & W^0 & 0 & 0 \\ 1 & -W^0 & 0 & 0 \\ 0 & 0 & 1 & W^1 \\ 0 & 0 & 1 & -W^1 \end{bmatrix} \begin{bmatrix} A_1(0,0) \\ A_1(0,1) \\ A_1(1,0) \\ A_1(1,1) \end{bmatrix}$$

The dual jump for this stage is $2^{r-i} = 2^{2-2} = 1$. Also, for $k = 0$, $p = 0$ as before. Now, one must shift k through the dual jump. This gives $k = 2 =$ binary (1,0). Shift this though $r - i = 2 - 2 = 0 \Rightarrow$ no shifts, and bit reverse to get $p =$ binary (0,1) = 1.

Unscrambling: In binary index form, this amounts to a simple bit reversal

$$\begin{bmatrix} A(0,0) \\ A(0,1) \\ A(1,0) \\ A(1,1) \end{bmatrix} = \begin{bmatrix} A_2(0,0) \\ A_2(1,0) \\ A_2(0,1) \\ A_2(1,1) \end{bmatrix}$$

The corresponding decimal assignments are

$$\begin{bmatrix} A(0) \\ A(1) \\ A(2) \\ A(3) \end{bmatrix} = \begin{bmatrix} A_2(0) \\ A_2(2) \\ A_2(1) \\ A_2(3) \end{bmatrix}$$

When real $x(t)$ sequence is used, note that half of the $X(f)$ sequence ($N/2$ points) is wasted because

$$X^*_{N-n} = X_n \quad \text{for real } x_m$$

Hence, one can make some gains in computational effort by converting a real-time sequence to an equivalent complex sequence prior to DFT.

D.3 DISCRETE CORRELATION AND CONVOLUTION

D.3.1 DISCRETE CORRELATION

The sampled data are formed according to

$$\begin{aligned} x_m &= x(m \cdot \Delta T) \quad &\text{for } m = 0, \ldots, M-1 \\ &= 0 \quad &\text{otherwise} \\ y_k &= y(k \cdot \Delta T) \quad &\text{for } k = 0, \ldots, K-1 \\ &= 0 \quad &\text{otherwise} \end{aligned} \quad \text{(D.29)}$$

Appendix D Digital Fourier Analysis and FFT

Using the trapezoidal rule, the sequence $\{z_n\}$ that approximates the sampled values $z(n \cdot \Delta T)$ of the correlation function of x and y can be computed, using

$$z(n \cdot \Delta T) \cong z_n = \frac{1}{N} \sum_{r=0}^{N-1} x_r y_{r+n} \tag{D.30}$$

in which $N > \max(M, K)$. It is noted that, in the summation, an upper limit greater than $\min(M-1, K-1-n)$ is redundant. Because the divisor is the constant value N rather than the actual number of terms in the summation, equation (D.30) represents a biased estimate of the mean lagged product. Nevertheless, it is convenient to use equation (D.30) in this analysis.

Discrete Correlation Theorem

A DFT result for discrete correlation is now established for discrete data. The inverse DFT equation is used in equation (D.30) in conjunction with the fact that $x_m = [x_m]^*$ for real x_m:

$$\begin{aligned} z_n &= \frac{1}{N} \sum_{r=0}^{N-1} \frac{1}{N\Delta T} \sum_{m=0}^{N-1} [X_m]^* \exp(-j2\pi mr/N) \frac{1}{N\Delta T} \sum_{k=0}^{N-1} Y_k \exp[j2\pi k(n+r)/N] \\ &= \frac{1}{N^3 \Delta T} \sum_{m=0}^{N-1} \sum_{k=0}^{N-1} [X_m]^* Y_k \exp(j2\pi kn/N) \sum_{r=0}^{N-1} \exp[j2\pi r(k-m)/N] \end{aligned} \tag{D.31}$$

The orthogonality condition is used in the last summation term of equation (D.31). Consequently,

$$Tz_n = \frac{1}{N\Delta T} \sum_{m=0}^{N-1} [X_m]^* Y_m \exp(j2\pi mn/N) \tag{D.32}$$

where $T = N \cdot \Delta T$. It follows that $T\{z_n\}$ is the inverse of DFT of $\{[X_m]^* Y_m\}$. Equation (D.32) is the discrete correlation theorem.

Discrete Parseval's theorem is given by

$$\Delta T \sum_{m=0}^{N-1} y_m^2 = \Delta F \sum_{n=0}^{N-1} |Y_n|^2 \tag{D.33}$$

Discrete Convolution Theorem

The convolution theorem equation of two signals $u(t)$ and $h(t)$, defined over the finite durations $(0, T_1)$ and $(0, T_2)$, respectively, can be computed, using a digital processor, to obtain $y(t)$. First, the sample step ΔT is chosen and the two sequences $\{u_m\}$ and $\{h_k\}$ of sampled data are formed according to

$$\begin{aligned} u_m &= u(m \cdot \Delta T) \quad &\text{for } m = 0, \ldots, M-1 \\ &= 0 \quad &\text{otherwise} \\ h_k &= h(k \cdot \Delta T) \quad &\text{for } k = 0, \ldots, K-1 \\ &= 0 \quad &\text{otherwise} \end{aligned} \tag{D.34}$$

in which $M = \text{integer}(T_1/\Delta T)$ and $K = \text{integer}(T_2/\Delta T)$. In order to eliminate the wraparound error, it is required that the number of samples of $y(t)$ be $N = M + K - 1$. The direct digital computation of convolution can be performed using the trapezoidal rule:

$$y(n \cdot \Delta T) = y_n = \Delta T \sum_{m=0}^{N-1} u_m h_{n-m} = \Delta T \sum_{m=0}^{N-1} u_{n-m} h_m \quad \text{for } n = 0, 1, \ldots, N-1 \quad \text{(D.35)}$$

In view of the zero terms in the two sequences $\{u_m\}$ and $\{h_k\}$ as given by equation (D.34), it is equally correct to make the lower and the upper limits of the first summation be $\max(0, n - k + 1)$ and $\min(n, M - 1)$, respectively. Similarly, the two limits in the second summation could be $\max(0, n - M + 1)$ and $\min(n, K - 1)$. In any event, by direct counting through summation of series, it can be shown that the computation of equation (D.35) needs KM real applications and $KM - N$ real additions. Alternatively, the discrete convolution result that is analogous to the continuous counterpart in the frequency domain can be used to evaluate equation (D.35) indirectly.

By substituting the inverse DFT equation in equation (D.35), one obtains

$$\begin{aligned} y_n &= \Delta T \sum_{m=0}^{N-1} \frac{1}{N\Delta T} \sum_{r=0}^{N-1} U_r \exp(j2\pi rm/N) \frac{1}{N\Delta T} \sum_{k=0}^{N-1} H_k \exp[j2\pi k(n-m)/N] \\ &= \frac{1}{N^2 \Delta T} \sum_{r=0}^{N-1} \sum_{k=0}^{N-1} U_r H_k \exp(j2\pi kn/N) \sum_{m=0}^{N-1} \exp[j2\pi(r-k)m/N] \end{aligned} \quad \text{(D.36)}$$

The orthogonality condition is used in the last summation. Consequently,

$$y_n = \frac{1}{N \cdot \Delta T} \sum_{r=0}^{N-1} U_r H_r \exp(j2\pi rn/N) \quad \text{(D.37)}$$

D.4 DIGITAL FOURIER ANALYSIS PROCEDURES

Proper interpretation of the DFT results is extremely important in digital Fourier analysis. For example, only the first $N/2 + 1$ points of the DFT array approximate the Fourier transform of the data signal. The remaining $N/2 - 1$ points correspond to the negative frequency spectrum and should be interpreted accordingly. The error caused by interpreting all N points in the DFT array as the positive frequency spectrum corresponding to the data signal is so great that the analysis would become worthless. In this section, some useful DFT procedures are outlined. Emphasis is placed on correct interpretation of the results. Some ways to reduce computation time and memory requirements in real-time applications are described.

D.4.1 FOURIER TRANSFORM USING DFT

Given an analog signal (continuous time) $x(t)$, the major steps for obtaining a suitable approximation to its Fourier transform $X(f)$, using digital Fourier analysis, are as follows:

1. Pick the sample step ΔT. Theoretically, $\Delta T = 1/(2 \times \text{highest frequency of interest})$. This value should be sufficiently small in order to reduce the aliasing distortion in the frequency domain.
2. Sample the signal up to time T, where $T = N \cdot \Delta T$ and $N = 2^r$. The duration $[0, T]$ of the sampled record must be sufficiently long in order to reduce the truncation error (leakage).
3. Obtain the DFT $\{\tilde{X}_n\}$ of the sampled data sequence $\{x_m\}$ using FFT.
4. A discrete approximation to the Fourier transform $X(f)$ is constructed from $\{\tilde{X}_n\}$ according to:

Appendix D Digital Fourier Analysis and FFT

$X(n \cdot \Delta F) \cong X_n$ for $n = 0, 1, \ldots, N/2$, and $X(-n \cdot \Delta F) \cong X_{N-n}$ for $n = 1, \ldots, N/2 - 1$,

where $\Delta F = 1/T$.

D.4.2 Inverse DFT Using DFT

The inverse DFT can be written as

$$[x_n]^* = \frac{1}{N\Delta T} \sum_{m=0}^{N-1} [X_m]^* \exp(-j2\pi mn/N) \qquad (D.38)$$

where [ ]* denotes the complex conjugation operation. It is observed that equation (D.38) is identical to the forward DFT equation except for a scaling factor. Consequently, the forward DFT algorithm can be used in the computation of the inverse DFT. The sampled data should be reorganized and complex-conjugated, however, before using DFT. Finally, the scaling factor should be accounted for so that the final results have the proper units.

Given the complex spectrum $X(f)$, which is the FIT of a real signal $x(t)$ with $x(t) = 0$ for $t < 0$, the main steps of determining a good approximation to the original signal using digital Fourier analysis are as follows:

1. Let F be the highest frequency of interest in $X(f)$, and let $[0, T]$ be the interval over which real signal $x(t)$ is required. The sample step $\Delta F = 1/T$. It is required that ΔF be sufficiently small (T sufficiently large) to reduce aliasing distortion in the time domain. Also, F should be sufficiently large to reduce truncation error. Furthermore, the number of samples $F/\Delta F = N = 2^r$, if radix-two FFT is used.
2. Sample $X(f)$ at intervals ΔF over the frequency interval $[-F/2, F/2]$ according to $X_n = X(n \cdot \Delta F)$ for $n = -N/2, \ldots, 0, \ldots, N/2$ and properly scale the data.
3. Form the sequence $\{\tilde{X}_n\}$ according to:

$$\tilde{X}_n = X_n \qquad \text{for } n = 0, 1, \ldots, N/2$$
$$= X_{n-N} \qquad \text{for } n = N/2 + 1, \ldots, N - 1$$

4. Form the complex conjugate sequence $\{[\tilde{X}_n]^*\}$.
5. Obtain the DFT of $\{[\tilde{X}_n]^*\}$ using FFT. This results in $\{[\tilde{x}_m]^*\}$, which has complex elements with negligible imaginary parts.
6. Construct:

$$x(m \cdot \Delta T) \cong real[\tilde{x}_m]^* \qquad \text{for } m = 0, 1, \ldots, N-1$$

D.4.3 Simultaneous DFT of Two Real Data Records

Considerable computational advantages can be realized when the DFTs $\{Y_m\}$ and $\{Z_m\}$ of two real sequences $\{y_n\}$ and $\{z_n\}$ are required simultaneously. The procedure given in this section achieves this using only a single DFT rather than two separate DFTs.

It is recalled that $\{x_n\}$ is generally a complex sequence. When a real sequence is used, half the storage requirement is wasted. Instead, the DFT of the complex sequence

$$\{x_n\} = \{y_n\} + j\{z_n\} \qquad (D.39)$$

is obtained using FFT. This results in $\{X_m\}$. It is evident from the DFT equation that

$$X_{N-m} = \Delta T \sum_{n=0}^{N-1} x_n \exp(j2\pi mn/N) \qquad (D.40)$$

recalling that $\exp(-j2\pi n) = 1$. Consequently,

$$[X_{N-m}]^* = \Delta T \sum_{n=0}^{N-1} [X_n]^* \exp(-j2\pi mn/N) \qquad (D.41)$$

Since $[x_n]^* = y_n - jz_n$, it is straightforward to observe from the equation and (D.41) that

$$Y_m = \frac{1}{2}\left(X_m + \{X_{N-m}\}^*\right) \qquad (D.42)$$

and

$$Z_m = \frac{1}{2j}\left(X_m - [X_{N-m}]^*\right) \qquad (D.43)$$

From the complex sequence $\{X_m\}$, the required complex sequences $\{Y_m\}$ and $\{Z_m\}$ are constructed according to equations (D.42) and (D.43).

D.4.4 Reduction of Computation Time for a Real Data Record

The DFT of a $2N$-element real sequence $[x_0, x_1, \ldots, x_{2N-1}]$ can be accomplished by means of a single DFT of an N-element complex sequence, using the concept discussed in the preceding section. From the DFT equation, one has

$$\begin{aligned}X_m &= \Delta T \sum_{n=0}^{2N-1} x_n \exp[-j2\pi mn/(2N)] \\ &= \Delta T \sum_{n=0}^{N-1} x_{2n} \exp[-j2\pi m(2n)/(2N)] \\ &\quad + \Delta T \sum_{n=0}^{N-1} x_{2n+1} \exp[-j2\pi m(2n+1)/(2N)]\end{aligned} \qquad (D.44)$$

Consequently,

$$X_m = \Delta T \sum_{n=0}^{N-1} x_{2n} \exp(-j2\pi mn/N) + \exp(-j\pi m/N)\Delta T \sum_{n=0}^{N-1} x_{2n+1} \exp(-j2\pi mn/N) \qquad (D.45)$$

Two real sequences, each having N elements, are defined by separating the even and the odd terms of the given sequence $\{x_n\}$ according to:

Appendix D Digital Fourier Analysis and FFT

$$y_n = x_{2n}$$
$$z_n = x_{2n+1} \quad n = 0, 1, \ldots, N-1 \quad (D.46)$$

The DFT sequences $\{Y_m\}$ and $\{Z_m\}$ of the two real sequences $\{y_n\}$ and $\{z_n\}$ are obtained using the procedure given in the preceding section. Finally, the required DFT sequence is obtained using equation (D.45):

$$X_m = Y_m + \exp(-j\pi m/N)Z_m \quad \text{for } m = 0, 1, \ldots, N-1 \quad (D.47)$$

It should be noted that only the first N terms of the transformed sequence are obtained by this method. This is not a drawback, however, because it is clear that due to aliasing distortion in the frequency domain, the remaining terms correspond to the negative frequencies of $X(f)$.

D.4.5 Convolution of Finite Duration Signals Using DFT

Direct computation of the convolution is possible using the trapezoidal rule. Also, from equation (D.37), it is clear that the required sequence $\{y_n\}$ is the inverse of DFT of $\{U_r H_r\}$, in which $\{U_r\}$ and $\{H_r\}$ are the DFTs of the N-point sequences $\{u_r\}$ and $\{h_r\}$, respectively. This gives rise to the following procedure for evaluating the convolution:

1. Determine $\{U_r\}$ and $\{H_r\}$ by the DFT of the N-point sequences $\{u_r\}$ and $\{h_r\}$, respectively.
2. Evaluate $\{y_n\}$ from the inverse DFT of $\{H_r U_r\}$.

If the slow DFT is used, the foregoing procedure requires $3N^2 + N$ complex multiplications and $3N(N-1)$ complex additions. If the FFT is employed, however, only $1.5N\ln_2 N + N$ complex multiplications and $3N\ln_2 N$ complex additions are necessary. For large N, this can amount to a considerable reduction in computer time. It can be shown that the trapezoidal rule is the most economical method for $N < 200$ (approximately). For larger values of N, the FFT method is recommended.

Wraparound Error

A direct consequence of the definition of the DFT equation is the N-term periodicity of the sequence $\{X_m\}$:

$$X_m = X_{m+iN} \quad \text{for } i = \pm 1, \pm 2, \ldots \quad (D.48)$$

Similarly, from the inverse DFT equation, it follows that the sequence $\{x_n\}$ has the N-term periodicity

$$x_n = x_{n+iN} \quad \text{for } i = \pm 1, \pm 2, \ldots \quad (D.49)$$

Accordingly, whenever a particular problem allows variation of the indices of X_m or x_n beyond their fundamental period $(0, N-1)$, the periodicity of the sequences should be properly accounted for, and the indiscriminate use of DFT should be avoided under such circumstances. An example for such a situation is the evaluation of the discrete convolution equation (D.35) using DFT.

The direct evaluation of equation (D.35) using the trapezoidal rule does not cause any discrepancy because the correct values as given by equation (D.34) are used in this case. When the DFT method is used, however, the N-term periodicity is assumed for the sequences $\{u_m\}$ and $\{h_k\}$. Since this is not true according to equation (D.34), the use of DFT can introduce a technical error into computation. It can be shown that, unless $N \geq M + K - 1$, the first $M + K - 1 - N$ terms in the N-point sequence $\{y_n\}$ do not represent the correct discrete convolution results.

In the first relation of equation (D.35), as m varies from 0 to $N - 1$, the highest value of m for which $u_m \neq 0$ is $M - 1$. The corresponding index of h is $n - M + 1$. Because of the N-term periodicity assumed in DFT, the terms in the sequence $\{h_k\}$ with indices ranging from $(-N)$ to $(-N + K - 1)$ are also non-zero; but if they are included in the discrete convolution, they lead to incorrect results because, in the correct sequence [equation (D.34)], these terms are 0. This is known as the *wraparound error*. It follows that, in order to avoid the discrepancy, one must require $n - M + 1 > -N + K - 1$. In other words, the condition $n > M + K - 2 - N$ must be satisfied to avoid the discrepancy. Since n ranges from 0 to N, the condition is satisfied if and only if $M + K - 2 - N \leq -1$. Consequently, it is required that $N \geq M + K - 1$ in order to avoid the wraparound error.

Data-Record Sectioning in Convolution

The result

$$\int_{-\infty}^{\infty} u(\tau + t_1)h(t - \tau + t_2)d\tau = \int_{-\infty}^{\infty} u(\tau')h(t + t_1 + t_2 - \tau')d\tau \tag{D.50}$$

is obtained using the change of variable $\tau' = \tau + t_1$. In view of the convolution equation, one obtains

$$\int_{-\infty}^{\infty} u(\tau + t_1)h(t - \tau + t_2)d\tau = y(t + t_1 + t_2) \tag{D.51}$$

From equation (D.51), it follows that, if the two convolving functions are shifted to the left through t_1 and t_2, the convolution shifts to the left through $t_1 + t_2$.

Suppose that the time history $u(t)$ is of short duration and that the nonnegligible portion of $h(t)$ represents a relatively long period. If proper sampling of $h(t)$ can exceed the available memory of the digital computer, the function $h(t)$ is sectioned into several portions of equal length T_2, and the convolution integral is computed for each section. Finally, the total convolution integral is obtained using these individual results. The concept behind this procedure is as follows:

$$h(t) = \sum_i h_i(t) \tag{D.52}$$

On substituting in the convolution equation, one obtains

$$y(t) = \sum_i y_i(t) \tag{D.53}$$

where

$$y_i(t) = \int_{-\infty}^{\infty} u(\tau)h_i(t - \tau)d\tau \tag{D.54}$$

However, $h_i(t) = 0$ over $0 \leq t < iT_2$. Because of these trailing zeros, the use of the DFT method becomes extremely inefficient for large i. To overcome this, each segment $h_i(t)$ is shifted to the left through iT_2, which results in a set of modified functions $h_i(t + iT_2)$ that do not contain the trailing zeros. The corresponding convolutions,

$$y_i(t + iT_2) = \int_{-\infty}^{\infty} u(\tau) h_i(t - \tau + iT_2) d\tau \tag{D.55}$$

can be evaluated very efficiently using FFT in the usual manner. Subsequently, the functions $y_i(t + iT_2)$ are shifted to the right through iT_2 to obtain $y_i(t)$. Finally, $y(t)$ is constructed by superposition [equation (D.53)]. It should be noted that evaluation of equation (D.55) using DFT or FFT is performed as described earlier. The major steps of the procedure are as follows:

1. Choose the sample step ΔT in the usual manner. Choose T_2 based on computer memory limitations or computational speed requirements. Section $h(t)$ at periods of T_2. Move each section to the origin and sample each section. A separate memory or storage segment can be used to store the sectioned and sampled data sequences $\{h_k\}_i$.
2. Sample $u(t)$ at ΔT. This results in the sequence $\{u_m\}$.
3. Using $N = \dfrac{(T_1 + T_2)}{\Delta T}$ as the period, obtain the discrete convolution $\{y_n\}_i$ of each pair $\{u_m\}$ and $\{h_k\}$.
4. Shift each sequence $\{y_n\}_i$ to the right through $iK = \dfrac{iT_2}{\Delta T}$ elements and superpose (add the overlapping elements).

Appendix E
Reliability Considerations for Multicomponent Units

In the practice of vibration (e.g., vibration monitoring, isolation, control, and testing), one depends on the proper operation of complex and multicomponent equipment. Equipment that has several components that are crucial to its operation, can have more than one mode of failure. Each failure mode of the overall system will depend on some combination of failure of the components. Component failure is governed by the laws of probability. Consider first some of the fundamentals of probability theory that are useful in the reliability or failure analysis of multicomponent units.

E.1 FAILURE ANALYSIS

E.1.1 Reliability

The probability that a component will perform satisfactorily over a specified time period t (component age) under given operating conditions is called *reliability*. It is denoted by R. Hence,

$$R(t) = \wp(\text{Survival}) \tag{E.1}$$

where $\wp$ denotes "the probability of."

E.1.2 Unreliability

The probability that the component will malfunction or fail during the time period t is called its *unreliability*, or its *probability of failure*. It is denoted by F. Hence,

$$F(t) = \wp(\text{Failure}) \tag{E.2}$$

Because it is known as a certainty that the component will either survive or fail during the specified time period t, one can write

$$R(t) + F(t) = 1 \tag{E.3}$$

The probability of survival of a component usually decreases with age. Consequently, the typical $R(t)$ is a monotonically decreasing function of t, as shown in Figure E.1. If it is known as a certainty that the component is good in the beginning, then $R(0) = 1$. Because of manufacturing defects, damage during shipping, etc., however, one usually has $R(0) \leq 1$. For a satisfactory component, $R(t)$ should not drop appreciably during its design life T_d. The drop is faster initially, however, because of infant mortality (again due to manufacturing defects and the like), and later on, as the component exceeds its design life because of old age (wear, fatigue, etc.).

It is clear from equation (E.3) that the unreliability curve is completely defined by the reliability curve. As shown in Figure E.1, transforming one to the other is a simple matter of reversing the axis.

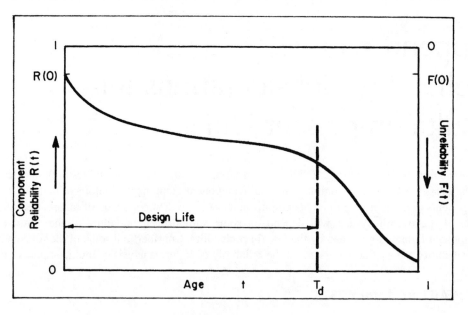

FIGURE E.1 A typical reliability (unreliability) curve.

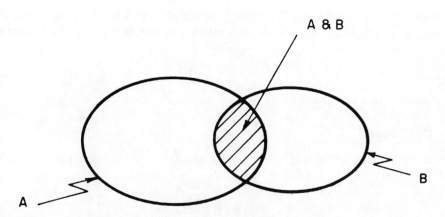

FIGURE E.2 Venn diagram illustrating the inclusion–exclusion formula.

E.1.3 Inclusion–Exclusion Formula

Consider two events, A and B, that are schematically represented by areas (as in Figure E.2). Each event consists of a set of outcomes. The total area covered by the two sets denoted by A and B is given by adding the area of A to the area of B and subtracting the common area.
This procedure can be expressed as

$$\wp(A \text{ or } B) = \wp(A) + \wp(B) - \wp(A \text{ and } B) \tag{E.4}$$

Example

Consider the rolling of a fair die. The set of total outcomes consists of six elements forming the space

$$S = \{1,\ 2,\ 3,\ 4,\ 5,\ 6\}$$

Each outcome has a probability of 1/6. Now consider the two events:

$$A = \{\text{Outcome is odd}\}$$

$$B = \{\text{Outcome is divisible by 3}\}$$

Then,

$$A = \{1, 3, 5\}$$

$$B = \{3, 6\}$$

Consequently,

$$A \text{ or } B = \{1, 3, 5, 6\}$$

$$A \text{ and } B = \{3\}$$

It follows that

$$\wp(A) = 3/6; \quad \wp(B) = 2/6; \quad \wp(A \text{ or } B) = 4/6; \quad \wp(A \text{ and } B) = 1/6$$

These values satisfy equation (E.4)

If the events A and B do not have common outcomes, they are said to be mutually exclusive. Then, the common area of intersection of sets A and B in Figure E.2 will be 0. Hence,

$$\wp(A \text{ and } B) = 0 \tag{E.5}$$

for mutually exclusive events.

E.2 BAYES' THEOREM

A simplified version of Bayes' theorem can be expressed as

$$\wp(A \text{ and } B) = \wp(A/B)\wp(B)$$
$$= \wp(B/A)\wp(A) \tag{E.6}$$

in which $\wp(A/B)$ denotes the conditional probability that event A occurs, given the condition that event B has occurred.

In the previous example of rolling a fair die, if it is known that event B has occurred, the outcome must be either 3 or 6. Then, the probability that event A would occur is simply the probability of picking 3 from the set $\{3, 6\}$. Hence, $\wp(A/B) = 1/2$. Similarly, $\wp(B/A) = 1/3$. It should be noted that equation (E.6) holds for this example.

E.2.1 PRODUCT RULE FOR INDEPENDENT EVENTS

If two events A and B are independent of each other, then the occurrence of event B has no effect whatsoever on determining whether event A occurs. Consequently,

$$\wp(A/B) = \wp(A) \tag{E.7}$$

for independent events. Then, it follows from equation (E.6) that

$$\wp(A \text{ and } B) = \wp(A)\wp(B) \tag{E.8}$$

for independent events. Equation (E.8) is the *product rule*, which is applicable to independent events.

It should be emphasized that although independence implies that the product rule holds, the converse is not necessarily true. In the example on rolling a fair die, $\wp(A/B) = \wp(A) = 1/2$. Suppose, however, that it is not a fair die and that the probabilities of the outcomes {1, 2, 3, 4, 5, 6} are {1/3, 1/6, 1/6, 0, 1/6, 1/6}. Then,

$$\wp(A) = 1/3 + 1/6 + 1/6 = 2/3$$

whereas,

$$\wp(A/B) = \frac{1/6}{1/6 + 1/6} = 1/2$$

This shows that A and B are not independent events in this sample.

Furthermore, $\wp(B) = 1/6$ and $\wp(A \text{ and } B) = 1/6$. It is seen that Bayes' theorem is satisfied by this example.

E.2.2 Failure Rate

The function $F(t)$ defined by equation (E.2) is the probability-distribution function of the random variable T denoting the time to failure. The rate functions can be defined as:

$$r(t) = \frac{dR(t)}{dt} \tag{E.9}$$

$$f(t) = \frac{dF(t)}{dt} \tag{E.10}$$

where

$$R(t) = \wp(T > t)$$
$$F(t) = \wp(T \leq t)$$

In equation (E.10), $f(t)$ is the probability-density function corresponding to the time to failure. It follows that:

$$\wp(\text{Component survived up to } t, \text{ failed within next duration } dt) =$$
$$\wp(\text{Failed within } t, t + dt) = dF(t) = f(t)dt \tag{E.11}$$

Also,

$$\wp(\text{Component survived up to } t) = R(t) \tag{E.12}$$

Define the function $\beta(t)$ such that:

$$\wp(\text{Failed within next duration } dt/\text{Survived up to } t) = \beta(t)dt \qquad (E.13)$$

By substituting equations (E.11) through (E.13) into equation (E.6), one obtains

$$f(t)dt = \beta(t)dt R(t)$$

or

$$\beta(t) = \frac{f(t)}{R(t)} = \frac{f(t)}{1 - F(t)} \qquad (E.14)$$

Now suppose that there are N components. If they all have survived up to t, then, on the average, $N\beta(t)dt$ components will fail during the next dt. Consequently, $N\beta(t)$ corresponds to the rate of failure for the collection of components at time t. For a single component ($N = 1$), the rate of failure is $\beta(t)$. For obvious reasons, $\beta(t)$ is sometimes termed *conditional failure*. Other names for this function include *intensity function* and *hazard function*, but *failure rate* is the most common name.

In view of equation (E.10), one can write equation (E.14) as a first-order linear, ordinary differential equation with variable parameters:

$$\frac{dF(t)}{dt} + \beta(t)F(t) = \beta(t) \qquad (E.15)$$

Assuming a good component initially, one has

$$F(0) = 0 \qquad (E.16)$$

The solution of equation (E.15) subject to equation (E.16) is

$$F(t) = 1 - \exp\left(-\int_0^t \beta(\tau)d\tau\right) \qquad (E.17)$$

where τ is a dummy variable. Then, from equation (E.3),

$$R(t) = \exp\left(-\int_0^t \beta(\tau)d\tau\right) \qquad (E.18)$$

It is observed from equation (E.18) that the reliability curve can be determined from the failure-rate curve, and the reverse.

A typical failure-rate curve for an engineering component is shown in Figure (E.3). It has a characteristic "bathtub" shape, which can be divided into three phases, as in the figure. These phases might not be so distinct in a real situation. The initial burn-in period is characterized by a sharp drop in the failure rate. Because of such reasons as poor workmanship, material defects, and poor handling during transportation, a high degree of failure can occur during a short initial period of design life. Following that, the failures typically will be due to random causes. The failure rate is approximately

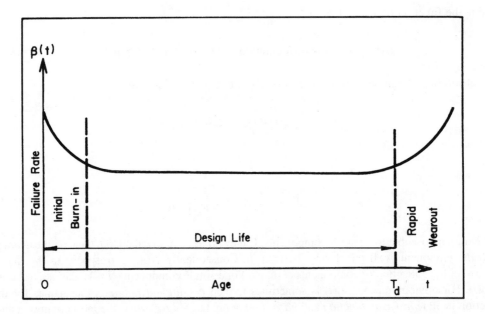

FIGURE E.3 A typical failure-rate curve.

constant in this region. Once the design life is exceeded (third phase), rapid failure can occur because of wearout, fatigue, and other types of cumulative damage, and eventual collapse will result.

It is frequently assumed that the failure rate is constant during the design life of a component. In this case, equation (E.18) gives the exponential reliability function:

$$R(t) = \exp(-\beta t) \tag{E.19}$$

This situation is represented in Figure E.4. This curve is not comparable to the general reliability curve shown in Figure E.1. As a result, the constant failure rate should not be used for relatively large durations of time (i.e., for a large segment of the design life) unless it has been verified by tests. For short durations, however, this approximation is normally used and it results in considerable analytical simplicity.

E.2.3 Product Rule for Reliability

For multicomponent equipment, if it is assumed that the failure of one component is independent of the failure of any other, the product rule given by equation (E.8) can be used to determine the overall reliability of the equipment. The reliability of an N-component object with independently failing components is given by:

$$R(t) = R_1(t) R_2(t) \ldots R_N(t) \tag{E.20}$$

where $R_i(t)$ is the reliability of the ith component. If there is no component redundancy, which is assumed in equation (E.20), none of the components should fail (i.e., $R_i(t) \neq 0$ for $i = 1, 2, \ldots, N$) for the object to operate properly (i.e., $R(t) \neq 0$). This follows from equation (E.20).

In vibration testing, a primary objective is to maximize the risk of component failure when subjected to the test environment (so that the probability of failure is less in the actual in-service environment). One way of achieving this is by maximizing the test-strength-measure function given by

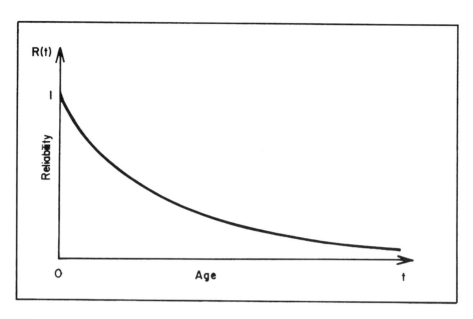

FIGURE E.4 Reliability curve under constant failure rate.

$$TS = \sum_{i=1}^{r} F_i(T)\Phi_i \quad \text{(E.21)}$$

in which $F_i(T)$ is the probability of failure (unreliability) of the ith component for the test duration T, and Φ_i is a dynamic-response measure at the location of the ith component. The parameters of optimization may be the input direction and the frequency of excitation for a given input intensity.

Regarding component redundancy, consider the simple situation of r_i identical subcomponents connected in parallel (r_ith-order redundancy) to form the ith component. The component failure requires the failure of all r_i subcomponents. The failure of one subcomponent is assumed to be independent of the failure state of other subcomponents. Then, the unreliability of the ith component can be expressed as

$$F_i = \left(F_{0i}\right)^{r_i} \quad \text{(E.22)}$$

in which F_{0i} is the unreliability of each subcomponent in the ith component. This simple model for redundancy might not be valid in some situations.

There are two basic types of redundancy: *active redundancy* and *standby redundancy*. In active redundancy, all redundant elements are permanently connected and active during the operation of the equipment. In standby redundancy, only one of the components in a redundant group is active during equipment operation. If that component fails, an identical second component will be automatically connected.

For standby redundancy, some form of switching mechanism is needed, which means that the reliability of the switching mechanism itself must be accounted for. Component aging is relatively less, however, and the failure of components within the redundant group is mutually independent. In active redundancy, there is no need for a switching mechanism; but the failure of one component in the redundant group can overload the rest, thereby increasing their probability of failure (unreliability). Consequently, component failure within the redundant group is not mutually independent in this case. Also, component aging is relatively high because the components are continuously active.

Answers to Numerical Problems

CHAPTER 2

2.22 $k_{eq} = 3EI/l^3$; $m_{eq} = 33m/140$
2.31 (d) 2.75×10^{-2} rad·s^{-1}, 0.564. Yes, at 2.27×10^{-2} rad·s^{-1}

CHAPTER 3

3.5 (c) 6.455×10^3 N·s·m^{-1}
3.7 (d) (i) $r > \sqrt{2}$, (ii) $r > 1.73$, 1.964, 2.871, 3.77, 7.075; (e) Yes
3.12 (c) 1.0, $\sqrt{2}$, $1/\sqrt{2}$
3.16 (c) 1.0, $\sqrt{2}$, $1/\sqrt{2}$
3.18 (a) $-0.55 \pm j\, 1.6424$

CHAPTER 4

4.1 b
4.2 c
4.3 1
4.6 0.31, $-72°$, 0.10, -0.29, -2.7 dB
4.14 (i) 10 samples/s; (ii) 102.4 s; (iii) 1024 complex values. First and last 512 give the same values; (iv) 512; (v) 0.01 Hz; (vi) 5 Hz; (vii) 400; (viii) 4 Hz
4.15 1.0, $1/\sqrt{0.875}$, $\sqrt{3}$, $1/\sqrt{0.3974}$, $\sqrt{(8/3)}$, ?
4.20 (ii) Meaningless; (iii) 2 decades; (iv) 1/2 octave; (v) No dimensions
4.21 1–80 Hz; 4–50 Hz; 1–200 Hz; 20 Hz–20 kHz; Problem dependent

CHAPTER 5

5.8 (b) $(0, 7/6)$, unstable; $(\pi, 5/6)$, unstable; $(\pi/3, 4/3)$, stable
5.14 (c) $\omega_0 = 1.0$ rad·s^{-1}, $\omega_1 = 0.765367$ rad·s^{-1}, $\omega_2 = 1.847759$ rad·s^{-1}; $\theta_1 = 0.25\cos\omega_1 t + 0.75\cos\omega_2 t$; $\theta_2 = \sqrt{2}/4 \cos\omega_1 t - 3\sqrt{2}/4 \cos\omega_2 t$

CHAPTER 6

6.12 (a) 8.0×10^{10} N·m^{-2}; (b) Steel
6.25 $0.8603336/l\sqrt{(G/\rho)}$, $\sin 0.8603336 x/l$

CHAPTER 7

7.7 (a) 0.08
7.8 (b) 0.045
7.9 (b) 0.08
7.13 0 to 2.06×10^7 N·m^{-1}
7.14 (b) (i) 50.6 rad·s^{-1}, 0.1; (ii) 1.5 cm; (iii) 50.1 rad·s^{-1}, 10.02 rad·s^{-1}

CHAPTER 8

- 8.4 (b) 0.99%
- 8.5 (d) 0.09; (e) 6.25×10^4 N·m^{-1} to 25.0×10^4 N·m^{-1}
- 8.11 1000 Hz
- 8.18 100.0 s
- 8.20 (a) 98.37% to 99.99%; (c) 170 Hz
- 8.26 144.0, 5.9%

CHAPTER 9

- 9.2 (b) 44.5 Mbits·s^{-1}
- 9.3 −7.5 V, −14 V
- 9.6 31.8 kHz, 5 μs
- 9.16 (a) 3600 rpm, 15
- 9.20 1%

CHAPTER 10

- 10.2 Directions 1, 2, 3, or 4
- 10.8 (a) 2000 Hz; (b) 80 Hz; (c) 50 Hz; (d) 2000 Hz; (e) 500 Hz; (f) 40 Hz
- 10.11 (i) $2^{1/3}$; (iii) Two; (iv) 1/2
- 10.15 (c) 4.6 g, 30.0 rad·s^{-1}; (d) 0.23 cm·s^{-1}; 653.3 rad·s^{-1}
- 10.17 (c) (i) 25 Hz, 80 g; 50 Hz, 160 g; (ii) 3.2 cm, 160 g; (iii) 3.2 cm, 3.0 m·s^{-1}; 29 g; 23 g

CHAPTER 11

- 11.5 (b) 20 dB/decade; (c) −20 dB/decade
- 11.6 (a) 0.356 Hz; (b) 0.355 Hz
- 11.8 0.023, 0.038
- 11.16 1.0 rad·s^{-1}, 0.2

CHAPTER 12

- 12.6 82.0 N·cm^{-1}, 6.1 cm
- 12.7 2.7×10^4 N·m^{-1}; 286.0 N·m^{-1}·s, 38.1 rad·s^{-1}
- 12.8 3.3×10^5 N·m^{-1}, 7.8×10^3 N·m^{-1}·s
- 12.9 3.3, 0.45
- 12.10 Mount 1
- 12.11 6.9 gm at −6.6°
- 12.12 (b) 8.9 gm at −23°
- 12.13 (b) Plane 1: 42.3 gm at 107°; Plane 2: 18.8 gm at 121°
- 12.14 (c) (iii) 0.4 kg, 2.6 kg
- 12.15 15.3 gm at 53.4°
- 12.23 0.2
- 12.24 (a) 2338.6 rpm; (b) 3.376 cm; (c) 17
- 12.28 119.2 rad·s^{-1}
- 12.29 (a) 2430 cycles·m^{-1}, 3330 cycles·m^{-1}; (b) 2616 to 2832 rpm; (c) 2.8×10^{-3} mm·N^{-1}, 18.4×10^{-3} mm·N^{-1}
- 12.30 3.2 kg and 1.7×10^5 N·m^{-1}
- 12.31 3.33 kg, 1.184×10^5 N·m^{-1}, 180.4 N·s·m^{-1}, 8.26/0.074
- 12.32 0.1
- 12.34 (b) 156 rad·s^{-1}, 2.75 kg, 1.085×10^4 N·m^{-1}, 58.7 N·m^{-1}·s

Index

A

A-type elements, 841
Absolute sum, 633
Absolute viscosity, 848
Absorber location, 837
AC bridge circuits, 547
AC bridge, 467, 502, 569
AC signal conditioning, 546
AC signal transmission, 546
AC tachometer, 498, 547
AC-coupled amplifier, 521
Accelerance, 117
Accelerated aging, 656
Accelerating conditions, 186
Acceleration limit, 734
Acceleration spectrum, 628
Acceleration transfer function, 111
Acceleration, 862
Accelerometer, 113, 137, 713, 750
    crystal, 452
    equivalent circuit, 499
    mounting techniques, 452
    null-balance, 451
    piezoelectric, 449
    sensitivity, 659
    servo force balance, 451
    strain gage, 451, 463, 500
    vibrating-wire, 451
Acceptance criteria, test, 639
Accumulator, fluid, 78
Accuracy, 426
Across variable, 118, 146, 203, 484, 840
Active bridge, 466
Active circuits, 509
Active coil, eddy current, 444
Active control of vibration, 5, 801, 805
Active control, saw blade vibration, 813
Active equipment, 638
Active filters, 531, 524, 541
Active gage, 466
Active mount, 496
    engine, 771
Active redundancy, 917
Active stiffness, 500
Active strain gage, 461
Active suspension, 393
Active test, 647
Active transducers, 421
Active vibration isolators, 495
Actuators, 731, 813
ADC conversion rate, 561
ADC performance characteristics, 560
ADC, *see* analog-to-digital conversion
Added inertia, 37
Addressing, MUX, 565
Adiabatic process, 847
Adjoint of a matrix, 852, 887, 893
Admissible set of transfer functions, 699
Aerodynamic forces, 779
Aeronautical engineering, 1
Aerospace engineering, 1
Aging level, 658
Aging, component, 917
Air cushion, 407
Aircraft, 145, 151
Aircraft, 151
    vibration, 299
Aliasing distortion, 164, 525, 904
    frequency domain, 165
    illustration, 170
    time domain, 167
Alignment shaker, 7
Amplification factor, 106, 375
Amplification, 399, 509
Amplifier, 13
    performance ratings, 519
    shaker, 411
    variable-gain, 418
Amplitude modulation, 539, 543
Amplitude servo-monitor, 417
Amplitude spectrum, 152
Amplitude, 27, 151
Analog control, 806
Analog multiplier, 545
Analog MUX, 565
Analog scanner, 565
Analog-to-digital conversion, 13, 548, 555, 590, 807
Analogies, 13
Analogies, 839, 840
Analysis bandwidth, 183
Analytical model, 17, 839
Analyzing filter bandwidth, 419
Angular coordinates, balancing, 830
Angular displacement sensing, 437
Angular frequency, 27, 152, 614
Angular momentum principle, 865
Angular momentum, 865, 867
    about centroid, 869
    about fixed point, 868
Angular motions, 144
Angular velocity
    definition, 858
    rigid-body, 859
Anti-aliasing filter, 167, 525
Antiresonance, 792
    shaker, 411
Arch, Jefferson memorial, 9
Archimedes principle, 36
Armature circuit, 141, 146
Assembly level, test object, 640
Assembly of subsystems, 785
Assembly plant, 260
Asymptotic stability, 42, 45

Attenuation, 738
Autocorrelation, 172, 618
Auto-transfer function, 696, 699
Autocovariance function, 619
Automated transit, 3
Automatic control, shaker, 415
Automatic scaling, 601
Automobile engine, balancing, 762, 764
Automobile, 78
    model, 263
    suspension, 10
Autonomous system, 849
Auxiliary differential equation, 82
Auxiliary variables, 840
Average value, 156
Averaging, 601
Axial and bending, self-compensation, 471
Axial tension, beam vibration, 319
Axial testing, damping, 388

B

Back emf, 411
Balanced bridge, 444, 459
Balancing experiment, 760
Balancing masses, 829
Balancing of reciprocating machines, steps, 770
Balancing of rotating machinery, 749
Balancing steps
    dynamic, 758
    static, 758
Balancing, 13
    approach, 750
    reciprocating machines, 762
Ballast circuit, 581
Ballast circuit, strain gage, 457
Band-limited data, 544
Bandpass filter, 182, 523, 534
    frequency tuned, 417
Band-reject filter, 538
Bandwidth method, damping measurement, 376
Bandwidth, 182, 423, 425
    amplifier, 519
    analysis, 183
    computational, 169
    filter, 172
    half-power, 183
    signal analyzer, 596
Bars, *see* rods
Bartlett window, 178
Basis of vector space, 214, 891
Bathtub curve, 915
Bayes' theorem, 913
Beam response, steady state, 313
Beam vibration control, 816
Bearing
    friction force, 773
    housing, 761
    monitoring, 606
Bearing planes, 760
Bearing-roller hammer, 547
Beat phenomenon, 654
    ultrasound, 481
Belt drive, 268, 503
Bending stiffness, 22
Bending system, 22

Bending vibration, 299
    axial tension, 319
    mode shapes, 320
    natural frequencies, 320
    thick beams, 322
Bernoulli, Daniel, 10
Bernoulli-Euler beam, 299, 302, 326, 716, 723, 817
    damped, 353
Bias current, op-amp, 520
Bias signals, op-amp, 521
Biharmonic operator, 335
Black box testing, 665, 666
Blade-passing frequency, 547
Block diagram, 94, 106, 146, 651
Bode plot, *see* Bode diagram
Body coordinates, 752
Body reference, 754, 828
Boom, 71
Bouncing back, test, 653
Boundary condition
    end mass, 326, 346
    end rotor, 346
    rotary inertia, 328
    beam vibration, 306
    cable, 272, 273
    dynamic, 306
    flexible, 273
    inertial, 273
    modal, rods, 286
    rods, 285
Boundary mass, torsion, 341
Boundary spring, torsion, 341
Boundary, 18
Boundary-layer effect, 356
Bounded-input-bounded-output stability, 45, 88
Boxcar window, 174
Break points, test psd, 671
Bridge circuit, 13, 568, 582, 584
    amplifiers, 572
Bridge completion
    impedance, 448
    resistor, 466
Bridge constant, 461
Bridge output, imbalance, 466
Bridge sensitivity, 459
Bridge vibration, 299
Bridge,
    active, 466
    balanced, 449, 459
    reference, 466
Broadband random testing, 639
Buffer amplifier, 514
Buffer size, 194
Building model, 249
Building vibration, 299
Bulk modulus, 80, 846
Bump test, 189, 652
Buoyancy, 37
Burn-in failure, 656
Butterworth filter, 529, 541

C

Cabinet-mounted equipment, test, 648
Cable capacitance, 447
    sensor, 490

Cable tension, 276
Cable, 268
    boundary conditions, 272, 273
    characteristic equation, 272
    characteristic roots, 272
    fixed ended, 282
    modal analysis, 271
    mode shape, 271
    natural frequency, 271
    vibration, equations, 283
Calibration constant, bridge, 460
Calibration curve, 426
Calibration records, 659
Calibration, 582
Cam-follower mechanism, 70
Canadarm, 2
Canonical space, 230
Cantilever, 723
    thin, vibration, 316
    with end mass, 724
Capacitance bridge, 447
Capacitive angular velocity sensor, 446
Capacitive coupling, 608
Capacitive sensors, 575
    displacement sensor, 445, 582
    rotation sensor, 446
Capacitor element, 843
Capacitor, 17, 24
Carrier frequency, 432, 539
Carrier input, 524
Carrier signal, 419, 539
Cart, 249
Cartesian coordinates, 859, 860
Cascade connection, 484
Cascade diagram, 187
Cascaded RC circuits, 535
Cathode ray tube, 597
Causal relationship, 839
Causal system, 174
Causality, 853, 855
Cause-effect relation, 839
Cayley-Hamilton equation, 851
Cayley-Hamilton theorem, 851
CCD, *see* charge-coupled device
Center of precession, 67
Centrifugal acceleration, 773
Centrifugal force, 750
Centrifugal pump, 72, 82, 246
Centrifuge, 250
Centripetal acceleration, 864
Centroid, 868
Cepstrum, 180
Channel address, MUX, 565
Channel selector, 718
Channel-select speed, MUX, 565
Characteristic equation, 39, 45, 49, 104, 106, 208, 238, 272, 697, 810, 851
    undamped, 133
Characteristic polynomial, 91, 104, 697, 851, 856
Characteristic roots, 272
Characteristics of packaging, 668
Charge amplifier, 113, 452, 453, 489, 807
Charge leakage, 452
Charge sensitivity, 450
Charge-coupled device camera, 602
Charging time constant, S/H, 563
Charging-discharging curve, 559
Chatter, 386

Chattering, 381
    of machine tools, 779
Chebyshev filter, 541
Circle fit method, 704, 721
    hysteretic damping, 706
    viscous damping, 704
Civil engineering, 1, 8
Clamped end, 306
Clough, 11
CMRR, 522
Code ambiguity, DAC, 554
Coefficient of friction, 380, 385
Cofactor
    matrix, 893
    of matrix element, 887
Coherence function, 175, 176
    ordinary, 620
Coherent output power, 185
Column vector, 880
Combustion
    force, 771
    load, IC engines, 770
    pressure, 763
Commercial EMA systems, 718
Common-mode
    error, 518, 581
    input voltage, 521
    gain, 518, 522
    output voltage, 520, 521
    rejection ratio, 521
    rejection, op amp, 487
    signals, 518
    voltage, op amp, 518, 487
Commutative property, 882
Commutator ripple, 443
Commutator, 440
Comparator, 555
Comparison of test excitations, 636
Comparison vibration representations, 634
Compatibility, 785
    condition, 30
    equations, 840
    motion, 244
Compensating coil, eddy current, 444
Compensating resistor, 582
Compensator, 807
Complete state feedback, 822
Complex conjugates, 40, 272, 622, 710, 808, 891, 905
Complex exponential, 39
Complex exponents, FFT, 901
Complex failures, test, 668
Complex modes, 329
Complex modulus of elasticity, 353
Complex representation of signal, 155
Complex response method, 91
Complex roots, filter, 536
Complex spectrum, 157, 905
Complex stiffness, 362
Complex systems, 702
Complex vector approach, 751
Compliance, 117
Component degradation, 179
Component failure, 180, 917
Component interconnection, 482
Component matching, 482
Component modification, 691, 779, 780, 787
Compound pendulum, 66
Compression molding, 498

Compressor circuit, shaker control, 416
Compressor, wind tunnel, 798
Computation reduction, 906
Computational bandwidth, 169
Computational speed, 909
Computer model, 839
Conditional failure, 915
Conditional probability, 913
Conditioning amplifier, 714
Conditioning monitoring, 691
Conduction, 845
Conductive coupling, 608
Conductive plastic, 428
Conductive resistance, 845
Conductivity, 845
Cone suspension, 10
Configuration space, 214
Configuration space, 223
Congruence transformation, 257
Conjugate matching, amplifier, 489
Connecting rod, 763
Connection rules,
        flexibility, 206
        stiffness, 205
Conservation equations, 840
Conservation of energy, 21
Conservative forces, 875
Conservative system, 31
Constant failure rate, 917
Constant-current bridge, 571
Constant-gain feedback, 812
Constant-parameter systems, 53, 850
Constitutive relation, 202, 841
        for momentum, 873
Constraints,
        holonomic, 198
        nonholonomic, 198
Continuity equations, 840
Continuity, assembly, 785
Continuous systems, 5, 12, 267
        damped, 328
Continuous-parameter models, 840
Contractor, 674
Control law, 807, 811
Control of vibration, 731
        beam, 816
Control sensor, 609, 642
        vibration testing, 659
Control system capabilities, test, 687
Control system, 400, 806
        shaker, 415
Control techniques, 807
Control variable, 808
Control, 691
        vibration, 13
Controllability, 820
Controlled variable, 808
Controller, 399, 806
Convection, 846
        heat transfer coefficient, 846
Convective resistance, 846
Conversion complete, 555
Conversion rate, ADC, 561, 563
Conversion start, 557
Conveyor motor, 835
Conveyors, 67
Convolution function, 190
Convolution integral, 54, 57, 62, 102, 232, 620, 852

Convolution theorem, 174
        discrete, 903
Convolution using DFT, 907
Cooley and Tukey, 895
Coordinates,
        generalized, 198
        incremental, 198
        natural, 230
Core, 432
Coriolis acceleration, 864
Correction, 427
Correlation theorem, 177
        discrete, 903
Correlation-function coefficient, 176, 620
Cos, 304
Cosh, 304
Cost function, 812
Coulomb friction, 237, 354, 381
Coulombs, 450
Coulomb, Charles, 10
Counter-rotating eccentric masses, 723
        shakers, 409
Counter-type ADC, 559, 562
Coupled system, 848
Coupling, flexible, 76
Covariance function, 175
Crandall, 11
Crank mass, 763
Crank pin, 763
Crank shaft, 763
Crank, 763
Critical components, test, 664
Critical interface requirements, test, 659
Critical speed, 775, 776
Critically damped motion, 43
Critically damped response, 107
Cross correlation, 172
Cross product, vector, 865
Cross-axis motions, 641
Cross-covariance function, 175, 620
Cross-product relation, 867
Cross-sensitivity, 425, 452, 464, 467
Cross-spectral density, 176
Cross-spectral density, 622
Cross-transfer function, 696, 699
Crosshead, 763
Crossing frequency, guideway, 296
CRT, see cathode ray tube
Cumulative damage, 916
        theory, 658
Cumulative fatigue, 639
Current amplifier, 513
Current sensors, 575
Current, 843
Curve shapers, 581, 585
Curve-fitting
        multi-dof, 709
        multi-resonance, 709
        single-resonance, 704
        steps, 713
        transfer functions, 702, 722
Curvilinear coordinates, 861
Curvilinear distance, 861
Customer, test, 674
Cutoff frequency, 165, 436
        filter, 529
        lower, 533
        upper, 533

Index

Cutting load, 391
Cyclic frequency, 27, 104, 152, 614
Cyclic integration, 372

D

D'Alembert's principle, 451, 763
D-type elements, 842
Da Vinci, 380
DAC error sources, 554
DAC, *see* digital-to-analog conversion
Damage criteria, 639
    test, 671
Damaging effects of whirling, 776
Damped continuous systems, 328
Damped natural frequency, 40, 94, 328
Damped oscillator, 90, 99
    characteristics, 50
Damped systems, 235
    properties, 238
    terminology, 238
Damper,
    broad-band effect, 802
    locations, 822
Damping capacity, 361, 366, 372
    per unit volume, 351, 357
Damping classification, 358
Damping coefficient, *see* damping constant
Damping constant, 38
    hysteretic, 366, 390
Damping effects, test, 643
Damping force vector, 359
Damping
    in bearings, 385
    in gears, 385
    in rotary devices, 385
Damping matrix, 236, 693
    inertial, 237
    stiffness, 237
Damping measurement, 42, 368
    methods, 379
    bandwidth method, 376
    hysteresis loop method, 371
    magnification factor method, 374
    of multi dof system, 375, 378
    step response method, 370
Damping mechanisms, 350
Damping models, 388
    multivariable, 359
Damping parameters, definitions, 367
Damping ratio, 39
    modal, 331
    equivalent, 361
Damping types, 349
Damping values, seismic applications, 380
Damping, 4, 12, 349
    Coulomb, 237
    momentum, 237
    quadratic, 80
    strain-rate, 237
    structural, 237
    viscoelastic, 237
Data acquisition board, 549
Data block length, 194
Data channels, 420
Data record sectioning, convolution, 908

Data valid signal, 559
Data-bus transfer, MUX, 565
DC bridge, 569
DC motor, 63
    armature-controlled, 141
DC tachometer, 441, 494
de Laval, 11
Dead weight, 407
Decades, 614
Decaying sine signal, 615
Decelerating conditions, 186
Decibels, 425
Decoding, 542
Deflection tolerances, 639
Degradation of components, 639
Degree of correlation, 620
Degrees of freedom, 197, 855, 875
Delay time, 422
Delay, 109
    dynamic, 111
Delta function, 53
Demodulation, 434, 539, 546, 587
    phase-sensitive, 498, 543
Demodulators, 13
Demultiplexing, 564
Den Hartog, 11, 797
Design approaches, test-based, 787
Design basis excitation, 616
Design development, 609, 637
Design example, beam control, 822
Design for vibration, 1, 731, 779, 784
Design life, 915
Design modification, 779
    vibration, 731
Design ratings, test object, 640
Design steps, isolator, 745
Design through modal testing, 779
Design, 13, 197, 691
    vibration isolation, 743
    strain gage torque sensor, 472
Design-development tests, 675
Determinant of a matrix, 884, 886, 893
    properties, 893
Deterministic error, 427
Deterministic signals, 157, 611
DFT, *see* discrete Fourier transform
Diagnosis, 544, 547, 691
Dielectric constant, 444
Dielectric properties, 447
Difference amplifier, 492, 516
Difference equation, 3, 567, 848
Differential amplifier, 486, 492, 516
Differential gain, 517, 519
    op-amp, 521
Differential transformer, 434, 547
Digital controller, 416
Digital counter, 590
Digital demultiplexing, 566
Digital filter, 567
Digital Fourier analysis, 895
    procedures, 904
Digital Fourier analyzer, 595
Digital MUX, 565
Digital oscilloscope, 600, 714, 761
    capabilities, 601
Digital plotter, 718
Digital signal analyzers, 594, 597
Digital spectral computation relations, 185

Digital spectrum analyzer, 714
Digital tachometer, 440
Digital transducers, 440, 590
Digital-to-analog conversion, 13, 549, 806
Digitizing speed, 170
Dimension of vector space, 891
Dimensional gaging, 443, 481
Diode,
        forward bias, 593
        reverse bias, 593
Dirac delta function, 53, 160, 281, 313, 338, 818
Direct sensitivity, 425
Direct-drive robot, 141
Direction of sensitivity,
        strain gage, 468
        test, 665
Directional measurements, 437
Discharge coefficient, 78
Discrete convolution, 904
Discrete correlation, 902
Discrete Fourier transform, 162
Discrete frequencies, 163
Discrete model, 197
Discrimination, signal, 539
Dish washer, 825
Displacement limit, 734
Displacement response vector, 692
Displacement sensor, 430
Displacement space, 223
Displacement spectrum, 627
Displacement tolerances, 638
Displacement transfer function, 111
Displacements,
        flexural, 201
        rotatory, 201
        translatory, 200
Display, 399
Distinct eigenvalues, 712, 852
Distributed components, 5
Distributed-parameter experiment, 716
Distributed-parameter systems, 5, 12, 267, 816
Distribution environment, test, 668
Distribution qualification, 668
Distribution spectra, test, 670
DOF, see degree-of-freedom
Dominant pole, 48
Dominant resonant frequencies, 834
Dominant time constant, 47
Doppler effect, 478
Double pendulum, 253
Drag coefficient, 356, 395
Drag force, 356, 395
Drawings, test object, 640
Drift, 546
        op-amp, 512
Drive frequency, 788
Drive motor, 714
Drive signal generator, 417
Driving-point transfer function, 696
Drop test, 414, 652, 653
Dry friction, 354
Dual jump, FFT, 901
Dual terms, FFT, 900
Dual-arm component, test, 665
Dual-slope ADC, 557, 562
Dummy elements, bridge, 569
Dummy gage, 466
Dummy weight tests, 648

Dummy weights, 645
Duration of exposure, 733, 735
Dwell frequency, 615
Dwell time interval, 615
Dynamic absorber, 132, 787
Dynamic balancing, 755
Dynamic boundary conditions, 306
Dynamic coupling, test, 662
Dynamic delay, 109
Dynamic flexibility, 117
Dynamic force, rotating, 752
Dynamic friction, 381
Dynamic inertia, 117
Dynamic instability, shaker, 415
Dynamic interaction, shaker-object, 413, 415
Dynamic model, 701, 839
Dynamic qualification, 656
Dynamic range, 417, 425, 542
Dynamic reliability, test, 664
Dynamic state, 223
Dynamic stiffness, 117, 130
Dynamic strain, 467
Dynamic symmetry, 209
Dynamic system, 18, 578, 848
Dynamic viscosity, 848
Dynamic-excitation equivalence, 657
Dynamometer, reaction type, 475

E

Eccentric loading, torsion, 296
Eccentric masses, 715
Eccentricity, 547
Eddy current sensor, 443, 575
Effective emmisivity, 846
Effective noise bandwidth, 182
Effective surface area, radiation, 846
Effort variable, 118
Eigenfrequencies, 208
Eigenfunctions, beam vibration, 304
Eigenvalue problem, 207, 223
        coordinate transformation, 870
Eigenvalues, 39, 45, 49, 104, 106, 695, 851
        of beam, 823
Eigenvectors, 208, 242
Eight-cylinder engine, balancing, 769
Elastic curve, 267
Elastic foundation, 345
Elastic potential energy, 20, 843
Elastomeric mounts, 825
Electrical circuit, 24, 81
Electrical elements, 843
Electrical engineering, 1
Electrodynamic shaker, 402, 685, 716
Electrohydraulic shaker, 402
Electromagnetic actuator, 806
Electromagnetic energy, 25, 844
Electromagnetic force, 411
Electromagnetic induction, 431
Electromagnetic noise, 524
Electromagnetic radiation, 478
Electromagnetic shaker, 402
Electromagnetic shaker, 410, 684
Electromagnetic torque, 475
Electrostatic energy, 25, 844
Elevated guideway, 3, 295

Elevator, 75
EMA components, 719
EMA steps, 692, 703
EMA system configuration, 719
EMA, *see* experimental modal analysis
Encoding, 542
Energy approach, beam bending, 324
Energy conservation, 21
Energy dissipation, 4, 12
Energy equivalence, 32
Energy reduction factor, 192
Energy storage, 17
    elements, 19, 244, 840
Energy, 17
Energy-dissipation elements, 840
Enforced response, 86
Engine box, 713
Engine unit, 713
Ensemble average, 171, 618
Environment, 18
Environmental conditions,
    instrumentation, 428
    test, 671
Equalization,
    frequency band, 609, 610, 624
    test signal, 658
Equalizer,
    random signal, 416
    test excitation, 416
Equidistant impulses, 163
Equilibrium state, 849
Equipment cabinets, 645
Equipment malfunction, 675
Equipment manufacturer, 674
Equipment supplier, 674
Equivalence for mechanical aging, 656
Equivalent damping model, 235
Equivalent damping ratios, 361
Equivalent end mass, cantilever, 724
Equivalent mass, 28
Equivalent series impedance, 505
Equivalent stiffness, 28
Equivalent viscous damping, 360
Ergodic hypothesis, 171, 611, 616, 618
Ergodic random signals, 171, 618
Error sources
    of EMA, 702
    of DAC, 554
Euclidean norm, 892
Euclidean space, 223
Euler's angles, 871
Euler's equations, 871
Euler's theorem, 857
Euler, Leonard, 10
Euler-Bernoulli beam model, 299
Excavator, 71
Excitation control, 416
Excitation controller, 417
Excitation frequency, 788
    nondimensional, 738
Excitation record, 616
Excitation signal, 610
Excitation system, 400
Excitation vector, 692
Excitation, 399
Excitation-intensity equivalence, 657
Exciter, *see* shaker
Exciters, 400, 401

vibration, 12
Experiment for rotor balancing, 760
Experimental modal analysis, 4, 691
Experimental model, 839
Experimental modeling, 691, 698
Experimental setup, 715, 717
    saw blade vibration, 817
    balancing, 762
Experimental vibration, 399
Exploratory testing, 650
Exponential decay, 158
Exposure to vibration, 733
External damping, 397
External trigger, 598
Extrinsic method,
    fiber optic, 478
    ultrasound, 481

F

Factors determining RRS, 680
Failure modes, test, 668
Failure rate, 914, 915
    curve, 916
Fan, 144
Fast Fourier transform, 13, 895
Fault detection, 544, 547
Fault diagnosis, 12
Feed-forward control, 806
Feed-forward path, 856
Feedback capacitance, op amp, 490
Feedback control, 806
    shaker, 408
    testing, 609
Feedback resistor, 582
    op-amp, 515
Feedback sensor, 146
Feedforward action, 111
Feedforward gain matrix, 108
Ferromagnetic core, 437
Ferromagnetic medium, 431
FFT analyzer, 595, 605, 718
Fiber-optic gyroscope, 478
Fiber-optic position error, 477
Fiber-optic proximity sensor, 477
Fiber-optic sensing, extrinsic, 477
Fiber-optic sensing, intrinsic, 477
Field coil, shaker, 410
Field, 889
Field-effect transistor, 512
Filter, 13
    bandwidth, 172, 625
    definitions, 541
    order, 525
    poles, 525
    recursive, 568
    types, 541
Filtering, 807
Finite element analysis, 813
Firing cycle, engine, 771
First order system, 44, 84
Fish processing, 35
    machine, 394
Fixed edge, membrane, 333
Fixed end, 306
Fixture dynamics, 641

Flaws in roller bearings, 547
Flexibility matrix, 202
Flexible boundary conditions, 273
Flexible coupling, 135
Flexible foundation, 345
Flexible mount, 66, 713, 746
Flexible shaft, 76, 143
Flexibly coupled system, 203
Flexural motion, *see* bending
Flexural system, 22, 201
Flexural vibration, 299
Floor response spectra, 646
Floor vibrations, 826
Floor-mounted equipment, 646
Flow rate, 846
Flow source, 78
Flow variable, 118
Flow-induced vibrations, 779
Fluid capacitance, 80
Fluid capacitor, 846
Fluid coupling, 97, 128
Fluid damping, 356, 358
Fluid elements, 846
Fluid inertor, 847
Fluid level oscillations, 444
Fluid resistance, 37, 848
Fluid resistor, 82
Fluid spring, 846
Fluid systems, 34
Fluorescent screen, 598
Flushing tank, 80
Flutter, 395
    of aircraft, 779
Flux linkage, 431
Flux variable, 118
Foil-type strain gage, 457
Fold catastrophe, 424, 578
Folding, 168
Force isolation, 735
Force rating, shaker, 404
Force sensors, 456, 462
Force source, 261
Force transmissibility, 124, 736
Force transmitted to support, 802
Force, 842
Force-current analogy, 841
Force-response time history, 720
Forced motion, 230
Forced oscillations, 2
Forced response, 17, 52, 86
    beam, 310
    concepts, 57
    continuous systems, 312
    equations, 233
Forced vibration, 230
    beam bending, 310
    beam, 313
    test, 652
Forcing excitation, 52
    vector, 692
Forcing function, harmonic, 85
Forward transform, 160
Foundation,
    damping modulus, 345
    elastic modulus, 345
    elastic, 345
Four-cylinder engine, 495
    balancing, 831

Four-stroke engine, 495
Fourier analysis, 13, 158, 895
Fourier coefficients, 161
Fourier integral transform, 160, 895
Fourier series expansion, 161, 895
Fourier spectrum, 605, 634
    method, 623
Fourier transform, 11, 110, 620
    properties, 164
    results, 173
    unification, 895
Fourier, Joseph, 11
Frahm absorber, 787
Free decay, 723
Free motion, 26
Free response, 17, 26, 51, 52, 233, 855
    continuous systems, 311
Free vibration, 848
Free-body diagram, 28, 114, 206, 854
Frequency analyzer, 595
Frequency content, 616, 733
    test, 655
Frequency control, 263
Frequency counter, 418
Frequency creation, 424, 578
Frequency domain, 3, 85, 153
    formulation, EMA, 692
    results, 110
Frequency modulation, 540, 591
    tape, 419
Frequency of oscillation, 27
Frequency ranges, test, 684
Frequency ratio, 113
Frequency response function, 92, 104
    test, 651
Frequency response of I-beam, 726
Frequency response, 85, 110
    of op-amp, 512
Frequency shifting property, 546
Frequency spectrum, 12, 140
Frequency transfer function, 92, 104, 484
Frequency uncertainty, 183
Frequency, 151
    angular, 27
    cyclic, 27
Frequency-domain model, 840
Frequency-domain specification, 733
Frequency-domain test, 715
Frequency-shifting theorem, 543
Frequency-to-voltage converter, 591
Frequency-tuned bandpass filter, 417
Fringes, 479
Full bridge, 568
Full-scale drift, 425
Full-scale reading, 426
Full-scale value, ADC, 561
Full-scale voltage, DAC, 549
Full-wave demodulation, 547
Functional failure, 639
Functional operability, 609, 638
    test, 674
Functional testing, 638
FVC, *see* frequency-to-voltage converter

G

Gage factor,

strain gage, 457, 502
    eddy current sensor, 444
Gain margin, 423
Gain scheduling, 582
Gain, op-amp, 454
Galloping of transmission lines, 395, 779
Gantry conveyor, 28
Gantry truck, 250
Gas pressure,
    balancing, 832
    IC engine, 763
Gas-force reactions, 771
Gaussian window, 174
Gear transmission, 73
Generalized coordinates, 198, 230, 281, 875
    beam, 307
    guideway, 297
    modal, 780
Generalized forces, 853
Generalized mass, 213
Generalized stiffness, 214
Generalized velocities, 853
Generic test, 646
Geometric connectivity, 840
Geometric factors, instrumentation, 428
Geometry file, EMA, 720
Glitches, DAC, 554
Gram-Schmidt orthogonalization, 892
Graphics terminal, 718
Gravitational potential energy, 18, 21
Gravity-type fluid capacitor, 82
Gray code, DAC, 554
Gray level of image, 602
Gray-box testing, 667
Ground loop, 493
Ground motion, seismic, 677
Ground points, 494
Ground potential, 494
Ground transit vehicles, 735
Ground wire, 494
Ground-based suspension, 127
Ground-loop noise, 516
Guideway span, 296
Guideway vibration, 299
Gyroscopic forces, 771
Gyroscopic sensors, 481

H

Half-bridge circuit, 572
    LVDT, 607
Half-power bandwidth, 99, 183, 426
    filter, 527, 536
Half-power level, 183
Half-power points, 95
Half-wave demodulation, 547
Hamilton's principle, 875, 876
Hamiltonian approach, 17
Hammer test, 134, 189, 414, 652, 716
Hamming window, 174
Handling loads, test, 668
Hanning window, 174
Hardware analyzer, 595
Hardware availability, test, 647
Hardware characteristics, shaker control, 688
Hardware control, 806
Hardware filters, 568

Hardware redundancy, 639
Harmonic component, 153
Harmonic excitations, 85, 750
Harmonic input, 105
Harmonic motion, 207
Harmonic response, 99
Harmonics, 424
Hazard function, 915
Heat generation, 803
Heat transfer rate, 845
Heat-sensitive paper, 595
Heave motion, 215, 263
Heavy spring, 32
Helical spoilers, 825
Helical spring, 71
Helicopter vibration, 8
Hermitian property, 167
Hermitian transpose, 891
Hertz, 104
Hidden-line display, 728
High-pass filter, 523, 531
Hold pulse, 563
Hold-down hardware, test, 660
Holonomic constraints, 198
Holonomic systems, 13, 198, 875
Homogeneous equation, 38
Homogeneous response, 86
Homogeneous solution, 38, 52, 99, 855
Hooke's law, 20, 29, 244, 843
Hooke, Robert, 10
Houdaille damper, 802, 803
Hydraulic actuator and valve, 792
Hydraulic pump, 143
Hydraulic servo, 792
Hydraulic shaker, 6, 402, 407
Hysteresis loop, 351
    damping measurement, 371
    for damping models, 355
Hysteresis, 424, 578
Hysteretic damping coefficient, 366
Hysteretic damping, 329, 352
    model, 347

I

I-beam vibration, 726
IC engine, *see* internal combustion engine
Identification,
    model, 691
    system parameter, 420
Identity matrix, 109, 884
Ignition cycle, 771
Image frame, CCD, 602
Imbalance output, 569
Imbalance, *see* unbalance
Impact test, 716
Impedance bridge, 443, 574
Impedance characteristics, 482
Impedance circuits, 127
Impedance head, 414, 483
Impedance test, 137
Impedance transformer, 487, 514
Impedance transforming amplifier, 452
Impedance, 117, 423, 425, 484, 505, 736
    of a damper, 118
    of a mass, 117
    of a spring, 117

Impedance-matching amplifiers, 483, 485
Impeller, 144
Impulse response function, 57, 102, 106, 173, 628
Impulse response, 53, 59
Impulses, equidistant, 162
Impulsive excitation, test, 652
In-line engines, 769
        balancing, 832
In-service functions, 610
In-service loading, 647
In-service mounting, 642
In-service operating conditions, 674
In-service orientation, test, 660
Inclusion-exclusion formula, 912
Incompressible fluid, 847
Incremental coordinates, 198, 875
Incremental variables, 81
Independent coordinates, 197, 875
Independent events, 913
Independent linearity, 426
Independent signals, 619
Induced voltage, 432
Induction motor, 144, 246, 260, 262
        three-phase, 258
Inductive coupling, 608
Inductor element, 844
Inductor, 17, 24
Inertia block, 744, 745, 825
Inertia effects, test, 643
Inertia element, 842
Inertia force, 120, 451
Inertia load of engine,
        primary component, 766
        secondary component, 766
Inertia matrix, 204, 692, 868
Inertia torque, 476
Inertia, 19
Inertial boundary conditions, 273, 326
Inertial damping matrix, 237, 360
Inertial frame, 864
Inertial shaker, 402, 408
Inertially coupled system, 206
Infant mortality, 656, 911
Influence coefficient method, 202, 205
Influence coefficients, 12
        complex, 752
        in balancing, 757
        damping, 236
        flexibility, 204
        inertia, 206
        stiffness, 203
Information acquisition, test, 639
Infrared source, 478
Initial condition response, 231
Initial conditions, 27, 40
        beam vibration, 308
        cable, 283
Initial-condition response, 52, 233
Initial-displacement excitation, 652
Inner product, vector, 891
Input gain matrix, 108, 241, 850
Input impedance, 423, 484, 510
Input port, 146
Input vector, 108, 241
Input-bias current, op amp, 487
Input-output board, 420, 549
Input-output differential equation, 75, 855
Instability in machine tools, 381, 386

Instrument accuracy, 426
Instrument calibration, 659
Instrument instability, 426
Instrument ratings, 425, 494
Instrumentation amplifier, 491, 572
Instrumentation, 399, 517
Instrumented robot, 722
Instruments, commercial, 13
Integral control, 423
Integrated circuit, 510
Integrating ADC, 557
Integration by parts, 309
Intensity function, product failure, 915
Intensity level, excitation, 612
Intensity reduction factor, test, 667
Intensity, test, 655
Interaction of test object and shaker, 640
Interconnection laws,
        for impedances, 118
        for receptance, 130, 133
Interconnection, component, 482
Interface board, 549
Interface conditions, assembly, 785
Interface damping, 380
Interface details, test object, 640
Interface dynamics, 641
Interface linkages, 785
        vibration test, 660
Interfacing devices, 400
Interference, 479
Interferometer, laser Doppler, 478
Internal combustion engines, 8
Internal damping, 329, 350, 358
Internal isolation, electrical, 493
Internal noise, op-amp, 520
Internal trigger, 598
International combustion engine, balancing, 763
Intersection of sets, 913
Intrinsic method,
        fiber optic, 478
        ultrasound, 480
Intrinsive coordinates, 861
Inverse DFT, 905
Inverse Fourier transform, 160
Inverse of a matrix, 888
Inverse transformation, 701
Inverted pendulum, 67
Inverting amplifier, 490, 515, 583
Inverting gain, op-amp, 513
Inverting input, 486, 510
Inverting lead, op-amp, 454
Iron Butcher, 394
Isolation level, 827
Isolation pads, 745
Isolation region, 738, 742
Isolation, electrical, 493
Isolation, vibration, 731
Isolator spring, 744
Isothermal process, 846

J

Jump phenomenon, 424, 578

K

K-orthogonality, 214, 218
Kelvin-Voigt damping kodel, 329, 351, 396, 817
Kinematic viscosity, 848
Kinematics, 857
Kinetic coenergy, definition, 873
Kinetic energy, 17, 19, 842
    beam bending, 324
    definition, 873
Kirchhoff, 11
Kronecker delta, 161

L

Laboratory experiments, 13, 713
    balancing, 761, 830
Ladder type DAC, 550, 551
Lag circuit, 526
Lagrange's equations, 31, 876
Lagrange, Joseph, 10
Lagrangian approach, 3, 13
    beam bending, 324
Lagrangian mechanics, 857, 873
Lagrangian, 875
    beam bending, 324
Laplace domain, 852
Laplace operator, 335
Laplace tables, 103
Laplace transform, 11, 102, 110, 161
Laplace variable, 91
Large test objects, 690
Laser Doppler interferometer, 478, 480
Laser, 478
Lateral vibration, 268
Lattice form, bridge arms, 568
Lead circuit, 532
Leakage inductance, 440
Leakage, in signal processing, 174, 178, 904
Least significant bit, 550
Least squares fit, 709
Level of coupling, test, 640
Level programming, 418
Liebnitz's rule, 62
Lift coefficient, 395
Lift force, 395
Light fringes, 479
Light-emitting diode, 478, 506
Limit cycles, 424, 578
Line noise, 516
Line structure, 267, 284
Line trigger, 598
Line-mounted equipment, 645, 677
Linear algebra, 849, 879
Linear calibration, capacitive sensor, 445
Linear dampers, 821
Linear dependence, vector, 890
Linear equations, solution, 893
Linear independence, modal vectors, 214
Linear interpolation, 580
Linear models, 208
Linear momentum, 865
    principle, 865, 866
Linear quadratic regulator, 811
Linear sensitivity, strain gage, 468
Linear system, 850

Linear viscous damping, 329
Linear viscous model, 235
Linear-variable differential transformer, 432
Linearity, 425
Linearization, 849
    by analog circuitry, 578, 581
    by hardware, 578, 581
    by software, 578, 580
Linearized equation, 23
Linearizing devices, 577
Linearly independent equations, 711
Linearly independent vectors, 242
Liquid column, 24
Liquid slosh, 24
Liquid tank, 24
Lissajous patterns, 598
Load cell, strain gage, 461, 503
Load cycles, 656
Load impedance, 440
Load resistance, 431
Load torque, 77
Load, op-amp, 515
Load-deflection of spring mounts, 827
Loading conditions, test, 674
Loading effect, 428, 483
Loading error, 440
    strain gage, 457
    filter, 533
Loading,
    electrical, 429
    mechanical, 429
Logarithmic decrement method, 41, 368, 653
Logarithmic decrement, 46, 369
    per radian, 368
Long-term drift, amplifier, 604
Longitudinal stiffness, 388
Longitudinal vibration of rods, 284
Loop adaptor, 419
Loss factor values of materials, 367
Loss factor, 366, 372, 387
Low-pass filter, 435, 523, 525
LQR, see linear quadratic regulator
Lumped-parameter approximation, 32
Lumped-parameter experiment, 713
Lumped-parameter model, 12, 197, 840
LVDT, 432
LVDT, signal conditioning, 434

M

M-normal modal vector, 693
M-orthogonality, 213, 218
Machine and support structure, 809
Machine tool, 139, 739
    vibration, 8
Machine vision, 602
Magnetic field, shaker, 410
Magnetic head, 419
Magnetic tape player, 417
Magnetic torque, 77, 135, 141, 265
Magnetic-induction sensors, 547
Magnetostrictive property, 480
Magnification factor, damping measurement, 374
Magnitude, 92
Magnitude, transfer function, 366
Main lobe, 184

Mainframe dynamics, test, 664
Malfunctions, 197, 638
Manual control, shaker, 415
Manufacturing engineering, 1, 8
Marginal stability, 45, 89
Mass eccentricity, 749
Mass element, 842
Mass matrix, 692
Mass, 19
Mass-spring-damper model, 808
Material damping, 329, 350, 397
Material fatigue, 656
Materials, test object, 640
Mathematical model, 839
Mathieu equation, 393
Matrices, 879
Matrix addition, 882
Matrix exponential, 242, 851
Matrix inverse, 884
Matrix multiplication, 883
Matrix properties, 889
Matrix Riccati equation, 812
Matrix subtraction, 882
Matrix transpose, 885
    property, 694
Matrix-eigenvalue problem, 242
Maxwell's principle of reciprocity, 122, 696
Maxwell bridge circuit, 606
Maxwell model, 352
Mean error, 427
Mean value, 156
Mean-squared amplitude spectrum, 156
Measurand, 421
Measurement accuracy, 426
    charge amplifier, 455
Measurement bandwidth, 435
Measurement error, 426
Measurement gain matrix, 852
Measurement matrix, 108
Measurement
    of acceleration spectrum, 113
    of damping, 42, 368
Measuring device, perfect, 422
Mechanical aging, 418, 636, 656
Mechanical circuit, 128
Mechanical elements, 842
Mechanical engineering, 1
Mechanical exciters, 408
Mechanical impedance, 116, 484
Mechanical mobility, 484
Mechanical system,
    translatory, 21
    vibration, 732
Mechanical wearout, 656
Mechanics, 857
Membrane analogy of torsion, 293
Membrane vibration, 332
Memory, 853
Menu-driven test system, 671
Metacenter, 37, 68
Metal-film capacitor, ultrasound, 480
Method of undetermined coefficients, 87
Microstrain, 467
Milling machine, 826
MIMO, *see* multi-input-multi-output
Miner's damage theory, 658
Minimal set of transfer functions, 699
Minor of matrix element, 887

Misalignment, 547
Missile vibration, 299
Mixed systems, 17, 839
Mobility function, 117, 484, 705, 721, 736
Modal analysis expressions, 216
Modal analysis processor, 718
Modal analysis systems, 727
Modal analysis, 3, 197, 241
    approaches, 226
    beam bending, 302
    experimental, 4, 691
Modal boundary conditions,
    beams, 304
    rods, 286
Modal control, 812
Modal damping matrix, 693
Modal damping ratio, 331, 693
Modal damping, 653
Modal data, 698
    extraction steps, 699
Modal decomposition, 236
Modal expansion, 819
Modal formulation,
    non-symmetric, 224
    transformed symmetric, 224
Modal interference, 380
Modal mass, 213
    matrix, 693
Modal matrix, 222, 230, 695
Modal motions, 207
Modal natural frequencies, cable, 274
Modal parameter extraction,
    multi-dof, 709
    single dof, 704
Modal parameters, 691
Modal response specification, 691, 779, 787
Modal series expansion, cable, 274
Modal stiffness, 214
    matrix, 693
Modal superposition, 281
Modal testing, 691, 714
Modal transformation, 233
Modal vector, 695
    normalized, 214
Modal vibrations, 197, 207
Mode shape file, EMA, 720
Mode shape of I-beam, 725
Mode shape vector, 210
Mode shapes, 3, 207, 654
    beam vibration, 304
    cable, 275
    normalized, 279
Model development, 840
Model identification, 4, 691
Model types, 208
Model,
    discrete, 197
    lumped-parameter, 197
Modeling process, 840
Modeling, 839
Modes, uncoupled, 214
Modification, proportional, 782
Modified Gram-Schmidt procedure, 892
Modulated wave, 88
Modulating signal, 539
Modulation theorem, 543
Modulation, 539
Modulators, 13

Modulus of elasticity, 300
    complex, 353
Modulus of rigidity, 717
Mohr's circle, 871
Moment of inertia, 22
    polar, 292
Momentum damping, 237
Monitoring channels, 654
Monochromatic light, 478
Monotonically decreasing function, 911
Monotonicity
    of ADC, 561
    of DAC, 555
Monte Carlo simulations, 671
Most significant bit, 550
Motion isolation, 735
Motion sickness, 734
Motion transducers, 431
Motion transmissibility, 124, 736
Motor torque, 77
Mounting details, test object, 640
Mounting fixtures, 400, 407
Mounting techniques, accelerometer, 452
Moving light beam, 430
Multi-component system, 131
Multi-degree-of-freedom system, 131
Multi-input-multi-output system, 852
Multi-test sequences, 667
Multicomponent equipment, 645
Multicomponent systems, 911
Multicylinder engines, balancing, 766, 832
Multifrequency signals, 616
Multifrequency testing, 679
Multiple impacts, test, 653
Multiple-mode excitation, 668
Multiplexer, 564
Multiplier circuit, 435
Multivariable systems, 110
Mutual-induction transducers, 431
MUX, *see* multiplexer

N

N-element data sequence, 899
N-type strain gages, 468
Narrow-band excitation, 645
Narrow-band filter, 416
Natural coordinate frames, 860
Natural coordinates, 230
Natural frequencies, 3, 207
    beam, 321
    damped, 40
    repeated, 209
    saw blade, 816
Natural frequency effects, test, 643
Natural modes, 197
Natural motion, 876
Natural oscillations, 2, 848
Natural response, 86
Negative damping constant, 386
Negative feedback, 811
Negative frequency spectrum, 904
Neutral axis, 300
Neutral equilibrium, 850
Newton's second law, 19, 206, 842
Newton, Sir Isaac, 10
Newtonian approach, 3, 13, 31

Newtonian mechanics, 857, 864
No load condition, shaker, 402
Node point, 221, 247, 257, 654, 784
    beam vibration, 318
Noise bandwidth, 184
Noise immunity, 546
Noise ripples, 592
Nomograph, 631
    specification, 734
Non-holonomic systems, 12
Non-homogeneous equation, 52
Non-inverting amplifier, 435, 490
Non-inverting gain, op-amp, 513
Non-inverting input, 486, 510
Nonanticipative transformation, 853
Noncircular section, torsional vibration, 293
Nonconservative forces, 875
Noncontact sensor, 437, 813
Noncontact actuators, 813
Nonharmonics, 424
Nonholonomic constraints, 198
Nonhomogeneous equation, 230, 711
Noninverting amplifier, 515
Nonlinear damping matrix, 258
Nonlinear damping torque, 258
Nonlinear model, 75, 208
Nonlinear resistance, fluid, 848
Nonlinear system, 23
Nonlinearities, 58, 555
    ADC, 561
    bridge, 571
    dynamic, 424, 578
    static, 424, 578
Nonlinearity error in differential transformers, 437
Nonlinearity percentage, bridge, 572
Nonsingular inertia matrix, 701
Nonsingular matrix, 884
Nonstationary random process, 616
Norm, 892
Normal mode motion, 243
Normal mode solution, cable, 275
Normal modes, 213, 729
    cable, 274
Normalization,
    beam modes, 307
    vector, 892
Normalized frequency, 113, 126
Normalized modal vector, 214
Normalized response spectra, 681
Notch filter, 443, 538
Notch frequency, 539
Nozzle, 78
Nuclear plant equipment, 646
Null balance, strain gage, 466
Null matrix, 882
Null space, 894
Null vector, 882
Null voltage, 433
Numerator polynomial, 108, 856
Numerical model, 839
Nyquist frequency, 165, 525, 561, 596, 897
Nyquist plot, 721

O

OBE, *see* operating basis earthquake
Octaves, 614

Offset compensation, 581
Offset current, op-amp, 513
Offset error, ADC, 561
Offset nonlinearity, 579
Offset signals, op-amp, 520
Offset voltage, op-amp, 513
Offset, 578
Offsetting circuitry, 581, 582
Offsetting voltage, 582
One-sided spectra, 156
One-third-octave band, 416
One-third-octaves, 614
Op-amp gain, open-loop, 510
Op-amp, 435
    ideal properties, 522
Open-loop gain, op-amp, 512
Operating basis earthquake, 676
Operating conditions, 77
Operating environment, 640
Operating frequency range, shaker, 402
Operating frequency, 140, 738
Operating interval, 423
Operating point, 392, 426, 849
Operating range, 137
Operating site, 640
Operating speed, 81
    compressor, 798
    steady-state, 144
Operating-basis earthquake, 646
Operation malfunction, 650
Operational amplifier definitions, 522
Operational amplifier, 485, 510
    ideal, 506
Optical coupling, 608
Optical encoder, 478, 713
Optical potentiometer, 430
Optimal absorber design, 797, 801, 837
Optimal control, 812
Optimal single damper, 823
Optimized two dampers, 823
Order analysis, 186
Order tracking, 188
Order, 856
    filter, 525
    of a system, 81
Ordinary coherence function, 176, 620
Origin, 48
Orthogonal matrix, 869
Orthogonal transformation, 869, 872
Orthogonal vectors, 892
Orthogonality condition, modes, 213
Orthogonality of modes, 12, 221
Orthogonality,
    beam bending, 327
    beam vibration, 310
    cable modes, 275, 276, 283
    damped beam, 331
    discrete Fourier transform, 163, 904
    end mass, 338
    Fourier series, 161
    Fourier transform, 160
    guideway modes, 298
    inertial boundary conditions, 326
    modes, beam, 308
    modes, rods, 286
    torsional modes, shaft, 293
Oscillation, 1
    forced, 2
    natural, 2
Oscillator characteristics, 50
Oscillator circuits, 575
    for LVDT, 434
Oscillator, 417
    inductance-capacitance, 445
    shaker control, 416, 418
    VFC, 589
Oscillatory mode, 234
Oscillatory systems, 242
Oscilloscopes, 13, 597
Output feedback control, 821
Output gain matrix, 108, 852
Output impedance, 423, 484, 510
Output port, 146
Output variables, 852
Output vector, 108
Output, 18
Overdamped filter, 531
Overdamped motion, 42
Overdamped response, 107
Overhead gantry truck, 250
Overhead transport device, 140
Overlapped processing, 184, 186
Overtesting, 667, 671
Owen bridge, 575
Oxidation, potentiometer, 429

P

P-type strain gages, 468
Paint pumping system, 260
Pancake motors, 749
Paper mill, 836
Parallel axis theorem, 869
Parallel element, generalized, 119
Parameter estimation, model, 691
Parameter file, EMA, 720
Parameters, 18
Parametric drift, 426
Parametric errors, DAC, 554
Parseval's theorem, 172, 176, 619
Partial fractions, 712
Particular integral, 52
Particular response, 86
Particular solution, 55, 99
Parzen window, 174
Pass-band frequencies, 537
Passenger compartment, 496
Passenger discomfort, 734
Passive control, 5
    vibration, 787, 801
Passive equipment, 638
Passive filters, 524, 541
Passive transducers, 421
Path equations, 840
Pause time, 616
PD control, *see* proportional-plus-derivative control
Peak acceleration, shaker, 403
Peak detection, 593
Peak displacement, shaker, 403
Peak magnitude, 113
Peak picking, 703, 708
    transfer functions, 709
Peak time, 422
Peak transmissibility, 740

Index

Peak velocity, shaker, 403
Peak-hold circuit, 593, 594
Peaks of frequency response functions, 130
Pendulous load, 250
Pendulum,
    compound, 66
    double, 253
    inverted, 67
    simple, 22, 250
Percentage overshoot, 422
Performance characteristics, ADC, 560
Performance curve, shaker, 402
Performance improvement, whirling, 778
Performance monitoring, test, 638
Performance specification, 12, 421
    frequency domain, 423
    time domain, 422
Period, 151
Periodic forcing function, 85
Periodic signals, 157, 162
Periodic solution, 27
Permanent-magnet transducers, 431, 438
Phase angle, 27, 92, 152
Phase lead, 98
    transfer function, 366
Phase margin, 423
Phase meter, 714
Phase modulation, 541
Phase sensitive demodulation, 498, 543
Phase shifter, 586
Phase, 153, 855
Phasing of excitations, test, 666, 668
Phasor notation, 757
Phasor, 27, 154
Photocopier, 69
Photodetector, 506
Photodiode, 506
Photointensity, 602
Photoresistive layer, 430
Phototransistor, 506
Physical model, 839
Physically non-realizable system, 120
Physical realizability, 855
Physically realizable systems, 103, 174
Picocoulomb, 450
Piezoelectric accelerometer, 685, 807
Piezoelectric actuator, 806
Piezoelectric element, 137
Piezoelectric transducers, 449, 489
Piezoresistive material, 467
Piezoresistive property, 457
Pinned beam, 305, 306
Pipeline segment, 244, 853
Piston pin, 763
Piston-cylinder, 407
Pitch motion, 216
Planar structures, 332
Planar system, 267
Plant builder, 674
Plant owner, 674
Plant, *see* process
Plotting, response spectra, 629
Pluck test, 414, 653, 654
Pneumatic mount, 744
Poincare, 11
Point transfer functions, 699
Poisson's ratio, 323, 334, 461
Poisson, Simeon-Dennis, 11

Polar coordinates, 860
    for whirling, 772
Polar moment of area, 470
Polar moment of inertia, 292
Pole assignment, *see* pole placement
Pole placement, 812
Pole, dominant, 48
Poles, 39, 45, 49, 104, 106, 695
    filter, 525
Polynomial equations, 580
Portable exciters, 690
Portable shakers, 655
Portion control, 35
Position feedback, 811
Positive frequency spectrum, 904
Postmultiplication, 883
Pot, 428
Potential energy, 17, 843, 875
    of beam bending, 324
    elastic, 20
    gravitational, 21
Potential variable, 118, 840
Potentiometer circuit, 572, 581
    strain gage, 457, 502
Potentiometer, 428
Power amplifier, 514, 717
    shaker, 610
Power rating, shaker, 405
Power spectral density, 172, 619, 624, 635
    of excitation, 416
Power spectrum, 156
Power supply, 714
    current-regulated, 571
    regulated, 428
    voltage-regulated, 571
Power, 95
PPD control, *see* proportional-plus-derivative control
Prandtl's membrane analogy, 293
Precision of estimates, 625
Precision, 426, 427
Preconditioning, test, 650
Premultiplication, 882, 883
Pressure boundary, 638
Pressure load, IC engines, 770
Pressure transducer testing, 646
Pressure, 846
Pressure-regulated system, 78
Pretest inspection, 636, 649
Pretest procedures, 636
Primary coil, potentiometer, 429
Primary component
    of engine inertia, 766
    balancing, 831
Primary windings, 432
Principal axes,
    dynamic, 661
    geometric, 661
Principal coordinates, 230
Principal directions, 869, 871
Principal moments of inertia, 870
Principal stress, 470
Principle of reciprocity, 122, 696
Principle of superposition, 52, 577
Probability density function, 914
Probability distribution function, 914
    joint, 619
Probability of failure, 911
Probability, 911

Process monitoring, 179
Process, 805
Product qualification, 5, 418, 609, 637, 668
Product rule, 913
Product rule, reliability, 916
Product testing, 12, 13
Production, 609
Program instructions, 580
Programmable gain, 516
Programmable unijunction transistor, 589
Programmed tests, 672
Propellers, vibration, 290
Proportional damping, 236, 239, 255, 329, 359, 697
Proportional output circuitry, 581, 584
Proportional-plus-derivative control, 811
Prototype, 17, 839
Proximity probe, 432
Proximity probe, 438
Proximity sensor,
    mutual induction, 438
    self-induction, 438
PSD, *see* power spectral density
Pseudo-random excitations, 635
Pulse waveforms, 418
Pulse-code modulation, 541
Pulse-frequency modulation, 541
Pulse-width modulation, 540
Pump, 72, 78
PUT, *see* programmable unijunction transistor
Pythagoras, 10

Q

Q-factor method, 95
Quadratic damping force, 80
Quadratic error function, 710
Quadrature error, 433
Qualification by symmetry, 638
Qualification criteria, 675
Qualification of products, 5
Qualification procedure, 675
Qualification testing, seismic, 673
Qualification, product, 13, 668
Quality assessment, 12
Quality assurance, 637
    test, 656, 675
Quality factor, 97
Quality of cut, 813
Quantization error, 543
    ADC, 560
Quartz crystal accelerometer, 499
Quasi-periodic signals, 157
Quefrency, 180

R

r-stage factorization, FFT, 899
Rack and pinion, 254
Rack and pinion, 261
Radial acceleration, 773
Radian frequency, 614
Radiation, 846
Radio frequency, 444
Radius of curvature, 300
Radius of gyration, 66

Radix-two FFT, 898, 901
Rail car, 21
Random environments, 635
Random error, 427
Random excitation, 611
Random process, 611, 618
Random signal generator, 417, 419, 617
Random signals, 12, 157, 171
Random testing, 635
Random vibration tests, 684
Rank, matrix, 893
Rankine, 11
Rate error, 437, 438
Rate gyro, 145, 481, 504
Rating parameters, 425
    for sensors, 483
Ratings of test apparatus, 640
Rational fraction form, 711
Rayleigh damping, 236
Rayleigh, 11
Reaction-type shakers, 409
Reactive transducer, 444
Read-and-write memory, 580
Real modes, 238, 729
Real-time digital analysis, 624
Real-time signal analyzer, 596
Receptance method, 130
Receptance relations, 149
Receptance, 117, 705
    cross, 134
    direct, 134
Reception properties, 133
Reciprocating engines, 832
Reciprocating machines, balancing, 762
Reciprocity property of transmissibility, 124
Record length, 625
Record length, effective, 186
Recording, 399
Rectangular matrix, 881
Rectangular membrane, 333
Rectangular plate, vibration, 334
Rectification, 434
Rectilinear biaxial test, 662
Rectilinear motion, 267
Rectilinear test, 661
Rectilinear triaxial test, 662
Rectilinear uniaxial test, 662
Recursive algorithm, 568
Recursive digital filters, 568
Redundancy, component, 917
Reference bridge, 466
Reference excitation, 610
Reference frame, 864
Reference voltage variation, DAC, 554
Refresh rate, picture, 602
Regulatory agency, 674
    testing, 636
Relays, 638
Reliability curve, 912, 916
Reliability, 13, 911
Reluctance, 431
Repeatable accuracy, 427
Repeated natural frequencies, 209
Repeated roots, 39, 44
Required input motion curve, 676
Required response spectrum, 404, 632
Reset action, 423
Residual flexibility, 214

Residue extraction, 711
Residue, 695, 698
Resistance bridge, 569
Resistive feedback, charge amplifier, 505
Resistively coupled transducer, 429
Resistivity, 456
Resistor, fluid, 82
Resolution of Fourier results, 184
Resolution, 425
    ADC, 560
    ordinary, 194
    potentiometer, 429
Resolvent matrix, 851
Resonance bandpass filter, 533
Resonance creation, absorbers, 799
Resonance search, 636, 650
Resonance, 1, 90, 93, 115
Resonant condition, 88
Resonant frequency, 1, 93, 99, 129, 423, 791
Resonant frequency, absorber, 796
    filter, 530
    with hysteretic damping, 364
Resonant peak, 95, 113, 129, 724
    of filter, 536
Response curve, 42
Response of simple oscillator, 46
Response spectrum, 402, 593, 625, 634, 724
    of excitation, 416
    plotting paper, 629
    utilization, 632
Response vector, 692
Response, 18, 52, 86
Response, dynamic, 609
Reversible failure, 639
Reviewer, test, 674
Reynold's number, 356
Riccati equation, 812
Ride comfort, 735
Ride quality, 3, 6, 134, 499
    specification, 9
Rigid base suspension, 124
Rigid body dynamics, 867
Rigid body modes, 12, 209, 210, 215, 234, 710
RIM, *see* required input motion
Rise time, 422
    of amplifier, 519
RMS amplitude spectrum, 155
RMS value, 156
RMS, *see* root-mean-square
Road disturbances, 111
Robot arm, 63
Robotic manipulator, 265
Robustness, 425
Rocket vibration, 299
Rods, longitudinal vibration, 284
Roll-off rate, filter, 527
Roll-up slope, filter, 533
Root-mean-square spectrum, 155
Root-mean-square value, 619
Rotary inertia boundary condition, 328
Rotating components, 749
Rotating frame, 859, 862
Rotating machinery,
    balancing, 749
    diagnosis, 544, 547
Rotating systems, 760
Rotating vector, 750, 752
Rotation, rigid-body, 858

Rotational mechanical elements, 841
Rotational system, 22
Rotatory inertia, 301, 322
Rotatory system, 22
Rotatory viscous damper, 97
Rotatory-variable differential transformer, 432
Rotors, vibration, 290
Rounding off, ADC, 561
Row vector, 880
RRS specification, 646, 658
    generation, 680
RRS, *see* required response spectrum
Rubber buffing machine, 258
RVDT, 432

S

S/H, *see* sample and hold
Safe-shutdown earthquake, 646, 676
Saint Venant theory of torsion, 293
Sallen-Key filter, 541
Sample and hold circuit, 555
Sample function, 171, 611, 618
Sample pulse, 563
Sample-and-hold circuit, 563
Sampled data, 164, 898
Sampling frequency, 525, 596
Sampling rate, 165, 170
    of data, 420
Sampling theorem, 165
Saturation nonlinearity, 525
Saturation, 424, 578
    of op-amp, 511
Saw blade frequencies, 816
Saw blade vibration, 813
Saw-tooth wave, 590
Sawing accuracy, 813
Scalars, 880
Scale change, 582
Scale factor drift, 426
Scale factor, DAC, 555
Scragg saw, 836
Second law for particle system, 866
Second moment of area, 300
Second order system, 44
Secondary coils, 432
Secondary component of engine inertia, 766
Secondary components, balancing, 831
Secondary windings, 432
Seismic mass, 137
Seismic qualification, 609, 673
Seismograph, 10
Self-compensation, bridge sensing, 471
Self-excited vibrations, 386, 779
Self-induction transducers, 431
Sensitivity axes, test, 683
Sensitivity drift, 426
Sensitivity, 425
    accelerometer, 452
    bridge, 459
    capacitive sensor, 447
    charge, 450
    circuit, 585
    strain gage, 457
    voltage, 450
Sensor locations, 640

Sensor sensitivity, 472
Sensor stiffness, 472
Sensors and transducers, 731
Sensors, 12, 399, 401, 807
    piezoelectric, 449
    rating parameters, 483
Separable solution, 283
    beam vibration, 303, 304
    beam with tension, 320
    cable, 271, 274
    membrane vibration, 333
Separation constant, 271
    beam vibration, 303
Sequence,
    data, 162
    spectral results, 162
Series element, generalized, 119
Series opposition connection, 435
Service functions, 638, 639
Servo balancing, bridge, 460
Servo-valve, 407
Sets, 879
Settling time, 422
    DAC, 554
Sewing robot, 261
Shaft deflection, whirling, 771
Shaft vibration, circular section, 290
Shaft-disk system, 774
Shafts, torsional vibration, 290
Shaker apparatus, 609
Shaker rating,
    force, 404
    power, 405
    stroke, 405
Shaker selection, 403
Shaker table, 407
Shaker test, 340, 654
Shaker, 400, 724
    alignment, 7
    control system, 415
    counterrotating mass, 409
    electrical model, 412
    feedback control, 408
    head, 410
    hydraulic, 6
    mechanical model, 412
    no-load condition, 402
    operation capabilities, 402
    performance curve, 402
    reaction-type, 409
Shannon's sampling theorem, 165
Shape factor, radiation, 846
Shear coefficient, beam bending, 322
Shear deformation, 322
Shear modulus, 70, 291, 323
Shear strain, 291
Shear strain, beam bending, 322
Shear stress, 291
Shielded cables, 494
Ship, 68
    motion, 199
    vibration, 8, 299
Shock absorbers, 825, 837
Shock and vibration, 732
Shock excitation, 732
Shock isolation, 732
Shock load, 732
Shock response, 593

Shutdown conditions, 186, 745, 799
Side frequencies, 544
Side lobes, 174, 178, 544
Signal acquisition, 400
Signal analysis, 12, 151
Signal analyzer, 594, 595
    bandwidth, 596
Signal classification, 160
Signal conditioning, 13, 399, 401, 509
    for LVDT, 434
Signal conversion, 509
Signal generating equipment, 417
Signal generator, 399, 418, 610
Signal modification, 400, 509
Signal modulation, 539
Signal processing, 165
Signal types, 157
    signal types, 617
Signal-to-noise ratio, 419, 702
Signum function, 380
Silicon-boron semiconductor gage, 469
Similarity law, test, 671
Similarity transformation, 257
Simple harmonic motion, 27
Simple integrator, 120
Simple oscillator, 18, 89, 111, 231
    damped, 38, 46, 694
    harmonic response, 99
    response spectrum, 632
    rods, 289
    undamped, 27
Simple pendulum, 22, 250
Simply-supported beam, 205, 306
Simulated excitations, 617
Simulation block diagram, 856
Simultaneous DFT of two signals, 905
Sin, 304
Sine beat with pauses, 616
Sine beat, 615
Sine decay, 615
Sine dwell, 106, 410, 614
Sine signal generator, 417
Sine sweep, 106, 158, 410, 612, 650
Sine test, 134, 635
Sine-beat signal, 419
Sine-random signal generator, 417, 716
Single cylinder engine, balancing, 763
Single degree-of-freedom test, 661
Single frequency testing, 677
Single-axis shaker, 411
Single-cylinder model, balancing, 831
Single-damper control, 823
Single-ended amplifier, 486, 516
Single-frequency signals, 612
Single-frequency testing, 645, 668
Single-plane balancing, 755, 828
Single-resonance curve fit functions, 722
Sinh, 304
Sinusoidal motion, 27
Sinusoidal oscillation, 27
Six-cylinder engine, balancing, 766, 832
Ski testing, 716
SkyTrain, 7
Slew rate,
    amplifier, 519
    op-amp, 522
Sliding end, 306
Slip ring and brush, 442, 467

Index

Slosh, liquid, 24
Small rotations, 858
Small-signal bandwidth, 519
SNR, *see* signal-to-noise ratio
Snubber, 21
Software analyzers, 596
Software control, 806
Software filters, 568
Sound, 480
Source alteration, 731
Sources of vibration, 732, 762
Space robot, 722
Space shuttle, 2
Space station, 2
    vibration, 299
Spacecraft vibration, 299
Span, vector space, 891
Spatial differential operator, 329
Spatial system, 267
Specific damping capacity, 366
Specific heat, 847
Specification,
    modal response, 782
    peak level, 733
    performance, 12
    ride quality, 9
    RMS value, 733
    severe discomfort vibration, 735
    vibration, 732
Spectral content, 669
Spectral lines, 158, 163, 184
    resolution, 702
Spectral map,
    speed, 187
    time, 187
Spectral raising, 617
Spectral suppressing, 616, 617
Spectrum analyzer, 13, 113, 595, 597
Spectrum shaper, 419
Spectrum,
    imaginary part, 161
    real part, 161
Speed controller, 714
Speed of response, 45, 50, 423, 473
    amplifier, 519
Speed spectral map, 187
Spherical polar coordinates, 860, 861
Splines, 580
Spool valve, two-stage, 407
Spray painting robot, 260
Spring element, 20, 843
Spring force, resolution, 72
Spring mounts, 825
Spring, heavy, 32
Square matrix, 881
Square-root of sum of the squares, 633
SRSS, *see* square root of sum of the squares
SSE, *see* safe shutdown earthquake
Stability curves, 394
Stability, 4, 45, 50, 423, 852
    of amplifier, 519
Stabilizer suspensions, 825
Stable equilibrium, 850
Stages of seismic qualification, 674
Stain-rate damping, 237
Standard deviation, 427
Standard linear solid model, 352
Standard linear solid model, 396

Standby redundancy, 917
Startup conditions, 186, 745, 799
State model, 392
State of capacitor element, 844
State of inductor element, 845
State of inertia element, 842
State of spring element, 843
State space approach, 240
State space model, 819
State space, 223
State trajectory, 853
State variable, 853
State vector, 108, 223, 241, 244, 392, 849, 853
State, definition, 853
State-space approach, 12
State-space model 66, 78, 108, 808, 848
    nonlinear, 75
State-transition matrix, 851
Static balancing, 749, 750
Static deflection, isolator, 743
Static equilibrium position, 56
Static friction, 381
Static gain, 375, 423
Static linearization, 578
Static modes, 12, 211, 213, 711
Static operating curve, 578
Statically coupled system, 203
Stationary random signals, 172, 618
Statistical error, 185
Statistically independent signals, 175
Stator, shaker, 411
Stead-state response, 94
Steady operating conditions, 77
Steady-state error removal, 581
Steady-state error, 142, 422, 473
Steady-state performance, 423
Steady-state response, 55, 98, 105
    beam, 313
Steady-state solution, 93
Steady-state value, 422
Steady-state whirling, 774
Steam generator, vibration, 835
Stefan-Boltzmann constant, 846
Step input, 232
Step response, 57, 60, 108
Steps of test procedure, 672
Stick, 71
Stick-slip vibration, 386, 394
Stiffening cross bars, 825
Stiffness damping matrix, 237, 360
Stiffness element, 843
Stiffness matrix, 202, 693
Stiffness,
    flexural, 201
    tensile/compressive, 200
    torque sensor, 472
    torsional, 201
Stochastic error, 427
Stochastic process, 611
Stochastic representation, 624
Stochastic signals, 157, 171, 611, 617
Stockbridge damper, transmission lines, 800
Stodola, 11
Strain capacity, 472
Strain gage accelerometer, 463
Strain gage force sensor, 470
Strain gage, 456
    active, 461, 466

dummy, 466
foil-type, 457
nonlinearity, 472
semiconductor, 467
Stress-strain relations, damping, 387
Stribeck damping, 804
Stribeck effect, 381
String, *see* cable
Strings, 268
Strobe frequency, 476, 716, 761
Strobe synchronization, 751, 753
Strobe-tacho, 761
Stroboscope tachometer, 761
Stroboscope, 476, 716
Stroke rating, shaker, 405
Stroke, 402
Strong response duration, 151
Strong-motion earthquakes, 681
Structural damping, 237, 354, 358, 380
Structural failure, 650
Structural integrity, 638, 639
    test, 674
Structural modification, 197
Structural strength, 638
Structural stress, 635
Subharmonics, 424
Subspace, 890
Substructuring, 197, 691, 779, 784, 787
Subsystem assembly, 785
Subsystems, design, 784
Successive-approximation ADC, 555, 559, 562
Summer type DAC, 550
Summing amplifier, 582
Superposition method, 82
Support motion excitation, 627
Support motion, 55, 109, 125, 127
Support pier, guideway, 296, 347
Support structure, vibration transmission, 789
Supporting structures, test, 660
Suspension bridges, 268
Suspension, automobile, 10
Sweep duration, 612
Sweep oscillator, 418
Sweep rate, 655
Sweep, up-down, 418
Symmetric matrix, 885
Symmetrical rectilinear testing, 664
Symmetry,
    dynamic, 209
    dynamic-axial, 663
    test object, 682
Synchronized strobe signal, 828
Synchronous whirl, 771, 775, 833
System boundary, 18
System matrix, 108, 241, 850
System of particles, 866
System order, 81
System parameters, 18
systematic error, 423, 427
Systems, 13

T

T-type elements, 841
Table lookup, 580
Tachometer, 146, 148

AC, induction, 442
AC, permanent-magnet, 441
mechanical, 482
Tangential-normal coordinates, 860, 861
Tape player, 419
Tape transport mechanism, 419
Taylor series expansion, 850
Temperature drift, amplifier, 519, 604
Temperature, 845
Temporal mean, 171, 618
Tensile test, damping, 388
Tension sensor, strain gage, 503
Test excitation characteristics, 655
Test fixture, 640
Test input characteristics, 610
Test laboratory capabilities, 645
Test laboratory, 674
Test nomenclature, 660
Test object data, 640
Test object movement, 674
Test object, 400, 610
Test package, 403
Test plan, 647
Test preliminaries, 675
Test procedure, 650, 671
Test program planning, 611, 636, 645
Test purpose, 647
Test report, 673
Test response spectrum, 404, 632
Test signal generation, 636, 669, 670
Test signals, 611
Test specimen, 662
Test spectrum equalization, 672
Test strength measure, 916
Test types, comparison, 669
Test-based design approaches, 787
Test-object mounting, 659
Testing instrumentation, 401
Testing objectives, 650
Thermal capacitance, 845
Thermal capacitor, 845
Thermal conduction blocks, 802
Thermal cycling, 670
Thermal elements, 845
Thermal resistance, 845
Thermal systems, 2
Thevenin's theorem, 505
Thick beams, bending vibration, 322
Thin beam, bending vibration, 299
Thin closed section, torsion, 295
Thin open section, torsion, 295
Thin plate, vibration, 334
Thin solid section, torsion, 293
Third-harmonic distortion, 419
Three-degree-of-freedom test, 661
Three-phase induction motor, 258
Through variable, 118, 146, 203, 484, 840
Time constant, 45, 49, 50
    amplifier, 519
    charge amplifier, 453
    dominant, 47, 426
    shaker, electrical, 411
    shaker, mechanical, 411
    piezoelectric sensor, 499
Time delay, excitation, 623
Time domain model extraction, 700
Time domain, 3
Time history, 151, 611

Index

Time lag, 154
Time lead, 154
Time reference, 154, 587
Time response, 850
Time signals, 634
Time spectral map, 187
Time-division MUX, 564
Time-domain model, 840, 848
Time-domain tests, 716
Time-scaling criterion, 671
Time-variant systems, 53
Timoshenko beam model, 322
Timoshenko, 11
Tooth-meshing frequency, 547
Torque motor, linear, 407
Torque sensing, 470
Torque sensor design formulas, 472
Torque sensor, 456, 501
    deflection type, 473
    reaction pipe, 474
Torque source, 144, 262
Torque, 865
Torsion member, 470
Torsion, thin section, 293
Torsional boundary conditions, 341
Torsional guideway transit, 295, 342
Torsional parameter, noncircular section 293, 294
Torsional rigidity, 292
Torsional stiffness, 292
Torsional system, 201
Torsional vibration,
    noncircular section, 293
    shafts, 290
Trace impurity, doping, 467
Trace of a matrix, 885
Tracking filter, 417, 522, 524
Tracking frequency, 523
Train vibration, 299
Train, 251
Transcendental equation,
    beam, 317
    rods, 288
Transducer impedance, 430
Transducers, 12, 807
Transfer function file, EMA, 720
Transfer function matrix, 108, 110, 692, 694, 880
Transfer functions, 3, 101, 856
    admissible set, 699, 700
    EMA, 721
    minimal set, 699, 700
    non-admissible set, 700
Transfer matrix, 852
Transform pair, 160
Transform techniques, 101
Transform, Fourier, 11
Transform, Laplace, 11
Transformation, congruence, 257
Transformation, similarity, 257
Transient exciters, 414
Transient motion sensing, 432
Transient signals, 157
Transit systems, 3
Translational transformation, 869
Translatory displacements, 200
Translatory mechanical system, 21
Translatory motions, 144
Transmissibility function, 738
Transmissibility ratio, 748

Transmissibility, 124
Transmissibility, magnitude, 126
Transmission lines, 8, 268
Transmission of random excitations, 620
Transmission wires, 799, 800
Transverse displacements, 201
Transverse motion, *see* bending
Transverse response, cable, 280
Transverse sensitivity, 452
Transverse vibration, 268
    beams, 299
    cables, 268
    membranes, 332
    plates, 334
Trapezoidal integration, 163
Trapezoidal rule, 903
Traveling wave, 270
Traveling-wave solution, 283
Trial mass, 750
Trial shipments, 670
Trial solution, 44
Trigger, external, 598
Trigger, internal, 598
Trigger, line, 598
Trivial solution, 273, 307, 890
TRS generation, 658
TRS, *see* test response spectrum
Truncation error, 174, 904
Tunable bandpass filter, 714
Tuned absorber, 789
Turbine rotor, whirling, 833
Turbulent effects, 356
Twin T circuit, 538
Two dof biaxial test, 662
Two dof triaxial test, 662
Two's complement representation, 557
Two-cylinder engine, balancing, 766
Two-damper control, 823
Two-degree-of-freedom test, 661
Two-dimensional members, 332
Two-plane balancing, 755, 828
Two-plate capacitor, 444
Two-port devices, 484
Two-port element, 118
Two-sided spectra, 156

U

Ultrasonic sensors, 480
Ultrasound probes, 480
Ultrasound waves, 480
Unbalance deflection, 771
Unbalance, 547
Uncorrelated excitations, test, 662, 663
Uncorrelated signals, 175, 619, 620
Uncoupled modes, 214
Undamped absorber design, 801
Undamped natural frequency, 693
Undamped oscillator, 88, 99
Underdamped motion, 40
Underdamped response, 106
Undermined coefficient method, 87
Unit vectors, 859, 892
Unity matrix, *see* identity matrix
Unmodeled signals, amplifier, 519
Unreliability curve, 912

Unreliability, 911
Unscrambling, FFT, 901
Unstable equilibrium, 850
Unstable response, 89
Unstable vibrations, 779
Useful frequency range, 423, 425, 435
    accelerometer, 452
Useful frequency, 188
Utilization, 609

V

V6 engine, 769
V8 engine, 769
Valve actuators, 638
Valve spring, 70
Valves, piezoelectric, 449
Vanes, vibration, 290
Variable cross-section, beam, 309
Variable speed drive, 410
Variable-capacitance transducers, 444
Variable-inductance sensors, 575
Variable-inductance transducers, 431
Variable-reluctance transducers, 431
Variable-tension cable, 283
Variational approach, 324, 346
Variational coordinates, 875
VCC, see voltage-to-current converter
Vector component, 881
Vector element, 881
Vector kinematics, 857
Vector mechanics, 857
Vector space, 853, 889
Vector, order, 881
Vectors, 879
Vehicle body, 250
Vehicle model, 253, 258, 728
Vehicle suspensions, 298
Velocity distribution, parabolic, 847
Velocity feedback gain, 811
Velocity feedback, 416, 811
Velocity limit, 734
Velocity profile, parabolic, 80
Velocity source, 261
velocity spectrum, 628
Velocity transfer function, 111
Velocity, 842
Vendor, 674
Venn diagram, 912
Ventilation system, whirling, 776
VFC, see voltage-to-frequency converter
Vibration absorber characteristics, 789
Vibration absorber equations,
    forcing excitation, 790
    support motion, 790
Vibration absorber, 835
    damped, 793
    rotary, 800
    undamped, 788
Vibration amplification, 799
Vibration control summary, 801
Vibration control,
    active, 5, 805
    passive, 5
Vibration dampers, 801
Vibration design, substructuring, 785

Vibration environment, 611
Vibration exciter, see shaker
Vibration exciters, 400
Vibration instrumentation, 399, 401
Vibration isolation formulas, 744
Vibration isolation, 122, 735, 788, 825
    flexible system, 745
Vibration isolator, 137, 826
    active, 495
Vibration level, 733
Vibration meter, 417, 714
Vibration mitigation, 736
Vibration monitoring, 606
Vibration specification, 732, 736
Vibration suppression, 132
Vibration test configurations, 662
Vibration test purpose, 637
Vibration test scheme, 610
Vibration testing, 5, 13, 113, 137, 419, 609
Vibration, membranes, 332
Virtual work, 324
Viscicoder, 595
Viscoelastic damping, 237, 329, 351
Viscoelastic parameter, 351, 352
Viscous damping constant, 38
Viscous damping, equivalent, 360
Visual inspection, 639
Voltage amplifier, 513
Voltage control, 263
Voltage follower, 487
Voltage ripple, 442
Voltage sensitivity, 450
Voltage source, 428
Voltage, 843
Voltage-controlled oscillator, 589
Voltage-sensitive switch, 589
Voltage-to-current converter, 592
Voltage-to-frequency converter, 589
Vortex shedding, 395

W

Washing machine, 825
Waterfall display, 187
Wave equation, 270
Wave generator, 591
Wave interference, 479
Wave speed, 270, 283
Waveform mixer, 658
Weighted type DAC, 550
Weighted-resistor DAC, 550
Weighting matrix, 812
Wheatstone bridge, 457, 569, 607
Whipping, 771
Whirling formulas, 777
Whirling measurement, 438
Whirling of shafts, 771
Whirling speed, 771
Whirling, 13
Whirling, steady-state, 774
White noise, 182, 419
White-box testing, 667
Wide band of frequencies, 802
Wide-band disturbances, 646
Wide-band random excitation, 407
Wien-bridge oscillator, 577

Index

Wind-induced vibrations, 779
Window functions, 174, 177, 178
Window selection, 180
Window, correction factor, 192
Wiper arm, 428
Wood cutting machine, 77
Wood cutting system, 391
Wood-cutting saw, 836
Work done, 19
Work, 875
World coordinates, 752
Wraparound error, 907
Wraparound error, convolution, 903

Y

Yaw motion, 145
Young's modulus, 300, 323

Z

Zero crossings, 151
Zero drift, 425
Zero initial conditions, 59, 103
Zero-detect unit, 558
Zero-initial-condition response, 233
Zero-input response, 233, 852
Zero-order hold, 564
Zero-period acceleration, 631, 633
Zero-period displacement, 631
Zero-period velocity, 631
Zero-state response, 52, 55, 852
Zeros, notch filter, 539
Zoom analysis, 597
Zooming, 601
ZPA, *see* zero-period acceleration